Heating, Ventilating, Air Conditioning, and Refrigeration

BILLY C. LANGLEY

PRENTICE HALL
Englewood Cliffs, New Jersey 07632

Library of Congress Cataloging-in-Publication Data

Langley, Billy C.
    Heating, ventilating, air conditioning, and refrigeration / Billy
C. Langley.—[4th ed.]
        p. cm.
    Rev. ed. of: Refrigeration and air conditioning. 3rd ed. c1986.
    Includes index.
    ISBN 0-13-385634-8
    1. Refrigeration and refrigerating machinery. 2. Air
conditioning. I. Langley, Billy C. Refrigeration and air
conditioning. II. Title.
TP492.L354 1990   89-33494
621.56—dc20   CIP

Editorial/production supervision: WordCrafters Editorial Services, Inc.
Interior design: Andrew Zutis
Cover design: Suzanne Behnke
Manufacturing buyer: Gina Brennan
Page layout: Gail Collis
Cover illustration: Lennox Industries, Inc.

Printed in the United States of America

10 9 8 7 6 5 4 3 2 1

ISBN 0-13-385634-8

Prentice-Hall International (UK) Limited, *London*
Prentice-Hall of Australia Pty. Limited, *Sydney*
Prentice-Hall Canada Inc., *Toronto*
Prentice-Hall Hispanoamericana, S.A., *Mexico*
Prentice-Hall of India Private Limited, *New Delhi*
Prentice-Hall of Japan, Inc., *Tokyo*
Simon & Schuster Asia Pte. Ltd., *Singapore*
Editora Prentice-Hall do Brasil, Ltda., *Rio de Janeiro*

Contents

Refrigeration Materials and Hand Tools 62

Compressors 91

Condensers and Receivers (124)

Evaporators 140

Flow Control Devices 149

9 Accessories 191

10 Refrigerants 207

11 Introduction to Electricity 221

12 Basic Solid-State Theory 256

13 Test Equipment 284

14 Electric Motor Theory 298

15 Electric Motor Controls 320

16 Domestic Refrigeration 336

17 Room Air Conditioning and Dehumidifiers   372

18 Commercial Refrigeration   383

19 Commercial Ice Machines   471

20 Air Conditioning (Heating)   506

26 Solar Energy   687

27 Starting and Operating a Business   703

Glossary   713

Index   732

Preface

Heating, Ventilating, Air Conditioning, and Refrigeration has been written to provide a thorough and practical overall study of the heating, ventilation, air conditioning, and refrigeration industry. The text begins with an entire chapter on safety and then presents the basic principles involved in air conditioning and refrigeration, followed by a discussion of the advanced theories and technology of modern systems. The book is intended to be used as a curriculum guide, a textbook, or a course for independent study. The practical fundamentals are covered, as are recommended service and installation procedures. *Heating, Ventilating, Air Conditioning, and Refrigeration* serves as a comprehensive textbook for the beginning student and a valuable reference for the experienced technician. The material in the text is organized to accommodate short survey courses, courses in private trade schools, or complete two-year college technical programs.

Heating, Ventilating, Air Conditioning, and Refrigeration is the accumulation of more than thirty-five years of field experience in the areas of research, service, installation, business, and teaching. It is written from the dual viewpoints of education and field experience using practical and everyday terminology and language, thus preparing the reader for the working environment. Some of the newer types of air conditioning and refrigeration systems and components are included.

Throughout the book, the author covers not only the theory necessary for reader and service technician alike, but presents examples to help reinforce that theory as well. Operating instructions from the various manufacturers who provide service and/or maintenance are incorporated wherever practical.

Approach, Philosophy, and Unique Features. While the material in *Heating, Ventilating, Air Conditioning, and Refrigeration* is based on both theory and experience, emphasis is constantly placed on the practical factors of the industry to aid the reader in making an easier entrance into the industry. Terminology used in the industry is incorporated throughout.

Troubleshooting procedures are covered wherever practical. The troubleshooting and service procedures are based on both theory and field experience. The newer technologies on some service procedures are presented at appropriate points in the text; they are inserted in such a way that they do not interfere with the reader's study of the material. The newer types of systems are included to help the reader realize the rapidly changing characteristics of the heating, ventilating, air conditioning, and refrigeration industry.

As each component is presented, its operation and purpose is discussed. Presentation in this manner helps the reader learn the purpose and operation of each component in the overall system without turning through page after page and breaking concentration.

Each chapter covers a specific area of the industry. Thus, after the reader has mastered the fundamentals, appropriate chapters may be consulted for information on specific areas. Each chapter ends with a set of review exercises, representing the minimum information the reader should understand from the chapter. Finally, a summary is included at the end of each chapter; it may also be used as a quick review for class-administered exams.

The text begins with an entire chapter on safety. Safety procedures are also integrated into the text as well as summarized at the end of each chapter. This emphasis on safety will make the reader more aware of safety procedures while doing the actual work.

A second color highlights key concepts and topics. Over 100 two-color illustrations show the reader the important details about the material being presented. Examples provide additional illustration of important and difficult material.

The Review Exercises at the end of each chapter provide an opportunity for self-evaluation before classroom testing. These Review Exercises provide immediate feedback on key concepts. An extensive glossary provides the reader with a single place for reference to the technical terms used in the heating, ventilation, air conditoning, and refrigeration industry.

Extensive Supplements Package. The supplements package for *Heating, Ventilating, Air Conditioning, and Refrigeration* is designed with two purposes in mind: to make the transition to using it as easy as possible, and to make the textbook and supplements a permanent part of the library for student and instructor alike.

A Student Air Conditioning and Refrigeration Troubleshooting Handbook is shrinkwrapped with every textbook (free of charge). The handbook is designed to aid the reader in learning the troubleshooting procedures used in the field. It contains a Troubleshooting Chart that is keyed to the Component Troubleshooting section of the handbook. The Component Troubleshooting section provides information on testing, checking, troubleshooting, and replacing specific components. It is much smaller than the textbook and can be carried on the service truck to serve as a handy reference until the reader is comfortable with the procedures used.

A Laboratory Manual contains over 100 Laboratory Exercises for the reader to perform, thus providing actual hands-on experience with specific tasks that are commonly performed in the field. There is a space for the instructor to initial the exercise; the space may also be used to suggest any additional information that should be studied to make the learning experience more complete.

A Standard Service Procedures Manual provides step-by-step, hands-on instructions for performing over 100 procedures that are used in the field.

A Transparency Master Manual contains more than 100 transparency masters of figures in the textbook to aid the instructor in making classroom presentations.

An Instructor's Resource Manual available upon adoption (free of charge) provides lesson plans for each chapter, with space for the instructor to make individual additions to the lesson plans. Answers to the Review Exercises at the end of each chapter in the textbook are included, along with the page in the textbook where the material is covered. Extra exam sets with answers are provided to make the testing process easier for the instructor.

Chapter Organization. Chapter 1 introduces the reader to safety procedures that should be observed while working on heating, ventilating, air conditioning, and refrigeration equipment. Safety procedures for equipment and personnel are included. Fundamentals of the industry are covered in Chapter 2. The reader is introduced first to the opportunities in the heating, ventilation, air conditioning, and refrigeration industry, and then to the theories upon which operation of heating, ventilation, air conditioning, and refrigeration systems are based.

Chapter 3 introduces the reader to the various types of refrigeration systems. The chapter begins with compression refrigeration systems and ends with thermoelectric refrigeration systems. Chapter 4 contains a description of the most popular hand tools and materials used in the heating, ventilation, air conditioning, and refrigeration industry. The reader learns about those tools and materials that are unique to the heating, ventilation, air conditioning, and refrigeration industry.

In Chapter 5 we describe compressor types, component construction, and other important details about compressor operation. Also included are the new scroll and discuss high-efficiency compressors. The chapter concludes with a section on compressor failure and prevention measures. Condensers and receivers are the subject of Chapter 6. The purpose and operation of both air-cooled and water-cooled condensers are covered. This chapter also provides information on condenser service. Chapter 7 introduces evaporator styles and types. Evaporator frosting and defrosting are covered, along with oil circulation. The effects of temperature differences on dehumidification are also covered.

Chapter 8 discusses the various types of refrigerant flow control devices and their purpose in the refrigerant circuit. The chapter concludes with installation and testing recommendations for thermostatic expansion valves and service hints. Accessories used in refrigeration systems are introduced in Chapter

9. Their purpose and operation are presented so that the reader will be aware of why they are desirable.

The most popular types of refrigerants used in the heating, ventilation, air conditioning, and refrigeration industry are covered in Chapter 10. Included is information on the handling of refrigerant cylinders and refrigerant tables. Electrical fundamentals are covered in Chapter 11; this coverage enables the reader to understand wiring diagrams and electrical circuits. Alternating and direct current are presented, along with magnetism and electromagnetism, to aid the reader in understanding how various components in the electric circuit operate.

Chapter 12 covers the basic solid-state theory used in some of the newer types of control circuits. Integrated circuits, printed circuit boards, and modular circuits are covered. The chapter includes information on transistors, diacs, triacs, optoelectronics, and fiber optics. It concludes with coverage on the types of control systems used in heating, ventilation, air conditioning, and refrigeration circuits. In Chapter 13 the reader is introduced to the most popular types of test instruments used in the service of heating, ventilation, air conditioning, and refrigeration systems and circuits. Their purpose, selection, care, and application are discussed so that the reader will be better informed about these types of instruments.

Electric motor theory and motor operation are presented in Chapter 14. The various types of motors used in heating, ventilation, air conditioning, and refrigeration systems are presented. Information on testing motor insulation, and motor speeds and capacitor connections are discussed. Chapter 15 describes the types and methods of controlling electric motors. Information covering full-voltage starting, star-delta and part-winding motor controllers is included in the chapter. Domestic refrigerator systems and cabinets are discussed in Chapter 16. The procedures used to charge domestic refrigerators and symptoms of an overcharge or a shortage of refrigerant are covered. A troubleshooting chart and domestic refrigeration electric circuits are presented. Chapter 17 covers room air conditioners and dehumidifiers. Refrigerant flow control application and sizing procedures are provided. The chapter also includes information on room unit compressors, and on special design features.

Commercial refrigeration cabinets and systems are the topic of Chapter 18. Heat recovery systems, refrigerant piping, and water coolers are covered. The chapter concludes with reasons for compressor failure, prevention of failure, and charging of commercial refrigeration systems. Chapter 19 introduces the reader to commercial ice machines. Both flaker and cuber types of units are covered. Water conditioning and troubleshooting procedures for electronic control circuits are included.

Chapter 20 provides information on standard and high-efficiency furnaces. The chapter introduces the reader to humidifier operation and heating controls, and concludes with solid-state electronic ignition systems. The various types of heat pump systems are described in Chapter 21. The reader is introduced to both standard heat pump systems and closed-loop earth-coupled water-source heat pump systems.

Chapter 22, "Air Conditioning (Cooling)," presents human comfort conditions, air distribution, and psychrometrics. Discussions of heat load calculations, duct system design, and air conditioning equipment conclude the chapter. In Chapter 23 the reader is introduced to systems using chilled and hot water for heating and cooling purposes. Start-up and maintenance procedures for chilled and hot water systems are included.

Chapter 24 describes both split and self-contained units used in commercial air conditioning applications. Maintenance procedures and thermal storage systems are also discussed. "Automotive Air Conditioning," Chapter 25, introduces the reader to the refrigeration system and the types of control circuits used in these systems. Installation and evacuation and charging procedures are covered.

Chapter 26 provides information on solar energy, including system design and operation and heat storage. In Chapter 27, "Starting and Operating a Business," information on the various forms of businesses is discussed. Included are discussions of human relations, picking a company name, and required licensing, advertising, and insurance needs. The chapter ends with tips on avoiding installation and service legal problems, and what to do in case of an accident.

□ ACKNOWLEDGMENTS □

A package containing as many items as are included with *Heating, Ventilating, Air Conditioning, and Refrigeration* would have been much more difficult to complete without the help of many people. Some of the people who provided technical assistance and other much needed support are:

Greg Burnell, Mariann Hutlak, Amy Ness, Ruth Semper. Also, many thanks to those with whom I did not have direct contact.

☐ To the Student ☐

Most students have asked themselves why is it necessary to learn a given subject, especially when the fundamentals of a subject are being covered, as they are in the first ten chapters of this textbook. Most students want to study immediately how to repair the equipment. However, knowledge of the fundamentals will not only make the job of repair much easier, but will make the diagnosis of problems possible.

It is especially important in the heating, ventilating, air conditioning, and refrigeration industry that fundamentals be understood fully. It is also important to review the fundamentals periodically, even after your formal training is completed. Remember, the more you know about heating, ventilation, air conditioning, and refrigeration equipment, the easier it is to service, maintain, and repair it. Also, the more you know, the faster you can get promotions or make more money in your own business. As the saying goes, "Someone can take away your material things but they cannot take away your knowledge. With knowledge you can get more material things. But without it little is possible."

The learning process takes place in several steps. First it is necessary to take an active part in the learning process. Getting involved requires self-discipline and the development of several habits. You must attend class regularly. While you are there, take part in class activities. Ask questions about information that is not clear to you. The Laboratory Exercises provide hands-on training for the procedure being covered. If you stand back and allow someone else to do all the exercises, you will be the loser and they will gain all the knowledge. You must also read that part of the textbook covering the material before the next class period.

It is best to set aside a particular time to start and a particular time to stop studying on a regular basis, preferably on a daily basis. Also, when a term that

you do not understand is encountered, take time to learn what it means. If you do not, the full meaning of the material will be almost impossible to understand. Learn the terminology. Every industry has a special terminology and you must know what the terms mean and how to use them in order to learn and to converse with others in the field.

Study the examples in the textbook. They are included to aid you in understanding the material. When they are ignored, the learning process is slowed down. The figures are used in the textbook to help you understand the material. When a figure is referred to, study it along with the text for a maximum learning experience and retention of the material.

Sometimes it is helpful to read the Review Exercises at the end of the chapter before studying the material. Do not try to memorize them. When you are reading through the text and find material related to the Review Exercises you will probably remember it. When you have finished studying a chapter, be sure to answer all of the Review Exercises. This procedure will allow you to review material that you do not understand before class time. The instructor can then help you with any problems you may have. The Review Exercises indicate the minimum amount of knowledge you should get from the chapter.

Take class notes; seldom do two instructors consider the same topics equally important. Frequently the instructor will put more emphasis on the areas that can be covered in his particular laboratory. This is perfectly satisfactory because learning several things well is usually better than learning a little bit about a lot of things. The laboratory is used to reinforce the theories learned in the classroom and the textbook. The class notes are an additional source of review for an exam.

Active learning requires a great deal of effort on your part, but it will be well worth the time and effort in the long run. Your understanding of each area covered in the textbook will permit easier learning of the following areas of the heating, ventilation, air conditioning, and refrigeration industry.

Be sure to ask questions about the areas that you do not fully understand. Remember, there are no dumb questions—only dumb answers. Continue to study and success will surely be yours.

Billy C. Langley

1 Safety

Performance Objectives

After completion of this chapter, the student should:

■ Be familiar with the safety procedures used in air conditioning and refrigeration work.

■ Be more aware of why safety procedures should be followed.

■ Know the basic procedures used in artificial respiration.

■ Recognize and practice safety procedures when working on refrigeration and air conditioning equipment.

■ Complete the Exercises with at least 80 percent accuracy.

A general feeling is that safety is a state of mind—that it is the attitude with which a person approaches his or her daily way of living while at work, play, or at home. Where safety is concerned, we are generally our own worst enemies, most of the time through a lack of knowledge, but more often because of failure to practice what we already know.

The use of safety aids is important in protecting ourselves from injury. Many potential tragedies have been avoided by wearing gloves, goggles, safety shoes, and other proper clothing. Safety is also promoted by using the proper tools, by posting safety rules in the workplace, and by the recognition of hazards; but the most important thing is the mental attitude that we have when going about our activities that causes us to look for the safe way of accomplishing these activities.

☐ Safety Procedures ☐

The old saying that "safety doesn't cost, it pays" is certainly true. It is most important that proper safety procedures be followed, especially by workers in the air conditioning and refrigeration industry. Since these people generally work alone, help may be difficult to obtain should a serious injury occur. Safety may be divided into three general categories: safety of the worker, safety of the equipment, and safety of the contents.

Safety of the worker is, of course, of prime importance. Machinery and materials can always be replaced, but a human life cannot. When the equipment and tools are used properly, there is very little danger to the worker. Always use the leg muscles when lifting heavy objects; never use the back muscles. Ask for help in lifting objects weighing more than 30 to 35 lb (13.6 to 15.9 kg). Keep the floor clear of water and oil. A slip while carrying heavy objects usually causes serious injury.

Air conditioning and refrigeration equipment use electricity as their source of power. Be sure that all electrical circuits are disconnected from the power source before working on them. The disconnect switch or circuit breaker will usually break the circuit. Never work a "hot" circuit.

Always wear safety goggles when working with refrigerants. If refrigerants should get in the eye, a frozen eye will result. When this happens, gently wash the eye with tap water and see a physician as soon as possible. Protect the skin from frostbite when working with refrigerants.

Safety of the equipment is also of great concern. Many components on refrigeration and air conditioning equipment are easily broken. Some of these components are very expensive and all of them take time to replace. Use the proper torque and sequence when tightening bolts and nuts. Before operating the equipment be sure that the fan blades and belts are clear of all objects.

Safety of the contents is a basic requirement and will depend to a great extent on the care given to the equipment during installation and servicing. The contents of a given installation must be kept at the desired temperature; this is the responsibility of the service technician. Technicians must know what the desired conditions are and maintain the equipment to produce these conditions. Most manufacturers provide tables and charts showing the desired operating conditions and the required control settings.

□ Welding Safety (Oxyacetylene)* □

Caution: Always read the instruction sheets provided with the equipment before operating the equipment.

Safety Precautions. Safe and effective use of flame tool equipment depends on the user fully understanding and carefully following practical, time-tested safety and operating instructions to prevent unnecessary injuries and costly property damage.

For adequate personal safety the user must be fully aware at all times that the torch flame can reach almost 6000 °F (3298 °C) and the workpiece can reach a high heat of almost 3000 °F (1649 °C), which may produce flying sparks, slag, fumes, and intense light rays, all of which can be harmful to the user or nearby persons without proper precautions and protection.

Practical proper "head-to-toe" protection includes hair and head coverage, welding- and cutting-type safety-tempered lens eye goggles (shade 5 is standard). Body coverage includes proper gloves and proper shoes. Avoid wearing anything flammable or clothing that has been exposed to flammables (oil, grease, solvents, etc.). Sparks and molten materials have a way of finding the unprotected areas, so be

*This section is used with the permission of Uniweld Products, Inc., Fort Lauderdale, Florida.

prepared before starting work by being properly protected.

Adequate ventilation must be provided, especially in confined areas, to remove harmful fumes and provide an adequate air supply for the user.

Important: Oxygen rapidly increases the burning rate of almost any ignited material, especially oil and grease; therefore, it must never be allowed to saturate a confined work area. Oxy–fuel or air–fuel concentrations in confined areas can also be hazardous and explosive. Never use a torch on any kind of container or pipe until it has been properly cleaned, purged, and vented.

Fire protection must be provided for the work area. The user must be fully aware of the impact of the torch flame, sparks, and molten materials on both close and surrounding areas [sparks can fly over 35 ft (10.64 m)]. Keep hoses clear of sparks and slag. Remove all flammables and carefully cover or shield with fireproof materials anything that could possibly catch fire, explode, or both.

Note: Concrete can chip explosively when overheated by slag and molten metal. Carefully check out the work area after the work has been completed for places where sparks or molten material could light and smolder. Douse doubtful areas with water or a fire extinguisher. Have a bucket of water and a proper fire extinguisher available in the work area at all times. A water bucket is also handy for leak-testing hoses and torches and for cooling the work.

Important: To avoid a dangerous pressure imbalance and cylinder contamination due to reverse flow of gases, do not allow cylinders (especially oxygen) to become empty in use. Check for adequate gas supplies before starting work. Also, install reverse-flow check valves in the welding hoses.

Checking Out the Equipment. New equipment in the original packaging has been inspected and tested and is ready for connecting. It is standard safety practice to check out and leak-test all connection points from the cylinder valves to the tips to assure safe, leaktight service before using the equipment. This checkup and leak testing is especially important after the equipment has been used and has been exposed to dust, dirt, oil, grease, and possible damage. The following guidelines for checking out, cleaning, and testing the equipment before using it are designed to cover normal working conditions. The proper procedures should be followed each time the equipment is set up and used.

Cylinders. The following procedures should be used when working with cylinders.

1. The cylinders must be secured to keep them from falling over.
2. Fuel gas cylinders must be used in the upright position.
3. Do not store cylinders and equipment in unventilated, confined spaces, closed vehicles, or near any source of heat or ignition.
4. The cylinder valves must be closed securely when not in use and when empty.
5. To avoid a dangerous pressure imbalance and cylinder contamination, do not allow cylinders, especially oxygen, to become completely empty in use. Check for adequate gas supplies before starting work. For added protection, install external-type reverse-flow check valves on regulators and/or torches. They are easy to install, test, and replace.
6. Valve protection caps should be replaced when the cylinders are empty, stored, or when the regulator is removed.
7. Do not expose cylinders to torch flames or electric arcs because they can seriously damage the cylinders.
8. Do not use a cylinder, full or empty, as a roller or support. The cylinder wall may be damaged and result in a rupture or an explosion.
9. Do not use a cylinder gas without a suitable pressure-reducing regulator.
10. Do not allow oil or grease to come in contact with cylinder valves, especially oxygen. Do not use if oil or grease is on the cylinder valve; return the cylinder to the supplier.
11. Do not use a cylinder with a leaking valve. Move it away from any source of ignition, preferably outdoors, and notify the supplier.
12. Handle all cylinders with care, and protect the valves from damage.

Regulators. The following procedures should be used when working with regulators.

1. Use the regulators for the gas or gases and services for which they were intended. Oxygen regulators must be used only for oxygen service.
2. Regulators and cylinder connections must be free of dirt, dust, grease, and oil. Remember that oxygen reacts violently with oil or grease. Inspect the regulator and cylinder valve connections for damaged threads, dirt, dust, oil, or grease.

Remove any dust and dirt with a clean cloth. Do not use the regulator if it is damaged or has oil or grease on it. Have your authorized repair station or Uniweld Products clean the regulator or repair any damage. Do not use cylinders with damaged or dirty valves. Return such cylinders to the supplier.

3. Do not crack fuel gas or oxygen cylinders near an open flame or any source of ignition. Make sure that you are in a well-ventilated area, and stand clear of the valve outlet when opening it.
4. Carefully crack (open briefly) the cylinder valves, one at a time, to clean the valves of dust, dirt, or foreign matter; then retighten.
5. Connect by attaching the inlet connection nut to the cylinder valve. Make sure that the threads engage properly. Tighten the nut with a wrench, but do not use excessive force. Excessive force could damage the nut and valve threads.
6. Turn the pressure-adjusting screw counterclockwise until the tension is fully released, to shut off the regulator. *Note:* The regulator should always be shut off when not in use to help in avoiding gas loss if the cylinder or the torch valves leak.
7. Do not stand directly in front of or in back of regulators when opening them. Open the cylinder valve slowly. Open fully only after the contents gauge stops moving. *Note:* Open the acetylene valves three-fourths to one turn only; all others are opened fully to seal the valve packing.

Welding Torches and Cutting Attachments. The following procedures should be used when connecting, testing, and operating the equipment.

1. Carefully check the torch handle head and valve hose connections, the cutting attachment head, and the hose and hose connections for damage, dirt, oil, or grease; clean, correct, or replace as necessary. Carefully check the welding tip, heating tip, and cutting-attachment cone end connection O-rings. There must be two O-rings in good condition to seal the gases properly. Do not use equipment with damaged or missing O-rings. If the hose is new or used and has been exposed to dirt and dust, blow it out safely in a well-ventilated area away from any sources of ignition before connecting it to the torch. Connect the hose to the regulator and use approximately 5 psi (34.45 kPa) regulator pressures for each gas.
2. Make all hose connections wrench-tight, except for the welding tip, and cutting-attachment

O-ring connections, which are firmly hand-tightened only. Cutting tips must be made wrench-tight (after being checked and cleaned, if necessary). Do not use damaged, plugged, or dirty tips. Cutting tip seats must fit leaktight in the cutting head.

3. Close all torch valves except the torch handle oxygen valve when the cutting attachment is connected. This valve must be fully open.

4. Turn the regulator adjusting screw clockwise until the proper delivery pressures for the tip size is obtained (see Table 1–1). *Warning:* Do not try to exceed the maximum pressure range of the delivery gauge because the gauge mechanism could be damaged. Do not use acetylene with a pressure of more than 15 psi (103.35 kPa). Above this pressure, acetylene becomes unstable.

5. Check for leaks. All connections from the gas cylinders to the welding or cutting tip must be leaktight and tested. Use a detergent solution or the water bucket test for torches and hoses. Fuel gas leaks can also be detected by their strong odor. Use the following leak detection procedure for each type of gas.

(a) Close the cylinder valve and turn the pressure-adjusting screw one turn counterclockwise. If the high-pressure gauge reading drops, there could be a leak in the cylinder valve to the regulator inlet connection. If the low-pressure gauge drops, there could be a leak in either the hose connection or the torch valve. *Note:* If the pressure on the high-pressure gauge drops and at the same time the pressure on the low-pressure gauge rises, there is a leak in the regulator seat. The regulator must be repaired by an authorized repair station or Uniweld.

(b) To check the welding or heating tip connection and torch handle, open the torch oxygen valve and plug the tip end. Check for leaks with a detergent solution or the water bucket test.

(c) To check the cutting-attachment connections, the torch handle oxygen valve must be fully

TABLE 1–1 Operational and Performance Data for Welding and Brazing Tips[a]

Welding Metal Thickness (in.)	Type 1, 4, 13, 17		Type 70		Type 08		Pressure (psig)[b]		Acetylene (scfh)[c]
	Size	Drill Size	Size	Drill Size	Size	Drill Size	Oxygen	Acetylene	
1/64	000	75	0	70	00	76	5	5	1
1/32	00	70	1	67	0	72	5	5	2
3/64	00	70	2	62	1	68	5	5	2
1/16	0	65	3	57	1	68	5	5	3
5/64	0	65	4	56	2	62	5	5	3
3/32	1	60	4	56	2	62	5	5	5
1/8	2	56	5	55	3	56	5	5	8
3/16	3	53	6	53	4	54	7	7	14
1/4	4	49	7	50	5	51	8	8	21
3/8	5	43	8	47	6	48	9	9	31
1/2	5	43	9	43	7	45	10	10	31
3/4	6	36	10	40	8	40	11	11	41
1	7	30			9	35	12	12	52
1 1/2	7	30			10	30	12	12	52
2	8	29					12	12	65
2 1/2	9	28					15	14	80
3	10	27					20	14	94
3 1/2	11	26					21	15	111
4	12	25					23	15	130

Source: Uniweld Products, Inc.

[a]Gas usage data are approximate, for estimating purposes. It will vary due to the materials, working conditions, and the operator's skill. The pressures on this chart are figured using 25 ft of hose. Additional lengths of hose require increased pressure, aproximately 3 psi per 25 ft of additional length. *Warning:* Acetylene pressure should never exceed 15 psig.

[b]Use same pressures for other fuel gases. Check valves require 2 to 3 psig increase in working pressure.

[c]Torch oxygen usage is 1.1 times acetylene with neutral flame—2 to 4 times for fuel gases.

Caution: The withdrawal rate of an individual acetylene cylinder should not exceed one-seventh of the cylinder contents per hour. Use an adequate-size cylinder or an adequate acetylene manifold system to supply the necessary volume for large usage. To avoid dangerous reverse flow of gases due to unbalanced pressures, *do not allow gas cylinder (especially oxygen) to become completely empty in use.*

SAFETY

open. Plug the cutting tip end and open the preheat oxygen valve and the cutting oxygen lever. Check for leaks using the water bucket test. *Note:* If the tube connections to the attachment body leak and cannot be corrected by tightening, the attachment must be repaired by an authorized repair station or Uniweld.

   (d) It is important to leak-test used hoses, especially any places that are frayed, worn, cut, burned, cracked, or damaged. Use the water bucket or detergent test. Repair or replace as necessary before using the hoses. *Do not* use the equipment until all the connections are leaktight. Always retighten, recheck, and retest any leaking connections.

6. Purge the hoses and equipment and check the pressures by briefly opening and closing the oxygen valve and then opening and closing the fuel gas valve. Be sure to vent the gases safely.

7. To light the welding or heating tips, open the fuel gas valve about one-half turn and quickly ignite the gas with a spark lighter (not matches). Adjust the fuel gas valve until excessive smoking stops. This is especially important for large tips and heating tips for adequate gas flow. Next, open the oxygen valve and adjust the flow until the white feathers disappear into a sharp cone (neutral flame).

8. *Important:* Correct fuel gas flows and pressures must be maintained at the tip to avoid flame backfiring (loud popping) or flashback (squealing) inside the tip. This can also be caused by plugged holes or valve seat damage. To stop the inside burning, quickly turn off the *oxygen valve first,* then the fuel gas valve. *Note:* The flame cannot burn inside the torch without oxygen. Cool the torch and check the tip and pressures before relighting.

Pressure Settings, Flame Adjustment, and Cutting Procedure for Cutting Attachments. Use the following steps when attempting these procedures.

1. Turn the torch oxygen handle valve fully open for a full supply of oxygen to the attachment (leave it full open when using the attachment). Close all other valves.

2. Set the correct regulator pressures for each gas used (see Table 1–2). For example, use not less than 5 psi (34.45 kPa) acetylene or fuel gas and 30 to 35 psi (206.7 to 241.15 kPa) oxygen for tip size 0 or 1 with approximately 25 ft (7.6 m) of hose. *Note:* The correct pressure settings are "working pressures" needed at the torch. Compensate for pressure drops by increasing the regulator pressure as necessary. Hoses over 25 ft (7.6 m) require an increased pressure of about 3 psi (20.67 kPa) per 25 ft (7.6 m) of hose.

3. Check the pressures with the desired gas flow conditions. Also purge the oxygen passages by briefly opening and closing the preheat oxygen valve and the cutting oxygen lever. Check the pressure and purge the fuel gas passages by briefly opening and closing the torch handle fuel gas valve. Vent the gases safely. *Note:* Purge the cutting oxygen before each torch lighting and use.

4. To light the cutting tip, open the torch handle fuel gas valve about one-half turn and quickly cup and light the tip with a spark lighter (not matches). Adjust the flame until it just leaves the tip end, then return the flame to the tip, but not with excessive smoke. This setting will maintain a stable preheat flame.

5. Open the preheat oxygen valve slowly until the white feathers disappear into the sharp blue-white flame cones. These are the correct neutral flame cones.

6. Press down and hold the cutting oxygen lever while slightly increasing the preheat oxygen to remove the feathers and restore the correct flame cones (Figure 1–1).

7. To start the cut, hold the preheat flame cones steady just above an edge until it glows bright orange, then press the cutting oxygen lever down slowly. If the metal is hot enough, the cutting action will start. To restart a cut, release the cutting lever and re-preheat the metal. Move the torch along steadily to avoid overheating and excess melting until the cut is completed.

8. *Important:* The correct fuel gas flow to the tip must be maintained to avoid flame backfiring (loud popping) or flashback (squealing) inside the attachment, which can also be caused by a

Picture shows correct flame cone shapes with cutting oxygen jet stream

FIGURE 1–1 Correct flame. (Courtesy of Uniweld Products, Inc.)

TABLE 1-2 Operational and Performance Data for Cutting Tips[a]

Metal Thickness (in.)	Tip Type 1-101 3-101 5-101	144 164	6290	Cutting Oxygen psig	scfh	Acetylene[b] psig	scfh	Speed (ipm)
1/8	000	00	000	20–25	25	5	5	28–32
1/4	00	0	00	20–25	36	5	6	25–30
3/8	0	1	00	25–30	60	5	8	24–28
1/2	0	1	0	30–36	65	5	10	20–24
3/4	1	2	1	30–40	85	5	13	17–20
1	2	2	1	36–50	140	6	16	15–20
1 1/2	2	3	2	40–50	160	7	18	12–17
2	3	3	3	40–56	180	9	22	12–15
2 1/2	3	4	3	45–56	230	10	26	10–13
3	4	5	4	45–60	280	10	30	9–12
4	5	5	4	45–60	360	12	34	9–11
5	5	6	4	50–65	420	12	38	7–9
6[c]	6	6	5	50–70	460	12	40	6–8
8[c]	6	7	5	55–75	600	14	44	5–8
10[c]	7	7	6	55–85	700	14	50	4–5
12[c]	8	8	6	55–96	900	14	56	3–5

[a]Check valves require 2 to 3 psig increase in working pressure.
[b]Use the same pressures for other fuel gases.
[c]Use 3/8-in. hose when using size 6 or larger tip or a cut of 6 in. or larger.

Cutting Tip Oxygen Pressure Guide:
    TYPE 1-101, 3-101, 5-101—use lower psig
    TYPE 144, 164—use middle psig
    TYPE 6290—use upper psig

Cutting Capacities (in.):
    Heavy-duty cutting torches: 12
    Heavy-duty cutting attachments: 8
    Medium-duty cutting attachments: 6
    Light-duty cutting attachments: 2

plugged, dirty, damaged, or loose cutting tip. To stop the inside burning, quickly turn off the preheat oxygen valve, then the torch fuel gas valve. *Note:* The flame cannot burn inside the torch without oxygen. Cool the cutting attachment and check the preheat gas flows and pressures. Before relighting, check the tip for plugged holes or seat damage. Clear and clean the holes with the proper-size tip cleaners. Do not use a tip with damaged seat or plugged holes. Contact your supplier if the problem continues. *Note:* Include the tip in use with the torch to be checked and repaired.

9. For piercing holes or cutting into confined spaces, the tip should be tilted over at an angle to avoid metal spatter from plugging the tip and blowing toward you when the cutting action starts. Press the lever down slowly, then return the tip upright and raise it slightly after the cut starts.

TABLE 1–2
Cutting Tip Orifice Sizes

| Cutting Tip Size | Type 1 101, 3 101, 5 101 | | Type 144, 164 | | | Type 6290A, 6290S | | |
	Cutting Oxygen	Preheat	Cutting Oxygen	Preheat 144	Preheat 164	Cutting Oxygen	Preheat A	Preheat S
000	71	74				68	72	71
00	67	74	68	65	66	64	72	71
0	60	74	62	63	65	60	72	69
1	56	71	56	60	61	56	72	65
2	53	68	54	59	60	52	72	59
3	50	66	52	57	59	48	69	58
4	45	64	49	56	57	42	69	56
5	39	61	45	56	56	35		55
6	31	58	41	55	56	30		55
7	28	56	34	55	55			
8	20	53	30	55	55			

Source: Uniweld Products, Inc.

10. Do not choke down or starve the preheat flames to try to reduce the heat for cutting thin sections. Always use the full correct flame with the tip tilted over so that the preheat flames hit the surface at an angle which may be increased or decreased as needed to vary the heat. Move the tip fast enough to avoid overheating and excess melting. (Use the correct small tip size for the thin metal, to avoid overheating and over-melting; see Table 1–3.)

11. To shut down when your operations are finished, shut off the oxygen valve first, then the torch fuel gas valve (shutting off the fuel gas valve first can cause backfire into the torch). Close the cylinder valves, then open the torch valves separately and allow all pressure to drain

TABLE 1–3 Operational and Performance Data for Heating Tips[a]

Heating: Oxygen/Acetylene	Tip Size	Pressure [psig] Acetylene	Pressure [psig] Oxygen	Consumption [scfh] Acetylene	Consumption [scfh] Oxygen
Type 11 tip	5	10	15	25	25
	6	10	20	45	50
	8	12	25	85	90
	10	15	40	100	120
	12	15	60	150	175
	15	15	60	225	250
Types 28 and 29 tips	2	10	12	10	12
	4	10	15	20	25
	6	12	20	45	50
	8	15	25	85	90

All Fuel Gases Not Acetylene	Tip Size	Pressure [psig] Oxygen	Pressure [psig] Fuel Gas	Consumption [scfh] Oxygen	Consumption [scfh] Fuel Gas
Type 45 tip	10	90	25	410	175
	15	110	35	750	300
	20	125	45	1050	450

Source: Uniweld Products, Inc.

[a]Use adequate pressures on large heating tips to avoid backfire and flashback from low gas flows; do *not* allow heating tips to come in contact with work. Check valves require 2 to 3 psig increase in working pressure.

Welding Safety (Oxyacetylene)

from each regulator [delivery gauges to 0 psi (0 kPa)]. Close the valves on the torch and then release all tension on the regulator pressure-adjusting screw by turning it counterclockwise until it turns freely. Closing the regulator helps to avoid gas loss if a cylinder or equipment valve leaks.

12. Use the correct Spiral Tip Cleaners to keep the cutting, welding, and heating tips in effective and efficient operating condition, ready for every job.

Instructions for Cutting, Welding, and Brazing. Use the following instructions when performing any of these procedures.

Cutting Steel. Position the preheat cones just above the work [about ⅛ in. (0.317 cm)] and hold it steady until the metal turns bright orange. Then slowly press the cutting lever and move the torch along steadily to avoid overheating and excess melting. To restart the cut, release the lever and repreheat the metal before pressing the lever again.

Piercing. For piercing holes or cutting into confined spaces, the tip should be tilted over to keep the metal spatter from plugging the tip. Return the tip upright after the cut is started.

Cutting Thin Steel. Tilt the tip over so that the preheat flames hit the surface at an angle that can be increased or decreased as needed to vary the heat. Move the tip fast enough to avoid overheating and excess melting of the metal. Use the correct tip size for the job and only 15 to 20 psi (105 to 140 kPa) cutting oxygen.

Welding Steel. *Note:* Oxyacetylene must be used for welding steel. See Table 1–1 for the proper tip size. Clean and assemble the parts with good fit-up. Position the flame cone just above the surface until a puddle forms and penetrates well into the joint. Then dip the rod into and out of the puddle to melt it as needed to fill the joint and move the puddle along the joint. When welding metal under ⅛ in. (0.317 cm), melt the rod ahead of the flame. Tilt the tip to decrease or increase the heat to the joint; also slightly withdraw the flame as needed to avoid overheating the puddle. For metal over ⅛ in. (0.317 cm), bevel the edge of the joints approximately 30° for good weld penetration.

Brazing. Brazing uses metals that melt and flow below the melting point of the base metal. Brazing temperatures range from 1150 to 1600°F (620 to 870°C). When brazing, the proper alloy filler rod and flux must be used. The joints must be clean and assembled with good fit-up. Be sure to provide proper ventilation to remove any harmful fumes. Heat the tip of the brazing rod and dip it into the flux (not necessary for prefluxed rods). Preheat the metal to a dull red color and apply the fluxed rod to the base metal and melt off a small amount of the rod. If it flows on the base metal, the heat is right; continue to heat and flow the rod on the joints. *Note:* Phos-copper brazing rod does not need fluxing when used on copper. Flux must be used on dissimilar metals. Do not use phos-copper on iron or steel, because a brittle joint can result.

Brazing Silver. Most silver alloys flow at lower temperatures [about 1200°F (612°C)] than brass alloys. The joints must be clean and have a good close fit-up [joints with maximum strength have about 0.002 to 0.003 in. (0.05 to 0.07 mm) fitting]. The joint must be coated with flux and the joint heated evenly. When the flux bubbles and melts clear, start applying the rod to the joint. If it flows, the heat is right and the alloy will flow quickly in and around the joint. Do *not* overheat the joint. Remove the flame when the alloy has flowed into the joint.

Soft Soldering. Most soft solders melt at around 400 to 500°F (205 to 260°C). The joints must be clean, have a good fit-up, and be well covered with soft solder flux. Heat the joint until the flux boils, then apply the solder. If it flows, the heat is right and the solder will flow quickly in and around the joint. Do *not* overheat the joint. Remove the flame when the solder has flowed into the joint.

Safety and Setup Checklist. Use the following steps to ensure a safe setup.

1. Review the previous information before using the equipment and use the precautions and procedures outlined.
2. *Safety precautions.* Provide for personel protection, ventilation, and fire protections and precautions.
3. *Cylinders.* Cylinders should be secured upright and should have adequate gas supply to avoid dangerous empty cylinder conditions. Provide for safe storage, with valves closed when not in use or empty. Use protective caps as needed.
4. *Regulators.* Check for clean, leaktight connections, turned off before opening cylinder valve slowly and closed after work is completed to avoid leaks.
5. *Torches and tips.* Make all connections wrench-

tight and leak-test before using the equipment. Use proper operating pressures and clean, efficient tips. Purge the oxygen before each torch use.

6. *Operations.* Follow proper lighting and shutdown procedures. Turn fuel gas *on first* and *off last.* Check for proper full neutral flame adjustment; proper tip tilt to decrease or increase the heat to the work for cutting, welding, or brazing; and proper speed to avoid excess heating and melting of thin material when cutting or welding. Purge the oxygen before each torch lighting and use.

7. *Shutdown procedure. Shut oxygen off first,* fuel gas off last (remember that to use the same procedure to stop flashback and to avoid backfire, *the flame cannot burn in the torch without oxygen*). Close the cylinder valves, drain off each gas separately, close the regulators and avoid leaks, close all torch valves, and store cylinders and equipment away from heat and ignition sources in a well-ventilated area.

8. Use the proper-size tip cleaners to keep the tips effective and efficient for every job.

Caution: The withdrawal rate of an individual acetylene cylinder should not exceed one-seventh (15%) of the cylinder contents per hour. Use a cylinder of adequate size or an adequate acetylene and oxygen manifold system to supply the necessary volume for large usage to avoid dangerous reverse flow of gases due to unbalanced pressures: *Do not allow gas cylinders (especially oxygen) to become completely empty while in use.*

☐ Safety in Handling Refrigerants ☐

In operating refrigeration machinery, it is necessary that only the type of refrigerant for which the machine was designed be used in it. When other types of gases are used in the system, it will not operate as it was designed. In some cases the pressures may become excessive.

Personal Protection. Most of the refrigerants used today are nontoxic and present no personal hazard unless there is a sufficient concentration to displace the oxygen in the air inside the space. Without oxygen the air will not support life. The refrigerant vapors are three to five times heavier than air and will therefore tend to collect in a strong concentration near the floor or in low places and displace the air. The stronger the concentration of refrigerant, the less air there is to breathe, and a person can actually suffocate.

When refrigerant is purged from a system there are three precautions that should be exercised: (1) make certain that it is vented out of doors; (2) be sure that the space is well ventilated; and (3) be sure that all refrigerant vapors are removed from the space before beginning work.

The deliberate inhalation of refrigerant vapors can cause a cardiac sensitization and death. This is an extremely dangerous act because death can occur without warning. When entering tanks or other places where there is the possibility of a strong concentration of refrigerant vapors, wear an air mask, and use the buddy system and a lifeline.

At times it is difficult to smell refrigerant vapors except in strong concentration; they have a faintly sweet odor that is difficult to detect. When you are working in an enclosed space where there has been a refrigerant leak or refrigerant has been purged into the air and you begin to feel a little dizzy, immediately get out of the space and into fresh air before it is too late.

Should refrigerant be drawn into the mouth, do not induce vomiting. When vomiting is induced, there is a possibility that the refrigerant can be drawn into the lungs. Should refrigerant be drawn into the lungs, get medical attention immediately. This condition could cause chemical pneumonitis, pulmonary edema, and hemorrhage.

Most of the refrigerants used in modern equipment are in vapor form at room temperature. These vapors have very little, if any, effect on the skin or the eyes. However, those refrigerants that remain a liquid at room temperature, such as Refrigerant-11 (R-11) and R-113, have a tendency to dissolve the protective fat from the skin, which will cause dryness and some irritation after prolonged contact or repeated contact with the liquid form. Any time there is the possibility of coming in contact with liquid refrigerant, protective clothing should be worn. Always wear protective eye goggles when working with refrigerants to prevent their entrance into the eye. Should liquid refrigerant get into the eye, flush the eye with water and get medical attention immediately. Liquid refrigerant will freeze the moisture in the eye and possibly cause blindness.

Should R-12, R-22, or any of the medium-temperature refrigerants come into contact with the skin, frostbite is probable. If treatment is started within 20 to 30 minutes after contact with a liquid refrigerant, soak the exposed area in lukewarm water. Never use hot or cold water. If the situation prevents starting treatment after 30 minutes has passed, apply a light coat of petroleum jelly to the area. If the frostbite occurred in a place where it would be difficult to apply the petroleum jelly, apply a light bandage to the affected area. Seek medical attention as soon as possible to reduce the chance of gangrene.

Handling Refrigerants under Pressure. Following are some of the precautions of handling refrigerants under pressure.

1. Do not overfill a refrigerant container. Should an overfilled container be subjected to temperatures greater than 125 °F (51.5 °C), the container may become "liquid full," causing an immediately dangerous hydrostatic pressure to build up inside the container. The container may then rupture.
2. Do not overheat or store a refrigerant container in the direct sunlight. There is the possibility of a buildup of hydrostatic pressure inside a correctly filled container when heated above 125 °F (51.5 °C).
3. Never connect a refrigerant container to the discharge side of a refrigeration system. The pressure inside the container may rise above the capacity of the relief device and possibly cause the container to rupture. This is very possible with the 14-oz (396.9-g) R-12 cans used to charge automobile air conditioners. It is possible that they will explode when connected to the discharge side of the system.
4. Cylinder abuse is the most common cause of pressure-related problems with refrigerants. Four recommendations are as follows:
   (a) Never heat cylinders above 125 °F (51.5 °C).
   (b) Never tamper with the valves or the safety devices.
   (c) Never refill disposable cylinders with anything or convert them into compressed air tanks. The empty cylinders should be vented and discarded. Reusing disposable cylinders is a dangerous practice because of the possibility of corrosion and weakening of the cylinder walls. The corrosion is greater when it is caused by water vapor in the refrigerant or in the compressed air. The greatest danger

is that the damage may not be visible until the cylinder explodes. It is then too late.
   (d) Disposable cylinders should always be stored in a dry place to prevent rusting and corrosion of the cylinder.

Handling Refrigerant Cylinders. The following are some of the more important procedures for handling refrigerant cylinders.

1. Make certain that the cylinder is properly connected to the system.
2. Open the valves very slowly to avoid rupturing a hose or ruining a gauge.
3. Store the cylinders in a dry place to protect them from rusting.
4. Do not tamper with the pressure-relief devices. To do so may cause the cylinder to rupture when subjected to excessive pressures.
5. Do not drop or otherwise abuse the cylinders.
6. Do not use disposable refrigerant cylinders for air tanks. The moisture in the air will cause the cylinder to corrode from the inside out. It may then rupture when least expected and cause serious injury.
7. Do not refill disposable cylinders. They are designed for one time use only, and any moisture in the refrigerant will cause corrosion on the inside of the cylinder.
8. Do not force connections. This weakens the connection that may blow out when subjected to pressure and cause serious injury.
9. Do not heat the cylinder above 125 °F (51.5 °C). To do so may cause the cylinder to become liquid full, and a hydrostatic pressure will build up rapidly and possibly rupture the cylinder.
10. Make certain that the refrigerant, the cylinder color, and the label match.
11. The caps that are provided for valve protection should always be on the valve except when the cylinder is in use.
12. Never mix refrigerant gas in a cylinder.
13. When a cylinder is emptied, immediately close the cylinder valve to prevent the entrance of moisture, air, and dirt into the cylinder.
14. Never use refrigerant cylinders for rollers, supports, or for any purpose other than to carry refrigerant.
15. Never attempt to repair or alter cylinders or valves.

Brazing or Welding on Refrigeration System Piping. The following are some of the precautions used when attempting these operations:

1. Remove all the refrigerant from the part of the system that is to be heated.
2. Be sure to purge all refrigerant out of doors.
3. Provide plenty of ventilation to the space. Use an auxiliary vent fan if necessary.
4. Purge the part of the system that is to be heated with nitrogen to force as much of the refrigerant vapor out of the system as possible.
5. Leave the system open during the welding or brazing process to prevent a buildup of pressure inside the system and decomposition of the refrigerant. Decomposed refrigerant may cause toxic and irritating compounds, such as hydrogen chloride and hydrogen fluoride.
6. Do not breathe the vapors produced when welding or brazing on refrigerant lines. These vapors are acidic and will irritate your nose and throat. Anyone that is exposed to these vapors should be taken to fresh air immediately and given medical attention.
7. Some silver brazing alloys contain cadmium. These alloys are very dangerous to use. Most silver brazing alloys contain large quantities of cadmium, and when thse alloys are melted they emit cadmium oxide fumes into the air. These fumes are very toxic and must be avoided. An increase in the temperature above the molten state will cause an increase in the amount of fumes produced. Because oxyacetylene gas burns at a temperature of about $6000\,°F\ (3298\,°C)$, higher concentrations of the cadmium oxide fumes are produced than with other types of welding and heating fuels.

Caution: You cannot smell cadmium oxide fumes, thus making them even more dangerous. A lethal dose does not necessarily need to be irritating enough to cause discomfort until after enough has been absorbed by a worker to place his or her life in immediate danger. Some of the more common symptoms of cadmium poisoning are headache, fever, chills, irritation of the throat, vomiting, weakness, and diarrhea. These symptoms may not appear for many hours after the exposure to the fume. The primary injury is to the respiratory system.

Those who breathe the air within 5 ft (1.52 m) of where the silver brazing is done are in the most dangerous area. When these materials are used in production work, precautions are taken to protect the worker from the fume. However, when a refrigeration or air conditioning technician is working in the field these precautions are not available. It is therefore up to technicians to protect themselves. Lunches should not be stored or eaten in the area where silver brazing with these materials is done. Workers should be sure to wash both their hands and face before eating, smoking, or leaving work.

The nice thing is that today no one must use these types of silver brazing materials. There are materials available that do not contain cadmium. However, technicians must make certain that the types they are using are safe. Several companies manufacture cadmium-free silver-brazing materials.

☐ Safety in Working with Heating Equipment ☐

Heating equipment is probably the most dangerous of all the equipment used in the industry, especially natural and liquefied petroleum (LP) gases. Oil is somewhat dangerous, but it is not quite as explosive as the gases. The danger is in the combustion process that provides heat for the building. Electricity is probably the least dangerous of all the heat sources. However, all these heat sources must be treated with respect. Therefore, service and installation technicians must know the combustion process and practice safety procedures to prevent loss of life, property, and equipment.

The following are some of the more common safety procedures to follow when working on heating equipment.

1. Do not tamper with the safety controls. If a problem arises with one, replace it. Tampering may cause delayed ignition, which could cause an explosion, or an overheated condition, which could cause a fire.
2. Do not light the pilot until the firebox has been cleared of all unburned fuel.
3. Follow the manufacturer's directions carefully. Much time and money has been spent in developing these directions, and they should be heeded.
4. Always follow the building safety codes effective in your locality.

5. Be sure the unit is properly grounded electrically.

6. Do not make adjustments without knowing what the results will be.

7. Do not alter the venting system in any way.

8. Never reduce the size of the vent pipe.

9. Increase the size of the vent pipe only when it is needed to provide proper ventilation of the products of combustion.

10. Be sure the proper ventilation air is supplied in the equipment room.

11. Never allow the unit to be operated when a bad heat exchanger is suspected. The products of combustion will enter the building and possibly cause death or illness of the occupants. A bad heat exchanger should be suspected when the occupants of a building complain of having a headache that is relieved when they leave the building for a period of time but returns after being in the building for a short time.

12. Never leave a furnace in operation if it is not functioning properly. To do so is inviting an explosion, fire, or other damage to the building and/or its occupants.

13. Always check the venting system and remove any obstructions before lighting the burners. If a furnace is fired with a restricted vent, the products of combustion will be forced into the building.

14. When fuel oil is used as the source of heat, it must be properly stored and piped to the unit without leaks in the piping.

15. Clean up all oil spills immediately.

16. All safety devices must be in excellent working condition and be properly adjusted. Never leave a safety device bypassed.

17. It is never a good idea to work on ''hot'' electrical systems. However, at times it is necessary to find a problem with the equipment. Take precautions to avoid becoming grounded and receiving an electrical shock.

18. The equipment and its components must be installed according to the local, state, and national electrical codes. The electrical circuits should be installed by a licensed electrician. If there is a local licensing code for heating, refrigeration, and air conditioning technicians, it must be adhered to. The equipment should have Underwriters' Laboratory (UL) approval.

19. Leak testing should be done with soap bubbles or a liquid leak detector. Never check for leaks with an open flame. If a leak is present, there will probably be a flame out of control.

20. Always allow the furnace to stand open for at least 5 minutes before lighting the pilot or the burner. This is to allow any accumulation of fuel to escape to the atmosphere without an explosion.

21. An accumulation of fuel oil or a gas odor is a signal that there is a problem and the technician should proceed with caution.

22. When working on a hot water or a steam heating system, remember that the system and its contents are under pressure. To rupture a line or a component will result in hot water or steam escaping from the system. If a person is standing near the leak, a severe burn is quite possible.

23. Always exercise care when working around heating equipment to prevent burns because almost all of the surfaces are hot when the equipment is operating.

24. When in doubt about the safe operation of a heating unit, check the vent products with one of the meters which are available for this purpose. Be safe, not sorry.

25. When servicing heating equipment, before leaving the job, check all safety devices to make certain that they are operating properly.

☐ First Aid for Victims of Electric Shock ☐

Air conditioning and refrigeration technicians are involved in the installation, service, and maintenance of electrical, electronic, and mechanical equipment in which dangerously high voltages are present. This work is often done in confined spaces. Among the hazards of this work are electrical shock, electrical fires, harmful gases which are sometimes generated by faulty electrical and electronic devices, and injuries which may be caused by the improper use of tools.

Because of these dangers, technicians should formulate safe and intelligent work habits since these are fully as important as knowledge of electronic equipment. One primary objective should be to learn to recognize and correct dangerous conditions and avoid unsafe acts. These persons should also know the proper procedures for dealing with fires of an electrical origin, for treating burns, and for giving artificial respiration to persons suffering from electric shock. In some cases, artificial respiration may have to be accompanied by cardiac massage to restore heart beat.

Electric Shock. Electric shock may cause burns of varying degree, the stoppage of breathing and unconsciousness, ventricular fibrillation or cardiac arrest, and death. If a 60-hertz alternating current is passsed through a person from hand to hand or from hand to foot, the effects when current is gradually increased from zero are as follows:

1. At about 1 milliampere (0.001 ampere) the shock will be felt.
2. At about 10 milliamperes (0.01 ampere) the shock will be severe enough to paralyze muscles and a person may be unable to release the conductor.
3. At about 100 milliamperes (0.1 ampere) the shock is usually fatal if it lasts for 1 second or more. *It is important to remember that, fundamentally, current, rather than voltage, is the criterion of shock intensity.*

It should be clearly understood that the resistance of the human body will vary. That is, if the skin is dry and unbroken, body resistance will be quite high, on the order of 300,000 to 500,000 ohms (Ω). However, if the skin becomes moist or broken, body resistance may drop to as slow as 300 Ω. Thus a potential as low as 30 volts (V) could cause a fatal electrical shock. Therefore, any circuit with a potential in excess of this value must be considered dangerous. *It should be noted that the intentional taking of an electrical shock should be avoided.*

Care of Shock Victims. Electric shock is a jarring, shaking sensation resulting from contact with electric circuits or from the effects of lightening. The victim usually experiences the sensation of a sudden blow, and if the voltage is sufficiently high, unconsciousness. Severe burns may appear on the skin at the place of contact; muscular spasm can occur, causing a person to clasp the apparatus or wire that caused the shock and be unable to turn loose. Electric shock can kill its victim by stopping the heart or by stopping breathing, or both. It may sometimes damage nerve tissue and result in a slow wasting away of muscles that may not become apparent until several weeks or months after the shock is received.

The following procedure is recommended for the rescue and care of electrical shock victims.

1. Remove the victim from electrical contact at once, but *do not endanger yourself*. This can be done by:
   (a) Throwing the switch, if it is nearby.
   (b) Using a dry stick, rope, leather belt, coat, blanket, or any other nonconductor of electricity.
   (c) Cutting the cable or wires to the apparatus, using insulated tools, while taking care to protect your eyes from the flash when the wires are cut.
2. Determine whether the victim is breathing. If so, keep the person lying down in a comfortable position. Loosen the clothing about his or her neck, chest, and abdomen so that the person can breathe freely. Watch the victim carefully and protect him or her from exposure to cold.
3. Keep the victim from moving about. After shock, the heart is very weak, and any sudden muscular effort or activity on the part of the patient may result in heart failure.
4. Do not give stimulants or opiates. Send at once for a medical person and do not leave until the patient has adequate medical care.
5. If the victim is not breathing, it will be necessary to apply artificial respiration without delay, even though the person may appear to be lifeless.

Artificial Respiration. Artificial respiration, the mechanical promotion of breathing, is used to resuscitate persons whose breathing has stopped as a result of electric shock, drowning, asphyxiation, strangulation, or the presence of a foreign body in the throat. To revive persons suffering from electrocution resulting in stoppage of breathing, begin artificial respiration immediately. Send another person for competent medical aid. If there is any serious bleeding, stop it first, but do not waste time on anything else. Seconds count—the longer you wait to begin, the less are the victim's chances.

The purpose of artificial respiration is to provide a method of air exchange until natural breathing is reestablished. Artificial respiration should be given only when natural breathing has stopped; it must not be given to any person who is breathing naturally without assistance. Do not assume that a person's breathing has stopped merely because he or she is unconscious or has been rescued from the water, from poisonous gas, or from direct contact wtih an electric wire. *Remember: Do not give artificial respiration to a person who is breathing naturally.*

To check for breathing, *look* for chest movement or put your ear next to the victim's nose and mouth to *listen* and *feel* for breath. If the victim needs assistance in breathing, first clear the mouth of any food particles, foreign objects, or loose dentures. Do this by sweeping the mouth clean with your fingers.

When the victim becomes unconscious, the muscles relax and the tongue may slide back into the airway blocking it (Figure 1–2). To open the airway in this situation, the head or jaw tilt technique is recommended.

Head Tilt. The head tilt technique is performed with the victim on his or her back and the rescuer kneeling at the person's side—head and back level with the victim. Place one hand on the victim's forehead and the other under the neck. Apply pressure on the forehead and at the same time lift the neck. Tilting the head in this manner opens the airway (Figure 1–3).

Jaw Tilt. A second technique for opening the airway is the jaw tilt. This technique is accomplished by kneeling by the top of the victim's head and placing your fingers behind the angles off the lower jaw. Forcefully bring the jaw forward while tilting the head backward. Separate the lips with your thumbs to allow breathing through the mouth as well as the nose. If no air is exchanged, start artificial respiration immediately.

Mouth-to-Mouth Respiration. If the victim is not breathing naturally, start artificial respiration using the mouth-to-mouth technique. This method is as follows:

1. Clean the mouth of all foreign material.
2. Place the victim on his or her back (face up) (Figure 1–4). Lift the person's neck and place a folded coat or blanket under the shoulders. Tilt the victim's head as far back as possible.
3. Assume a position at the side of the victim's head. Grasp the jaw with the thumbs in one side of the mouth and pull it forward (Figure 1–5). Maintain this position to keep the air passage open.
4. Pinch the victim's nostrils shut with the thumb and forefinger (Figure 1–6). Take a deep breath and place your mouth over the person's mouth and your thumb, creating a tight seal. Blow into

FIGURE 1–4 Placement of victim.

FIGURE 1–2 Blocked airway.

FIGURE 1–5 Opening of air passage.

FIGURE 1–3 Head tilt open airway.

FIGURE 1–6 Inflation.

SAFETY

FIGURE 1-7 **Exhalation.**

the victim's mouth until you see his or her chest rise.

5. Remove your mouth and let the victim exhale passively (Figure 1-7). Inflate the lungs at a rate of about 12 times per minute for adults and 20 times per minute for children. Continue until the victim begins to breathe normally.

Keep the following procedures in mind:

1. Keep the victim sufficiently covered so that he or she will not suffer from exposure.
2. Do not allow bystanders to crowd around the victim and interfere with the first-aid treatment.
3. Do not attempt to give the victim anything to drink while the person is unconscious, since this may cause choking.
4. Artificial respiration must be continued for at least 4 hours, unless the natural breathing is restored or a medical person declares the person dead. Some people have been saved after as much as 8 hours of artificial respiration.
5. To assist shallow natural breathing, blow in at the moment the victim inhales and take your mouth off quickly when the person exhales.

Continue artificial respiration until the victim breathes naturally or professional medical aid arrives!

Mouth-to-Nose Respiration. Mouth-to-nose respiration is effective when the victim has extensive facial or dental injuries or is very young, as it permits an effective air seal. To administer this method, place the heal of one hand on the victim's forehead and use the other to lift the person's jaw. After sealing the victim's lips, take a deep breath, place your lips over the victim's nose, and blow. Observe the chest for movement and place your ear next to the victim's nose to listen, or feel, air exchange. Again you must continue your efforts at the rate of 12 ventilations per minute, or one breath

every 5 seconds, until the victim can breathe without help.

Cardiac Arrest (Loss of Heartbeat). If the victim of an electrical shock has no heartbeat (no detectable pulse at wrist or neck, and heart cannot be heard), the person has suffered a cardiac arrest. In this case, severe brain damage may occur if circulation is not restored within 4 minutes. Closed cardiac massage is the recommended procedure to be used in such cases. However, a person should not attempt to perform closed cardiac massage unless he or she has been instructed in the proper procedure by competent personnel (a doctor or someone certified in such treatment) and then only as a last resort, since serious or even fatal injury may result if improperly applied.

☐ Safety in Working with Electricity and Electronics ☐

Working on Energized Circuits. Insofar as practical, a technician should not undertake working on energized circuits and equipment. However, if it should become necessary, as when making operational adjustments, one should carefully observe the following safety precautions:

1. Ensure that you have adequate illumination. If you are to perform the job properly and safely, you must be able to see clearly.
2. Ensure that you are insulated from ground by an approved type of rubber mat, or layers of dry canvas and/or wood.
3. Where practical, use only one hand, keeping the other behind you or in your pocket.
4. If the system voltage exceeds 150 V, rubber gloves should be worn.
5. An assistant should be stationed near the main switch or circuit breaker, so that the equipment may be deenergized immediately in case of an emergency.
6. A person qualified in first aid for electrical shock should be standing by during the entire operation.
7. Do *not* work alone.
8. Do *not* work on any type of electrical apparatus when wearing wet clothing or if the hands are wet.
9. Do *not* wear loose or flapping clothing.

10. The use of thin-soled shoes and shoes with metal plates or hob nails is prohibited.

11. Flammable articles, such as celluloid cap visors, should not be worn.

12. All rings, wristwatches, bracelets, and similar metal items should be removed before working on the equipment. Also, ensure that clothing does not contain exposed metal fasteners such as zippers, snaps, buttons, and pins.

13. Do not tamper with interlock switches; that is, do not defeat their purpose by shorting them or blocking them open.

14. Ensure that the equipment is properly grounded before energizing.

15. Deenergize the equipment before attaching alligator clips to any circuit.

16. Use only approved meters and other indicating devices to check for the presence of voltage.

17. Observe the following procedures when measuring voltages in excess of 300 V:
    (a) Turn off the equipment power.
    (b) Short circuit or ground the terminals of all components capable of retaining a charge.
    (c) Connect the meter leads to the points to be measured.
    (d) Remove any terminal grounds connected previously.
    (e) Turn on the power and observe the voltage reading.
    (f) Turn off the power.
    (g) Short circuit or ground all components capable of retaining a charge.
    (h) Disconnect all meter leads.

18. On all circuits where the voltage is in excess of 30 V and where metal is present, the worker should be insulated from accidental ground by the use of an approved insulating material. The insulating material should have the following characteristics:
    (a) It should be dry, without holes, and should not contain conducting materials.
    (b) The voltage rating for which it is made should be clearly marked on the material, and the proper material should be used so that adequate protection from the voltage can be possible.
    (c) Dry wood may be used, or as an alternative, several layers of dry canvas, sheets of phenolic insulating material, or suitable rubber mats.
    (d) Care should be exercised to ensure that

moisture, dust, metal chips, and so on, which may collect on insulating materials is removed at once. Small deposits of such materials can become electrical hazards.
    (e) All insulating materials on machinery and in the area should be kept free of oil, grease, carbon dust, and so on, since such deposits destroy insulation.

Using a Safety Shorting Probe. Always assume that there is voltage present when working with circuits having high capacitance even when the circuit has been disconnected from its power source. Therefore, capacitors in such circuits should be discharged individually, using an approved type of shorting probe. High-capacity capacitors may retain their charge for a considerable period of time after having been disconnected from the power source.

When using the safety shorting probe, always be sure first to connect the test clip to a good ground (if necessary, scrape the paint of the grounding metal to make good contact). Then hold the safety shorting probe by the insulated handle and touch the probe end of the shorting rod to the point to be shorted out. The probe end is fastened so that it can be hooked over the part or terminal to provide a constant connection by the weight of the handle alone. Always take care not to touch any of the metal parts of the safety shorting probe while touching the probe to the exposed "hot" terminal. It pays to be safe; use the safety shorting probe with care.

Another method of discharging a system or a component is to use a resistor with at least 20,000 Ω resistance equipped with pieces of wire connected to the ends. Touch the wire ends to the points to be discharged. Be careful not to touch the shorting wires while they are in contact with the exposed terminals to prevent electrical shock.

Working on Deenergized Circuits. When any piece of electronic equipment is to be repaired or serviced, certain general safety precautions should be observed. They are as follows:

1. Remember that electric and electronic circuits often have more than one source of power. Take time to study the schematics or wiring diagrams of the entire system to ensure that all sources of power have been disconnected.

2. If pertinent, inform all persons regarding the circuit on which the work will be performed.

3. Use only one hand while turning switches either on or off.

4. Safety switches, such as interlocks, overload relays, and fuses should never be altered or disconnected except for replacement. In addition, they should never be changed or modified in any way without knowledge of the danger involved.

5. Fuses should be removed and replaced only after the circuits have been deenergized. When a fuse blows, the replacement should be of the same type and have the same current and voltage ratings. A fuse puller should be used to remove and replace cartridge-type fuses.

6. All circuit breakers and switches from which power could possibly be supplied should be secured (locked if possible) in the open or off (safety) position and tagged.

7. After the work has been completed, the tag, or tags, should be removed *only* by the person who signed it when the work began.

8. Keep clothing, hands, and feet dry if at all possible. When it is necessary to work in wet or damp locations, use a dry platform or wooden stool to sit or stand on, and place a rubber mat or other nonconductive material on top of the wood. Use insulated tools and insulated flashlights of the molded type when required to work on exposed parts.

Electrical Fires. Fire is a hazard. The three general classes of fires are A, B, and C. Class A fires involve wood, paper, cotton and wood fabrics, rubbish, and the like. Class B fires involve oil, grease, gasoline and aircraft fuels, paints, and oil-soaked materials. Class C fires involve insulation and other combustible materials in electrical and electronic equipment.

Electrical or electronic equipment fires result from overheating, short circuits (parts failure), friction (static electricity), or radio-frequency arcs. Also, any equipment may be ignited by exposure to nearby class A or B fires. Since class C fires involve electrical circuits, electrical shock is an added hazardous condition. Thus, whenever possible, any electrical equipment exposed to a class A or class B fire, or actually ignited by such a fire, should be deenergized immediately. If the equipment cannot be deenergized completely, protective measures must be enforced to guard against electrical shock. In addition, extinguishing agents other than gases will contaminate delicate instruments, contacts, and similar electrical

devices. Therefore, carbon dioxide (CO_2) is the preferred extinguishing agent for electrical fires because it does not conduct electricity and rapidly evaporates, leaving little or no residue. Thus its use reduces the possibility of electrical shock to personnel and damage to delicate equipment as a result of contamination.

A dry chemical type of extinguishing agent, composed chiefly of potassium bicarbonate (purple-K), is suitable for electrical fires because it is also a nonconductor and therefore protects against electrical shock. However, damage to electrical or electronic parts may result from the use of this agent. The dry chemical extinguisher is similar in appearance to the CO_2 extinguisher.

A solid stream of water must never be used to extinguish electrical fires in energized equipment because water usually contains minerals that make it conductive. (Pure distilled water is not a good electrical conductor and may be used in an emergency on small electrical fires.) If circumstances demand the use of fresh water, fog produced by a special hose nozzle (fog head or tip) may be used in electrical or electronic equipment spaces. However, even though the fog is a fine diffusion or mist of water particles, with very little conductivity, there is still some danger of electric shock, unless the equipment is completely deenergized. Also, fog condensation on electrical equipment frequently damages the components, and this damage must be corrected after the fire. Exercise care when using the fire hose. Pressure at the fireplug may be as much as 100 psi, and an unrestrained fire hose may result in whiplash.

Foam is not recommended for electrical fires because of equipment damage and the possible shock hazard to personnel; however, if necessary, foam may be used only on *deenergized* circuits. When a blanket of foam is applied to a burning substance, the foam smothers the fire; that is, it cuts off the air supply to the burning substance. Thus the supply of oxygen necessary to support combustion is isolated from the substance, and the fire will be extinguished.

The following general procedure is used for fighting an electrical fire:

1. Promptly deenergize the circuit or equipment affected.

2. Sound an alarm. Give the fire location and state what is burning. If possible, report the extent of

the fire, that is, what its effects are upon the surrounding area.

3. Control or extinguish the fire using a CO_2 fire extinguisher.
4. Avoid prolonged exposure to high concentrations of carbon dioxide in confined spaces since there is danger of suffocation unless special breathing apparatus is available.
5. Administer artificial respiration and oxygen to personnel overcome by carbon dioxide fumes and keep the patient warm.

Using Cleaning Solvents. Only nonvolatile solvents should be used to clean electrical or electronic apparatus. In addition to the potential hazard of accidental fires, many cleaning solvents are capable of damaging the human respiratory system in cases of prolonged inhalation. The following is a list of "do nots" when using cleaning solvents:

1. Do *not* work alone in a poorly ventilated compartment.
2. Do *not* breathe directly the vapor of any cleaning solvent for a long time.
3. Do *not* spray cleaning solvents on electrical windings or insulation.
4. Do *not* apply solvents to warm or hot equipment, since this increases the toxicity hazard.

The following reminders are positive safety steps to be taken when cleaning operations are underway.

1. Use a blower or canvas wind chute to blow air into a compartment in which a cleaning solvent is being used.
2. Open all possible windows, doors, and so on, and place wind scoops in them.
3. Place a fire extinguisher close by, ready for use.
4. Use water compounds in lieu of other solvents where feasible.
5. Wear rubber gloves to prevent direct contact.
6. Use goggles when a solvent is being sprayed.
7. Hold the nozzle close to the object being sprayed.

☐ Safety in Working with Hoisting Equipment ☐

Often, air conditioning and refrigeration personnel require the use of cranes and other hoisting equipment to get equipment onto the roof of a building or to move the equipment from one place to another. Although this type of equipment is very useful, there are many safety procedures that must be followed to prevent accidents. The following is a general list to be followed. This list should in no way replace any instructions that are provided by the manufacturer of the equipment. Always follow any safety instructions provided by the manufacturer.

1. Never use a crane under electric power lines. To do so may allow them to come into contact with the lines, causing a high-voltage electrical short through the equipment which could possibly electrocute anyone touching the equipment while being grounded.
2. Always check for the center of gravity before hoisting any piece of equipment. If the weight is not centered on the hoist, it may slip and fall, damaging the equipment and possibly injuring persons standing nearby.
3. Always check the total equipment weight before attempting to hoist it to make certain that the hoisting equipment can lift it adequately.
4. Never move a loaded hoist, crane, or chain fall until everyone is clear of the area underneath the equipment and the path through which the hoisting equipment is to travel.
5. Do not use eyebolts to hoist an entire piece of equipment. The eyebolt may be designed for lifting only a portion of the assembly and may not be adequate for the entire assembly.
6. Do not use rigging equipment that needs repair or is not designed to lift the objects being lifted.
7. Do not use rigging equipment on slippery or otherwise hazardous surfaces where slipping may be a possibility.
8. Do not use ladders that do not have the proper angle from top to bottom.
9. Use only approved methods to rig equipment to be hoisted.
10. Never climb over a unit. Use the proper catwalk or platforms when crossing over equipment.
11. Always be aware of where fellow workers are when moving equipment with hoisting equipment.
12. Use lifting lugs only in accordance with the accompanying instructions.
13. Always post signs indicating that heavy equipment is being moved or hoisted so that other people will stay clear of the area.
14. Always make a final check of all connections and fastenings before hoisting equipment.

☐ Safety in Working with Reciprocating Compressors ☐

When working on reciprocating compressors there are certain precautions that should be taken to prevent personal injury, damage to the equipment, or damage to the property. The following is a basic list of the procedures that should be taken to prevent an accident when working with these units.

1. Never work on the electrical wiring to a compressor until all electrical power is off.
2. Never attach an ohmmeter to any electrical wiring, connections, or terminals until the electric power is turned off. To do so will ruin the meter and possibly cause an injury to the user.
3. Do not purge the refrigerant from the system through a cut, loosened connection, or a broken refrigerant pipe. Always use the gauge manifold so that the flow rate can be controlled to prevent purging oil from the system and possible property damage.
4. Do not turn on the electric power to the compressor until the terminal box cover is in place. This is to prevent possible injury if a compressor terminal should be blown out.
5. Pump the system down to about 2 psig (13.79 kPa) before loosening any bolts or connections to the compressor. This is to prevent possible injury or damage from a refrigerant blowout.
6. Do not attempt to operate a compressor with the discharge service valve closed. To do so could possibly cause damage to the compressor.
7. Open, tag, and lock all electrical disconnect switches while working on any of the electrical circuits, components, or connections.
8. All the compressors in a multiple-compressor system must be valved off before a single compressor can be purged.
9. Never use a hermetic or semihermetic compressor for a vacuum pump. To do so may cause the motor winding to become overheated and short out or cause a terminal to blow out, causing serious injury or property damage.
10. Never weld on a compressor having pressure inside it. To do so may cause a refrigerant and oil blowout, causing serious injury or property damage.
11. Never use oxygen for pressure testing a refrigeration system. The oxygen and compressor oil may combine to cause an explosive mixture.

☐ Safety in Working with Air-Handling Equipment ☐

Air conditioning and refrigeration personnel are constantly involved in working with air-handling equipment. There are certain procedures that should be followed to prevent personal injury, property damage, or damage to the equipment. The following is a list of these safety procedures. This list does not replace the safety instructions provided by the equipment manufacturer.

1. Do not enter the air-handling section while the unit is running. The fan sometimes has a great amount of force that could pull a person into the blades.
2. Do not work on electric circuits, heating elements, or the connections until the electricity is turned off to the unit. The disconnect switch should be tagged, locked, and the fuses removed to prevent turning on equipment accidentally.
3. Never operate belt-driven equipment without the belt guards in place. This is to prevent the possibility of getting clothing or limbs caught in the belt.
4. Make certain that all personnel are clear of the equipment before steam cleaning the coil to prevent possible personal injury to the personnel.
5. Do not work on large fans without securing the sheave with a rope or a strap to prevent it from freewheeling and causing personal injury.
6. Do not reach into a fan or equipment cabinet while the fan is running. The hand or fingers may be caught in the blades.
7. To prevent accidental crushing of the hands or fingers in the dampers, do not work on the air control dampers until the operators are disconnected.
8. To prevent using all the oxygen, always make certain that there is plenty of ventilation when welding or cutting inside an air-handling unit.
9. Do not attempt to remove large access doors of roof-mounted units during high winds without sufficient help.
10. Make certain that rooftop units are properly grounded to prevent accidental shock or property damage should an electric short occur.

☐ Personal Safety Protection ☐

In addition to the safety precautions listed previously, there are other personal precautions that should be observed in special situations that may be encountered in everyday operations involving air-conditioning and refrigeration equipment. Use the following rule when in doubt: If something looks dangerous, it probably is.

1. Always wear a hard hat or other head protection when there is the possibility of falling objects.
2. Do not work on live electrical circuits on a wet surface, while wearing wet shoes or when barefooted.
3. Never touch electrical connections with wet or sweaty hands.
4. Always wear gloves when handling components from a system having a burned-out compressor motor winding. The refrigerant and oil from this type of system contains acid, which could possibly cause acid burns to the one handling them.
5. Always wear safety shoes when working around, lifting, or moving heavy objects.
6. Always wear hearing protection equipment when working in areas having noise levels greater than 90 dB.
7. Never wear rings or other jewelry when working around moving belts, moving equipment, electrical wiring, or electrical equipment.
8. Never wear rings, jewelry, long ties, loose clothing, or gloves while working around moving belts and machinery.
9. Always wear safety goggles equipped with side shields when working in manufacturing plants or at construction sites.
10. Always wear goggles and gloves when handling chemicals.
11. Always keep the work area clean of debris and free of liquid spills.

☐ EXERCISES ☐

1. Why is it especially important for workers in the air conditioning and refrigeration industry to practice safety?
2. What should be done before operating any piece of equipment?
3. What temperature can the acetylene torch flame reach?
4. What is the standard shade of eye protection used when welding?
5. What must be done to a container or pipe before using a welding torch on them?
6. What must be done with welding hoses when welding or cutting?
7. Name two uses for a bucket of water in the welding area.
8. What should be done before using new welding equipment?
9. In what position must a fuel gas cylinder be used?
10. Should the fittings and valves on welding equipment be lubricated with oil or grease?
11. What should be done when a cylinder valve becomes dirty?
12. Where should the operator be standing when opening the valve on a welding cylinder?
13. How far should acetylene cylinder valves be opened?
14. How far should oxygen cylinder valves be opened?
15. How many O-rings are required in a welding torch or cutting attachment?
16. What are the recommended gas pressures at the torch for welding equipment?
17. What size wrench is needed to tighten the welding tip and cutting-attachment O-ring connection?
18. What is the maximum acetylene pressure that can safely be used on welding equipment?
19. Is it desirable to field repair welding regulators?
20. Describe the procedure for lighting a welding torch.
21. Is it safe to light the welding torch with a match or a cigarette lighter?
22. What must be done to prevent flame backfiring or flashback inside the welding tip?
23. Each 25 ft (7.6 m) of welding hose requires how much pressure increase?
24. What is the first step in the welding unit shutdown procedure?
25. What is the maximum gas withdrawal rate of a single welding cylinder?
26. Is it proper procedure to use more than one type of refrigerant in a system?
27. What is the greatest personal hazard when using refrigerants in a confined space?
28. Name three precautions that should be observed when purging refrigerant from a system.
29. What could be the result of inhaling refrigerant vapors?
30. Describe the odor of refrigerant vapors when in heavy concentration.
31. What will direct contact of liquid refrigerant with the skin cause other than frostbite?

32. Should hot water be used on refrigerant frostbite?
33. What could possibly happen if the temperature of an overfilled refrigerant cylinder were exposed to temperatures over 125 °F (51.5 °C)?
34. Why should a refrigerant container never be connected to the high side of the refrigeration system?
35. Why is reusing disposable cylinders a bad practice?
36. Why should a refrigeration system be left open during a welding operation on the tubing?
37. What causes cadmium oxide fumes?
38. Are cadmium oxide fumes dangerous to breathe?
39. What type of equipment is probably the most dangerous to work with?
40. What complaint is a good indication of a bad heat exchanger in a gas heating unit?
41. Is it good practice to leave heating controls bypassed?
42. With what should natural or LP gases be leak tested?
43. Name the possible results of an electric shock.
44. At what amperage is an electrical shock usually fatal?
45. What is the most dangerous to people, electrical current or voltage?
46. What is the first step to follow when a shock victim is discovered?
47. What is artificial respiration?
48. How long should artificial respiration be given to a shock victim?
49. How long after the heart stops beating is brain damage likely?
50. Is it a good idea to work alone when working on live electrical circuits?
51. What should be assumed when working on circuits having high capacitance?
52. What is the purpose of a shorting probe?
53. What are the three general types of fires?
54. What type of fire extinguishing agent is best for use on electrical fires?
55. What is the first step to follow when an electrical fire is present?
56. What two checks should be made before trying to lift a piece of equipment?
57. Why should eyebolts never be used to lift an entire piece of equipment?
58. When working on a compressor, what should be done before turning on the electricity?
59. Is it good practice to operate belt-driven equipment without the belt guards in place?
60. What personal protection should be used when working with chemicals?

2 Fundamentals of Refrigeration

Performance Objectives

After completion of this chapter, the student should:
- Understand the basic refrigeration cycle.
- Recognize and practice safety precautions while working on refrigeration systems.
- Complete the Exercises with at least 80 percent accuracy.
- Complete the Laboratory Exercises with at least 80 percent accuracy.

The modern way of life depends on refrigeration equipment for the preservation of food and for human comfort. Many industrial and commercial processes, such as printing, textile manufacturing, food storage, and so forth, are heavily dependent on the proper conditioning of air for efficient, economical operation of the necessary equipment.

The air conditioning and refrigeration field affords many opportunities to those interested in designing, selling, manufacturing, installing, and servicing its equipment. There is a heavy demand for people who are willing to become proficient in any of these areas. But in order to become proficient, the basic principles must be thoroughly understood.

The air conditioning and refrigeration industry encompasses many fields, such as chemistry and physics. Mechanical ability is also of great importance in order to design, install, or service these systems. Customer relations is another important area of competence for success in this field.

Opportunities in Air Conditioning and Refrigeration

Without a doubt, the use of refrigeration and air conditioning can be dated back as far as the history of humankind. In the early stages of cooling, however, water and ice were used to lower the temperature of food and the surroundings. Only in the last century has mechanical refrigeration been developed and used on a widespread basis. Most fruits and vegetables are refrigerated directly after being harvested. The sanitary handling and preservation of meats is also accomplished through adequate cooling and freezing methods.

Because of the ever-expanding uses for refrigeration and air conditioning equipment, there is always a growing demand for personnel to design, sell, install, service, and maintain this equipment. The Manpower Survey Report, which is published by the Air Conditioning and Refrigeration Institute (ARI), recently estimated that for every 1 million units of installed equipment, the following personnel would be needed:

1 graduate engineer
2 salespeople
2 technicians/master mechanics
2 technician helpers
11 refrigeration, air conditioning, and heating mechanics
7 sheet-metal mechanics
1 sheet-metal helper

The industry can generally be divided into the following categories: domestic refrigeration, commercial refrigeration, industrial refrigeration, residential air conditioning, commercial air conditioning, industrial air conditioning, and wholesale supply houses. It is common for technicians to be qualified in more than one of these areas. The different categories overlap somewhat, but they generally include specific types of equipment: The domestic refrigeration category includes home refrigerators, freezers, and sometimes window air conditioning units; the commercial refrigeration category includes systems used in grocery stores and supermarkets, beverage coolers, water fountains, and truck refrigeration systems; the industrial refrigeration category includes equipment for packing plants, cold storage, processing plants, and ice-skating rinks; residential air conditioning generally includes air conditioning systems up to about 5 tons installed in residences and some commercial buildings; commercial air conditioning equipment includes air conditioning units used in supermarkets, office buildings, and cold storage plants; industrial air conditioning includes equipment installations in large manufacturing plants that may involve intricate treatment of the inside air; the wholesale supply house employee is involved in the selling and supplying of equipment and parts needed for the installation, service, and maintenance of the equipment.

The people who work in this industry have many responsibilities. The kind of work they perform is also varied. Design engineers, because they are responsible for designing the equipment or the plans for installation of the equipment, are required to have a good knowledge of all the different categories in the industry. They may be involved in designing a large industrial system one day, and the next day designing a small residential air conditioning installation. Design engineers must be familiar with the different code requirements and have a good knowledge of fluid flow, heat transfer, static pressure, and so forth.

The salesperson must be versatile for the same reasons. He or she must be able to sell equipment in all the categories, estimate material and labor costs, and in many installations oversee the actual installation.

Technicians are used in all phases and categories of the industry. Some technicians prefer to work in research and development; they work in a laboratory and perform many tests on the equipment.

Service technicians drive service trucks and perform service operations on equipment. These people must know how several categories of equipment function, be familiar with the pricing of invoices, the material and labor costs, and so forth. They must be able to supervise service helpers and at times other service technicians. They are responsible for customer relations, the truck and its contents, and sometimes the collecting of money.

The installation technician usually drives the truck to and from the installation site. This person must know how to make solder and silver-soldered joints on copper tubing, as well as knowing how to make and install ductwork. Often, he or she is involved in the installation and connecting of electric wiring for the control circuits. Because installation technicians are responsible for the work of several helpers and often that of other technicians, they should be familiar with the local code requirements in order to properly install the equipment.

Service technician helpers, usually considered to be trainees, work with several service technicians. They should have a good knowledge of the basics of refrigeration and air conditioning so that advancement can be made rapidly and easily. Most often, helpers are involved in jobs that require two people; however, they can sometimes complete less complex jobs alone.

Installation technician helpers work with the installation technician during the installation of the refrigerant piping, duct fabrication, and during the installation of the duct system, equipment, and control wiring. These people are usually considered to be trainees and should have good knowledge of the soldering and silver-soldering process and be able to make less complicated duct system components and install them without the help of the installation technician. To advance to the installation technician status, they must be able to complete a duct system, install the refrigeration piping, show a basic knowledge of how to install the control wiring, and be able to install the equipment.

The supply house employees generally start out in the warehouse to learn the operations and the different types of parts that are carried by the wholesale institution. They advance to the inside sales section and get the parts needed by the technicians and their helpers. Often, advancement is made to the outside sales area. The outside sales force makes calls on the service and installation companies who purchase parts for use in their business. In some cases the outside salesperson is asked to make system designs so that his equipment will be properly sized and to help in selling the equipment to the service or installation company.

The business owners in the air conditioning and refrigeration industry strive to make a profit; their employees also must make a living wage to exist. Obviously, it is not good business to pay an employee more than he or she makes for the company. Thus, to advance, people must try to make more money for their employer. This task is much easier with knowledge. The more employees know and apply, the more money they will make for the company and as a result the more they will earn. Therefore, knowledge and the application of that knowledge is the key to success in this industry. It is difficult for a person to gain sufficient knowledge to satisfactorily advance through on-the-job training alone; however, reading, studying, and additional training make the road to advancement much easier. People who utilize their leisure by studying at home are usually increasing their ability.

☐ Brief History of Refrigeration ☐

Through the ages people have used some means of cooling foods. First, they cooled foods by lowering them into wells or storing them in caves. Following that, natural ice was used. This ice was cut in the winter and stored underground for use in warm months.

The Egyptians used another means of cooling water. They found that they could keep foods cool by storing them in clay containers. This method was quite successful because the water slowly seeped through the porous walls and dried on the outside wall. This drying action, known as evaporation, caused the jar and its contents to cool. This same principle of evaporation of a liquid is the basis of modern mechanical refrigeration.

Mold, yeast, and bacteria require warmth in order to grow and multiply. Their growth and multiplication can, therefore, be retarded by cooling the food to a sufficiently low temperature. This reduction in temperature may be done by modern refrigeration equipment. Refrigeration is actually the most valuable method of protecting foods against spoilage. It is the only method that preserves food in its original state and does not appreciably affect its flavor, appearance, or nutritive value.

Before the existence of refrigeration, most food had to be eaten shortly after being obtained. Therefore, people who lived in one area would have difficulty enjoying foods that were plentiful in another area. Also, foods that were plentiful in one season of the year could not be enjoyed in another season.

It was discovered many years ago that several benefits are derived from cooling food. All food, except frozen foods that are properly prepared and protected, should be kept below 50°F (10°C) and above 32°F (0°C) to prevent spoilage by either microbic growth or by freezing. The zone between 50°F (10°C) and 32°F (0°C) is recognized as the *food safety zone* and is commonly referred to as *safety zone refrigeration*.

The icebox came into existence when natural ice became available. People who lived in warm, dry sections of the country, however, could not afford natural ice because transportation was expensive. It was also discovered that some of the natural ice was contaminated and could not be used safely.

More than 150 years ago, an English scientist changed ammonia gas to a liquid by applying pressure and lowering the temperature. As the pressure was released, the ammonia liquid boiled off rapidly and changed back to a gas. As the liquid boiled off, heat was absorbed from the surroundings. This discovery proved to be of tremendous importance in the development of modern refrigeration equipment.

The first commercial ice machine was introduced in 1825. The ice produced by the machine was cleaner and purer than natural ice. The ice also could be produced regardless of weather conditions.

By the early 1900s industrial refrigeration using the mechanical cycle had been developed, and meat packers, butcher shops, breweries, and most of the other industries were beginning to make use of the mechanical refrigeration units that were available.

When the electric industry began to grow and when homes were beginning to be wired for its use, household refrigerators became more popular and began replacing the common window and standing iceboxes. These types of iceboxes required a block of ice every day to function properly. The new refrigerator did not have this requirement.

The interest and demand for household refrigerators was aided by the design and develop-

*The metric equivalents of units in the English measuring system are included in this book where applicable, to aid in the transition from the English measuring system to the metric measuring system.

ment of fractional horsepower motors, which were used to operate the compressors used in these refrigerators. These units began being produced in large numbers in the early 1920s and have become a necessity for all rather than a luxury for the rich.

With the development of the modern apartment house and the increased demand for more and more cooling, the need for refrigeration equipment became more apparent.

Home food preservation today is not the only use of these systems, but commercial food preservation has become one of the most important present-day applications of refrigeration units. The commercial preservation and transportation of food is so common today that it would be difficult to imagine living without this industry.

Approximately three-fourths of the food used in homes today is produced, packaged, shipped, stored, and preserved through the use of refrigeration equipment. There are literally millions of tons of food stored in warehouses that are cooled by a refrigeration unit. Also, there are millions of tons more that are stored in frozen-food warehouses, in individual locker plants, and in packaging and processing plants.

The commonplace storage and transportation of all types of perishable products would be absolutely impossible without the use of different types of refrigeration and air conditioning equipment.

Due to the refrigeration process, we are not limited to the enjoyment of fruits and vegetables and other produce that are locally grown at any given time of the year; we can also have foods grown and processed in other parts of the country on a year-round basis.

The evolution of refrigeration has improved the economy of almost all areas because it is a means of preserving products while being shipped to customers. It has played a large role in the development of agricultural regions because of the greater demand for products. The dairy- and livestock-producing areas have also enjoyed the growth brought about by the use of refrigeration.

From the previous discussion we can define refrigeration as *a process of removing heat from an enclosed space or material and maintaining that space or material at a temperature lower than its surroundings.* As heat is removed, a space or material becomes colder. The more heat is removed, the colder the object becomes. *Cold,* therefore, is a relative term signifying a condition of lower temperature or less heat.

☐ Common Elements ☐

There are more than 100 basic elements that make up everything around us. There are 92 natural elements; the rest are synthetic. Everything in nature on or around the earth, moon, sun, stars, and the human body is made up of these basic elements. In most cases two or more elements are combined to make a substance. A discussion of some of the most common of these elements follows.

Aluminum, Cadmium, Chromium, Copper, Gold, Iron, Lead, Nickel, Silver, Tin, Tungsten, and Zinc. These elements are known as metals and are generally used by themselves. However, some of the elements are sometimes found in compounds, or mixtures, called alloys. These elements normally exist as solids.

Calcium, Potassium, Silicon, Sodium, and Sulfur. These elements generally exist in solid form. They are found in many materials. They are almost always found in chemical combination with one or more of the other elements.

Carbon. This is the principal element found in coal, cloth, gasoline, natural gas, oil, and paper. It is also found in many gases such as carbon dioxide, methyl chloride, and the fluorocarbon refrigerants used in air conditioning and refrigeration units. Carbon exists as a solid at atmospheric pressures and temperatures.

Nitrogen. Air is made up of several gases. Nitrogen constitutes about 78% of the air in the atmosphere. It is very important for the proper growth and life of plants. Nitrogen is a gas.

Oxygen. About 21% of air is oxygen. It is a gas and is essential to all animal and human life. Because oxygen is an active element, it combines readily with most of the other elements to form oxides or more complex chemicals.

Air. The principal ingredient in air are nitrogen (78%) and oxygen (21%). The remaining 1% consists of other gases such as argon, carbon dioxide, helium, hydrogen, krypton, ozone, and xenon.

Hydrogen. This element is found in many compounds, water being the most important. It is especially important in acids, fuels, and oils. It is rarely found alone in nature. Hydrogen is an extremely light gas. When hydrogen is burned, water is formed. Burning is the process of properly combining hydrogen and oxygen.

Atoms. Each element in nature is made up of billions of tiny particles known as atoms. An atom is the smallest particle of which an element is made up that still maintains the characteristics of that element. An atom is so small that it cannot be seen even with a very powerful microscope. An atom, for our purpose, is considered to be invisible and unchangeable. An atom cannot be divided (broken up) by ordinary means. Atoms of all elements are different. That is, iron is composed of iron atoms, and hydrogen is composed of hydrogen atoms.

Scientists know many things about atoms. How they know these things is outside the scope of this text; we must accept these as true in order to understand the subject being studied.

Molecules. The molecule is the next larger particle of material. It consists of one or more kinds of atoms. When a molecule contains atoms of only one kind, it is said to be a molecule of that element; two or more elements combined are molecules of a chemical compound. Usually a molecule of an element contains only one atom. However, a molecule can contain several atoms of the same kind. Example: A molecule of iron contains only one iron atom while a molecule of sulfur usually contains eight sulfur atoms.

A small piece of any element consists of billions of molecules. Each of these molecules consists of one or more atoms of that element.

Chemical Compounds. The molecule of a chemical compound consists of two or more atoms of different elements. For example, carbon monoxide (CO) is a simple compound with one atom of each element. This combining of different elements causes the material to become entirely different. The new material does not resemble either of the elements that make it up. For example, a molecule of water contains two atoms of hydrogen and one atom of oxygen.

The refrigerants used in air conditioning and refrigeration units are examples of chemical compounds. A molecule of R-12, a colorless gas, consist of one carbon atom, two chlorine atoms, and two fluorine atoms. A molecule of R-22 consists of one carbon atom, one hydrogen atom, one chlorine atom, and two fluorine atoms.

Many of the substances used in our daily living are chemical compounds. Some of these substances are table salt, baking soda, and calcium chloride.

Molecular Motion. Scientists have found that all matter is made up of small particles called molecules. These molecules may exist in three states: solids, li-

quids, and gases (in this text, the terms *gas* and *vapor* will be used interchangeably). Molecules can be broken down into atoms. Atoms are studied in more detail in Chapter 11.

However, we will study the theory of molecular movement and action as it is involved in air conditioning and refrigeration work. A molecule is the smallest particle to which a compound can be reduced before breaking down into its original elements. For instance, water is made up of two elements, hydrogen and oxygen. The movement or vibration of these molecules determines the amount of heat present in a given body. This heat is caused by the friction of the molecules rubbing against each other. The attraction of these molecules to each other is reduced as the temperature increases. When a substance is cooled to absolute zero, all molecular motion stops. At this temperature the substance contains no heat.

Molecules vary in weight, shape, and size. They tend to cling together to form a substance. The substance will assume the character of the combining molecules. Because molecules are capable of moving around, the substance form will be, to a degree, dependent on the space between them. The molecules in a solid have less space between them than either a liquid or a gas. A liquid has more space between the molecules than a solid and less than a gas. A gas has more space between the molecules than either a solid or a liquid. Many substances can be made to exist in any one of these three forms, depending on the temperature and pressure. Water is a very common example of this type of substance.

Solids. In solids, the vibrating rate of the molecules is very low. Therefore, the attraction of the molecules to each other is very strong and a solid must have support or it will fall. The force on a solid is upward [Figure 2–1(a)].

Liquids. In liquids, the vibrating motion of the molecules is faster than in solids. Therefore, the attraction of the molecules to each other is less and a liquid must be kept in a container of some type. A liquid will take the shape of the container [Figure 2–1(b)]. The higher the temperature of the molecules, the faster they vibrate. The warmer molecules will move upward in the container toward the surface of the liquid because they are less attracted to each other and require more space. Therefore, they become lighter and rise upward.

The force exerted by a liquid is toward the sides and to the bottom of the container. The force will be greater on the bottom than on the sides because

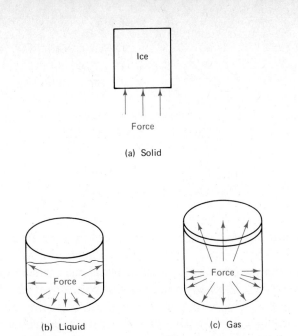

(a) Solid

(b) Liquid   (c) Gas

FIGURE 2-1   **Three states of a substance.**

of the weight of the liquid. As the surface of the liquid is approached, the force will decrease because of the reduced weight of the liquid.

Gases. In gases, the vibrating motion of the molecules is even faster than in liquids. Therefore, the attraction of the molecules to each other is very small and a gas (vapor) must be kept in a closed container or it will escape into the atmosphere. A gas will take the shape of the container on all sides [Figure 2-1(c)]. The molecules of a gas have little or no attraction for any substance. Actually they bounce off each other as well as molecules of other substances.

The force exerted by a gas (vapor) is equal in all directions. Most gases (vapors) are lighter than air. Therefore, they tend to float upward, causing a weightless condition.

With the proper regulation of temperature any substance can be made to remain in any of the three forms: solid, liquid, or gas. Also, any substance can be made to change from one form to another by the proper use of temperature and pressure. This change in form is known as the *change of state.*

Change of State. The addition of heat to a substance may cause, in addition to a rise in the temperature of that substance, a change of state of that substance. That is, an addition of heat may cause a substance to change from a solid to a liquid, or from a liquid to a gas (vapor). There are three states of any substance. The states of water are: ice, water, and steam, that is, solid, liquid, and gas (vapor).

□ **Heat** □

There are two terms that should not be confused. They are heat and temperature. *Heat* is considered as the measure of quantity; *temperature,* the measure of degree or intensity. For instance, if we have a 1-gal container of water and a 2-gal container of water and both are boiling, the 2 gal of water will contain twice as much heat as the 1 gal, even though they both will be at 212 °F (or the same temperature).

Temperature is measured in degrees with a thermometer. Heat is measured in Btu [British thermal unit(s)]. A Btu is defined as the amount of heat required to raise the temperature of one pound of pure water one degree Fahrenheit.

Methods of Heat Transfer. It is important to know that heat always flows from a warmer object to a colder object. The rate of heat flow depends on the temperature difference between the two objects. For example, consider two objects lying side by side in an insulated box. One of the objects weighs 1 lb and has a temperature of 400 °F (204 °C), while the second object weighs 1000 lb (454.54 kg) and has a temperature of 390 °F (199 °C). The heat content of the larger object will be far greater than that of the smaller object. However, because the temperatures are different, heat will flow from the smaller object to the larger object until their temperatures are the same. The three ways that heat travels are: (1) conduction, (2) convection, and (3) radiation.

Conduction. Conduction is the flow of heat through an object [Figure 2-2(a)]. For heat transfer to take place between two objects, the objects must touch [Figure 2-2(b)]. This is a very efficient method of heat transfer. If you have ever heated one end

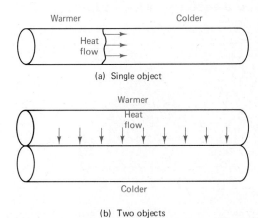

(a) Single object

(b) Two objects

FIGURE 2-2   **Heat transfer by conduction.**

of a piece of metal and then touched the other end, you felt the heat that had been conducted from the heated end of the metal.

Convection. Convection is the flow of heat by use of a fluid, either gas or liquid. The fluids most commonly used with this method are air and water. The heated fluids are less dense and rise, while the cooler fluids are more dense and fall, thus creating a continuous movement of the fluid. Another example of convection is the heating furnace. Air is heated in the furnace and blown into a room to heat the objects in the room by convection.

Radiation. Radiation is the transfer of heat by wave motion. These waves can be light waves or radio frequency waves. The form of radiation that we are most familiar with is the sun's rays. When heat is transferred by radiation, the air between the objects is not heated, as can be noticed when a person steps from the shade into the direct sunlight. The air temperature is about the same in either place. However, you feel warmer in the sunlight (Figure 2–3). This is because of the heat being conducted by the rays of the sun. There is little radiation at low temperatures and at small temperature differences. Therefore, heat transfer by radiation is of little importance in actual refrigeration applications. However, if the refrigerated space is located in the direct rays of the sun, the cabinet will absorb heat. This heat absorption from direct sun rays can be a major factor in the calculation of the heat load of a refrigerated space.

Heat will travel in a combination of these processes in a normal refrigeration application. The ability of a piece of refrigeration equipment to transfer heat is known as the *overall rate* of heat transfer. As was learned earlier, heat transfer cannot take place without a temperature difference. Different materials have different abilities to conduct heat. Metal is a good conductor of heat.

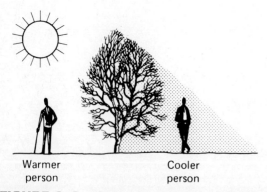

FIGURE 2–3   **Heat transfer by radiation.**

FIGURE 2–4   **Sensible heat.**

Sensible Heat. Heat added to or removed from a substance, resulting in a change in temperature but no change of state, is called sensible heat. The word *sensible* as applied to heat refers to that heat which can be sensed with a thermometer. An example of sensible heat is when the temperature of water is raised from 42 to 212 °F (5.6 to 100 °C). There was a change in temperature of 170 °F (94.4 °C). This change is sensible heat. It can be measured with a thermometer (Figure 2–4).

Latent Heat. Heat added to or removed from a substance during a change of state, but no change in temperature is called latent heat. The types of latent heat are:

1. *Latent heat of fusion:* amount of heat that must be added to a solid to change it to a liquid at a constant temperature. Latent heat of fusion is also equal to the heat that must be removed from a liquid to change this liquid to a solid at a constant temperature.
2. *Latent heat of condensation:* amount of heat that must be removed from a vapor to condense it to a liquid at a constant temperature.
3. *Latent heat of vaporization:* amount of heat that must be added to a liquid to change it to a vapor at a constant pressure and a constant temperature.
4. *Latent heat of sublimation:* amount of heat that must be added to a substance to change it from a solid to a vapor (gas), with no evidence of going through the liquid state. This process is not possible in all substances. The most common example of this is *dry ice.* The latent heat of sublimation is equal to the sum of the latent heat of fusion and the latent heat of vaporization.

Figure 2–5 shows graphically what happens when 1 lb of ice is heated from 0 to 220 °F (−17.8 to 104 °C) and then cooled back to 0 °F. The arrows

| Btu | 16 | 144 | 180 | 970 | 4 | = 1310 Btu |
| Kcal | 4.03 | 36.28 | 45.36 | 244.44 | 1 | 331.11 kcal |

FIGURE 2–5 Relationship of heat and temperature when changing ice to steam.

in the figure indicate that the process is reversible. That is, the water may be changed to steam by adding heat, or the steam may be condensed back to water by removing heat.

Let us follow the process starting in the lower left-hand corner. The first step in the process shows that when the ice is heated from 0 to 32 °F (−17.8 to 0 °C) 16 Btu (4032 cal) of heat are absorbed by the ice. These 16 Btu are sensible heat because they did not change the state of the ice.

When 144 Btu more (36.28 kcal) are added to the ice, the temperature is still 32 °F (0 °C) but the ice has all melted. Since the ice has changed to water—solid to liquid—these 144 Btu are termed *latent heat of fusion.*

When 180 Btu (45.36 kcal) more are added to the water, the temperature is caused to rise to 212 °F (100 °C), which is the boiling point of water at sea level and atmospheric pressure. Since these 180 Btu changed the temperature but did not change the state of the water, it is termed *sensible heat.*

When 970 Btu (244.44 kcal) more are added to the water at 212 °F (100 °C), it is changed into steam at 212 °F. Since there is no change in temperature, but a change of state—liquid to vapor—these 970 Btu are termed *latent heat of vaporization.*

When 4 Btu (1 kcal) more are added to the steam at 212 °F (100 °C), the temperature is raised from 212 to 220 °F (100 to 104 °C). Since this heat resulted in a change in temperature but no change of state, it is termed *sensible heat.*

If the arrow is followed backward, it easily can be seen that this process is reversible. Exactly the same amount of heat must be given up by the substance as was absorbed by it.

Specific Heat. The amount of heat required to raise the temperature of 1 lb of any substance one degree Fahrenheit is called specific heat. Specific heat is also the ratio between the quantity of heat required to change the temperature of a substance one degree Fahrenheit and the amount of heat required to change an equal amount of water one degree Fahrenheit.

From the definition of a Btu given earlier, it can be seen that the specific heat of water must be 1 Btu per pound. The specific heat values of some of the more popular foods are given in Table 2–1. If you will note, after foods are frozen their specific heat values drop considerably. It may be assumed that the specific heat is a little more than one-half of what it was before the foods were frozen.

TABLE 2–1 Specific Heat [Btu [cal]] of Foods

Food	Unfrozen		Frozen	
Veal	0.70	[176.4]	0.39	[98.28]
Beef	0.68	[171.36]	0.38	[95.76]
Pork	0.57	[143.64]	0.30	[75.6]
Fish	0.82	[206.64]	0.43	[108.36]
Poultry	0.80	[201.6]	0.42	[105.84]
Eggs	0.76	[191.52]	0.40	[100.8]
Butter	0.55	[138.6]	0.33	[83.16]
Cheese	0.64	[161.28]	0.37	[93.24]
Whole Milk	0.92	[231.84]	0.47	[118.44]

Superheat. Heat added to a vapor after the vapor is no longer in contact with its liquid is called superheat. For example; When enough heat is added to a liquid to cause all the liquid to vaporize, any additional heat added to the vapor is termed *superheat* (Figure 2–6).

Heat Measurement. As was explained earlier, temperature is the measure of degree or intensity. Heat quantity is the amount of heat present. As the temperature of a body is increased, a definite amount of heat is absorbed. The greater the rise in

Heat

FIGURE 2-6 Explanation of superheat.

temperature, the greater the amount of heat absorbed by the body. Heat will flow from one body to another only when there is a temperature difference. In every case the heat will flow from the warmer body to the cooler body. The greater the temperature difference, the faster the flow of heat.

Water has a greater heat capacity than almost any other substance, and, therefore, is used as a basis in the definition of the Btu. When warming 1 pound of pure water 1 degree Fahrenheit, the water absorbs 1 Btu. When the water is cooled 1 degree Fahrenheit, it will give up the same amount of heat—1 Btu.

Heat calculation is done to determine the amount of heat in Btu that is gained or given up by a substance while it is changing temperature.

The calculation of heat gain or loss involves the use of the following equation:

$$\text{Btu (cal)} = \Delta t \times W \times \text{SH} \qquad (2\text{-}1)$$

where

Δt = change in temperature, [°F(°C)]
W = weight of the substance [lb (kg)]
SH = specific heat of the substance

EXAMPLE 1

If we are to heat 60 lb (27.2169 kg) of water from 50 to 120 °F (10 to 49 °C), how many Btu is required? The specific heat of water is taken as 1.

Solution
Use Eq. 2-1:

$$
\begin{aligned}
\text{Btu} &= \Delta t \times W \times \text{SH} \\
&= (120 - 50) \times 60 \times 1 \\
&= 70 \times 60 \times 1 \\
&= 4200 \\
\text{cal} &= \Delta t \times W \times \text{SH} \\
&= (49 - 10) \times 27.21 \text{ (kg)} \times 1 \\
&= 39 \times 27.21 \times 1 \\
&= 1.061 \text{ kcal}
\end{aligned}
$$

EXAMPLE 2

Now let us cool this same 60 lb of water from 120 to 50 °F (49 to 10 °C) and determine how many Btu (cal) is to be removed.

Solution
$$
\begin{aligned}
\text{Btu} &= \Delta t \times W \times \text{SH} \\
&= (120 - 50) \times 60 \times 1 \\
&= 70 \times 60 \times 1 \\
&= 4200 \\
\text{cal} &= \Delta t \times W \times \text{SH} \\
&= (49 - 10) \times 27.21 \text{ (kg)} \times 1 \\
&= 39 \times 27.21 \text{ (kg)} \times 1 \\
&= 1.061 \text{ kcal}
\end{aligned}
$$

Notice particularly that in these two examples the amount of heat involved is the same. In the first case the heat is applied; in the second case the heat is removed.

In refrigeration, the total weight (W) of the material to be stored in the refrigerator must first be calculated or estimated and then multiplied by the number of degrees of difference between the original temperature and the storage temperature. This value must then be multiplied by the specific heat of the items that are being stored. This will give us the number of Btu that must be removed to lower the temperature of the stored items.

Whenever there are different materials or foods to be stored or refrigerated, each item's temperature should be calculated separately and added together. If the quantities of the different materials are approximately equal in weight, the average specific heat may be used in the calculation.

Heat of Compression. Heat of compression occurs when a vapor is compressed mechanically. As the vapor is compressed, the temperature rises. There is practically no increase in the amount of heat in the vapor; only the amount of heat caused by the friction in the compressor is added to the vapor. The temperature is raised because the gas molecules are forced closer together causing more friction, which in turn causes the increase in temperature.

This amount of heat is the heat energy equivalent to the work that was done during the cycle of the compression refrigeration system. Theoretically, this is the discharge temperature of the gas, assuming of course that the saturated vapor enters the refrigerant cycle. In actual operation the discharge temperature may be from 20 to 35° higher than that predicted. On an operating system this can be checked by fastening a thermometer or preferably a thermocouple sensing probe, to the outlet of the compressor discharge service valve.

A part of this additional heat is dissipated through the compressor cylinder walls. Therefore, a lot depends upon the design of the compressor, the conditions under which it is operating, and the balance between the heat gain and the heat loss of the refrigerant.

In practice the low-temperature, low-pressure vapor is drawn from the evaporator in a refrigeration system to the compressor, where its volume is reduced and its temperature is increased to a point above the temperature of the air (or water) that cools the condenser. The heat will be removed from the vapor by the surrounding air or water, causing a change of state to occur. The vapor will be changed to a liquid.

☐ Temperature ☐

As stated earlier, temperature is the measure of degree or intensity of heat. The device used to measure temperature is a thermometer. There are two types of thermometers—Fahrenheit and Celsius (centigrade). In the United States the Fahrenheit thermometer is most often used, while in Europe the Celsius is in common usage.

Temperature Scales. In 1724, Fahrenheit, after whom the thermometer scale is named, placed his thermometer into a mixture of ice and snow and thought he had reached the lowest temperature possible and marked that point "zero." However, not long after that it was discovered that he had made a mistake. Since that time, scientists have produced temperatures very nearly down to absolute zero, which on the Fahrenheit scale is 460° below zero. However, the name and arrangement of the Fahrenheit scale still remains in daily use even

though it is not best as a scientific scale for measuring temperature.

The Celsius thermometer used throughout Europe is the scale commonly used for scientific work in the United States. This thermometer has a scale on which the freezing point of water is marked at 0 °C and the boiling point of water is marked at 100 °C. This provides a metric thermometer with 100 equal spaces or degrees between the freezing and boiling points of water.

Fahrenheit thermometers use 32 °F as the freezing temperature of water and 212 °F as the boiling temperature of water. There is a difference of 180 divisions between these two points (Figure 2–7). The scale is divided evenly into degrees Fahrenheit. These temperature measurements were proven when taken at sea level because a different atmospheric pressure will change the boiling and freezing temperatures.

Celsius thermometers use 0 °C as the freezing temperature of water and 100 °C as the boiling temperature of water. There is a difference of 100° on the Celsius thermometer between these two points. The scale is divided into degrees Celsius.

Measurement of temperature is obtained by placing the thermometer in the area to be tested and reading the scale directly.

There is a great difference in temperature between any certain number of degrees on the different thermometer scales. It is often necessary when doing laboratory work to change one type of thermometer reading to the other. Since there are 180° difference between the freezing and the boiling temperatures on the Fahrenheit scale and only 100° difference on the Celsius scale, we can see that a Celsius degree is 1.8 times larger than the Fahrenheit degree. Because of the recent popularity of metric scales in the United States, it may be necessary to convert from one scale to the other.

To convert Celsius degrees to Fahrenheit degrees, multiply the number of Celsius degrees by 1.8 and then add 32. The 32 is added because the Fahrenheit zero is 32° below the freezing temperature of water.

$$\text{Fahrenheit temperature} = (°C \times 1.8) + 32 \quad (2\text{–}2)$$

EXAMPLE 1

Change 90 °C to Fahrenheit.

Solution

$$\begin{aligned}
\text{Fahrenheit temperature} &= (°C \times 1.8) + 32 \\
&= (90 \times 1.8) + 32 \\
&= 162 + 32 \\
&= 194 °F
\end{aligned}$$

FIGURE 2-7 **Comparison of the Fahrenheit and Celsius thermometer scales.**

To convert Fahrenheit degrees to Celsius degrees, subtract 32 from the Fahrenheit reading and then divide by 1.8.

$$\text{Celsius temperature} = \frac{°F - 32}{1.8} \qquad (2\text{-}3)$$

EXAMPLE 2
Change 40°F to Celsius.

Solution

$$\begin{aligned}
\text{Celsius temperature} &= \frac{°F - 32}{1.8} \\
&= \frac{40 - 32}{1.8} \\
&= \frac{8}{1.8} \\
&= 4.4°C
\end{aligned}$$

Absolute zero is the lowest possible temperature that can be achieved with any substance. This is the temperature at which all molecular movement in a substance stops. On the Fahrenheit scale absolute zero is −460°F. On the Celsius scale absolute zero is −273°C. It is practically impossible to reach absolute zero.

Critical Temperature. Critical temperature is the highest temperature at which a gas may be liquefied, regardless of how much pressure is applied to it. In refrigeration applications the condensing tempera-

ture must be kept below the critical temperature. If the critical temperature is reached in the condenser, the refrigeration process will stop.

Saturation Temperature. Saturation temperature is a condition of both pressure and temperature at which both liquid and vapor can exist in the same container simultaneously. A saturated liquid or gas is at its boiling temperature. The saturation temperature increases with an increase in pressure and decreases with a decrease in pressure.

☐ Enthalpy ☐

The calculations necessary for some refrigeration and air conditioning estimating involve the use of enthalpy. Enthalpy is simply the total heat contained within a body—both sensible and latent. It is measured in Btu/lb (kcal/kg) of that substance. In theory, enthalpy is measured from absolute zero of the particular scale being used, −460°F and −273°C. However, the large figures involved when using absolute zero as a reference point have caused other reference points to be set. The zero reference point for water is 32°F and for refrigerants is −40°F. Once the zero reference point is established, any enthalpy above that point is positive and any enthalpy below that point is negative.

The following examples show how to calculate enthalpy using letters as designations:

$$E = W \times \text{SH} (\Delta t) + \text{LH} \qquad (2\text{-}4)$$

$$E = W \times \text{SH} (\Delta t) \qquad (2\text{-}5)$$

where

E = enthalpy

W = weight of substance

SH = specific heat of substance

Δt = change in temperature

LH = latent heat of substance

EXAMPLE 1

Calculate the enthalpy of 1 lb (0.454 kg) of steam at 212°F (100°C). [Use 32°F (−0°C) as the reference point.] Use Eq. 2-4 because latent heat is involved.

Solution

$$
\begin{aligned}
E &= W \times \text{SH} (\Delta t) + \text{LH} \\
&= 1 \times (212 - 32) + 970 \\
&= 1 \times 1 (180) + 970 \\
&= 180 + 970 \\
&= 1150 \text{ Btu/lb} \\
E &= W \times \text{SH} (\Delta t) + \text{LH} \\
&= 0.454 \times 1000 (100 - 0) + 244.44 \text{ kcal} \\
&= 45400 \text{ kcal} + 244.44 \text{ kcal} \\
&= 45644.44 \text{ kcal/kg}
\end{aligned}
$$

EXAMPLE 2

Calculate the enthalpy of 1 lb of water at 80°F (26.7°C). [Use 32°F (0°C) as the reference point.] Use Eq. 2-5 because the latent heat is not included without a change of state.

Solution

$$
\begin{aligned}
E &= W \times \text{SH} (\Delta t) \\
&= 1 \times 1 (80 - 32) \\
&= 1 \times 1 (48) \\
&= 1 \times 48 \\
&= 48 \text{ Btu/lb} \\
E &= W \times \text{SH} (\Delta t) \\
&= 0.454 \times 1000 (26.7 - 0) \\
&= 454 \times 26.7 \\
&= 12121.8 \text{ kcal/kg}
\end{aligned}
$$

□ Pressure □

Because refrigeration systems depend basically on pressure differences inside the system, a basic understanding of pressure and the laws that govern it is very important to the designer and technician.

Pressure is defined as the weight or force per unit area and is generally expressed in pounds per square inch (psi) or pounds per square foot. The normal atmospheric pressure at sea level is 14.7 psi.

All substances exert pressure on the materials that support them. A book exerts pressure on the table. A liquid exerts pressure on the bottom and sides of its container, and a gas exerts pressure on all the surfaces of its container, such as a ballon.

If we had a solid cube of 1 in. in all dimensions that weighed 1 lb, it would exert a pressure of 1 psi on a table top when it is placed on it.

The liquid in a container maintains a greater pressure on the bottom and sides of its container as the liquid level is raised. However, gases do not always exert a constant pressure on the container because the amount of pressure is determined by the temperature and the quantity of gas inside the container.

The air around us exerts pressure, which is called *atmospheric pressure*. All liquids have a definite boiling temperature at atmospheric pressure. If the pressure over a liquid is increased, the boiling temperature will also be increased. If the pressure over a liquid is lowered, the boiling temperature will also be lowered. All liquids have a definite boiling temperature for each pressure. This is one of the basic principles used in refrigeration work.

Atmospheric Pressure. The pressure exerted on the earth by the atmosphere above us is called atmospheric pressure. At any given point, this atmospheric pressure is relatively constant, except for the changes caused by the weather. As a basic reference for comparison, the atmospheric pressure at sea level has been universally accepted as being 14.7 psi. This pressure is equal to that exerted by a column of mercury 29.92 in. (75 cm) high.

The depth of the atmosphere is less at altitudes above sea level; therefore the atmospheric pressure is less on a mountain. For example, at 5000 ft elevation the atmospheric pressure is only 12.2 psia (83.81 kPa).

Absolute Pressure. The pressure measured from a perfect vacuum is called absolute pressure. The atmospheric pressure and absolute pressure are the same. Atmospheric pressure is 14.7 psia (101.32 kPa). Absolute pressure is normally expressed in terms of pounds per square inch absolute (psia). Absolute pressure is equal to gauge pressure plus atmospheric pressure. To find absolute pressure, add 14.7 to the pressure gauge reading.

Gauge pressure is zero pounds per square inch gauge (psig) when the gauge is not connected to a source of pressure. Pressures below 0 psig are negative gauge readings and are commonly referred to as *inches of vacuum*. Refrigeration compound gauges are calibrated in inches of mercury (in. Hg) for readings below atmospheric pressure. Since 14.7 psi is equal to 29.92 in. Hg, 1 psi is equal to 2 in. Hg:

$$\frac{29.92}{14.7} = 2.03 \text{ in. Hg}$$

It should be remembered that gauge pressures are only relative to absolute pressures (Table 2–2).

Pressure Measurement. Measurement of low pressures requires a unit of measurement smaller than the pound or the inch of mercury. The *micron* is commonly used for measuring low pressures. A micron is a metric measurement of length and is used in measuring the vacuum in a refrigeration system. It is considered as being absolute pressure.

One micron (μm) is equal to 1/1000 of a millimeter. There are 25.4 mm in an inch. Therefore, 1 μm is equal to 1/25,400 of an inch. A system that has been evacuated to 500 μm would have an absolute pressure of 0.02 in. Hg. At standard conditions this would be equal to a vacuum reading of 29.90 in. Hg, which is impossible to read on a refrigeration compound gauge.

Effects of Temperature. Effects of temperature on the pressure of gases is of great importance in the refrigeration industry. It must be thoroughly understood before a good knowledge of the refrigeration cycle can be obtained. There are several laws that deal with the effects of temperature on the pressure of a vapor within a confined space. These laws are as follows:

Charles' law of gases (vapors) states that "with the pressure constant, the volume of a vapor is directly proportional to its absolute temperature." In mathematical form this is stated as

$$\frac{V_1}{V_2} = \frac{T_1}{T_2} \tag{2-6}$$

where

V_1 = old volume

V_2 = new volume

T_1 = old temperature

T_2 = new temperature

In practical applications this can be proven by use of a properly fitted piston within a cylinder [Figure 2–8(a)]. In this example the cylinder is fitted with a sliding piston. The cylinder is full of vapor at atmospheric pressure. Heat is applied to the cylinder, causing the temperature to rise. Because the piston is easily moved, the volume of vapor increases but the pressure remains constant at atmospheric pressure [Figure 2–8(b)]. On the other hand, if the gas is cooled, the volume of vapor will become smaller [Figure 2–8(c)]. If it were possible to cool

TABLE 2-2 Comparison of Atmospheric and Absolute Pressures at Various Altitudes

Altitude [ft [m]]	Pressure psia [kPa]	Pressure in. Hg [cm Hg]	Boiling Point of Water [°F [°C]]	Refrigerant Boiling Point [°F [°C]] R-12	Refrigerant Boiling Point [°F [°C]] R-22	Refrigerant Boiling Point [°F [°C]] R-502
0 [0]	14.7 [1.03]	29.92 [75.9]	212 [100]	−22 [−30]	−41 [−40.5]	−50 [−45.6]
1000 [304]	14.2 [0.998]	28.85 [73.2]	210 [99]	−23 [−30.5]	−43 [−41.7]	−51 [−46]
2000 [608]	13.7 [0.96]	27.82 [70.6]	208 [97.44]	−25 [−31.7]	−44 [−42.2]	−53 [−47]
3000 [912]	13.2 [0.93]	26.81 [68.1]	206 [96.3]	−26 [−32.2]	−45 [−42.8]	−54 [−47.8]
4000 [1216]	12.7 [0.89]	25.84 [65.6]	205 [95.8]	−28 [−33.3]	−47 [−43.9]	−56 [−48.9]
5000 [1520]	12.2 [0.857]	24.89 [63.2]	203 [94.7]	−29 [−33.9]	−48 [−44.4]	−57 [−49.5]

FIGURE 2-8   Application of Charles' law.

(a)   (b)
Old pressure   New pressure
and volume   and volume

FIGURE 2-9   Application of Boyle's law.

the vapor to absolute zero temperature, $-460\,°F$ $(-273.15\,°C)$, the volume would be zero because there would be no molecular movement at this temperature.

Boyle's law of gases (vapors) states that "with the temperature constant, the volume of gas [vapor] is inversely proportional to its absolute pressure." In mathematical form this is stated as

$$P_1V_1 = P_2V_2 \qquad (2-7)$$

where

P_1 = old pressure

P_2 = new pressure

V_1 = old volume

V_2 = new volume

As before, this can be proven by use of a cylinder with a properly fitted piston [Figure 2–9(a)]. By taking simultaneous pressure and volume readings as the vapor is slowly compressed within the cylinder so that no temperature increase will be experienced, each side of Eq. 2-7 will always be equal. This is possible because a decrease in volume will always

be accompanied by an increase in pressure [Figure 2–9(b)].

Dalton's law of partial pressures states that "gases occupying a common volume each fill that volume and behave as if the other gases were not present." This law, along with the combination of Charles' law and Boyle's law, forms the basis for deriving the psychrometric properties of air.

A practical application of Dalton's law is: The total pressure in a cylinder of compressed air, which is a mixture of oxygen, nitrogen, water vapor, and carbon dioxide, is found by adding together the pressures exerted by each of the individual vapors.

Pascal's law states that "the pressure applied upon a confined fluid is transmitted equally in all directions." A practical application of Pascal's law is shown with a cylinder of liquid and a properly fitted piston in Figure 2–10. The piston has a cross-sectional area of 1 in². With 100 psig (788 kPa) pressure applied to the piston, the pressure gauges show that the pressure exerted in all directions is equal.

This is the basic principle used in most hydraulic and pneumatic systems.

The *general gas (vapor) law* is made by combining Boyle's law and Charles' law of gases (vapor).

FIGURE 2-10   Application of Pascal's law.

The general gas (vapor) law is sometimes expressed as

$$\frac{P_1 V_1}{T_1} = \frac{P_2 V_2}{T_2} \qquad (2\text{-}8)$$

A more useful form of the equation is

$$PV = MRT \qquad (2\text{-}9)$$

where

P = absolute pressure of the vapor (lb/ft²)

V = volume of the given quantity of vapor (ft³)

M = weight of the given amount of vapor (lb)

R = universal gas constant of 1545.3 divided by the molecular weight of the vapor

T = absolute temperature of the vapor

The general gas (vapor) law may be used to study changes in the conditions of a vapor as long as absolute temperature and absolute pressure are used. Gauge pressures cannot be used. This law is used in calculating the psychrometric properties of air in air conditioning work.

EXAMPLE

Calculate the volume of 2 lb (907.2 g) of dry air at 20 lb gauge pressure (238.39 kPa) and 80 °F (26.7 °C).

Solution

$$R = \frac{\text{universal gas content}}{\text{molecular weight of the vapor}}$$
$$= \frac{1545.3}{28.97}$$
$$= 53.4$$
$$P = 20 \div 15 = 35 \times 144$$
$$= 5040$$

Use Eq. 2–9.

$$PV = MRT$$
$$V = \frac{MRT}{P}$$
$$= \frac{2 \text{ lb} \times 53.4 \times (80 + 460)}{(20 + 15) \times 144}$$
$$= \frac{57{,}672}{5040}$$
$$= 11.44 \text{ ft}^3$$

Pressure–Temperature Relationship. Of vital importance in refrigeration equipment design and servicing is the relationship of pressure to temperature. The temperature at which a liquid will boil is dependent on the pressure applied to it, and the pressure at which it will boil is dependent on its temperature. From this it can be seen that for each pressure exerted on a liquid there is also a definite temperature at which it will boil, providing an uncontaminated liquid is being measured.

In practice, because all liquids react in the same manner, pressure provides us with a convenient means of regulating the temperature inside a refrigerated space. When an evaporator is part of a refrigeration system that is isolated from the atmosphere, pressure can be applied to the inside of the evaporator which is equal to the boiling pressure of the liquid at the desired cooling temperature. The liquid will boil at that temperature and, as long as heat is being absorbed by the liquid, refrigeration is being accomplished.

This process is also reversible. When the pressure over a vapor is increased enough to cause the temperature of the vapor to be higher than the surrounding medium, heat will be given up and condensation of the gas will occur. This is the principle used in the condenser of a refrigeration system.

Critical pressure of a liquid is the pressure at or above which the liquid will remain a liquid. With this pressure applied, the liquid cannot be changed to a vapor by adding heat.

☐ Cooling ☐

As stated earlier in this chapter, cooling is merely the removal of heat from a substance. This cooling can be accomplished in several ways. However, we will discuss only the evaporation and expansion methods at this time.

Evaporation. The process that causes water left in an open container to disappear is called evaporation. Evaporation of water depends on two things—temperature and moisture in the air. When water is left in an open container in the summertime, it dries rapidly because of the high temperature. If the temperature drops, the rate of evaporation will decrease. If the temperature goes up, the rate of evaporation will increase. Even at temperatures of −40 °F (−40 °C) evaporation takes place.

If the air is nearly saturated with moisture, it will absorb additional moisture very slowly. If the air is dry, it will absorb moisture very rapidly. During a hot spell when the air is muggy or humid, evaporation of the perspiration from the human body is slow because the saturated air absorbs moisture very slowly.

When water evaporates, a sufficient amount of heat will be absorbed in order to supply the latent heat of vaporization. This heat is absorbed from the water itself or from any object (or air) in contact with the water.

The cold, clammy feeling experienced by a person wearing a wet bathing suit is caused by evaporation of the water and absorption of heat from the material of the suit and from the skin of the person.

Expansion. The process that causes a cooling effect by compressing a vapor, then rapidly reducing the pressure on the vapor is called expansion. When a vapor is compressed, heat is generated in an amount equivalent to the amount of work done in compressing the vapor. This may be demonstrated by the bicycle pump that gets hot due to the heat developed by compressing air into the tire. The greater the compression, the higher the temperature of the compressed air.

The cooling action that takes place when compressed air is allowed to expand is just the reverse of the compression effect. After a long ride over hot roads, the air in an automobile tire becomes very hot and the pressure within the tire is increased. If the valve stem were opened and the hot air allowed to escape, the air would become cool as it was released due to the expansion and resulting reduction in the pressure of the air.

Some of the early refrigeration systems used this principle. The air was compressed and then cooled by passing it through a water-cooled coil. The water was used to absorb and carry away heat from the compressed air.

When the air was allowed to expand, the temperature dropped in relation to the amount of heat that was removed by the water while it was in the compressed state.

There are three steps involved in the expansion refrigeration cycle:

1. Air or gas is compressed to a high pressure.
2. The heat produced by compression is removed.
3. The air or vapor is expanded, causing a reduction in temperature through the absorption of heat by the air.

Subcooling. Subcooling occurs when a liquid is at a temperature lower than the saturation temperature corresponding to its pressure. Water at any temperature below its boiling temperature [212 °F (100 °C) at sea level] is said to be subcooled.

In a refrigeration system this subcooling may occur while liquid refrigerant is temporarily stored in the condenser or receiver. Some of the heat may also be dissipated to the ambient temperature while passing through the liquid line on route to the flow control device. The use of a subcooler will pay for itself through the increased capacity and efficiency of the refrigeration system.

▢ Humidity ▢

The term *humidity*, in the general sense, refers to the moisture content of the air. Humidity also can be defined as the water vapor within a given space. The actual amount of moisture, or water vapor, that air can hold depends on the pressure and temperature of the air. As the temperature of the air is lowered, it can hold less moisture. As the temperature is increased, the amount of moisture it can hold at a given temperature is also increased. This amount of moisture is usually expressed as a percentage figure and is in direct relation to the amount of moisture the air can absorb at that temperature.

Humidity can be expressed in two ways—absolute humidity and relative humidity. *Absolute humidity* is all the moisture present in a given quantity of dry air. It can be expressed as grains of water per pound of dry air or as pounds of water vapor per pound of dry air. *Relative humidity* indicates the amount of moisture, in percentage form, actually present in the air compared to the maximum amount the air could hold under the same conditions. When air contains all the moisture it can hold at any given temperature, it is saturated. When this condition occurs, the relative humidity is 100%. If the air contains only half of the moisture it can hold at a given temperature, the relative humidity is then 50%.

When air is at the saturation point, 100% relative humidity, and the temperature is lowered further, a certain amount of moisture will drop out in the form of water droplets. These droplets appear as ice on the evaporator in a refrigerator or as condensation in an air conditioner. A more common example of this is the water droplets that form on the outside of a glass containing a cold drink (Figure 2-11).

FIGURE 2-11   Air at 100% relative humidity.

As the air surrounding the glass comes in contact with the cold surface, the temperature of the air in contact with the glass is reduced sufficiently to cause condensation. This condensation is in the form of water droplets.

Effects of Humidification. Effects of humidification have a tremendous outcome on the operation of refrigeration and air conditioning installations. In refrigeration applications, when the relative humidity is too low, the food will have a tendency to dry out and become useless. If the relative humidity is too high, however, the formation of mold will occur and cause the food to spoil.

In an air-conditioned building, when the relative humidity is low, the structure will dry out and cracking will appear. Also, the occupants of the building will feel cold and excessive drafts will be noticed. On the other hand, if the relative humidity is high, the occupants will feel warm and clammy. Condensation may appear on the windows during the winter months, causing the structure to rot and mildew.

☐ Dehydration ☐

There has been a great amount of money spent on researching the effects of heat and noncondensables in a refrigeration system. Even now, many of the effects are still a mystery. We do know that their presence can result in many forms of damage in a refrigeration system such as sludging, corrosion, oil breakdown, carbon formations, and copper plating. These contaminants are usually the cause of compressor failure.

Dehydration, also known as evacuation, is the removal of air, moisture, and noncondensables from a refrigeration system. Dehydration protects the refrigeration system as much as possible from contaminants, and causes it to operate as efficiently as possible with a minimum amount of equipment failure.

Methods of dehydration are many and varied. As

stated above, two methods used to cause a liquid to boil are lowering the pressure exerted on it and applying heat to the liquid. Some of the ways to eliminate moisture from a refrigeration system by the boiling process are:

1. Move the system to a higher elevation where the ambient temperature is high enough to boil water at the existing pressure.
2. Apply heat to the system, causing the moisture inside it to boil.
3. Use a vacuum pump to lower the pressure inside the refrigeration system so that the ambient temperature will boil the moisture.

In practice, the first two choices are impractical. Therefore, the vacuum pump method is the most desirable means of removing moisture and noncondensables from a system. To accomplish effective dehydration, the refrigeration system must be evacuated to at least 500 μm.

Before attempting to charge a system with refrigerant, it must be evacuated with some type of vacuum pump. This is especially true since manufacturers of refrigeration and air conditioning units have changed most of their equipment to air-cooled condensers. This change has resulted in higher operating head pressures and temperatures and higher condensing pressures, especially when hermetic compressors are used. Also, the greater use of low-temperature refrigeration equipment results in higher compressor ratios. When single-stage compressors are used, still higher discharge temperatures are present.

These higher head temperatures make it even more important to remove all of the air, as well as the moisture, from the system to a point below the critical point. To keep the possibility of moisture damage to a minimum, the moisture inside the system must be reduced to 5 ppm (parts per million) in the refrigeration system.

The process of removing the air from the system can be referred to as removing the noncondensables, or "degassing" the unit. The worst contaminant is the oxygen in the remaining air. Oxygen is one of nature's most chemically active elements, and its rate of reaction with the refrigeration oil in the system increases very rapidly with any increase in temperature above 200°F.

Because of this, the most important factor is removing all of the oxygen from the system, or at least to a very minimum. In the process of remov-

FUNDAMENTALS OF REFRIGERATION

ing all of the oxygen, the system will also become adequately dehydrated. Filter-driers are installed in the refrigerant lines to pick up any contaminants, such as soldering flux, any moisture that may exist in the oil and refrigerant charge, and materials of construction.

In order to appreciate the importance of reducing the amount of oxygen remaining in the unit to an extremely small amount, the rapid rate of chemical reaction oxygen has with oil in the system must be realized, especially when accompanied with higher temperatures. Unfortunately, this chemical reaction is not in direct proportion to the temperature rise. Rather, for each 18°F rise in the temperature above 200°F, the rate of chemical reaction of the oil with oxygen doubles!

Because of this basic law, we must take every precaution to ensure that the final oxygen content of the system is so low that it cannot support the oxidation of the oil after the system has been charged with refrigerant and oil. It should be noted that the chemical reaction between the oxygen and oil at a temperature of approximately 440°F is about 10,000 times greater than it is at 200°F (Figure 2–12). Much higher localized temperatures are more easily attained, and with corresponding astronomical increases in the rate of reaction, when sluggish valve action occurs in the discharge plate area of an air-cooled compressor.

As the oil begins the breakdown process, deposits will build up in the discharge valve area. Sticky and leaky valves also create a restriction to the flow and high velocities, resulting in localized temperatures that are high enough to start the chemical reaction with, and the breakdown of, the refrigerant. This action results in an acid condition, which invites a vicious maintenance cycle or costly compressor motor burnouts.

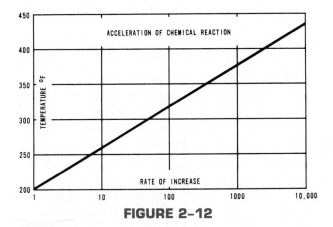

FIGURE 2–12

With the use of a high-vacuum pump and a proper vaccum gauge, a service technician can be certain that the refrigeration system has been properly evacuated to prevent the possible early breakdown of the lubricating oil and the refrigerant.

There are basically two adequate methods of evacuating a refrigeration system with a vacuum pump. One is to use a vacuum pump that is capable of creating a vacuum high enough to do the complete job by evacuation alone. The second method is to initially evacuate the system to about 27 in. of vacuum, then charge the system with an anhydrous nitrogen or a dry refrigerant charge having a moisture content of only 5 or 10 ppm. This dry gas is allowed to remain inside the system for at least one hour so that it can diffuse throughout the system to absorb as much of the moisture as its saturation point will allow. This moisture-laden gas is then released to a pressure that is just above atmospheric pressure. The system is then evacuated again and recharged. When this process has been done three times, the system is finally evacuated by using a high-vacuum pump. The high-vacuum pump allows the use of a simple electronic high-vacuum gauge for checking the final condition of the system being evacuated. This method is commonly called the "triple" evacuation or "blotter method."

The choice of which one of the methods to use is generally determined by experience on similar systems and by the most timesaving and economical steps to use. There are many variables involved, including the volume of the suction and discharge lines, convenience and available time for the procedure, and the amount of moisture inside the system being serviced.

Also, while we are on the subject of adequate evacuation, a thought that is quite commonly associated with air conditioning systems should be discussed. Due to the fact that some air conditioning systems operate at temperatures above freezing, some engineers do not think that it is necessary to evacuate the refrigerant system thoroughly. This process does not permit a long, uninterrupted life of the system. The refrigerant system on air conditioning units must be sufficiently evacuated to adequately reduce the oxygen content remaining inside the system.

A low-type vacuum pump is more popular than a high-vacuum pump because it costs much less. However, it does have many limitations, and in reality a low-vacuum pump is the most costly of the two when modern specifications are considered. One of its major limitations is that it cannot be used for adequately evacuating a system by vacuum alone. It must instead be used only on the multiple-evacuation methods. Even then, the effectiveness of proper evacuation is left to unreliable mathematical assumptions.

The low-vacuum pumps used are of the reciprocating compressor type or the hermetic compressor type, which is used as a vacuum pump. Either of these types are not capable of creating a vacuum much higher than about 28 in. We will refer to these types as compressor pumps for simplicity in our discussions.

However, when a high-vacuum pump is used, it may be used to evacuate a system adequately by: (1) straight vacuum alone, (2) multiple-evacuation methods, or (3) a combination of these two methods.

It is quite common for equipment manufacturers to require that a refrigeration system be completely evacuated to total absolute pressures of less than 1 mm Hg. It is easier to use the next smaller unit of length in the metric system when discussing pressures of less than 1 mm Hg. When 1 mm is divided into 1000 divisions, a unit of length called the micron is obtained, which is very well suited for our use in measuring a vacuum (Figure 2–13).

The compound gauge used on refrigeration manifolds are not capable of measuring this type of pressure. For example, the following is a comparison between micrometer readings and mercury readings:

Microns	Inches of Mercury (Gauge Reading)
0	29.921
200	29.91
500	29.90
2,000	29.84
4,000	29.76
20,000	29.13

FIGURE 2–13 Comparison of millimeters and microns. (Courtesy of Airserco Manufacturing Company.)

It should be obvious that it would be impossible to read a compound gauge to this accuracy.

□ Final Equilibrium Pressure □

The use of test manifolds will permit a distinctive and important test in the process of making a refrigeration system meet the warranty requirements of the manufacturer. This test is known as the pressure-rise test and is used for determining the final equilibrium pressure that has been reached by the system evacuation.

Manufacturers are fully aware that to merely attain an instant low pressure, that which is shown on a total absolute pressure gauge, is not a satisfactory criterion to show that the entire system has been adequately evacuated.

The most important requirement is to know the final equilibrium pressure of the complete system, after it has been evacuated and before it is charged with refrigerant. To determine this fact, close a valve in the system that will isolate the system from the vacuum pump; then watch the gauge. This procedure will hold the system under a vacuum. The reading of the gauge after the needle levels off indicates the equilibrium pressure of the moisture remaining in the system. The amount of pressure rise over a period of time, such as from 2 to 10 min, must be within the specified limits of the manufacturer.

To meet the requirements of modern manufacturers, both the air and moisture must be removed from the refrigeration system to a point much lower than was acceptable just a few years ago. It is no longer advisable to leave to chance that the evacuation process has been completed adequately.

□ High-Vacuum Pump □

The thoroughness of the evacuation process and the universal application of a high-vacuum pump is accepted as the most sound investment for field and shop use in refrigeration service work. So that the great value of a high-vacuum pump can be appreciated, let us review the fundamentals involved in the operation of the vacuum pump and its economical evacuation. To remove some of the mystery that seems to accompany the word *vacuum*, we will discuss how a vacuum pump pumps.

It is no problem for us to understand how a liquid pump operates because we can see and measure the substance being pumped and the amount of liquid being discharged. It is a different story, however, when a vacuum pump is operating; there is no visible sign that the work is actually being accomplished. We must depend on the gauge indication. The selection of the type of gauge to use is just about as important as the type of vacuum pump.

In our everyday lives we have learned that everything in nature seeks an equilibrium of pressure. As an example, when an automobile tire blows out, the air rushes from inside the tire to the atmosphere where the pressure is lower. This difference in pressure causes the air to seek an equilibrium of pressure. Any time there is a greater pressure in one area than in another this differential in pressure exists. When an automobile tire has a slow leak, a greater amount of time is required for the equilibrium to take place. When we remember this differential in pressure, we can see the importance of a high-vacuum pump, as well as the use of a large-diameter connecting line.

The purpose of a vacuum pump is to remove contaminants from inside a closed system, thus reducing the pressure inside the system. To cause the mixture of gases to flow from inside the system to the pump, the pump must create a sufficiently lower pressure inside its cylinder than the one inside the system being evacuated.

When a pump reaches the point where the pressure differential is no longer great enough to cause this flow, a state of equilibrium is present, the flow of gases stops, and the pressure inside the system is no longer decreased. When this state of equilibrium is reached, it is referred to as the "blank-off pressure" of the pump.

An advantage of having a high-vacuum pump in good condition is that a very low blank-off pressure can be reached, thus creating a large differential in pressure. This large pressure differential determines that the high-vacuum pump has a faster pump-down than a low-vacuum pump.

The most economical investment in modern evacuation equipment is not in the purchase of a high-vacuum pump alone but the investment in two additional timesaving accessories. These two accessories are a two-valved high-vacuum test manifold, and an electronic high-vacuum gauge having a range from 0 to at least 20 mm Hg. The combination of these two components is applicable to all pumping units that are used either in the field or in the shop or for small service stations.

A vacuum system made of these components will save many hours by continuously providing information about the progress of the system pump-down. These types of instruments will give confidence that the evacuation meets the modern requirements and furnishes complete quality control of the process.

The High-Vacuum Pump versus the Low-Vacuum Pump. The following are characteristics of the high-vacuum pump:

1. The pump dehydrates and degases by high vacuum alone or in combination with multiple evacuation and blotting.
2. Fast pump-down is possible.
3. The pump uses an electronic high-vacuum gauge to supply continuous visual indication of the progress of the dehydrating and degassing cycle from start to finish.
4. The two-valved test manifold allows the high-vacuum gauge (a) to be used as a reliable leak and moisture indicator, (b) to quickly show the blank-off pressure of the pump, and (c) to show the final equilibrium, or balanced pressure inside the system.

The following are characteristics of a low-vacuum pump:

1. It is limited to only the multiple-evacuation and blotter method. The system must be heated to dehydrate by vacuum alone.
2. Pump-down is slow.
3. The pump does not create a sufficient vacuum to use a high-vacuum gauge; therefore, questionable mathematical computations must be used to determine when the system has been adequately dehydrated and degassed.
4. The pump does not create a vacuum high enough to use a high-vacuum test manifold and high-vacuum gauge; because of this, it cannot detect small leaks nor will it indicate moisture inside the system.

Types of High-Vacuum Pumps for Servicing Refrigeration Systems. There are two classifications of high-vacuum pumps: (1) the sliding-vane type, and (2) the cam-and-piston type. Either of these types is well suited for dry high-vacuum requirements; but when dehydrating a refrigeration system for field or shop work, moisture is usually included along with other contaminants in the installation. Therefore, it is important to select a design that will permit easy field maintenance of the pump without the use of special tools or skills. The cam-and-piston type falls into this category. It has heavier bearings and sturdier parts, which allow it to be used for heavy and continuous industrial work as well as laboratory service.

Also, the cam-and-piston type of pump will retain its positive-displacement pumping action indefinitely, a great advantage of this type of pump. The spring-actuated sliding-vane type loses its pumping action if the vanes should fail to slide for some reason.

The delays for sending a pump back to the factory for repairs are far more costly than the more expensive pump. Therefore, the cam-and-piston pump that can be repaired by the user should be the choice.

When selecting the size of the pump for your requirements, it should be remembered that it is the length and diameter, or volume, of the line to be evacuated that govern the pump size rather than the horsepower or the Btu rating of the unit.

Size of the Vacuum Line. Regardless of how short or how large the diameter of the connecting line from the pump to the unit being evacuated, the line offers some resistance to the flow of gas. The amount of resistance is in direct proportion to the length of the line and to the fourth power of the inside diameter. To shorten the pump-down time, use a short line with as large a diameter as is feasible; never use tubing as small as ¼ in. Use at least ⅜-in. tubing for small pumps because under a vacuum the carrying capacity is 16 times greater than the ¼-in. tubing.

When subjected to a high vacuum, the leakage rate of a conventional neoprene refrigerant charging hose is generally too great to make a successful important pressure-rise test. It is therefore recommended that high-vacuum seamless metal hoses be used. These are available in a standard length of 6 ft and a ⅜ in. interior diameter and with a carrying speed 16 times greater than ¼-in. rubber hose.

High-Vacuum Oil. The important and distinctive characteristic of an oil for a rotary high-vacuum pump service is that it be refined from a high-quality paraffin-base crude oil and be processed to have a vapor pressure at 100 °F not greater than 0.005 mm Hg total absolute pressure. Its viscosity at 100 °F should be approximately 300 SSU.

This oil, because it is from a paraffin base, is suitable for continuous use in high-vacuum systems because it has sealing properties that are more stable than oils from other types of crudes. Refrigeration oil, because it is refined from napathanic crude oils to achieve a low pour point, does not have the equal qualities or stability for high-vacuum pump service. Because of this, an oil should be selected that will provide the type of performance desired from this type of pump.

Remember, a vacuum pump cannot create a total absolute pressure less than the vapor pressure of its sealing oil!

The oil must be changed on a regular basis so that a high-vacuum can be obtained. Remember that the pump is removing moisture from the system and some of this moisture remains in the vacuum pump oil. The pump cannot produce the desired vacuum when moisture is in the oil because the moisture boils out of the oil rather than being removed from the system.

☐ Energy ☐

Energy may be defined as the capacity to do work. Energy may be in the form of *heat* energy, *electrical* energy, or *mechanical* energy. Energy cannot be destroyed. It may, however, be converted from one form to another.

In practice, if we are to remove heat from a product in a refrigerator by mechanical means, the heat that is removed constitutes a load on the machine that is used to remove this heat. It is necessary, therefore, that we have some understanding of elementary physics.

☐ Force ☐

Force is defined as that which tends to, or actually does, produce motion. Force may be exerted in any direction. In the simple definition of force direction is not considered. The common unit by which force is measured is the pound.

☐ Motion ☐

Motion is the movement or constantly changing position of a body. Motion is commonly expressed in feet. The simple definition of motion does not involve the factor of time.

☐ Velocity ☐

Velocity is the rate of change of position, or the speed of movement per unit of time. Velocity measurements are expressed in feet per minute (ft/min or fpm) or feet per second (ft/sec).

☐ Work ☐

Work is the force applied multiplied by the distance through which this force acts. The unit of work is the foot-pound (ft-lb). A foot-pound is the work done by a force of 1 pound through a distance of 1 foot.

$$\text{Work} = \text{force} \times \text{distance} \qquad (2\text{--}10)$$

EXAMPLE
If we are to move a 20-lb (9.07-kg) weight through a distance of 6 ft (1.8 m), how much work would be done?

Solution

$$
\begin{aligned}
W &= F \times D & W &= F \times D \\
&= 20 \text{ lb} \times 6 \text{ ft} & &= 9.07 \text{ kg} \times 1.82 \text{ m} \\
&= 120 \text{ ft-lb} & &= 16.34 \text{ m-kg}
\end{aligned}
$$

☐ Power ☐

Power is the time rate of doing work. It is work done divided by time. This gives us the amount of work done in a certain amount of time.

$$\text{Power} = \frac{\text{work}}{\text{time}} \qquad (2\text{--}11)$$

EXAMPLE
A weight of 600 lb (272.16 kg) has been moved through a distance of 10 ft (3.04 m) in 10 min. How much power was used?

Solution

$$
\begin{aligned}
W &= F \times D \\
&= 600 \text{ lb} \times 10 \text{ ft} \\
&= 6000 \text{ ft-lb} \\
W &= F \times D \\
&= 272.16 \text{ kg} \times 3.04 \text{ m} \\
&= 872.37 \text{ m-kg}
\end{aligned}
$$

then

$$
\begin{aligned}
P &= \frac{W}{T} & P &= \frac{W}{T} \\
&= \frac{6000 \text{ ft-lb}}{10 \text{ min}} & &= \frac{872.37 \text{ m-kg}}{10 \text{ min}} \\
&= 600 \text{ ft-lb/min} & &= 87.237 \text{ m-kg}
\end{aligned}
$$

This problem is the same regardless of the direction of the applied force, whether it is applied vertically or horizontally; nor does it matter whether the resistance is caused by friction, gravity, or gas pressure as in the electric refrigeration compressor.

☐ Horsepower ☐

In calculation, the foot-pound is much too small a unit for ordinary work. Therefore, a larger unit called the horsepower (hp) is used. One hp is equal to 33,000 ft-lb/min. Notice that time is involved in calculating horsepower.

$$\text{Horsepower} = \frac{\text{weight} \times \text{distance}}{\text{time} \times 33{,}000} \qquad (2\text{--}12)$$

EXAMPLE
A weight of 10,000 lb is to be moved 20 ft/min. What is the horsepower required?

Solution

$$
\begin{aligned}
\text{Horsepower} &= \frac{\text{weight} \times \text{distance}}{\text{time} \times 33{,}000} \\
&= \frac{10{,}000 \times 20}{1 \times 33{,}000} \\
&= \frac{200{,}000}{33{,}000} \\
&= 6.06 \text{ hp}
\end{aligned}
$$

☐ Mechanical Equivalent of Heat ☐

The mechanical equivalent of heat is the heat produced by the expenditure of a given amount of mechanical energy. This has been determined by accurate scientific experiments. If the heat energy represented by one Btu could be changed to mechanical energy without energy loss, it would represent 778 ft-lb of work. Thus, we have established a definite relationship between the Btu and the ft-lb. This relationship is represented as:

1 Btu of heat energy is equivalent to 778 ft-lb.
1 ft-lb of mechanical energy is equivalent to 1/778 Btu or 0.00128 Btu.

To obtain the heat equivalent in Btu from any number of ft-lb, divide by 778.

$$Btu = \frac{foot\text{-}pounds}{778} \qquad (2\text{-}13)$$

To obtain ft-lb of work done from the heat in Btu, multiply by 778.

$$W = Btu \times 778 \qquad (2\text{-}14)$$

☐ Density ☐

Density may be defined as the weight per unit volume of a substance; it is ordinarily taken as the weight per cubic foot.

Under ordinary circumstances, most substances expand when heated and contract when cooled. However, a great many liquids vary from this rule. For example, when water is cooled, it contracts until the temperature is lowered to 39°F (3.89°C). At this temperature the water starts to expand. Therefore, the water is at its greatest density at 39°F. As the temperature is lowered to the freezing temperature, the water expands further and ice is formed at 32°F (0°C). Because of this, the density of ice is less than that of water.

☐ Basic Refrigeration System ☐

The basic components of a refrigeration system are compressor, condenser, flow control device, evaporator, connecting tubing, and the refrigerant (Figure 2-14). The compressor is known as the heart of the system. It causes the refrigerant to circulate in the system. The refrigerant is pushed by the compressor to the condenser where both sensible and la-

FIGURE 2-14 Basic refrigeration system.

tent heat are removed. The liquefied refrigerant then goes to the flow control device where the pressure is reduced, allowing the refrigerant to expand and absorb heat from within the refrigerated cabinet. This low-pressure, heat-laden refrigerant vapor is then drawn to the compressor where the cycle is repeated. The refrigerant systems will be taken up in greater detail in a later chapter.

☐ SUMMARY ☐

☐ The growth and multiplication of mold, yeast, and bacteria can be retarded by cooling.
☐ Refrigeration is the only process that preserves food in its original state.
☐ All food, except frozen foods, should be kept between 50 and 35°F (10 and 1.7°C).
☐ Refrigeration is the process of removing heat from an enclosed space or material and maintaining that space or material at a temperature lower than its surroundings.
☐ All matter is made up of molecules and may exist in three states—solid, liquid, and gas.
☐ The movement of molecules determines the amount of heat present in a body.
☐ In solids, the movement of the molecules is slow.
☐ In liquids, the molecular movement is faster than in solids.
☐ In gases, the molecular movement is faster than in either solids or liquids.
☐ The force exerted on a solid is always upward.
☐ The force exerted by a liquid is mainly downward.
☐ The force exerted by a gas (vapor) is in all directions.
☐ The addition or removal of a sufficient quantity of heat will result in a change of state of a substance.
☐ The Btu is the quantity of heat.
☐ A Btu is the amount of heat required to raise the temperature of 1 pound of pure water 1 degree Fahrenheit.
☐ The three methods of heat transfer are: (1) conduction, (2) convection, and (3) radiation.

- [ ] Sensible heat is the heat added to or removed from a substance resulting in a change of temperature but no change of state.
- [ ] Latent heat is the heat added to or removed from a substance during a change of state but with no change in temperature.
- [ ] Specific heat is the amount of heat required to raise the temperature of 1 pound of any substance 1 degree Fahrenheit.
- [ ] Heat of compression occurs when a gas is compressed mechanically.
- [ ] Temperature is a measure of degree or intensity of heat.
- [ ] Critical temperature is the temperature above which a gas (vapor) will never liquefy.
- [ ] Fahrenheit thermometers use 32°F as the freezing temperature of water and 212°F as the boiling temperature of water.
- [ ] Celsius thermometers use 0°C as the freezing temperature of water and 100°C as the boiling temperature of water.
- [ ] For conversion from one type of thermometer scale to the other use these formulas:

$$°F = (°C \times 1.8) + 32 \qquad °C = \frac{°F - 32}{1.8}$$

- [ ] At absolute zero there is no molecular movement because no heat is present.
- [ ] Enthalpy is the total heat contained within a body.
- [ ] Atmospheric pressure is the pressure exerted on the earth by the atmosphere.
- [ ] Absolute pressure is the pressure measured from a perfect vacuum.
- [ ] To find the absolute pressure add 14.7 to the gauge pressure reading.
- [ ] Gauge pressures are relative to absolute pressures.
- [ ] One micron is equal to 1/1000 of a millimeter or 1/25,400 of an inch.
- [ ] Charles' law of gases (vapors) states that "with the pressure constant, the volume of gas [vapor] is directly proportional to its absolute temperature."
- [ ] Boyle's law of gases (vapors) states that "with the temperature constant, the volume of gas [vapor] is inversely proportional to its absolute pressure."
- [ ] Dalton's law of partial pressures states that "gases [vapors] occupying a common volume each fill that volume and behave as if the other gases [vapors] were not present."
- [ ] Pascal's law states that "the pressure applied upon a confined fluid is transmitted equally in all directions."
- [ ] The temperature at which a liquid will boil is dependent on the pressure applied to it.
- [ ] The critical pressure of a liquid is the pressure at or above which the liquid will remain a liquid.

- [ ] Evaporation of water depends on the temperature of the air and the amount of moisture in the air.
- [ ] Expansion is the process that causes a cooling effect by compressing a vapor, then rapidly reducing the pressure on the vapor.
- [ ] Humidity is the term generally used when referring to the moisture in the air; the actual amount of moisture the air can hold depends on the pressure and temperature of the air.
- [ ] Absolute humidity is all the moisture present in a given quantity of dry air.
- [ ] Relative humidity indicates the amount of moisture, in percentage form, actually present in the air compared to the maximum amount that the air could hold under the same conditions.
- [ ] Dehydration, also known as evacuation, is the removal of air, moisture, and noncondensables from a refrigeration system.
- [ ] Dehydration is best accomplished by use of a vacuum pump.
- [ ] Effective dehydration requires a vacuum of at least 500 microns.
- [ ] Force is that which tends, or actually does, produce motion.
- [ ] Motion is the movement or constantly changing position of a body.
- [ ] Velocity is the rate of change of position, or the speed of movement per unit of time, measured in feet per second.
- [ ] Work is the force applied multiplied by the distance through which this force acts.
- [ ] Power is the rate of doing work. One horsepower is equal to 33,000 ft-lb/min.
- [ ] The mechanical equivalent of heat is the heat produced by the expenditure of a given amount of mechanical energy.
- [ ] The six components of a basic refrigeration system are: compressor, condenser, flow control device, evaporator, connecting tubing, and refrigerant.

◻ EXERCISES ◻

1. In what three states does matter exist?
2. In what direction is the force exerted by a gas (vapor)?
3. What causes a change of state of a substance?
4. How is heat measured?
5. What are the three methods of heat transfer?
6. Define *sensible heat*.
7. Name the heat that is added to or removed from a substance during a change of state but no change of temperature.
8. Define *specific heat*.
9. When does the heat of compression occur?

10. How is temperature measured?
11. What is the unit of measurement used to measure heat?
12. Define *critical temperature*.
13. What happens to the molecular structure of a substance at absolute temperature?
14. What is the term applied to the total heat contained within a body?
15. What is atmospheric pressure at sea level?
16. Give the formula for obtaining absolute pressure.
17. If the pressure is reduced over a liquid, what will happen to its boiling temperature?
18. On what two things does the evaporation process depend?
19. Briefly define the cooling by expansion process.
20. How is relative humidity expressed?
21. What is the purpose of dehydration?
22. What is the most practical method of dehydration?
23. Convert 32 °F to Celsius.
24. Convert 25 °C to Fahrenheit.
25. What are the six important parts of a basic refrigeration system?
26. Define *refrigeration*.

3 Compression Refrigeration Systems

Performance Objectives

After completion of this chapter, the student should:
■ Have a basic understanding of how compression refrigeration systems operate.
■ Be familiar with the various types of compression refrigeration systems.
■ Recognize and practice safety procedures while working on refrigeration and air conditioning equipment.
■ Complete the Exercises with at least 80 percent accuracy.
■ Complete the Laboratory Exercises with at least 80 percent accuracy.

There are two types of refrigeration systems in use today. The *compression system,* which is the more popular of the two, and the *absorption system*, which is used in highly specialized applications. The compression system is sometimes called a mechanical system. We will deal only with the compression system in this text.

As was stated earlier, refrigeration is the process of removing heat from an enclosed space or material and keeping that space or material at a temperature lower than its surroundings.

Food is refrigerated in an ordinary icebox because the ice melts. When the ice melts, it absorbs the latent heat of fusion from the foods stored in the refrigerator. Heat is also absorbed through the walls of the icebox. In this operation heat is taken from a warm body and space and added to the cooler body—the ice.

A liquid may be made to boil by the application of heat. If a vacuum is created above the liquid, it may be made to boil at a lower temperature. If the liquid, or refrigerant in our case, exists as a vapor at atmospheric pressure and temperature, it will immediately begin to boil if it is left in an open container, absorbing heat from its surroundings and any material which is in contact with it.

Consequently, any liquid, such as a refrigerant, that will boil at a temperature below the freezing temperature of water will make ice cubes and keep food cool in a mechanical refrigerator.

☐ Refrigeration by Evaporation ☐

When we think of anything boiling, we naturally think of it as being hot. However, that is not true in all cases. Just because water boils at 212 °F (100 °C) does not mean that all other substances will boil at the same temperature.

The simplest way to produce refrigeration by evaporation is to place a bottle containing a refrigerant, such as R-12, inside an insulated cabinet with an opening at the top (Figure 3–1). The liquid R-12 immediately begins to absorb heat from its surroundings. When the liquid has absorbed enough heat to raise its temperature above −21.66 °F (−30 °C), it starts to boil vigorously. The temperature of the liquid remains at −21.66 °F because it is absorbing the latent heat of vaporization as it changes to a vapor. These things take place at a constant temperature of −21.66 °F and an atmospheric pressure of 14.7 psig (202.57 kPa).

This process will continue as long as the liquid R-12 in the bottle lasts, and the bottle is left open so that the R-12 will evaporate. The liquid refrigerant also continues to absorb heat from the inside of the

FIGURE 3-1 Refrigeration by evaporation.

FIGURE 3-2 Elementary refrigerator.

box, lowering the temperature inside the box. The arrows in Figure 3-1 indicate the circulation of the air inside the box. Notice that the air circulation is exactly the same as in an ordinary icebox. The air being cooled by the refrigerant container drops because of its own weight. The air that is warmed at the sides of the box by the heat passing through the insulation rises.

We can apply these basic principles to the development of an elementary refrigerator cabinet as illustrated in Figure 3-2. Notice that the cabinet has no machinery connected to it. It merely contains a bottle of liquid refrigerant with a vent that allows the vapors resulting from the boiling of the liquid refrigerant to escape. As the boiling liquid absorbs heat from the inside of the refrigerator, it lowers the temperature. Even though refrigeration is accomplished, this process is not practical because refrigeration will stop when all the liquid refrigerant in the bottle has evaporated. Furthermore, the expense and trouble of always supplying new liquid

refrigerant makes this unit too costly and bothersome for practical use.

☐ Compression System Principles ☐

To eliminate these objections, some means must be provided to capture and reuse the refrigerant vapors. This is where we can put pressure to good use in a compression refrigeration system. With pressure, we can compress these refrigerant vapors, thereby concentrating the heat they contain. When heat is concentrated in a vapor, we increase the intensity, or the degree, of the heat. In short, we increase the temperature because temperature is only a measurement of heat intensity. The most important part is that we have made the vapors hotter without actually adding any additional quantity of heat (Figure 3-3).

Now we have covered the scientific ground rules that apply to mechanical refrigeration. They are

FIGURE 3-3 Compression principle. (Courtesy of Frigidaire Division, General Motors Corp.)

probably still a little hazy; however, it is easy enough to remember these main points:

1. All liquids absorb a lot of heat without getting any warmer when they boil into a vapor.
2. Pressure can be used to make vapors condense back into liquid so they can be used again. With that amount of scientific knowledge, we can build a refrigerator.

☐ Elementary Compression Refrigeration System ☐

A bottle of liquid refrigerant can be placed in an insulated box. We know that it will boil and absorb heat from the box and its contents because it boils at such a low temperature at atmospheric pressure.

The vapors that result because of the boiling liquid can be piped to the outside of the cabinet, providing a way to carry the heat outside the cabinet (Figure 3–4). When the heat-laden vapor reaches the outside, we can compress it with a pump. When enough pressure is applied, we can squeeze the heat out of the cold vapor even in a warm room. An ordinary radiator will help to get rid of the heat.

By removing the heat and making the refrigerant vapor into a liquid, it is the same as it was in the bottle. We can now run another pipe to the inside of the box and return the refrigerant to the bottle to be used over again.

☐ The Basic Refrigeration Cycle ☐

Mechanical refrigeration is achieved by constantly circulating, evaporating, and condensing a fixed supply of refrigerant inside a closed system. Evaporation occurs at a low pressure and low temperature. Condensation occurs at a high temperature and high pressure. Thus it is possible to transfer heat from an area of low temperature (refrigerator cabinet) to an area of high temperature (the kitchen).

If we begin the refrigeration cycle at the evaporator inlet (1) in Figure 3–5, the low-pressure liquid expands, absorbs heat, and evaporates, changing to a low-pressure vapor at the outlet of the evaporator (2). The compressor (3) pumps this gas from the evaporator and increases its pressure. The high-pressure gas is then discharged to the condenser (4). In the condenser some of the heat is removed from the gas, causing it to condense and become a warm, high-pressure liquid.

The next device encountered by the refrigerant as it travels through the system is the liquid line strainer-drier (5), which prevents plugging of the flow control device by trapping scale, dirt, and moisture. The flow of refrigerant into the evaporator is controlled by a pressure differential across the flow control device.

As the warm, high-pressure liquid refrigerant enters the evaporator, it is subjected to a much lower pressure due to the combined suction of the com-

Refrigerant as a liquid

Refrigerant as a vapor

Pump

Refrigerant laden with heat from the milk bottle

Heat traveling from a warm object to a colder object

Heat from milk bottle being discharged into surrounding air by condenser

Condenser

Refrigerant on its way to pick up more heat from milk bottle

FIGURE 3–4 Elementary compression refrigeration system. (Courtesy of Frigidaire Division, General Motors Corp.)

FIGURE 3-5 **Compression refrigeration system.**

FIGURE 3-6 **High- and low-side diagram.**

pressor and the pressure drop across the flow control device. Thus the refrigerant tends to expand and evaporate. In order to evaporate, the liquid must absorb heat from the air passing over the evaporator.

Eventually, the desired temperature is reached and a temperature control will stop the compressor. As the temperature of the air passing over the evaporator rises, the temperature control starts the compressor again. After the compressor starts, the cycle is continued.

All refrigeration systems operate at two definite pressure levels. The dividing line passes through the compressor discharge valve on one end and the orifice of the flow control device on the other end (Figure 3-6). The compressor discharge valve and the orifice of the flow control device are the dividing points between the high side and the low side of the system.

The high side of the system includes all the components operating at or above the condensing pressure. In the diagram the high side includes the discharge side of the compressor, the discharge service valve, condenser, liquid receiver, liquid line shutoff valve, dryer, liquid line strainer, and the liquid line. In practice, the complete compressor is considered as being on the high side of the system.

The low side of the system includes all components operating at or below the evaporating pressure. In the diagram the low side includes the low-pressure side of the flow control device, evaporator, suction line, suction service shutoff valve, compressor crankcase, and the piston and suction valves of the

compressor. In practice, the flow control device is considered as being on the low side of the system.

Although there are many types of compression refrigeration systems, the cycle in all of them is the same. We can summarize the cycle of a compression refrigeration system as follows:

1. The refrigerant vapor is compressed by the compressor and discharged into the condenser.
2. The compressed vapor is cooled and condensed to a liquid in the condenser.
3. The liquid refrigerant is passed through the flow control device to the evaporator or cooling coil.
4. Due to the reduced pressure in the evaporator, vaporization takes place. Refrigeration results from the absorption of heat during vaporization in the cooling coils.
5. The vapor is drawn into the compressor and the cycle is repeated.

It is extremely important to analyze completely every system and understand the intended function of each component before attempting to determine the cause of a malfunction or failure.

□ Types of Refrigeration Systems □

While there are five general types of refrigeration systems based on the type of flow control device, there are two more classifications based on the condition of the refrigerant in the evaporator. The two latter classifications are the *flooded* and *dry* systems.

Flooded Systems. Flooded systems operate with a definite liquid refrigerant level in the evaporator.

FIGURE 3-7 Flooded system evaporator.

FIGURE 3-8 Dry system evaporator.

This liquid refrigerant level is maintained in the evaporator through the action of the refrigerant flow control device as shown in Figure 3-7. There are several advantages of the flooded system over the dry system. A few of these advantages are: higher efficiency (system operates at higher average suction pressures with resultant shorter operating time), lower operating costs, less cycling (starts and stops), higher rate of heat transfer, and closer control of temperature.

More liquid on the low-pressure side of the system, as in the flooded system, provides a greater area of wetted surface and allows a higher rate of heat transfer through the evaporator walls and tubing. The refrigerant flow control devices which provide a flooded evaporator are the *low-side float, high-side float,* and the *capillary tube* or *restrictor.*

Dry Systems. Dry systems have an almost all-refrigerant vapor condition in the evaporator. The

flow control is accomplished by either an automatic expansion valve or a thermostatic expansion valve. The refrigerant passing through the valve is partially evaporated immediately after passing the orifice. The fine, suspended droplets of liquid refrigerant are completely evaporated as they flow through the balance of the evaporator (Figure 3-8). Dry-type evaporators are generally made from one continuous length of tubing. The continuous length of tubing ensures better control of the evaporation of the liquid refrigerant when the flow control device is properly adjusted.

Low-Side Float Systems. Low-side float systems have a pool of liquid refrigerant in the evaporator (Figure 3-9). In operation, the refrigerant gas is compressed by the compressor. From the compressor the high-temperature, high-pressure vapor passes directly to the condenser where it is cooled and liquefied, changing to a warm, high-pressure liquid. The warm,

FIGURE 3-9 Low-side float system.

Types of Refrigeration Systems

high-pressure liquid then flows into the liquid receiver. The warm, high-pressure vapor fills the upper portion of the liquid receiver. The warm, high-pressure liquid sinks to the bottom. The pressure of the vapor forces the warm, high-pressure liquid through the liquid line to the seat of the low-side float.

As the level of the low-pressure, low-temperature liquid in the float chamber falls, the float also lowers and pulls the needle from the seat of the valve. This permits the warm, high-pressure liquid to flow into the float chamber where the pressure and temperature are reduced. The purpose of the float is to maintain a constant level of liquid refrigerant in the evaporator. The liquid refrigerant fills the tubes of the evaporator and a part of the float chamber. As heat is absorbed by the liquid refrigerant, evaporation occurs and the resulting vapor collects in the upper portion of the float chamber. The low-pressure vapor is then drawn through the suction line to the compressor, and the cycle is repeated.

The refrigerant charge used on the low-side float flooded system is not critical because any excess or overcharge of refrigerant will remain in the liquid receiver. This is an important point to remember because there are other systems in which the refrigerant charge must be accurate to within a few ounces of the recommended charge or operating troubles may result.

High-Side Float Systems. High-side float systems have a majority of the liquid refrigerant in the low-pressure side of the system. The running cycle begins with the refrigerant being compresed in the compressor cylinder. It then passes through the compressor discharge valve, located above the cylinder, to the compressor discharge chamber. The high-pressure, high-temperature gas then flows to the condenser where it is cooled and liquefied, changing to a warm, high-pressure liquid. The warm, high-pressure liquid then flows to the liquid receiver where it falls to the bottom and any warm, high-pressure vapor present rises to the top. As the liquid level in the bottom of the receiver rises, it causes the float to rise also (Figure 3–10). As the float rises, it pulls the needle from its seat, opening the valve which allows the warm, high-pressure refrigerant to flow into the liquid line. In this system the high-side float is located inside and is a part of the receiver. There are other designs of the high-side float system that have the float located in a small, separate chamber of its own away from the receiver.

FIGURE 3–10 High-side float system.

COMPRESSION REFRIGERATION SYSTEMS

At the inlet of the evaporator the warm, high-pressure liquid flows through a pressure-reducing valve called a *liquid temperature valve*. This valve is made of a weighted needle valve. When the pressure of the warm, high-pressure liquid is great enough to overcome the weight, it will force the valve open. The liquid temperature valve will remain open until the pressure of the warm, high-pressure liquid in the liquid line drops below the pressure required to keep the valve open. The weight will then close the valve, stopping the flow of refrigerant. The refrigerant pressure is reduced further inside the evaporator. As the liquid refrigerant flows down into the evaporator coils, evaporation occurs. The low-pressure liquid absorbs heat from its surroundings and changes to a low-pressure vapor. The low-pressure vapor is then drawn through the suction line to the compressor where the cycle is repeated.

In the high-side float system, the flow of liquid refrigerant is controlled by both the high-side float in the liquid receiver and the liquid temperature valve at the evaporator. The liquid temperature valve automatically regulates the flow of the refrigerant from the liquid line into the evaporator. The valve is caused to open when the difference in pressure between the refrigerant in the liquid line and the refrigerant in the evaporator reaches a certain point. The valve immediately closes again as soon as the difference in pressure is reduced sufficiently.

The refrigerant charge in the high-side float system is very critical because the high-side float in the receiver opens and liquid refrigerant is pushed into the liquid line when the liquid refrigerant level in the receiver reaches a certain point. This causes any excess or overcharge of refrigerant in the system to pass over to the evaporator. Any excess of refrigerant would then raise the liquid refrigerant level in the evaporator, which could possibly result in *floodback*. Floodback occurs when liquid refrigerant flows into the suction line, causing frost to appear and possible damage to the compressor. If the refrigerant charge is insufficient, the liquid refrigerant level in the receiver may never rise enough to cause the float valve to open and allow refrigerant to pass into the liquid line. This condition would result in a starved evaporator and a loss of refrigeration. Therefore, the refrigerant charge in the high-side float system must be kept within a few ounces of the manufacturer's recommended charge.

Capillary Tube or Restrictor Systems. Capillary tube systems continue to be more widely adopted by manufacturers for both refrigeration and air conditioning applications because of their low cost, simplicity, and trouble-free operation. They also lend themselves more readily to hermetically sealed systems.

The evaporator in the system using this type of refrigerant flow control is also a flooded type. The major difference is the way in which the refrigerant is fed into the evaporator (Figure 3–11). In this type of system the capillary tube is between the strainer-drier and the evaporator. The liquid refrigerant is fed into the evaporator through a small tube. The amount of refrigerant allowed to flow is dependent mainly on the length and inside diameter of the tube.

Capillary tube systems require an accurate charge of refrigerant for several reasons. First, any overcharge of refrigerant will be in the low side of the system, resulting in floodback. Second, when the compressor stops, all the liquid refrigerant on the high side of the system tends to run over into the evaporator until the pressures balance throughout the entire system. This indicates that if there is an overcharge of refrigerant in the system, the liquid will flood the evaporator on the off cycle and cause floodback when the compressor starts the next cycle.

Automatic Expansion Valve Systems. These systems use a pressure-operated refrigerant flow control device. The purpose of the expansion valve is to control the flow of refrigerant into the evaporator. When this control is properly adjusted, it will keep the evaporator fully refrigerated.

In operation, the refrigerant gas is compressed in the compressor where the temperature and pressure are both raised. From the compressor the high-temperature, high-pressure gas passes to the condenser where the vapor is cooled and liquefied, changing to a warm, high-pressure liquid. The warm, high-pressure liquid then flows to the liquid receiver. The warm, high-pressure vapor fills the top part of the receiver and the warm, high-pressure liquid falls to the bottom. The pressure of the vapor forces the warm, high-pressure liquid through the liquid line to the expansion valve (Figure 3–12). The warm, high-pressure liquid passes through the automatic expansion valve into the evaporator. As the refrigerant passes through the expansion valve, both the pressure and temperature are reduced and the liquid refrigerant is *atomized,* broken into fine droplets of liquid. The expansion valve regulates the flow of refrigerant to the evaporator in response to the refrigerant pressure in the evaporator. In the

FIGURE 3–11   Capillary tube system.

FIGURE 3–12   Automatic expansion valve system.

evaporator, the low-temperature, low-pressure liquid refrigerant changes to low-temperature, low-pressure vapor as heat from the surroundings is absorbed. Notice that the liquid refrigerant does not collect in a pool inside the evaporator on a dry-type system. Instead, it passes through the evaporator as a mist and vapor combination. As compared to the flooded-type systems, there is relatively little liquid

COMPRESSION REFRIGERATION SYSTEMS

refrigerant in the evaporator. The low-pressure vapor is drawn through the suction line to the compressor and the cycle is repeated.

Thermostatic Expansion Valve Systems. These systems are also included in the dry-type system classification. Thermostatic control is used in combination with evaporator pressure, as in the automatic expansion valve systems, to provide more accurate control of the flow of refrigerant into the evaporator (Figure 3–13). While the evaporator operates basically as a dry system with the thermostatic expansion valve, control of refrigerant flow is much more positive and the operating efficiency is improved, closely paralleling that of the flooded system.

Operation of the thermostatic expansion valve is the same as the automatic expansion valve system, with one exception. The evaporator outlet temperature is used along with evaporator pressure. This additional control is accomplished by placing a temperature-sensitive bulb at or near the evaporator outlet.

Compound Refrigeration Systems. Compound refrigeration uses two or more compressors connected in series (Figure 3–14). When compressors are connected in series, the first compressor discharges through an oil separator and an intercooler, then into the suction of the second com-

pressor. The refrigerant vapor is desuperheated in the intercooler. The second compressor then discharges into the system condenser. The refrigerant vapor is condensed in the condenser and passes on to the liquid receiver.

The liquid refrigerant then flows to the flow control device, usually a thermostatic expansion valve (TXV), and into the evaporator. The refrigerant then absorbs heat and is changed to a vapor. The refrigerant vapor then flows back to the first compressor, and the cycle is repeated.

Compound refrigeration systems are used to increase capacity when operating at such low temperatures that one compressor cannot effectively meet the requirements. These systems require compressor motors that have a high starting torque.

Multiple-Evaporator Systems. These systems are used on commercial refrigeration systems where one condensing unit (compressor and condenser) is capable of operating satisfactorily under varying load conditions and is connected to two or more evaporators (Figure 3–15). The evaporators may be operating at the same temperature or there may be considerable temperature difference between them. A separate flow control device is used on each evaporator. An evaporator pressure regulating (EPR) valve is installed in the suction line to each evaporator. This valve maintains an evaporator pressure at or above a desired pressure. The com-

FIGURE 3–13 Thermostatic expansion valve system.

FIGURE 3-14 Compound refrigeration system.

pressor may be controlled electrically by a thermostat at each evaporator or a pressure control sensing the common suction pressure of the evaporators.

In operation, the refrigerant is discharged from the compressor into the condenser. The vapor is condensed in the condenser, and the liquid refrigerant flows to the liquid line manifold. The liquid refrigerant is divided in the manifold, and part goes to each flow control device. The liquid absorbs heat in the evaporators and changes to a vapor. The vapor pressure is controlled by the EPR valve at the evaporator outlet. The suction vapor flows down each suction line to the suction line manifold. In the manifold the suction vapor is directed into a common suction line. The suction vapor then enters a

suction accumulator that prevents liquid refrigerant from returning to the compressor. The suction vapor then enters the compressor suction valve, and the cycle is repeated.

Cascade Refrigeration Systems. These systems are generally used in industrial or commercial applications where cooling the objects to $-50\,°F$ ($-46\,°C$) or lower is required.

Too or more refrigeration systems are connected together so that the evaporator of one cools the condenser of the other (Figure 3–16). Both systems are in operation at the same time. The evaporator of the second system supplies the desired cooling. Each unit is a complete system in itself. Both systems start

FIGURE 3-15 Multiple-evaporator refrigeration system.

and stop at the same time by a temperature control bulb sensing the evaporator temperature of the second unit.

Modulating Refrigeration Systems. Modulating systems are used on installations that have varying load conditions, such as air conditioning systems. These systems were developed to more closely match the unit to the heat load. This is accomplished by using two or more compressors connected in parallel, with each compressor being operated by its own motor control (Figure 3-17). This control is usually a low-pressure control sensing the refrigerant suction pressure.

In operation, when the system is started, full capacity is required and all the compressors are operating. As the temperature inside the space is reduced, the evaporator load is reduced, resulting in a lower suction pressure. At a predetermined suc-

tion pressure, one compressor will stop operating. If the heat load is such that the suction pressure continues to fall, another compressor will stop operating. The system will continue operating in this manner until the system and load are balanced or the maximum system capacity reduction has been accomplished. When the temperature is increased inside the space, the suction pressure will also increase to the point that each compressor will be restarted until maximum system capacity has been reached.

Thermoelectric Refrigeration Systems. Thermoelectric refrigeration is based on the Peltier effect, which was discovered in about 1835. This method of producing refrigeration makes use of electron movement rather than using a refrigerant such as that used in the compression refrigeration systems (Figure 3-18). In practice this principle makes use of a simple thermoelectric couple. The heat is re-

FIGURE 3-16 Cascade refrigeration system.

moved from inside the insulated space and is transferred to a heat exchanger outside the space by the couple. The electrons move the heat from where it is not wanted to where it is not objectionable. The fins on the evaporator inside the space increase the flow of heat, and the fins on the heat exchanger outside the space give off the heat to the ambient air.

The two semiconductors, P and N, have different energy levels which cause the electrons to move from one place to another and take the heat with them. The P and the N refer to different types of semiconductor material and not positive and negative as may be thought. The types of materials used in making the P and N components have a great effect on their energy levels and the efficiency of the device on which they are used. These couples can be connected together in series to provide more capacity. When they are connected in this manner they are called a module (Figure 3-19).

The thermostat inside the space controls the temperature through the transformer-rectifier, whose purpose is to control the flow of dc current to the modules. The 120-V ac current is stepped down to 20 V dc in the rectifier. These units can also be designed to operate on 12 V dc, making them very popular for use as portable units used in automobiles, for taking on picnics, and for camping uses.

These units are not very efficient. The amount of refrigeration obtained for the amount of energy used is lower than for the conventional compression refrigeration system. These units have the capability of being used for either heating or cooling of the insulated space. Reversal of the dc current will cause the hot and cold surfaces to reverse, producing the heating or cooling effect desired. Some units have a reversal switch built in for this convenience.

FIGURE 3-17 Modulation refrigeration system.

FIGURE 3-18 Thermoelectric couple used for refrigerating a space.

FIGURE 3-19 Couples connected together to increase capacity.

Types of Refrigeration Systems

▣ Safety Procedures ▣

It should be remembered that when working with refrigeration equipment some of the components are very hot and will cause burns, while other components are very cold and can cause frostbite. Be sure to exercise caution when working near these components. Never put the tongue to an evaporator to determine if it is cold. Always use caution when working around rotating parts to prevent clothing and limbs from being caught by them. Be sure to prevent moisture from entering the system.

▣ SUMMARY ▣

☐ The two types of systems in use today are the compression system and the absorption system.

☐ Food is cooled in an ordinary icebox because the ice absorbs heat while melting.

☐ Any liquid that will boil at a temperature below the freezing temperature of water will keep food cool and make ice cubes.

☐ The simplest way to produce refrigeration is to place an open container of refrigerant inside an insulated cabinet and allow the refrigerant to escape.

☐ The boiling temperature of R-12 at 14.7 psig (202.57 kPa) is $-21.66\,°F$ $(-30\,°C)$. The temperature of the evaporator will cause air to circulate inside an insulated cabinet.

☐ A compressor increases the pressure of a vapor, which increases the temperature of the vapor.

☐ Only the heat of compression is added to the vapor during the compression process.

☐ The basic points to remember about a refrigeration system are: (1) all liquids absorb a lot of heat without getting any warmer when they boil into a vapor, and (2) pressure can be used to make vapors condense back into a liquid.

☐ Mechanical refrigeration is achieved by constantly circulating, evaporating, and condensing a fixed supply of refrigerant inside a closed system.

☐ It is possible to transfer heat from a cooler area to a warmer area because heat absorption takes place at a low pressure and low temperature, while condensation occurs at a high pressure and high temperature.

☐ In a basic refrigeration system, the compressor draws the refrigerant from the evaporator, increases its pressure and temperature, and discharges it to the condenser. The gas is cooled and changed to a liquid in the condenser. From the condenser, the refrigerant flows to the flow control device where the refrigerant flow to the evaporator is controlled. On passing through the flow control device, the pressure of the refrigerant is lowered. The refrigerant absorbs heat in the evaporator and boils into a vapor, which is removed from the evaporator by the compressor, and the cycle is repeated.

☐ The dividing points between the high side and the low side of the system are the compressor discharge valve and the orifice of the flow control device.

☐ The high side of the system includes all the components operating at or above condensing pressure.

☐ The low side of the system includes all components operating at or below the evaporating pressure.

☐ The flooded and dry systems are classified based on the condition of the refrigerant in the evaporator.

☐ Flooded systems have a pool of liquid refrigerant in the evaporator.

☐ Dry systems have only droplets of liquid in the evaporator.

☐ Flooded systems use the low-side float, high-side float, and capillary tube or restrictor as their flow control devices.

☐ Dry-type systems use the automatic expansion valve and the thermostatic expansion valve as their flow control devices.

☐ Thermoelectric refrigeration is based on the Peltier effect, which was discovered about 1835.

☐ The electrons move the heat from where it is not wanted to a place where it is not objectionable.

☐ The two semiconductors, P and N, have different energy levels, which cause the electrons to move from one place to another and take the heat with them.

☐ The thermostat inside the space controls the temperature through a transformer-rectifier, whose purpose is to control the flow of dc current to the modules.

☐ These units have the capability of being used for either heating or cooling of the insulated space.

▣ EXERCISES ▣

1. What method, other than the application of heat, can be used to cause a liquid to boil?
2. Why does the air circulate inside a refrigerated cabinet that does not have a fan in it?
3. How is the heat concentrated in a refrigerant vapor?
4. Is any heat actually added to a refrigerant vapor during the compression process?
5. What are the two main points to be remembered as far as basic ground rules are concerned about the refrigeration system?
6. Why do vapors form inside the evaporator of a refrigeration system?

7. What two things make it possible for us to transfer heat from a cooler area to a hotter area?

8. What does the compressor in a refrigeration system do?

9. What happens to the refrigerant in the condenser?

10. What happens to the refrigerant in the evaporator?

11. What are the dividing points between the high side and the low side of a refrigeration system?

12. What components are located in the high side of the refrigeration system?

13. Summarize the cycle of a compression system.

14. Name the two classifications of refrigeration systems that consider the condition of the refrigerant in the evaporator.

15. Name the five general types of flow control devices used on compression refrigeration systems.

16. What does the term *critical charge* mean?

17. What are the types of flow control devices that require a critical charge?

18. What is the difference between the automatic expansion valve and the thermostatic expansion valve?

19. Does the automatic expansion valve or the thermostatic expansion valve provide better control of the refrigerant?

20. Are the expansion valve systems considered flooded or dry-type systems?

21. Into what does the first compressor discharge on a compound refrigeration system?

22. How is the refrigerant flow controlled into the evaporator on multiple-evaporator systems?

23. What type of system is used in industrial or commercial applications where temperatures of $-50\,^\circ$F ($-46\,^\circ$C) or lower are required?

24. In a cascade refrigeration system, what does the first evaporator cool?

25. Why were modulating refrigeration systems developed?

4 Refrigeration Materials and Hand Tools

Performance Objectives

After completion of this chapter, the student should:
- Have a thorough understanding of the various types of piping and tubing used in refrigeration and air conditioning systems.
- Know how the various types of fittings are used in piping refrigeration systems.
- Know how to work with refrigerant tubing.
- Be able to name the basic hand tools and explain how they are used.
- Be able to work with the different types of materials used in refrigeration systems.
- Recognize and practice safety procedures while working on refrigeration and air conditioning systems.
- Complete the Exercises with at least 80 percent accuracy.
- Complete the Laboratory Exercises with at least 80 percent accuracy.

□ Piping and Tubing □

A complete refrigeration unit consists of several parts that function to vaporize and condense the refrigerant as it circulates through the system. When these parts are located at a distance from each other where there can be no direct connection from one to the other, the refrigerant must have some means of travel. This is accomplished by the use of piping and tubing that is only a hollow bar used to transfer liquids and vapors from one part of the system to another. There is no definite ruling that defines the difference between piping and tubing. However, piping is ordinarily understood to have walls of considerable thickness, while tubing has comparatively thin walls.

Welded and brazed pipe and tubing is made from several different metals. Copper, brass, iron, and steel are the most popular for use in refrigeration and air conditioning installations. Pipe made from iron or steel is formed into a cylindrical shape from a flat strip so that the two edges meet and are butt-welded together. Sometimes, the edges are beveled and lap-welded together. The lap-welded tubing is stronger and better adapted for bending. Copper and brass pipe may be made in a similar manner. However, instead of welding, a process called brazing is used to join the edges. The brazed type of pipe or tubing is not suitable for use when bending is required.

Seamless pipe and tubing is a variety of pipe and tubing that is better for refrigeration use than the welded or brazed types. The piping and tubing used in refrigeration installations are generally of the seamless type made by forcing a solid bar of hot metal over a mandrel. This is known as the extrusion process, and the tubing has no seam. This type of pipe and tubing is much stronger and better adapted to bending and forming than the welded or brazed types.

□ Seamless Copper Tubing □

This type of tubing is more commonly used on refrigeration systems using refrigerants other than ammonia. Copper is never used when ammonia is the desired refrigerant because it will quickly corrode and disintegrate if left in contact with the vapor any length of time. It is used successfully with all

FIGURE 4–1 Roll of soft-drawn copper tubing.

the Freon group of refrigerants.* The heat conductivity of copper is the greatest of all the available commercial metals. It is quite difficult to thread copper tubing because of its softness. Therefore, soldered connections, which make use of the maximum strength of the tube in construction of the joint, are used. In addition to its good conductivity of heat, soft copper tubing is easy to bend, which eliminates many elbows and joints.

Copper tubing is available in either soft-drawn or hard-drawn types. Soft-drawn copper tubing is available in 25-, 50-, and 100-ft (7.6-, 15.2-, and 30.4-m) rolls of ⅛ to 1⅜ in. (3.175 to 34.93 mm) outside diameter (OD). The most popular length is the 50-ft (15.2-m) roll (Figure 4–1). Soft-drawn copper tubing also is available in straight lengths in larger sizes. Soft-drawn tubing of ⅝ in. and smaller is easy to work with because it is fairly flexible. It can be formed by hand. However, care must be taken to avoid kinks in the tubing which will restrict the flow of refrigerant. Soft-drawn copper may be joined by use of soft soldering, silver soldering, flaring techniques, and epoxy procedures.

Hard-drawn tubing is available in 20-ft (6.08-m) lengths and from ¼ to 6⅛ in. (6.5 to 155.57 mm) OD. This tubing cannot be easily formed by hand. It may kink or collapse, resulting in a restriction that must be removed from the system. Minor bends can be achieved with this tubing, however, if the area to be bent is heated to a cherry red and then allowed to cool in the air before bending is attempted. This

process is known as *annealing*. Hard-drawn tubing is especially useful when long lengths, which cannot be supported, must be used. It is also used when the appearance of the installation is important. Annealing is also required if flare-type fittings are to be used. Hard-drawn copper may be joined by use of soft soldering, silver soldering, and epoxy methods.

Safe Working Pressures. Safe working pressures of copper tubing will vary with the size of tubing and the temperature of the material inside the tube (Table 4–1). When the temperature of the material (refrigerant in this case) increases, the safe working pressure of the tube decreases.

The safe working pressures for hard-drawn copper tubing are different from those for soft-drawn tubing (Table 4–2). Care must be used to ensure that these pressures are not exceeded.

TABLE 4–1 Safe Working Internal Pressures (psi) of Soft-Drawn ACR Tubing

Tube OD (in.)	Wall Thickness (in.)	Weight per Foot (lb)	Temperature (°F)			
			150	250	350	400
¼	0.030	0.0804	1230	1130	970	720
⅜	0.032	0.134	860	790	670	500
½	0.032	0.182	630	580	490	370
⅝	0.035	0.251	540	500	430	320
¾	0.035	0.305	440	400	350	260
⅞	0.045	0.455	500	460	390	300
1⅛	0.050	0.655	430	400	340	250
1⅜	0.055	0.884	390	360	300	230
1⅝	0.060	1.140	370	340	280	220

TABLE 4–2 Safe Working Internal Pressures (psi) of Hard-Drawn ACR Tubing

Tube OD (in.)	Wall Thickness (in.)	Weight per Foot (lb)	Temperature (°F)			
			150	250	350	400
⅜	0.030	0.126	900	870	570	380
½	0.035	0.198	800	770	500	330
⅝	0.040	0.285	740	720	470	310
¾	0.042	0.362	650	630	410	270
⅞	0.045	0.455	590	570	370	250
1⅛	0.050	0.655	510	490	320	210
1⅜	0.055	0.884	460	440	290	190
1⅝	0.060	1.14	430	420	270	180
2⅛	0.070	1.75	370	360	230	150
3⅛	0.090	3.33	330	320	210	140
3⅝	0.100	4.29	320	310	200	130
4⅛	0.110	5.38	300	290	190	120
5⅛	0.123	7.61	280	270	180	120

*Freon is a trade name for a family of synthetic chemical refrigerants manufactured by DuPont.

Classification. Seamless copper tubing is made in four general groups as follows:

1. SAE soft tubing, thin wall
2. Type K soft or hard tubing, extra-thick wall
3. Type L soft or hard tubing, thick wall
4. Type M soft or hard tubing, standard wall

The SAE and type L are the most popular for use in refrigeration work and are usually designated as air conditioning and refrigeration (ACR) tubing. Type M tubing is used for condensate drain lines and uses where the operating pressures do not exceed 150 psig (1131.49 kPa).

ACR tubing is cleaned, dried, and capped to prevent contamination by moisture and other foreign particles in accordance with ASTM 3280 and ANSI-B9.1-1971 refrigeration industry standards. When a piece of tubing is not used, the ends should be sealed to keep it clean and dry for later use on refrigeration systems. In addition to cleaning and dehydrating, hard-drawn tubing is pressurized with dry nitrogen and reusable plugs are inserted in the ends (Figure 4–2).

FIGURE 4–2 Dehydrated, pressurized ACR tubing.

Cutting Copper Tubing. Many times during the installation and service of air conditioning and refrigeration systems it becomes necessary to cut the copper tubing. The tubing must be cut properly to allow leakproof joints to be made and to prevent restrictions to the refrigerant flow.

Basically, there are two methods used to cut copper tubing: (1) hand-held tubing cutters and (2) the hacksaw and a sawing fixture (Figure 4–3). The hand-held cutters are available in sizes to cut tubing from ⅛ in. (3.18 mm) OD, to approximately 4⅛ in. (104.78 mm) OD. The hand-held cutter is placed on the tubing at the desired length and the cutter blade is brought into contact with the tubing (Figure 4–4). The final adjustment is made by turning the knob approximately ¼ turn (Figure 4–5). The cut is then made by rotating the cutter around the tube. After the completion of each turn around the tube, tighten the cutter knob another ¼ turn until the tube is completely cut through. There will be a burr inside the tubing that will be a resistance to the refrigerant flow. This burr must be removed with the reamer blade on the cutter (Figure 4–6).

The hacksaw method of cutting tubing is generally used on the larger size hard-drawn tubing. The saw should be used with a sawing fixture so that a square end can be accomplished on the tubing (Figure 4–7).

To obtain the best cut possible, the hacksaw blade should have at least 32 teeth per inch (25.40 mm). The tubing should be positioned so that the saw filings will not fall into the tubing to be used. Be sure

Retract Reamer

FIGURE 4–3 Typical tubing cutters. (Courtesy of Gould, Inc., Fluid Components Division.)

(a)

(b)

FIGURE 4–4 Tubing cutter application.

REFRIGERATION MATERIALS AND HAND TOOLS

FIGURE 4–5 Tubing cutter adjustment.

Retractable
locking
reamer

FIGURE 4–6 Tubing cutter reaming blade.

FIGURE 4–7 Cutting tube with a hacksaw and sawing fixture. (Courtesy of Gould, Inc., Fluid Handling Division.)

to remove any saw filings that accidentally enter the tubing to be used.

□ Seamless Steel Tubing □

This type of tubing is manufactured by the extrusion process. It may be either hard drawn or soft drawn. The hard-drawn tubes are the hardest and may be softened and toughened by the annealing process. The soft drawn is best for use where bending is to be done. Steel tubing is sized according to the outside diameter. The wall thickness is made to standard gauge numbers. The seamless tube is more

expensive than the welded type, but it will withstand a great deal more pressure. It is usually obtained in straight lengths of 20 ft (6.08 m). The most common method of joining steel tubing is soft soldering and silver soldering.

Steel tubing is being used almost exclusively in the manufacture of condensers for domestic refrigerators. The heat-conducting capabilities of steel are fewer than those of copper. However, the steel is closely fastened to the shell of the refrigerator, which in effect increases the area of the condenser. This increased area increases the heat transfer to closely parallel the heat transfer capabilities of an equal length of copper tubing. The difference in cost between copper and steel tubing will more than offset the cost of fastening the steel tubing to the refrigerator shell.

□ Aluminum Tubing □

This type is being used extensively in the manufacture of evaporators for domestic refrigerators. The heat transfer capabilities of aluminum are not as good as those of copper. Again, however, the cost of aluminum as compared to copper will more than offset any added cost in manufacturing. Aluminum tubing is easily formed by hand, eliminating almost entirely the need for fittings. Soft soldering and special welding alloys are used to join aluminum tubing. Sometimes epoxy is used in the joining process.

□ Stainless Steel Tubing □

This tubing, like the others, is available in the usual ACR copper tube sizes. Stainless steel is strong and has a high resistance to corrosion. Stainless steel tubing 304 is the most popular type for refrigeration work. This tubing has a low carbon, nickel, and chromium content. Some food processing requires this type of tubing. Joining is commonly accomplished by use of soft-soldering techniques.

□ Flexible Tubing □

Automotive air conditioning systems and some transport refrigeration systems make use of flexible rubber tubing (Figure 4–8). The vibration caused by

FIGURE 4-8 Flexible refrigeration tubing (hose). (Courtesy of Refrigeration and Air-Conditioning Division, Parker Hannifin.)

the constant movement of the transport vehicle would work-harden metal tubing and allow the system to leak. Flexible tubing usually has neoprene liners that are covered by rayon braids and bonded to provide leak-free, long-lasting service. It is usually purchased in bulk lengths and cut to meet specific needs. The fittings are easily installed on the tubing to provide flexibility in use and design.

☐ Fittings ☐

There are several types of fittings in refrigeration work today. The type used depends largely on the tubing material used and the degree of permanency of connections desired. The most popular are the wrot (sweat), flare, compression, and O-ring and hose fittings.

Wrot Fittings. These are more commonly referred to as sweat fittings. They are made to be used with either soft solder or silver solder. They are designed to make the necessary turns and connections with a minimum amount of pressure drop in the pipe (Figure 4-9). In use, the tubing must be cut squarely so that a proper fit will be possible at the stop in the fitting (Figure 4-10). If this joint is not properly made, excess solder may flow through and get inside the refrigeration system. Inside the system, the solder will cause extra friction to the refrigerant and possible damage to other components.

There is a wrot fitting to fill almost any need. The service and installation technician should be familiar with them so that he will be able to make repairs to the system or accomplish the efficient replacement of parts.

Joining Wrot Fittings. The joining together of tubing is a very important operation to the installation and service technician. Because it is a common job, many people tend to be lax about using the proper procedures.

There are six basic steps that must be followed to produce strong leaktight solder joints. The following steps are used in soft soldering and silver solder-

FIGURE 4-9 Wrot copper fittings.

FIGURE 4-10 Proper joint between tube and fitting.

ing operations: (1) proper cutting of the tubing, (2) good fit with proper clearance, (3) cleaning the tube and fitting, (4) proper fluxing, (5) proper heating and flowing of the solder, and (6) final cleaning.

The following is an explanation of the six basic steps:

1. *Proper cutting of the tubing* (Figure 4-11). Cut the tube to the desired length using a tubing cutter or hacksaw. Be sure that the end is cut square. When using a hacksaw, be sure to tilt the end of the tube downward so the saw filings will fall from the tube. Remove the burrs from inside the tube with a reamer. A half-round file may also be used. Tilt the tube downward so the shavings will fall from the tube (Figure 4-12).

FIGURE 4–11 Cutting tube. (Courtesy of Mueller Brass Company.)

FIGURE 4–12 Deburring. (Courtesy of Mueller Brass Company.)

FIGURE 4–13 Sanding. (Courtesy of Mueller Brass Company.)

FIGURE 4–14 Wire brushing. (Courtesy of Mueller Brass Company.)

2. *Good fit with proper clearance.* Place the fitting on the tube end and chcck to be sure there is a good close fit with uniform clearance all the way around. Use a sizing tool to reshape out-of-round tubing. This will bring the damaged tubing back to the proper size and shape.

3. *Cleaning the tube and fittings* (Figure 4–13). The tube should be cleaned until all oil, grease, rust, or oxidation is removed and the metal has a shiny new appearance. Use a piece of sand cloth to polish the metal. A cleaning solvent may be used to remove any contamination from the metal. Apply the solvent with a brush or rag and rub briskly. Do not allow any of the solvent to remain inside the tube or fitting. To do so will only add to the contaminants inside the refrigeration system.

Clean the inside of the fitting with a clean wire brush (Figure 4–14). Do not use steel wool or emery cloth. The shreds of the steel wool may enter the system while the emery cloth may contain oil and abrasives—all of which must be kept out of the system. The cleaning should be done just before the soldering process so that oxidation will be kept to a minimum. Do not touch the cleaned surfaces with your hand or a dirty rag after cleaning.

4. *Proper fluxing of the joint.* The purpose of the flux is to keep the metal clean after it has been cleaned in step 3. Flux prevents air from coming in contact with the heated metal, which causes the surface to oxidize. Soldering flux may be obtained in both paste and liquid form. Both types produce satisfactory results. The steps listed below should be followed when applying flux to a joint.

(a) Always select the proper flux for the job. Use soft solder flux on soft solder jobs and silver brazing flux on silver brazing joints. Never use the wrong flux.

(b) Stir the flux before applying it to the joint (Figure 4–15). When flux stands, especially in hot weather, the chemicals tend to settle to the bottom.

FIGURE 4–15 Stirring flux. (Courtesy of Mueller Brass Company.)

Fittings

(c) Insert the tube part way into the fitting.

(d) Apply the flux with a brush. Brush the flux all around the outside of the joint. Never apply the flux with your fingers because perspiration and oils may prevent the solder from sticking to the tube. Do not use too much because it may enter the system and cause damage.

(e) Insert the tube all the way into the fitting. If possible, rotate the fitting or tubing to spread the flux uniformly over the complete joint (Figure 4–16).

5. *Proper heating and flowing of the solder.* To make a soft solder joint, apply the flame to the fitting and direct the flame so the heat will also warm the tube (Figure 4–17). Do not allow the flame to touch the flux because the flux will burn and the joint will have to be recleaned and refluxed. Heat the heaviest part of the fitting first, then move toward the joint to be soldered. Move the heat around the fitting so the total joint will be uniformly heated. Occasionally, remove the flame and touch the opening between the fitting and tube with the solder (Figure 4–18). When the joint is hot enough, the tube will melt the solder.

FIGURE 4–18   Applying solder. (Courtesy of Mueller Brass Company.)

FIGURE 4–19   Large-size work. (Courtesy of Mueller Brass Company.)

FIGURE 4–16   Joining. (Courtesy of Mueller Brass Company.)

FIGURE 4–17   Applying flame. (Courtesy of Mueller Brass Company.)

Do not melt the solder with the torch flame because a cold, weak joint will result. Capillary attraction will draw the solder into the joint. When a fillet of solder appears where the tube and fitting join, the joint is complete. On large fittings of 2 in. (50.80 mm), it is desirable to heat the tube with a special Y-shaped torch tip (Figure 4–19). Two torches may be used if a Y-shaped torch tip is not available. Also, on larger tubing each fitting should be tapped with a small mallet at two or three points around the joint while feeding the solder. This action helps to settle the joint and releases any trapped gases that may interfere with the flow of solder.

To silver solder a joint, a soft oxyacetylene flame will provide the best type of heat. Adjust the flame to a slightly reducing flame—use less oxygen. Start heating the tube about ½ to 1 in. (12.70 to 25.40 mm) away from the fitting. Heat the tube evenly all the way around. When the flux has turned to a clear liquid, move the heat to the fitting (Figure 4–20). Move the heat back and forth from tube to fitting, keeping it directed toward the tube. Do not allow the heat to remain

FIGURE 4–20   Brazing horizontal joints.

FIGURE 4–21   Smoothing solder. (Courtesy of Mueller Brass Company.)

FIGURE 4–22   Flare tube fittings.

in one spot too long because this can cause over-heating.

When the flux turns to a clear liquid, pull the flame back from the joint a little and apply the silver solder to the opening where the tube and fitting join. When the joint is properly heated, the silver solder will flow into the joint. When silver soldering larger joints, it is best to complete the process on small segments at a time overlapping the alloy at each segment.

Apply only enough solder to fill the joint. If too much solder is used, it will either be wasted or will flow into the system and possibly cause damage. A piece of solder or silver solder the length of the tube diameter is sufficient to make a leak-tight joint.

6. *Final cleaning of the joint.* To clean a joint and make a professional looking job, wipe the joint clean with a clean cloth while the solder is still melted (Figure 4–21).

Flare Fittings.   The 45° flare is used in refrigeration piping work (Figure 4–22). These fittings are usually more easily installed because of the type of tools needed in making the connection. In practice, the flare nut is placed over the tube, which is then flared and clamped to another fitting. The flare on the tube should be the same size as the chamfer (beveled edge) on the fitting. The nut is then screwed onto the mating part, squeezing the tubing between the parts

FIGURE 4–23   Flare connection.

(Figure 4–23). A drop of refrigeration oil should be put on each of the mating parts to prevent twisting of the tubing as the connection is tightened. The seal between the flare fittings is accomplished by the tubing between the parts. Therefore, if the fitting is tightened too much, the flare may either be cut off or the tubing squeezed too much to allow a proper seal to be made. Usually, hand tight plus one turn is sufficient. If flare fittings are used where the temperature varies between the on and off cycles of the unit, a leak will sometimes occur. This leak is the result of expansion and contraction caused by the warming and cooling of the metal. The expansion and contraction will cause the flare nut to loosen and will require retightening before refrigerant is added to the system.

Joining Flare Fittings. There are eight basic steps that must be followed to produce strong, leaktight flare joints. These steps are: (1) properly cut the tubing, (2) install the flare nut on tubing, (3) place flare block over tube, (4) install flare yoke, (5) make the flare, (6) remove the yoke, (7) remove the flare block, and (8) check flare size.

The following is an explanation of the eight basic steps.

1. *Properly cut the tubing* (Figure 4-24). Cut the tube to the desired length using a tubing cutter or hacksaw. Be sure that the end is cut square. When using a hacksaw, be sure to tilt the end of the tube downward so that the saw filings will fall from the tube. Remove the burrs from inside the tube with a reamer. A half-round file may also be used. Tilt the tube downward so that the shavings will fall from the tube (Figure 4-25).

2. *Install the flare nut on the tube* (Figure 4-26). Slide the flare nut onto the tubing with the threaded end toward the end to be flared. Be sure that there is enough room to install the flare nut, flare block, and yoke before bending or completing any fittings close to the flared end of the tube.

3. *Place the tube in the flaring block* (Figure 4-27). Insert the tube through the flaring block with the proper amount extending through to make a proper flare. Be sure to insert the tube through the proper hole. When too much tube is extended, a large flare will result and possibly split the tube. When too little tube is extended, too small a flare will result, which will probably not make a leaktight seal. Tighten the block on the tube to prevent slipping during the flaring process.

4. *Install the yoke on the flaring block* (Figure 4-28). Install the yoke on the flaring block with the flaring cone over the tube end. Put one or two drops of refrigeration oil on the cone to allow it to rotate inside the tube more freely.

5. *Make the flare* (Figure 4-29). When the yoke is

FIGURE 4-24 Cutting tube. (Courtesy of Mueller Brass Company.)

FIGURE 4-25 Deburring. (Courtesy of Mueller Brass Company.)

FIGURE 4-26 Flare nut on tubing.

FIGURE 4-27 Placing tube in flare block. (Courtesy of Gould, Inc., Valve and Fittings Division.)

FIGURE 4-28 Installing yoke on flaring tool. (Courtesy of Gould, Inc., Valve and Fittings Division.)

FIGURE 4-29 Making flare on tubing. (Courtesy of Gould, Inc., Valve and Fittings Division.)

FIGURE 4-32 O-ring fitting.

in place, turn the feed screw five or six turns after the cone touches the tubing. Do not overtighten the yoke because the metal making the flare will become dead (nonresilient) and will prevent a leaktight connection.

6. *Remove the yoke from the flare block.* When the flare is completed, turn the feed screw in the opposite direction until the cone is free of the tube. Backing off the cone automatically burnishes the flare to a highly polished finish.

7. *Remove the flaring block from the tube.* When the yoke has been removed from the flaring block, loosen the clamp screw enough so that the die can be separated enough to allow removal of the block from the tube. Remove the block.

8. *Check the flare size* (Figure 4-30). Slide the threaded side of the flare nut over the flare. The flare should just pass through the threads and seat on the chamfer in the flare nut.

Compression Fittings. These types of fittings are becoming popular for making quick connections to condensing units with refrigerant lines (Figure 4-31). In practice, hard-drawn tubing is preferred for use with these fittings. Soft-drawn tubing may be used if it is perfectly round and no kinks are close to the fitting. These fittings will usually provide a leaktight

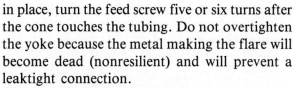

FIGURE 4-30 Checking flare size.

FIGURE 4-31 Compression fitting. (Courtesy of Dearborn Division, Addison Products Co.)

joint when tightened hand tight plus 1½ turns. To prevent damage to the tubing, be careful not to tighten these fittings excessively. If the tubing is dented enough to allow refrigerant leakage, then the tubing must be cut off past that point and the fitting reassembled and tightened.

Hose (Barbed) Fittings. These types of fittings are commonly used on automotive and transport refrigeration systems. They are designed to allow connections to be made between a neoprene hose and the screwed fittings of system components. In practice, the threaded connection is made, then the hose is slipped over the barbed end and a clamp is tightened on the hose. Usually, once the clamp has been tightened, the hose will have to be cut off the fitting. These fittings provide a gastight seal and are satisfactory for refrigeration work.

O-Ring Fittings. These types of fittings are used on some service valves and special connections. These fittings use mating parts with an O-ring placed in a groove to provide a seal (Figure 4-32). In practice, caution must be taken to ensure that the O-ring is inside the groove provided for it. Usually, hand tight plus ½ turn will ensure a leaktight fit. Care must be taken when making soldered connections close to these fittings because heat will destroy the O-ring, rendering it useless. If the tubing must be heated close to these fittings, a damp cloth can be wrapped around the O-ring fitting to keep it cool. The cloth must be kept damp.

▢ Working with Tubing ▢

A few of the more common tasks of working with tubing deserve mention here.

Bending Tubing. The use of tubing benders is recommended when making a bend (turn) in tubing.

(a)   (b)

FIGURE 4-33 **(a)Tubing bender (courtesy of Robinair Manufacturing Company); (b) recommended technique in bending tubing by hand (Courtesy of Gould, Inc., Valve and Fittings Division.)**

Bending tubing eliminates elbows and soldered or flared connections. The tubing is measured and marked where the bend is desired. It is then placed in the bender of the proper size and the handles are pulled together to form the desired angle [Figure 4-33(a)]. The bend should exceed the desired angle by about 3 to 5° because when the tubing is removed from the bender, it will straighten some. It is almost impossible to completely straighten a piece of tubing that has been bent with a bender. Therefore, be sure to put the bend in the right place the first time.

Hand bending is another method of forming tubing. When making a hand bend, start with a larger radius and work the tubing into the desired bend by reducing the radius gradually [Figure 4-33(b)]. Do not attempt to make a small radius on the first bending of the tube because the tube may be kinked. The minimum bending radius, when hand bending tubing, is about five times the tube diameter, while larger tubing may require about 10 times the tube diameter.

Changing Tube Sizes. Many times an installation and service technician will be required to change sizes of refrigerant piping. The proper way to accomplish this is to use a reducing fitting. However, many times he will not have all the necessary fittings. This problem can be overcome very easily in most cases, especially when changing the size of tubing. The tubing used in refrigeration work is sized in $\frac{1}{8}$ in. (3.175 mm). This allows one size of tube to be placed inside the next larger size of tube. A joint made this way is not within the acceptable tightness tolerance to be soft soldered. However, silver solder will work very satisfactorily. This type of joint is commonly used in service and installation procedures.

FIGURE 4-34   Silver solder line repair.

Line Repair. Line repair is sometimes needed when a cracked line occurs and the desired fittings are not available. The service technician may repair the line satisfactorily and get the system operating again by using this method. First, be certain that no pressure exists in the system because injury may result if the pressure builds enough to blow hot oil and metal from the line. Next, clean the joint with sand cloth and apply silver solder flux. *Note:* This procedure should not be attempted with soft solder.

Heat the line to a dull cherry red and apply silver solder until the area around the crack is slightly built up (Figure 4-34). Be sure not to overheat the tube. This type of repair will last indefinitely if properly done.

Equivalent Length of Pipe. The use of equivalent length of pipe is very important in refrigerant piping procedures. The number of fittings in a refrigeration system should be kept to a minimum. Every turn encountered by the refrigerant flowing through the tubing is an added point of resistance. Excessive resistance will reduce the equipment capacity. Every elbow gives an added resistance (pressure drop) equal to a given length of that size tubing (Table 4-3). Pressure drop and line sizing tables are usually designed on the basis of a given pressure drop for each 100 ft (30.4 m) of straight pipe. Therefore, the use of an equivalent length of pipe allows the data to be used directly.

TABLE 4-3 Equivalent Length (ft) of Straight Pipe for Valves and Fittings

OD Line Size (in.)	Globe Valve	Angle Valve	90° Elbow	45° Elbow	Tee Line	Tee Branch
½	9	5	0.9	0.4	0.6	2.0
⅝	12	6	1.0	0.5	0.8	2.5
⅞	15	8	1.5	0.7	1.0	3.5
1⅛	22	12	1.8	0.9	1.5	4.5
1⅜	28	15	2.4	1.2	1.8	6.0
1⅝	35	17	2.8	1.4	2.0	7.0
2⅛	45	22	3.9	1.8	3.0	10.0
2⅝	51	26	4.6	2.2	3.5	12.0
3⅛	65	34	5.5	2.7	4.5	15.0
3⅝	80	40	6.5	3.0	5.0	17.0

When accurate calculations of pressure drop are required, the equivalent length for each fitting should be calculated. In practice, an experienced system designer may be capable of making an accurate overall percentage allowance unless the piping arrangement is very complicated. For long runs of piping (100 ft or more) a 20 to 30% allowance of the actual length may be adequate. For short runs an allowance as high as 50 to 75% of the lineal length may be required. Judgment and experience are necessary in making a good estimate. These estimates should occasionally be checked with actual calculations to ensure reasonable accuracy.

☐ Hand Tools ☐

In addition to the commonly used mechanic's hand tools and the refrigeration test instruments, there are some specialized hand tools used by service and installation technicians. The most popular of these tools will be covered at this time, although all of Chapter 13 is devoted to refrigeration test instruments.

Tubing Cutters. Tubing cutters are used to cut tubing with a square end, enabling a service technician to make leaktight connections. In practice, as shown in Figure 4–4(a), the cutter is placed around the tube and the cutter wheel is then tightened on the copper tubing [Figure 4–4(b)]. Care should be taken not to tighten the cutter wheel too much because the tubing will be flattened and a leaktight joint will be almost impossible to make. After the tubing is touched with the cutter wheel, the cutter handle should not be turned more than ½ turn (Figure 4–5).

Make a complete revolution of the tube with the cutter, then tighten the cutter handle no more than ½ turn again. Repeat this procedure until the tubing has been completely separated. To obtain a square end on the tubing, the tubing must be placed on the rollers so that the cutter wheel will be centered on the tube (Figure 4–35).

After the tubing has been completely separated, there will be a burr on the inside of the tubing. This burr must be removed or it will cause a restriction to the refrigerant flow. Most tubing cutters have an attached reaming blade (Figure 4–36). This blade is rotated outward and then inserted into the tubing end and rotated to remove the burr. Care should be taken to prevent the loose metal chips from falling into the tubing or they will clog the strainers and filters, which will need to be removed for cleaning or replacement.

Roller type with flare cut-off groove

FIGURE 4-35   Tubing cutter rollers.

Retractable locking reamer

FIGURE 4-36   Tubing cutter reaming blade.

Tubing cutters are also equipped with a flare cutoff groove in the rollers so that bad flares can be easily cut from the tubing (Figure 4–37). This permits easy removal of the flare with a minimum amount of wasted tube and prevents threading.

Flaring Tools. Flaring tools (Figure 4–38) are used to make a flare on the end of tubing so that flare fittings can be used to make leaktight connections. When a flare is to be made, insert the tubing into the flare block so that approximately ¼ in. (6.5 mm) extends above the flare block [Figure 4–39(a)]. Swing the clamp into place against the end die and tighten it sufficiently to prevent slippage of the tube in the block [Figure 4–39(b)]. Slide the yoke over the tubing and turn the feed screw 5 to 6 turns after the cone touches the tubing [Figure 4–39(c)]. When the flare is completed, reverse the flaring process to remove the tube from the flaring block. Backing off of the cone automatically burnishes the flare to a highly polished finish. Flares made in this type of flaring block are stronger because the flare is made above the flaring block die, not against it, and the original wall thickness is maintained at the base of the flare. There is no chance of "washing out" the flare. Flaring tools can be used for re-rounding and sizing bent tubing.

Swaging Tools. Swaging tools (Figure 4–40) are used to enlarge one end of a piece of tubing so that two pieces the same size can be soldered together, eliminating the use of a coupling. Each size tubing requires a separate swaging tool. Swaging tools are used with the flaring block part of the flaring tool. The tubing is inserted into the proper-size die with enough extending out to allow the swaging process to be completed with the block tightened securely. Do not extend the tubing too much past the block because the tubing may bend and ruin the swage. If, however, too little tubing is extended, it may be cut when the swaging tool reaches the flaring block. After the tubing is properly placed in the block, the swaging tool is placed in the tube and struck with a hammer. Between each blow with the hammer, rotate the swaging tool to prevent its sticking in the tubing. Should it stick, a few light sideways licks with the hammer will loosen the swaging tool and allow its removal from the tube.

Tubing Reamers. Tubing reamers are used to remove burrs from refrigeration tubing (Figure 4–41). After the tubing has been cut, the reamer is placed over the tube and rotated, removing the burrs from either the inside or the outside of the tube. Then the reamer is turned around and placed over the tube again and rotated to remove the burrs. Since these are multisized tools, one tool will fit many sizes of tubing.

FIGURE 4–37 Tubing cutter with flare-removing rollers. (Courtesy of Gould, Inc., Valve and Fittings Division.)

FIGURE 4–38 Flaring tool. (Courtesy of Gould, Inc., Valve and Fittings Division.)

Tube height gauge
Slot in yoke is used for tube height gauge.

Secure clamping
Sliding dies with lever clamping action.

Slip-on yoke
Yoke slips over top of bar, then locks into position with slight turn.

(a)   (b)   (c)

FIGURE 4–39 Use of flaring tool. (Courtesy of Gould, Inc., Valve and Fittings Division.)

FIGURE 4–40　Swaging tool. (Courtesy of Robinair Manufacturing Company.)

FIGURE 4–41　Tubing reamer. (Courtesy of Gould, Inc., Valve and Fittings Division.)

FIGURE 4–42　Pinch-off tool. (Courtesy of Gould, Inc., Fluid Components Division.)

FIGURE 4–43　Positive seal made by pinch-off tool. (Courtesy of Robinair Manufacturing Company.)

FIGURE 4–44　Tubing bender. (Courtesy of Gould, Inc., Valve and Fittings Division.)

FIGURE 4–45　Spring-type tubing bender. (Courtesy of Gould, Inc., Valve and Fittings Division.)

Pinch-Off Tools. Pinch-off tools are used to block the escape of refrigerant from a sealed system through the process tube after service operations have been performed. After the service operations are finished, the pinch-off tool (Figure 4–42) is placed on the process tube of the system and tightened. The tool closes on the tube and causes a positive seal. The service tools are then removed and the process tube is sealed with silver solder. (Figure 4–43). The pinch-off tool is then removed.

Tubing Benders. Tubing benders are used to make bends in tubing. Any angle of bend (turn) can be made with these tools. Their use eliminates elbows and many soldered or flared connections. The tubing is measured and marked where the bend is desired. It is then placed in a bender of the proper size and the handles are pulled together to form the desired angle (Figure 4–44). The bend should exceed the desired angle by about 3 to 5° because when the tubing is removed from the bender it will straighten some. It is almost impossible to completely straighten a piece of tubing that has been bent with a bender.

Another type of bender sometimes used is the spring bender (Figure 4–45). The spring is placed over the tubing and the bend is formed by hand. The bend should exceed the desired angle by about 5° and be restraightened to the desired angle so the bender can be slipped from the tube. Care should be taken when using this type of bender because if too sharp a bend is attempted the tubing may crimp and make removal of the spring difficult.

FIGURE 4-46 **Service valve adapter kit.**

Hermetic Service Valve Kits. These kits are used to perform service operations involving the refrigerant and oil in sealed systems (Figure 4–46). In use, the proper adapter is installed on the unit and the gauge manifold is then attached to the adapter. This allows checking of the refrigerant pressures and charging procedures to be completed without cutting the tubing or making sweat connections.

Reversible Ratchets. These ratchets are used to turn the stem on service valves to allow service operations to be completed. The most common size is ¼ in. (6.5 mm). A ³⁄₁₆-in. (4.7625-mm) square is provided on the end with a ½-in. (12.7-mm) six-point socket (Figure 4–47). There are a variety of sockets available to fit almost any size of valve stem that will enable the service technician to connect the refrigeration gauges to the system (Figure 4–48). Also available are a variety of packing gland nut sockets that can be used with the reversible ratchet (Figure 4–49). These are used to tighten the packing glands on certain valves to prevent leakage of refrigerant.

Gauge Manifolds. Probably the most often used tool in the service technician's tool chest is the gauge manifold. A gauge manifold is comprised of a compound gauge, a pressure gauge, and the valve manifold (Figure 4–50). All the service operations performed involving the refrigerant, lubricating oil, and evacuation can be accomplished by use of this tool.

FIGURE 4-47 **Reversible ratchet. (Courtesy of Bonney.)**

FIGURE 4-48 **Valve stem socket. (Courtesy of Bonney.)**

FIGURE 4-49 **Packing gland socket. (Courtesy of Bonney.)**

Compound Gauges. These gauges (Figure 4–51) are used to read pressures both above atmospheric and below atmospheric (vacuum). In practice, they are used to determine pressures in the low side of the system. The outside scale is in pressure (psig).

FIGURE 4-50 Refrigeration gauge manifold. (Courtesy of Robinair Manufacturing Company.)

FIGURE 4-52 Bourdon tube principles.

FIGURE 4-51 Compound gauge. (Courtesy of Marshalltown Instruments.)

FIGURE 4-53 Pressure gauge. (Courtesy of Marshalltown Instruments.)

The inside scales are the corresponding temperatures for different refrigerants.

Gauges operate from the action of a *Bourdon tube* (Figure 4-52). When pressure inside the Bourdon tube is increased, the element tends to straighten. As the pressure is decreased, the element tends to curve again. A Bourdon tube is a flattened metal tube sealed at one end, curved, and soldered to the pressure fitting on the other end. The movement of the element will pull a link that is attached to the pointer through a series of gears. This movement will be shown by the gauge hand, or pointer.

Compound retard gauges have a retarder that permits accurate readings within a given range. In refrigeration work this range would be between 0 and 100 psig. These gauges can be recognized by the change in graduations at pressures higher than those usually encountered.

Pressure Gauges. Pressure gauges are used to determine pressures on the high side of the system (Figure 4-53). The outside scale is calibrated in psig, and the inside scales indicate the corresponding temperatures of the different types of refrigerants. Some pressure gauges are not designed to operate

Special washer allows
stem to close without turning
. . . no twisting or tearing seat

Glass-impregnated
Teflon* seat

Stem moves in and out
without turning

Flow

Chrome-plated
investment casting

Hot forged
brass body

Special "O" rings
provide positive seal

Two full turns allow
maximum flow

FIGURE 4–54 Valve manifold. (Courtesy of Robinair Manufacturing Company.)

at pressures below atmospheric pressure. Therefore, caution should be used during evacuation procedures to prevent damage to these gauges.

Valve Manifolds.   These manifolds provide openings through which the various service operations are performed on the refrigeration system (Figure 4–54). The proper manipulation of the hand valves will permit almost any function. When the valves are screwed all the way in, the gauges will indicate the pressure on the corresponding line. The center-line is usually connected to a vacuum pump, a refrigerant cylinder, or an oil container.

Every system operates with specific operating pressures on both the high and low sides at any given ambient temperature. Because of this fact, the technician can use the gauges to determine whether or not the system is functioning properly for the ambient temperatures in which it is operating. Equipment manufacturers provide operating manuals that specify the proper operating pressures for the particular system at any given ambient temperature.

The operating pressures and temperatures vary inside a system depending on what the problem may be. A higher-than-normal pressure can exist on both the high and low sides of the refrigeration system. Also, there can be an extremely low side pressure and a higher-than-normal high-side pressure because of a restriction in the refrigerant system.

The proper use of the service manifold can help in determining what is wrong with the system. By taking these pressures properly, the technician can make a more accurate determination as to what the problem may be. However, it is necessary that the technician have a thorough understanding of the design of the manifold and how its features will aid in diagnosing problems before the full capabilities of this tool can be realized.

The valve body is manufactured from forged brass that will last many years when properly cared for. There are ports designed so that the desired procedures can be completed with a minimum amount of time (Figure 4–54). Both the high- and the low-side gauges are open to their related port at all times. The position of the manifold hand valves control the use of the center port. The center port is used for making connections to equipment separate from the system, such as refrigerant cylinder, vacuum pump, and other equipment.

Color-coded gauges, hand valves, and charging hoses may also be used to aid in connecting the proper gauge to the desired part of the system. The following colors are usually recommended:

White or yellow: charging or purging line (center hose)
Blue: low-pressure side of the system
Red: high-pressure side of the system

It should be standard procedure to have the hoses either connected to the system or to the hose holder to help keep the lines clean and free of moisture. The ends of the hoses should never be left open to the atmosphere. This will prevent the entrance of contaminants into the hoses and eventually into a refrigeration system.

Using the Gauge Manifold. The following list illustrates some of the more typical procedures that are followed when using the gauge manifold. These procedures are not intended to replace or conflict with any equipment manufacturer's recommendations for specific test procedures.

1. To determine the operating pressures of the system (Figure 4–55):
   (a) Backseat the system service valves C and D.
   (b) Close the gauge manifold hand valves A and B.
   (c) Connect the proper-colored hoses to the desired system service valves.
   (d) Crack the system service valves off the backseat just enough to allow the pressures to be shown on the gauges, but not enough so that the gauge needles wiggle excessively when the unit is running.
   (e) Slightly loosen the hoses at the gauge manifold and purge the hoses for a few seconds to remove any contaminants from them.
   (f) Start the unit and allow it to run for at least 5 min to stabilize the system pressures.
   (g) Read the operating pressures and the corresponding refrigerant saturation temperatures on the gauges.

FIGURE 4–55 Connecting the gauge manifold to the system.

2. To purge refrigerant from the system (Figure 4–55):
   (a) Turn off the system.
   (b) Close the gauge manifold hand valves A and B.
   (c) Connect the proper-colored hose to the correct system service valves C and D.
   (d) Crack the system service valves C and D off the backseat.
   (e) Crack open the gauge manifold hand valves A and B and allow vapor to escape slowly through the center hose. Use caution to prevent the escape of oil and/or liquid refrigerant from the system.
   (f) If only a part of the refrigerant is to be removed, close the gauge manifold hand valves A and B after some of it has been removed. Check the operation of the system to determine if the proper refrigerant charge remains in the system.
   (g) If all the refrigerant is to be purged, the gauges will indicate 0 psig (0 kPa) when the procedure is complete.

3. To pump refrigerant from the system (Figure 4–56):
   (a) Turn off the system.
   (b) Close the gauge manifold hand valves A and B.
   (c) Connect the proper-colored hoses to the system service valves C and D.
   (d) Crack the system service valves off the backseat.
   (e) Connect the center hose to the pump-out unit but do not tighten this connection.
   (f) Crack open the gauge manifold hand valves A and B and slowly allow refrigerant vapor to escape through the center hose at the pump-out unit for a few seconds to remove contaminants, then tighten the connection.
   (g) Connect a charging hose to both the pump-out unit and a clean, dry refrigerant cylinder of the proper type which is large enough to hold the complete charge of refrigerant. Or have enough extra cylinders available so that the operation can be completed. Do not tighten the connection at the refrigerant cylinder.
   (h) Purge the contaminants from the line between the pump-out unit and the refrigerant cylinder by opening the outlet valve on the pump-out unit. After purging the line for a few seconds, tighten the connection at the cylinder.

FIGURE 4–56 **Connections for pumping refrigerant from a system.**

FIGURE 4–57 **Charging refrigerant vapor through the system suction service valve.**

(i) Place the refrigerant cylinder on a scale so that the amount of refrigerant pumped into the cylinder can be determined. Do not over-fill the cylinder. The amount of refrigerant in pounds that the cylinder can safely hold is marked on the cylinder.

(j) Start the pump-out unit and pump the refrigerant from the system and into the cylinder.

(k) All of the refrigerant has been removed when the gauges read 0 psig (0 kPa) and the operation is complete. Do not lower the pressure inside the system below 0 psig (0 kPa) to prevent the entrance of contaminants during the service operation.

4. To charge refrigerant vapor through the system suction service valve (Figure 4–57):

(a) Close the gauge manifold hand valves *A* and *B*.

(b) Connect the color-coded charging hoses to the proper system service valves.

(c) Crack the system service valves *C* and *D* off the backseat.

(d) Connect the center charging hose to a cylinder of the proper-type refrigerant. Do not tighten this connection.

(e) Crack open the gauge manifold hand valves *A* and *B*. Let the refrigerant escape through the loose connection at the cylinder for a few seconds to purge any contaminants from the hose. Tighten the connection.

(f) Start the unit if it is not running.

(g) Crack open the low-side gauge manifold hand valve *A* and meter the correct amount

of refrigerant into the system by opening and closing the valve. Observe the discharge pressure and if a sudden increase in the discharge pressure occurs stop the unit and determine and correct the cause for the sudden increase in pressure.

5. To evacuate and vapor charge a system (Figure 4–58):

(a) Connect the color-coded charging hoses to the proper system service valves.

(b) Crack the system service valves off the backseat.

(c) Remove the refrigerant from the system by either purging or pumping the refrigerant into a cylinder of the proper type.

FIGURE 4-58 **Evacuating and vapor charging a system.**

(d) Connect the center charging hose to the micron meter (Chapter 12) and the vacuum pump at valve *E* and a cylinder of refrigerant at valve *F*.

(e) Open the gauge manifold hand valve *A* all the way and close the hand valve *B*.

(f) Backseat the system service valve *D*.

(g) Move the system service valves to the midway position.

(h) Fully open valve *E* at the vacuum pump.

(i) Start the vacuum pump.

(j) Observe the micron meter and when the desired vacuum is reached, usually 500 μm, close off valve *E*.

(k) Stop the vacuum pump.

(l) Break the vacuum by opening valve *F* at the refrigerant cylinder and allow vapor to enter the system.

(m) When the system pressure rises above 0 psig (0 kPa), crack the system service valve *D* off the backseat until pressure is shown on the pressure gauge.

(n) Start the refrigeration unit. *Caution:* Do not start a hermetic or semihermetic compressor while it is in a vacuum and empty of refrigerant to prevent possible damage to the motor windings.

(o) Meter in the correct amount of refrigerant by opening and closing the manifold gauge hand valve *A*. Observe the discharge pressure and if it starts to rise suddenly, stop charging the unit and determine and correct the reason for the sudden rise in pressure.

6. To charge compressor oil through the system suction service valve (Figure 4–59):

FIGURE 4-59 **Charging oil into a compressor suction service valve.**

(a) Close the gauge manifold hand valves *A* and *B*.

(b) Connect the color-coded charging hoses to the proper system service valves.

(c) Crack the system service valves off the backseat.

(d) Crack open the gauge manifold hand valve *A* to purge the hoses, then close the valve.

(e) Place the center charging hose in a container with enough oil in it to make certain that the hose end remains under the oil at all times.

(f) Front seat the system suction service valve with the system operating.

(g) When the low-side gauge indicates a vacuum of about 10 in. (254 mm), stop the unit.

Caution: Hermetic and semihermetic compressors should never be operated in a vacuum over 18 in. (458 mm) to avoid possible damage to the motor windings.

(h) Turn off the unit.

(i) Meter in the correct amount of the proper grade refrigeration oil by opening and closing the gauge manifold hand valve *A*.

7. To check compressor discharge valves (Figure 4–60):

(a) Turn off the system.

(b) Connect the color-coded charging hoses to the proper system service valves *C* and *D*; do not tighten these connections.

(c) Crack the system service valves off the backseat and purge the hoses. Tighten the hose connections.

(d) Either pump the system down, pump the refrigerant from the system, or purge the refrigerant from the system.

(e) Front seat the system service valves *C* and *D*.

(f) Connect the center charging hose to a cylinder of the proper type refrigerant at valve *E*.

(g) Open the cylinder valve *E*.

(h) Open the high-side gauge manifold hand valve *B* and admit refrigerant vapor only into the compressor discharge service valve.

(i) Close the gauge manifold hand valve *B*.

(j) Observe the pressure on both the high- and low-side gauges. The pressures should remain the same.

(k) Wait 5 or 10 min and reread the high- and low-side pressures. There will be no, or very little, change in the pressures if the valves are good. A drop in high-side pressure and a rise in the low-side pressure indicates bad compressor discharge valves. The high-side pressure will drop more than the rise in the low-side pressure because of the difference in volume between the two areas of the compressor.

8. To check compressor discharge valves with refrigerant in the system (Figure 4–61):

(a) Close the gauge manifold hand valves *A* and *B*.

(b) Connect the color-coded charging hoses to the proper system service valves *C* and *D*.

(c) Crack the system service valves off the backseat.

(d) Purge the charging hoses by slightly opening the gauge manifold hand valves *A* and

FIGURE 4–60 Checking compressor discharge valves.

FIGURE 4–61 Checking compressor discharge valves with refrigerant in the system.

B and allowing refrigerant vapor to escape through the center charging hose.

(e) Close the gauge manifold hand valves *A* and *B*.

(f) Front seat the system suction service valve *C*.

(g) Pump a vacuum of 5 or 10 in. (127 or 254 mm) with the compressor.

(h) Stop the unit.

(i) Front seat the system discharge service valve *D*.

(j) Observe the pressures on the gauges.

(k) Wait 5 or 10 min and reread the high- and

low-side pressures. There will be no, or little, change in pressures if the valves are good. A drop in the high-side pressure and a rise in the low-side pressure indicates bad compressor discharge valves. The high-side pressure will drop more than the rise in the low-side pressure because of the difference in volume between the two areas of the compressor.

9. To check compressor suction valves (Figure 4–62):
   (a) Stop the compressor.
   (b) Close the gauge manifold hand valves *A* and *B*.
   (c) Connect the color-coded hoses to the proper system service valves *C* and *D*.
   (d) Crack the system discharge service valve *D* off the backseat.
   (e) Front seat the system service valve *C*.
   (f) Crack open the gauge manifold hand valves *A* and *B* and purge the charging hoses through the center hose, then close the gauge manifold hand valves.
   (g) Start the unit and run it until a vacuum of 10 to 15 in. (254 to 381 mm) on the low side is obtained.
   (h) Stop the compressor and observe the low-pressure gauge.
   (i) Wait 5 or 10 min and reread the low-pressure gauge. There should be no, or very little, increase in the low-side pressure. If more than 5 in. (0.197 mm) loss of vacuum is indicated,

FIGURE 4–63 Charging hose. (Courtesy of Robinair Manufacturing Company.)

start the compressor and run it until 15 in. (381 mm) vacuum is indicated in the low side. Stop the compressor and wait 5 to 10 min and reread the low-pressure gauge. If 5 in. (0.197 mm) or more loss of vacuum is indicated, the compressor suction valves are bad.

Charging Hoses. These hoses are flexible (Figure 4–63) and are used to connect the gauge manifold to the system. Charging hoses are equipped with ¼-in. (6.5-mm) flare connections on each end; some automotive units require a different size. One end usually has a valve core depressing attachment for attaching the gauges to *schrader valves.** Charging hoses may be purchased in a variety of colors, which facilitate making the connections to the unit and are designed for a working pressure of 500 psi (3450.00 kPa) and an average bursting pressure of 2000 psi (13,800.00 kPa).

Pocket Thermometers. An asset to the service technician is the pocket thermometer. Through their use much time can be saved. Operating temperatures of a unit can be checked upon completion of the service work to determine whether or not the unit is operating properly. The two types currently used are the mercury and the bimetal thermometer. The bimetal type (Figure 4–64) is the most popular because of its durability. Pocket thermometers can be purchased with a convenient shock-resistant carrying tube which can be clipped into the shirt pocket for ready use. Pocket thermometers are accurate to within 1%.

Capillary Tube Cleaners. These cleaners are extremely useful when servicing domestic refrigerators and freezers. The capillary tube flow control device used on these units has such a small inside diameter

Low pressure   High pressure

A →   ← B

C → ⊗   ⊗ ← D

Suction →   → Discharge

Compressor

FIGURE 4–62 Checking compressor suction valves.

---

*A schrader valve is a type of service valve used on systems not provided with a means for servicing.

FIGURE 4-64 Various types of thermometers. (Courtesy of Marshalltown Instruments.)

FIGURE 4-65 Portable welding unit. (Courtesy of Union Carbide Corp.)

that it can become clogged easily. The capillary tube cleaner is connected to the plugged cap tube and up to 15,000 psi (103,500 kPa) of hydraulic pressure may be applied to force chips, filings, flux, oil, etc. from the cap tube into the evaporator, where it either remains harmlessly or circulates to the strainer or filter. A high grade of refrigeration oil is used to provide the hydraulic pressure.

Welding Units. These units are a must for the service technician. The most popular type is the portable model shown in Figure 4-65. These ultra-portable units are convenient, economical, and easy to use. Each one includes a complete apparatus for handling a wide range of repair and maintenance work. A flame temperature near 6000°F (3298°C) can be obtained with these units.

There are two types of acetylene tanks and one oxygen tank used with these units (Figure 4-66). Types MC and B acetylene tanks are commonly used with the R oxygen tank. The disposable propane (DP) tank is used for lower-temperature requirements. The specifications for these tanks are shown in Table 4-4.

FIGURE 4-66 Acetylene and oxygen tanks. (Courtesy of Union Carbide Corp.)

TABLE 4-4 Tank Specifications

Style	Capacity	Height (in.)	Diameter (in.)	Weight, Full
Disposable propane (DP)	1 pt, 10.7 fl. oz	10 ½	2¾	2 lb, 2 oz
Type MC acetylene	10 ft³	14	4	8 lb
Type B acetylene	40 ft³	23	6¼	26 lb
Type R oxygen	20 ft³	14	5³⁄₁₆	13½ lb

Source: Union Carbide Corp.

FIGURE 4-67 Chargefaster. (Courtesy of Watsco, Inc.)

Chargefaster Units. These units (Figure 4-67) are a tremendous time saver for the service technician. Tests have shown that when charging refrigeration systems the service technician does not need to know how much refrigerant is needed because the chargefaster method provides this information. In practice, the chargefaster is connected to the low side of the gauge manifold. Liquid refrigerant can then be charged directly into the low side of any refrigeration system safely and quickly. As the liquid goes through the chargefaster, it is converted to a saturated vapor at a rate of flow less than the compressor capacity.

Since the rerigerant leaves the cylinder in liquid form and expansion takes place at the outlet of the chargefaster unit, there is no pressure drop in the charging cylinder. Therefore, heating the cylinder is neither necessary nor desirable.

Because chargefaster contains a Magni-Chek* valve assembly which automatically permits the flow

*Magni-Chek is the brand name of a valve used to prevent the backward flow of fluids.

FIGURE 4-68 Charging cylinder. (Courtesy of Robinair Manufacturing Company.)

to bypass when drawing a vacuum through the gauge manifold, there is no restriction or slowdown of evacuation of the system.

Charging Cylinders. These cylinders are used by the service technician to charge the correct amount of refrigerant into a system in a short period of time. In practice, the charging cylinder is loaded with the manufacturer's recommended amount of refrigerant for a particular unit (Figure 4-68). While the system is evacuated, the complete refrigerant charge is

FIGURE 4-69 Fitting brushes. (Courtesy of Schaefer Corp.)

FIGURE 4-70 Soft solder. (Courtesy of J.W. Harris Co., Inc.)

dumped into it. Some of these units have a built-in electric heater to speed up the dumping process. These cylinders should never have a direct flame applied to them, nor should they be stored in direct sunlight or with liquid refrigerant in them.

Fitting Brushes. Fitting brushes are used to clean the inside of fittings that are to be soldered. These brushes (Figure 4-69) can be obtained in a size for each fitting size. In practice, the brush fits tightly into the size fitting for which it has been designed. The brush is then rotated by hand to clean the fitting. If too large a brush is used or if it is rotated in more than one direction, it will be ruined.

□ Materials □

Various materials are peculiar to the refrigeration and air conditioning industry. Because there are so many in use, we will cover those that are most commonly used.

Soft Solder. Soft solder is a low-temperature method of joining metals together. It is an adhesion process. That is, the materials being joined are not melted; only the solder is melted. Soft solder usually is a mixture of tin and lead. This includes tin and antimony; 95% tin and 5% antimony (95/5), which has a melting point of 460 °F (238 °C); and 50/50, which has a melting point of 421 °F (235.76 °C). Both temperatures are well below the melting point of copper, which is 1984 °F (585 °C). Soft solder can be purchased in 1-lb spools of either ¹⁄₁₆ in. (1.5875 mm) or ⅛ in. (3.175 mm) diameter at a refrigeration supply house (Figure 4-70). Soft solder's low melting temperature, low tensile strength, and

inability to withstand high vibration make it unsuitable for most modern refrigeration applications. This is especially so with systems using R-22 as the refrigerant. The compressor discharge temperature is sometimes very high, and when combined with the vibration of the compressor, leaks develop quite easily. Therefore, a stronger solder is needed.

Silver Solder (Stay-Silv 15). This solder is the most common agent used in silver brazing joints in modern refrigeration equipment. It is sometimes referred to as silver brazing. It is available in rods ⅛ in. (3.175 mm) × ⅛ in. (3.175 mm) × 20 in. (50.8 cm); .050 in. (12.7 mm) × ⅛ in. (3.175 mm) × 20 in. (50.8 cm); and 1, 3, 5, and 25 troy ounces and ¹⁄₁₆ in. (1.5875 mm) in diameter.

To silver braze copper to steel or to braze dissimilar metals, the recommended product is Stay-Silv 45 alloy, which has a melting temperature of 1145 °F. Also available are the Safety-Silv cadmium-free silver solders 1200 (56% silver), Safety-Silv 1370 (45% silver), and Safety-Silv 1350 (40% silver). Their melting temperatures are according to their designated alloy number as listed (Figure 4-71). These solders have strong tensile strength and are not affected by vibration. Some types of solder contain their own flux, eliminating the need for extra flux on the joint. This is especially true on copper tubing.

Almost all alloys used today for brazing operations contain cadmium, a product that gives off poisonous fumes when heated. When using one of these alloys make certain that there is plenty of ventilation.

It is possible for a worker to breathe enough cadmium in about 2½ to 3 hr of soldering time to be lethal. Most of the brazing alloys in use today have

FIGURE 4-71 Silver solder. (Courtesy of J.W. Harris Co., Inc.)

large quantities of cadmium (15 to 25%) in their makeup. When these alloys are in the molten state, they will emit cadmium oxide fumes to the atmosphere. These fumes are toxic. When the temperature is increased above the molten state, the quantity of fumes produced increases. When oxyacetylene welding gas (which burns at about 6000°F) is used, higher concentrations of cadmium oxide fumes are produced. Air-acetylene, which burns at about 4000°F, and air-natural gas or a resistance oven produces a temperature of about 3500°F, all of which produce less fumes than the oxyacetylene process.

These fumes are insidious; you cannot smell them. A lethal dose does not necessarily become irritating to the worker until after he has absorbed a sufficient amount to be in immediate danger of his life. Some of the symptoms of cadmium poisoning are headache, fever, irritation of the throat, vomiting, nausea, chills, weakness, and diarrhea—none of which may appear until some hours have passed after exposure. The primary injury is to the respiratory passages. An area within 5 ft of the workpiece is the most dangerous. In production work the installation of glass shields between the workers and the workpiece help in diverting some of the fumes in a

well-ventilated room. The use of hooded exhaust systems in the immediate area can be very effective. It is considered necessary to practice good housekeeping and personal hygiene when brazing is in process. Storing of lunches or eating should not be permitted in the area where silver brazing is done. The workers should always wash their hands and faces before eating, smoking, or before leaving work.

The threshold limit value for cadmium oxide fumes is 0.1 mg/m³ of air for daily 8-hr exposures. This value represents the maximum tolerance under which most workers can be exposed without adverse effects. In practice, control measures should be sought that will achieve an amount less than the maximum.

In modern times there is no need to use an alloy that contains cadmium for the normal silver brazing operations. Most manufacturers make a cadmium-free silver brazing alloy. It is recommended that one of these be used when at all possible.

Flux. Flux is used when making sweat joints to aid in making leaktight connections. It is available in both paste and liquid forms (Figure 4-72).

Caution: Flux has an acid content and should not be allowed on clothing, near the eyes, or on open cuts. To remove, wash with soap and water or see a physician.

There is a flux designed specifically for each type of solder that cannot be used for any other type of solder. Flux should be put on the joint sparingly. Do not over-flux a joint or allow flux to enter the refrigeration system because the acid will be harmful to the motor windings of the compressor and other internal components.

Sand Cloth. Sand cloth is used mainly to clean the surfaces to be soldered. It is available in rolls at the refrigeration supply house (Figure 4-73). Short strips

FIGURE 4-72 Soldering flux. (Courtesy of J.W. Harris Co., Inc.)

Materials

FIGURE 4–73 Sand cloth.

are torn from the roll and wrapped around the tube and each end is pulled alternately until the tube is shiny bright. Caution should be used when cleaning tubing to prevent the particles from entering the tube. Perhaps the best way to accomplish this is to turn the end of the tube being cleaned downward so that the particles will fall away from it.

☐ Safety Procedures ☐

Always practice safety and know the proper steps required for completing a job. The following are only a few of the safety points to be remembered.

1. Never use a screwdriver for a pry bar. The blade is hardened to retard wear and it is, therefore, brittle to sideways pressure. It may break and cause injury to the user and equipment.
2. Always grind the "mushroom" from the head of chisels and punches. When struck with a hammer, this mushroom may break and the particles may cause injury to the user or to a bystander.
3. Never use a file without a handle. The tang may cause injury to the hands.
4. Always wear goggles when drilling. Chips may break away and enter the eye, causing serious injury and possibly loss of sight.
5. Never use oxygen when pressure testing a refrigeration system. Oxygen and oil together form an explosive mixture.
6. Never heat a refrigeration pipe containing a refrigerant to a cherry red condition. The pressure that is built up inside the pipe could cause a "blowout" and serious injury could result.
7. Never use carbon tetrachloride for any purpose because it is toxic and harmful to the skin.
8. Always provide plenty of ventilation when working with silver brazing materials. These materials contain cadmium, the fumes of which are poisonous.
9. Never use emery cloth for cleaning refrigeration tubing. The grit is extremely hard and could cause damage to the system if allowed to enter.

10. Practice care when using epoxy bonding materials because they may irritate the skin.
11. Never fill refrigerant cylinders to more than 85% of their rated capacity. Hydrostatic pressure may cause them to burst when they become warm.
12. Always pull on a wrench, never push.
13. Always use the proper size wrench to prevent injury to the user and equipment.
14. Always "crack" cylinder service valves before opening all the way. This will allow for control of the gas if a dangerous situation occurs.
15. Always protect an open refrigeration system from moisture. Keep all refrigeration components clean and dry.

☐ SUMMARY ☐

☐ In a refrigeration system, the piping is used to convey the refrigerant from one system component to another.

☐ The types of tubing used are welded or brazed and seamless tubing. Seamless tubing is the stronger of the two.

☐ Copper tubing is never used when ammonia is the desired refrigerant. It is used for condensers and evaporators because of its availability and heat-conducting capabilities.

☐ In refrigeration work, copper tubing is measured by its outside diameter (OD). It is rated as ACR for refrigeration applications.

☐ Two types of copper tubing used in refrigeration work are soft and hard drawn.

☐ Copper tubing is manufactured according to ASTM B88 and ANSI-B9.1-1971 refrigeration industry specifications.

☐ For refrigeration work, tubing is cleaned, dehydrated, and capped.

☐ Steel tubing is being used almost extensively in modern domestic refrigerators as condensers.

☐ Aluminum tubing is being used in domestic refrigerator evaporators.

☐ Flexible tubing is used on automotive air conditioning and transport refrigeration.

☐ The fittings most commonly used in refrigeration work are wrot (sweat), flare, compression, and hose. Each fitting develops a resistance to refrigerant flow that is rated by an equivalent length of pipe.

☐ Flare fittings are susceptible to leaking when used in places of varying temperatures.

- Compression fittings are becoming popular for making quick connections to condensing units with refrigerant lines.
- Hard-drawn tubing is preferred for use with compression fittings.
- One size of tubing can satisfactorily be inserted into another piece of tubing the next size larger and silver soldered.
- Tubing cutters are used to cut a square end on tubing to enable the technician to make a leaktight joint. To prevent damage to the tubing, the cutter should not be overtightened.
- Flaring tools are used to make flares on tubing so that threaded flare fittings can be used.
- The flaring tool should not be overtightened, ruining the flare.
- A drop of refrigeration oil should be put on both faces of the flare to prevent twisting of the tube when tightening the connection.
- The flare fitting should be tightened hand tight plus one turn.
- Swaging tools are used to enlarge one end of a piece of tubing so that two pieces the same size can be soldered together. Care should be exercised to prevent enlarging the swage or cutting it with the swaging tool.
- Tubing reamers are used to remove burrs from tubing to reduce friction to the refrigerant. Care should be taken to prevent metal chips from entering the tubing. One size of reamer fits several sizes of pipe.
- Pinch-off tools are used to aid in removing gauges from sealed systems after repairs have been made. The process tube is silver soldered before the pinch-off tool is removed.
- Hermetic service valve kits are used to gain entrance into hermetically sealed systems without cutting and soldering lines. An adapter is available for almost any hermetic system.
- Reversible ratchets are used to turn the stem of service valves. They are available in practically any size.
- Adapters are available for use with the ¼-in. (6.5-mm) size to make it more flexible in use.
- Gauge manifolds are the most used tools in the service technician's tool chest. They are used for checking pressures, adding refrigerant, or any process involving refrigerants, oil, and evacuation.
- Compound gauges are used to read pressures above and below atmospheric.
- Pressure gauges are used to read pressures above atmospheric only.
- Gauges use the Bourdon tube principle.
- Charging hoses are used to connect the gauge manifold to the system.
- Pocket thermometers are an asset to the technician because the system can be checked after completion of repairs to determine whether or not it is operating properly. The most popular type is the bimetal thermometer because of its durability.
- Capillary tube cleaners are used to clean foreign material from small-diameter tubes with hydraulic pressure. A high grade of refrigeration oil is used to produce the hydraulic pressure.
- Portable welding units are very popular for work on tubing.
- A temperature of 6000 °F (3298 °C) can be obtained through proper adjustment of the oxygen and acetylene.
- There are two acetylene tanks and one oxygen tank available for use in portable units.
- Chargefaster units enable the technician to charge liquid refrigerant quickly and accurately into any size system without damage to the unit. The unit can also be left in the line during evacuation procedures.
- Charging cylinders are used to aid the technician in charging the correct amount of refrigerant into a system. These cylinders should never be heated with an open flame or stored in direct sunlight or with refrigerant left in them.
- Fitting brushes are used to clean the insides of fittings. There is a size for each size tubing. Care should be taken not to damage the brush.
- Soft solder is a low-temperature method of joining metals together by adhesion.
- Fittings to be soft soldered must be cleaned to a shiny bright and flux applied. Excessive temperature must be avoided.
- A piece of soft solder the length of the diameter of the tube is sufficient quantity to make a leaktight joint.
- Apply only enough heat to melt the solder. It has a melting point of 460 °F (238 °C) for 95/5 and 421 °F (235.76 °C) for 50/50.
- Silver solder is the most popular method of soldering joints in modern refrigeration systems. It has a high melting point and tensile strength. It is not affected by vibration. When heated, it gives off a poisonous vapor from the cadmium in it.
- Flux is used in making soldered joints.
- Each type of solder requires a different flux. Care should be taken to prevent its entering the refrigeration system. It should not be allowed on clothing or skin because of its acid content.
- Repair of cracked refrigerant lines can be accomplished by applying silver solder to the area of the crack until a slight buildup is apparent. Do not use soft solder for this procedure. Be sure the refrigerant is purged from the system to prevent a refrigerant-oil blowout and possible human injury.

1. What are the two types of tubing and pipe mentioned in this text?
2. Why is seamless tubing used in refrigeration work?
3. Why is copper tubing not used with ammonia refrigerant?
4. What are the two types of tubing used as refrigerant lines?
5. How is tubing that is used for refrigeration work measured (OD or ID)?
6. What should be done to hard-drawn tubing before attempting to bend it?
7. What two factors vary the safe working pressures of copper tubing?
8. What are the four classifications of copper tubing?
9. What three things are done to ACR tubing during the manufacturing process?
10. What should be done to copper tubing that is not used?
11. What part of the refrigeration system is more commonly made of steel?
12. What is done to increase the heat transfer capabilities of steel?
13. Where is aluminum tubing being used in refrigeration systems?
14. What is the most popular type of stainless steel tubing used in refrigeration work?
15. What type of tubing is more popular in automotive and transport refrigeration work?
16. What problem does flexible tubing eliminate?
17. What are the most popular types of tube fittings used in refrigeration work?
18. To what does the term *equivalent length of pipe* refer?
19. What percentage of the actual length of pipe is allowable on 100 ft (30.4 m) of pipe or more?
20. What is another name for wrot fittings?
21. What is the angle in degrees of the flare fittings used in refrigeration work?
22. What should be done to prevent twisting of the tubing when tightening a flare fitting?
23. What causes flare fittings to leak after being used awhile?
24. What caution should be taken when assembling an O-ring fitting?
25. Where are compression fittings becoming popular in refrigeration work?
26. Why should tubing cutters be used to cut tubing?
27. How much should the handle on the tubing cutter be tightened after each complete revolution of the tubing?
28. For what is the flaring tool used?
29. What tool is used to enlarge the end of tubing so that two pieces of the same size may be soldered together?
30. What is the purpose of the pinch-off tool?
31. What does a tubing bender eliminate?
32. On what type of systems are hermetic service valve kits used?
33. For what purpose are reversible ratchets used?
34. What is the most popular tool in the service technician's tool chest?
35. On what principle do refrigeration gauges operate?
36. For what purpose is a gauge manifold used?
37. What does the expression 95/5 indicate when referring to soft solder?
38. What is the main criterion used when selecting the type of silver solder to be used?
39. Can any soldering flux be used on all types of soldered joints?
40. What is the length of solder required when making a sweat joint?
41. What precaution should be taken when silver soldering a joint?
42. What are the two methods used to cut copper tubing?
43. How should the tubing be positioned when cutting tubing with a hacksaw?
44. What should be used on the mating parts of a flare fitting to prevent the tube from twisting when a fitting is being tightened?
45. What does the bending of refrigeration tubing eliminate?

5 Compressors

Performance Objectives

After completion of this chapter, the student should:

- Be able to name the different types of compressors that are used for refrigeration and air conditioning systems.
- Know how a reciprocating compressor operates.
- Know how centrifugal compressors operate.
- Know how screw compressors operate.
- Be able to explain how compressor shaft seals prevent leaks.
- Be familiar with the various methods of cooling a compressor.
- Know how the scroll compressor operates.
- Know how the discus compressor operates.
- Complete the Exercises with at least 80 percent accuracy.
- Complete the Laboratory Exercises with at least 80 percent accuracy.

All compression refrigeration systems use a compressor of some type to circulate the refrigerant through the system. The compressor has two functions. First, it draws the refrigerant vapor from the evaporator and then lowers the pressure of the refrigerant in the evaporator to the desired evaporating temperature. Second, the compressor raises the pressure of the refrigerant vapor in the condenser high enough so that the saturation temperature is higher than the temperature of the cooling medium used to cool the condenser and condense the refrigerant.

Compressor Types

Several different types of refrigeration compressors are in use today. The newer designs include the screw, discus, and the compliant scroll compressors. The most popular types are the reciprocating, rotary, and centrifugal. The most popular of these is the reciprocating compressor.

Reciprocating Compressors. These compressors are used in the smaller horsepower sizes for commercial refrigeration, domestic refrigeration, and industrial applications.

Reciprocating compressors are quite similar in design to the automobile engine. A piston is driven from a crankshaft and makes alternating suction and compression strokes inside a cylinder, which is equipped with suction and discharge valves (Figure 5-1).

As the piston is pulled down inside the cylinder, refrigerant vapor is drawn from the evaporator into

FIGURE 5-1 **Reciprocating compressor. (Courtesy of Frigidaire Division, General Motors Corp.)**

the cylinder of the compressor. A reedlike flapper valve in the cylinder head acts as a check valve. It allows the refrigerant vapor to be drawn into the cylinder (Figure 5–2). As soon as the refrigerant vapor pressure inside the cylinder is equal to the refrigerant vapor pressure in the suction line, the spring action of the reed closes the suction valve.

When the piston has reached the bottom of its stroke, it has drawn in all the refrigerant vapor possible. The piston then starts on its upstroke, pushing the entrapped refrigerant vapor ahead of it. The vapor cannot return to the evaporator because the suction valve is closed, blocking its path.

However, there is another reedlike flapper valve in the cylinder head. This one is arranged so that vapor can get out of the cylinder. All the vapor that is forced through this valve is directed along a path that leads to the condenser. The discharge valve, like the suction valve, will allow the refrigerant to flow in only one direction. It lets the vapor out of the compressor cylinder but blocks the way for it to return to the cylinder (Figure 5–3). When the piston

FIGURE 5–2 **Suction stroke of a reciprocating compressor. (Courtesy of Frigidaire Division, General Motors Corp.)**

FIGURE 5–3 **Compressor discharge stroke. (Courtesy of Frigidaire Division, General Motors Corp.)**

has reached the top of its stroke, it is ready to start down again and repeat the cycle by drawing more refrigerant vapor into the cylinder.

Because the reciprocating compressor is a positive-displacement pump, it is suitable for small displacement volumes, and it is quite efficient at high condensing pressures and high compression ratios. Other advantages of the reciprocating compressor are its adaptability to a number of different refrigerants (liquid refrigerant may easily be run through connecting piping because of the high pressure created by the compressor), its durability, basic simplicity of design, and relatively low cost.

The reciprocating motion of the compressor can be accomplished in several ways. The most popular means of producing this motion in modern refrigeration equipment is by the *crank throw* type crankshaft to which the piston is joined by a connecting rod, again, much like the automobile engine (Figure 5–4).

The other popular type of reciprocating compressor is the *eccentric disc* type. The eccentric disc uses a straight shaft, which can be very inexpensively machined, and requires no special forging like the crank throw type (Figure 5–5). The piston is connected to the eccentric disc on the crankshaft by an eccentric strap instead of a connecting rod. The eccentric disc is fastened to the shaft by an eccentric lock bolt or setscrew. The strap fits over the eccentric disc without being bolted.

The difference in shaft construction of the crank throw and the eccentric disc type of compressors is shown in Figures 5–6 and 5–7.

Discus Compressors. The discus compressor designed and manufactured by Copeland Refrigeration Corporation is a newer design of the reciprocating compressor. The purpose of this new design is to increase the efficiency of the reciprocating-type compressor. This new design includes a redesigned piston and valve plate which allows the dome of the piston to enter into the valve plate. The valve plate is much thicker to allow the entrance of the higher piston dome and to allow the returning refrigerant to enter the cylinder through the center of the valve plate rather than going into the cylinder head and then through the valve plate.

The raised dome on the piston reduces the amount of reexpansion gas left in the cylinder after the compression stroke of the piston. The valve plate design helps the compressor to operate cooler and, therefore, provide a longer compressor life. The discus

FIGURE 5-4 Crank-type reciprocating compressor.

valve plate and the standard valve plate are not interchangeable. The gaskets are also somewhat different and should not be interchanged.

Rotary Compressors. These are much more simple in construction than the reciprocating types. In some respects, the very essence of their simplicity makes it a little more difficult to understand how they work

(Figure 5-8). Actually, the only moving parts in a rotary compressor consist of a steel ring, an eccentric or cam, and a sliding barrier (Figure 5-9).

Both the ring and the cam are housed in a steel cylinder. The steel ring is a little smaller in diameter than the cylinder. It is situated off-center so that one point on the outer circumference of the ring is always in contact with the wall of the cylinder (Figure 5-10).

1.	Base	30.	Lock Nut
2.	Base Gasket	31.	Lock Washer
9.	Crankshaft	32.	Oil Plug
12.	Crankshaft Thrust Ball	33.	Oil Plug Gasket
14.	Crankshaft Thrust Plug	34.	Piston
15.	Crankshaft Thrust Plug Gasket	35.	Piston Pin
16.	Cylinder Block	36.	Piston Pin Button
17.	Cylinder Head	37.	Piston Ring
19.	Discharge Valve	39.	Seal
20.	Discharge Valve Guide	40.	Seal Bellows
21.	Discharge Valve Plate	41.	Seal Cover
22.	Discharge Valve Plate Gasket, Lower	42.	Seal Cover Gasket
23.	Discharge Valve Plate Gasket, Upper	43.	Seal Guide
24.	Discharge Valve Safety Spring	44.	Seal Spring
25.	Discharge Valve Safety Spring Retainer	45.	Set Screw
27.	Discharge Valve Spring	46.	Suction Valve
28.	Eccentric Disc	47.	Suction Valve Guide
29.	Eccentric Strap and Rod	48.	Woodruff Key

FIGURE 5–5 Eccentric disc-type reciprocating compressor.

This, of course, leaves an open, crescent-shaped space on the opposite side between the ring and the cylinder wall.

An electric motor rotates the cam. As the cam rotates, it carries the ring around with it, imparting a peculiar rolling motion to the ring (Figure 5–11). The ring literally rolls on its outer rim around the wall of the cylinder.

If a port is installed in the cylinder wall and used as an entry to permit refrigerant vapor from the evaporator to flow into the crescent-shaped space (Figure 5–12), and if the cam is rotated just a fraction of a complete turn, the ring will almost immediately cover up the port. As soon as this happens,

the refrigerant vapor is trapped in the crescent-shaped space. There is no place for it to escape.

However, if another port is installed near the other end of the crescent-shaped space, the entrapped vapor will have a way to escape. A piece of pipe is connected from the discharge port to the condenser. As the cam continues to rotate and roll the ring around in the cylinder, it pushes the crescent-shaped wedge of refrigerant ahead of it, compressing the vapor as it is forced out through the discharge port to the condenser.

As can be seen in Figure 5–12, the ring cannot roll very far before it uncovers both the suction and discharge ports. Obviously, there must be some means

FIGURE 5-6   Piston and crankshaft details.

FIGURE 5-7   Eccentric disc details.

of directing the refrigerant vapor out through the discharge port while at the same time blocking its passage to the suction port.

The most simple and effective way to block the passage is to place a barrier between the two ports. This barrier must be flexible or sliding because one end of it is constantly moving back and forth as the ring rolls around in the cylinder (Figure 5-13). To

FIGURE 5-8   Rotary compressor. (Courtesy of Frigidaire Division, General Motors Corp.)

FIGURE 5-9   Rotary compressor parts. (Courtesy of Frigidaire Division, General Motors Corp.)

FIGURE 5-10   Internal view of a rotary compressor. (Courtesy of Frigidaire Division, General Motors Corp.)

FIGURE 5-11   Rotary compressor motion. (Courtesy of Frigidaire Division, General Motors Corp.)

FIGURE 5-12   Rotary compressor compression phase. (Courtesy of Frigidaire Division, General Motors Corp.)

FIGURE 5-13   Sliding barrier. (Courtesy of Frigidaire Division, General Motors Corp.)

install a barrier, a slot must be cut in the cylinder wall deep enough so that the barrier can slide all the way in. A spring is placed behind the barrier to push it out and hold it snugly against the ring no matter what position the ring is in (Figure 5-14). As the ring rolls around in the cylinder, the movable barrier will follow every movement.

With the movable barrier in place, the refrigerant vapor entrapped in the crescent-shaped space can go only one way as the ring rolls around and pushes the vapor ahead of it (Figure 5-15). The only way it can go out is through the discharge port because the movable barrier blocks its passage to the inlet port.

Barrier Held Against Ring by Spring

Barrier

Ring

Cam

Bottom End of Motor Shaft
Rotates in a Smooth Manner

Cam rotating as a part of the
motor shaft causes the ring to roll
in a circle larger than itself. In
doing so, it is always in contact
with both the cylinder wall and
the barrier.

FIGURE 5–14   Barrier in cylinder. (Courtesy of Frigidaire Division, General Motors Corp.)

Compression and intake
phases half completed

To condenser

Refrigerant vapor in compression

Refrigerant vapor from freezer

FIGURE 5–15   Rotary compressor compression cycle. (Courtesy of Frigidaire Division, General Motors Corp.)

The foregoing discussion describes the simple cycle of operation of a rotary-type compressor. As the ring rolls around in the cylinder, its point of contact literally *runs in a circle* around the cylinder wall. All refrigerant ahead of that point of contact is pushed toward the movable barrier, which deflects it out through the discharge port (Figure 5–16). Meanwhile, once the point of contact has passed the suction port, a fresh charge of refrigerant vapor is drawn from the evaporator to the compressor.

As with the reciprocating-type compressor, there are numerous variations of design to be found in different makes of rotary compressors. None of them, though, is more simple or has fewer moving parts than the rotary compressor just described.

Rotary-type compressors are popular in domestic refrigeration and have recently been used in small air conditioning units. These compressors must be manufactured in large volumes to be economically

Finish of compression phase

To condenser

Refrigerant vapor in compression

Refrigerant vapor from freezer

FIGURE 5–16   Finish of compression phase of a rotary compressor. (Courtesy of Frigidaire Division, General Motors Corp.)

produced because of the precision machining required to provide the desired performance. A check valve is usually needed in either the suction line or the discharge line to prevent refrigerant backup from the condenser to the evaporator during the off cycle. Liquid floodback of the refrigerant to the compressor must be eliminated because the suction line enters the compression chamber directly.

Rotary compressors are suited for applications where large volumes of vapor are to be circulated and where a low compression ratio is desired. They are positive-displacement pumps.

Screw Compressors. These are positive-displacement compressors. They are available in sizes of about 100 to 700 tons. They are generally used on chilled-water systems. Systems using screw compressors will perform satisfactorily over a wide range of condensing temperatures (Figure 5–17).

In operation, the refrigerant vapor is drawn in to fill the space between the lobes on the screws (Figure 5–18). As the screws are rotated, the space between the lobes moves past the inlet port, sealing the space. As the screws are rotated farther, the space between the lobes is progressively reduced, causing the vapor to be compressed. The vapor is discharged as the discharge port is opened by the rotating screw.

These compressors will operate smoothly when the capacity is reduced as low as 10%. Capacity control is accomplished by recirculating the refrigerant vapor inside the compressor. Screw compressors are not presently being used on small air conditioning and refrigeration units.

Compressor Types

FIGURE 5-17 Screw compressor. (Courtesy of York, Division of Borg-Warner Corporation.)

FIGURE 5-18 Screw compressor compression cycle. (Courtesy of York, Division of Borg-Warner Corporation.)

FIGURE 5-19 Scroll compressor cutaway. (Courtesy of Lennox Industries, Inc.)

Compliant Scroll Compressor. The scroll compressor design is simple, efficient, and requires few moving parts (Figure 5-19). It is designed to operate with increased efficiency and economy. The scrolls are located in the top of the compressor housing and the motor is located in the bottom of the housing. The oil level is immediately below the motor.

The scroll is a simple compression concept that is centered around the unique spiral shape of the scroll and its inherent properties (Figure 5-20). The two identical scrolls are mated together forming concentric spiral shapes (Figure 5-21). One scroll remains stationary, while the other is allowed to "orbit" (Figure 5-22). Note that the orbiting scroll does not rotate or turn but merely orbits the stationary scroll.

The clockwise orbiting scroll draws gas into the outer crescent-shaped gas pocket created by the two scrolls [Figure 5-22(a)]. The centrifugal action of the orbiting scroll seals off the flanks of the scrolls [Figure 5-22(b)]. As the orbiting motion continues, the gas is forced toward the center of the scroll and the gas pocket becomes smaller and further com-

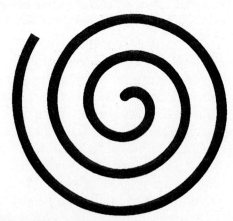

FIGURE 5-20 Scroll form. (Courtesy of Lennox Industries, Inc.)

pressed [Figure 5-22(c)]. When the compressed gas reaches the center [Figure 5-22(d)], it is discharged vertically into a chamber and discharge port in the top of the compressor (Figure 5-19).

The discharge pressure forcing down on the top scroll helps seal off the upper and lower edges (tips) of the scroll (Figure 5-21). During a single orbit,

FIGURE 5-21   Cross section of the scrolls. (Courtesy of Lennox Industries, Inc.)

several pockets of gas are compressed simultaneously, providing smooth continuous compression, thus reducing the amount of compressor vibration.

The scroll compressor can tolerate the return of liquid refrigerant. If liquid enters the scrolls, the orbiting scroll is allowed to separate from the stationary scroll. The liquid refrigerant is worked toward the center of the scroll and is discharged. If the compressor should ever need to be replaced, conventional cleanup practices must be used. *Note:* The

top of a scroll compressor housing may be hot since it is in constant contact with the discharge gas.

Each scroll compressor is equipped with a discharge temperature sensor located on the outside top of the compressor housing (Figure 5–23). The sensor is a single-pole single-throw (SPST) thermostatic switch which opens when the discharge temperature exceeds 280 ± 8 °F (120 ± 4.48 °C) temperature rise. When the switch opens, the circuit to the compressor contactor and the timed "off" control is deenergized and the unit stops. The switch automatically resets when the compressor temperature drops below 130 ± 14 °F (54.44 ± 7.84 °C).

Access to the sensor can be obtained by prying up on the snap plug on top of the compressor (Figure 5–23). Make sure to reseal the snap plug securely after replacement. The sensor terminals are located inside the compressor terminal box (Figure 5–24).

Two-Stage Compressors.   Two-stage compressors are used on applications operating in the range −30 to −80 °F (−34.4 to −62.2 °C). Two-stage compression may be accomplished in two ways: (1) two compressors with one discharging into the suction of the other, and (2) one compressor with multiple cylinders. Because of the problem of maintaining proper oil levels in the crankcase when two compressors

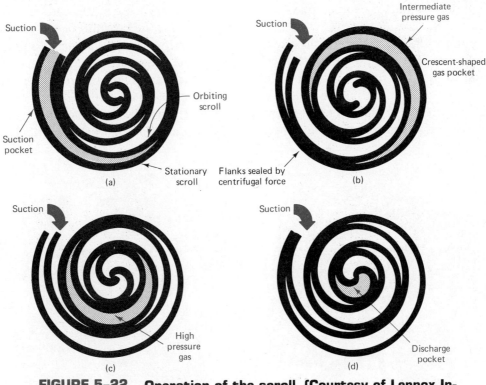

FIGURE 5-22   Operation of the scroll. (Courtesy of Lennox Industries, Inc.)

FIGURE 5-23 **Scroll compressor discharge temperature sensor. (Courtesy of Lennox Industries, Inc.)**

FIGURE 5-24 **Scroll compressor terminal box. (Courtesy of Lennox Industries, Inc.)**

are used, it is more satisfactory to use one compressor (Figure 5-25).

Two-stage compressors are divided internally to provide the two stages. The suction vapor enters the first-stage cylinders from the suction line; then it is compressed and discharged and metered into the interstage manifold where it is used to cool the compressor motor. The superheated refrigerant vapor then enters the suction ports of the second-stage cylinders. The vapor is then discharged to the condenser.

Centrifugal Compressors. These are not positive-displacement pumps like those previously discussed. These compressors produce a pressure through centrifugal effect alone. Compression is caused by whirling the mass of gaseous refrigerant at a high rate of speed, causing it to be thrown outward by centrifugal force into a channel where it is caught.

This action can be related to a ball being whirled at the end of a string. The higher the speed of rotation, the greater the centrifugal pull as the ball tries to get away. If the string were cut during rotation, the ball would fly off into space. In the centrifugal compressor, the molecules of gas are treated much the same way. By whirling them at a high rate of speed, they are thrown off the outer edge of the rotating wheel. Instead of flying off into space, however, they are caught in a channel where they are compressed by the following stream of molecules. In the reciprocating and rotary compressors, the refrigerant molecules are actually squeezed together inside the cylinder by the positive action of the piston. Compression is produced and maintained by the action of the suction and discharge valves. In the centrifugal compressor there are no valves, in the ordinary sense of the word, and the only wearing parts are the bearings at the ends of the shaft.

A centrifugal compressor is similar to a water pump, but it operates at a much higher speed than a water pump (Figure 5-26). A centrifugal compressor consists of a wheel made of rotors or impellers. Each rotor is in a separate compartment or stage (Figure 5-27). In this particular compressor

FIGURE 5-25 **Typical two-staged compressors. (Courtesy of Copeland Refrigeration Corp.)**

FIGURE 5-26 Cross-sectional view of a centrifugal compressor.

FIGURE 5-27 Longitudinal section of a centrifugal compressor.

there are five stages, and the compressor is called a five-stage compressor. Each rotor has a series of vanes that are an integral part of the rotor. In operation, the refrigerant vapor enters through the suction inlet and flows through the suction passages into the first rotor, through openings near the shaft. The centrifugal force developed by the high speed of the rotor discharges the vapor from the periphery of the rotor at a higher pressure than it had at the suction passages. The refrigerant vapor compressed in the first rotor discharges into the space labeled C in Figure 5-27 between the first and second rotors. It passes back to the center of the compressor and enters the second rotor through the openings near the shaft. The rotating motion of the second rotor again forces the refrigerant outward, and it passes into passage D. In this manner the refrigerant

follows a zigzag path through the centrifugal compressor until it is finally discharged through the discharge port marked G. From there it enters the discharge line and flows to the condenser.

Each rotor, or impeller, is progressively smaller in diameter and smaller in thickness, since the same amount of refrigerant vapor passes through each rotor, but in a more highly compressed state from stage to stage. The passages labeled $C, D, E, F,$ and G are also progressively smaller. If the rotors and the passages were not made progressively smaller, the compressed vapor would reexpand to fill the space and the compression would be nullified.

The refrigerant passages $C, D, E, F,$ and G are formed of metal labyrinths, which prevent leakage of the refrigerant between the various stages of compression. The labyrinths are constructed so that they come comparatively close to the rotors without actually touching them. The clearance between these two parts is not at all the close clearance that is required in the rotary compressor. Lubrication is not needed at any place on the centrifugal compressor except at the end bearings of the shaft.

Since these end bearings are the only internal friction surfaces, the refrigerant vapor compressed by a centrifugal compressor is free from oil, giving it the advantage of preventing an accumulation of oil on the inside surfaces of the condenser and evaporator. This situation greatly improves heat transfer.

Centrifugal compressors are designed to operate at relatively high speeds. They are quite efficient and are well suited for large-capacity refrigerating plants. They are best used on systems of 250 tons up to 3000 tons. They are efficient at a wide range of operating temperatures from 50 to 120 °F (10 to 49 °C). Because they are not positive-displacement compressors, they are quite flexible under varying load conditions and operate at good efficiencies even when the demand is less than 40% of their designed capacity.

Reciprocating and centrifugal compressors are also typed according to their compressor body style. There are three styles of compressors: open, semi-hermetic, and hermetic. These may be further divided into belt driven and direct driven.

Open-Type Compressors. This type was used extensively in the early phases of the refrigeration industry. A separate motor was used, and the power was transmitted to the compressor by a belt and a set of pulleys (Figure 5-28). The pistons and cylinders were enclosed within a case with the crankshaft extending through the body with the attached pulley. In this type, a crankshaft seal prevented the loss of refrigerant and oil around the shaft.

FIGURE 5-28 Open-type belt-driven compressor. (Courtesy of Tecumseh Products Company.)

This type of compressor is very flexible in that the speed can be changed to accommodate different refrigerants as well as to provide different capacities. Repairs are fairly easy to make and the replacement of worn parts is easily accomplished.

Even though open-type compressors were widely used at one time, they have many inherent disadvantages. They are bulky and heavy because of

FIGURE 5-29 Semihermetic compressor. (Courtesy of Copeland Refrigeration Corp.)

the cast-iron body. They are expensive to manufacture. The shaft seals are vulnerable to leaks, and shaft alignment is difficult on direct-drive models. They are more noisy because of the belts and other external components. The belts wear out and need to be replaced. Because of these disadvantages, the open-type compressor is being replaced with the semihermetic or the hermetic-type compressors. The open-type compressor is only used in a few specialized applications, such as automotive air conditioning, transport refrigeration, large commercial and industrial units, and ammonia systems.

2-CYLINDER COPELAMETIC

FIGURE 5-30 Cross-sectional view of semihermetic compressor. (Courtesy of Copeland Refrigeration Corp.)

COMPRESSORS

Semihermetic Compressors. These are driven by an electric motor that is mounted directly on the compressor crankshaft (Figure 5–29). The working parts of both the motor and compressor are sealed within a common housing. The troublesome belts and shaft seal are eliminated. The motors can be sized specifically for the load to be handled, resulting in a compact, economical, efficient, and basically maintenance-free design.

The removable heads, stator covers, bottom plates, and housing covers allow access for easy field repairs in the event of compressor damage (Figure 5–30). The cast iron used in the manufacture of open and semi-hermetic-type compressors must be of the close-grained type to prevent refrigerant leakage through the pores of the metal.

Hermetic Compressors. Sometimes referred to as welded hermetic motor-compressors, hermetic compressors are produced in an effort to further decrease the size and the cost of manufacturing refrigeration units. They are widely used in smaller horsepower self-contained equipment. They are popular in up to about the 5-ton size in air conditioning applications. The motor and compressor are mounted on a common shaft as in the semihermetic units. However, the body is a formed metal shell hermetically sealed by welding (Figure 5–31). No internal field

FIGURE 5–31 Hermetic compressor. (Courtesy of Copeland Refrigeration Corp.)

repairs can be performed on this type of compressor since the only means of access is by cutting open the compressor shell.

The major internal parts of a hermetic compressor are listed in the same sequence as that of the refrigerant vapor flow through the compressor (Figure 5–32). First, the suction is drawn into the

FIGURE 5–32 Internal view of hermetic compressor. (Courtesy of Tecumseh Products Company.)

compressor shell, then to and through the electric motor, which provides power to the crankshaft. The crankshaft revolves in its bearings, driving the piston in the cylinder. The crankshaft is designed to carry oil from the oil pump in the bottom of the compressor housing to all bearing surfaces. Refrigerant vapor surrounds the compressor crankcase and motor as it is drawn through the compressor shell and into the cylinder through the suction muffler and suction valves. As the vapor is compressed by the moving piston, it is released through the discharge valves, discharge muffler, and compressor discharge tube.

☐ Compressor Component Construction ☐

The various components that are assembled together to build a compressor must be manufactured to certain specifications. If these specifications are not met, the quality of the compressor will be lowered. Most of these specifications are met at the factory; however, the service technician must be acquainted with them so that repairs to a compressor can be efficiently and economically completed.

Cylinders. Cylinders must be manufactured of a good grade of close-grained hard cast iron (Figure 5–33). In the factory, the cylinder is usually subjected to a slow running in process for a considerable period of time. This process enables the inside of the cylinder to attain a smooth, hardened, polished surface; in other words, the cylinder is very carefully broken in. Careful breaking in of the compressor will enable it to wear a long time. To prevent vapor leakage past the piston, it is necessary that the cylinder bore be very carefully honed to a perfect circular shape.

Pistons. Pistons are made from a hollow cylinder casting closed on one end and provided with a piston pin (Figure 5–34). The piston pin is used to connect the piston to the connecting rod. The diameter of the piston is a few thousandths of an inch smaller than the bore of the cylinder. This allows space for expansion and allows the lubricating oil to enter between the cylinder wall and piston. Rings are provided on the piston to prevent excessive oil from passing into the cylinder. The lubricating oil serves a dual purpose of lubricating the movement of the piston and sealing the pressure of the compressed refrigerant to prevent its leaking between the piston and the cylinder wall.

FIGURE 5–33   Compressor cylinder. [Courtesy of Copeland Refrigeration Corp.]

FIGURE 5–34   Compressor piston.

Valves. Valves used in most modern refrigeration compressors are termed *flapper valves*. They are used to control the flow of refrigerant through a compressor. The proper operation of a compressor depends greatly on their condition and operation. There are two such valves for each cylinder in a compressor. They are termed *suction* and *discharge* valves. The suction valve controls the flow of refrigerant vapor into the compressor, while the discharge valve controls the flow of refrigerant from the compressor.

The flapper valve consists of a thin flat strip of special valve steel. This type of valve operates successfully at extremely high compressor speeds and does not hammer down the valve seat like heavier type valves. It has very little valve drag due to its extremely light weight, and as a result the volumetric efficiency of the compressor is greater than that of a compressor using a heavier type valve with a coil spring. At high compressor speeds, heavy valves prove extremely inefficient.

The purpose of all compressor suction valves is to open and admit the refrigerant vapor to the cylinder chamber on the suction stroke of the piston. On the compression stroke, the suction valve closes and traps the refrigerant vapor within the cylinder chamber so that it can be compressed to a higher

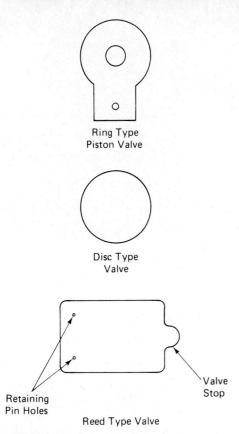

Ring Type
Piston Valve

Disc Type
Valve

Retaining
Pin Holes

Valve
Stop

Reed Type Valve

FIGURE 5–35 **Suction valve types.**

Piston Head
Plate

Spiral
Spring

Valve Ring

Valve Seat

Valve
Retainer Pin

Piston

Suction Valve Guide
Retainer

Suction Valve Guide

Suction Valve

FIGURE 5–36 **Suction valve locations. (Courtesy of Copeland Refrigeration Corp.)**

pressure and temperature. Suction valves are operated automatically by the differences in pressures within the compressor. In most types a spring is not used. The valve consists of a thin piece of steel that may take the shape of a ring, disc, or reed (Figure 5-35). In each case, whatever the shape may be, the thin metal covers a port when it is closed and is lifted from the port when open.

Suction valves may be placed in the head of the piston or located on the valve plate of the compressor. When the suction valve is located in the piston, it is commonly known as a piston valve (Figure 5-36).

Every type of valve must have some means of limiting its motion. If the valve movement is too great, it may not close with sufficient rapidity on the compression stroke to seal properly. In addition to the different types of valves, the means whereby valve motion is limited also vary. In the piston valve of the ring type the piston head plate, or cage, is screwed tightly to the piston and keeps the various parts of the valve in place (Figure 5-37). The valve seat consists of a circular raised edge on top of the piston. This edge is machined to a smooth surface and lapped. There is also another lapped circular edge or seat concentrically located within the valve

Top View of Assembled
Piston

Valve Cover

Valve Retainer } Retainer
Valve Spring } Assembly

Valve Disc

Piston

Parts Shown in Order of Assembly

FIGURE 5–37 **Suction valve stops.**

seat. Between these two lapped circular seats is the opening through which the refrigerant vapor passes when the valve ring is raised on the downstroke of the piston. The valve ring has an anchoring tongue extension with a hole which slips over the valve retainer pin.

The valve ring has a rather limited up and down motion since it can only move between the edge of the valve seat and the underside of the piston head plate. Its motion is opposed by a small spiral spring fastened to the piston head. The primary purpose of this spring is to prevent the valve from opening and closing abruptly, as this would introduce a clicking noise during operation of the compressor. In addition to its noise-eliminating action, the spring also helps keep the valve closed when the compressor is idle. This reduces the possibility that any refrigerant that has leaked past the discharge valve will enter the crankcase.

Not all suction valves are located in the head of the piston. Many compressors have the suction valves located in the compressor cylinder head. The valves are mounted on a head plate; which also carries the discharge valves. Suction valves and discharge valves move in opposite directions. The plate is known as a valve plate and is installed between the compressor cylinder proper and the cylinder head (Figure 5–38). The advantage of such construction is that the cool refrigerant vapor returning from the evaporator helps to cool the cylinder head. Because the vapor does not pass through the compressor crankcase, the possibility of pumping oil is reduced. Suction valves located in the valve plates have much the same construction as those located in the piston.

The reed-type suction valve is made of very thin spring steel and is a spring in itself. It is flexed, that is, bent downward, by the pressure of the incoming vapor (Figure 5–39). The reed constantly tends to hold itself closed except when its spring tension is overcome by the pressure of the suction vapor.

The purpose of discharge valves is to allow the compressed refrigerant vapor to be discharged to the condenser and to maintain a gastight seal between the high and low sides of the system when the compressor is at rest. This prevents the high-side condensed refrigerant from vaporizing and passing into the low side. The discharge valve in effect is a check valve since it checks the compressed refrigerant from reexpanding back into the cylinder chamber.

Discharge valves are almost always located in the

FIGURE 5–38   Valve plate location. (Courtesy of Copeland Refrigeration Corp.)

FIGURE 5–39   Reed-type suction valve.

cylinder head on the valve plate. There are several designs and types: the ring type; the disc type; and the reed type, or flapper valve as it is sometimes called. Most discharge valves are designed with an auxiliary safety discharge, which comes into play if a quantity of oil or liquid refrigerant enters the cylinder chamber. Oil and liquid refrigerant are incompressible and precautions must be taken to allow them to pass safely out of the cylinder chamber.

There are many designs of the flapper-type discharge valve. The bridged flapper valve is the most common, and this discussion will be limited to it. The bridged flapper discharge valve is so called because the valve bridges across the opening between the two valve guides (Figure 5–40). The valve guides consist of specially shaped machine screws which are screwed into the valve plate. The discharge valve proper consists of a steel reed secured at both ends by the valve guides. Directly above the reed is a valve spring stop. The valve spring stop has a slight upward bend or curvature in the middle so that the thin reed can be forced sufficiently upward by vapor pressure from the cylinder chamber to allow the compressed vapor to pass. Should a quantity of oil be trapped in the cylinder chamber, it will force the

FIGURE 5-40 Bridged-flapper-type discharge valve.

FIGURE 5-41 Rotary compressor discharge flapper valve.

valve up against the valve spring stop. The valve spring stop will be forced upward, compressing the valve springs. This action allows the oil to escape without permanently deforming or breaking the reed, since the amount of flexing of the reed is limited by the valve spring stop. The valve springs constitute a safety discharge feature and are often known as safety springs. The valve guides are wired together at the top by a locking wire to prevent possible loosening.

A thin film of oil between the valve and its seat maintains the actual seal in all cases regardless of the type of valve. When the oil charge in the crankcase is correct, a sufficient amount of oil flows through the system with the refrigerant to lubricate the valves and effect a gastight seal. It is therefore important that there be sufficient oil in the crankcase of the compressor. A compressor loses its efficiency when the discharge or suction valves leak.

To check the condition of the valves, first stop the flow of refrigerant to the suction side of the compressor. This is usually done by *front seating* (turning all the way in) the suction service valve with the gauge manifold attached. Run the compressor for a few minutes. The compressor should pump a vacuum of approximately 20 in. Hg. If it does not, stop the compressor for a few minutes, then start the compressor and see if it will pump the desired vacuum. If not, the suction valves are leaking and must be replaced.

To check the discharge valves, attach the gauge manifold, close off the flow of refrigerant to the suction side of the compressor, and run the compressor until it pumps as deep a vacuum as possible. Stop the compressor and observe the reading on the compound gauge. The pressure should not increase over 3 or 4 in. Hg. If it does, try pumping another vacuum and again observe the pressure reading on the compound gauge. If the pressure still increases, the discharge valve is leaking. One step more will relieve

any doubt about the condition of the discharge valve. While the vacuum remains on the compressor, front seat the discharge service valve with the compressor stopped. If the pressure reading on the compound gauge stops rising, the discharge flapper valve is positively bad and must be replaced.

The discharge valve on a rotary compressor prevents the compressed refrigerant from reentering the cylinder chamber after it has been discharged from the compressor. The valve seat is ground and polished or lapped to a mirror finish. The valve is a reed or flapper type made of spring steel and provided with a stop spring which returns the valve to its seat with each discharge (Figure 5–41). The valve cover and the muffler reduce the noise resulting from the discharge pulsations. The valve reed is assembled to the plate by means of a screw and lock washer. Not all types of rotary compressor discharge valves use a valve stop spring, since on some rotary compressors the valve is not a positive type and may allow some refrigerant to leak back into the compressor housing. A check valve is used to prevent the leakage of refrigerant from the compressor housing into the evaporator.

Connecting Rods. The connecting rod is the link that connects the piston to the crankshaft. It is provided with bearings on each end to take the wear and thrust at the piston pin and at the crankshaft (Figure 5–42). The wearing surfaces are provided with close-fitting bushings or bearings made of a bronze alloy or babbit.

Crankshafts. Crankshafts are made of forged steel. The crankshaft must be sufficiently heavy to resist deflection. A crankshaft that deflects interferes with the smooth up-and-down motion of the piston, which results in excessive vibration (whipping). In a two-cylinder compressor the crank throws are usually located at opposite sides of the shaft to balance each other. This enables the movement of the pistons to balance and also reduces vibration to

FIGURE 5-42 Connecting rod.

a great extent. Crankshafts are usually counter-balanced to prevent vibration.

The amount of vibration in a compressor is directly related to the number of cylinders it contains. A single-cylinder compressor will vibrate more than a compressor of the same capacity with two cylinders. Similarly, a two-cylinder compressor will vibrate more than a compressor of the same capacity with four cylinders. This is because the total volume is divided equally among the cylinders, reducing the quantity of refrigerant pumped by each cylinder and allowing a more constant flow of the discharged refrigerant.

□ Compressor Speed □

Early refrigeration compressors were designed to operate at relatively low speeds, usually less than 1000 rpm. Semihermetic and hermetic motor-compressors were designed to operate at 1750 rpm and made use of the standard four-pole electric motors. The increasing demand for lighter weight and more compact air conditioning equipment has been instrumental in the development of hermetic motor-compressors equipped with two-pole electric motors that operate at 3450 rpm.

Certain special applications such as automotive air conditioning, aircraft, or military air conditioning equipment make use of even higher speed compressors. For the normal commercial and domestic applications, however, the existing 60-hertz electric power supply will generally limit compressor speeds to the currently available 1750 and 3450 rpm.

Compressors operating at higher speeds tend to have lubrication problems and a shorter life. These factors, together with cost, size, and weight, must be considered in compressor design and application.

□ Lubrication □

The importance of maintaining the proper oil level cannot be overemphasized. Operating a compressor with too much oil in it will possibly do more damage than operating it with too little in it. If the compressor uses an oil pump, the oil safety control will stop the compressor if the oil level drops to a dangerously low level and prevent damage to it. An excessive amount of oil can possibly cause oil to return to the compressor in a large quantity after a defrost period of a commercial refrigeration unit. This condition is known as liquid slugging and can cause severe damage. The excess oil will circulate through the system and greatly reduce the efficiency by placing an insulating coat of oil on the heat transfer surfaces. This condition causes high discharge pressures and low suction pressures.

The normal oil level should be maintained at or slightly above the center of the *oil sight glass*, sometimes called the bull's-eye (Figure 5–43). The oil levels are different for compressors equipped with oil pumps and those without oil pumps. The following are the recommended levels:

- *Compressors with oil pumps.* On most compressors equipped with oil pumps, the oil level should be maintained at $\frac{5}{8}$ of the sight glass or less. Any oil above the $\frac{5}{8}$ sight glass level is only moved out into the refrigeration system to cause the problems mentioned above. It is generally recommended that the oil be maintained at $\frac{1}{3}$ to $\frac{1}{2}$ sight glass level.
- *Compressors without oil pumps.* The oil level in these compressors should be maintained at $\frac{1}{2}$ to $\frac{2}{3}$ sight glass. The level should never drop below $\frac{1}{2}$ sight glass.

The oil return on air-cooled compressors is usually very slow and the level should be checked quite frequently to assure adequate lubrication. However, the oil return to suction gas-cooled compressors is usual-

FIGURE 5-43 Oil sight glass.

ly very fast, and therefore they do not require as much attention as the air-cooled models.

It should be remembered that when excess oil is taken from a system, the area vacated by the oil must be filled with refrigerant. Because of this, it may be necessary to add more refrigerant to the system, which could cause more oil to return to the compressor because of the higher velocity of the suction gas.

Only a good grade of refrigeration oil should be used to lubricate refrigeration compressors. Refrigeration oil is made of a high grade of mineral oil. It is wax free and has additives to reduce foaming. At the factory, special precautions are taken to remove any moisture and foreign particles from it. After a container of oil has been opened, it should be tightly closed and precautions taken to prevent moisture and dirt from getting into the oil.

A perfect oil for use under all operation conditions encountered in refrigeration systems has not yet been developed. The different oils now available each have good and bad characteristics. These characteristics must be balanced against the system operating requirements.

Some of the essential qualities of refrigeration lubricants are:

1. They must remain stable at high operating temperatures.
2. They must remain fluid (viscosity) at low operating temperatures.
3. They must be free of moisture.
4. They must not contain wax, which will be deposited on the system components at low operating temperatures.
5. They must not react chemically with the refrigerant, metals, air, and/or other contaminants, or motor insulation when used with hermetic or semihermetic compressors.

Refrigeration lubricants can be separated into three main categories: (1) naphthene base, (2) paraffin base, or (3) a mixture of these two. These different-based lubricants are possible because the crude oil is found in different parts of the world.

Characteristics of Lubricating Oil

Flash Point. The flash point is the temperature at which oil vapor will flash into fire when exposed to an open flame. At this temperature the oil becomes unstable and tends to break down into its components. There is generally no danger of refrigeration lubricants reaching this temperature.

However, it must be avoided by using the proper lubricant.

Fire Point. Fire point is reached when the lubricant temperature is increased beyond the flash point of the oil vapor and the oil continues to burn.

Corrosion Tendency. Corrosion tendency is a measure of how much damage is possible to the metal components of a refrigeration system. A good lubricating oil should show only a minimum corrosive tendency when a strip of highly polished copper is immersed in a sample of the lubricant heated to 200°F (93.33°C) or higher for a period of 3 to 4 hr. When the strip is removed from the oil sample and there is pitting or more than a slight discoloring, the oil is corrosive and should not be used.

Pour Point. Pour point is a measure of the lowest temperature at which the oil will flow. A low pour point indicates that the oil will not congeal at the lowest temperatures encountered in the system at the design operating conditions.

Floc Point. Floc point is the temperature at which a mixture of refrigerant and oil becomes cloudy because of the separation of wax from the oil. The expended particles of wax will form into small balls. The wax will collect in the colder areas of the refrigeration system, reducing heat transfer and perhaps restricting the flow control device. The system efficiency will also be decreased.

Dielectric Strength. Dielectric strength is a measure of the resistance to the flow of an electric current. This is especially important in hermetic and semihermetic compressors because the refrigerant and oil returning from the evaporator flow across the motor windings. At this point, a path for electric current to flow would result in a damaged motor winding.

Viscosity. Viscosity is a measure of the resistance to flow of the oil. It simply reflects how thick or thin it is. When determining the viscosity, a measured sample of liquid at a specified temperature is allowed to flow through a calibrated orifice. The amount of time required, in seconds, for the sample of oil to flow through the orifice determines its viscosity.

Oil of the specific viscosity should be used for each temperature application. Each manufacturer has a recommended viscosity for each type of equipment application and, for best results, their suggestions should be followed.

Methods of Lubrication. There are two basic methods used to lubricate the compressor parts: (1)

splash, and (2) *forced*. Almost all compressors use a combination of these two.

Splash Lubrication. This is the simplest method of lubricating the compressor parts. The crankcase oil is splashed onto the moving parts by the rotating crankshaft. This method supplies oil to the cylinder walls and bearing surfaces. The splash method of compressor lubrication was satisfactory for the slower speed compressors. It became inadequate when the large modern high-speed compressors were designed because of their higher bearing and friction surface temperatures. Lubricating oil is not only used to reduce the friction between moving parts but to carry away some of the heat produced by this friction.

Forced Lubrication. This is provided on almost all modern refrigeration compressors. There are several ways to force lubricate a compressor. On the smaller size compressors, up to about 3 hp, the oil is forced to the desired points by providing rifled passageways. When this type of system is used, the oil is provided to the passageway by splashing. As the compressor parts move, the oil is forced to the bearing surface in the same manner that a projectile is forced to spin in a gun barrel.

In the larger horsepower models, over 3 hp, a more positive type of lubrication system is used. Compressor lubrication is provided by means of a positive-displacement oil pump. The pump is mounted on the rear bearing housing and is driven from a slot in the crankshaft into which the flat end of the oil pump drive shaft is fitted (Figure 5–44).

Oil is forced through a hole in the crankshaft to the compressor bearings and connecting rods. A spring-loaded ball check valve serves as a pressure-relief device, allowing oil to bypass directly to the compressor crankcase if the oil pressure rises above its setting (Figure 5–45).

Because the pump intake is directly connected to the compressor crankcase, the oil pump inlet pressure will always be equal to the crankcase pressure, and the oil pump outlet pressure will be the sum of the crankcase pressure plus the oil pump pressure. Therefore, the net oil pressure is always equal to the pump outlet pressure minus the crankcase pressure. When the compressor is operating with the suction pressure in a vacuum, the crankcase pressure is negative and must be added to the pump outlet pressure to determine the net oil pump pressure. A typical compound gauge is calibrated in inches of mercury for vacuum readings, and 2 in. Hg is approximately equal to 1 psi. (See Table 5–1.)

FIGURE 5–44 Oil pump location.

Remove schrader type valve cap for checking oil pressure

Oil pressure port
Remove seal cap when installing oil pressure safety switch

Identification stamped on oil pump

Spring loaded bypass assembly
Not adjustable
Do not disassemble

Oil pump mounting bolts
For replacing oil pump

Face plate attaching bolts
Do not remove

FIGURE 5–45 Location of a spring-loaded bypass assembly. (Courtesy of Copeland Refrigeration Corp.)

TABLE 5–1 Calculating Compressor Oil Pump Pressure

Crankcase Pressure	Pump Outlet Pressure [psi (kPa)]	Net Oil Pump Pressure [psi (kPa)]
50 psi [444.49 kPa]	90 [719.28]	40 [375.79]
8 in. vacuum [equivalent to a pressure reading of 4 psig [99.31 kPa]]	36 [348.3]	40 [375.79]

In normal operation, the net oil pressure will vary depending on the size of the compressor, the temperature and viscosity of the oil, and the amount of clearance in the compressor bearings. Net oil pressures of 30 to 40 psi (307.09 to 375.79 kPa) are normal, but adequate lubrication will be maintained at pressures down to 10 psi (169.69 kPa). The bypass valve is set at the factory to prevent the net oil pressure from exceeding 60 psi (513.19 kPa).

The oil pump may be operated in either direction; the reversing action is accomplished by a friction plate that shifts the inlet and outlet ports. After prolonged operation in one direction, however, wear, corrosion, varnish formation, or burrs may develop on the reversing plate which can prevent the pump from reversing. Therefore, on installations where the compressors have been in service for some time, care must be taken to maintain the original rotation of the motor if for any reason the electrical connections are disturbed.

The presence of liquid refrigerant in the compressor crankcase can materially affect the operation of the oil pump. Violent foaming on startup can result in a loss of oil from the crankcase, and a resulting loss of oil pressure until some of the oil returns to the crankcase. If liquid refrigerant or a refrigerant-rich mixture of oil and refrigerant is drawn into the oil pump, the resulting flash gas may result in large variations and possibly a loss of oil pressure. The crankcase pressure may vary from the suction pressure because liquid refrigerant in the crankcase can pressurize the crankcase for short intervals. Therefore, the oil pressure safety control low-pressure connection should always be connected to the crankcase.

◻ Acid Level ◻

When acid is present in a refrigeration system, the system oil acts as a scavenger and picks it up. Field and laboratory experience have shown that the measurement of the acid concentration of an oil sample taken from a compressor is a reliable method of determining the extent of contamination in the system.

Any acid in the system will have an adverse effect on the overall performance of the valves, filters, driers, and the related devices and can cause that system to fail completely.

Periodic checks of the oil will determine when the oil should be changed from the amount of contamination contained in it. Also, periodic checks of the oil during the cleanup period after a compressor burnout will indicate when the acid contamination has been reduced to a satisfactory and safe operating level.

There are several methods used for checking the acid level in the oil. The one used will depend on the preference of the user. Be sure to follow the manufacturer's recommended procedure for the one being used.

◻ Compressor Shaft Seals ◻

On the conventional open-type compressor, where the motor and compressor are two separate components, the shaft of the compressor extends out through the compressor crankcase. Some method of sealing around the rotating shaft must be used. This seal must positively prevent leaks both when the compressor is operating and when it is at rest. The seal must not allow vapor or oil to escape out of the compressor, nor must it allow air and moisture to leak into the compressor.

When the compressor is operating, the oil is splashed around the interior of the crankcase and is conducted to and lubricates the seal faces. It is the oil film between the seal faces that does the actual sealing. Therefore, it is important that this oil film never be permitted to dry out as this would immediately cause the seal to leak. A dry seal is also apt to squeal when the compressor is operating and wear out quickly.

Some refrigerants dissolve considerable oil. When an open-type compressor with a shaft seal is allowed to stand idle for a considerable length of time, the refrigerant may dissolve the oil from the seal faces. This type of compressor should not be allowed to stand idle for long periods of time unless it has been *pumped down*, that is, unless the refrigerant has been pumped into and stored in the liquid receiver and condenser with the compressor suction and discharge service valves closed.

Caution: Never attempt to operate a compressor with the discharge service valve closed because considerable damage could be done to the compressor.

Packing Gland Seals. These seals were used in early model compressors. This type of shaft seal is still used on large-capacity ammonia compressors and compressors that operate at slow speeds. The packing gland seal guards the drive shaft against loss of refrigerant on compressors operating above atmospheric pressure. On compressors operating below atmospheric pressure, it prevents air from being drawn into the system. The early type of packing consisted of a composition made of asbestos, fiber, and graphite. This material was placed in the housing surrounding the shaft and held in tightly by means of a metal packing gland (Figure 5–46). The compressor shaft extends through the compressor crankcase and is sealed by rings of semimetalic graphite impregnated packing. The packing gland is

FIGURE 5-46 Packing-gland-type shaft seal.

FIGURE 5-47 Stationary unbalanced bellows seal.

FIGURE 5-48 Unbalanced bellows seal installed on a compressor.

drawn down tightly by means of the takeup bolts. When the bolts are tightened, the packing is compressed, forming a seal between the shaft and the housing.

Packing glands have the disadvantage that they must be frequently tightened to prevent leakage; even so, there is a tendency for loss of refrigerant and oil. Another difficulty is that too tight a packing gland might interfere with proper starting of the compressor. Also, the pressure required to effect a tight seal causes considerable wear on the packing, which is softer than the shaft, and makes frequent replacement of the packing necessary.

Stationary Unbalanced Bellows Seals. These seals use one end of the bellows as a flange while the other end has a seal nose (Figure 5-47). The spring is of a smaller diameter than the bellows, which makes this seal the internal spring type. The seal nose and the flange are solidly soldered to the bellows. When the bellows seal is assembled on the compressor, the seal nose makes contact and rubs against the shoulder on the crankshaft (Figure 5-48). The bellows flange is clamped between the compressor body and the removable seal cover plate. This forms the second gastight seal by means of a gasket. The seal cover plate is secured to the compressor body by six cap screws. The shoulder of the shaft is case hardened, ground, and lapped to a very smooth surface. The seal nose also must be ground and lapped and free from any surface marks or defects. Only if both the face of the shoulder on the shaft and the nose of the seal are absolutely smooth will the oil be able to maintain a perfect seal.

Diaphragm-Type Seals. These are also known as the *balanseal*. The diaphragm is soldered to the seal nose

and clamped between the compressor body and the cover plate with a gasket. The seal nose does not ride directly on the shoulder of the crankshaft. Instead, it rubs against a hardened seal ring collar that rotates with the shaft. In case of wear the collar, as well as the seal nose, are easily replaceable (Figure 5-49).

The diaphragm maintains a practically constant pressure between the seal nose and the hardened seal collar regardless of the crankcase pressure. This is accomplished by a balancing effect created by a fulcrum located at the proper point outside the diaphragm. The exaggerated sections 1, 2, and 3 in Figure 5-49 indicate how the diaphragm acts and how the pressure between the two sealing surfaces is equalized when affected by varying pressures in

FIGURE 5-49 Diaphragm-type seal.

FIGURE 5-50 Rotary seal components.

FIGURE 5-51 Rotary seal installed.

the crankcase. Section 1 indicates the position of the diaphragm when the seal is first assembled and the crankcase is at atmospheric pressure. The fulcrum presses against the diaphragm; the pressure in turn is transmitted by the diaphragm to the seal nose, causing it to press against the hardened steel collar. When the pressure in the crankcase is increased, the diaphragm is caused to bulge outward as shown in section 2. This outward bulging action of the diaphragm working across the fulcrum causes the seal nose to be pressed more tightly against the seal collar. This compensates for the additional pressure in the crankcase which, of course, would act directly on the seal nose and tend to push it away. The additional pressure exerted by the diaphragm would cause these two pressures to equalize; thus, the friction between the two sealing surfaces would be practically unchanged.

In section 3 the appearance of the diaphragm shows that the pressure within the crankcase has decreased; that is, it has fallen below atmospheric pressure. The decreased pressure within the crankcase has caused the diaphragm to bulge inward. This action releases the pressure on the fulcrum and reduces the diaphragm pressure on the seal nose. Since the vapor pressure against the seal nose has also decreased, it can readily be seen that the pressures are equalized; thus, the friction between the two sealing faces has not been materially changed.

Rotary Seals. Rotary seals are widely used today. A rotary seal is available for replacing the bellows-type seals. In these seals the bellows rotates with the shaft. In the simplest adaptation of the rotary seal

principle, the seal assembly consists of seal face, friction ring, retaining shell, friction ring band, spring, and spring holder (Figure 5-50).

In operation, the entire seal assembly rotates with the shaft (Figure 5-51). A running seal joint is formed by contact between the surface of the seal face (1) within a mating stationary seal seat (A). These surfaces are finished to an extremely high degree of flatness and smoothness. The seal face (loosely encircling the shaft) is held tightly against the stationary seat by the pressure of the spring (5) bolstered by any internal vapor or liquid pressure; the mating is so perfect that no liquid or vapor can pass.

The back of the seal face rests against the front surface of the resilient friction ring (2), which fits frictionally tight on the shaft surface and prevents leakage at that point. The friction ring band (4) fits around the hub end of the friction ring and restricts its expansion, thus controlling and maintaining the tightness of the ring's fit on the shaft.

A retaining shell (3), fitting snugly over the flanged portion of the friction ring and loosely encircling the seal face, is provided with keys that engage in corresponding lateral grooves in the outer periphery of the seal face. This keying arrangement combines with the resiliency and snug fit of the friction ring both on the shaft and in the retaining shell to effect a positive, yet simple and flexible, driving connection between the shaft and the seal assembly unit.

The spring, encircling the shaft, is compressed to a predetermined length between the back wall of the retaining shell and the spring holder (6), which may rest against any part of the shaft or any member fastened to the shaft. The pressure developed holds the seal face in perfect sealing contact with the stationary sealseat, and also provides the force to form a gasket joint between the friction ring and the back of the seal face.

It should be noted that just one part, the relatively soft and elastic friction ring, is tightly fastened to the shaft. Consequently, the seal face can freely adjust its running lane to compensate for any shaft vibration, deflection, lateral play, or slight misalignment that might tend to break the running seal joint.

The end plate is stationary and forms the stationary surface of the seal (Figure 5-52). Thus, a seal is created between the rotating surfaces on the seal nose and the stationary seal surface of the end plate. A second seal is created by the friction ring, preventing leakage along the shaft. A third seal is formed by the gasket, which is placed under the end plate. The friction ring is made of a synthetic rubber substance, neoprene, which is not affected by refrigerant vapor or oil.

☐ Compressor Displacement ☐

The displacement of a compressor is the volume displaced by the pistons. Some manufacturers rate the displacement of their compressors in terms of cubic feet per hour, while others rate their compressors in terms of cubic inch displacement per revolution or in cubic feet per minute. For comparative purposes (see Table 5-2), compressor displacement may be calculated by the following formulas:

$$\text{cfm} = \frac{\pi \times D^2 \times L \times \text{rpm} \times N}{4 \times 1728}$$

$$\text{cfh} = \frac{\pi \times D^2 \times L \times \text{rpm} \times N \times 60}{4 \times 1728}$$

$$\text{in.}^3/\text{rev.} = \frac{\pi \times D^2 \times L \times N}{4}$$

where

cfm	=	cubic feet per minute
cfh	=	cubic feet per hour
in³/rev.	=	cubic inch displacement per revolution
π	=	3.1416
D	=	cylinder bore in inches
L	=	length of stroke in inches
N	=	number of cylinders
rpm	=	revolutions per minute
1728	=	cubic inches per cubic foot
$\pi D^2/4$	=	area of a circle

FIGURE 5-52 Rotary seal.

TABLE 5-2 Conversion Factors

Unit	1750 rpm	3500 rpm
cfh	60 × cfm	60 × cfm
cfh	60.78 × in.³/rev.	121.5 × in.³/rev.
cfm	1.013 × in.³/rev.	2.025 × in.³/rev.
in.³/rev.	0.01645 × cfh	0.00823 × cfh

☐ Compression Ratio ☐

Lack of knowledge is a tremendous factor in equipment failures because few service technicians really know how to calculate compression ratio. Many have no idea of the dangers involved in designing a system with a suction pressure that is too low. Surveys show that fewer than 10% of service technicians are able to calculate the compression ratio of a given system. A system operating with a high compression ratio may have a discharge temperature as much as 150°F (66°C) higher than it should be.

The rate of chemical reaction caused by reactive materials such as refrigerant oil, cellulose, oxygen, moisture, acid, heat, and pressure approximately doubles for each 18°F (−7.8°C) rise in discharge temperature. It is only natural that a system running at abnormally high head temperatures will develop more problems in a shorter period of time than one operating with normal discharge temperatures. Sound refrigeration design and common sense seem to indicate that the relationship between the head pressure and the suction pressure should be well within the accepted industry bounds of a 10:1 compression ratio.

Compression ratio is defined as *the absolute head pressure divided by the absolute suction pressure.* Most service technicians realize that an ordinary compound gauge does not indicate atmospheric pressure but reads zero when it is not connected to a system under pressure. Most service technicians know that to find the absolute head pressure or absolute suction pressure at zero pounds gauge pressure or above, 15 psi (103.05 kPa) must be added to the gauge reading.

The difficulty in computing the absolute suction pressure is when the suction pressure is below atmospheric pressure. This is done by subtracting the vacuum reading in inches from 30 in. (77 cm) and dividing the result by 2.

$$\text{Compression ratio} = \frac{\text{absolute head pressure}}{\text{absolute suction pressure}}$$

Case 1. For 0 psig pressure or above:

Absolute head pressure = gauge reading + 15 psi (103.05 kPa)

Absolute suction pressure = gauge reading + 15 psi (103.05 kPa)

Case 2. When the low-side reading is in a vacuum:

Absolute head pressure
= gauge reading + 15 psi (103.05 kPa)

Absolute suction pressure
$$= \frac{30 - \text{compound gauge reading in inches}}{2}$$

The calculation of compression ratio can be illustrated by the following examples:

EXAMPLE 1

Head pressure = 160 psi (1200.19 kPa)

Suction pressure = 10 psi (169.69 kPa)

$$\text{Compression ratio} = \frac{\text{absolute head pressure}}{\text{absolute suction pressure}}$$

$$= \frac{160 \ (1200.19 \text{ kPa})}{10 \ (169.69 \text{ kPa})} +$$

$$15 \ (103.05 \text{ kPa})$$

$$= \frac{175 \ (1303.24 \text{ kPa})}{25 \ (272.74 \text{ kPa})} = 7{:}1$$

EXAMPLE 2

Head pressure = 160 psi (1200.19 kPa)

Suction pressure = 10 in. vacuum (68.42 kPa)

Absolute
head pressure = 160 (1200.19 kPa) +
15 (103.05 kPa)

= 175 psia (1303.24 kPa)

Absolute
suction pressure $= \dfrac{30 - 10 \ (206.1 \text{ kPa} - 68.7 \text{ kPa})}{2}$

$$= \frac{20}{2}$$

= 10 psia (68.7 kPa)

$$\text{Compression ratio} = \frac{\text{absolute head pressure}}{\text{absolute suction pressure}}$$

$$= \frac{175 \text{ psi} \ (1303.24 \text{ kPa})}{10 \text{ psi} \ (68.7 \text{ kPa})}$$

= 17.5:1

In the two preceding examples, we can see what a great influence suction pressure has on the compression ratio of a system. A change in the head pressure does not produce such a dramatic effect.

If the head pressure in both examples was 185 lb (1374 kPa), the compression ratio in Example 1 would become 8:1, and in Example 2 it would be 20:1.

Example 2 falls somewhat in the class of one of our more serious troublemakers, which is the R-12 system operating on a coil of approximately −35 °F (−37.2 °C), with a suction pressure of 8 in. of vacuum (75.51 kPa) or lower. It is interesting to compare two systems using different types of refrigerants, one system operating with R-12 and the other with R-22. At the −35 °F (−37.2 °C) evaporation, the R-22 system would have a 10.9:1 compression ratio, while the R-12 system would have a 17.4:1 compression ratio. The R-22 system is a borderline case, but the R-12 system is far out of the safe range and would run very hot with all the accompanying problems.

A competent service technician can do several things to reduce the effects of excess heat, reactive materials, and catalysts. These steps include a deep vacuum on the system, proper dehydration of the system, acid removal, and refrigerant filteration. When these things are done properly, the system will remain operating for a longer period of time, even under adverse operating conditions.

☐ Clearance Volume ☐

As mentioned previously, the volumetric efficiency of a compressor will vary with the compressor design. If the valves seat properly, the most important factor affecting compressor efficiency is the clearance volume.

At the end of the compression stroke, there still remains some clearance space, which is essential if the piston is not to hit the valve plate (Figure 5-53). There is also a great deal more space in the discharge valve ports located in the valve plate, since the discharge valves are on top of the valve plate. This residual space, which is unswept by the piston at the end of its stroke, is termed *clearance volume*. It remains filled with hot, compressed vapor at the end of the compression stroke.

When the piston starts down on the suction stroke, the residual high-pressure vapor expands as its pressure is reduced. No vapor from the suction line can enter the cylinder chamber until the pressure in the cylinder has been reduced below the suction line pressure. (Figure 5-54). Thus the first part of the suction stroke is actually lost from a capacity stand-

FIGURE 5-53 **Compressor clearance volume.**

FIGURE 5-54 **Pressure in a compressor cylinder.**

point. As the compression ratio increases, a greater percentage of the suction stroke is occupied by the residual gas.

With high suction pressures, the compression ratio is low and the clearance volume is not critical from a capacity standpoint. An additional clearance volume is also helpful in reducing the compressor noise level. Lower gas velocities through the discharge ports reduce both wear and operating power requirements. Most air conditioning compressor

116 _____

COMPRESSORS

valve plates are designed with a greater clearance volume by increasing the diameter of the discharge ports.

On low-temperature applications, it is often necessary to reduce the clearance volume to obtain the desired capacity. Low-temperature valve plates, which have smaller discharge port sizes to reduce the clearance volume, are used on most low-temperature compressors.

□ Compressor Cooling □

Air-cooled compressors require an adequate flow of cooling air over the compressor body to prevent the compressor from overheating. The airflow from the fan must be discharged directly on the motor-compressor. Air drawn through the compressor compartment usually will not cool the compressor adequately.

Water-cooled compressors are provided with a water jacket or wrapped with a copper water coil. The water must be circulated through the cooling circuit when the compressor is in operation.

Refrigerant-cooled motor-compressors are designed so that suction vapor flows around and through the motor to help cool it. At evaporating

FIGURE 5-55 Oil location. (Courtesy of Tecumseh Products Company.)

Oil Cooler Tubes

temperatures below 0°F (−17.8°C), additional motor cooling by means of airflow is necessary because the decreasing density of the refrigerant gas reduces its cooling ability.

Another method of compressor cooling is to cool the oil. Oil coolers are found only on low-temperature refrigeration models. External tubes to the oil cooler connect to a coil or hairpin bend within the compressor oil sump (Figure 5–55). This coil or hairpin is not open inside the compressor. Its only function is to cool the compressor oil sump. The oil cooler tubes are most generally connected to an individually separated tubing circuit in the air-cooled condenser.

□ Heat Pump Compressors □

Heat pump compressors should have a maximum compression ratio of 7.5:1. Compressor motors used on these types of units generally run somewhat hotter because of the low refrigerant density at the lower operating suction pressures. The design of these compressors takes this into account and reroutes the refrigerant through the motor windings. In some units a nondirectional gas flow is used that also reduces the temperature. This design allows a liquid and gas separation to produce the required better cooling.

Heat pump compressors should use a low-foaming white lubricating oil. The old type of refrigeration oil is not generally considered satisfactory for heat pump compressors. The amount of refrigerant in the oil is also important because a mixture of 14% liquid refrigerant to oil starts oil breakdown. Usually, 10% is considered maximum.

Larger connection rod and bearings are used in these compressors. The bearing loads are reduced by 20 to 41%. Ringed pistons are used rather than lapped pistons. This step provides better cylinder and piston clearance. Also, an improved motor winding protector is used to provide better protection during the high compression ratio operation.

□ Factors Controlling Compressor Output □

There are many factors that control the pressure a compressor is able to pump. Briefly, these factors are the design and condition of the compressor, the type of refrigerant used, and the conditions under which the compressor must operate. The following list includes the main factors that control the pressures attained in the compressor and, as a result, control the output of the compressor:

1. *Type of refrigerant.* Examples are Freon, ammonia, etc.
2. *Compression ratio.* This is the ratio between absolute head pressure and absolute suction pressure.
3. *Volumetric efficiency (VE).* This is the ratio of the volume of refrigerant vapor actually drawn into the cylinder to the volume of the piston displacement. Piston displacement (PD) is the volume of space in a cylinder swept through by the piston.

$$PD = \frac{\pi \times D^2 \times L}{4}$$

$$VE = \frac{actual\ volume}{calculated\ volume}$$

4. *Cooling system efficiency.* An efficient cooling system increases the volumetric efficiency by lowering the refrigerant temperature, thus reducing expansion of the refrigerant. This allows a greater amount of gas by weight to enter the cylinder.
5. *Cylinder cooling system.* The amount of compression heat removed from the refrigerant by the cylinder cooling system will also increase the volumetric efficiency.
6. *Presssure of refrigerant in the suction line.* The lower the pressure of the refrigerant in the suction line, the smaller will be the volume of refrigerant drawn into the cylinder on the suction stroke.
7. *Compressor speed.* At a high compressor speed the valves have a tendency to "drag" due to their weight and inertia, reducing the quantity of vapor taken in on each stroke.
8. *Type and size of valves.* Size of the valve openings and the positiveness or speed of the valve action affect the volume of vapor compressed.
9. *Friction of the refrigerant vapor.* The friction of the vapor against the turns in the tubing and the openings to the compressor decreases the speed of the vapor entering the cylinder, which reduces the quantity per stroke.
10. *Mechanical condition of the compressor.* Leaky piston rings, leaky valves, and worn bearings decrease the compressor output.
11. *Lubricants.* Proper lubrication decreases friction and seals the piston and valves against leakage, thus increasing the compressor output.

One of the basic difficulties in preventing compressor failure is the inability to determine what actually causes the failure. The compressor is the functioning heart of the refrigeration system, and regardless of the nature of a system malfunction, the compressor must ultimtely suffer the consequences. Since the compressor is the component that fails, there often is a tendency to blame any failure on the compressor without determining the origin of the malfunction. In far too many cases, the actual cause of failure is not discovered and corrected, resulting in recurring failures that could have been prevented.

If service technicians are to help in eliminating the causes of compressor failure, they must thoroughly understand both the operation of the system and the possible causes of failure that might occur. They must be on the alert for any signs of system malfunction.

□ Compressor Failure □

Compressor failures have plagued this industry since the introduction of hermetic and semihermetic compressors. Even more frustrating for manufacturers is the repeat compressor failure rate. It is a well-known fact that replacement compressors fail at a rate that is several times greater than the original compressor failure rate. More than 30% of the replacement compressors returned failed because of service and system-related problems. The remainder either were good or had failed because of electrical or mechanical problems. The electrical failures include single-phasing, motor overheating, and some types of control problems. Mechanical failures include bent rods, scored shafts, overheating, and broken valves.

From this it can be seen that the job of a service technician is more than just changing out a bad compressor. The chances are very good that the replacement will fail unless the system is properly cleaned up.

Many compressors fail because of one or more of the following conditions: (1) slugging, (2) liquid flooding, (3) loss of lubricating oil, (4) contamination, (5) bad piping practices, (6) improper thermostatic expansion valve superheat setting, (7) flooded starts, and (8) overheating.

Slugging. Slugging usually occurs on compressor startup and lasts for only a short time. However, it can occur while the compressor is running during a rapid change in the system operating conditions. It is associated with a clattering noise much like an automobile engine under a heavy strain. This noise is created by the compressing of the liquid refrigerant and/or oil. Compressors are designed to compress vapor only and not liquid. When liquid is compressed by the compressor a hydraulic pressure that may exceed 1000 psig (6890 kPa) is created in the cylinder.

Slugging can result in blown head and/or valve plate gaskets, broken valve reeds, broken pistons, and damaged piston wrist pins. If any of these conditions are found, the conditions that cause slugging should be checked and if found corrected.

Causes of Slugging. Refrigerant can condense in any part of the system that is colder than the compressor during the off cycle. When the system next starts, the liquid refrigerant will return to the compressor in a large slug.

Oil trapping in the system is also a cause of compressor slugging. Oil trapping occurs when the suction gas velocity is so low that the oil is allowed to settle in low places and is not brought back to the compressor on a continuous basis. This is a typical problem when suction risers are not properly sized. This problem is more apt to occur on systems that use compressor unloading equipment and when the compressor is operated for long periods unloaded.

Liquid Flooding. Liquid flooding is the constant flow of liquid refrigerant droplets to the compressor. When liquid refrigerant enters the compressor, it goes into the crankcase, where the oil is diluted. Liquid refrigerant is an excellent cleaning agent and will wash oil from the compressor surfaces. In most cases the oil will foam, reducing its lubricating value and resulting in overheating of the bearing surfaces. This condition is more likely on air-cooled compressors than on refrigerant-cooled compressors.

In cases of severe liquid floodback, damage to the pistons, valves, and rings is probably due to the lack of lubrication. Also, there may be broken parts which show very little wear. Usually, liquid flooding is indicated by more wear on the parts farthest from the oil pump.

Causes of Liquid Flooding. Liquid flooding is generally caused by poor refrigerant control. Poor refrigerant control may be the result of an oversized expansion valve, too low a superheat setting, or a loose or poor contact of the external feeler bulb on the suction line or in the wrong location. When a capillary tube is used, an overcharge of refrigerant could be the cause, or a low load on the evaporator due to low airflow, dirty filters, a dirty blower, or a loose indoor fan belt. In any case the problem must be found and corrected. The proper system superheat at the compressor will help this problem if all the other causes are removed.

Loss of Lubricating Oil. The loss of lubricating oil will result in damage to the compressor. The wearing surfaces will be gauled, overheated, and as a result, ruined. The compressor motor will be overheated and will possibly burn out. As the quantity of oil drops in the crankcase, the temperature of the remaining oil increases. Compressor oil starts vaporizing at 310 to 320 °F (136.66 to 142.22 °C), which reduces the amount of lubrication the cylinders receive, resulting in excessive ring and cylinder wear. The oil breaks down completely at 350 °F (158.89 °C), causing contaminants and a complete loss of lubricating qualities.

Causes for Loss of Lubricating Oil. In addition to poor line sizing and design, slugging, and liquid floodback, there are basically three reasons for the loss of lubricating oil from a compressor: (1) refrigerant leak; (2) long period of operation with low suction gas velocities, with long periods of operation without defrosting; and (3) temporary liquid flooding after a defrost cycle.

Contamination. The contaminants in a system include air, moisture, dirt, and other foreign matter. These contaminants cause several system malfunctions, such as high discharge pressure, poor system performance, moisture freezing in the flow control device, oil contamination, and acid.

Causes for Contamination. Contaminants enter the refrigeration system during poor installation or service procedures. Care must be taken to prevent their entrance into the system if the compressor is to last for its designed lifetime expectancy. When installing a system, the refrigerant lines must be kept clean and capped until they are connected to the indoor and outdoor components. The charging hoses on the gauge manifold must be kept clean, capped, and purged to remove any contaminants before charging a system with refrigerant. Use only clean, dry refrigerant when leak testing or charging a system.

Bad Piping Practices. Bad piping practices can also contribute to oil leaving the compressor. When unnecessary traps are allowed to remain in the system, oil will settle in them and cause the compressor to operate short of oil. Sizing of the suction line is extremely important for proper oil return. If an accumulator is used, it must be of the proper size for the system. Accumulators should be capable of holding approximately one-half of the total refrigerant charge on smaller systems. It is generally desirable to check with the accumulator and equipment manufacturers for their recommendations. Also, an accumulator will not protect the compressor from refrigerant floodback during the off cycle. Chapter 18 provides a more in-depth coverage of proper piping practices.

Improper Thermostatic Expansion Valve Superheat Setting. When checking and adjusting thermostatic expansion valve superheat settings, the procedure should be to start at the evaporator coil. If there is more than one expansion valve, they must all be set to maintain the same superheat setting. If they are set to maintain different superheat settings, oil logging is quite possible, especially on low-temperature applications. The recommended TXV superheat settings are:

High-temperature applications: 8 to 12°F (4.48 to 6.72°C),

Medium-temperature applications: 5 to 8°F (2.8 to 4.48°C),

Low-temperature applications: 2 to 5°F (1.12 to 4.48°C).

Flooded Starts. A flooded start generally is caused by refrigerant migration to the compressor during the off cycle. The refrigerant moves into the compressor crankcase and mixes with the oil when the compressor is colder than the remainder of the system. When the compressor starts, the liquid refrigerant quickly boils off, causing the oil to foam.

The amount of refrigerant absorbed by the oil is determined by the oil temperature and the pressure in the crankcase. The lower the temperature and the higher the pressure, the more refrigerant will be absorbed by the oil. Under certain conditions the refrigerant–oil mixture will separate and stratify, with the liquid refrigerant settling to the bottom of the crankcase where the oil pump will pick it up first. When the compressor starts, the oil pump forces the liquid refrigerant into the bearings and any oil that is there will be washed away.

At the same time, the pressure inside the crankcase drops rapidly, causing the liquid to flash into a vapor, causing the oil to foam. This foaming can cause an increased pressure in the oil passages, resulting in a reduction of oil to the bearings. Also, the foaming action may cause a mixture of liquid refrigerant and oil to enter the cylinder, resulting in a slugging condition.

Overheating. Compressors are designed to withstand the heat generated by compression of the refrigerant vapor, the heat given off by the motor windings, and the heat caused by bearings and other surfaces which cause friction heat inside the compressor. However, they are not designed to tolerate excessive temperatures caused by high compression ratios, high discharge pressures, high discharge temperatures, and shortage of refrigerant. The temperature measured within 6 in. (15.24 cm) of the compressor discharge service valve should never be allowed above 275°F (172.77°C). A temperature of 275°F at this point in the system indicates a temperature between 300 and 325°F (131.11 and 145°C) at the discharge valve at the compressor valve plate. At this high temperature, breakdown of the lubricating oil begins and the compressor will soon fail due to a lack of lubrication. A preferred discharge line temperature would be 225°F (126°C) or less.

It has been determined in laboratory tests that for every 18°F (10.0°C) temperature rise in the discharge vapor above the normal operating conditions the chemical reaction doubles between the refrigerant and moisture, acids and oxides, acid and the oil, and the refrigerant and the oil inside the system. This kind of reaction causes the compressor to start the failing process very quickly. It must be remembered that these contaminants are circulating throughout the system and will cause problems in other areas also.

There are two basic reasons that will cause a compressor to overheat: (1) electrical problems, and (2) refrigerant problems.

Overheating Due to Electrical Problems. Overheating in single-phase motors could be the result of problems in the starting circuit. Generally, these problems are due to the use of wrong starting capacitors, wrong run capacitors, and/or a wrong starting relay.

Another cause of motor winding overheating is short cycling. When the compressor is allowed to cycle rapidly on refrigerant controls, temperature controls, and safety devices, the motor windings will become overheated, causing damage to the motor winding insulation and early failure of the motor.

In operation, every time the motor starts it draws locked rotor (LR) amperage for a brief period of time. Each start of the compressor requires several minutes of normal operating time for the refrigerant and oil to remove the heat generated in the winding by the locked rotor current draw. Therefore, short cycling of the compressor causes the motor to become overheated because all the heat generated during the previous locked rotor condition (startup) was not completely removed.

Three-phase motor overheating can be caused by unbalanced voltage, unbalanced current, and single phasing. Either/or both an unbalanced voltage condition and an unbalanced current condition can go undetected for a long period of time and can cause many problems due to overheating. The maximum amount of current or voltage unbalance from winding to winding is 2%. This is significant because the temperature rise of the winding, in percentage, is equal to an amount double to the square of the voltage unbalance. For example, a 2% voltage unbalance will cause an 8% increase in the temperature of the motor winding; a 3% voltage unbalance will cause an 18% increase in winding temperature; and a 5% increase in voltage unbalance will cause a 50% increase in winding temperature. A motor cannot withstand this kind of temperature for very long before being ruined.

An unbalanced voltage can cause an unbalanced current, but an unbalanced current does not always indicate an unbalanced voltage. For instance, in a three-phase system a dirty or pitted starter contact or a loose connection could cause a current unbalance by causing a higher resistance in that one leg. Because electricity follows the path of least resistance, more current will flow down the other two legs. Therefore, the problem is actually in the leg that has the least amount of current flowing through it. The current increase in the other two legs causes more than normal heat to be generated in those two windings. The percent of current unbalance is determined in the same manner as a voltage unbalance.

Single phasing of a three-phase motor occurs when one leg of the three-phase is lost. This will cause a higher-than-normal current draw and overheating in the other two windings. This condition is a more advanced condition of the one described above with the high resistance in one leg. However, the motor will fail more quickly when single phasing exists. The two windings will overheat and burn while the winding on the lost leg will not be affected.

If a motor is running when the single phase oc-curs, it may continue running. The problem is that the other two legs will try to carry the total load and their current draw will increase to about 1½ times their normal running current. On the other hand, if the compressor is fully loaded and one phase is lost, the overcurrent draw of the motor will cause the overload protection to trip and stop the motor. Should the motor be operating with a partial load, the current draw may not reach the cutout point of the overload protectors and the motor will continue to run. When this situation occurs, the windings in use will become overheated.

In most cases, if the motor stops when a single-phase condition is present, it cannot restart and will repeatedly trip the overload protectors if they are of the automatic reset type. In this case the motor will usually fail because of the heat generated in trying to start.

Voltage that is either too high or too low will also cause the motor to overheat. For a compressor that is rated for 230 V only, the normal operating voltage limits are 10%, or 207 V for a minimum and 253 V for a maximum. For a compressor rated for 208/230 V, the limits are 10% below the 208 V and 10% above the 230 V, or 187 V minimum and 253 V maximum.

Overheating Due to Refrigerant Problems. The improper setting of the refrigerant controls is the most common cause of compressor overheating from refrigerant problems. These controls are the flow control device, evaporator pressure regulator, compressor unloaders, pressure controls, hot gas bypass, or the use of devices that are not necessary, which could cause the suction vapor to the compressor to become superheated. An improperly insulated suction line is another reason for an elevated suction vapor temperature.

Because hermetic and semihermetic compressors and motors are cooled by the returning suction vapor, it must not be allowed to become overheated. Be sure to keep the unit operating within the ratings recommended by the manufacturer.

High compression ratios are another cause of compressor overheating. The compression ratio is the ratio of the absolute discharge pressure to the absolute suction pressure. Most compressor manufacturers will publish the maximum and the minimum compression ratios for their compressor.

High compression ratios can be caused by a dirty evaporator, dirty filters, low load on the evaporator, or a shortage of refrigerant. However, a high discharge pressure would more than likely be the cause of high compression ratios. A high discharge pressure can be caused by a dirty condenser, faulty condenser fan motor, high ambient temperature, or noncondensables in the system. When water-cooled equipment is used, the high discharge pressure may be due to a scaled-up condenser, reduced water flow from the pump, stopped up strainer, or a high condenser inlet water temperature.

High discharge temperatures can also cause other problems, such as seized or worn pistons and scored cylinder walls. The heat can cause the formation of certain contaminants in the system. These contaminants cause a reaction between the refrigerant and the oil, causing the oil to break down at around 300°F (131.11°C). When the oil breaks down, a sludge is formed, releasing carbon inside the system. Under these conditions proper lubrication of the moving parts would be almost impossible.

Also, a high discharge temperature is usually accompanied by a copper plating condition on all the steel parts in the system. As this copper plating builds up on the bearing surfaces, the clearance needed for the oil is reduced, the bearing becomes tight, and friction is increased. All of these conditions result in higher compressor temperatures, which reduce the life span of the compressor and the motor.

Also, to help keep the compressor cool during periods of low ambient temperature when it may be desirable to cycle the condenser fan, place an auxiliary fan to blow air across the compressor. This is especially important at night during mild weather conditions when the compressor is already operating under light-load conditions.

▣ Safety Procedures ▣

Refrigeration and air conditioning compressors are sometimes very heavy. Caution should therefore be exercised when removing or installing these components.

1. Never lift heavy objects with the back muscles; use the leg muscles.
2. Get help when lifting heavy objects.
3. Always protect open compressors and open refrigerant lines from dirt and moisture.
4. Clean up any oil that spills from the compressor. Slipping on oily surfaces can cause serious personal injury.
5. Use caution to prevent oil from a burnt compressor from contacting the skin. This oil is usually acidic and can cause irritation.

6. Be sure to tighten all bolts and nuts with the proper torque.
7. Always be sure that electrical power is disconnected before touching the terminals with hands or metal tools.
8. Be sure to tighten all bolts in the proper sequence to prevent ruining the parts.
9. Always wear goggles when charging or discharging the system refrigerant.
10. Exercise care when handling compressors that have been operating. There may be hot components that can cause burns.
11. Never close the discharge service valve without a pressure-relief port available.

▣ SUMMARY ▣

☐ The two functions of a compressor are: (1) to draw the refrigerant from the evaporator, lowering the pressure to the desired evaporating temperature, and (2) to raise the refrigerant vapor pressure high enough so that the saturation temperature is higher than the temperature of the condenser cooling medium.
☐ The most popular types of compressor are: (1) reciprocating, (2) rotary, and (3) centrifugal.
☐ The most popular ways of producing reciprocating motion are: (1) crank throw crankshaft, and (2) eccentric disc crankshaft.
☐ Connecting rods link the crankshaft and the piston together. Vibration in a compressor is in direct relation to the number of its cylinders.
☐ Only a good grade of refrigeration oil should be used to lubricate a compressor.
☐ The methods used to lubricate compressors are: (1) splash, (2) forced, and (3) a combination of the two methods. Oil pumps are used on compressors of 3 hp and up.
☐ Liquid refrigerant in the oil will cause erratic operation of the oil pump.
☐ Shaft seals are used on open-type compressors to prevent leaking between the compressor shaft and body.
☐ The types of shaft seals are: (1) packing gland, (2) stationary unbalanced bellows, (3) diaphragm, and (4) rotary.
☐ The displacement of a compressor is the volume displaced by the number of pistons.
☐ Compression ratio is the absolute head pressure divided by the absolute suction pressure.
☐ Accepted industry bounds of compression ratio are 10:1 or less. Air-cooled compressors must be provided with an adequate flow of air.
☐ Refrigerant-cooled compressors depends on suction vapor to cool the compressor motor windings.

☐ Oil coolers use a small portion of the condenser to cool the compressor oil.

☐ The major factors controlling compressor output are: (1) type of refrigerant, (2) compression ratio, (3) volumetric effiency, (4) cooling system efficiency, (5) cylinder cooling system, (6) pressure of the refrigerant in the suction line, (7) compressor speed, (8) type and size of valves, (9) friction of the refrigerant vapor, (10) mechanical condition of compressor, and (11) the lubricant.

☐ EXERCISES ☐

1. What are the two functions of a compressor?
2. What are the most popular types of reciprocating compressors?
3. What are the purposes of suction valves in a compressor?
4. What are the most popular ways of producing the reciprocating motion in a compressor?
5. What is the difference between a connecting rod and an eccentric strap?
6. What are the moving parts in a rotary compresssor?
7. What is the purpose of the barrier in a rotary compressor?
8. Where are most rotary compressors used?
9. Are centrifugal compressors positive-displacement-type pumps?
10. How is the number of stages determined in a centrifugal compressor?
11. Where is lubrication needed in a centrifugal compressor?
12. Where are centrifugal compressors best suited for use?
13. What is the difference between open, semihermetic, and hermetic compressors?
14. What are the different types of compressor suction and discharge valves?
15. What actually does the sealing on the shaft seals and valves?
16. What is the name of the compressor component where the valves are located?
17. Describe the procedure used to check operation of the suction valves.
18. Describe the procedure used to check operation of the discharge valves.
19. Where should the oil level be maintained in a compressor?
20. What should the range of the net oil pump pressure be on a compressor?
21. Why is oil foaming undesirable in a compressor?
22. Name the different types of compressor shaft seals.
23. What is meant by compressor displacement?
24. Define *compression ratio*.
25. Why is a clearance volume necessary?
26. How does the gas confined in the clearance pocket affect the compressor efficiency?
27. Why do compressors require cooling?
28. How will a lower-than-normal suction pressure affect the compressor efficiency?
29. What type of compressor requires oil coolers?
30. How can a service technician help to eliminate compressor failures?

6 Condensers and Receivers

Performance Objectives

After completion of this chapter, the student should:

■ Know the purpose of a condenser.

■ Be able to explain the basic concept of condenser theory.

■ Be able to list the various types of condensers.

■ Know what condenser capacity depends on.

■ Understand condensing temperature difference.

■ Know the purpose of liquid receiver safety devices.

■ Complete the Exercises with at least 80 percent accuracy.

■ Complete the Laboratory Exercises with at least 80 percent accuracy.

As the refrigerant leaves the compressor, it is still a vapor. It is now quite hot and ready to give up the heat it picked up in the evaporator along with the heat of compression. One of the easiest ways to remove this heat from the refrigerant vapor is to send it through a radiatorlike device known as a condenser. As heat is given off by the high-temperature, high-pressure refrigerant vapor, its temperature falls to the saturation point and the refrigerant vapor condenses to a liquid refrigerant, hence the name "condenser."

☐ Purpose of the Condenser ☐

The purpose of the condenser on a refrigeration unit is to remove heat from the compressed refrigerant vapor, changing it to liquid form. First, the sensible heat must be removed from the compressed refrigerant vapor until its temperature has been lowered to the condensing temperature. Then, as the latent heat of condensation is removed from the refrigerant, it gradually changes (condenses) to a liquid.

From the foregoing discussion, it can be said that a condenser is a device used for removing heat from the refrigeration system.

A basic concept in condenser theory is that the amount of heat given up by the refrigerant must equal the heat gained by the cooling medium. An example of this is shown in Figure 6–1, a temperature–Btu chart. The vertical scale is in degrees Fahrenheit, and the horizontal scale indicates the heat content of the refrigerant in Btu. All the refrigerant in the area to the right of line 1 is a vapor. The refrigerant to the left of line 2 is in the liquid state. The area between lines 1 and 2 represents a vapor–liquid mixture. In this example, Refrigerant-12 at 100 °F (37.78 °C) and 116.9 psig (906.15 kPa) is used.

When the refrigerant vapor enters the condenser from the compressor, it has a temperature of 120 °F (48.89 °C) at point A. The vapor has a 20 °F (−6.67 °C) superheat. As the vapor comes in contact with the tubes, the heat is given up to the condensing medium. This results in the removal of temperature only because the vapor is superheated. The temperature is reduced from 120 to 100 °F, represented by the line from point A to point B. The vapor, at this point, has reached the saturation temperature corresponding to its pressure. It is now ready to change to a liquid (condense). As the heat removal continues, all the vapor condenses to a liquid at point C. Notice that between points B and C there has been no change in temperature; only latent heat has been removed. All the refrigerant is

FIGURE 6-1 Temperature–Btu chart.

now a liquid, and any further cooling will produce a subcooled liquid. There would normally be some subcooling taking place in an operating condenser. However, in this example the liquid refrigerant is subcooled an additional 20 °F (11.2 °C) to point *D*. The refrigerant liquid leaves the condenser at 80 °F (26.67 °C).

It should be noted that the amount of sensible heat removed is small compared to the amount of latent heat removed. The total amount of sensible heat removed is 8.4 Btu/lb (8860.3 J/0.4536 kg). That is, from point *A* to point *B* the amount of sensible heat removed is 3.5 Btu/lb (3691.8 J/0.4536 kg), and from point *C* to point *D* it is 4.9 Btu/lb (5168.5 J/0.4536 kg). Compare this to 57.3 Btu/lb (60,440 J/0.4536 kg) of latent heat removed between points *B* and *C*. Thus most of the heat removed from the refrigerant in the condenser is latent heat.

In operation, the first heat to be removed from the refrigerant by the condenser is sensible heat in cooling the vapor down to the condensing temperature, or desuperheating the refrigerant vapor. This process takes from about 25 to 30% of the condenser coil (Figure 6–2). The next function of the condenser is to change the refrigerant vapor into a liquid. This process takes from 60 to 70% of the condenser coil. The third function of the condenser is to subcool the refrigerant liquid. This

process takes from 5 to 10% of the condenser coil. The heat of rejection of the condenser coil must be equal to the system capacity plus the power input to the compressor motor.

☐ Types of Condensers ☐

There are three types of condensers in use on modern refrigeration systems. They are: (1) air-cooled, (2) water-cooled, and (3) evaporative condensers. The air-cooled condenser uses air as the condensing medium, the water-cooled condenser uses water, and the evaporative condenser uses a combination of air and water.

Air-Cooled Condensers. These are the most popular type used on small commercial and domestic refrigeration and air conditioning units. They are constructed of the tube and external fin design, which dissipates heat to the ambient air. Except for very small domestic units, which depend on gravity air circulation, heat transfer is efficiently accomplished by forcing large quantities of air through a compact condenser assembly (Figure 6–3), usually about 1000 cfm per ton of refrigeration.

When natural-draft, static-type condensers are used, air circulates over the condenser by convection. When the air contacts the warm condenser surface, it absorbs heat and rises. As the warm air rises, cooler air replaces it and more heat is absorbed (Figure 6–4).

There are two types of natural-draft air-cooled condensers: the tube and fin type shown in Figure 6–4 and the plate type illustrated in Figure 6–5. In the plate-type condenser, the plates are pressed together to form a condenser coil and seam-welded together.

Natural-draft air-cooled condensers have very limited use. They are not capable of dissipating large quantities of heat because of the slow air movement over their surfaces. Because of this, relatively large surfaces are required, but they are economically manufactured, require very little maintenance, and are commonly used in domestic refrigerators.

Condenser capacity can be increased by forcing air over the surfaces. This is normally done by adding a fan to increase the airflow (Figure 6–6). The propeller-type fan or the centrifugal type may be used to move the air. The type of fan used depends

FIGURE 6-2 Percent of condenser used for different functions.

FIGURE 6–3 Air-cooled condensing unit. (Courtesy of Tecumseh Products Company.)

on design factors such as air resistance, noise level, space requirements, etc.

Some of the earlier air-cooled condensers were of the bare-tube-type construction. However, the bare tube type was inefficient and gave way to the tube and fin type of condenser. The forced-air type of air-cooled condenser is more practical for larger capacities than the natural-draft-type condenser. The major limiting factors for forced-draft air-cooled condensers are economics and available space.

Air-cooled condensers are easy to install, inexpensive to maintain, require no water, and present no danger of freezing in cold weather. A sufficient quantity of fresh air is necessary, however, and the fan may create noise problems in large installations. In very hot regions, the relatively high ambient air temperature may result in higher than normal condensing pressures. Nevertheless, if the condenser sur-

face is properly sized, air-cooled condensers may be used satisfactorily in all climatic regions.

When space permits, condensers may be manufactured with a single row of tubing. In an attempt to obtain a compact size, however, condensers are usually constructed with a comparatively small face area and several rows of tubing in depth. When the air is forced through the condenser, it absorbs heat and the air temperature rises. Therefore, the efficiency of each succeeding row in the coil is decreased. Coils eight rows in depth are frequently used.

Draw-through fans, the type that pulls the air through the condenser, produce a more uniform airflow through the condenser than the blow-through type (Figure 6–7). Because even air distribution will increase the condenser efficiency, the draw-through fans are preferred.

Most air-cooled refrigeration systems operated

FIGURE 6-4   Natural-draft condenser.

FIGURE 6-5   Plate-type condenser.

FIGURE 6-6   Forced-draft air-cooled condenser.

with the condenser in low-ambient-temperature conditions are susceptible to compressor damage due to abnormally low head pressures unless some means of maintaining normal head pressures is provided. This is true of roof-mounted refrigeration or air con-

FIGURE 6-7   Draw-through and blow-through condensers.

ditioning systems exposed to low outside ambient temperatures. The capacity of refrigerant flow control devices is dependent on the pressure difference across the device. Since these devices are selected for the desired capacity with normal operating pressures, abnormally low head pressure, which reduces the pressure difference across the flow control device, may result in insufficient refrigerant flow. This reduced refrigerant flow may cause erratic refrigerant feeding to the evaporator, which may result in frosting of the evaporator coil on air conditioning units. The lower refrigerant velocity, and possibly lower evaporator pressure, will permit oil to settle out and be trapped in the evaporator, sometimes causing a shortage of oil in the compressor crankcase.

The low limit of the head pressure depends on the required pressure drop across the flow control device. For normal air conditioning applications the head pressure should be maintained above a condensing temperature of 90 °F (32.2 °C). This, in effect, corresponds to a normal lower limit of about 60 °F (15.6 °C) ambient air. Since air conditioning is not normally required at these lower temperatures, condenser head pressure control may not always be necessary. However, for those applications for which operation is required below 60 °F ambient air temperature, three methods of condenser head pressure control are available to meet specific job requirements: (1) fan control, (2) damper control, and (3) refrigerant flooding the condenser.

Fan control is an automatic winter control method that will maintain a condensing pressure within reasonable limits by cycling one fan on a two-fan unit, two fans in sequence on a three-fan unit, or three fans on a four-fan unit in response to the outside air temperature entering the condensing coil.

The damper control method of controlling head pressure utilizes the principle of modulating the air volume flow through a section of the condenser coil by use of a face and bypass damper section located on the leaving-air side of the coil (Figure 6-8).

The damper control linkage is arranged so that when the face damper is completely closed, the bypass damper blades are in the wide-open position. This is the normal position for full effective dampering (Figure 6-9). In this position, the fan blows the air against the completely closed damper blades, out through the wide open bypass sections, back into the fan section, and then the air is recirculated. In this position very little, if any, air passes through the condenser coil.

The liquid flooding of the condenser type of head pressure control is based on liquid refrigerant flooding back into the condenser and thereby cutting down its effective condensing capacity.

Liquid flooding of the condenser system is an improved head pressure control incorporating two modulating valves. A check valve is also recommended to prevent migration to the condenser during the off cycle (Figure 6-10). Operation of this system is independent of any difference in elevation between the condenser and the receiver. It therefore permits the receiver and condenser to be more conveniently located to suit specific job conditions.

The main or liquid line valve is normally closed and opens on a pressure rise in the condenser. This valve is located in the liquid line between the condenser and receiver (Figure 6-11). The secondary vapor or bypass valve is normally open and closes on a pressure rise. This valve is located in the line between the liquid valve and the receiver.

System operation is as follows: on system startup, the bypass valve is normally open and the main or liquid valve is closed. Hot vapor moves from the compressor, some going into the condenser, and some going through the bypass circuit and the open valve into the receiver to maintain or build up pressure in the receiver as the liquid leaves. As the compressor continues to run, hot vapor condenses in the condenser and raises the liquid level since the main valve on the leaving side of the condenser is still closed. As the liquid level rises, the condensing capacity of the condenser decreases. As a result, the head pressure rises. The bypass vapor maintains or raises the pressure in the receiver. As the pressure in the condenser rises to the control point of the valve, the liquid valve starts modulating toward the open position, permitting liquid to leave the condenser and flow into the receiver. At the same time, the bypass valve starts modulating toward the closed

Bypass Damper

Face Damper

FIGURE 6-8 Damper control location. (Courtesy of McQuay Division, McQuay-Perfex, Inc.)

DAMPERTROL C head pressure control

DAMPERTROL C IN OPEN POSITION

DAMPERTROL C IN BY PASS POSITION

DAMPERTROL C head pressure control selections. Show air flow in open (normal position) and closed (winter start) position.

FIGURE 6-9 Damper control operation. (Courtesy of McQuay Division, McQuay-Perfex, Inc.)

FIGURE 6-10 Valve arrangement for liquid flooding of the condenser. (Courtesy of McQuay Division, McQuay-Perfex, Inc.)

FIGURE 6-11 Liquid valve location. (Courtesy of McQuay Division, McQuay-Perfex, Inc.)

position, limiting the hot vapor flow into the receiver. The modulating action of the two valves maintains the proper liquid level in the condenser to maintain proper head pressure.

On multifan condensers the liquid flooding of the condenser system is ordinarily used to extend the range of a fan cycling system. The use of fan cycling in conjunction with liquid flooding of the condenser reduces the amount of refrigerant charge required for a specific minimum design temperature condition.

Water-Cooled Condensers. When adequate low-cost condensing water is available, water-cooled condensers are often desirable because lower condensing pressure and better control of the discharge pressure is possible. Water, especially when obtained from underground sources, is usually much colder than daytime ambient air temperatures. If cooling towers are used, the condensing water can be cooled to a point closely approaching the ambient wet-bulb temperature. This allows the continuous recirculation of condensing water and reduces the water consumption to a minimum.

For a condenser to properly do what it is designed to do, there must be a sufficient amount of cooling medium passing through the condenser coil. An example of the amount of water required for a water-

lb Refrig. X Heat Lost/lb =
lb Water X Spec. Heat X Temp. Rise

FIGURE 6-12 Heat balance of water-cooled condenser. (Courtesy of Carrier Air-Conditioning Company.)

cooled condenser is: When 10 lb (4.536 kg) of refrigerant and 43.8 lb (19.87 kg) of water pass through a condenser, a temperature rise of 15°F (8.4°C) of the water represents the amount of heat gained as being equal to the heat lost by the refrigerant (Figure 6-12). Thus each pound of refrigerant that passes through the condenser gives up 65.7 Btu (69,530.36 J). Since there was 10 lb of refrigerant passing through the condenser, the total amount of heat lost by the refrigerant would be 10 times 65.7 or 657 Btu. At the same time, 43.8 lb of water is passing through the condenser with a temperature rise of 15°F (8.4°C). The amount of heat gained by the water is then 43.8 × specific heat of water × 15° temperature rise, or 657 Btu (695,303.60 J). In the operation of the unit, some heat is lost through radiation. This amount of heat is small and is not generally considered during these calculations. The concept of heat balance is valid for all types of condensers.

Because of the excellent heat transfer characteristics of water, water-cooled condensers are usually quite compact. There are several types of construction used for water-cooled condensers. The principal classifications of water-cooled condensers are: (1) shell and tube, (2) tube in tube, and (3) evaporative (Figure 6-13).

Shell-and-Tube and Shell-and-Coil Condensers. Insofar as external appearance is concerned, these condensers (sometimes called condenser-receivers), are similar to liquid receivers. The shell-and-tube condenser acts as both condenser and liquid receiver. It is constructed of a cylindrical tank having a refrigerant inlet and outlet. Installed within this tank is a water coil that also has inlet and outlet connections (Figure 6-14). The shell is made of steel. The ends may be either welded in place or bolted in place. Cleaning the condenser is easier when the

Shell and Tube

Tube in Tube

FIGURE 6-13 Water-cooled condensers.

refrigerant passes from the compressor through the condenser refrigerant inlet to the interior of the shell, where it comes into contact with the cool walls of the water coil and the cooler walls of the shell. The majority of heat is conducted through the copper coil walls to the water that carries it away. Some of the heat is conducted through the shell walls to the ambient air, which also aids in cooling the refrigerant. As heat is removed, the refrigerant condenses and the liquid collects in the bottom of the shell (Figure 6-17). The refrigerant outlet tube extends from the refrigerant outlet almost to the bottom of the receiver interior in order to be below the liquid level in the receiver. The liquid line to the evaporator connects to the receiver refrigerant outlet.

Shell-and-tube condensers may be either of horizontal or vertical construction. The horizontal type is most commonly used in the latest designs since it is installed underneath the base of the compressor and occupies space that would otherwise be wasted. Vertical condensers require space on top of the base, which generally increases the floor area required for the condensing unit (Figure 6-18).

Tube-in-Tube Condensers. These condensers are also commonly used on water-cooled refrigeration equipment. They are made of two copper tubes, one inside the other. For example, ⅝-in. (15.87-mm)-OD copper tubing could be used for the outside tube and ⁷⁄₁₆-in. (11.1125-mm)-OD copper tubing could be placed inside the larger tube (Figure 6-19).

By using special fittings at the ends, connections can be made so that water may be circulated through the inner tube with the refrigerant flowing through the space between the two tubes. The heat is transferred from the compressed refrigerant through the thin walls of the internal tube into the cooling water which removes the heat. The arrows in Figure 6-19 show the main direction of heat flow, which is from the compressed vapor to the water. However, a por-

ends are bolted in place (Figure 6-15). The water coil is usually made of copper tubing. To increase the efficiency of the condenser, most manufacturers have added fins to the copper tubing. The cooling water flows through the coil from the inlet end to the outlet. The outlet may be piped to a wastewater line or run to an open drain. To reduce waste, however, the cooling water is usually returned to a cooling tower and recirculated (Figure 6-16).

In operation, the compressed high-temperature

FIGURE 6-14 Shell-and-coil condenser.

FIGURE 6–15   Multipass water-cooled condenser with bolted-in-place ends.

FIGURE 6–16   Cooling water circuit.

tion of the heat from the refrigerant is dissipated into the surrounding air, which increases the cooling capacity of the condenser.

Tube-in-tube condensers are arranged in coils so that the compressed refrigerant enters at the top and flows downward as it liquefies in much the same manner as in an air-cooled condenser. The cooling water enters at the bottom and flows upward. Thus the liquid refrigerant and the cool water flow in opposite directions or in a counterflow direction (Figure 6–20).

A more efficient method used for tube-in-tube condenser makes use of a *counterswirl* design (Figure 6–21). The design gives superior heat transfer per-

FIGURE 6–17   Liquid refrigerant level in a shell-and-coil condenser.

Types of Condensers

FIGURE 6-18 Shell-and-tube condenser location. (Courtesy of Copeland Refrigeration Corp.)

High pressure vapor
from compressor

Water

Outer tube   Inner tube

FIGURE 6-19 Cross section of a tube-in-tube condenser.

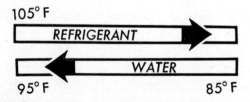

105°F

REFRIGERANT

WATER

95°F   85°F

FIGURE 6-20 Typical counterflow path. (Courtesy of Packless Industries.)

REFRIGERANT

WATER

FIGURE 6-21 Counterswirl design principle. (Courtesy of Packless Industries.)

formance. The tube construction provides for excellent mechanical stability. The water flow path is turbulent, and provides a scrubbing action to maintain cleaner surfaces. The construction method also exhibits very high system pressure resistance.

The pressure from city water mains is sometimes used to circulate the water through the condenser. In very large systems, the water is often recirculated to effect economy in water costs, particularly where water is expensive or difficult to obtain in unlimited quantities. In such cases, an evaporative condenser of some type is used.

Evaporative Condensers. Evaporative condensers are frequently used where lower condensing temperatures are desired than those obtainable with air-cooled condensers, and where the available water supply may not be adequate for heavy water usage.

A simple evaporative condenser consists of an air-cooled condenser over which water is allowed to flow. The air blast from the fan causes the water to evaporate from the surface of the condenser fins, absorbing heat from the refrigerant in the evaporation process. The water that is exposed to the airflow evaporates rapidly. The latent heat required for the evaporating process is obtained by a reduction in sensible heat and, therefore, a reduction in the temperature of the remaining water. An evaporative condenser can reduce the water temperature to a point closely approaching the wet-bulb temperature of the air (Figure 6-22).

The hot refrigerant vapor is piped through a spray chamber where it is cooled by evaporation of the

Air Outlet

Fan

Water In

Spray Header

Refrigerant
Inlet

Refrigerant
Outlet

Air Inlet

Water Pan

Water Make-up

FIGURE 6-22 Evaporative condenser.

water coming in contact with the refrigerant tubing. Since the cooling is accomplished by evaporation of water, water consumption is only a fraction of that used in conventional water-cooled applications in which the water is discharged to a drain after it is used. Because of this, evaporative condensers are used in hot, arid regions.

☐ Counterflow of Cooling Water ☐

A water-cooled condenser should cool the refrigerant to nearly the same temperature as the cooling water entering the condenser. If a condenser is to do this, it must operate on the counterflow principle (Figure 6–23). The hot refrigerant vapor enters the condenser at the top, and as it gradually liquefies it flows to the bottom. But the water enters at the bottom of the condenser and flows in the opposite direction to the top. When the counterflow method is used, the entering heated refrigerant is cooled by the warm water leaving the condenser, and upon leaving the condenser is further cooled by the cold entering water. Thus, the liquid refrigerant is cooled to very nearly the temperature of the cold entering water. If both the refrigerant and the cooling water were to travel in the same direction, the heated entering refrigerant would first make contact with the coldest water and the leaving liquid refrigerant would contact the warm leaving water. Therefore, maximum temperature difference between the two fluids would not be maintained and the flow of heat from the refrigerant to the water would be substantially reduced.

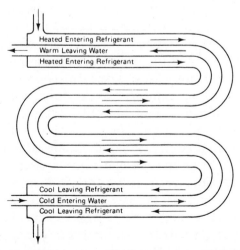

FIGURE 6–23 Counterflow principle.

☐ Condenser Capacity ☐

The heat transfer capacity of a condenser depends on several factors:

1. Surface area of the condenser.
2. Contact between the refrigerant and the internal surface of the condenser.
3. Temperature difference between the cooling medium and the refrigerant vapor.
4. The velocity of the refrigerant gas in the condenser tubes. Within the normal commercial operating range, the greater the velocity, the better the heat transfer factor, and the greater the capacity.
5. Rate of flow of the cooling medium over or through the condenser. Heat transfer increases with velocity for both air and water, and in the case of air, it also increases with density.
6. Material of which the condenser is made. Since heat transfer differs with different materials, more efficient metals will increase capacity.
7. Cleanliness of the heat transfer surface. Dirt, scale, or corrosion can reduce the heat transfer rate.
8. Rate of replacement of the condensed refrigerant by uncooled refrigerant.

For a given condenser, the physical characteristics are fixed, and the primary variable is the temperature difference between the refrigerant gas and the condensing medium.

☐ Condenser Precooler ☐

The purpose of the condenser precooler is to reduce the temperature of the air to the condenser. The reduced air temperature reduces the head pressure, extending the life of the unit and all its components, especially the compressor. It also helps reduce the electrical use.

The basic operating principle is the same as that of an evaporative cooler. Most precoolers cool the air through water evaporation (Figure 6–24). The precooler is mounted on the air inlet side of the condensing coil, causing the air to pass through the precooler before entering the condenser coil (Figure 6–25). Water is fed to the precooler and distributed across a feed tray. It then flows down the evaporative

FIGURE 6-24   **Precooler operation.**

FIGURE 6-25   **Precooler installation.**

pad so that the entire pad is wetted. Air is drawn across this wetted surface before entering the condensing coil. The evaporation of the water flowing over the pad cools the air. The lower the humidity and the higher the ambient air temperature, the greater the cooling effect.

They make use of a pulse-feed water system to deliver water in timed sequence rather than a continuous flow. This is electronically controlled, based on the outdoor ambient temperature, to provide only enough water to wet the pad thoroughly, therefore minimizing the amount of water used.

☐ Condensing Temperature ☐

The condensing temperature is the temperature at which the refrigerant vapor becomes a liquid. This temperature should not be confused with the temperature of the cooling medium, since the con-

densing temperature must always be higher in order for heat transfer to occur.

In order to condense the refrigerant vapor flowing into the condenser, heat must flow from the condenser at the same rate at which heat is introduced by the refrigerant vapor entering the condenser. As mentioned before, the only way in which the capacity of the condenser can be increased under a given set of conditions is by an increase in the temperature difference through the condenser walls.

Since a reciprocating compressor is a positive-displacement pump, the pressure in the condenser will continue to increase until such time as the temperature difference between the cooling medium and the refrigerant condensing temperature is sufficiently great to transfer the necessary amount of heat. With a large condenser, this temperature difference may be very small. With a small condenser, in the event that the air or water flow to the condenser has been blocked, the necessary temperature difference may be very large, resulting in dangerously high discharge pressures. It is, therefore, essential that the condenser be operating properly any time a refrigeration unit is in operation.

The condensing temperature and, therefore, the condensing pressure are determined by the capacity of the condenser, the temperature of the cooling medium, and the heat content of the refrigerant vapor being discharged from the compressor, which in turn are determined by the volume, density, and temperature of the discharge vapor.

☐ Condensing Temperature Difference ☐

A condenser is normally selected for a sytem by sizing it to handle the compressor load at a desired temperature difference between the condensing temperature and the expected temperature of the cooling medium. Most air-cooled condensers are selected to operate on temperature differences (usually termed TD) of 20 to 30 °F (11.1 to 16.7 °C) at design conditions. Higher and lower TDs are sometimes used on special applications. Standard production air-cooled condensing units are often designed with one condenser for a wide range of applications. In order to cover as wide a range as possible, the TD at high suction pressures may be from 30 to 40 °F (16.7 to 22.2 °C), while at low evaporating temperatures the TD is often no more than 4 to 10 °F

(2.2 to 5.6 °C). The design condensing temperature on water-cooled units is normally determined by the temperature of the water supply and the water flow rate available, and may vary from 90 to 120 °F (32.2 to 49 °C).

Because the condenser capacity must be greater than the evaporator capacity by an amount equal to the heat of compression and the motor efficiency loss, the condenser manufacturers may rate condensers in terms of evaporator capacity, or they may recommend a factor to allow for the heat of compression in selecting the proper-size condenser.

Condensers that are rated with higher temperature differences result in higher equipment operating costs and can possibly cause compressor failure due to high compression ratios because of the increased discharge pressure. Condenser temperature differences are calculated by the following formula:

Condenser TD = condensing temperature − ambient temperature

The condensing temperature is determined by taking the discharge pressure and converting this reading into the saturation temperature with the pressure–temperature chart. The ambient temperature is taken where the air enters the condensing coil, out of the direct rays of the sun.

Liquid flooding of the condenser effectively reduces the surface area for effective condensing of the refrigerant. Liquid flooding of the condenser is usually done to increase the amount of subcooling for the liquid. Liquid subcooling gains about ½% of system capacity for each degree Fahrenheit of subcooling obtained. Do not subcool the liquid enough to reduce the head pressure below the required operating limits. If an external subcooler is not used and a large amount of subcooling is obtained, the discharge pressure will be increased because there is less effective condenser area for condensing the refrigerant. This raising of the head pressure results in operating costs higher than before the subcooling process began.

To achieve subcooling in a system using a receiver tank, the tank must be full of refrigerant. When liquid and a vapor are together there is zero subcooling because the mixture is at the saturation point. When the receiver tank is full of refrigerant to provide subcooling of the liquid, it is no longer usable for pumping the unit down or storing the refrigerant because it is already full.

Systems that operate with a low discharge pressure must have matched evaporator, condenser, and expansion valves. If these components are not matched, the system will not operate as desired.

☐ Noncondensable Gases ☐

Air is composed primarily of nitrogen and oxygen. Both elements remain in a gaseous form at all temperatures and pressures encountered in commercial refrigeration and air conditioning systems. Therefore, although these gases can be liquefied under extremely high pressures and extremely low temperatures, they may be considered as noncondensable in a refrigeration system.

Scientists have discovered that one of the basic laws of nature is the fact that in a combination of gases, each gas exerts its own pressure independently of others, and the total pressure existing in a system is the total of all the gases present. This is known as Dalton's law (see Chapter 2). A second basic characteristic of a gas is that if the space in which it is enclosed remains constant, so that it cannot expand, its pressure will vary directly with the temperature. This is known as Charles' law. Therefore, if air is sealed in a system with a refrigerant, the nitrogen and oxygen will each add its pressures to the system pressure, and this pressure will increase as the temperature rises.

Since the air is noncondensable, it will usually remain trapped in the top of the condenser and the receiver. During operation, the compressor discharge pressure will be a combination of the refrigerant condensing pressure plus the pressure exerted by the nitrogen and oxygen. The amount of pressure above normal condensing pressure that may result will depend on the amount of trapped air, but it can easily reach 40 to 50 psig (375.79 to 444.49 kPa) or more. Any time a system is operating with abnormally high discharge pressure, air in the system is a prime suspect.

☐ Cleaning the Condenser ☐

Since the purpose of a condensor is to dissipate heat, it can readily be seen that a layer of dirt, lint, or scale will act as an insulator and prevent proper heat dissipation while causing the head pressure to increase. The surface of the condenser should be cleaned regularly by the service technician in order to maintain the greatest efficiency possible. An oil deposit often collects on the condenser and attracts a layer of dust on air-cooled condensers. This oil and dust must be removed. On water-cooled condenser tubes a formation of scale, much like that on the inside of a tea kettle, will appear. This scale reduces the heat transfer and must be removed. Since different parts of the country have different minerals in the water, different types of cleaners must be used. Local chemical companies should be consulted to determine the proper type and method to use.

☐ Location of the Condenser ☐

Since an air-cooled condenser requires a free flow of air over its surface, it is important that it be located in a well-ventilated room so that cool air will replace heated air. Insufficient quantities of cool air will result in high operating discharge pressures. The amount of air required for an air-cooled condenser is approximately 1000 cfm per ton of refrigeration. If the ambient temperature is unusually high more air may be required for the condenser to do the job for which it was intended.

Water-cooled condensers should not be located where the temperature may drop below the freezing point of water. Freezing may crack the compressor cylinder or split the condenser. The water tubes of a shell-and-tube condenser may burst when frozen and allow water to get into the refrigeration system. A burst water tube requires a thorough dehydrating of the entire system, including the compressor, condenser, receiver, and all the refrigerant lines.

For most satisfactory operation of the system, the temperature around a water-cooled condensing unit in use should preferably never fall below 60°F (15.6°C). If the unit is to be shut down for the winter, as a condensing unit on an air conditioning system, the cooling water should be drained out of the entire system.

The vertical type of water-cooled condenser is easily drained by opening a drain valve or removing a water drain usually provided at the bottom especially for this purpose. On the horizontal type of water-cooled condenser, however, simply disconnecting the water tubes at the ends is not sufficient to drain the coils because the bottom of each coil will not drain. The water must be removed by either blasting compressed air through the water coil to blow it out or by completely detaching the condenser from the base and raising one end to allow the water to drain out. Both the inlet and outlet must be disconnected to vent the coil; otherwise, the water will not drain properly. When the water coil is made of continuous tubing, the easiest way to drain it is to blow it out with compressed air. This method also requires that the water connections be completely disconnected.

☐ Water Flow Control Valves ☐

The flow of water to a condenser is controlled by means of an automatic water valve that prevents waste and maintains a constant discharge pressure on wastewater-type systems. On some of the larger systems, a cooling tower is used to control the discharge pressure.

The most popular type of water flow control valve is a bellows-type pressure-operated valve (Figure 6–26). The bellows of this type valve is connected to the discharge side of the refrigeration system, usually at the cylinder head. Consequently, the valve is actuated by the variation in the head or high side pressure, which is transmitted to the water valve through the high-pressure control tube. This valve is designed to open gradually as the discharge pressure increases. A flow of water is provided in proportion to the operating requirements. The pressure of the refrigerant operates against a metallic bellows that permits more water to flow when the pressure increases. This has the effect of automatically furnishing more cooling capacity and reducing the pressure in the condenser. When the condenser or high side pressure has been reduced, the valve is automatically restricted, reducing the flow of water in accordance with the needs of the condenser. During warm weather, the pressure in a refrigeration system is higher than in cool weather. The higher pressure automatically opens the water valve wider and supplies more water.

The water valve is usually installed in the inlet water line to the condenser (Figure 6–27). The direction of water flow should always be as indicated on the valve, usually by an arrow or "in" marked on the valve.

FIGURE 6-28   Liquid receiver tank location.

FIGURE 6-29   Extended tube in a receiver tank.

FIGURE 6-26   Water flow control valve. (Courtesy of Singer Controls Division.)

FIGURE 6-27   Location of water flow control valve.

▢ Liquid Receivers ▢

Some types of refrigeration units have sufficient space within the condenser to contain the entire refrigerant charge of the system. However, when the condenser does not have this added space, a receiver tank is usually provided. The amount of refrigerant required for proper operation of the unit will determine whether or not a receiver is used. When proper system operation requires approximately 8 lb (3.63 kg) or more of refrigerant, a receiver is generally used.

In most cases, the bottom connection from the condenser discharges into the top of the liquid receiver tank, while the liquid line connects to the bottom of the tank at the liquid line shutoff valve (Figure 6-28). If the liquid line connects to the top of the receiver tank, a tube is brought to the bottom of the receiver (Figure 6-29). To maintain normal operation, it is important that the level of the refrigerant never fall below the bottom end of the tube. On the other hand, an excess of refrigerant will not allow sufficient space for the gas above the liquid. This condition may also interfere with proper operation by causing excessive discharge pressures. The size or capacity of a liquid receiver is determined by the total quantity of refrigerant the system requires to operate properly. These factors are determined when the unit is designed.

▢ Liquid Receiver Safety Devices ▢

Household units have a relatively small refrigerant charge. Therefore, city ordinances do not require safety devices for such units. In larger refrigeration and air conditioning units, however, local city codes and ordinances require safety protection for liquid receivers to prevent possible bursting due to excessive pressures or fire. Safety devices take the form

Refrigerant Outlet — Fusible Plug — Refrigerant Inlet

Receiver Tank

FIGURE 6–30 Fusible plug location.

of a fusible metal plug generally installed in the end of the receiver (Figure 6–30). Should the liquid receiver be exposed to high temperature, as in the case of fire, the liquid receiver would not explode. Instead, the fusible plug would melt and the refrigerant would escape harmlessly. Large liquid receivers holding a considerable quantity of refrigerant are sometimes required by ordinance to provide a vent pipe connection to the fusible plug. In the event that the fusible plug should melt, the gas would be released to the outside of the building through the vent pipe.

During service operations, the service technician may have occasion to apply heat to a liquid receiver. When so doing, care should be taken that the liquid receiver is not heated to a point that will melt the fusible plug. Fusible plug melting points are commonly 165 and 212 °F (74 and 100 °C). It is very easy to accidentally heat a liquid receiver to these temperatures. If this should happen, it would be necessary to replace the fusible plug as well as the refrigerant charge.

Common practice on most installations is to charge the system until the liquid receiver is approximately one-third full. When the receiver is of the type having a liquid outlet tube extending to the bottom of the receiver, the liquid level must never fall below the lower end of the tube. In vertical receivers, the refrigerant level should extend about 3 in. (76.2 mm) above the lower end of the outlet tube, while in horizontal receivers 1 in. (25.4 mm) above is sufficient.

☐ Safety Procedures ☐

Most of the liquid refrigerant in a refrigeration system is located in condensers and receivers. The condensers are made with aluminum fins for more rapid transfer of heat. Receivers are made with safety devices incorporated. Care must be exercised when working with these devices.

1. Be careful of the aluminum fins on condensers. If the hand should slip, they may cause a serious cut, which is difficult to heal.
2. Do not bend the fins over. This prevents proper air circulation and heat transfer from the condenser.
3. Do not apply heat to a receiver tank that contains refrigerant because hydrostatic pressure may build up, resulting in a possible explosion.
4. Never cap the safety relief devices on receiver tanks.
5. Never plug or use the wrong type of solder on safety devices on a receiver tank.
6. Never fill a receiver tank or condenser to more than 85% of its rated capacity.

☐ SUMMARY ☐

☐ A condenser is a device used for removing heat from the refrigeration system.

☐ Types of condensers are: (1) air cooled, (2) water cooled, and (3) evaporative.

☐ Air-cooled condensers are popular on small commercial and domestic refrigeration and air conditioning units.

☐ Types of air-cooled condensers are: (1) forced draft and (2) natural draft.

☐ The major limiting factors for forced-draft air-cooled condensers are economics and available space.

☐ Most air cooled refrigeration systems operated with the condenser in low ambient temperature conditions are susceptible to compressor damage due to abnormally low head pressures, unless some means of maintaining normal head pressures is provided.

☐ Types of head pressure control for air-cooled condensers are: (1) fan control, (2) damper control, and (3) refrigerant flooding of the condenser.

☐ When adequate low-cost condensing water is available, water-cooled condensers are often desirable because of the lower condensing pressure and possibility of better control of the discharge pressure.

☐ Because of the excellent heat transfer characteristics of water, water-cooled condensers are usually quite compact.

☐ Water-cooled condensers are classified as: (1) shell and tube, (2) tube in tube, and (3) evaporative.

☐ The shell-and-tube condenser acts as both a condenser and liquid receiver.

☐ Shell-and-tube condensers are made in both horizontal and vertical models.

☐ Tube-in-tube condensers are made of two copper tubes, one inside the other.

☐ Tube-in-tube condensers transfer heat to both the surrounding air and the water.

☐ Evaporative condensers are used where lower condensing temperatures are desired than those obtainable with air-cooled condensers, and where the available water supply may not be adequate for heavy water usage.

☐ A simple evaporative condenser consists of an air-cooled condenser over which water is allowed to flow.

☐ A water-cooled condenser should cool the refrigerant to nearly the same temperature as the cooling water entering the condenser.

☐ Water-cooled condensers should operate on the counterflow principle with the refrigerant and water flowing in opposite directions.

☐ For a given condenser, the physical characteristics are fixed, and the primary variable is the temperature difference between the refrigerant vapor and the cooling medium.

☐ The condensing temperature is the temperature at which the refrigerant vapor is condensing to a liquid.

☐ The only way the capacity of a condenser can be increased under a given set of conditions is by an increase in the temperature difference through the condenser walls.

☐ The condensing temperature and, therefore, the condensing pressure are determined by the capacity of the condenser, the temperature of the cooling medium, and the heat content of the refrigerant vapor being discharged from the compressor, which in turn are determined by the volume, density, and temperature of the discharged vapor.

☐ A condenser is normally selected for a system by sizing it to handle the compressor load at a desired temperature difference between the condensing temperature and the expected temperature of the cooling medium.

☐ Since air is noncondensable, it will usually remain trapped in the top of the condenser and the receiver.

☐ To maintain the greatest efficiency possible, the surface of the condenser should be cleaned regularly by the service technician.

☐ On wastewater-type systems, the flow of water to a condenser is controlled by means of an automatic water valve that prevents waste and maintains a constant discharge pressure.

☐ The most popular type of water flow control valve is a bellows-type pressure-operated valve.

☐ To maintain proper operation of the unit, the liquid level in a liquid receiver should never fall below the pickup tube.

☐ EXERCISES ☐

1. What is the purpose of a condenser in a refrigeration system?
2. What are the three general types of condensers?
3. What methods are used to prevent damage to a compressor when it is operated at low ambient temperatures with an air-cooled condenser?
4. Why are water-cooled condensers sometimes more desirable than air-cooled condensers?
5. How are water-cooled condensers classified?
6. Which water-cooled condenser serves as both condenser and receiver?
7. What two sources are used as cooling media for water-cooled condensers?
8. On what principle do water-cooled condensers operate?
9. How are condensers selected?
10. Why will air usually be trapped in the condenser or the receiver?
11. What factors determine the condensing pressure?
12. How can the capacity of a condenser be increased?
13. What are the purposes of an automatic water valve?
14. Where are evaporative condensers desirable?
15. Where is the latent heat obtained for cooling in an evaporative condenser?
16. What factors determine condenser capacity?
17. Define condensing temperature.
18. How can the capacity of a condenser be increased?
19. What is condenser TD?
20. Does the condenser or evaporator have the greater capacity?
21. Why must noncondensable gases be removed from a refrigeration system?
22. What effect will a dirty condenser have on the head pressure?
23. Where should the condenser be located?
24. What is the purpose of the water flow control valve?
25. How does a water flow control valve operate?
26. What is the purpose of the liquid receiver?
27. What is the lowest liquid refrigerant level allowed in a receiver tank?
28. What is the purpose of the fusible plug?

7 Evaporators

Performance Objectives

After completion of this chapter, the student should:

- ■ Be able to name the two basic types of evaporators.
- ■ Be able to list the various styles of evaporators.
- ■ Be able to explain the process of heat transfer in evaporators.
- ■ Know which type of evaporator will transfer the most heat.
- ■ Explain the effects that temperature difference has on dehumidification.
- ■ Know the importance of oil circulation through evaporators.
- ■ Complete the Exercises with at least 80 percent accuracy.
- ■ Complete the Laboratory Exercises with at least 80 percent accuracy.

The evaporator is an extremely important part of the refrigeration system because it performs the actual cooling. The evaporator may be defined as *a device used for absorbing heat into the refrigeration system*. The evaporator is placed inside the area to be cooled. The heat is absorbed because of the vaporization of the refrigerant inside the evaporator tubes.

☐ Types of Evaporators ☐

There are two basic types of evaporators: (1) the dry or direct expansion type, and (2) the flooded type (Figure 7-1).

FIGURE 7-1 Types of evaporators.

Dry or Direct Expansion Evaporators. These types of evaporators use a continuous piece of tubing where the refrigerant is metered into one end by the flow control device and the suction line is connected to the other end. The refrigerant flows through the tubing while it vaporizes and absorbs heat. The refrigerant does not collect in a pool as in the flooded system. Also, there is no separation line between the

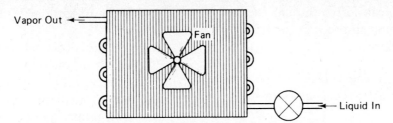

FIGURE 7–2 **Blower coil evaporator.**

Vapor out →

Liquid in →

FIGURE 7–3 **Bare coil evaporator.**

FIGURE 7–4 **Gravity evaporator.**

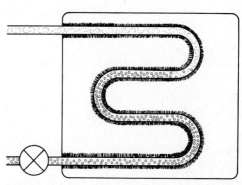

FIGURE 7–5 **Flat plate evaporator.**

liquid refrigerant and the gaseous refrigerant anywhere in the evaporator.

Flooded Evaporators. Flooded evaporators provide for the recirculation of liquid refrigerant by the addition of a *separation chamber*. The liquid refrigerant enters the separation chamber through the flow control device and flows down to the bottom of the separation chamber and the evaporator inlet. As the liquid refrigerant flows through the evaporator, it boils and absorbs heat. As the refrigerant leaves the evaporator, any liquid refrigerant present is separated from the vapor in the separation chamber and recirculated.

Controlling the liquid refrigerant level and recirculating the unevaporated liquid refrigerant assure that virtually all the inside surface of the evaporator is in contact with liquid refrigerant under any load conditions in a flooded evaporator. The evaporators shown in Figures 7–2 through 7–5 are of the single-pass type, but multiple-pass evaporators are more commonly used for greater efficiency and economy.

☐ Evaporator Styles ☐

Evaporators are made in many different shapes and styles to fill specific needs. Until the application is known, no one style can be considered superior.

The most common style is the *blower coil* or forced convection evaporator, in which the liquid refrigerant evaporates inside of finned tubes, absorbing heat from the air that is blown through the coil by a fan (Figure 7–2).

Specific applications may, however, use *bare coils* with no fins (Figure 7–3), *gravity coils* with natural-convection airflow (Figure 7–4), *flat plate surface* (Figure 7–5), or other specialized types of heat transfer surfaces.

☐ Heat Transfer in Evaporators ☐

The transfer of heat from the space being refrigerated to the refrigerant inside of the evaporator takes place in two steps:

1. The heat from the space being refrigerated must be absorbed by the metal of the evaporator.

2. The heat must be transferred through the metal of the evaporator and absorbed by the refrigerant inside the evaporator.

The transfer of heat from the air in the refrigerated space to the metal of the evaporator depends on several factors:

1. The nature of the exposed surface of the evaporator: bright or dull, rough or smooth
2. The temperature difference between the evaporator surface and the air surrounding it
3. The velocity of the heat-carrying air currents within the area being refrigerated
4. The conductivity of the metal from which the evaporator is constructed
5. The thickness of any frost coating on the evaporator

The difference in temperature between the air and the evaporator surface is quite important. If this temperature difference is small, the rate of heat transfer will be low.

The velocity of the heat-carrying air currents within the area being cooled depends on the design and construction of the evaporator. As the air in contact with the evaporator is cooled, it must be replaced with warm air. If the air circulation is restricted, the heat from the refrigerated space will not be properly transferred to the cooling unit, and the space will not be maintained at the desired temperature. If the air is allowed to circulate properly, the heat transferred to the air will be increased from four to six times over the rate of heat transfer to still air. In most installations, forced air circulation is induced over the surfaces of an evaporator, increasing the cooling capacity more than 20 times over natural-draft installations.

The heat transfer from the evaporator surface to the refrigerant is important because the refrigerant is what really does the work. The rate of heat transfer from the evaporator surface to the refrigerant depends on such factors as:

1. The surface area of the evaporator.
2. The temperature difference between the evaporating refrigerant and the medium being cooled.
3. The velocity of the refrigerant in the evaporator tubes. (In the normal commercial range, the higher the velocity, the greater the heat transfer rate.)

4. The ratio of primary surface to secondary surface.
5. The condition of the refrigerant (dry or flooded).
6. Freedom from oil film.
7. Rate of removal of the vaporized refrigerant.
8. The type of medium being cooled. Heat flows almost five times more effectively from a liquid to the evaporator than from air to the evaporator.
9. Dew-point temperature of the entering air. If the evaporator temperature is below the dew-point temperature of the entering air, latent heat as well as sensible heat will be removed.

The cooling capacity of the evaporator is proportionate to the exposed surface area. If the exposed area is small, the capacity of the evaporator will also be small. It is for this reason that fins have been added to the coils.

The ratio of primary surface to secondary surface simply means the ratio of tube area to fin area of the evaporator. Evaporators are manufactured for many different applications and the proper-size evaporator for any particular installation must be determined through calculation.

If the refrigeration unit is not functioning properly and the refrigerant vapor is not removed as rapidly as it forms in the evaporator, the pressure will increase in the evaporator and reduce the evaporation process. The increase in pressure will cause the boiling point of the refrigerant to rise and the corresponding temperature of the evaporator will rise and reduce the differential between the evaporator temperature and the temperature of the circulating air. This will effectively slow down the cooling rate of the unit.

In the final estimate of the capacity of a particular evaporator, it is this difference between the temperature of the refrigerant inside the evaporator and the temperature in the refrigerated space which is used. The difference in temperature is called *temperature differential*. For the ordinary household refrigerator, the refrigerant evaporating temperature is normally $5\,°F$ $(-15\,°C)$. If, then, we were to assume an average refrigerator temperature of $45\,°F$ $(7.2\,°C)$, we would have a temperature difference of $45\,° - 5\,°$ or $40\,°F$ $(22.2\,°C)$ between the temperature of the refrigerant in the evaporator and the temperature of the refrigerated space.

When the refrigerant enters the evaporator approximately 30 to 35% of the coil surface is used to produce flash gas. That is, to cool the incoming

FIGURE 7-6 **Percent of evaporator used for different functions.**

liquid refrigerant down to the temperature of the liquid already in the evaporator. About 50 to 65% is used to remove heat from the air passing through the coil, or for useful work. The remaining 5 to 10% is used for superheating the vapor before it leaves the evaporator coil (Figure 7-6).

☐ Evaporator Conductivity ☐

Metallic substances have a much higher rate of conductivity than nonmetallic substances. For this reason, metals are used as heat conductors, while the nonmetallic substances such as cork and glass are used as insulators since they do not readily conduct heat. The ability of metals to conduct heat is expressed by the *unit conductivity factor*. The unit conductivity factor is equal to the number of Btu (calories) that will pass through 1 ft² (0.093 m²) of 1-in. (2.54-cm)-thick material in 1 hr for each 1°F (1.8°C) difference in temperature between the two surfaces. The symbol used for this factor is k (Table 7-1).

From Table 7-1 it can be concluded that copper has a substantially higher rate of conductivity when compared with the other commercial metals. It also may be seen that steel and iron have merely a fraction of the conductivity of copper. However, since the metal used in the evaporator is relatively thin, iron and steel can be used successfully in evaporators. The maximum heat flow is obtained by using copper, but because of the constant film resistance of the surface of the metals, the heat transfer of a copper evaporator is only 10 to 20% greater than that of a steel evaporator.

Many of the nonmetallic substances such as cork and rock wool listed in Table 7-1 are used extensively as insulators in the refrigeration and air conditioning industry because of their low heat conductivity factor.

TABLE 7-1 Heat Conductivity

Substance	Conductivity Factor k [Btu (kcal)]
Copper	2680 [675.36]
Aluminum	1475 [371.7]
Iron	350–423 [88.2–106.6]
Steel	310 [78.12]
Concrete [average]	5.8 [1.46]
Porcelain	10 [2.52]
Water	3.85–5 [0.97–1.26]
Wood [with grain]	2.5 [0.63]
Wood [across grain]	1.0 [0.25]
Lubricating oil	1.2 [0.3]
Asbestos	0.94 [0.24]
Cork	0.3 [0.07]
Rock wool	0.26 [0.06]
Air	0.16 [0.04]

☐ Calculation of Heat Transferred ☐

Evaporator coils which have a higher number of rows of tubing are preferable in air conditioning work because they remove more humidity from the air passing through them. In refrigeration work, a lower number of rows in the coil is preferable because they remove less humidity. That is, thin coils are used to remove mostly sensible heat, while thicker coils are used to remove more latent heat. The total heat that may be transferred by any coil, provided that a steady temperature differential exists, is determined by the formula

$$Q = U \times A \times \Delta t \qquad (7-1)$$

where

Q = total Btu (cal) transferred or given up per hour

U = overall heat transfer coefficient (Btu/hr) (ft²)(°F Δt), R_t value ($R_t = R_1 + R_2 + R_3$ for various components)

A = area of surface in ft² (m²) through which the heat is flowing

Note: Calculating the U factor is usually quite involved.

The U factors for most materials have been tabulated and placed in tables for convenience (see Table 22-1). A more complete listing of U factors can be found in the American Society of Heating, Ventilating and Refrigerating Engineers' Society *Guide and Data Book*. The guide provides the most accurate information available on this subject.

When a heat load is to be calculated, each type of construction must be considered and calculated separately.

EXAMPLE:

An evaporator is capable of absorbing 2 Btu/ft² (504 cal/0.093 m²) per hour for each 1 °F (0.555 °C) of temperature difference. Therefore, the value of $U = 2$ (504 cal). The total surface area of the evaporator is $A = 5$ ft² (0.465 m²). The temperature in the refrigerated space = 45 °F (7.2 °C) and the refrigerant temperature = 5 °F (−15 °C). Calculate the heat transferred.

Solution

$Q = 2 \times 5 \times (45 - 5)$
$\quad = 10 \times 40$
$\quad = 400$ Btu/hr
$Q = 504 \times 0.465 \times [7.2 - (-15)]$
$\quad = 234.36 \times 22.2$
$\quad = 5202.79$ kcal/hr

From Eq. 7–1, it can be readily seen how the capacity of a coil (Q) is affected by the conductivity (U), the exposed surface (A), and the temperature differential (Δt). If any one of these factors is increased, the capacity of the coil will be increased.

The heat gained through any type of material will vary with the type of material, the area that is exposed to the different temperatures, the type and thickness of insulation, if any, and the temperature difference between the two temperatures.

Thermal conductivity varies directly with time, area, and temperature difference and is expressed in Btu/hr per square foot of area per degrees Fahrenheit of temperature difference per inch of thickness.

The amount of heat that may be transferred through any material is also subject to or affected by the surface resistance to heat flow, which is determined by the type of surface (rough or smooth), its position (vertical or horizontal), its reflective properties, and the rate of airflow over the surface.

Many laboratories have done extensive testing to determine the accurate values for heat transfer through all types of materials. Certain materials, such as insulation, have a high resistance to the flow of heat, while others do not offer such resistance to this flow.

In an effort to help in calculating heat loss, the refrigeration and air conditioning industry has developed a standard measuring term called the resistance (R) factor. The R factor is the amount of resistance to the flow of heat produced by a given type of material for a specified thickness of that material, or of an air space, an air film, or of an entire assembly of materials. This value is expressed as degrees Fahrenheit of temperature difference per Btu per hour per square foot. A high R value means that that material has a low heat flow through it. The resistances of several materials, of a wall for instance, may be added together to calculate the total resistance of that material or wall by the formula $R_t = R_1 + R_2 + R_3$. These R values are placed in tabular form for convenience (Table 7–2).

☐ Rate of Heat Transfer ☐

A flooded-type system will transfer approximately 50% more heat than a dry-type system. The refrigerant in a dry system exists as a mist (in the vaporous state) and, therefore, does not make as much contact with the inside of the evaporator tubes as does the liquid refrigerant in a flooded system. A flooded-type evaporator, which uses gravity air circulation within the food compartment, will absorb approximately 3 Btu (756 cal) per ft² (0.093 m²) per hour for every 1 °F (0.555 °C) difference in temperature. The dry system will not absorb as much, approximately 1½ to 2 Btu (378 to 504 cal) per ft² per hour for every 1 °F difference in temperature.

☐ Evaporator Design Factors ☐

Any pressure drop that occurs in an evaporator results in a reduction in system capacity due to the lower refrigerant pressure at the outlet of the evaporator coil. When a reduction in suction pressure exists, the specific volume of the refrigerant gas returning to the compressor increases, and the weight of the refrigerant vapor pumped by the compressor decreases. Therefore, the length of tubing in an evaporator must be kept to a minimum. When large-capacity evaporators are required, it is necessary to use several refrigerant circuits (Figure 7–7).

There are other factors that must also be con-

TABLE 7-2 Typical Heat Transmission Coefficients

Material	Density (lb/ft³)	Mean Temp. (°F)	Conductivity, k	Conductance, C	Resistance, R Per Inch	Resistance, R Overall
Insulating Materials						
Mineral wool blanket	0.5	75	0.32		3.12	
Fiberglass blanket	0.5	75	0.32		3.12	
Corkboard	6.5–8.0	0	0.25		4.0	
Glass fiberboard	9.5–11.0	−16	0.21		4.76	
Expanded urethane, R-11		0	0.17		5.88	
Expanded polystyrene	1.0	0	0.24		4.17	
Mineral wool board	15.0	0	0.25		4.0	
Insulating roof deck, 2 in.		75		0.18		5.56
Mineral wool loose fill	2.0–5.0	0	0.23		4.35	
Perlite, expanded	5.0–8.0	0	0.32		3.12	
Masonry Materials						
Concrete, sand and gravel	140		12.0		0.08	
Brick, common	120	75	5.0		0.20	
Brick, face	130	75	9.0		0.11	
Hollow tile, two-cell, 6 in.		75		0.66		1.52
Concrete block, sand and gravel, 8 in.		75		0.90		1.11
Concrete block, cinder, 8 in.		75		0.58		1.72
Gypsum plaster, sand	105	75	5.6		0.18	

Liquid in

Liquid in

Multiple circuit

FIGURE 7-7 Evaporator circuits.

sidered in evaporator design. If the evaporator tubing is too large, the refrigerant gas velocities may become so low that oil will accumulate in the tubing and will not be returned to the compressor. The only means of assuring satisfactory oil circulation is by maintaining adequate refrigerant vapor velocities. The heat transfer ability of the tubing also may be decreased greatly if the refrigerant velocities are not sufficient to scrub the interior walls of the tubing and keep it clear of an oil film. Since the goals of low pressure drop and high refrigerant velocities are directly opposed by the resistance of the tubing, the final evaporator design must be a compromise.

Pressure drops of approximately 1 to 2 psi (107.86 to 114.73 kPa) through the evaporator are acceptable on most medium- and high-temperature applications. Low-temperature applications commonly operate with ½ to 1 psi (104.42 to 107.86 kPa) pressure drop in the evaporator.

☐ Temperature Difference and Dehumidification ☐

In the study of evaporators, one property of air takes on a new importance. That property is the dew-point temperature of the air. When air is cooled, the temperature at which moisture will start to condense is known as the dew-point temperature.

Since for a given installation the physical characteristics are fixed, the primary variable is the temperature difference (TD) between the evaporating refrigerant and the medium being cooled. The colder the refrigerant is with respect to the temperature of the air entering the evaporator section, the greater the capacity of the evaporator.

Temperature differences of 5 to 20 °F (2.8 to 11.1 °C) are commonly used. Usually for best economy, the TD should be kept as low as possible because operation of the compressor will be more efficient at higher suction pressures.

The amount of moisture condensed out of the air is in direct relation to the temperature of the evaporator. An evaporator operating with too great a TD will tend to produce a low-humidity condition in the refrigerated space. In the storage of leafy vegetables, meats, fruits, and other perishable items, low humidity will result in excessive dehydration and damage to the product. For perishable commodities that require a very high relative humidity (RH) (approximately 90% RH), a TD from 8 to 12 °F (4.4 to 6.7 °C) is recommended. For relative humidities slightly lower (approximately 80% RH), a TD from 12 to 16 °F (6.7 to 8.8 °C) is normally adequate.

When the air temperature difference through an evaporator is increased, the capacity of the evaporator is increased. This is because the air temperature will drop more as it passes through the coil due to the lower refrigerant temperature inside the coil. However, this increase in evaporator temperature difference causes a decrease in compressor capacity. This decrease is because the lower temperature difference is accompanied with a lower-than-normal suction pressure which reduces the capacity of the compressor.

☐ Evaporator Frosting ☐

The formation of frost on the surface of the evaporator causes a considerable loss of heat transfer. First, the evaporator must cool the air below the condensation temperature. The moisture then gives up the latent heat of condensation as it condenses. The moisture then must be cooled to the freezing point. At this temperature, the moisture gives up the latent heat of fusion as it changes to frost. The frost must then be cooled down to the

temperature of the evaporator. A complete calculation will show that over 1200 Btu (302.4 kcal) must be absorbed by the evaporator to form 1 lb (0.4536 kg) of frost on its surface. When the evaporator is defrosted, only a small amount of this heat is regained; thus, it can be readily seen that frosting on an evaporator must be kept to a minimum.

☐ Defrosting of Evaporators ☐

Ice and frost will accumulate on evaporators that operate at or below freezing temperatures. Air flow through the evaporator will eventually be blocked unless the frost is removed. To allow continuous operation on refrigeration applications where frost accumulation can occur, periodic defrost cycles are necessary.

If the air returning to the evaporator is well above 32 °F (0 °C), defrosting can be accomplished by allowing the fan to continue to operate while the compressor is shut down, either for a preset time period or until the evaporator temperature rises a few degrees above 32 °F (the melting temperature of frost).

For low-temperature applications, some source of heat must be supplied to melt the ice. Electric defrost systems utilize electric heater coils or rods in the evaporator (Figure 7-8). Proprietary systems that use water for defrosting are available. Hot-gas defrosting is widely used, with the discharge vapor from the compressor bypassing the condenser and discharging directly into the evaporator inlet. In hot-gas defrost systems, the heat of compression or some source of stored heat provides the heat for defrost-

FIGURE 7-8 Electric defrost heaters.

ing. Adequate protective devices such as reevaporators or suction accumulators must be provided if necessary to prevent liquid refrigerant from returning to the compressor. Other systems may utilize reverse-cycle defrosting. In such a system, the flow of refrigerant is reversed to convert the evaporator into a condenser temporarily until the defrost period is complete. A drain pan heater is required on low-temperature applications to prevent refreezing of the melted condensate in the evaporator drain pan.

The refrigeration cycle on electric defrost systems is typically designed to operate 20 out of 24 hr a day. That is, the system is designed to be in the defrost cycle for a total of 4 hr per day. These 4 hr are scheduled for brief periods of time several times during the 24-hr period.

The refrigeration cycle on air defrost systems are typically designed to operate 16, 18, or 20 hr out of 24 hr a day. That is, the system may be in defrost for either 8, 6, or 4 hr out of the 24. The number of defrost periods and the length of each period will depend on the normal operating characteristics of the overall system.

□ Oil Circulation □

The fact that refrigeration compressors require lubrication allows some of this oil to circulate in the refrigeration system. The refrigerant vapor, in passing through the compressor, comes into direct contact with the oil on the cylinder walls, pistons, and in the crankcase. A small portion of this oil is carried along with the refrigerant vapor. A small amount of oil will pass through the system without any trouble. However, a large quantity of oil circulating in the system will allow oil logging of the evaporator, plus enough oil may leave the compressor to prevent proper lubrication of the moving parts.

Some compressors, during normal operation, pump relatively large amounts of oil. Unless some type of provision is made to return this oil to the compressor crankcase, the compressor may be damaged. The correct installation of the properly designed refrigerant lines with oil traps is needed.

Any oil allowed to collect in the circuits and passages of the evaporator will cause the evaporator temperature to rise, allowing less work to be done. The oil will take up space, which is needed for the evaporization of refrigerant, causing a decrease in efficiency.

Oil trapping in the evaporator may be due to inadequate defrost periods. If this is found to be the problem, increase the number of defrost periods or the length of each period to ensure proper oil return to the compressor.

□ Safety Procedures □

The evaporators on a refrigeration unit are on the low-pressure side of the system. Many times the low-pressure side is below atmospheric pressure. Therefore, when cleaning the evaporator, care must be taken not to cause leaks in the refrigeration system.

1. When defrosting an evaporator do not use sharp instruments that may puncture the metal.
2. Never block the airflow over an evaporator since damage to the compressor might result.
3. Care should be taken not to bend the evaporator fins.
4. Be careful when working around tube and fin evaporators because the fins can cause serious cuts on the hands and arms.
5. Always provide plenty of ventilation when working with coil cleaning solvents.
6. Always provide plenty of ventilation when purging refrigerant from a system.
7. Never heat a refrigerant line close to a flow control device without first wrapping a wet cloth around the flow control device.
8. Never unscrew any of the fittings when the evaporator is covered with condensation.
9. Never clean an aluminum evaporator with a caustic cleaner.

□ SUMMARY □

□ An evaporator is a device used for absorbing heat into the refrigeration system.
□ The two basic types of evaporators are: (1) dry or expansion, and (2) flooded.
□ The most common style of evaporator is the blower coil style.
□ The styles of evaporators are: (1) finned tube, (2) bare coil, and (3) flat plate coil.
□ Heat transfer in evaporators takes place in two steps: (1) heat from the air is absorbed by the metal, and (2) the heat is transferred through the metal to the refrigerant.

- The rate of heat transfer will be small if the TD is small.
- If air is allowed to circulate over an evaporator, the heat transfer will be four to six times as great as when still airflow is used.
- When forced-air circulation is used, the heat transfer is more than 20 times as great as natural-draft airflow.
- The refrigerant inside the evaporator is what actually does the work.
- The cooling capacity of an evaporator is proportionate to the exposed surface area.
- The ratio of primary surface to secondary surface is the ratio of tube area to fin area of the evaporator.
- The difference between the refrigerant temperature and the temperature of the refrigerated space is called the TD.
- Metallic substances are used for heat conductors, while nonmetallic substances are used as insulators.
- The ability of metals to conduct heat is expressed by the unit conductivity factor whose symbol is k.
- Copper has the highest conductivity factor (k) of the commercial metals.
- The formula for determining the total heat transferred is: $Q = U \times A \times \Delta t$.
- A flooded-type system will transfer approximately 50% more heat than a dry-type system.
- Any pressure drop that occurs in an evaporator results in a reduction in system capacity.
- When large-capacity evaporators are required, it is necessary to use several refrigerant circuits.
- The refrigerant velocity must be kept as high as possible to increase efficiency and affect oil return to the compressor.
- Pressure drops of more than 2 psi (114.73 kPa) should be avoided.
- The amount of moisture condensed out of air is in direct relation to the evaporator temperature.
- For perishable commodities that require 90% relative humidity, a TD of 8 to 12 °F (4.4 to 6.6 °C) is recommended.
- The formation of frost on an evaporator results in a great loss of heat.
- Defrost cycles are necessary on systems that require the evaporator temperature to be below 32 °F (0 °C).
- The types of defrost systems are: electric, water spray, and hot-gas defrost.

☐ EXERCISES ☐

1. Why is an evaporator so important?
2. Define *evaporator*.
3. What are the two basic types of evaporators?
4. Which type of evaporator is the most efficient?
5. What does the actual cooling in an evaporator?
6. List four styles of evaporators.
7. What two steps take place in the transfer of heat in the evaporator?
8. Will the heat transfer be greater with a high or a low TD?
9. How much is the heat transfer increased if forced air is used rather than still air over an evaporator?
10. What is proportionate to the cooling capacity of an evaporator?
11. What does the ratio of primary surface to secondary surface mean?
12. What will be the result if the refrigerant vapor is not properly removed from the evaporator?
13. Why are metallic substances rather than nonmetallic substances used in an evaporator?
14. Write the formula for determining the total heat transferred by a refrigeration unit.
15. Which type of system will transfer the greatest amount of heat?
16. How does a pressure drop affect the performance of an evaporator?
17. Why must the refrigerant velocity be kept in an evaporator?
18. What pressure drops through a high-temperature evaporator are acceptable?
19. What air property is important when studying evaporator operation?
20. What will an evaporator operating with a high TD tend to produce?
21. What will a heavy coating of frost on an evaporator cause?
22. What can be used to reduce frost on an evaporator operating at temperatures below 32 °F (0 °C)?
23. What are three types of defrost systems?
24. What device should be used to prevent liquid from returning to the compressor?
25. What is used to prevent moisture from freezing in the evaporator drain pan?

8 Flow Control Devices

Performance Objectives

After completion of this chapter, the student should:
- Know the purpose of flow control devices.
- Understand the theory of operation of flow control devices.
- Demonstrate how to adjust the different types of flow control devices.
- Complete the Exercises with at least 80 percent accuracy.
- Complete the Laboratory Exercises with at least 80 percent accuracy.

In modern refrigeration and air conditioning practice, a wide variety of refrigerant flow control devices are used to obtain efficient economic operation. Some of the smaller systems use a manual control or the simple on–off automatic control that requires very few flow control devices. However, large systems that use more elaborate automatic controls may have a multitude of controls, and the proper operation of each one is essential to the satisfactory performance of the system.

In order to adjust a control for its most efficient performance or recognize the effect of a particular malfunction, it is necessary that the function, operation, and application of each refrigerant flow control be thoroughly understood.

□ Purpose □

Though the flow control device is one of the more difficult refrigeration system components to understand, its basic function is extremely simple. The flow control device is used to control the flow of liquid refrigerant to the evaporator. The flow control device may be operated by different forces, such as temperature, pressure, or a combination of the two, but its only function is to control the flow of liquid refrigerant to the evaporator.

The proper control of refrigerant flow is also important for the following reasons:

1. The proper operation of the evaporator depends on the correct amount of liquid refrigerant and its swirling pattern through the evaporator tubes. Too much or too little refrigerant will provide less efficiency during the heat transfer process. To obtain the best heat transfer, the inside surface of the evaporator tubes must be completely wetted by the refrigerant, except in the last section of the evaporator, which is used to superheat the exiting refrigerant.

2. All the liquid refrigerant must be evaporated in the evaporator; otherwise, liquid refrigerant may return to the compressor and cause damage to the compressor valves and bearings. This is commonly known as *compressor slugging* and must be avoided.

□ Theory of Operation □

In order to understand how the flow control device accomplishes its job, a detailed explanation of the process is required. Figure 8–1 shows what happens to the refrigerant as it passes through the flow control device.

All the area to the left of line 2 in Figure 8–1

149

FIGURE 8–1 **Operation of flow control device.**

represents the refrigerant in the liquid state. The area between lines 1 and 2 represents a vapor–liquid mixture. The area to the right of line 1 represents the refrigerant in the vaporous state.

For this illustration, liquid R-12 enters the flow control device at point A under a condition of 117.1 psi (905.47 kPa) and 100 °F (37.8 °C). As it passes through the flow control device, the refrigerant pressure drops to 37 psi (355.18 kPa). The temperature at that pressure is 40 °F (4.4 °C). This is represented by point B. At point D, a mixture of vapor and liquid exists. This vapor has been formed by some of the liquid being evaporated in order to remove sensible heat from the remaining liquid to 40 °F (4.4 °C). This evaporating gas is termed *flash gas.*

Figure 8–1 also shows how the percentage (21.5%) of flash gas is obtained. As line 2 is the saturated liquid line, at C, all the refrigerant is in the liquid state at a temperature of 40 °F. At this condition, each pound (453.6 g) of R-12 would contain 17.3 Btu (4.36 kcal). As line 1 is the saturated vapor line, at D, all the refrigerant is in a saturated vapor state at a temperature of 40 °F. At this point, the refrigerant would contain 81.4 Btu/lb (44.77 kcal/kg).

For all practical purposes it can be said that there is no loss or gain in total heat of the refrigerant as it passes through the flow control device. Therefore, since 1 lb (453.6 g) of liquid refrigerant contains 31.1 Btu (7.84 kcal) as it enters the flow control device at point A, it will also contain 31.1 Btu as it leaves the flow control device.

The total heat that 1 lb of refrigerant could absorb between points C and D is 64.1 Btu (16.15 kcal). As shown in Figure 8–1, 13.8 Btu (3.48 kcal), represented by line CD, have been used in cooling the remaining liquid refrigerant from 100 °F (37.8 °C) to 40 °F (4.4 °C). The figure shows that 13.8 Btu (3.48 kcal) divided by 64.1 Btu (16.15 kal) = 21.5%. Therefore, 21.5% of the liquid has been evaporated or changed to flash gas.

Actually, there is a slight loss in the heat of the refrigerant because of the heat picked up from the valve, refrigerant lines to the evaporator, distributor, etc. But for all practical purposes the amount lost is too insignificant to be considered.

Flash gas may be caused by many factors in the refrigeration system. Figure 8–2 is the same as Figure 8–1 except that the line A'B' has been added. This

FIGURE 8–2 **Flash gas.**

line shows the effect of an increased compression ratio on flash gas. Assuming the low-side pressure will remain the same, the compression ratio has been increased by increasing the head pressure.

The line *CB* represents the amount of flash gas in Figure 8–1. The line *CB'* represents the new quantity of flash gas with a higher compression ratio. This is one very important reason why the compression ratio should be kept as low as possible. With the theory of the metering device, the full cycle of compression, condensation, expansion, and evaporation has been explained.

Figure 8–3 shows the entire cycle imposed on one temperature–Btu chart. As in the previous figures, all the area to the left of line 2 represents the refrigerant in the liquid state; all the area between lines 2 and 1 represents a vapor–liquid mixture; and all the area to the right of line 1 represents the refrigerant in a gaseous state.

For a brief review of the cycle, assume point *A* of Figure 8–3 to be the suction vapor entering the compressor. From point *A* to point *B*, the vapor is compressed in the compressor. Notice that not only does the temperature increase but there is an increase in the heat content of the vapor. This increase in heat is the result of the actual work done to compress the vapor. This increase in heat is called the *heat of compression*. From point *B* to point B_1, the superheat of the vapor is removed and it is cooled to the saturated vapor line or condensing temperature. From point B_1 to point B_2, the vapor is condensed. From point B_2 to point *C*, the condensed liquid is subcooled in the condenser. This heat, which is removed from the system, is known as the *heat of rejection*.

From point *C* to point *D*, the liquid passes through the flow control device. Although there is a change in temperature that corresponds to the pressure, and

a partial change of state, the amount of heat in the refrigerant remains the same.

The remaining part of the cycle is heat absorption. From point *D* to point D_1, the absorption is accomplished entirely by the evaporation of the refrigerant. This is latent heat since it results in a change of state of the refrigerant. From point D_1 to point *E*, the heat absorption results in a superheating of the refrigerant vapor before it leaves the evaporator at point *E*. This quantity of heat, which is absorbed from point *D* to point *E*, is known as the *net refrigerating effect* and is the actual work done by the refrigeration system. From point *E* to point *A*, a small amount of heat is absorbed into the system through the suction line, resulting in additional superheat to the refrigerant.

This cycle is the basis for all compression-refrigeration systems. If it is understood, a complete analysis of any compression-refrigeration system can be made.

☐ Types of Flow Control Devices ☐

There are seven types of refrigerant flow control devices: (1) hand-operated expansion valve, (2) low-side float, (3) high-side float, (4) automatic expansion valve, (5) thermostatic expansion valve, (6) capillary tube, and (7) thermal-electric expansion valve. The hand-operated expansion valve, low-side float, and high-side float have been replaced with the other more efficient types of flow control devices. Therefore, only the automatic expansion valve, thermostatic expansion valve, capillary tube, and the thermal-electric expansion valve will be discussed.

FIGURE 8–3 Refrigeration cycle.

□ Automatic (Constant-Pressure) Expansion Valve □

The constant-pressure or automatic expansion valve is the forerunner of the thermostatic expansion valve. The demand for energy conservation at minimum cost, however, has created a new interest in this simple, reliable control valve.

The automatic expansion valve is generally so termed because it opens and closes automatically without the aid of an external mechanical device. It maintains a nearly constant refrigerant pressure in the evaporator and on the low side of the system. Since the opening and closing of the automatic expansion valve is controlled by the refrigerant pressure in the low side of the system, this type of valve will not compensate for varying conditions in the high side or low side of the system, or for variation in load requirements.

Function. Automatic (constant pressure) expansion valves are basically pressure-regulating devices that respond to the refrigerant pressure of the valve outlet. They are installed at the evaporator inlet as a device to control refrigerant flow (Figure 8–4). The valve meters the refrigerant to maintain a constant evaporator pressure during machine operation.

This type of valve consists of a diaphragm, a control spring, and the basic valve needle (ball) and seat (Figure 8–5). The control spring above the diaphragm exerts a force to move the diaphragm downward, opening the valve (F1). On the opposite side of the diaphragm, the opposing force is developed by the low-side (evaporator) pressure (F2). This force tends to close the valve.

During the off cycle of the machine, the valve will be closed as the evaporator pressure builds up and overcomes the control spring pressure. When the machine starts operating, the compressor quickly reduces the low-side pressure. When the low-side pressure reaches the pressure of the control spring, the expansion valve is ready to open. It actually opens when the low-side (evaporator) pressure is reduced just below the control spring setting.

This is the valve opening point. The valve, however, must open further to satisfy the compressor capacity at the operating pressure. This occurs as the operation of the compressor further reduces the low-side pressure. The low-side pressure continues to drop, and the valve continues to open until a point is reached where liquid refrigerant is entering the

FIGURE 8–4 Location of automatic expansion valve. (Courtesy of Singer Controls Division.)

FIGURE 8–5 Automatic expansion valve. (Courtesy of Singer Controls Division.)

evaporator coil and vaporizing at a rate equal to the compressor pumping capacity. The evaporator pressure is now stabilized, and the system will operate at this point for the balance of the running cycle.

Adjustment. Automatic expansion valves feature a manual adjustment. The adjusting screw increases or decreases the tension of the control spring above the diaphragm, changing the valve opening point (Figure 8–6). The valve can be adjusted to open at a predetermined pressure within the range of the control spring. The valve operating point will be a few psi below the opening point. The exact differential is determined by the compressor's pumping capacity. It is the operating pressure that will be observed on a low-side pressure gauge attached to a system. A change in valve outlet pressure of 1 psi (6.87 kPa)

FIGURE 8-6 Automatic expansion valve adjustment. (Courtesy of Singer Controls Division.)

will move the valve stem approximately 0.001 in. (0.0025 cm).

The expansion valve should not be adjusted until the refrigeration unit has been operating from 24 to 48 hr. By this time, the refrigerant and the oil in the system should be properly distributed and the evaporator should be cold. When making adjustments, turn the adjusting screw one-quarter turn at a time and wait 15 min between each adjustment. When checking the operation of an expansion valve, the refrigeration unit should be operating continuously.

Features. Automatic expansion valves, because of their pressure-regulating performance, offer a number of operating features that are very useful on a variety of applications. The performance characteristics of the automatic expansion valve are discussed in the following sections.

Protection against Evaporator Icing. The use of automatic expansion valves eliminates the hazard of frost and ice accumulation on evaporator coils and prevents the freezing of product loads—for example, water in a water-chilling system. Unless protected during periods of low load, air conditioners accumulate a volume of frost, which retards air circulation over the evaporator coil and vastly reduces system capacity.

Automatic expansion valves maintain a constant low-side pressure and, therefore, a constant evaporator temperature. Adjusting the valve for an evaporator temperature just above the freezing point of water will completely eliminate the possibility of frost formation, regardless of ambient temperatures, heat load, or the length of time the unit is operated.

Other applications that require similar protection are drinking water coolers, soda fountain water coolers, photo developing tanks, and various industrial liquid chillers and fluid coolers.

Control of Relative Humidity. Adjusting the valve to maintain an evaporator temperature just above freezing not only prevents frost formation but maintains the low evaporator temperature required for maximum moisture removal—the means of controlling humidity in conditioned air. Humidity control is essential in conditioning air for human comfort.

Motor Overload Protection. Due to the close control of the system low-side pressure afforded by an automatic expansion valve, the possibility of excessive current draw by the motor operating the unit is completely avoided. Motors require this protection at high heat load conditions. With automatic expansion valves, the low-side (suction) pressure is constant and does not vary as the heat load fluctuates. Therefore, there are no variations in motor current requirements. Condensing unit current consumption is also automatically maintained within a safe maximum as dictated by the electrical wiring to the unit.

Lower-cost motors and wiring can be used with this flow control device. Motor horsepower is established on the basis of system load at normal ambient temperatures. Extreme ambients will not tax the unit because the evaporator pressure is always under control. Reserve motor capacity is not required and its cost is eliminated. Similarly, condenser surface area can be minimized.

Simplification of Field Service. The use of automatic expansion valves simplifies the field service of refrigeration and air conditioning units. Air conditioning unit manufacturers generally use valves with a fixed adjustment. The valves are set and sealed at the ideal operating condition for a particular unit for a refrigerant temperature above freezing. Common field difficulties such as stoppage of capillary tubes are eliminated.

The charging of expansion-valve-controlled air

conditioning systems with the proper weight of refrigerant is greatly simplified. A major problem in charging capillary tube systems is adjusting the refrigerant charge to the ambient temperature at the time of charging. It is very easy to overcharge a capillary tube system, which leads to operating difficulties when the ambient temperature increases.

If an expansion valve system is overcharged, the valve will automatically adjust the flow of refrigerant during the operating phase of the cycle to feed the evaporator properly. Any reserves of liquid are kept in the bottom of the condenser. Ambient temperature variations have very little effect upon the performance of an expansion-valve-equipped system.

High Capacity in a Small-Size Control. The overall expansion valve size has been reduced to the point where a physically small control device provides substantial refrigerating capacity. Valve bodies now machined from bar stock are in contrast to the bulkily forged bodies once used. Most modern valves are the bleed type; the bleed slot or bypass adds in large measure to the valve capacity. Other plus features for the reduced valve size are improved flow patterns through the valve, smaller but improved valve diaphragms, and silver solder construction.

The Right Valve for Water Coolers. On water cooler applications, automatic expansion valves are adjusted for minimum water temperatures. At the same time, the valve will provide absolute insurance against freeze-up of the water in the cooler, due to the fact that the valve maintains a constant, predetermined evaporator pressure and temperature. The valve is always adjusted for a refrigerant temperature above 32°F (0°C). It also assures no freezing at the valve orifice due to moisture in the system. A capillary tube cannot provide this assurance.

An Ideal Valve for Low-Starting Torque Motors. The bleed-type valve is used with split-phase and other low-starting torque motors commonly used in condensing units today. This expansion valve permits system unloading from high to low side during the system off cycle. All the advantages of automatic expansion valve operation are realized during the operating phase of the cycle.

A Low-Capacity Bypass Valve. Automatic expansion valves function as high- to low-side pressure regulators. Installed as a bypass and responding to the outlet pressure, the valve will open when the low-side pressure is reduced to the valve set point. The system's low-side pressure will be maintained at the required minimum as machine capacity is effectively reduced.

Constant Low-Side Pressure. The flow of refrigerant to the evaporator is maintained by the automatic expansion valve at a rate equivalent to the pumping capacity of the compressor. The fact that automatic expansion valves will open and remain open at a point where the refrigerant flow exactly meets compressor capacity accounts for the constant low-side pressure (and, therefore, evaporator temperature) maintained during the operating phase of the refrigeration cycle.

The balance in a refrigeration system using an automatic expansion valve is, therefore, between the expansion valve and the condensing unit. Since an automatic expansion valve is a differential-type control, the differential between the opening point and the operating point is automatically established by the running of the compressor. This differential provides the required valve movement to allow the flow of liquid refrigerant at the rate required by the compressor. Automatic expansion valves are ideally suited to refrigeration applications in which control of the evaporator temperature is required.

Bleed-Type Valves for Off-Cycle Unloading. Split-phase and other low-starting torque motors used in condensing units require an expansion valve that provides system unloading from the high to the low side during the off cycle.

Bleed-type valves permit pressures in the refrigeration system to reach a balance or near balance point during the off phase of the cycle. At the start of the next running phase, the motor starts under practically no load.

The bleed-type (or slotted orifice) valve is a standard automatic expansion valve with the addition of a small slot in the valve orifice (seat) to prevent complete valve close-off at the end of the machine's on cycle. Due to the bleed slot, the valve does not close completely when the unit stops operating, but permits refrigerant to continue to flow at a reduced rate (Figure 8–7).

Bleed slot

FIGURE 8–7 Bleed slot. (Courtesy of Singer Controls Division.)

Expansion valve bleeds add to the total valve capacity. Small *nicked seat* bleeds increase the valve capacity due to the flow of refrigerant through the nicks (openings) provided. The larger *drill bypass* bleeds are also open at all times to pass refrigerant when a pressure drop exists across the valve. Bleeds provide a fixed capacity depending on: (1) bleed size, (2) liquid density, and (3) the existing pressure drop.

Selecting the Proper-Size Bleed. Some care must be taken in selecting an expansion valve with the proper-size bleed for a particular application. Generally, the valve selected should feature the smallest bleed slot that will provide the required unloading (balancing of pressure) during the minimum length of machine off cycle. This is important because the bleed slots must not interfere with normal valve operation during the running phase of the cycle. A valve with a large bleed slot used with a relatively small condensing unit would cause difficulty at low suction temperatures, due to the bleed slot's closely matching the machine pumping capacity. Suction pressures below this point could not be reached by adjusting the automatic expansion valve.

To be certain that the bleed slot is not oversized for an application, turn the adjusting stem out to a setting lower than that at which the system will normally operate. Operate the unit and check the low-side pressure with a gauge. If the suction pressure drops down to the valve setting, or at least to a point below the normal operating pressure, the user can be sure the size of the bleed will not prevent the machine from operating at the desired point.

The rate of off-cycle unloading is an especially important consideration when machines may be required to start after a comparatively short off cycle. The larger the bleed and the greater the initial pressure drop across the valve, the faster the unloading rate.

Elevation Change—Effect upon Valve Setting. Automatic expansion valves are designed to admit atmospheric pressure to the valve diaphragm on the control spring side. The spring plus atmospheric pressure move the valve in the opening direction. The spring force can be varied by adjusting the screw, but any substantial change in altitude after a valve has been adjusted will alter the low-side pressure that the valve will maintain.

FIGURE 8-8   AXV as bypass valve. (Courtesy of Singer Controls Division.)

AXVs as Bypass Valves. Automatic expansion valves (AXVs) respond to valve outlet pressure. When this pressure is reduced to the valve opening point setting, the control spring above the diaphragm moves in the valve opening direction. This type of response is required for bypass valves as well as for flow control devices. Automatic expansion valves are, therefore, ideal as small-capacity bypass or capacity control valves.

Installed between the system's high and low sides, the bypass valve will open at preset outlet (low side) pressure and bypass high-pressure discharge gas to the low side of the system. Compressor capacity is modulated to prevent evaporator temperatures from dropping below a predetermined set point. In addition to capacity modulation the automatic expansion valve, when applied as a hot-gas bypass valve, can provide evaporator coil freeze-up protection regardless of the type of flow control device selected to meter liquid refrigerant to the evaporator (Figure 8-8).

Factors Affecting Valve Capacity. The factors that affect capacity of an automatic expansion valve are:

1. Orifice size
2. Amount of needle movement
3. Pressure drop across the valve
4. Type of refrigerant used
5. Condensing temperature or pressure
6. Size of the bleed slot
7. Evaporating temperature or pressure
8. Liquid subcooling

□ Thermostatic Expansion Valves □

The most commonly used device for controlling the flow of liquid refrigerant into the evaporator is the thermostatic expansion valve (Figure 8–9). An orifice in the valve meters the flow into the evaporator. The rate of flow is modulated as required by a needle-type plunger and seat that varies the orifice opening.

Function. The thermostatic expansion valve is a precision device designed to meter the flow of refrigerant into an evaporator in exact proportion to the rate of evaporation of liquid refrigerant in the evaporator, thereby preventing the return of liquid refrigerant to the compressor. By being responsive to the temperature of the refrigerant vapor leaving the evaporator and the pressure in the evaporator, the thermostatic expansion valve can control the refrigerant vapor leaving the evaporator at a predetermined superheat.

A vapor is said to be superheated whenever its temperature is higher than the saturation temperature corresponding to its pressure. The amount of

the superheat is, of course, the temperature increase above the saturation temperature at the existing pressure.

Consider a refrigeration evaporator operating with R-12 as the refrigerant at 37 psi (355.18 kPa) suction pressure (Figure 8–10). The R-12 saturation temperature at 37 psi is 40°F (4.4°C). As long as any liquid refrigerant exists at this pressure, the refrigerant temperature will remain at 40°F.

As the refrigerant moves along in the evaporator, the liquid boils off into a vapor and the amount of liquid decreases. At point *A* in Figure 8–10 all of the liquid has evaporated due to the absorption of a quantity of heat from the surrounding atmosphere, which is equal to the latent heat of vaporization of the refrigerant. The refrigerant vapor continues along in the evaporator and remains at the same pressure (37 psi); however, its temperature increases due to the continued absorption of heat from the surrounding atmosphere. By the time the refrigerant vapor reaches the end of the evaporator at point *B*, its temperature is 50°F (10°C). This refrigerant vapor is now superheated and the amount of superheat is, of course, 50°F − 40°F, or 10°F (5.6°C). The degrees to which the refrigerant vapor is superheated is a function of the amount of refrigerant being fed into the evaporator and the load to which the evaporator is exposed.

Operation. Three forces govern the operation of a thermostatic expansion valve (Figure 8–11). These forces are: (1) the pressure created by the remote bulb and power assembly (P_1), (2) the evaporator pressure (P_2), and (3) the equivalent pressure of the superheat spring (P_3).

The remote bulb and power assembly is a closed system and in the following discussion it is assumed that the remote bulb and power assembly charge is the same refrigerant as that used in the system.

The pressure within the remote bulb and power assembly, P_1 in Figure 8–11, corresponds to the saturation pressure of the refrigerant temperature

FIGURE 8–9   Thermostatic expansion valve. (Courtesy of Alco Controls Division, Emerson Electric Co.)

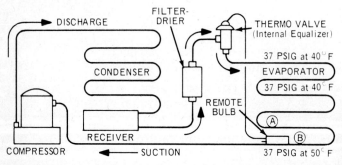

FIGURE 8–10   Basic refrigeration schematic. (Courtesy of Alco Controls Division, Emerson Electric Co.)

FIGURE 8-11 Thermostatic expansion valve basic forces. (Courtesy of Alco Controls Division, Emerson Electric Co.)

Superheat setting too low

Normal superheat setting

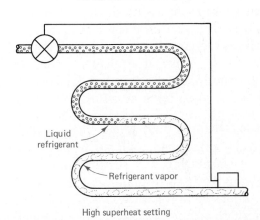

High superheat setting

FIGURE 8-12 Superheat setting and evaporator capacity.

leaving the evaporator, and moves the valve pin in the opening direction. Opposed to this force on the underside of the diaphragm, and acting in the closing direction, is the force exerted by the evaporator pressure (P_2), along with the pressure exerted by the superheat spring (P_3). The valve will assume a stable control position when these three forces are in equilibrium, that is, when $P_1 = P_2 + P_3$. As the temperature of the refrigerant vapor at the evaporator outlet increases above the saturation temperature that corresponds to the evaporator pressure, it becomes superheated. The pressure thus generated in the remote bulb and power assembly increases above the combined pressures of the evaporator pressure and the superheat spring, causing the valve pin to move in the opening direction. Conversely, as the temperature of the refrigerant vapor leaving the evaporator decreases, the pressure in the remote bulb and power assembly also decreases and the combined evaporator and spring pressures cause the valve pin to move in the closing direction.

The factory superheat setting of thermostatic expansion valves is made with the valve pin just starting to move away from the seat. These valves are so designed that an increase in superheat of the refrigerant vapor leaving the evaporator, usually 4°F (−15.6°C) beyond the factory setting, is necessary for the valve pin to open to its full open position. For example, if the factory setting is 10°F (−12.2°C) superheat, the operating superheat at the rated open position (full-load rating of the valve) will be 14°F (−10°C) superheat. If the system is operating at half load, with 50% compressor capacity reduction, the valve will operate at about 12°F (−11.1°C) superheat. It is important that internally adjustable type

thermostatic expansion valves be ordered with the correct factory superheat setting. It is also recommended that an externally adjustable thermostatic expansion valve be used in a pilot-model test to determine the correct factory superheat setting before ordering the internally adjustable type valve.

As the operating superheat setting is raised, the evaporator capacity decreases, since more of the evaporator surface is required to produce the superheat necessary to open the valve (Figure 8-12). It is obvious, then, that it is most important to adjust the operating superheat correctly. It is of vital im-

Thermostatic Expansion Valves **157**

portance that a minimum change in superheat is required to move the valve pin to the full open position because this provides savings in both the initial evaporator cost and the cost of operation. Accurate and sensitive control of the liquid refrigerant flow into the evaporator is necessary to provide maximum evaporator capacity under all load conditions.

Adjustment. Each thermostatic expansion valve is adjusted at the factory before shipment. This factory setting will be correct and no further adjustment is required for the majority of applications. When the application or operating conditions require a different valve setting, the valve may be adjusted to obtain the required superheat.

Some expansion valves are nonadjustable for use on original equipment manufacturer's units (Figure 8–13). These valves are set at a superheat predetermined by the manufacturer's laboratory tests and cannot be adjusted in the field.

Most nonadjustable models are modifications of standard adjustable-type valves. This is done by using a solid bottom cap instead of one equipped with an adjusting stem and seal cap (Figure 8–14). Adjustable bottom cap assemblies are available for converting some nonadjustable valves to the adjustable type. However, this is rarely required. If symptoms indicate that a valve adjustment is needed, carefully check the other possible causes of incorrect superheat before attempting an adjustment.

How to Determine Superheat. Thermostatic expansion valve performance cannot be analyzed properly by measuring the suction pressure or by observing the frost formation on the suction line. The initial step in correctly determining whether or not a thermostatic expansion valve is functioning properly is to measure the superheat. The following four steps show how to calculate the superheat setting:

1. Measure the temperature of the suction line at the point where the bulb is clamped.
2. Obtain the suction pressure that exists in the suction line at the bulb location by either of the following methods:
   (a) If the valve is externally equalized, a gauge in the external equalizer line will indicate the desired pressure directly and accurately.
   (b) Read the gauge pressure at the suction valve of the compressor. To the pressure reading add the estimated pressure drop through the

FIGURE 8–13 Nonadjustable thermostatic expansion valve. (Courtesy of Sporlan Valve Company.)

FIGURE 8–14 Adjustable and nonadjustable thermostatic expansion valves. (Courtesy of Singer Controls Division.)

TABLE 8-1 Temperature–Pressure Chart[a]

°F	R-12	R-13	R-22	R-500	R-502	R-717 (Ammonia)
−100	**27.0**	7.5	**25.0**		**23.3**	**27.4**
−95	**26.4**	10.9	**24.1**		**22.1**	**26.8**
−90	**25.7**	14.2	**23.0**		**20.7**	**26.1**
−85	**25.0**	18.2	**21.7**		**19.0**	**25.3**
−80	**24.1**	22.2	**20.2**		**17.1**	**24.3**
−75	**23.0**	27.1	**18.5**		**15.0**	**23.2**
−70	**21.8**	32.0	**16.6**		**12.6**	**21.9**
−65	**20.5**	37.7	**14.4**		**10.0**	**20.4**
−60	**19.0**	43.5	**12.0**		**7.0**	**18.6**
−55	**17.3**	50.0	**9.2**		**3.6**	**16.6**
−50	**15.4**	57.0	**6.2**		**0.0**	**14.3**
−45	**13.3**	64.6	**2.7**		2.1	**11.7**
−40	**11.0**	72.7	0.5	**7.9**	4.3	**8.7**
−35	**8.4**	81.5	2.6	**4.8**	6.7	**5.4**
−30	**5.5**	91.0	4.9	**1.4**	9.4	**1.6**
−28	**4.3**	94.9	5.9	0.0	10.6	0.0
−26	**3.0**	98.9	6.9	0.7	11.7	0.8
−24	**1.6**	103.0	7.9	1.5	13.0	1.7
−22	**0.3**	107.3	9.0	2.3	14.2	2.6
−20	0.6	111.7	10.1	3.1	15.5	3.6
−18	1.3	116.2	11.3	4.0	16.9	4.6
−16	2.1	120.8	12.5	4.9	18.3	5.6
−14	2.8	125.7	13.8	5.8	19.7	6.7
−12	3.7	130.5	15.1	6.8	21.3	7.9
−10	4.5	135.4	16.5	7.8	22.8	9.0
−8	5.4	140.5	17.9	8.8	24.4	10.3
−6	6.3	145.7	19.3	9.9	26.0	11.6
−4	7.2	151.1	20.8	11.0	27.7	12.9
−2	8.2	156.5	22.4	12.1	29.5	14.3
0	9.1	162.1	24.0	13.3	31.2	15.7
2	10.2	167.9	25.6	14.5	33.1	17.2
4	11.2	173.7	27.3	15.7	35.0	18.8
6	12.3	179.8	29.1	17.0	37.0	20.4
8	13.5	185.9	30.9	18.4	39.1	22.1
10	14.6	192.1	32.8	19.8	41.1	23.8
12	15.8	198.6	34.7	21.2	43.3	25.6
14	17.1	205.2	36.7	22.7	45.5	27.5
16	18.4	211.9	38.7	24.2	47.8	29.4
18	19.7	218.8	40.9	25.7	50.1	31.4
20	21.0	225.7	43.0	27.3	52.5	33.5
22	22.4	233.0	45.3	29.0	55.0	35.7
24	23.9	240.3	47.6	30.7	57.5	37.9
26	25.4	247.8	49.9	32.5	60.1	40.2
28	26.9	255.5	52.4	34.3	62.8	42.6
30	28.5	263.2	54.9	36.1	65.4	45.0
32	30.1	271.3	57.5	38.0	68.3	47.6
34	31.7	279.5	60.1	40.0	71.2	50.2
36	33.4	287.8	62.8	42.0	74.1	52.9
38	35.2	296.3	65.6	44.1	77.2	55.7
40	37.0	304.9	68.5	46.2	80.2	58.6
45	41.7	327.5	76.0	51.9	88.3	66.3
50	46.7	351.2	84.0	57.8	96.9	74.5
55	52.0	376.1	92.6	64.2	106.0	83.4
60	57.7	402.3	101.6	71.0	115.6	92.9
65	63.8	429.8	111.2	78.2	125.8	103.1
70	70.2	458.7	121.4	85.8	136.6	114.1
75	77.0	489.0	132.2	93.9	148.0	125.8
80	84.2	520.8	143.6	102.5	159.9	138.3
85	91.8		155.7	111.5	172.5	151.7
90	99.8		168.4	121.2	185.8	165.9
95	108.3		181.8	131.2	199.7	181.1
100	117.2		195.9	141.9	214.4	197.2
105	126.6		210.8	153.1	229.7	214.2
110	136.4		226.4	164.9	245.8	232.3
115	146.8		242.7	177.3	262.6	251.5
120	157.7		259.9	190.3	280.3	271.7
125	169.1		277.9	203.9	298.7	293.1
130	181.0		296.8	218.2	318.0	315.0
135	193.5		316.6	233.2	338.1	335.0
140	206.6		337.3	248.8	359.1	365.0
145	220.6		358.9	265.2	381.1	390.0
150	234.6		381.5	282.3	403.9	420.0
155	249.9		405.2	300.1	427.8	450.0
160	265.12		429.8	318.7	452.6	490.0

[a]Bold figures, in. Hg vacuum; light figures, psig.

suction line between the bulb location and the compressor suction valve. The sum of the gauge reading and the estimated pressure drop will equal the approximate line pressure at the bulb.

3. Convert the pressure obtained in step 2(a) or 2(b) above to the saturated evaporator temperature by using a temperature–pressure chart as shown in Table 8–1.

4. Subtract the two temperatures obtained in steps 1 and 3. The difference is the superheat.

Figure 8–15 illustrates a typical example of superheat measurement on an air conditioning system using R-12 as the refrigerant. The temperature of the suction line at the bulb location is read at 51°F (10.6°C). The suction pressure at the compressor is 35 psi (341.44 kPa), and the estimated pressure drop is 2 psi (13.74 kPa). The total suction pressure is 35 psi + 2 psi = 37 psi (355.18 kPa). This is equivalent to a 40°F (4.4°C) saturation temperature; therefore, 40°F − 51°F = 11°F (6.2°C) superheat setting.

Notice that subtracting the difference between the temperature at the inlet of the evaporator and the temperature at the outlet of the evaporator is not an accurate measure of superheat. This method is not recommended because any evaporator pressure drop will result in an erroneous superheat indication.

How to Change the Superheat Setting. To reduce the superheat setting, turn the adjusting stem *counterclockwise*. To increase the superheat setting, turn the adjusting stem *clockwise*. When adjusting the superheat, make no more than one turn of the stem at a time and observe the change in superheat closely to prevent overshooting the desired setting. As much as 30 min may be required for the new balance to take place after an adjustment is made.

If in doubt about the correct superheat setting for a particular system, consult the equipment manufacturer. As a general rule, the proper superheat setting will depend on the amount of temperature difference (TD) between the refrigerant temperature and the temperature of the air or other substance being cooled. Where high TDs exist, such as in air conditioning applications, the superheat setting can be as high as 15°F (−9.4°C) without noticeable loss in evaporator capacity. Where low TDs exist, such as in low-temperature blower coil applications, a superheat setting of 10°F (−12.2°C) or below is usually recommended for maximum evaporator capacity.

For the correct valve setting on factory-built equipment, the manufacturer's recommendations should be followed. Some manufacturers specify the superheat directly; others may recommend valve adjustment to provide a given suction pressure at certain operating conditions, or until a certain frost line is observed. Such recommendations, however they are stated, represent the results of extensive laboratory testing to determine the best possible operation.

Bulb Location and Installation. The location of the remote bulb is extremely important, and in some cases determines the success or failure of the refrigeration plant. For satisfactory expansion valve control, *good thermal contact* between the bulb and the suction line is essential. The bulb should be securely fastened with two bulb straps to a clean, straight section of the suction line.

Installation of the bulb to a horizontal run of suction line is preferred. If a vertical installation cannot be avoided, the bulb should be mounted so that the capillary tubing comes from the top (Figure 8–16).

FIGURE 8–15 Determination of superheat. (Courtesy of Sporlan Valve Company.)

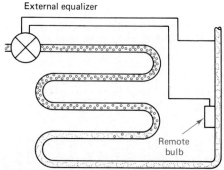

FIGURE 8–16 Remote-bulb installation on vertical tubing.

To install, clean the suction line thoroughly before clamping the remote bulb in place. When a steel suction line is used, it is advisable to paint the line with aluminum paint to minimize future corrosion and faulty remote-bulb contact with the line. On lines under ⅞ in. (22.225 mm) OD, the remote bulb may be installed on top of the line. On ⅞ in. OD and larger, the remote bulb should be installed at about the 4 or 8 o'clock position (Figure 8–17).

If it is necessary to protect the remote bulb from the effects of an airstream after it is clamped to the line, use a material such as sponge rubber, which will not absorb water when evaporator temperatures are above 32°F (0°C). Below 32°F cork or similar material sealed against moisture is suggested to prevent ice clogging at the remote-bulb location. The use of hair felt is not recommended. When the remote-bulb location is below the water or brine level of a submerged coil, use a waterproofing material or pitch that does not require heating above 120°F (49°C) when applying it in order to protect the remote bulb and tubing.

Accepted principles of good suction line piping should be followed to provide a bulb location that will give the best possible valve control. Never locate the bulb in a trap or pocket in the suction line. Liquid refrigerant or a mixture of liquid refrigerant and oil boiling out of the trap will falsely influence the temperature of the bulb and result in poor valve control.

Recommended suction line piping includes a horizontal line leaving the evaporator to which the thermostatic expansion valve bulb is attached. This line is pitched slightly downward, and when a vertical riser follows, a short trap is placed immediately ahead of the vertical line (Figure 8–18). The trap will collect any liquid refrigerant or oil passing through the suction line and prevent it from influencing the bulb temperature.

On multiple-evaporator installations, the piping should be arranged so that the refrigerant flow from any valve cannot affect the remote bulb of another. Approved piping practices including the proper use of traps ensure individual control for each valve without the influence of refrigerant and oil flow from other evaporators (Figure 8–19).

Recommended procedure for suction line piping when the evaporator is located above the compressor is to extend a vertical riser to the top of the evaporator to prevent refrigerant draining by gravity into the compressor during the off cycle (Figure 8–20). When a pumpdown control is used, the suction line may turn down immediately without a trap.

Remote-Bulb and Power Assembly Charges. Expansion valve charges are grouped into the following general classifications:

1. *Liquid charge.* This is usually the same as the system refrigerant. Valve control is not normally good. The charge tends to allow liquid re-

FIGURE 8–18 Evaporator suction piping. (Courtesy of Sporlan Valve Company.)

FIGURE 8–19 Evaporator suction piping on multiple evaporators. (Courtesy of Sporlan Valve Company.)

FIGURE 8–17 Remote-bulb installation on horizontal tubing. (Courtesy of Alco Controls Division, Emerson Electric Co.)

COMPRESSOR BELOW EVAPORATOR

Without pump down

Pump down control

FIGURE 8-20 Evaporator suction piping with compressor below evaporator. (Courtesy of Sporlan Valve Company.)

frigerant floodback to the compressor on system startup.

2. *Gas charge.* This is usually the same as the system refrigerant. Valve control is generally good. The charge will not operate in cross-ambient conditions and will condense at the coldest point. The valve will lose control if the power element or remote-bulb tubing becomes colder than the remote bulb.

3. *Liquid–vapor cross charge.* A volatile liquid refrigerant (not necessarily the same as the system refrigerant) is combined with a noncondensable gas. The charge does not lose control under cross-ambient conditions. The charge shifts slightly in control point.

4. *Cross-ambient vapor charge.* This is an all-purpose charge that replaces all three of the previous charge types, providing outstanding control under all conditions from −40 to 50°F (−40 to 10°C).

5. *Adsorption charge.* In adsorption, solids hold large quantities of gas not by taking them into the body of the solid, as in absorption, but by gathering them and holding them on the surface of the solid without chemical reaction. The vapor penetrates into the cracks and furrows of the solid, allowing considerably greater capacity than possible with absorption.

The advantage of an adsorption charge is that in a fixed volume, the quantity of vapor adsorbed varies with the temperature and the system. So it can be used to exert operating pressures as a function of temperature.

Typical adsorbants include charcoal, silica gel, and activated alumina.

Applications. The following is a list of the more practical applications of the more popular of the charges listed above:

1. *Adsorption charge:* air conditioning, medium-temperature commercial units, and chillers
2. *Liquid–vapor cross charge:* medium- and low-temperature commercial units, transport refrigeration, and ice makers
3. *Gas charge:* air conditioning and water chillers

Thermostatic Expansion Valve Equalizer. The operation of the thermostatic expansion valve is dependent on the relationship of the three fundamental pressures. That is, the bulb pressure acting on top of the diaphragm must always equal the sum of the evaporator (or suction pressure) and the spring pressure applied to the evaporator side of the diaphragm.

On an internally equalized valve, the pressure at the valve outlet (or evaporator inlet) is transmitted to the evaporator side of the diaphragm via a passageway within the valve or through clearance around the push rods (Figure 8-21).

On an externally equalized expansion valve the evaporator side of the diaphragm is isolated from the valve outlet pressure by packing around the push rods. The suction pressure is transmitted to the evaporator side of the diaphragm by a line usually connected between the suction line near the evaporator outlet (preferably downstream of the bulb) and an external fitting on the valve (Figure 8-22).

While internally equalized valves may be used with evaporators that have a low pressure drop, valves with an external equalizer *must* be used when there is an appreciable drop between the valve outlet and the remote-bulb location.

An internally equalized valve feeding an evaporator which, for purpose of illustration, has no pressure drop is shown in Figure 8-23. The pressure at the valve outlet and at the remote-bulb location is 27 psi (286.48 kPa). Therefore, the evaporator side of the valve diaphragm senses the evaporator pressure, which is also 27 psi and acts to close the valve. A spring pressure of 7 psi (48.09 kPa) also assists the evaporator pressure in attempting to close the valve. The valve consequently adjusts its refrigerant flow rate until the suction line vapor becomes sufficiently superheated to create a remote-bulb temperature of 37°F (2.8°C), which develops a pressure of 34 psi (334.57 kPa), balancing the evaporator spring. The resulting superheat is 9°F (5°C).

FIGURE 8-21 Internally equalized expansion valve. (Courtesy of Sporlan Valve Company.)

FIGURE 8-22 Externally equalized expansion valve. (Courtesy of Sporlan Valve Company.)

If this same internally equalized valve with the same spring adjustment is installed on an evaporator of equivalent nominal capacity but with a 6 psi (41.22 kPa) pressure drop, the operating superheat will increase to 15 °F (8.3 °C) (Figure 8–24). Now the valve senses a comparatively high pressure of 33 psi (327.70 kPa) at the valve outlet. The total closing pressure is 33 psi (327.70 kPa + 7 psi (48.09 kPa) = 40 psi (375.79 kPa). Since the bulb pressure must equal the total closing pressure, the valve reduces the flow of refrigerant to create the necessary superheat and bulb pressure. Consequently, excessive evaporator pressure drop causes an internally equalized valve to operate at an abnormally high superheat, and a serious loss of evaporator capacity results.

The problem of improper valve control may be corrected by using a thermostatic expansion valve

Closing pressure . . . . = 27 + 7 = 34 psi
  (Evaporator inlet pressure plus spring pressure)

Bulb pressure necessary to open valve . . . . 34 psi

Bulb temperature equivalent to 34 psi . . . . 37°F

Saturated temperature equivalent to evaporator outlet pressure 28°F
  Superheat . . . . 9°F
    Bulb temperature minus saturated evaporator temperature

FIGURE 8-23 Internally equalized valve illustration. (Courtesy of Sporlan Valve Company.)

Closing pressure . . . . = 33 + 7 = 40 psi
  (evaporator inlet pressure plus spring pressure)

Bulb pressure necessary to open valve . . . . 40 psi

Bulb temperature equivalent to 40 psi . . . . 43°F

Saturated temperature equivalent to evaporator outlet pressure 28°F
  Superheat . . . . 15°F
    Bulb temperature minus saturated evaporator temperature

FIGURE 8-24 Thermostatic expansion valve with coil pressure drop. (Courtesy of Sporlan Valve Company.)

Thermostatic Expansion Valves

Closing pressure . . . . = 27 + 7 = 34 psi
(suction pressure at bulb plus spring pressure)

Bulb pressure necessary to open valve . . . . 34 psi

Bulb temperature equivalent to 34 psi . . . . 37°F

Saturated temperature equivalent to evaporator outlet pressure 28°F
Superheat . . . . 9°F
Bulb temperature minus saturated evaporator temperature

FIGURE 8–25 Externally equalized expansion valve installation. [Courtesy of Sporlan Valve Company.]

with an external equalizer (Figure 8–25). This is the same system but with an externally equalized expansion valve. The suction pressure at the remote-bulb location is transmitted to the evaporator side of the diaphragm via the external equalizer line. The valve operation is now identical to that shown in Figure 8–23, and the superheat returns to 9°F (5 °C).

When to Use the External Equalizer. An internally equalized valve can tolerate less evaporator pressure drop at lower evaporator temperatures (Table 8–2). There are, of course, applications that may satisfactorily employ the internal equalizer when a higher pressure drop is present, but this should usually be verified by laboratory tests. The general recommendations given in Table 8–2 are suitable for most field-installed systems. Use an external equalizer when the pressure drop between the valve outlet and the remote-bulb location exceeds the values shown in Table 8–2.

When the expansion valve is equipped with an external equalizer connection, it must be connected—never capped—or the valve may flood, starve, or regulate erratically.

There is no operational disadvantage in using an external equalizer even if the evaporator has a low pressure drop.

TABLE 8–2 Maximum Pressure Drop [psi] for Which Internally Equalized Valves May Be Used[a]

| Refrigerant | Evaporating Temperature [°F] | | | | |
	40	20	0	−20	−40
R-12	2	1.5	1	0.75	0.5
R-22	3	2	1.5	1.0	0.75
R-500	2	1.5	1	0.75	0.5
R-502	3	2.5	1.75	1.25	1.0
R-717 [ammonia]	3	2	1.5	1.0	0.75

Source: Sporlan Valve Company.

[a]The external equalizer must be used on evaporators that employ a pressure drop type of refrigerant distributor.

Pressure-Limiting Thermostatic Expansion Valve. Standard thermostatic expansion valves are either pressure limiting or nonpressure limiting. Non-pressure-limiting valves are generally liquid charged. Various liquid charges are used depending on the system refrigerant, application temperature, and other factors. Pressure-limiting valves provide motor overload protection at times of high heat load. The pressure-limiting feature is commonly called *maximum operating pressure* (MOP) (Figure 8–26).

Present-day condensing units are designed to operate within a certain range of suction pressures. Operation of high-side equipment above the recommended maximum pressure imposes an overload on condensing units that may result in eventual damage to the condensing unit motor. To prevent suction pressures from soaring during unit operation, a pressure-limiting thermostatic expansion valve should be employed.

FIGURE 8–26 Maximum operating pressure expansion valve. [Courtesy of Sporlan Valve Company.]

A pressure-limiting valve will perform as a standard thermostatic expansion valve when operating at suction pressures within the range for which the condensing unit is designed. When the maximum operating pressure is reached, the valve's pressure-limiting feature takes control from the power element and prevents further increase in the flow of refrigerant to the evaporator. The system will continue to operate at this maximum suction pressure until the overload condition has passed. At this time, the suction pressure will drop below the maximum point, and the valve will again function as a standard thermostatic expansion valve.

The danger of floodback to the compressor at system start-up is also reduced with a pressure-limiting valve. The power element pressure above the diaphragm will not open the valve until the evaporator pressure is reduced below the pressure limit by the compressor. When the valve opens, it will operate on the valve's maximum operating pressure to minimize the initial surge of liquid refrigerant to the warm evaporator coil.

Most thermostatic expansion valves use two types of construction to provide the pressure-limiting feature: (1) limited liquid charge, and (2) liquid charge with a mechanical pressure limit.

Limited Liquid (Gas) Charge. Valves with a gas charge pressure limit are charged with a limited amount of refrigerant—a limited liquid charge. When the valve pressure limit has been reached, all the charge will have vaporized. As the evaporator temperature continues to rise, the power element vapor is superheated, with no appreciable increase in power element pressure. As a result, the pressure developed above the diaphragm is limited to a fixed maximum, and the valve prevents further increase in refrigerant flow at this point. Evaporator maximum operating pressure is thus established.

Gas-charged pressure-limiting valves are common in air conditioning and other high-temperature applications. However, they have one limitation. The thermal bulb must always be colder than the capillary tubing or valve head. If not, the thermal charge will condense at the coldest point and the remote thermal bulb will lose control.

Mechanical-Type Pressure-Limiting Valves. A thermostatic expansion valve with a mechanical pressure limit contains a standard liquid charge in the power element. The bulb always contains liquid and is always in control.

The pressure-limiting feature is accomplished by means of a special double diaphragm (Figure 8–27).

FIGURE 8–27 Mechanical pressure-limiting expansion valve. (Courtesy of Singer Controls Division.)

At all pressures below the pressure-limit setting, the valve functions as a standard thermostatic expansion valve. The lower diaphragm operates the push pins which in turn control the valve needle.

When the valve pressure-limit setting has been reached, the increased pressure between the diaphragms offsets the spring force above the upper diaphragm. The upper diaphragm rises and the two diaphragms (now interlocked) act as one. The valve now performs as a constant-pressure expansion device and meters liquid refrigerant to maintain the fixed evaporator pressure. When the overload condition has passed, the upper diaphragm moves down to a stop, and the lower diaphragm again assumes control. Normal thermostatic expansion valve operation is now resumed.

Distributors. When the refrigeration load is such that large evaporators are required, multiple refrigerant circuits are necessary to avoid an excessive pressure drop through the evaporator. To ensure uniform refrigerant feed from the expansion valve to each of the various circuits, a refrigerant distributor is normally used (Figure 8–28). The refrigerant distributor is mounted on the evaporator with the necessary tubing (Figure 8–29).

As liquid refrigerant is fed through the expansion valve, a portion of the liquid flashes into vapor in

FIGURE 8-28 Refrigerant distributor. (Courtesy of Singer Controls Division.)

Refrigerant Distributor

FIGURE 8-29 Mounted refrigerant distributor. (Courtesy of Singer Controls Division.)

FIGURE 8-30 Refrigerant distributor circuits. (Courtesy of Sporlan Valve Company.)

give satisfactory performance when properly applied in accordance with the manufacturer's instructions.

☐ Balanced-Ported Expansion Valves ☐

In a conventional expansion valve with the refrigerant flow direction such that the inlet pressure is applied under the valve pin, as the refrigerant pressure increases, the valve tends to move in the opening direction (Figure 8-31). Conversely, when the inlet refrigerant pressure falls, the valve tends to move in a closing direction. This change in valve pin position with a change in the inlet pressure is called unbalance. It results in an erratic variation from the original superheat setting. The same phenomenon occurs when the outlet (evaporator) pressure varies as load conditions vary—with the same results.

The balanced-ported, sometimes called double-ported, valves have been designed with two ports or "double ported." The inlet is so designed as to create a "counterflow" against the double-ported balance-ported valves and thus eliminate any unbalance across the two ports (Figure 8-32). Refrigerant flow through the upper port enters the upper radial holes of the cage assembly, moves upward and across the upper seat, down through the internal passage of the spool, and out the holes in the bottom of the spool. The pressure drop across this port exerts a force in a closing (upward) direction.

Refrigerant flowing through the bottom port enters the lower radial holes of the cage seat assembly and moves downward through the port formed by the cage seat and the valve spool. High-pressure liquid refrigerant acts downward on the spool, the

order to reduce the liquid temperature to the evaporator temperature (flash gas). This combination of liquid and flash gas is fed into the distributor from the expansion valve and is then distributed evenly through small feeder tubes, the number of feeders depending upon the construction of the distributor and the number of circuits required to provide proper refrigerant velocity in the evaporator.

Without the distributor, the refrigerant flow would split into separate vapor and liquid layers, resulting in the starving of some evaporator circuits. To avoid variations in circuit feeding, extreme care must be taken to ensure that feeder tubes are equal in length, so that equal resistance is offered by each circuit.

There are two different approaches in the design of a distributor. A high-pressure drop distributor depends on the turbulence created by an orifice to achieve good refrigerant distribution (Figure 8-30). A low-pressure drop distributor depends on a counterflow pattern to provide the proper distribution of the refrigerant flow. Both types of distributors

FLOW CONTROL DEVICES

makes it possible for the new double-ported cage assemblies to modulate over a much wider load range than is possible with the single-ported standard-type expansion valves. The reverse flow valves provide satisfactory control at loads less than 15% of the nominal capacity. Their performance is superior. Actual field performance has proven the superiority of double-ported valves and their ability to reduce hunting to a very minimum.

Adjustment. Since it is the purpose of the expansion valve to feed the evaporator fully, they must be adjusted to maintain this desired condition. Before the expansion valve can be adjusted properly, the unit must have been operating long enough for the interior of the unit to reach the normal operating temperature. Also, the evaporator must be free of heavy frost accumulation. At this point the valve can be adjusted as follows: Fasten two thermometer sensing elements to the evaporator coil, one at the inlet of the coil and one under the feeler bulb strap at the outlet of the coil (Figure 8-33). During normal operation these valves will "hunt" to some degree. The valve is properly adjusted when the maximum difference between the two probes is 3 to 5°F (1.68 to 2.80°C) during the hunting portion of the valve cycle. During a portion of the hunting cycle, when the valve is adjusted in this manner, the temperature difference between the two thermometer sensing elements will be less than 3°F (1.68°C).

FIGURE 8-31   Standard TEV in unbalanced condition, pin being forced open. (Courtesy of Alco Controls Division, Emerson Electric Co.)

Equal areas

FIGURE 8-32   Balanced-port TEV forces balanced (Courtesy of Alco Controls Division, Emerson Electric Co.)

pressure drops across the spool, and the seat exerts a force in an opening direction.

Since the effective port area of both the upper and lower cage ports is almost the same, the net force unbalanced across them is negligible. This feature

Expansion valve
Probe
Thermometer
Evaporator
Expansion valve sensor
Probe
Strap

FIGURE 8-33   Adjusting a balanced-port expansion valve.

FIGURE 8–34 **Bidirectional expansion valve. (Courtesy of Alco Controls Division, Emerson Electric Co.)**

Adjustments of only ½ turn of the valve stem should be made at a time. Then wait 15 min before checking the temperature and/or making further adjustments. Do not set the valve for a superheat higher than is required because a high valve superheat setting can cause a heavy frost accumulation on the coil, which would require a premature defrost cycle.

FIGURE 8–35 **Capillary tubing.**

Bidirectional Expansion Valves

These valves are designed for use on heat pump systems. They will allow the refrigerant to flow in either direction; thus one valve eliminates the need for two flow control devices and two check valves in the system, thereby simplifying the piping (Figure 8–34).

The superheat is controlled by a single sensing bulb and external equalizer connection placed on the common suction line, where it leaves the reversing valve close to the compressor. They are generally of the nonadjustable superheat type, which prevents unauthorized adjustments. They are designed to control the superheat properly over a wide range of applications and load conditions.

Bidirectional expansion valves will protect the compressor from high compression ratios by decreasing the superheat as the operating temperatures drop. They will also provide a constant superheat through the complete operating evaporator temperature range in the cooling mode.

Capillary Tubes

The capillary tube is the simplest type of refrigerant flow control device used in modern refrigeration systems. However, its application is limited to single cabinets having their own condensing unit and cannot be used in multitemperature or multiple-cabinet systems. A capillary tube is a small-diameter tube through which the refrigerant flows to the evaporator (Figure 8–35). The capillary tube is not a true valve since it is not adjustable and cannot be readily regulated. It is used only on flooded systems and allows the liquid refrigerant to flow into the evaporator at a predetermined rate, which is determined by the size of the refrigeration machine and the load it must carry. The capillary tube acts in exactly the same manner as a small-diameter water pipe that holds back water, allowing a higher pressure to be built up behind the water column with only a small rate of flow. In the same manner the small-diameter capillary tube holds back the liquid refrigerant, enabling a high pressure to be built up in the condensing unit during operation of the unit, at the same time permitting the liquid refrigerant to flow slowly into the evaporator.

The capillary tube is placed between the liquid line and the evaporator coil (Figure 8–36). The inside diameter of the capillary tube varies with the type of refrigerant, the unit capacity, and the length of the tube.

In operation, the capillary tube uses the princi-

FLOW CONTROL DEVICES

FIGURE 8–36   Location of capillary tube.

FIGURE 8–37   Capillary tube and strainer. (Courtesy of Henry Valve Company.)

ple of restriction to accomplish its purpose. It is the pressure differences between the high and the low sides of the refrigeration system that determine the flow of refrigerant. A capillary tube is chosen that will allow just the right amount of refrigerant into the evaporator to replenish that which has evaporated as the compressor is operating. Therefore, the pressure is reduced from the high pressure in the condenser to the low pressure inside the evaporator. There is no change in the liquid itself except for a slight drop in pressure for about the first two-thirds of the length of the capillary tube. At this point, the liquid begins to change to a vapor. At the end of the tube approximately 10 to 20% of the liquid has vaporized. It is this increased volume of vapor that causes most of the pressure drop to occur in the last one-third of the tube.

Since the orifice is fixed, the rate of feed is relatively inflexible. Under conditions of constant load, and constant discharge and suction pressures, the capillary tube performs very satisfactorily. However, changes in the evaporator load or fluctuations in head pressure can result in under- or overfeeding of the evaporator with refrigerant.

When the condensing unit stops, the condenser pressure and the evaporator pressure gradually equalize as the liquid refrigerant continues to flow through the capillary tube. Eventually, the compressor is able to start under balanced pressure conditions. This allows the use of low-starting torque motors, a big advantage of capillary tubes. However, it is necessary that the tube be free from obstructions because if it becomes clogged, the refrigeration system will not function.

Because of the small bore in the capillary tube, it is essential that the system be kept free from dirt and foreign matter. Usually, a filter is placed before the capillary tube to prevent dirt from plugging this

tube (Figure 8–37). If the capillary tube becomes plugged, the evaporator will defrost, the unit will run continuously, or the thermal overload may cut out. The discharge pressure will become very high unless the liquid receiver or condenser has capacity enough to take the entire charge of refrigerant. In many cases where the capillary tube is used, it is connected directly to the outlet end of the condenser and no liquid receiver is used. This means that most of the refrigerant is in the evaporator. In this case, it should be evident that a plugged capillary tube will result in an excessively high head pressure.

The refrigerant charge is critical in capillary tube systems because normally there is no receiver to store the excess refrigerant. Too much refrigerant will cause high head pressures, compressor motor overloading, and possible liquid floodback to the compressor during the off cycle. Too little refrigerant will allow vapor to enter the capillary tube causing a loss in system capacity.

Due to its basic simplicity, the elimination of the need for a receiver, and the low-starting torque motor requirement, a capillary tube system is the least expensive of all flow control systems.

For the service technician to be able to handle capillary tube installations properly, there are two basic facts that must be understood: cleanliness and size.

Cleanliness. Every precaution possible should be used to prevent chips, dirt, flux, moisture, filings, and other types of contaminants from entering the system when making repairs. Most capillary tubes have very small holes and it does not take much to plug them.

Size. In the normal installation some variation from the ideal length can be tolerated, but it is important to understand those factors that can be controlled by the service technician. Thus the size of the cap tube is fairly critical. Unlike orifices such as expansion valve seats, capillary tubes are dependent on their length as well as their diameter to determine their total restriction.

A change in the diameter on a percentage basis can change the flow more than an equal change in

FIGURE 8-38 Capillary tube selection—R-22 high temperature. (Courtesy of Copeland Refrigeration Corp.)

the length. As an example, changing the diameter by 0.005 in. as between 0.026 in. ID and 0.031 in. ID can double the flow through the tube for a given pressure differential.

The length of the capillary tube is the one factor that the service engineer has the easiest and most immediate control over. The best known method of changing the restriction of a capillary tube is to change the length. Cutting off a part of the tube reduces its restriction, while adding to its length increases its restriction.

The proper size of capillary tube is difficult to calculate accurately and can best be determined by actual test on the system. Once the size is determined, the proper capillary tube can be applied to identical systems, so it is well adapted to production-type units. Figures 8–38 to 8–42 give tentative selection data for capillary tubes.

FIGURE 8-39 Capillary tube selection—R-22 medium temperature. (Courtesy of Copeland Refrigeration Corp.)

☐ Aeroquip Flow Control Device ☐

This is a primary refrigeration expansion device. When properly sized, it can be used to replace other types of flow control devices. The Aeroquip flow control device is made up of three different components: the field connector, the restrictor, and the distributor (Figure 8–43). The field connector attaches to the liquid line. The restrictor fits inside and meters the refrigerant flow to the evaporator coil. The restrictor can easily be replaced in the field, providing the possibility to fine tune the system. The distributor provides refrigerant flow to the coil and is available in a wide variety of outlet configurations to match the system requirements (Figure 8–44).

Medium Temperature

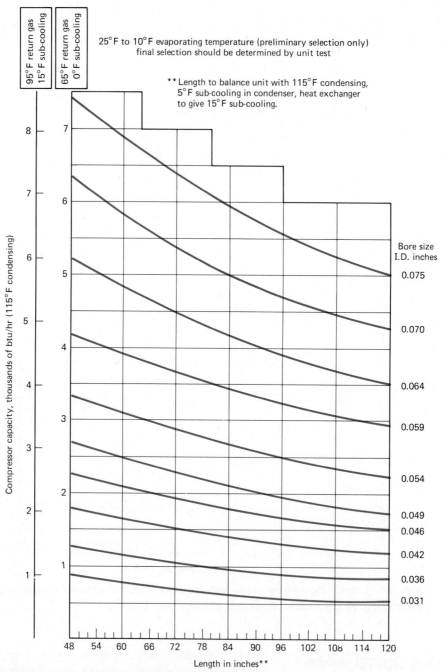

**FIGURE 8-40 Capillary tube selection—R-12 medium temperature.
(Courtesy of Copeland Refrigeration Corp.)**

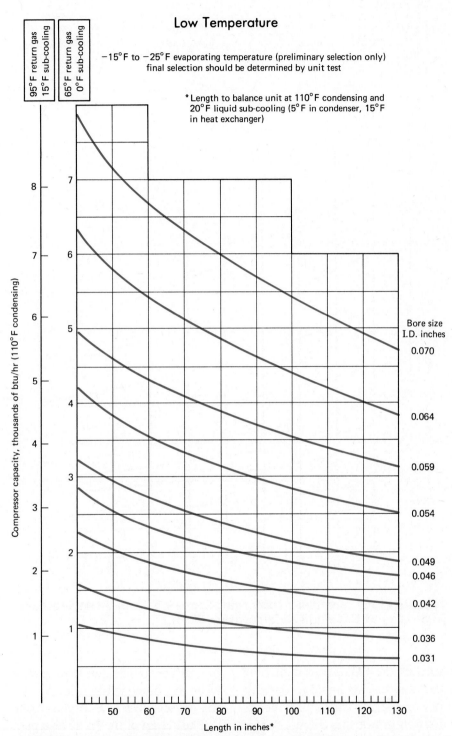

FIGURE 8-41 Capillary tube selection—R-12 low temperature. (Courtesy of Copeland Refrigeration Corp.)

FIGURE 8-42 Capillary tube selection—R-502 low temperature. (Courtesy of Copeland Refrigeration Corp.)

Operation. The heart of the FD20 flow control is the field-replaceable restrictor. As liquid refrigerant flows through the device, the restrictor orifice meters the flow into the distributor housing (Figure 8-45). This direction of flow results in a proper pressure drop to vaporize the refrigerant as it travels through the coils. The rate at which the refrigerant is vaporized is determined by the restrictor orifice size. The restrictor can be replaced quickly with another one

of a different size and flow rate, allowing a system to be fine tuned in the field. A wide variety of restrictor orifice sizes is available from Aeroquip.

The design of the FD20 also permits a reverse flow of refrigerant during the heating cycle in heat pump applications. As the liquid refrigerant flows in the reverse direction through the device, the restrictor moves back to a free-flow position, offering no, or very little, restriction to the flow of refrigerant. This

Field Connector Restrictor Distributor

FIGURE 8-43 AeroQuip flow control device components. (Courtesy of AeroQuip.)

reverse flow capability eliminates the need for check valves and their related plumbing in heat pump applications (Figure 8-46).

Selecting the Right Restrictor. The FD20 flow control is designed to accommodate a wide range of restrictor sizes. This allows a selection of the distributor housing/restrictor/field connector combinations which is best for the specific application. It allows mix-matching units, plus it offers the flexibility to fine tune the system in the field to meet changing needs and conditions.

Figures 8-47 to 8-49 are used to select the proper restrictors for the application. Use the following procedure in making the selection:

FIGURE 8-44 AeroQuip distributor configurations. (Courtesy of AeroQuip.)

FIGURE 8-45 Refrigerant flow during cooling cycle. (Courtesy of AeroQuip.)

FIGURE 8-46 Refrigerant flow during heat pump heating cycle. (Courtesy of AeroQuip.)

FIGURE 8–47   10°F (5.6°C) subcooling/40°F (4.4°C) evaporator. (Courtesy of AeroQuip.)

1. Determine the amount of subcooling desired.
2. Select the chart closest to the desired subcooling.
3. Determine the design inlet pressure (P).
4. From the selected chart, choose the line with the appropriate pressure.
5. Determine the desired mass flow.
6. Draw an imaginary straight line from the desired mass flow to the pressure line chosen, then drop the imaginary line down to the orifice diameter axis.
7. The orifice diameter can be read directly to the nearest thousandth, for example: 0.075 in.

Flow Control Assembly Procedures. Use the following recommended procedures when assembling the flow control device.

Factory Installation

1. Mount the flow control distributor housing on the unit with the appropriate mounting flange.
2. Insert the restrictor fully into the flow control distributor housing with the rounded "bullet" nose toward the unit as shown. Make sure that the restrictor stays inserted in the distributor housing before assembling the field connector.
3. Thread the flow control assembly together by hand to ensure proper mating of the threads, then

tighten until the components "seat." They are seated when a definite resistance is felt.
4. Tighten the assembly to 10 to 12 ft/lb of torque. Use a suitable torque wrench with a properly sized wrench or socket.
5. If no torque wrench is available, use a marker or ink pen to scribe a line lengthwise from the body hex to the bulkhead. Then tighten an additional ¼ turn. The misalignment of the line will show the amount the assembly has been tightened. This final ¼ turn is necessary to ensure a leakproof joint.
6. Package the unit to prevent contamination and/or damage during shipment and storage.

Field Replacement of the Restrictor. Refer to Figure 8–50. *Note:* Do not connect line sets! If the restrictor needs to be changed, change the restrictor first.

1. Remove the charge or pressure from the unit (if necessary). *Caution:* Always follow the equipment manufacturer's instructions. Extreme care must be taken in removing the charge.
2. Disassemble the flow control assembly by turning the field connector body hex.
3. If the existing restrictor has not dropped out of

**FIGURE 8-48 15°F (8.3°C) subcooling/40°F (4.4°C) evaporator.
[Courtesy of AeroQuip.]**

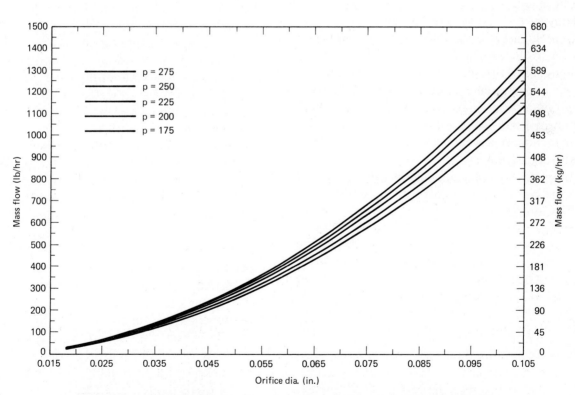

**FIGURE 8-49 20°F (11°C) subcooling/40°F (4.4°C) evaporator.
[Courtesy of AeroQuip.]**

BRAZE STYLE

PRECHARGE COUPLING STYLE

BULLET NOSE

RESTRICTOR

MOUNTING FLANGE

DISTRIBUTOR HOUSING

FIGURE 8-50 Factory installation of the flow control assembly. (Courtesy of AeroQuip.)

the body when disassembled, remove it by using a pin or paper clip. Discard the original restrictor.

4. Insert the new restrictor fully into the distributor housing with the rounded "bullet" nose toward the unit as shown. (Be sure that the restrictor stays inserted in the body while attaching the field connector.)

5. Thread the assembly halves together by hand to ensure proper mating of the parts. Tighten until the components "seat." They are seated when a definite resistance is felt.

6. Using a marker or pen, mark a line lengthwise from the connector hex to the bulkhead. Then tighten an additional ⅙ turn (or one flat). The misalignment of the line will show the amount the assembly has been tightened. This final ⅙ turn is necessary to ensure that a leakproof joint has been made.

7. Connect the liquid line to the flow control assembly either by a 5780 coupling connection (refer to the instructions under "Flow Control Liquid Line Installation"), by brazing directly to the copper tubing, or by attaching a flared tubing connection (Figure 8-51). *Note:* The field connector body hex must be held during the flare connection to prevent overtightening of the flow control assembly.

8. Complete the piping and installation of the unit according to the equipment manufacturer's installation instructions.

Caution when using 5780 series coupling or flare connection: If the liquid line is ever disconnected, the flow control assembly connection may also be loosened. If this occurs, care must be taken to prevent loss of the restrictor. If loosened, repeat steps 5 and 6 to ensure a leakproof joint.

Flow Control Liquid Line Installation Instructions

1. Review the equipment manufacturer's specifications and verify that the restrictor is properly sized for the unit being installed. If not, replace the restrictor according to the section "Field Replacement of the Restrictor."

FIGURE 8-51 5780 series AeroQuip refrigeration coupling. (Courtesy of AeroQuip.)

2. Route the suction and liquid line between the indoor and outdoor units.

3. For precharged coupling liquid line:
   (a) Remove the restrictor caps and plugs.
   (b) If unnecessary, carefully wipe the coupling seats and threaded surfaces with a clean cloth to prevent the inclusion of dirt or any foreign material into the system.
   (c) Lubricate the male half of the diaphragm and synthetic rubber seal with refrigerant oil. Thread the coupling halves together by hand to ensure proper mating of the threads. Use properly sized wrenches (on the line set coupling body hex and the union nut) and tighten until the coupling bodies "seat."
   (d) Using a marker or ink pen, mark a line lengthwise from the coupling union nut to the bulkhead. Then tighten an additional ¼ turn; the misalignment of the line will show the amount the coupling has been tightened. This final ¼ turn is necessary to ensure a leakproof joint is made. If a torque wrench is used, tighten to 10 to 12 ft/lb.
   (e) Finish the installation checkout of the unit according to the equipment manufacturer's installation instructions.

4. For sweat-style liquid line:
   (a) Cut the tubing square. Make sure that it is round and free of burrs at the connecting ends. Clean the tubing to prevent contaminants from entering the system.
   (b) Wrap a wet rag around the brass body of the flow control assembly.
   (c) Sparingly apply paste flux to the copper tube and insert it into the stub. Excessive or liquid flux can run inside the assembly and cause corrosion. No flux is necessary if a low or zero silver braze alloy is used.
   (d) After brazing, quench the joint with a wet rag to cool and remove any flux residue.
   (e) Evacuate, purge, or charge the lines and unit per the equipment manufacturer's installation instructions.
       Note: If a flow control sweat connector is not factory installed, refer to the factory installation procedure for proper assembly instructions.
5. For flare-style liquid line:
   (a) Connect the line in accordance with the equipment manufacturer's specifications for assembly and tightening the flare connection. *Note:* The flow control field connector body hex must be held during tightening of the flare connection to prevent overtightening of the flow control assembly.
   (b) Evacuate, purge, or charge the lines and unit per the equipment manufacturer's installation instructions.
6. Complete the piping and installation of the unit according to the equipment manufacturer's installation instructions.

Caution when using the 4780 series coupling or flare connection: If the liquid line is ever disconnected, the flow control assembly connection may also be loosened. If this occurs, care must be taken to avoid loss of the restrictor. If loosened, repeat steps 5 and 6 under "Field Replacement of the Restrictor" to ensure a leakproof joint.

□ Thermal-Electric Expansion Valves □

Conventional-type expansion valves have remained unchanged in basic design and function for more than 30 years. They have inherent limitations; their functional scope is narrow; and they do not permit

FIGURE 8-52 Thermal-electric expansion valve and thermistor. (Courtesy of Singer Controls Division.)

a refrigeration system to operate at maximum efficiency. Manufacturers of refrigeration and air conditioning equipment are constantly improving the reliability of their products, increasing their performance, and at the same time lowering the cost. They try new ideas, develop new systems, improve performance, modernize, and revolutionize equipment in order to stay competitive.

The thermal-electric expansion valve is the only refrigerant flow control device in step with this thinking (Figure 8-52). It is the only refrigerant control compatible with modern electronic sensing and monitoring devices. It is the only flow control device offered to the refrigeration industry in over 30 years that permits entirely new and fresh approaches in the field of refrigerant metering and system modulation.

Operation. The electric expansion valve, which is part of a valve and sensor package, is a simple heat-motor-operated needle valve infinitely positionable in response to voltage input (Figure 8-53). An increase in voltage opens the valve and increases the flow of refrigerant. A decrease in voltage reduces the flow or closes the valve altogether.

The changes in voltage necessary to modulate this control can be established by using various temperature, pressure, or liquid refrigerant sensors (Figure 8-54). Therefore, the function of the valve can be changed instantly by merely switching from one sensing device to another.

The valve opens to full capacity at 4.1 W and is designed to operate at 24 V with either ac or dc power.

FIGURE 8–53 Thermal-electric expansion valve cross section (Courtesy of Singer Controls Division.)

FIGURE 8–54 Liquid sensing thermistor. (Courtesy of Singer Controls Division.)

The valve does not depend on refrigerant temperature or pressure to create modulation. It responds only to voltage input. Therefore, the same valve works equally well with any common refrigerant except ammonia. It does not require external equalizer lines regardless of coil pressure drop and does not require a charged bulb and capillary lines.

The same valve will work just as well on low-temperature freezer applications as it does on unitary air conditioners. It can be used to control coil superheat [and 0°F (−17.8°C) superheat at that] or coil pressure, or both. The capacity of this valve, as with all expansion valves, remains a function of fluid flow and refrigerant effect.

0°F (−17.8°C) Superheat Control. The thermal-electric expansion valve is capable of maintaining saturation conditions throughout the entire evaporator at all loads while permitting the suction line

to superheat the refrigerant in order to prevent flooding the compressor (Figure 8–55).

A small negative coefficient thermistor is directly exposed to the refrigerant in the suction line by mounting it in a ¼-in. (6.5 mm) male pipe fitting requiring only a short female receptable.* The thermistor is placed in series electrically with the expansion valve heater, making the current input to the heater a function of thermistor resistance, which in turn is a function of refrigerant conditions.

Exposure to vaporous or superheated refrigerant permits the thermistor to self-heat, lowering its resistance, and increasing the current input to the expansion valve heater. The expansion valve responds by modulating in an opening direction, increasing the flow of refrigerant to the evaporator. This process continues until saturated suction line conditions occur, permitting wet refrigerant to contact the thermistor. Liquid refrigerant or wet refrigerant vapor will immediately cool the thermistor, increasing its resistance, which causes the expansion valve to modulate in the closing direction.

The termination point of refrigerant saturation can be controlled by the location of the thermistor, and the suction line becomes a logical spot for this transition to occur. However, the ability of the thermistor to sense liquid and meter the refrigerant accordingly permits consideration of new approaches to superheat control that are difficult, if not impossible, with conventional valves. These new approaches are:

1. The 0°F (−17.8°C) superheat point can be shifted in the evaporator by using more than one thermistor probe and the desired point of control can be switched-in on demand. This procedure could be used to size the evaporator to the compressor on systems using automatic compressor unloading.
2. The thermistor can be used to control the liquid level in a suction accumulator or riser, preventing flooded or semiflooded evaporator control with dry suction vapor returning to the compressor.
3. Evaporators can be split into independent sections using two or more electric valves and thermistors to control a specific series of passes or surface area. Each valve and thermistor will operate completely independent of the others.

*A thermistor is a semiconductor whose electrical resistance varies with the temperature.

FIGURE 8-55   Refrigeration system with thermal-electric expansion valve. (Courtesy of Singer Controls Division.)

Evaporator surface can be added to, or subtracted from, the total coil by simply switching the desired electric valves and their controlled passes off or on. Evaporator capacity, evaporator temperature, and moisture removal can be changed or regulated with this simple control arrangement.

Constant-Pressure Control Electric Expansion Valve. The electric expansion valve can be used to control evaporator pressure by placing the valve heater in electrical series with a pressure or

temperature switch, sensing evaporator pressure or temperature, respectively (Figure 8–56). A slow make-and-break switch, which opens on a rise in pressure, is suitable for this function because the electrical load is light, has low voltage, and is non-inductive. The principle of control is extremely simple.

For example, a switch set to control 70 psi (581.89 kPa) closes when the evaporator drops below this pressure, energizing the expansion valve. The resultant increase in refrigerant flow will raise the evaporator pressure, opening the pressure switch and

FIGURE 8-56   Installation of electric valve as constant-pressure control (Courtesy of Singer Controls Division.)

Thermal-Electric Expansion Valves

allowing the expansion valve to modulate toward closing again. The slow, modulating nature of the heat motor expansion valve permits the system to hover at the pressure setting of the switch.

Pressure switches can be used in this manner to limit both maximum and minimum system pressure. Switches can be placed anywhere in the system to close the expansion valve or limit its capacity when the desired levels of pressure or temperature are reached. The control of maximum head pressure is a good example of this type of superimposed function.

0 °F (– 17.8 °C) Superheat and Pressure-Limit Control. The thermal-electric expansion valve will provide both superheat control and pressure limit if the liquid-sensing thermistor is used in conjunction with pressure switches. In Figure 8–57 a high-pressure limit switch is placed in series with the expansion valve heater, preventing further increase in capacity when its pressure setting is exceeded. A low-pressure limit switch is placed in parallel with the thermistor, shunting it when the system pressure drops below its setting. (Both switches open on a rise in pressure.) This arrangement results in system operation at 0 °F (– 17.8 °C) superheat between the high- and low-pressure limits. The pressure switches

convert the expansion valve from a superheat control to a constant-pressure control automatically.

Liquid Line Shutoff Control. A switch, whether automatic or manual, can be used to close the expansion valve, thus allowing the system to pump down. Being a heat motor, the valve will open slightly in ambients above room temperature, slowly bleeding off refrigerant. Below room ambient, the valve will remain closed.

☐ Installation and Testing Recommendations ☐

Mounting Position. The valve should be installed such that the thermal head is within 45 ° of an upright position to ensure maximum capacity.

Thermistor Suction Line Adapter. The thermistor should be located flush, or slightly less, with the inside wall of the suction line. A $\frac{5}{16}$-in. (7.94-mm) dimension will correctly locate the thermistor assembly with the inside wall of the suction line (Figure 8–58). Projection of the female adapter fit-

CONSTANT SUPERHEAT CONTROL (WITH HIGH AND LOW PRESSURE LIMIT)

FIGURE 8–57 Superheat and pressure-limit control with electric valve. (Courtesy of Singer Controls Division.)

FIGURE 8-58 Correct location of thermistor on line. (Courtesy of Singer Controls Division.)

INCORRECT   CORRECT

FIGURE 8-59 Incorrect location of thermistor on line. (Courtesy of Singer Controls Division.)

FIGURE 8-60 Proper location of thermistor on line. (Courtesy of Singer Controls Division.)

GRAVITY SIDE PREFERABLE

FIGURE 8-61 Thermistor located on slanting line. (Courtesy of Singer Controls Division.)

ting, as well as the thermistor assembly, into the suction line should be avoided (Figure 8-59).

Thermistor Location and Sensing Positions. The liquid-sensing thermistor assembly can be used in any suction line with a diameter of ½ in. (12.7 mm) or larger. It will work on both vertical and horizontal suction lines, but it should never be located where liquid refrigerant is likely to accumulate or trap off, for instance, in a bottom U-bend connecting two vertical risers (Figure 8-60).

Because suction refrigerant flow depends on many factors, including suction line size, suction vapor velocity, elbow reducers, etc., it is impossible to establish firm rules regarding the best location of the liquid-sensing thermistor. The following observations have been made on a wide variety of applications and have proven to be useful guideposts for locating thermistor assemblies in suction lines:

1. High-velocity suction locations are preferable over low-velocity locations.
2. Smaller-diameter suction locations are preferable over larger-diameter locations.
3. Unless the flow pattern around an elbow or reducer is well known or specifically designed for the liquid-sensing thermistor assembly, it is best

to stay at least 6 in. (152.4 mm) away when locating downstream of them.
4. Vertical suction lines make excellent locations, but trapping should be avoided (Figure 8-60). If the line is at some angle other than vertical, the lower or gravity side is preferable (Figure 8-61).
5. Thermistor location on horizontal suction lines is more common, and in some cases (large-diameter, low-velocity suction lines) permits adjustments to be made in suction vapor saturation (Figure 8-62). The best thermistor-sensing positions in horizontal suction lines are generally between 4 and 8 o'clock in the lower half of the suction line, although successful applications have been made with the thermistor in all axial locations. As a rule, the thermistor should be located as high axially as possible on horizontal suction lines, using tolerable suction gas wetness as the limiting factor.
6. The most important rule regarding the location of the liquid-sensing thermistor on any applications is simply this: *make sure that liquid or wet refrigerant gas can come into good contact with the thermistor at all loads.* Once a location has been established on a given application, subsequent units will show amazingly consistent repeated performance.

Installation and Testing Recommendations

FIGURE 8-62 **Thermistor located on horizontal lines. (Courtesy of Singer Controls Division.)**

higher ambients. The thermistor, however, generally compensates for these changes automatically.

Effect of Ambient during Off-Cycle Periods. During off-cycle periods, the valve will bleed refrigerant if exposed to ambients above 70 °F (21.1 °C) and will close off relatively tightly if exposed to ambients of less than 70 °F. In short, the valve is simply calibrated to close at 70 °F with no electrical energy applied.

Off-Cycle System Unloading. The valve can be left energized 100% of the time, in which case it will completely and rapidly unload the system during off-cycle periods. The valve can be deenergized during off-cycle periods with the compressor, in which case it will partially unload the system, depending on the amount of charge and the ambient temperature.

Effects of Ambients and Blowers during Running Periods. Exposure to fan or blower air movement has almost no effect on the valve and it will operate properly in all ambients from −40 to 150 °F (−40 to 66 °C). The valve has a normal tendency to lose capacity at lower ambients and gain capacity at

Testing for Proper Operation. The valve and thermistor should be wired in series as shown in Figure 8-63. For pressure switch wiring circuits refer to Figures 8-56 and 8-57. All wiring should be made in accordance with UL low-voltage class 2 codes.

A voltmeter placed across the terminals of the elec-

FIGURE 8-63 **Wiring diagram of thermistor and valve. (Courtesy of Singer Controls Division.)**

tric valve becomes a handy tool for analyzing proper valve selection and system operation as follows:

1. A properly sized valve will cause the voltmeter to fluctuate as it modulates refrigerant flow during normal operation. Each time wet refrigerant vapor hits the thermistor, the voltmeter reading should drop; each time the suction vapor dries out, it should rise. The voltmeter may read as high as 20 V and as low as 8 V during normal operation.
2. When the system is operating near or at peak load, the voltmeter reading should remain at higher (15 to 20 V) levels longer and may even hover at some fixed point between 15 and 20 V.
3. When the system is operating at lower loads, the voltmeter reading should remain at the lower (8 to 14 V) levels longer and may even hover at some fixed point between 8 and 14 V.
4. If the valve remains at 17 to 20 V for long periods (over 3 min) without satisfying the system (suction line remains superheated), it is too small.
5. If the valve remains at 8 to 10 V for longer periods (over 3 min) and the system is not at minimum-load conditions, the valve is too large.
6. If either the valve or thermistor is suspected of being defective, a simple ohmmeter check can be used to prove their electrical continuity.

Connect the ohmmeter to the leads of the electric expansion valve or the thermistor assembly, whichever is the case, and compare their resistances against the manufacturer's specifications (Table 8-3). The valve should have 70 Ω resistance. The valve should check within 5% of the nominal valve listed and the thermistor within 40%.

Note: Both the valve and thermistor, but particularly the thermistor, must be at room temperature when making this check. The reading on the thermistor must be observed within 2 sec after connecting it to the ohmmeter. (The slight current input created by the ohmmeter causes the thermistor to self-heat and gradually drop in resistance.)

Service hints for the thermostatic expansion valve are summarized in Table 8-4.

☐ Safety Procedures ☐

The flow control device is normally a trouble-free device. It is, however, easily damaged. Care should be exercised to prevent damage when working on or around these devices.

1. Never apply an open flame to an expansion valve, as damage to the gaskets might result.
2. Do not hammer an expansion valve to free it. If it is stuck, it should be replaced.
3. Never unscrew any connections while refrigerant is in the system. An explosion might occur, resulting in serious personal injury.
4. Always use two wrenches when removing a valve from the system to prevent twisting of the lines.
5. Wear safety goggles when purging refrigerant from the system.
6. Keep the floor clear of moisture and oil.
7. Do not suddenly open an adjustment on an expansion valve fully because flooding of the compressor might result.
8. Always provide plenty of ventilation when soldering or welding on a flow control device.
9. Never try to weld or solder a line while refrigerant pressure is present.
10. Never leave the control bulb on a thermostatic expansion valve loose because flooding of the compressor may result.

TABLE 8-3 Liquid-Sensing Thermistor Specifications[a]

Typical Applications	Thermistor Resistance at 77°F (Ω)	Size	Number
Air conditioners, heat pumps, meat cases, chillers, ice makers	50	¼ MPT	189150
Low-temperature display cases, ice cream freezers, cascade systems, low-temperature blower coils	19	¼ MPT	189485
Special applications with extremely high load changes	100	¼ MPT	189519

Source: Singer Controls Division.
[a]Fitting material, brass; net weight, ¾ oz; shipping weight, 2 oz.

TABLE 8-4 Thermostatic Expansion Valve Service Hints

Probable Cause	Remedy

High Discharge Pressure

Probable Cause	Remedy
Insufficient cooling water due to inadequate supply or faulty water valve.	Start pump and open water valves. Adjust, repair or replace any defective equipment.
Condenser or liquid receiver too small.	Replace with correct-size condenser or liquid receiver.
Cooling water above design temperature.	Increase supply of water by adjusting water valve, replacing with a larger valve, etc.
Air or noncondensable gases in condenser.	Purge and recharge system.
Overcharge of refrigerant.	Bleed to proper charge.
Condenser dirty.	Clean condenser.
Insufficient cooling air circulation over air cooled condenser.	Properly locate condenser to freely dispel hot discharge air. Tighten or replace slipping belts or pulleys and be sure blower motor is of proper size.

Low Suction Pressure . . . High Superheat

A. Expansion valve limiting flow

Probable Cause	Remedy
Inlet pressure too low from excessive vertical lift, undersize liquid line or excessive low condensing temperature. Resulting pressure difference across valve too small.	Increase head pressure. If liquid line is too small, replace with proper size.
Gas in liquid line . . . due to pressure drop in the line or insufficient refrigerant charge. If there is no sight glass in the liquid line, a characteristic whistling noise will be heard at the expansion valve.	Locate cause of liquid line flash gas and correct by use of any or all of the following methods: 1. Add charge. 2. Clean strainers, replace filter driers. 3. Check for proper line size. 4. Increase head pressure or decrease temperature to ensure solid liquid refrigerant at valve inlet.
Valve restricted by pressure drop through evaporator.	Change to an expansion valve having an external equalizer.
External equalizer line plugged or external equalizer connector capped without providing a new valve cage or body with internal equalizer.	If external equalizer is plugged, repair or replace. Otherwise, replace with valve having correct equalizer.
Moisture, wax, oil, or dirt plugging valve orifice. Ice formation or wax at valve seat may be indicated by sudden rise in suction pressure after shutdown and system has warmed up.	Wax and oil indicate wrong type of oil is being used. Purge and recharge system, using proper oil. Install a filter-drier to prevent moisture and dirt from plugging valve orifice.
Valve orifice too small.	Replace with proper valve.
Superheat adjustment too high.	See "How to Change the Superheat Setting."
Power assembly failure or partial loss of charge.	Replace power assembly (if possible) or replace valve.
Gas charged remote bulb of valve has lost control due to remote bulb tubing or power head being colder than the remote bulb.	Replace with a cross-ambient power assembly. See "Remote Bulb and Power Assembly Charges."
Filter screen clogged.	Clean all filter screens.
Wrong type of oil.	Purge and recharge system and use proper oil.

B. Restriction in system other than expansion valve. (Usually, but not necessarily, indicated by frost or lower than normal temperatures at point of restriction.)

Probable Cause	Remedy
Strainers clogged or too small.	Remove and clean strainers. Check manufacturer's catalog to make sure that correct stainer was installed. Add a filter-drier to system.
A solenoid valve not operating properly, or undersized.	If valve is undersized, check manufacturer's catalog for proper size and conditions that would cause malfunction.
King valve at liquid receiver too small or not fully opened. Hand valve stem failure or valve too small or not fully opened. Discharge or suction service valve on compressor restricted or not fully opened.	Repair or replace faulty valve if it cannot be fully opened. Replace any undersized valve with one of correct size.

TABLE 8–4 [cont.]

Probable Cause	Remedy
Plugged lines.	Clean, repair, or replace lines.
Liquid line too small	Install proper size liquid line.
Suction line too small.	Install proper size suction line.
Wrong type oil in system, blocking refrigerant flow.	Purge and recharge system and use proper oil.

Low Suction Pressure . . . Low Superheat

Probable Cause	Remedy
Poor distribution in evaporator, causing liquid to short circuit through favored passes and throttling valve before all passes receive sufficient refrigerants.	Clamp power assembly remote bulb to free draining suction line. Clean suction line thoroughly before clamping bulb in place. Install a refrigerant distributor. Balance evaporator load distribution.
Compressor oversize or running too fast due to wrong size pulley.	Reduce speed of compressor by installing proper size pulley or provide compressor capacity control.
Uneven or inadequate evaporator loading due to poor air distribution or brine flow.	Balance evaporator load distribution by providing correct air or brine distribution.
Evaporator too small . . . often indicated by excessive ice formation.	Replace with proper-size evaporator.
Excessive accumulation of oil in evaporator.	Alter suction piping to provide proper oil return or install oil separator, if required.

High Suction Pressure . . . High Superheat

Probable Cause	Remedy
Unbalanced system having an oversized evaporator, and undersized compressor and a high load on the evaporator. Load in excess of design conditions.	Balance system components for load requirements.
Compressor undersized.	Replace with proper-size compressor.
Evaporator too large.	Replace with proper-size evaporator.
Compressor discharge valves leaking.	Repair or replace valve.

High Suction Pressure . . . Low Superheat

Probable Cause	Remedy
Compressor undersized.	Replace with proper-size compressor.
Valve superheat setting too low.	See "How to Change the Superheat Setting."
Gas in liquid line with oversized expansion valve.	Replace with proper-size expansion valve. Correct cause of flash gas.
Compressor discharge valves leaking.	Repair or replace discharge valves.
Pin and seat of expansion valve wire drawn, eroded, or held open by foreign material, resulting in liquid floodback.	Clean or replace damaged parts or replace valve. Install a filter-drier to remove foreign material from system.
Ruptured diaphragm or bellows in a constant-pressure (automatic) expansion valve, resulting in liquid floodback.	Replace valve power assembly.
External equalizer line plugged, or external equalizer connection capped without providing a new valve cage or body with internal equalizer.	If external equalizer is plugged, repair or replace. Otherwise, replace with valve having correct equalizer.
Moisture freezing valve in open position.	Apply hot rags to valve to melt ice. Install a filter-drier to ensure a moisture-free system.

Fluctuating Suction Pressure

Probable Cause	Remedy
Incorrect superheat adjustment.	See "Measuring and Adjusting Operating Superheat."
Trapped suction line.	Install P trap to provide a free draining suction line.
Improper remote bulb location or installation.	Clamp remote bulb to free draining suction line. Clean suction line thoroughly before clamping bulb in place.
Floodback of liquid refrigerant caused by poorly designed liquid distribution device or uneven evaporator loading. Improperly mounted evaporator.	Replace faulty distributor with a refrigerant distributor. If evaporator loading is uneven, install proper load distribution devices to balance air velocity evenly over evaporator coils. Remount evaporator lines to provide proper angle.

TABLE 8-4 [cont.]

Probable Cause	Remedy
External equalizer lines tapped at a common point although there is more than one expansion valve on same system.	Each valve must have its own separate equalizer line going directly to an appropriate location on evaporator outlet to ensure proper operational response of each individual valve.
Faulty condensing water regulator, causing change in pressure drop across valve.	Replace condensing water regulator.
Evaporative condenser cycling, causing radical change in pressure difference across expansion valve. Cycling of blowers or brine pumps.	Check spray nozzles, coil surface, control circuits, thermostat overloads, etc. Repair or replace any defective equipment. Clean clogged nozzles, coil surface, etc.
Restricted external equalizer line.	Repair or replace with correct size.

Fluctuating Discharge Pressure

Probable Cause	Remedy
Faulty condensing water regulating valve.	Replace condensing water regulating valve.
Insufficient charge . . . usually accompanied by corresponding fluctuation in suction pressure.	Add charge to system.
Cycling of evaporative condenser.	Check spray nozzles, coil surface, control circuits, thermostat overloads, etc. Repair or replace any defective equipment. Clean clogged nozzles, coil surface, etc.
Inadequate and fluctuating supply of cooling water to condenser.	Check water regulating valve and repair or replace if defective. Check for water circuit restrictions.
Cooling fan for condenser cycling.	Determine cause for cycling fan, and correct.
Fluctuating discharge pressure controls on low ambient air-cooled condenser.	Adjust, repair, or replace controls.

Source: Alco Controls Division, Emerson Electric Co.

▣ SUMMARY ▣

☐ The purpose of the flow control device is to control the flow of refrigerant to the evaporator.

☐ Flow control devices may be operated by forces such as temperature, pressure, or a combination of the two.

☐ Proper operation of the evaporator depends on the correct amount of liquid refrigerant and its swirling pattern through the evaporator tubes.

☐ All the liquid refrigerant must be evaporated in the evaporator or slugging may occur.

☐ As liquid refrigerant passes through the flow control device flash gas occurs.

☐ Flash gas is equal to approximately 20% of the total liquid passing through the flow control device.

☐ Flash gas cools the remaining liquid refrigerant to the evaporator temperature.

☐ Demand for energy conservation at a minimum cost has created a new interest in the automatic expansion valve.

☐ The automatic expansion valve controls the flow of refrigerant without an external mechanical device.

☐ The automatic expansion valve maintains a nearly constant evaporator pressure.

☐ Opening and closing of the automatic expansion valve is controlled by the refrigerant pressure in the low side of the system.

☐ The automatic expansion valve will not compensate for varying conditions.

☐ The evaporator pressure is stabilized when the liquid refrigerant evaporating in the evaporator is equal to the compressor's pumping capacity.

☐ The expansion valve should not be adjusted until the system has operated 24 or 48 hr.

☐ Automatic expansion valves may be used for the following performance functions: (1) protection against evaporator freezing, (2) control of relative humidity, (3) motor overload protection, (4) simplification of field service, (5) high capacity in a small size control, (6) the right valve for water coolers, (7) ideal valve for low-starting torque motors, and (8) low-capacity bypass valve.

☐ Bleed-type valves permit pressures in the refrigeration system to reach a balance or near balance point during the off cycle.

☐ The bleed-type (or slotted orifice) valve is a standard automatic valve with the addition of a small slot in the valve orifice to prevent complete valve close-off at the end of the running cycle.

☐ Bleeds provide a fixed valve capacity depending on (1) bleed size, (2) liquid density, and (3) the existing pressure drop.

☐ The valve must be selected with the smallest size bleed slot that will provide the required unloading during the minimum length of off cycle.

☐ Any substantial change in altitude after an automatic

expansion valve has been adjusted will alter the low side pressure that the valve will maintain.

☐ Automatic expansion valves can be applied as bypass or capacity control valves.

☐ Thermostatic expansion valves are the most commonly used devices for controlling the flow of liquid refrigerant into the evaporator.

☐ Thermostatic expansion valves are designed to meter the flow of refrigerant into an evaporator in exact proportion to the rate of evaporation of the liquid refrigerant in the evaporator.

☐ Thermostatic expansion valves are responsive to the temperature of the refrigerant gas leaving the evaporator and the refrigerant pressure inside the evaporator.

☐ The pressures that cause the thermostatic expansion valve to operate are: (1) pressure in the remote bulb and power assembly, (2) evaporator pressure, and (3) equivalent pressure of the superheat spring.

☐ The thermostatic expansion valve is said to be in equilibrium when $P_1 = P_2 + P_3$.

☐ An increase in the superheat of the refrigerant leaving the evaporator will cause the expansion valve to open.

☐ An increase in the evaporator pressure will cause the valve to close.

☐ As the superheat setting of a thermostatic expansion valve is increased, the evaporator capacity is decreased.

☐ The initial step in determining whether or not a thermostatic expansion valve is functioning properly is to measure the superheat.

☐ Determining the difference in temperature at the inlet and outlet of the evaporator is not an accurate measure of superheat.

☐ To reduce the superheat setting, turn the adjusting stem counterclockwise.

☐ As a general rule, the proper superheat setting will depend on the TD between the refrigerant temperature and the temperature of the medium being cooled.

☐ Good thermal contact between the remote bulb and the tube is essential for the successful operation of the unit.

☐ Expansion valve charges are grouped into the following general classifications: (1) liquid charge, (2) gas charge, (3) liquid–vapor cross charge, and (4) cross-ambient vapor charge.

☐ External equalized thermostatic expansion valves must be used on evaporators with an appreciable pressure drop between the valve outlet and the remote-bulb location.

☐ External equalizer connections must be made—*never cap these connections.*

☐ Maximum operating pressure (MOP) expansion valves provide motor overload protection at times of high heat load.

☐ When the MOP of the equipment is reached, the valve pressure-limit feature takes control from the power element and prevents further flow of refrigerant into the evaporator.

☐ MOP is provided by use of two features: (1) limited liquid charge, and (2) liquid charge with mechanical pressure limit.

☐ When large evaporators are needed, distributors are required to reduce the pressure drop through the evaporator.

☐ The capillary tube is the simplest type of refrigerant flow control device.

☐ The capillary tube is limited to single cabinet and single temperature applications.

☐ The capillary tube system is a critical charge type system.

☐ The electric expansion valve is simply a heat-motor-operated needle valve.

☐ The required voltage change can be caused by use of various temperature, pressure, or liquid refrigerant sensors.

☐ The electric expansion valve opens fully at 4.1 W of power.

☐ The electric expansion valve does not depend on refrigerant temperature or pressure to create modulation.

☐ The thermistor is located on the suction line allowing full refrigeration of the evaporator.

☐ The electric expansion valve will operate with a 0 °F (−17.8 °C) superheat on the coil.

☐ A voltmeter placed across the terminals of the electric valve becomes a handy tool for analyzing proper valve selection and system operation.

☐ EXERCISES ☐

1. What is the purpose of any flow control device?
2. Why must all the liquid refrigerant be evaporated in the evaporator?
3. What is flash gas?
4. What percentage of the total refrigerant becomes flash gas?
5. What is the net refrigerating effect?
6. What forces operate an automatic expansion valve?
7. Does the automatic expansion valve open or close during the off cycle of the system?
8. What is known as the valve opening point of an automatic expansion valve?
9. Is it correct to install an expansion valve and adjust it immediately?
10. How will an altitude change affect an automatic expansion valve?
11. What are the eight performance characteristics of an automatic expansion valve?

12. At what rate equivalent is the refrigerant admitted to the evaporator?
13. Why are bleed-type valves used?
14. What are the two basic types of bleed valve seats?
15. Explain how to test whether or not the bleed-type expansion valve is the correct size.
16. How can an automatic expansion valve be used as a bypass valve?
17. What is the thermostatic expansion valve designed to do?
18. What are the forces that operate the thermostatic expansion valve?
19. When is the thermostatic expansion valve in equilibrium?
20. Explain how to check the superheat of a thermostatic expansion valve.
21. Explain how a nonadjustable thermostatic expansion valve can be adjusted.
22. What does clockwise turning of the adjusting stem of a thermostatic expansion valve do to the superheat setting?
23. Where should the remote bulb be mounted?

24. What is the purpose of the cross-ambient vapor charge on a thermostatic expansion valve?
25. Should the external equalizer connection on a thermostatic expansion valve be capped off?
26. Where is the external equalizer line connected into the suction line?
27. When is an external equalizer used?
28. What does MOP mean and how is it used?
29. What two methods are used to provide pressure limiting on a thermostatic expansion valve?
30. For what purpose are refrigerant distributors used?
31. What is a capillary tube?
32. Which flow control device discussed in this book is used as a critical charged system?
33. What opens and closes the thermal-electric expansion valve?
34. What is the lowest superheat that can be obtained with a thermal-electric expansion valve?
35. What is the purpose of the thermistor used on the thermal-electric expansion valve?
36. Where should the thermistor be located for best results?

9 Accessories

Performance Objectives

After completion of this chapter, the student should:

- Exhibit a thorough understanding of the use and selection of accumulators.
- Be able to explain how and where driers are used and tell how to select them.
- Know the purpose of strainers.
- Know where to install and how to use moisture-liquid indicators.
- Be able to explain the purpose of an oil separator.
- Understand the use of heat exchangers.
- Know the importance of discharge mufflers.
- Understand the purpose of compressor service valves.
- Be able to discuss the operation of solenoid valves.
- Explain the purpose of check valves.
- Explain the purpose of water regulating valves.
- Complete the Exercises with at least 80 percent accuracy.
- Complete the Laboratory Exercises with at least 80 percent accuracy.

A number of accessory items are used in refrigeration systems for specific purposes. Their requirement in a particular system will depend on the application of the system. An accessory will allow the basic refrigeration system to have conveniences or reach a degree of performance that could not possibly be reached by the basic system. Therefore, an accessory may be defined as an article or device that adds to the convenience or the effectiveness of the system, but is not essential.

☐ Accumulators ☐

Compressors are designed to compress vapors, not liquids. Many systems, especially low-temperature systems, are subject to the return of excessive quantities of liquid refrigerant to the compressor. Liquid refrigerant returning to the compressor dilutes the oil, washes out the bearings, and in some cases causes complete loss of oil in the compressor crankcase. This condition is known as oil pumping or slugging and results in broken valve reeds, pistons, rods, crankshafts, etc. The purpose of the accumulator is to act as a reservoir to temporarily hold the excess oil–refrigerant and return it at a rate the compressor can safely handle.

Proper installation of a suction accumulator in the suction line just before the compressor eliminates damage (Figure 9–1). If the accumulator is correctly sized, relatively large quantities of liquid refrigerant may return through the suction line and the suction accumulator will prevent damage to the compressor. The liquid refrigerant is temporarily held in the suction accumulator (Figure 9–2) and is metered back to the compressor along with any oil, at a controlled rate through the metering orifice. Therefore, damage to the compressor is prevented and the compressor immediately and quietly goes to work.

Selection of Accumulator. An accumulator must be selected for a low refrigerant pressure drop, proper oil-return, and for adequate volume to ensure

Discharge Line
to Condenser ◄— — Suction Line to Evaporator —►

Compressor   Accumulator

FIGURE 9-1 Accumulator installation.

↑ TO COMPRESSOR   ↓ FROM COIL

PATENTED INLET DEFLECTOR

Will Not Trap Oil
·······················
Protects the Compressor

LEEK PRUF FUSIBLE PLUG

METERING ORIFICE

FIGURE 9-2 Internal view of suction accumulator. (Courtesy of Refrigeration Research, Inc.)

storage of at least 50% of the total system charge. The suction accumulator should not necessarily be selected to have the same size inlet and outlet as the compressor suction line.

◻ Driers ◻

Most authorities agree that moisture is the single most detrimental factor in a refrigeration system. A unit can stand only a small amount of water. The maximum amount of moisture allowed in a refrigeration system is 5 parts per million (ppm). For this reason, the majority of refrigeration and air conditioning systems, both field and factory assembled, contain driers (Figure 9-3). Several designs of heat pumps have two driers.

FIGURE 9-3 Liquid line drier. (Courtesy of Alco Controls Division, Emerson Electric Co.)

Refrigerant flow control device

Liquid line drier

To evaporator

Condenser   Compressor

FIGURE 9-4 Liquid line drier location.

Liquid line driers (Figure 9-4) have been used in refrigeration systems for many years. Their primary function, as their name implies, is to remove moisture from the refrigerant. Moisture in a refrigeration system is a contributing factor in the formation of acids, sludge, and corrosion. Hydrochloric and hydrofluoric acids are formed by the interaction of halocarbon refrigerants and small amounts of moisture. Moisture in a refrigeration system can cause a multitude of difficulties varying from expansion valve freeze-up to possible hermetic motor compressor burnout. In large amounts, water can directly produce mechanical malfunctions. It is important to remove quickly and completely any moisture and any acid or sludge that may have formed from a refrigeration system.

Over the years refrigeration systems have become more complex and driers are required to perform functions other than drying. Modern liquid line driers are capable of performing the following functions:

1. Moisture removal
2. Filtering
3. Acid removal

Selection of Liquid Line Drier. The following factors influence the selection of the correct size of drier:

1. Type and amount of refrigerant
2. Refrigeration system tonnage
3. Line size
4. Allowable pressure drop

When the refrigerant, line size, and application are known, the drier is generally selected on the basis of recommended capacities, which take into account both drying and refrigerant flow capacity. The flow and moisture capacity information is published by the liquid line drier manufacturer in tabular form for all popular refrigerants (Table 9–1).

It should be taken into consideration that the pressure drop through a drier, either liquid or suction line, will have a very bad effect on the overall operation of the system. A pressure drop of 1 psig (6.89 kPa) may cause up to 6000 Btu (6 kcal) loss of system capacity. When a pressure drop occurs in the drier, it should be replaced, not removed, with one of the proper type. It may be desirable to install one of greater capacity than the original because there are contaminants in the system.

Driers that are to be installed on low-temperature systems using R-502 as the refrigerant should have a medium that will remove wax to prevent it from clogging the flow control device and other system components. This wax is a natural component of the compressor lubricating oil.

Types of Liquid Line Driers. There are two general types of liquid line driers in use: (1) the straight-through sealed type (Figure 9–5) and (2) the angle-replaceable core (replaceable-cartridge) type (Figure 9–6).

Another type of liquid line drier has been especially developed for use on heat pump systems. This filter-drier provides system protection in both heating and cooling cycles with one filter-drier. These driers are equipped with internal check valves that allow refrigerant flow and filtration in either direc-

TABLE 9–1 Filter-Drier Selection

Catalog Number Flare	Catalog Number OD Solder	Size of Connection (in.)	Core Filter Area (in.²)	Shell Diameter	Overall Length Flare	Overall Length OD Solder	Weight (lb)	Recommended Tonnage[a] R-12	Recommended Tonnage[a] R-22
H032	H032-S	¼	11	1⅝	4¼	3¹³/₁₆	⅜	¾	¾
H052	—	¼	17	2⁵/₁₆	4¹¹/₁₆		¾	1	1
H053	—	⅜			5³/₁₆		⅞	1	1
H082	H082-S	¼			5⁹/₁₆	5⅛	1⅛	1	1
H083	H083-S	⅜	24	2⁵/₁₆	6¹/₁₆	5⅛	1⅛	2	2
H084	H084-S	½			6¼	5⅜	1¼	2	2
H162	—	¼			6⁵/₁₆	—	1½	2	2
H163	H163-S	⅜			6¾	5¹⁵/₁₆	1½	3	3
H164	H164-S	½	36	2⅞	7	6¹/₁₆	1¾	4	4
H165	H165-S	⅝			7¼	6⁵/₁₆	1¾	5	5
H303	—	⅜			9¹¹/₁₆	—	3⅜	4	5
H304	H304-S	½			9⅞	9	3⅜	7½	7½
H305	H305-S	⅝	57	3	10³/₁₆	9¼	3⅜	10	10
—	H307-S	⅞			—	9⅞	3¼	10	15
H414		½			9¹⁵/₁₆	—	4⅜	10	10
H415	H415-S	⅝	71	3½	10¼	9⁵/₁₆	4½	10	15
—	H417-S	⅞			—	9¹⁵/₁₆	4½	15	20
—	H4190S	1⅛			—	9⅞	4½	15	20
—	H607-S	⅞						20	25
—	H609-S	1⅛	106	3	—	16⁹/₁₆	6	25	30
—	H755-S	⅝			—	15¹/₁₆		15	20
—	H757-S	⅞	123	3½	—	15¹¹/₁₆	8¼	25	30

Source: Henry Valve Company.

[a]Recommended tonnage rating based on both drying and flow capacity is shown in table. For Refrigerants-500 and 502 use data shown for Refrigerant-12.

Driers

tion. Thus, the external check valves normally used on heat pump systems are eliminated (Figure 9–7). These units reduce system complexity and cost while providing effective removal of moisture, acid, and solid contaminants.

FIGURE 9–5   Straight-through liquid line drier. (Courtesy of Henry Valve Company.)

Dimension "D" is the minimum space required to remove the filter-drier core from the shell.

FIGURE 9–6   Angle-replaceable core drier. (Courtesy of Henry Valve Company.)

FIGURE 9–7   Heat pump filter-drier. (Courtesy of Watsco, Inc.)

☐ Suction Line Filter ☐

The proper suction line filter (Figure 9–8) will protect the refrigeration compressor by collecting all foreign matter and preventing it from entering the compressor where it could damage the highly machined and polished internal working parts.

The suction line filter should be installed in a vertical position as close as possible to the compressor prior to the system startup (Figure 9–9). When installed in this manner, it will remove foreign particles that are present even when the best installation procedures are carefully followed.

Selection of Suction Line Filters.   The following factors influence the selection of the correct size of filter:

1. Type of refrigerant
2. Suction line size
3. Allowable pressure drop
4. Application (low temperature, commercial refrigeration, or air conditioning)

When the refrigerant type, line size, and application are known, the filter is selected on the basis of refrigerant flow capacity expressed in tons. Flow capacity information is published by the suction line filter manufacturer in tabular form for all popular refrigerants (Table 9–2).

Dimension "D" is the minimum space required to remove the filter-drier core from the shell.

FIGURE 9–8   Suction line filter. (Courtesy of Henry Valve Company.)

Discharge Line to Condenser

Suction Line Filter

Suction Line to Evaporator

FIGURE 9–9   Suction line filter installation.

ACCESSORIES

TABLE 9-2 Suction Line Filter Selection

Catalog Number	Size of Connection (in.)	Shell Diameter (in.)	Filter Area (in.ª)
853A-⅝	⅝		
853A-⅞	⅞	2⅞	25
853-1⅛	1⅛		
853-1⅜	1⅜	3	57
853-1⅝	1⅝		
843-2⅛	2⅛	4¾	94
853-2⅝	2⅝		

Source: Henry Valve Company.

FIGURE 9-10 Straight-through suction line filter. (Courtesy of Henry Valve Company.)

Replaceable fluted core

FIGURE 9-11 Angle-replaceable core suction line filter. (Courtesy of Henry Valve Company.)

Types of Suction Line Filters. There are two general types of suction line filters in use: (1) the straight-through sealed type (Figure 9-10) and (2) the angle-replaceable core type (Figure 9-11).

☐ Suction Line Filter-Drier ☐

The suction line filter-drier offers all the advantages of a suction line filter plus the ability, through a blend of activated alumina and molecular sieve desiccants (drying agents), to remove acids and moisture

FIGURE 9-12 Suction line filter-drier. (Courtesy of Henry Valve Company.)

from the refrigerant vapor in the suction line along with any foreign matter present (Figure 9-12). Acid is best removed in the gaseous state. Therefore, liquid line driers are not used for removing it from the system. Activated alumina has a high affinity for acids and is a very popular desiccant for suction line filter-driers. The acid is not actually removed or neutralized, it is simply held in the pores of the desiccant. A large filter area and traverse flow passageway for the refrigerant vapor permits installation in the suction line with a minimum loss of refrigeration capacity due to pressure drop. Only clean and dry refrigerant is returned to the compressor.

☐ Strainers ☐

Strainers remove foreign matter such as dirt and metal chips from liquid or suction lines. This foreign matter could clog the small orifices of expansion and solenoid valves and also enter the compressor.

Types of Strainers. There are three general types of strainers: (1) straight-through sealed type (Figure 9-13), (2) cleanable angle type (Figure 9-14), and (3) cleanable Y type (Figure 9-15).

FIGURE 9-13 Straight-through sealed filter. (Courtesy of Henry Valve Company.)

FIGURE 9-14 Cleanable angle filter. (Courtesy of Henry Valve Company.)

INLET

FIGURE 9-15 Cleanable Y filter. [Courtesy of Henry Valve Company.]

FIGURE 9-16 Moisture-liquid indicator. [Courtesy of Sporlan Valve Company.]

Refrigerant flow control device

Liquid line drier

Moisture-liquid indicator

Suction line to evaporator

Condenser

Compressor

FIGURE 9-17 Location of moisture–liquid indicator.

☐ Moisture-Liquid Indicators ☐

The installation of a moisture–liquid indicator is actually the most inexpensive preventive maintenance item that can be placed in the refrigeration system (Figure 9-16). The liquid refrigerant is readily visible through the glass port or the indicator, and the presence of bubbles warns the operator or service technician of a shortage of refrigerant, a restriction in the liquid line, or improper subcooling of the refrigerant.

Purpose of the Moisture Indicator. The purpose of the moisture indicator is to reveal the presence of any moisture in the refrigerant that could be harmful to the system.

All moisture indicators have a moisture-sensing element of some type. This moisture-sensing element generally consists of a porous filter paper or cloth impregnated with an anhydrous cobalt salt. Cobalt salts are unique in that they have the capability of changing color with the addition or release of a very small amount of moisture. By changing colors, the moisture content of the refrigerant is indicated. Drier placement is required when a "wet" system is indicated.

To ensure accuracy, the indicator must always be installed in the liquid line as shown in Figure 9-17. Only when a system is in operation with the refrigerant flowing will the sensitized element indicate the correct condition of the refrigerant.

The sensitized element is protected from discoloration by dirt and sludge; however, contaminants, excessive amounts of water, or high temperatures can permanently discolor or damage the element. If the sensitized element is damaged, the indicator must be replaced.

☐ Oil Separator ☐

In any refrigeration system, refrigerant and oil are always present. Refrigerant is required for cooling. Oil is required for lubrication of the compressor. Refrigerant and oil are miscible in different percentages depending on the type of refrigerant, the temperature, and the pressure. Certain amounts of oil will always leave the compressor crankcase with the refrigerant.

The use of an oil separator in refrigeration systems has become quite widespread (Figure 9-18). In fact, a separator is considered by most engineers to be an essential item of installation on low-temperature units and on large air conditioning units up to 150 tons. The overall efficiency of a system is much improved where an oil separator is used, particularly in open-top display cases in supermarkets where we have to deal with evaporating temperatures of -30 to $-40\,°F$ (-34.4 to $-40\,°C$). This is equally true on industrial, laboratory, and environmental equipment operating to $-100\,°F$ ($-73.3\,°C$) and lower.

FIGURE 9-18 Oil separator. (Courtesy of AC&R Components, Inc.)

It is noted that most manufacturers require oil separators on their two-stage compressors.

Purpose of the Oil Separator. Originally, the purpose of an oil separator was to maintain the correct level of oil in the compressor crankcase, but this has long been overshadowed by the benefits it provides in preventing free circulation of even small amounts of oil in the refrigeration of air conditioning system.

The prime function of an oil separator, however, is to separate oil from the refrigerant and return it to the compressor crankcase before it can enter the other components of the system. Remember, oil is not a refrigerant; it belongs in the crankcase as a lubricant.

How the Oil Separator Functions. The oil separator is installed in the discharge line of the unit between the compressor and the condenser (Figure 9-19). The vapors containing the oil enter the oil separator inlet in the form of fog and pass through the inlet baffling. In so doing, they are forced to change direction several times on the surfaces of the baffle. A good percentage of the oil is in finer particles and these can only be removed by causing them to collide with one another to form heavier particles. This can only be achieved by changing the velocity of the vapor and oil fog mixture. This change in velocity occurs as the mixture passes into the shell of the oil separator, which is designed to reduce the velocity. As this reduction occurs, the oil particles have more momentum than the refrigerant vapor, causing them to collide with each other and impinge on the surfaces of the shell wall.

The refrigerant vapor then passes out through the outlet screens where there is a final scavenging action as the vapors speed up to their original velocity and the oil-free vapor then goes on to the condenser. The oil, thus separated, drips to the bottom of the oil

Oil return
line

Compressor   Oil separator   Condenser

Remember
Oil is Not a Refrigerant.
Oil Belongs in the Crankcase
to Lubricate

FIGURE 9-19 Oil separator location. (Courtesy of AC&R Components, Inc.)

separator where there is an adequate reservoir (sump) to permit any foreign matter, sludge, etc., to settle out and be trapped out of circulation. A magnet is installed in the bottom of the sump to collect any metallic particles, and stop them from being circulated in the system. Oil will collect until a sufficient level is reached to raise the float and open the oil return needle valve. The oil is then returned to the crankcase of the compressor.

The float-operated oil return needle valve is located high in the sump to allow clean oil to be automatically returned to the compressor crankcase.

Only a small amount of oil is required to actuate the float mechanism, so only a small amount of oil is absent from the crankcase at any time when an oil separator is used in the system.

Experiments with glass oil separators have shown that oil cannot be separated completely by baffles alone. There must also be a velocity drop to determine the ratio of the volume of the shell to the cross-sectional area of the discharge line. There is relatively little pressure drop through a properly designed and sized oil separator.

Selecting the Size of an Oil Separator. Although oil separator catalogs show their capacity in tons of refrigeration or by horsepower, the actual tonnage or Btu capacity of a system may vary widely from the horsepower size of the compressor. This is due to the speed of the compressor and the density of the suction vapor. It is a fact that the actual capacity of a compressor is dependent on the suction pressure. On any given size compressor, the higher the suction pressure, the higher the actual capacity—the lower the suction pressure, the lower the actual capacity of the compressor.

The sizing of the suction line and hot-vapor discharge lines is based largely on obtaining the desired velocity of the gases in the system. Velocity drop in the vapor and oil mixture as it passes through the oil separator is a major factor in oil separation. Therefore, the relationship of the volume of the oil separator connections to the volume of the shell and the baffling is important. The larger the capacity of the compressor, the larger the volume of the shell required, even if the connections remain the same. Since there must be minimal pressure drop, the connections must be able to carry the vapors at the same pressure as the discharge line. Obviously, the connections must be the same size or larger than the discharge line.

Heat Exchangers

Any device that brings into contact two substances of different temperatures for the purpose of heating or cooling one of these substances is called a heat exchanger. In refrigeration, the heat exchanger brings the liquid refrigerant and the suction vapor into thermal contact with each other.

Refrigerant in the liquid line is ordinarily slightly above room temperature. Suction vapor leaves the evaporator at a temperature near that maintained in the unit. Lowering the temperature of the liquid refrigerant before it reaches the flow control device increases the overall efficiency of the refrigeration unit by reducing the quantity of flash gas. During the heat exchange, the liquid is subcooled while any liquid present in the suction line is evaporated.

While it is sometimes possible to solder together lengths of the liquid and suction line to effect heat exchange (Figure 9–20), a carefully designed heat exchange allows complete and more reliable use of evaporator surface. System heat exchangers do not increase system capacity because what is taken from the system is put back into it. They do, however, increase the efficiency by reducing the amount of flash gas at the inlet of the evaporator.

Types of Heat Exchangers. There are two styles of small, highly efficient heat exchangers. One is a fountain-type heat exchanger, which is a double-tube device in which the liquid refrigerant passes through the space between the inner and outer tubes. A per-

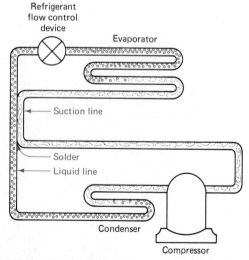

FIGURE 9-20 Liquid and suction lines soldered together.

FIGURE 9-21 Fountain-type heat exchanger.

FIGURE 9-22 Coil-type heat exchanger.

forated internal tube causes the cool suction gas to impinge at high velocity on the walls of the inside tube, resulting in a high rate of heat exchange (Figure 9-21). A second type is a coil-type heat exchanger, which has a coil made from small diameter tubing within a larger copper tube shell (Figure 9-22).

While both types give excellent heat exchange, the fountain type is recommended if the device must be located where condensation may be a problem.

Because the warm liquid refrigerant is in the outer shell, with the gas traveling through the center tube, atmospheric condensation with resultant dripping is not likely to occur.

Selection of Heat Exchangers. Field practices vary so greatly that capacity ratings applied to heat exchangers in laboratory tests would have no practical significance. However, recommendations can be made on the basis of maximum evaporator capacity, Btu per hour, using only the liquid and suction line sizes (Table 9-3).

☐ Vibration Eliminators ☐

In order to prevent the transmission of noise and vibration from the compressor through the refrigeration piping, vibration eliminators are frequently installed in both the suction and discharge lines (Figure 9-23).

On small units, where small-diameter soft copper tubing is used for the refrigerant lines, a coil of tubing may provide adequate protection against vibra-

TABLE 9-3 Recommended Maximum Evaporator Capacities [Btu/hr] for Heat Exchangers at Selected Suction Temperatures

Heat Exchanger Type	Part No.	R-12		R-22		R-502[a]	
		5°F	40°F	5°F	40°F	5°F	40°F
Fountain	A-13730	4,000	5,000	5,000	6,000	4,000	9,000
	A-14893	6,000	8,000	8,000	10,000	6,000	12,000
	A-14894	8,000	11,000	11,000	13,000	8,000	18,000
	A-14895	12,000	18,000	18,000	21,000	21,000	27,000
	A-14896	18,000	21,000	21,000	27,000	18,000	36,000
	A-14897	27,000	36,000	36,000	42,000	27,000	48,000
Coil	A-14891	4,000	5,000	5,000	6,000	4,000	9,000
	A-14890	5,000	6,000	6,000	7,000	5,000	11,000
	A-14892	6,000	8,000	8,000	10,000	6,000	12,000

[a]Refrigerant-502 has a higher density than R-22—and to minimize pressure drop, it is recommended that the evaporator capacities to R-12 be used.

FIGURE 9-23 Vibration eliminator location.

FIGURE 9-24 Metallic vibration eliminator. (Courtesy of Packless Industries.)

tion. On larger compressors, flexible metallic hose is most frequently used (Figure 9-24).

☐ Discharge Mufflers ☐

On systems where noise transmission must be reduced to a minimum, or where compressor pulsation might create vibration problems, a discharge muffler is frequently installed to dampen and reduce compressor discharge noise. The muffler is basically a shell with baffle plates, with the internal volume required primarily dependent on the compressor displacement, although the frequency and intensity of the sound waves are also factors in muffler design (Figure 9-25).

Purpose of the Discharge Muffler. The purpose of the muffler is to dampen or remove the hot-vapor pulsations set up by a reciprocating compressor.

Every reciprocating compressor will create some hot-gas pulsations. Although great effort is made to minimize pulsations in the system and compressor design, gas pulsation can be severe enough to create two closely related problems: (1) noise that although irritating to users of the equipment, does not necessarily have a harmful effect on the system, and (2) vibration that can result in refrigerant line breakage. Frequently, these two problems appear simultaneously.

Discharge Muffler Location. The muffler is installed in the discharge line as close to the compressor as possible. In the case of hermetic compressors, the muffler is frequently installed in the compressor shell itself.

Because the normal construction of a muffler is within a shell, natural trap is formed. Mufflers will trap oil easily and may trap liquid refrigerant. If a muffler is used, it should be installed in either a downward or a horizontal line as shown in Figure 9-26.

FIGURE 9-25 Discharge muffler.

Discharge Muffler

Discharge Line to Condenser

FIGURE 9-26 Discharge muffler location.

Selection of Discharge Muffler. The selection of a muffler can be a difficult engineering problem. Improper field selection of a muffler will sometimes lead to increased rather than decreased vibration.

☐ Crankcase Heaters ☐

When a compressor is installed in a location where it will be exposed to ambient temperatures colder than the evaporator, refrigerant migration to the crankcase can be aggravated by the resulting pressure difference between the evaporator and the compressor during the off cycle. To protect against the possibility of refrigerant migration, crankcase heaters are often used to keep the oil in the crankcase at a temperature high enough so that any liquid refrigerant that enters the crankcase will evaporate and create a pressure sufficient to prevent a large-scale refrigerant migration to the compressor.

If liquid refrigerant does migrate to the compressor, a reduction in suction pressure on compressor start-up will cause the liquid to boil (evaporate). When the refrigerant boils, the oil–refrigerant mixture will foam. Some of this foam will leave the crankcase and pass into the cylinder. This liquid causes a condition known as *slugging*, which will damage the compressor valves. The refrigerant also reduces the lubricating effectiveness of the oil.

Types of Crankcase Heater. Crankcase heaters are made in two types: (1) insert type (Figure 9-27), and (2) externally mounted (Figure 9-28).

FIGURE 9-27 **Insert-type crankcase heater.**

FIGURE 9-28 **External-type crankcase heater.**

Selection of Crankcase Heater. The crankcase heater is a low-wattage resistance element, and may be energized continuously, or on the compressor off cycle, and must be carefully selected to avoid overheating of the compressor oil. Compressor manufacturers generally have a list of recommended crankcase heaters for any particular model.

□ Compressor Service Valves □

Open-type and semihermetic compressors are usually fitted with compressor service valves, one at the suction port and one at the discharge port of the compressor. These service valves have no operating function, but they are indispensable when service operations are to be performed on any part of the refrigeration system.

Compressor service valves are of the back-seating type, so constructed that the stem forms a seal against a seat whether the stem is full-forward (front-seated) or full-backward (back-seated). A valve packing is depended on only when the stem is in the intermediate position. In one style of construction,

FIGURE 9-29 **Double-port compressor service valve.**

FIGURE 9-30 **Single-port compressor service valve.**

FIGURE 9-31 **Adjustable compressor service valve.**

the front seat, including the gauge connection, is threaded and soldered to the body after the stem has been assembled (Figure 9-29).

When the valve is full open (the normal position when the compressor is running), the gauge and charging port plug or cap may be removed without loss of refrigerant and a charging line or pressure gauge may be attached (Figure 9-30). It is also possible to repack the valve without interruption of service.

Compressor valves are machined to fit standard dimensions and many special flange dimensions are used by the several compressor manufacturers (Figure 9-31).

The valve bodies are fabricated of forged brass or cast iron. The valve openings are designed for unrestricted refrigerant flow. Each valve must be able to withstand 300 psi (2161.99° kPa) without leaking refrigerant.

☐ Solenoid Valves ☐

Solenoid valves are widely used in all types of refrigeration work. They are used as electrically operated line stop valves and perform in the same manner as hand shut-off valves. However, being electrically operated, they may be conveniently operated in remote locations by any suitable electric control device. The automatic control of the flow of refrigerants, brine, or water frequently depends on the use of solenoid valves. These valves are very popular for use when a pumpdown cycle of the system is required.

Solenoid Valve Operation. A magnet in which the lines of force are produced by an electric current is called an electromagnet. This type of magnet is important to automatic control design because the magnetic field can be created or destroyed by turning an electric current on or off. A solenoid valve is a simple form of electromagnet consisting of a coil of insulated copper wire or other suitable conductor which, when energized by the flow of an electric current, produces a magnetic field that will attract magnetic materials such as iron and many of its alloys. In this way an armature (frequently called a plunger) can be drawn up into the core of a solenoid coil (Figure 9–32). By attaching a stem and pin to this plunger, we are able to open or close a valve port by energizing or deenergizing the solenoid

coil. This magnetic principle constitutes the basis of design of all solenoid valves.

Types of Solenoid Valves. Solenoid valves can be divided into two general types: (1) direct acting, and (2) pilot operated. In the direct-acting type, the pull of the solenoid coil opens the valve port directly by lifting the pin out of the valve seat (Figure 9–33).

Since this type of valve depends solely on the power of the solenoid for operation, its port size for a given operating pressure differential is limited by

FIGURE 9–33   Direct-acting solenoid valve. (Courtesy of Alco Controls Division, Emerson Electric Co.)

FIGURE 9–32   Solenoid coil and plunger. (Courtesy of Alco Controls Division, Emerson Electric Co.)

FIGURE 9–34   Pilot-operated solenoid valve. (Courtesy of Alco Controls Division, Emerson Electric Co.)

the practical limitations of a solenoid size. Therefore, larger solenoid valves—type 2—are usually of the pilot-operated design (Figure 9–34).

In this type, the solenoid plunger does not open the main port directly, but merely opens the pilot port *A*. The pressure trapped on top of piston *B* is released through the pilot port, thus creating a pressure imbalance across the piston. The pressure underneath is now greater than the pressure above and the piston moves upward, thus opening the main port *C*. To close, the plunger drops and closes the pilot port *A*. In the next step, the pressure above and the pressure below the piston equalize again, then the piston drops and closes the main valve. The pressure difference across the valve, acting upon the area of the valve seat, holds the piston in a tightly closed position.

Selection of Solenoid Valve. The following performance characteristics are to be considered when selecting a solenoid valve for application: (1) capacity, (2) maximum operating pressure differential (MOPD), (3) electrical rating, (4) seat leak, and (5) special features.

1. Select a valve that meets the required flow capacity. Additional costs (when too large) or restricted refrigerant flow (when too small) are involved when a solenoid valve is improperly selected.
2. The MOPD should be the extreme pressure difference that the valve will operate against and not an average value. A valve failing to open due to occasional excessive pressures causes a constant "in-rush" current, and consequently coil burnouts.
3. Choose a valve with the proper voltage and frequency rating. Installation in communities where the electrical voltage varies beyond the rated electrical input limits of the coil will result in valve failure. When the voltage is too high, the coil will eventually burn out due to excessive current draw. The valve may fail to open when the voltage is too low and again cause the coil to burn out, due to the in-rush current.
4. It is advantageous to select a *metal-to-metal* seat valve when a seat leak is permissible. Soft seat valves are not as dependable throughout the life span due to the resilient property of the seat material. Continuous operations eventually cause an impression (set) in the seating material conforming to the seating surface that could lead to excessive seat leak. A metal-to-metal seat has a

minimum amount of wear and is uniform. This type has the inherent property of maintaining constant or decreasing seat leak with use. Foreign particles on the seating surface will cause excessive seat leak and sometimes are mistaken for a seat breakdown.

5. Fitting sizes, metering, materials of construction, and any other special features are also to be considered when selecting a valve for a particular application.

□ Check Valves □

Frequently, refrigeration systems are designed in which refrigerant liquid or vapor must flow to several components but must never flow backward in a given line. In such installations, a check valve is needed (Figure 9–35).

Purpose of Check Valves. As its name implies, a check valve checks or prevents the flow of refrigerant in one direction, while allowing free flow in the other direction. For example, if two evaporators are controlled by a single condensing unit, a check valve should be placed in the line from the lower-temperature evaporator to prevent the suction vapor from the higher-temperature evaporator from entering the lower-temperature evaporator.

Check valves are designed to permit flow in only one direction. When the fluid flows in the direction of the arrow on the valve, the force of the fluid will force the valve gate from its seat, allowing the fluid to flow through. Should the fluid attempt to flow in the wrong direction, the gate will close and stop the flow. Check valves are useful in installations where the liquid refrigerant may have a tendency to return to the compressor through the discharge line during the shutdown period. They are also used in

FIGURE 9–35 Check valve. [Courtesy of Henry Valve Company.]

heat pump systems, or reverse-cycle systems, to help in directing the flow of the refrigerant during the different cycles of operation.

These valves have many different uses and are considered to be relatively trouble free. They do create a pressure drop in the refrigerant line and, therefore, must be used with caution and replaced with the proper replacement.

Most check valves are designed to eliminate chattering and to give maximum refrigerant flow when the unit is operating with a spring tension sufficient to overcome the weight of the valve disc (Figure 9–36). Check valves must be mounted in the proper position.

Another type of check valve that does not use springs is the Magni-check valve (Figure 9–37). The Magni-chek valve disc is positioned against the brass seat by an alinco magnet. There are no springs to create excessive back pressure, and the valve can be installed in any position without affecting valve performance.

A fraction of an ounce of pressure moves the polished steel disc, allowing smooth refrigerant flow. The permanent magnet draws the disc to the closed position when the system pressure balances.

☐ Water-Regulating Valves ☐

On some applications it is desirable to use water-cooled condensing units. In order to save water and cause the unit to operate more efficiently, water-regulating valves are used (Figure 9–38).

On water-cooled condensers, a modulating water-regulating valve is normally used to economize on water usage and to control condensing pressures within reasonable limits. Water valves may be either pressure actuated or temperature actuated. They act to throttle water flow as required by the condenser. Proper sizing of these valves is important because oversizing will cause a fluctuating head pressure, while undersizing will cause a high head pressure.

FIGURE 9–36 Spring-loaded check valve. (Courtesy of Henry Valve Company.)

FLARE CONNECTED

SOLDER CONNECTED

SWEDISH STEEL DISC (IN OPEN POSITION)   BRASS BODY

PERMANENT ALNICO MAGNET

FIGURE 9–37 Magni-check valve. (Courtesy of Watsco, Inc.)

FIGURE 9–38 Pressure-actuated water-regulating valve. (Courtesy of Penn Controls, Inc.)

☐ Safety Procedures ☐

Accessories are devices that are used to cause a refrigeration system to function more efficiently and economically. Care should be exercised when working on these devices to prevent upsetting their value to the system.

1. Always obtain help when lifting heavy objects.
2. Never make adjustments that are not absolutely necessary.
3. Always replace a component rather than completely removing it from the system; removal of a component may reduce the efficiency of the equipment.
4. Never touch components on the compressor discharge line while the compressor is running because they are very hot.
5. Keep hands, feet, and clothing away from belts and revolving or moving parts.
6. Do not touch a compressor crankcase heater; severe burns may result.
7. Never force a service valve because the stem might be broken or the valve might rupture, causing a loss of refrigerant and possible personal injury.
8. Never heat a suction accumulator with an open flame; the working pressure might be exceeded, resulting in an explosion.
9. Never replace a fusible plug in a receiver with a pipe plug.
10. Never put weight on vibration eliminators; they are not designed to withstand excessive sideways force.

☐ SUMMARY ☐

☐ Broken compressor valves, rods, pistons, crankshafts, etc., are usually the result of oil slugging.

☐ Accumulators can prevent slugging by trapping liquid refrigerant in the suction line.

☐ Accumulators are installed in the suction line of the refrigeration system.

☐ Accumulators should be selected for a low refrigerant pressure drop, proper oil return, and adequate volume to store at least 50% of the refrigerant charge.

☐ Liquid line driers are used to remove moisture from the refrigerant.

☐ Modern liquid line driers perform the following functions: (1) moisture removal, (2) acid removal, and (3) filtering.

☐ Selection of liquid line driers should be influenced by: (1) type and amount of refrigerant, (2) refrigeration system tonnage, (3) line size, and (4) allowable pressure drop.

☐ The two types of liquid line driers in use are: (1) straight-through, and (2) angle-replaceable core.

☐ The proper suction line filter will protect the compressor by trapping foreign matter before it enters the compressor.

☐ The following factors influence the selection of suction line filters: (1) type of refrigerant, (2) suction line size, (3) allowable pressure drop, and (4) application.

☐ The two general types of suction line filters are: (1) straight-through, and (2) angle-replaceable core.

☐ The suction line filter drier offers all the advantages of a suction line filter plus the ability to remove acids and moisture.

☐ Strainers remove foreign matter such as dirt, metal chips, etc., from the refrigerant lines.

☐ The three general types of strainers are: (1) straight-through, (2) cleanable angle type, and (3) cleanable Y type.

☐ Moisture-liquid indicators are installed in the liquid line and used to determine the amount of refrigerant and moisture in the refrigeration system.

☐ The oil separator is installed in the discharge line between the compressor and the condenser.

☐ The purpose of an oil separator is to maintain the correct oil level in the compressor.

☐ The prime function of the oil separator is to separate oil from the refrigerant and return it to the compressor before it can enter the other system components.

☐ The oil separator reduces the velocity of the refrigerant and oil, which allows the heavier oil to fall out and collect in the bottom of the oil separator, from where it is fed back into the compressor.

☐ In selecting an oil separator, one thing that should be considered is that the larger the capacity of the compressor, the larger the volume of the shell required.

☐ Any device that brings into contact two substances of different temperatures for the purpose of heating or cooling is called a heat exchanger.

☐ Heat exchangers are used to increase the capacity of refrigeration sytems by bringing the liquid refrigerant into contact with the suction vapor.

☐ Vibration eliminators are used to prevent the transmission of noise and vibration from the compressor through the refrigeration piping.

☐ Discharge mufflers are used to reduce noise transmissions to a minimum, or where compressor pulsations might create vibration problems.

☐ The discharge muffler is located in the discharge line as close to the compressor as possible.

☐ Improper field selection of a discharge muffler will

sometimes lead to increased rather than decreased vibration.

☐ Crankcase heaters are used to prevent the migration of refrigerant to the compressor when installed in a colder location than the evaporator.

☐ Crankcase heaters are low-wattage resistance elements, and may be energized continuously, or on the compressor off cycle, and must be carefully selected to avoid overheating of the compressor oil.

☐ Compressor service valves are used to aid in service operations that involve the refrigerant or the oil.

☐ Solenoid valves are used in refrigeration work to perform the same function as hand shutoff valves.

☐ Two types of solenoid valves are: (1) direct acting, and (2) pilot operated.

☐ Check valves prevent the flow of refrigerant in one direction while allowing free flow in the other direction.

☐ Water-regulating valves are used to control the head pressure and conserve water on some water-cooled condensers.

☐ EXERCISES ☐

1. What does liquid refrigerant cause in a compressor crankcase?
2. What are accumulators designed to do?
3. What would a broken valve in a compressor indicate?
4. Where should an accumulator be installed?
5. How much refrigerant should an accumulator hold?
6. How much moisture can a refrigeration system tolerate?
7. What is the purpose of a drier?
8. What three functions are modern driers capable of performing?
9. What four factors influence the selection of a drier?
10. What are the two general types of liquid line driers?

11. How does a suction line filter protect the compressor?
12. What four factors influence the selection of suction line filters?
13. What are the two general types of suction line filters?
14. Why does the suction line filter-drier offer more advantages than the suction line filter?
15. What is the purpose of a refrigerant strainer?
16. Where is the moisture-liquid indicator installed?
17. State the purpose of a moisture indicator.
18. Can an excessive amount of moisture discolor a moisture indicator?
19. How does an oil separator function?
20. What is the purpose of an oil separator?
21. Why is an oil separator needed on some installations?
22. How are oil separators selected?
23. Define *heat exchanger*.
24. How do heat exchangers work?
25. How does a heat exchanger increase the capacity of a refrigeration system?
26. How are heat exchangers selected?
27. What are vibration eliminators?
28. Where are vibration eliminators installed?
29. What is the purpose of a discharge muffler?
30. What will the improper sizing of a discharge muffler cause?
31. Where is a crankcase heater installed?
32. What two types of crankcase heaters are used in refrigeration work?
33. How do crankcase heaters operate?
34. What is meant when a compressor service valve is said to be back seated?
35. What is the operating force in a solenoid valve?
36. Name two types of solenoid valves.
37. What performance characteristics are to be considered when selecting a solenoid valve?
38. What is the purpose of a check valve?
39. Where are check valves used?
40. What is the purpose of a water-regulating valve?

10 Refrigerants

Performance Objectives

After completion of this chapter, the student should:
- Be able to explain the effects of pressure on the boiling point of a refrigerant.
- Understand the term condensing pressure.
- Understand the importance of the vaporizing pressure of a refrigerant.
- Understand the use of latent heat of vaporization as applied to refrigeration systems.
- Explain refrigerant-oil relationships.
- Discuss how to handle refrigerant cylinders.
- Complete the Exercises with at least 80 percent accuracy.
- Complete the Laboratory Exercises with at least 80 percent accuracy.

The vapors or liquids used in refrigeration systems are known as refrigerants. The evaporation of the refrigerant within the evaporator extracts heat from the surrounding objects. The various parts of the refrigeration unit compress and condense the refrigerant so that it can be used over and over again. Even though there are many different types of refrigerants, only the more common fluorocarbons will be considered here.

Definition. For practical purposes, a *refrigerant* is a fluid that absorbs heat by evaporating at a low temperature and pressure and gives up that heat by condensing at a higher temperature and pressure.

☐ Characteristics of Refrigerants ☐

Most of the commonly used refrigerants exist in a vaporous state under ordinary atmospheric pressures and temperatures. To change these vapors to liquid form, it is necessary to compress and cool them as is done by the condensing unit on a refrigeration system. A fluid is a liquid, gas, or vapor. The words *gas* and *vapor* are ordinarily used interchangeably, although to be perfectly technical, perhaps we should explain that a gas near its condensation point is called a vapor. All fluids have both a liquid and gaseous state. Some fluids have high boiling points, which means that they exist as a gas only when heated to a high temperature or when placed under a vacuum. Fluids that have low boiling points are in the form of vapor at ordinary room temperatures and pressures. Many of the common refrigerants such as those in the Freon group are in this category. If these vapors are to be liquefied, they must be compressed and cooled, or condensed.

Water is a fluid that exists as a liquid at atmospheric pressure and temperature. The boiling point of water under atmospheric pressure at sea level is 212 °F (100 °C). If the water is left in an open basin, it will evaporate very slowly. If heat is applied to the water and its temperature raised to its boiling point, it will then evaporate, or boil, very rapidly. The water will change to the gaseous form of water known as steam or water vapor. If the water is boiled in an open container, its temperature will not rise above 212 °F. All the heat supplied by the flame is used to boil off or vaporize the water.

If a liquid refrigerant is similarly placed in an open container, it will immediately begin to boil vigorously and vaporize, but at a very low temperature. Liquid Refrigerant-12, under atmospheric pressure, will boil at $-21.6°F$ ($-29.7°C$). It will absorb sufficient heat from the container and the surrounding air to enable it to boil. It would not be necessary to heat it with a flame as was done with the water.

A vaporizing refrigerant will absorb heat equal to the amount of energy necessary to change the physical form of the refrigerant from the liquid state to the gaseous state. Each refrigerant will absorb an amount of heat per pound of refrigerant equal to its latent heat of vaporization.

Applications of Different Refrigerants. In larger industrial refrigeration systems such as cold storage warehouses and ice manufacturing plants, ammonia is used almost exclusively. Under conditions where the odor of ammonia gas would be undesirable or where there are certain dangers from fire or explosion as on ships, carbon dioxide is used. Both ammonia and carbon dioxide require high condenser pressures and are therefore not suitable for use in household compression-type units. Ammonia is sometimes used in absorption units, but no compressor is used in this type of system.

The units for use in household systems and small commercial units are designed to use either R-12 or R-22. Some of the smaller commercial units are being designed to use R-500 or R-502. R-11 is used extensively in large centrifugal air conditioning units. These refrigerants are safe for ordinary use. They do not require high pressures and are quite suitable in many other ways.

☐ Effects of Pressure on Boiling Point ☐

The boiling point of any liquid may be raised or lowered in accordance with the amount of pressure applied to the liquid. The greater the pressure, the higher the boiling point; the lower the pressure, the lower the boiling point. Thus, a liquid can be caused to boil at a low temperature by placing it in a partial vacuum.

Certain refrigerants require high pressure when used in a refrigeration unit. R-112 (Freon-112) is an example of such a refrigerant (see Table 10-1). The energy consumed by a refrigeration unit using R-112 is required chiefly on the compression side for compressing the vaporous refrigerant. R-14 and R-502 are examples of refrigerants that do not require such high pressures. R-12, R-22, R-500, and R-502 are used in most household and smaller commercial refrigeration units.

TABLE 10-1 Boiling Point of Freon Fluorocarbon Refrigerants

Product	Formula	Molecular Weight	Boiling Point [°F (°C)]
Freon-14	CF_4	88.0	-198.4 $[-128.0]$
Freon-23	CHF_3	70.0	-115.7 $[-82.1]$
Freon-13	$CClF_3$	104.5	-114.6 $[-81.4]$
Freon-116	CF_3-CF_3	138.0	-108.8 $[-78.2]$
Freon-13B1	$CBrF_3$	148.9	-72.0 $[-57.8]$
Freon-502	$CHClF_2/CClF_2-CF_3$ [48.8/51.2% by weight]	121.2	-50.1 $[-45.6]$
Freon-22	$CHClF_2$	86.5	-41.4 $[-40.8]$
Freon-115	$CClF_2-CF_3$	154.5	-37.7 $[-38.7]$
Freon-12	CCl_2F_2	120.9	-21.6 $[-29.8]$
Freon-C318	C_4F_8 [cyclic]	200.0	21.5 $[-5.8]$
Freon-114	$CClF_2-CClF_2$	170.9	38.4 [3.6]
Freon-21	$CHCl_2F$	102.9	48.1 [8.9]
Freon-11	CCl_3F	137.4	74.8 [23.8]
Freon-114B2	$CBrF_2-CBrF_2$	259.9	117.5 [47.5]
Freon-113	$CCl_2F-CClF_2$	187.4	117.6 [47.6]
Freon-112	CCl_2F-CCl_2F	203.9	199.0 [98.2]

Source: Freon Products Division, E.I. DuPont de Nemours & Co., Inc.

☐ Critical Temperature ☐

The critical temperature of a gas is the temperature above which the gas cannot be liquefied regardless of the amount of pressure applied. If the vapor is heated sufficiently to raise its temperature above its critical temperature, the vibrations of the molecules in the vapor become so intense that pressure cannot bring them into sufficient contact with each other to exist as a liquid.

Refrigerants used in refrigeration units change from the liquid state to the vaporous state and back again while going through the refrigeration cycle. For this reason, a refrigerant must be used below its critical temperature in order to obtain the liquefaction phase of the refrigeration cycle (see Table 10–2).

When the temperature of the vapor is reduced, the amount of pressure required to effect liquefaction decreases. From this we can deduce the fact that for every temperature below the critical temperature, there is a corresponding pressure that will cause liquefaction.

There are graphs and tables that indicate the relationship between a temperature and the corresponding pressure that will cause liquefaction of the refrigerant vapor (Figure 10–1). The temperature in degrees Fahrenheit is indicated on the bottom. The temperature in degrees Celsius is indicated on the top of the chart. Absolute pressure (psia) is on the left, while the gauge pressures (psig) and inches of mercury vacuum are on the right side of the chart. The metric equivalent values are not listed because the United States has not yet confirmed which of the SI units will be used. For example, if we have R-12 at 80 °F, what will its saturation pressure be? Find 80 °F on the bottom line of the chart. Follow this line vertically, until the curve for R-12 is intersected. Read to the left for absolute pressure and

TABLE 10–2 Critical Temperatures of Fluorocarbon Refrigerants

Refrigerant	Critical Temperature [°F [°C]]
R-11	388.4 [198.0]
R-12	233.6 [122.0]
R-22	204.8 [96.0]
R-502	194.1 [90.1]

to the right for gauge pressure. Find R-12 at this temperature to have 99.2 psia and 84.2 psig. This is the pressure required to liquefy the refrigerant at 80 °F temperature.

☐ Standard Conditions ☐

The capacity of any refrigeration unit will vary as the refrigerant temperatures vary on the high- and low-pressure sides of the system. The latent heat of the refrigerant, its condensing pressure, and its vaporizing pressure will also vary as the temperature of the refrigerant varies. For purposes of comparing different refrigerants and refrigeration units, certain standards have been developed. The refrigeration industry has set forth conditions known as *standard conditions*. These conditions are established with the following temperatures at various locations in the refrigeration cycle: a temperature in the evaporator of 5 °F (− 15 °C); a temperature in the saturated portion of the condenser of 86 °F (30 °C); a liquid temperature at the flow control device of 77 °F (25 °C); and a suction gas temperature of 14 °F (− 10 °C).

Now if we use the following factors in comparing any two refrigerants, with these temperatures as a basis, we have a true comparison and can arrive at correct conclusions.

Condensing Pressure. The condensing pressure will depend on the temperature at which the vapor will liquefy. In refrigeration work, it is desirable to avoid high condensing pressures if at all possible. Therefore, the condensing medium (air or water) must be as cool as possible. Ordinarily, a water-cooled condenser will operate at a lower condensing temperature and pressure than an air-cooled condenser. Because of this, there is some difference in the operating pressure for these two types of condensers.

In general, it may be assumed that the condensing temperature and pressure of an air-cooled unit will be approximately 25 to 35 °F (14 to 19.6 °C) higher than ambient temperatures. The actual temperature and pressure will, however, depend on the efficiency of the condenser itself, the location of the condenser, whether or not sufficient air circulation is obtained, and the cleanliness of the condenser surface. When water-cooled condensers are

FIGURE 10-1 Pressure-temperature relationships of Freon compounds.

used, the condenser temperature is generally lower than the ambient temperature. Therefore, the condensing pressure of a water-cooled condenser is lower than the condensing pressure of an air-cooled condenser.

The saturation pressures shown in Figure 10-1 should not be considered as operating head pressures for a refrigeration unit. The operating head pressures will be higher than the saturation pressures because of the addition of the heat of compression. If condensation is to take place in the condenser, the temperature of the refrigerant inside must be higher than the temperature of the cooling medium used for the condenser.

Vaporizing Pressure. The vaporizing pressure of a refrigerant is important because the refrigerant must evaporate without requiring too low a suction pressure. A temperature of 5°F (−15°C) is considered the temperature of most domestic refrigerator evaporators. This is the same temperature as that set for the standard conditions used for comparison of different refrigerants and refrigeration units. In general, a refrigerant is desired that has an evaporating pressure at or near atmospheric pressure (see Table 10-3). A refrigerant that requires a vacuum to produce evaporation is not practical under ordinary conditions because of the tendency for air to leak into the system. The air will not condense and will cause the condenser pressure to become very high. This high pressure will reduce the efficiency of the refrigeration unit. Refrigerants that have a vaporizing pressure above atmospheric pressure do not allow air to be drawn into the system through a leak.

It should be noted that the pressure in the evaporator and the low side of the system will be the same. Also, that the temperature of the evaporating refrigerant will correspond to the temperature shown on the pressure–temperature chart, the pressure in the evaporator, or the pressure in the low side of the system (see Table 10-4).

Latent Heat of Vaporization. The amount of heat in Btu (cal) required to change a liquid to a vapor, where the change takes place at a constant temperature, is known as the latent heat of vaporization. This definition can now be developed so as to apply to 1 lb of refrigerant, where the vaporization takes place at atmospheric pressure and the liquid is at a temperature equal to its boiling point when the operation begins. Thus the latent heat of vaporization of a liquid is the amount of heat in Btu (cal) required to vaporize 1 lb of the liquid at atmospheric pressure, with the liquid at its boiling point when the operation begins. To convert 1 lb of water at 212°F (100°C) into steam at the same temperature and pressure, the water must absorb 970 Btu/lb (533.5 kcal/kg). This quantity of heat is the total latent heat of 1 lb of water at atmospheric pressure.

Any refrigerant, when evaporating in the evaporator, must absorb heat from within the cooled space exactly equal to its latent heat of vaporization. When a refrigerant has a high latent heat, it will absorb more heat per pound of liquid than a refrigerant with a lower latent heat of vaporization. Thus if a refrigerant with a high latent heat of vaporization is used, a smaller compressor, condenser, and evaporator can be used (see Table 10-5).

The latent heat of vaporization of a liquid will vary with the pressure and the corresponding temperature at which the vaporization occurs. When lower temperatures and pressures are encountered, the latent heat of vaporization increases.

□ Types of Refrigerants □

There are many types of commercially available refrigerants. Those used in early refrigeration units were ammonia, sulfur dioxide, methyl chloride, propane, and ethane. Ammonia is still used in large refrigeration equipment; however, all the others have been discontinued due to the fact that they are either toxic, dangerous, or have other undesirable characteristics. Specialized refrigerants have been developed for ultra low temperature applications or for large centrifugal compressors. For normal commercial refrigeration and air conditioning systems which use reciprocating compressors, R-12, R-22,

TABLE 10-3 Vaporizing Pressures [in. Hg (kPa)] of Refrigerants at 5°F (−15°C)

Refrigerant	Pressure
R-11	−24 [41.22]
R-12	11.8 [182.05]
R-22	28.1 [294.04]
R-500	16.4 [213.66]
R-502	36 [348.31]

TABLE 10–4 Pressure–Temperature Relation Chart[a]

Temperature (°F)	Refrigerant R-12	R-22	R-500	R-502	R-717
−60	*19.0*	*12.0*	*17.0*	*7.2*	*18.6*
−55	*17.3*	*9.2*	*15.0*	*8.8*	*16.6*
−50	*15.4*	*6.2*	*12.8*	*8.2*	*14.3*
−45	*13.3*	*2.7*	*10.4*	*1.9*	*11.7*
−40	*11.0*	**0.5**	*7.6*	**4.1**	*8.7*
−35	*8.4*	**2.6**	*4.6*	**6.5**	*5.4*
−30	*5.5*	**4.9**	*1.2*	**9.2**	*1.6*
−25	*2.3*	**7.4**	**1.2**	**12.1**	**1.3**
−20	**0.6**	**10.1**	**3.2**	**15.3**	**3.6**
−18	**1.3**	**11.3**	**4.1**	**16.7**	**4.6**
−16	**2.0**	**12.5**	**5.0**	**18.1**	**5.6**
−14	**2.8**	**13.8**	**5.9**	**19.5**	**6.7**
−12	**3.6**	**15.1**	**6.8**	**21.0**	**7.9**
−10	**4.5**	**16.5**	**7.8**	**22.6**	**9.0**
−8	**5.4**	**17.9**	**8.8**	**24.2**	**10.3**
−6	**6.3**	**19.3**	**9.9**	**25.8**	**11.6**
−4	**7.2**	**20.8**	**11.0**	**27.5**	**12.9**
−2	**8.2**	**22.4**	**12.1**	**29.3**	**14.3**
0	**9.2**	**24.0**	**13.3**	**31.1**	**15.7**
1	**9.7**	**24.8**	**13.9**	**32.0**	**16.5**
2	**10.2**	**25.6**	**14.5**	**32.9**	**17.2**
3	**10.7**	**26.4**	**15.1**	**33.9**	**18.0**
4	**11.2**	**27.3**	**15.7**	**34.9**	**18.8**
5	**11.8**	**28.2**	**16.4**	**35.8**	**19.6**
6	**12.3**	**29.1**	**17.0**	**36.8**	**20.4**
7	**12.9**	**30.0**	**17.7**	**37.9**	**21.2**
8	**13.5**	**30.9**	**18.4**	**38.9**	**22.1**
9	**14.0**	**31.8**	**19.0**	**39.9**	**22.9**
10	**14.6**	**32.8**	**19.7**	**41.0**	**23.8**
11	**15.2**	**33.7**	**20.4**	**42.1**	**24.7**

Temperature (°F)	Refrigerant R-12	R-22	R-500	R-502	R-717
12	**15.8**	**34.7**	**21.2**	**43.2**	**25.6**
13	**16.4**	**35.7**	**21.9**	**44.3**	**26.5**
14	**17.1**	**36.7**	**22.6**	**45.4**	**27.5**
15	**17.7**	**37.7**	**23.4**	**46.5**	**28.4**
16	**18.4**	**38.7**	**24.1**	**47.7**	**29.4**
17	**19.0**	**39.8**	**24.9**	**48.8**	**30.4**
18	**19.7**	**40.8**	**25.7**	**50.0**	**31.4**
19	**20.4**	**41.9**	**26.5**	**51.2**	**32.5**
20	**21.0**	**43.0**	**27.3**	**52.4**	**33.5**
21	**21.7**	**44.1**	**28.1**	**53.7**	**34.6**
22	**22.4**	**45.3**	**28.9**	**54.9**	**35.7**
23	**23.2**	**46.4**	**29.8**	**56.2**	**36.8**
24	**23.9**	**47.6**	**30.6**	**57.5**	**37.9**
25	**24.6**	**48.8**	**31.5**	**58.8**	**39.0**
26	**25.4**	**49.9**	**32.4**	**60.1**	**40.2**
27	**26.1**	**51.2**	**33.2**	**61.5**	**41.4**
28	**26.9**	**52.4**	**34.2**	**62.8**	**42.6**
29	**27.7**	**53.6**	**35.1**	**64.2**	**43.8**
30	**28.4**	**54.9**	**36.0**	**65.6**	**45.0**
31	**29.2**	**56.2**	**36.9**	**67.0**	**46.3**
32	**30.1**	**57.5**	**37.9**	**68.4**	**47.6**
33	**30.9**	**58.8**	**38.9**	**69.9**	**48.9**
34	**31.7**	**60.1**	**39.9**	**71.3**	**50.2**
35	**32.6**	**61.5**	**40.9**	**72.8**	**51.6**
36	**33.4**	**62.8**	**41.9**	**74.3**	**52.9**
37	**34.3**	**64.2**	**42.9**	**75.8**	**54.3**
38	**35.2**	**65.6**	**43.9**	**77.4**	**55.7**
39	**36.1**	**67.1**	**45.0**	**79.0**	**57.2**
40	**37.0**	**68.5**	**46.1**	**80.5**	**58.6**
41	**37.9**	**70.0**	**47.1**	**82.1**	**60.1**

Temperature (°F)	Refrigerant R-12	R-22	R-500	R-502	R-717
42	**38.8**	**71.4**	**48.2**	**83.8**	**61.6**
43	**39.8**	**73.0**	**49.4**	**85.4**	**63.1**
44	**40.7**	**74.5**	**50.5**	**87.0**	**64.7**
45	**41.7**	**76.0**	**51.6**	**88.7**	**66.3**
46	**42.6**	**77.6**	**52.8**	**90.4**	**67.9**
47	**43.6**	**79.2**	**54.0**	**92.1**	**69.5**
48	**44.6**	**80.8**	**55.1**	**93.9**	**71.1**
49	**45.7**	**82.4**	**56.3**	**95.6**	**72.8**
50	**46.7**	**84.0**	**57.6**	**97.4**	**74.5**
55	**52.0**	**92.6**	**63.9**	**106.6**	**83.4**
60	**57.7**	**101.6**	**70.6**	**116.4**	**92.9**
65	**63.8**	**111.2**	**77.8**	**126.7**	**103.1**
70	**70.2**	**121.4**	**85.4**	**137.6**	**114.1**
75	**77.0**	**132.2**	**93.5**	**149.1**	**125.8**
80	**84.2**	**143.6**	**102.0**	**161.2**	**138.3**
85	**91.8**	**155.7**	**111.0**	**174.0**	**151.7**
90	**99.8**	**168.4**	**120.6**	**187.4**	**165.9**
95	**108.2**	**181.8**	**130.6**	**201.4**	**181.1**
100	**117.2**	**195.9**	**141.2**	**216.2**	**197.2**
105	**126.6**	**210.8**	**152.4**	**231.7**	**214.2**
110	**136.4**	**226.4**	**164.1**	**247.9**	**232.3**
115	**146.8**	**242.7**	**176.5**	**264.9**	**251.5**
120	**157.6**	**259.9**	**189.4**	**282.7**	**271.7**
125	**169.1**	**277.9**	**203.0**	**301.4**	**293.1**
130	**181.0**	**296.8**	**217.2**	**320.8**	
135	**193.5**	**316.6**	**232.1**	**341.2**	
140	**206.6**	**337.2**	**247.7**	**362.6**	
145	**220.3**	**358.9**	**264.0**	**385.0**	
150	**234.6**	**381.5**	**281.1**	**408.4**	
155	**249.5**	**405.1**	**298.9**	**432.9**	

Source: Sporlan Valve Company.

[a]Italic figures, in. Hg, vacuum; bold figures, psig.

TABLE 10-5 Latent Heat of Vaporization [Btu/lb [kcal/kg]] at 5°F [–15°C]

Refrigerant	Latent Heat
R-11	83.459 [45.902]
R-12	68.204 [37.512]
R-22	93.206 [51.263]
R-500	82.45 [45.34]
R-502 at 40°F [4.4°C]	63.1 [34.7]

R-500, and R-502 are presently used almost exclusively. R-11 is used in centrifugal compressors.

R-11 Trichlorofluoromethane (CCL_3F).

This refrigerant is a synthetic chemical product that is used as a refrigerant. It is considered stable, nonflammable, and has a very low toxicity rating. It operates with a low-side pressure of 24 in. of vacuum (20.61 kPa) at 5°F (–15°C), and a high-side pressure of 18.3 psia (125.72 kPa) at 86°F (30°C). It has a latent heat value at 5°F of 83.459 Btu/lb (45.9 kcal/kg) (see Table 10-6). The boiling point at atmospheric pressure is 74.87°F (23°C).

R-11 may be used as a cleaning agent to aid in cleaning systems that have been contaminated by motor compressor burnout or moisture. This process is very expensive and should be used only as a last resort. A thorough evacuation and replacement of driers in conjunction with oil sample testing is recommended to clean the contaminated systems. Leak detection is accomplished by use of a soap solution, a halide torch, or an electronic leak detector. The cylinder color code for R-11 is orange.

R-12 Dichlorodifluoromethane (CCL_2F_2).

R-12 is a refrigerant used in nearly all facets of refrigeration and air conditioning. It has a boiling point of –21.6°F (–30°C) at atmospheric pressure. Therefore, it exists as a vapor at ordinary room temperatures and atmospheric pressure. R-12 will liquefy under a pressure of 76 psi (623.11 kPa) at 75°F (23.9°C) (see Table 10-7).

R-12 has very little odor, but in large concentrations a faint, sweet odor may be detected. Its critical temperature is 233.6°F (112°C). It is a colorless gas and liquid. It is nontoxic, nonflammable, and nonirritating, and is noncorrosive to any of the ordinary metals even in the presence of water. It is stable under all conditions and temperatures normally encountered in refrigeration. R-12 liquid will dissolve lubricating oil in all proportions; the oil in return will absorb refrigerant vapor. There is no separation or oil blanket formed in the evaporator to interfere with proper vaporization.

R-12 is only slightly soluble in water. It is this characteristic that prevents corrosion of metals in the presence of water. Leaks may be detected with a halide torch, an electronic leak detector, or a soap-

TABLE 10-6 R-11 Refrigerant Standard Ton Characteristics

Evaporator pressure at 5°F [psia]	2.9373
Condenser pressure at 86°F [psia]	18.186
Compression ratio [86°F/5°F]	6.19
Latent heat of vaporization at 5°F [Btu/lb]	83.459
Net refrigerating effect [Btu/lb]	66.796
Refrigerant circulated per ton of refrigeration [lb/min]	2.9942
Saturated liquid volume at 86°F [ft³/lb]	0.010942
Liquid circulated per ton of refrigeration [in.³/min]	56.614
Saturated vapor density at 5°F [lb/ft³]	0.081933
Saturated vapor density at 86°F [lb/ft³]	0.44668
Compressor displacement per ton of refrigeration [ft³/min]	36.544
Refrigeration per cubic foot of compressor displacement [Btu]	5.473
Heat of compression [Btu/lb]	13.292
Temperature of compressor discharge [°F]	110.9
Coefficient of performance	5.025
Horsepower per ton	0.9383

Source: Freon Products Division, E.I. DuPont de Nemours & Co., Inc.

TABLE 10-7 R-12 Refrigerant Standard Ton Characteristics

Evaporator pressure at 5°F [psia]	26.483
Condenser pressure at 86°F [psia]	108.04
Compression ratio [86°F/5°F]	4.08
Latent heat of vaporization at 5°F [Btu/lb]	68.204
Net refrigerating effect [Btu/lb]	50.035
Refrigerant circulated per ton of refrigeration [lb/min]	3.9972
Saturated liquid volume at 86°F [ft³/lb]	0.012396
Liquid circulated per ton of refrigeration [in.³/min]	85.621
Saturated vapor density at 5°F [lb/ft³]	0.68588
Saturated vapor density at 86°F [lb/ft³]	2.6556
Compressor displacement per ton of refrigeration [ft³/min]	5.8279
Refrigeration per cubic foot of compressor displacement [Btu]	34.318
Heat of compression [Btu/lb]	10.636
Temperature of compressor discharge [°F]	100.84
Coefficient of performance	4.704
Horsepower per ton	1.0023

Source: Freon Products Division, E.I. DuPont de Nemours & Co., Inc.

bubble solution. The halide torch imparts a blue-green tinge to the flame in the presence of R-12. The cylinder color code for R-12 is white.

R-22 Monochlorodifluoromethane (CHCLF₂).

R-22 has been developed for refrigeration installations that have a low evaporating temperature. In most physical characteristics it is similar to R-12. However, its saturation pressures are higher than those for R-12 for the same temperatures. It has a latent heat of vaporization at 5 °F (−15 °C) of 93.21 Btu/lb (51.26 kcal/kg). It has a lower specific volume than R-12 and, therefore, a greater refrigeration capacity for a given volume of saturated refrigerant vapor. The boiling point is −41.36 °F (−41 °C) at atmospheric pressure. R-22 has an operating head pressure at 86 °F (30 °C) of 172.8 psia, or 158 psi (1186.45 kPa) (see Table 10–8). Its evaporator pressure at 5 °F is 42.88 psia, or 28 psi (294.58 kPa). R-22 is nontoxic, nonflammable, noncorrosive, and nonirritating. It is a stable refrigerant under normal operating conditions. Because of its low evaporating temperatures and high compression ratios, however, the temperature and compressed R-22 vapor may become so high that damage to the compressor could result under ultralow-temperature applications.

Where size of equipment and economy are a factor, such as in packaged air conditioning units, R-22

TABLE 10–8 R-22 Refrigerant Standard Ton Characteristics

Evaporator pressure at 5 °F [psia]	42.888
Condenser pressure at 86 °F [psia]	172.87
Compression ratio [86 °F/5 °F]	4.03
Latent heat of vaporization at 5 °F [Btu/lb]	93.206
Net refrigerating effect [Btu/lb]	70.027
Refrigerant circulated per ton of refrigeration [lb/min]	2.8560
Saturated liquid volume at 86 °F [ft³/lb]	0.013647
Liquid circulated per ton of refrigeration [in.³/min]	67.351
Saturated vapor density at 5 °F [lb/ft³]	0.80422
Saturated vapor density at 86 °F [lb/ft³]	3.1622
Compressor displacement per ton of refrigeration [ft³/min]	3.5512
Refrigeration per cubic foot of compressor displacement [Btu]	56.32
Heat of compression [Btu/lb]	15.022
Temperature of compressor discharge [°F]	128.4
Coefficient of performance	4.662
Horsepower per ton	1.0114

Source: Freon Products Division, E.I. DuPont de Nemours & Co., Inc.

is commonly used. If the evaporating temperature of R-22 of −40 °F (−40 °C) is reached, the oil will begin to separate from the refrigerant and form a film on the surface that will interfere with proper evaporation. Leaks can be detected with a halide torch, soap-bubble solution, or with an electronic leak detector. The cylinder color code for R-22 is green.

R-500 Refrigerant (CCL₂F₂/CH₃CHF₂).

This refrigerant is an azeotropic mixture of 26.2% R-152a and 73.8% R-12. An azeotrope is the scientific name given to a specific mixture of different compounds in which the resulting mixture has different characteristics than either of the components, and that can evaporate and condense without a change in its composition. The vapor–pressure temperature curve is fairly constant and is different from either of the compounds used in making it. For the same size compressor, it offers approximately 20% more refrigeration capacity than R-12. Refrigerant-500 has an evaporator pressure of 31.21 psia, or 19.7 psi (214.41 kPa) at 5 °F (−15 °C). Its condensing pressure is 127.6 psia, or 113 psi (876.61 kPa) at 86 °F (30 °C). The boiling point is −28 °F (−33 °C) at atmospheric pressure, and its latent heat value is 82.45 Btu/lb (45.35 kcal/kg) at 5 °F.

R-500 is used in both commercial and industrial applications and only with reciprocating compressors. It also can be used when greater refrigeration capacity is required than that possible with R-12.

It is fairly soluble with oil and its solubility with water is very critical. Therefore, proper dehydration procedures and the use of driers is strongly recommended. Leak detection is accomplished with a halide leak detector, soap-bubble solution, or an electronic leak detector. The cylinder color code for R-500 is yellow.

R-502 Refrigerant (CHCLF₂/CCLF₂CF₃).

R-502 is an azeotropic mixture of 48.8% R-22 and 51.2% of R-115. It is similar to R-12 and R-22 in most physical characteristics. The latent heat of vaporization is 72.5 Btu/lb (39.87 kcal/kg). Its refrigerating capacity is comparable to that of R-22, except at lower temperatures it will usually be greater. The boiling point is −49.76 °F (−46 °C) at atmospheric pressure.

The evaporator pressure of R-502 is 30 psia, 15 psi (206.1 kPa) at −20 °F (−28.9 °C) evaporating temperature. The condenser pressure is 297.4 psia, or 282.4 psi (2043.14 kPa) at a condensing temperature of 120 °F (49 °C) (see Table 10–9).

TABLE 10-9 R-502 Refrigerant Standard Ton Characteristics

Evaporating temperature [°F]	20
Return gas temperature [°F]	65
Condensing temperature [°F]	120
Evaporator pressure [psia]	30.01
Condenser pressure [psia]	297.4
Compression ratio	9.91
Net refrigerating effect [Btu/lb][a]	45.26
Refrigerant circulated per ton [lb/min]	4.418
Saturated liquid volume at 120°F [ft³/lb]	0.01472
Liquid circulated per ton [in.³/min]	112.4
Vapor density at 65°F [lb/ft³][b]	0.6165
Compressor displacement per ton [ft³/min]	7.167
Refrigeration capacity [Btu/ft³]	27.90
Heat of compression [Btu/lb]	22.52
Temperature of compressor discharge [°F]	226.0
Coefficient of performance	2.010
Horsepower per ton	2.346

Source: Freon Products Division, E.I. DuPont de Nemours & Co., Inc.

[a]Enthalpy of vapor at 65°F and evaporator pressure—enthalpy of liquid at 120°F.

[b]It is assumed that vapor enters the compressor cylinder at a temperature of 65°F.

As with R-22, a compressor with a lower displacement may be used for performance equal to R-12. Because of the excellent low-temperature characteristics, R-502 is well suited for low-temperature refrigeration applications. R-502 is recommended for all single-stage installations with an evaporating temperature of 0°F (−17.8°C) or below. It also is very satisfactory for use in two-stage systems providing they are extra low temperature applications, and is becoming very popular for use in the medium temperature range.

It is nonflammable, noncorrosive, nontoxic, and stable under all normal situations. Leak detection is accomplished by use of halide leak detectors, soap-bubble solution, or electronic leak detectors. Oil return is not a problem with this refrigerant down to −40°F (−40°C) evaporator temperatures. However, oil separators are sometimes used when lower temperatures are necessary. The cylinder color code for R-502 is orchid.

☐ Refrigerant-Oil Relationships ☐

When reciprocating compressors are used, oil and refrigerant mix continuously. Oils used in refrigeration are soluble in liquid refrigerants, and at room temperatures will mix completely. *Miscibility* is the term used to express the ability of liquid refrigerant to mix with oil.

Any oil circulating in a refrigeration system is exposed to both very high and very low temperatures. Because of the critical nature of lubrication under these extreme conditions, and the damage that can be done to the system by wax or other impurities that may be in the oil, only highly refined oils specifically prepared for use in refrigeration systems can be used.

In general, naphthenic oils are more soluble in refrigerants than paraffinic oils. Separation of the oil and refrigerant into separate layers can take place with either type of oil. However, naphthenic oils separate at somewhat lower temperatures. This separation does not necessarily affect the lubricating properties of the oil, but it could cause problems in properly supplying the oil to the working parts.

Because oil must pass through the compressor cylinders to provide lubrication of moving parts, a small amount of oil is always circulating with the refrigerant. Oil and refrigerant vapor do not mix readily. Therefore, oil can be properly circulated through the system only if the refrigerant vapor velocities are kept high enough to sweep the oil along the piping. If the velocities are not kept high enough, the oil will tend to lie on the bottom of the refrigeration tubing, which will decrease the heat transfer and, perhaps, cause a shortage of oil in the compressor. As the refrigerant evaporating temperatures become lower, the oil separation becomes more critical. The viscosity of the oil increases with a decrease in temperature. Therefore, the proper design of refrigerant piping is necessary for proper oil return.

One of the basic characteristics of a refrigerant and oil mixture inside a sealed system is that liquid refrigerant is attracted by oil. This liquid refrigerant will vaporize and migrate through the system to the compressor crankcase, even though no pressure differential exists to cause the movement. When the refrigerant reaches the compressor crankcase, it will condense back into a liquid. This refrigerant migration will continue until the compressor oil is saturated with liquid refrigerant.

The excess of liquid refrigerant in the compressor crankcase will result in violent foaming and boiling action of the oil. This may cause all of the oil to leave the compressor crankcase, and may, in turn, result in lubrication problems. Therefore, some provisions, must be made to prevent the accumulation of excess liquid refrigerant in the compressor crankcase. One such provision is a crankcase heater.

The refrigerants R-22 and R-502 are much less soluble in oil than is R-12. Therefore, the proper piping and system design for these two refrigerants is more critical with regard to oil return.

□ Refrigerant Tables □

In order to accurately determine the operating performance of a refrigeration system, very precise and accurate information is required concerning the various properties of refrigerants at any pressure and temperature that may be considered. Refrigerant manufacturers have calculated and compiled the data in the form of tables listing the thermodynamic properties of each refrigerant.

An excerpt from an R-12 saturation table, which lists five major saturation properties of this refrigerant is shown in Table 10–10. Pressure, volume, and density were previously discussed.

Enthalpy is a term used in thermodynamics to describe the heat content of a substance. When discussing refrigerants, enthalpy is expressed in terms of Btu/lb of refrigerant. An arbitrary base of saturated liquid at −40°F (−40°C) has been accepted as the standard zero value. More simply stated, the enthalpy of any liquid refrigerant is zero for liquid at −40°F. Liquid refrigerant below that temperature is considered to have a negative enthalpy, while at all temperatures above −40°F the refrigerant has a positive enthalpy value.

The difference in enthalpy values in different parts of the refrigeration system is commonly used to rate the performance of a refrigeration unit. If the heat content per pound of a refrigerant entering and leaving a cooling coil can be determined, and the flow rate of the refrigerant is known, the cooling capabilities of that coil can be calculated.

Entropy (see Table 10–10) is best described as a mathematical ratio used in thermodynamics. Thermodynamics is used in the solving of complex refrigeration engineering problems, but is seldom used in commercial refrigeration and air conditioning applications. Therefore, a discussion of it is not pertinent in this book.

□ Pocket Pressure-Temperature Charts □

Small pocket-sized tables that list the saturation temperatures and pressures of common refrigerants are available from expansion valve manufacturers, refrigerant manufacturers, and refrigeration supply stores (see Table 10–10).

These charts are a ready reference and an invaluable tool for the service technician. They are used to check the performance of a refrigeration system. The suction and discharge pressures can be readily checked by using gauges. These pressures indicate the evaporating and condensing temperatures of the refrigerant.

□ Handling of Refrigerant Cylinders □

The pressure created by a liquid refrigerant in a sealed container is equal to its saturation pressure at that liquid temperature as long as there is space above the liquid for the vapor. However, if the refrigerant cylinder is overfilled, or if the cylinder is gradually and uniformly overheated, the liquid refrigerant will expand until the cylinder becomes full of liquid. When this occurs, hydrostatic pressure builds up rapidly, producing pressures well above saturation pressures. Figure 10–2 illustrates the

FIGURE 10–2 Hydrostatic pressure with R-12. [Courtesy of Copeland Refrigeration Corp.]

TABLE 10-10 Saturation Properties—Temperature Table of R-12

TEMP. °F	PRESSURE		VOLUME cu ft/lb		DENSITY lb/cu ft		ENTHALPY Btu/lb			ENTROPY Btu/(lb)(R)		TEMP. °F
	PSIA	PSIG	LIQUID v_f	VAPOR v_g	LIQUID $1/v_f$	VAPOR $1/v_g$	LIQUID h_f	LATENT h_{fg}	VAPOR h_g	LIQUID s_f	VAPOR s_g	
−40	9.3076	10.9709*	0.010564	3.8750	94.661	0.25806	0	72.913	72.913	0	0.17373	−40
−39	9.5530	10.4712*	0.010575	3.7823	94.565	0.26439	0.2107	72.812	73.023	0.000500	0.17357	−39
−38	9.8035	9.9611*	0.010586	3.6922	94.469	0.27084	0.4215	72.712	73.134	0.001000	0.17343	−38
−37	10.059	9.441*	0.010596	3.6047	94.372	0.27741	0.6324	72.611	73.243	0.001498	0.17328	−37
−36	10.320	8.909*	0.010607	3.5198	94.275	0.28411	0.8434	72.511	73.354	0.001995	0.17313	−36
−35	10.586	8.367*	0.010618	3.4373	94.178	0.29093	1.0546	72.409	73.464	0.002492	0.17299	−35
−34	10.858	7.814*	0.010629	3.3571	94.081	0.29788	1.2659	72.309	73.575	0.002988	0.17285	−34
−33	11.135	7.250*	0.010640	3.2792	93.983	0.30495	1.4772	72.208	73.685	0.003482	0.17271	−33
−32	11.417	6.675*	0.010651	3.2035	93.886	0.31216	1.6887	72.106	73.795	0.003976	0.17257	−32
−31	11.706	6.088*	0.010662	3.1300	93.788	0.31949	1.9003	72.004	73.904	0.004469	0.17243	−31
−30	11.999	5.490*	0.010674	3.0585	93.690	0.32696	2.1120	71.903	74.015	0.004961	0.17229	−30
−29	12.299	4.880*	0.010685	2.9890	93.592	0.33457	2.3239	71.801	74.125	0.005452	0.17216	−29
−28	12.604	4.259*	0.010696	2.9214	93.493	0.34231	2.5358	71.698	74.234	0.005942	0.17203	−28
−27	12.916	3.625*	0.010707	2.8556	93.395	0.35018	2.7479	71.596	74.344	0.006431	0.17189	−27
−26	13.233	2.979*	0.010719	2.7917	93.296	0.35820	2.9601	71.494	74.454	0.006919	0.17177	−26
−25	13.556	2.320*	0.010730	2.7295	93.197	0.36636	3.1724	71.391	74.563	0.007407	0.17164	−25
−24	13.886	1.649*	0.010741	2.6691	93.098	0.37466	3.3848	71.288	74.673	0.007894	0.17151	−24
−23	14.222	0.966*	0.010753	2.6102	92.999	0.38311	3.5973	71.185	74.782	0.008379	0.17139	−23
−22	14.564	0.270*	0.010764	2.5529	92.899	0.39171	3.8100	71.081	74.891	0.008864	0.17126	−22
−21	14.912	0.216	0.010776	2.4972	92.799	0.40045	4.0228	70.978	75.001	0.009348	0.17114	−21
−20	15.267	0.571	0.010788	2.4429	92.699	0.40934	4.2357	70.874	75.110	0.009831	0.17102	−20
−19	15.628	0.932	0.010799	2.3901	92.599	0.41839	4.4487	70.770	75.219	0.010314	0.17090	−19
−18	15.996	1.300	0.010811	2.3387	92.499	0.42758	4.6618	70.666	75.328	0.010795	0.17078	−18
−17	16.371	1.675	0.010823	2.2886	92.399	0.43694	4.8751	70.561	75.436	0.011276	0.17066	−17
−16	16.753	2.057	0.010834	2.2399	92.298	0.44645	5.0885	70.456	75.545	0.011755	0.17055	−16
−15	17.141	2.445	0.010846	2.1924	92.197	0.45612	5.3020	70.352	75.654	0.012234	0.17043	−15
−14	17.536	2.840	0.010858	2.1461	92.096	0.46595	5.5157	70.246	75.762	0.012712	0.17032	−14
−13	17.939	3.243	0.010870	2.1011	91.995	0.47595	5.7295	70.141	75.871	0.013190	0.17021	−13
−12	18.348	3.652	0.010882	2.0572	91.893	0.48611	5.9434	70.036	75.979	0.013666	0.17010	−12
−11	18.765	4.069	0.010894	2.0144	91.791	0.49643	6.1574	69.930	76.087	0.014142	0.16999	−11
−10	19.189	4.493	0.010906	1.9727	91.689	0.50693	6.3716	69.824	76.196	0.014617	0.16989	−10
−9	19.621	4.925	0.010919	1.9320	91.587	0.51759	6.5859	69.718	76.304	0.015091	0.16978	−9
−8	20.059	5.363	0.010931	1.8924	91.485	0.52843	6.8003	69.611	76.411	0.015564	0.16967	−8
−7	20.506	5.810	0.010943	1.8538	91.382	0.53944	7.0149	69.505	76.520	0.016037	0.16957	−7
−6	20.960	6.264	0.010955	1.8161	91.280	0.55063	7.2296	69.397	76.627	0.016508	0.16947	−6
−5	21.422	6.726	0.010968	1.7794	91.177	0.56199	7.4444	69.291	76.735	0.016979	0.16937	−5
−4	21.891	7.195	0.010980	1.7436	91.074	0.57354	7.6594	69.183	76.842	0.017449	0.16927	−4
−3	22.369	7.673	0.010993	1.7086	90.970	0.58526	7.8745	69.075	76.950	0.017919	0.16917	−3
−2	22.854	8.158	0.011005	1.6745	90.867	0.59718	8.0898	68.967	77.057	0.018388	0.16907	−2
−1	23.348	8.652	0.011018	1.6413	90.763	0.60927	8.3052	68.859	77.164	0.018855	0.16897	−1
0	23.849	9.153	0.011030	1.6089	90.659	0.62156	8.5207	68.750	77.271	0.019323	0.16888	0
1	24.359	9.663	0.011043	1.5772	90.554	0.63404	8.7364	68.642	77.378	0.019789	0.16878	1
2	24.878	10.182	0.011056	1.5463	90.450	0.64670	8.9522	68.533	77.485	0.020255	0.16869	2
3	25.404	10.708	0.011069	1.5161	90.345	0.65957	9.1682	68.424	77.592	0.020719	0.16860	3
4	25.939	11.243	0.011082	1.4867	90.240	0.67263	9.3843	68.314	77.698	0.021184	0.16851	4
5	26.483	11.787	0.011094	1.4580	90.135	0.68588	9.6005	68.204	77.805	0.021647	0.16842	5
6	27.036	12.340	0.011107	1.4299	90.030	0.69934	9.8169	68.094	77.911	0.022110	0.16833	6
7	27.597	12.901	0.011121	1.4025	89.924	0.71300	10.033	67.984	78.017	0.022572	0.16824	7
8	28.167	13.471	0.011134	1.3758	89.818	0.72687	10.250	67.873	78.123	0.023033	0.16815	8
9	28.747	14.051	0.011147	1.3496	89.712	0.74094	10.467	67.762	78.229	0.023494	0.16807	9
10	29.335	14.639	0.011160	1.3241	89.606	0.75523	10.684	67.651	78.335	0.023954	0.16798	10
11	29.932	15.236	0.011173	1.2992	89.499	0.76972	10.901	67.539	78.440	0.024413	0.16790	11
12	30.539	15.843	0.011187	1.2748	89.392	0.78443	11.118	67.428	78.546	0.024871	0.16782	12
13	31.155	16.459	0.011200	1.2510	89.285	0.79935	11.336	67.315	78.651	0.025329	0.16774	13
14	31.780	17.084	0.011214	1.2278	89.178	0.81449	11.554	67.203	78.757	0.025786	0.16765	14
15	32.415	17.719	0.011227	1.2050	89.070	0.82986	11.771	67.090	78.861	0.026243	0.16758	15

Source: Freon Products Division, E.I. DuPont de Nemours & Co., Inc.

*Inches of mercury below one atmosphere.

pressure–temperature relationship of a liquid refrigerant before and after the cylinder becomes full of the expanded liquid under gradual and uniform overheating. The true pressure–temperature relationship exists up to the point where expansion room is no longer available in the cylinder.

The extremely dangerous pressures that can result under such circumstances can cause the rupture of the refrigerant cylinder. Under uniform heating conditions the cylinder can rupture at approximately 1300 psi (8931.0 kPa). If, however, heat is applied with a welding torch, the area of the cylinder wall where the heat is applied may be weakened and the danger of rupture increased.

The Interstate Commerce Commission's regulations prescribe that a liquefied vapor container shall not be full of liquid when heated to 131 °F (55 °C). If cylinders are filled in compliance with this regulation, liquid refrigerant may completely fill the cylinder because of the expansion of the liquid at temperatures above 131 °F. Fusible metal plugs are designed to protect the refrigerant cylinder in case of fire. However, they will not protect the cylinder from a gradual uniform overheating. Fusible plugs begin to soften at 157 °F (69.5 °C), but the hydrostatic pressure created at this temperature far exceeds the cylinder test pressure.

The following safety rules should always be followed when handling compressed gas cylinders:

1. Never heat a cylinder above 125 °F (51.5 °C).
2. Never store refrigerant cylinders in the direct sunlight.
3. Never apply flame directly to a cylinder.
4. Never place an electric resistance heater in direct contact with a refrigerant cylinder.
5. Do not drop, dent, or otherwise abuse cylinders.
6. When refilling small cylinders, never exceed the weight stamped on the refrigerant cylinder.
7. Always keep the valve cap and head cap in place when the cylinder is not in use.
8. Always open all cylinder valves slowly.
9. Secure all cylinders in an upright position to a stationary object with a strap or chain when they are not mounted in a suitable stand.
10. Never force connections.
11. Do not tamper with safety devices.
12. Do not alter cylinders.
13. Protect cylinder from rusting during storage.

□ Lubricants □

The common fluorocarbon refrigerants in use today are miscible with the lubricating oil in amounts depending on viscosity, temperature, suction pressure, etc. In a properly designed refrigeration system, the oil is absorbed in the refrigerant and will travel through the system with the refrigerant. The lubricating oil should be a straight-run refined mineral oil, free from moisture, sediment, acid, soap, or any substance not derived from petroleum. The use of such compounds as ethylene glycol, glycerine, or castor oil is not recommended because these substances are all hygroscopic—that is, they all absorb moisture and become gummy. They also may produce a sticky sludge in the compressor crankcase when the system is idle or when the oil becomes heated. The lubricating oil must be thoroughly dehydrated to avoid the freezing of moisture in the refrigerant flow control device and to prevent possible emulsification of the oil and the formation of sludge.

□ Safety Procedures □

Many of the more common refrigerants do not have a disagreeable odor. It might therefore be possible to work in an area where there is a considerable amount of refrigerant vapor. Many refrigerants are heavier than air and will replace the air within the room. This is dangerous because a person must have at least 19.1% oxygen in the air breathed; otherwise, he might become unconscious.

Extreme care should be exercised when handling any of the hydrocarbon refrigerants in the presence of an open flame. When these refrigerants are subjected to an open flame, phosgene gas is produced. This gas is very toxic and will cause many health problems when breathed.

1. Be sure that the room is thoroughly ventilated before repairing a refrigerant leak.
2. Check the type of refrigerant before charging the system with refrigerant.
3. Be sure to wear goggles when charging or purging a system.

4. Be sure to keep liquid refrigerant out of the eyes. If liquid refrigerant should enter the eyes, flush them gently with tap water to prevent freezing, and see a physician immediately.
5. Do not breathe the refrigerant vapor from a system that has a burned compressor.
6. Keep burned refrigerant off the skin to prevent acid burns.
7. Never apply an open flame to a refrigerant cylinder; an explosion and serious personal injury could result.
8. Always protect refrigerants from being contaminated with moisture, air, or other substances.
9. Do not allow oil from a burned compressor to touch the skin, as it is acidic and may burn the skin.
10. If liquid refrigerant comes in contact with the skin, flush the skin with water and treat the affected area for frostbite.

◻ SUMMARY ◻

◻ Most of the commonly used refrigerants are in the gaseous state at atmospheric pressures and temperature.

◻ The words *gas* and *vapor* are ordinarily used interchangeably.

◻ All fluids have both a liquid and a gaseous state.

◻ A vaporizing refrigerant will absorb heat equal to the amount of energy necessary to change the liquid to vapor.

◻ Each refrigerant will absorb an amount of heat per pound of refrigerant equal to its latent heat of vaporization.

◻ The greater the pressure exerted on a refrigerant, the greater the boiling point.

◻ A liquid can be caused to boil by placing it in a partial vacuum.

◻ The critical temperature of a vapor is the temperature above which the vapor cannot be liquefied, regardless of the amount of pressure applied.

◻ The condensing pressure will depend on the temperature at which the vapor will liquefy.

◻ Water-cooled condensers will operate at lower condensing temperatures and pressures than air-cooled condensers.

◻ The vaporizing pressure of a refrigerant is important because it must evaporate without requiring a low suction pressure.

◻ The amount of heat in Btu (cal) required to change a liquid to a vapor (the change taking place at a constant temperature) is known as the latent heat of vaporization.

◻ Any refrigerant when evaporating in the evaporator must absorb heat from within the cooled space exactly equal to its latent heat of vaporization.

◻ The latent heat of vaporization of a liquid will vary with the pressure and the corresponding temperature at which the vaporization occurs.

◻ The fluorocarbon refrigerants may be tested for leaks with a halide torch, a soap-bubble solution, or an electronic leak detector.

◻ All fluorocarbon refrigerants are miscible with oil.

◻ *Miscibility* is the term used to express the ability of liquid refrigerant to mix with oil.

◻ In general, naphthenic oils are more soluble in refrigerants than paraffinic oils.

◻ Because oil must pass through the compressor cylinders to provide lubrication of moving parts, a small amount of oil is always circulating with the refrigerant.

◻ Excess liquid refrigerant in the compressor crankcase may result in improper lubrication of the compressor because of the foaming and boiling action on start-up.

◻ Proper piping and system design are required for systems using R-22 and R-502 to ensure proper oil return to the compressor.

◻ *Enthalpy* is a term used in thermodynamics to describe the heat content of a substance.

◻ Only the proper amount of refrigerant should be put in a refrigerant cylinder to prevent explosions resulting from hydrostatic pressure.

◻ Refrigeration lubricating oils should be straight-run refined mineral oil, free from moisture, sediment, acid, soap, or any substance not derived from petroleum.

◻ EXERCISES ◻

1. In what state do most refrigerants exist at atmospheric pressure and temperature?
2. How much heat will a liquid refrigerant absorb per pound in changing from a liquid to a gas?
3. How will an increase in pressure on a liquid refrigerant affect its boiling point?
4. What does the term *critical temperature* of a refrigerant mean?
5. What happens to the condensing pressure of a refrigerant when the temperature of the gas is reduced?

6. List the standard conditions established by the refrigeration industry to compare refrigerants and refrigeration units.
7. On what does the condensing pressure depend?
8. Will a water-cooled condenser or an air-cooled condenser operate with a lower condensing temperature?
9. How can the approximate condensing temperature and pressure of an air-cooled condenser be determined?
10. Why is the vaporizing pressure of a refrigerant important?
11. What will air in a refrigeration system cause?
12. Define *latent heat of vaporization* of a refrigerant.
13. What methods of leak detection are used for testing fluorocarbon refrigerants?
14. What does the term *miscibility* mean?
15. Why is a small amount of oil always circulating in a refrigeration system?
16. What precautions should be observed on systems using R-22 and R-502?
17. What does the term *enthalpy* indicate?
18. What precautions should be observed when filling a refrigerant cylinder?
19. Why is it not safe to heat a refrigerant cylinder with a welding torch?
20. Why must moisture be kept out of refrigeration oil?

11 Introduction to Electricity

After completion of this chapter, the student should:

- Have a working knowledge of the electron principle.
- Be able to name the sources of electricity.
- Know what electrical pressure (electromotive force) is.
- Explain the concept of rate of flow of electricity.
- Have a working knowledge of the various types of electrical circuits.
- Understand electromagnetism.
- Have a working knowledge of transformers.
- Complete the Exercises with at least 80 percent accuracy.
- Complete the Laboratory Exercises with at least 80 percent accuracy.

Human beings have known about electrical phenomena for centuries. The word *electric* is derived from the Greek word *elektron* which means amber. Amber was the first substance to show identifiable electrical characteristics as far back as 600 B.C. However, we did not learn to harness this powerful and mysterious force until the nineteenth century. Thomas A. Edison began the new "electric era" when he invented the incandescent light bulb in 1879. After that time, practical applications for electrical energy multiplied at a rapid pace.

☐ The Electron Principle ☐

The basis of the electron is the fact that all known matter or substance regardless of its form—solid, liquid, or gas—is fundamentally composed of very small particles called *atoms*. These atoms, in turn, are composed of even smaller particles called *electrons* and *protons*. Both electrons and protons have an electrical charge. The electrons are negatively charged and the protons are positively charged. Thus, everything on earth, including human beings, is basically electrical in nature.

Further research has revealed a number of other interesting and important facts regarding the construction of an atom. For example, it was found that a positively charged proton is almost 2000 times heavier than a negatively charged electron. Also, each atom has a core or center called a *nucleus*. The nucleus contains the heavier protons along with uncharged particles called *neutrons*. A neutron is made up of one electron and one proton. The electrical charges of these particles have been neutralized. Because of the combined weight of the protons and neutrons, the nucleus is relatively stationary as compared to the electrons, which are known to be rotating at high speeds in orbits or shells around the nucleus. Except for the speed, this phenomenon compares with the earth and other planets rotating in orbits around their nucleus, the sun.

This principle is the very foundation upon which electricity is based. In order to more fully understand electricity, it is necessary to learn more about the makeup of matter. It is this makeup that helps us to understand why different substances or materials are used in the construction of electrical components and devices.

☐ Nature of Matter ☐

Scientists tell us that everything is made up of matter. Matter is generally defined as anything that occupies space or has mass. However, some matter differs from other matter, and observable characteristics will not always allow identification of a substance. Thus, it is necessary to complete a chemical analysis to determine the nature and behavior of some substances in order to identify them. Almost all materials are made up of a combination of various kinds of matter or mixtures of matter.

It must be understood, in the study of electron theory, that some types of elementary matter may exist that are neither mixtures nor compounds, and that any further effort to break this matter into parts by chemical decomposition will produce no change in any of the characteristics of the matter. When matter has been divided into its simplest components, these components are known as *elements*.

Elements. Even though matter is basically comprised of atoms, in a larger sense it is composed of one or more elements. An element may be defined as a substance containing only one kind of atom.

There are at the present time 106 known elements. However, only 90 are considered to be natural elements. The remaining 16 are *synthetic elements,* which have been produced by man through atomic research. As a point of interest, 11 of the 90 natural elements are gases, 2 are liquids, and the remaining 77 are solids at normal temperature and pressure.

Very few people realize the role that elements play in our daily lives. The food we eat, the clothes we wear, and even our bodies are made up basically of elements (Figure 11-1). A human body is made up basically of 15 elements, 6 of which vary from 1 to 65% of the whole for a total of 99%. The other 9 elements combine to produce the remaining 1%.

Because each of the 106 elements is made up of only one kind of atom, it follows that there are 106 kinds of atoms. It is known, however, that the electrons and protons in one atom are identical to those in all other atoms. The difference between atoms lies in the number of electrons and protons each con-

FIGURE 11-1 **Elements of a human body. (Courtesy of Frigidaire Division, General Motors Corp.)**

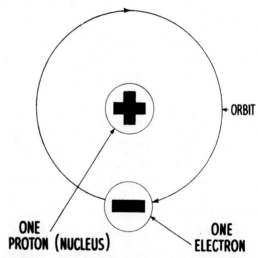

FIGURE 11-2 **Hydrogen atom. (Courtesy of Frigidaire Division, General Motors Corp.)**

tains and the arrangement of the electrons in orbit around the nucleus.

The structure of the hydrogen atom is the simplest of all atoms as shown in Figure 11-2. It has one electron rotating in orbit around one proton as the nucleus.

Note: The nucleus of the hydrogen atom also contains one neutron. In most instances, the number of neutrons of any atom is equal to the number of protons and to the number of electrons in orbit.

The hydrogen atom is relatively simple, but as the number of electrons and protons increases, the struc-

FIGURE 11–3   Oxygen atom. (Courtesy of Frigidaire Division, General Motors Corp.)

NUCLEUS
8 PROTONS

ORBITS
8 ELECTRONS

TABLE 11–1   Some Commonly Known Elements

Element	Atomic Number	Atomic Weight
Hydrogen[a]	1	1.008
Carbon	6	12.01
Nitrogen[a]	7	14.008
Oxygen[a]	8	16.00
Aluminum	13	26.97
Iron	26	55.85
Nickel	28	58.69
Copper	29	63.57
Silver	47	107.88
Tin	50	118.70
Gold	79	197.0
Mercury[b]	80	200.61
Lead	82	207.21
Uranium	92	238.07

Source: Frigidaire Division, General Motors Corp.
[a]Gas.
[b]Liquid.

ture of the atom becomes more complex. Helium has 2 protons in the nucleus and 2 electrons in orbit around it. Carbon has 6 protons in the nucleus and 6 electrons in orbit. This establishes the pattern for all atoms; the number of protons in the nucleus is always equal to the number of electrons in orbit.

An oxygen atom has 8 protons and 8 electrons (Figure 11–3). The electrons are divided into two groups with 2 electrons in an orbit near the nucleus and the other 6 in an orbit at a distance farther from the nucleus. As the number of electrons in the elements increases, the number of orbits and their spacing also increases. For example, the 92 electrons in a uranium atom are divided into 7 orbits differently spaced around the 92 protons in the nucleus. All elements have an atomic number and weight (Table 11–1). The number is based on the number of electrons and protons in the element.

FIGURE 11–4   Molecule of water. (Courtesy of Frigidaire Division, General Motors Corp.)

OXYGEN ATOM

HYDROGEN ATOMS

Compounds. Except for the elements, all other known matter may be defined as substances that are made up of two or more elements. In a compound, two or more elements combine chemically to form a distinct substance with the same proportion of elements always present in its composition.

When an atom of one kind of element is combined with an atom of another kind, it forms a larger particle called a *molecule*. A molecule may be defined as the smallest particle of a compound that is of the same chemical composition as the compound itself.

The chemical formula for water, H_2O, is known to almost everyone. The formula indicates that the two elements, hydrogen and oxygen, are chemically

combined. There are 2 atoms of hydrogen for every atom of oxygen. The smallest particle of water must contain 2 hydrogen atoms and 1 oxygen atom in order to retain the same chemical properties as water itself (Figure 11–4). The atoms are locked together due to the crossing orbits of the rapidly rotating electrons. Of course, it would require millions of these molecules to form a single drop of water. All other compounds are made up in the same general manner.

Many compounds are composed of more than two elements. These compounds would include all of the refrigerants commonly used in refrigeration and air conditioning. For example, a molecule of R-12 contains 1 carbon atom, 2 chlorine atoms, and 2 fluorine

Nature of Matter

atoms locked together by the rotating electrons. The chemical name for R-12 is dichlorodifluoromethane.

Electron Movement. The relationship between the electrons and protons of an atom is the basis for the well-known facts in electricity that positive charges repel each other, negative charges repel each other, and positive charges attract negative charges. This attraction of the protons in the nucleus for the surrounding electrons is the force that holds the atom together. The strength of this attraction varies widely in different atoms. As the distance from the nucleus to the electron orbit increases, the force of attraction between the nucleus and the electrons decreases proportionately.

Even though the atoms are locked together by the overlapping orbits, some of the electrons are able to move from one atom to another because of the weakness of attraction to their nucleus. These electrons are called *free electrons*. The remaining electrons are close to the nucleus and the attraction is too strong for them to leave their orbit. These electrons are called *planetary* or *bound electrons*. According to the electron principle, electricity is the movement of free electrons from one atom and molecule to another. The forces that cause these electrons to move will be explained later. Because of its free electrons (Figure 11–5) and other desirable characteristics, carbon is used as a conductor for electricity on many applications.

Any substance with a great number of free electrons would make a good conductor of electricity. Such substances as platinum, gold, silver, and tin each have many free electrons and make excellent conductors. However, their cost and other characteristics make them impractical for commercial use. In spite of its cost, however, silver is used extensively for electrical contacts in switches and contactors to ensure good electrical performance and give long, trouble-free service. Other substances such as copper, brass, aluminum, iron, tungsten, and nichrome are widely used to either conduct electricity from one place to another or in the manufacturing of the multitude of electrical devices in use today.

On the other hand, there are many substances in which the force of attraction between the protons and the electrons is so strong that there are very few free electrons. These substances are good insulators because they have low conductance. A few examples of good insulators are glass, slate, wood, mica, and plastic. Insulation is used on wires to protect accidental short circuits and to avoid electrical contact between the conductors and any supporting devices. It is apparent that some materials resist the flow of electrons, while others offer less resistance to the flow of electrons. Because of this, resistance must always be considered as an important characteristic of an electrical circuit.

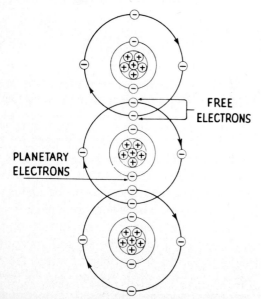

FIGURE 11–5 Combining of carbon atoms. (Courtesy of Frigidaire Division, General Motors Corp.)

FREE ELECTRONS

PLANETARY ELECTRONS

☐ Sources of Electricity ☐

The source of all energy is the sun and electricity is a form of energy. This means that we are continually being bombarded with untold billions of electrons and protons from the sun. This energy is absorbed and stored by the earth and stored by the earth and everything on it. Except for the splitting of atoms, energy can neither be created nor destroyed. Therefore, it can be said that electricity cannot be made by machines or devices. Although we say that batteries and generators produce electricity, they actually are devices that make it possible for the ever-present free electrons in a substance to be moved from one place to another. There are six sources of electricity: (1) friction, (2) chemical action, (3) magnetism, (4) heat, (5) pressure, and (6) light (Figure 11–6). Each of these sources of electricity has significance in modern everyday life.

PRESSURE

LIGHT

FRICTION

CHEMICAL ACTION

MAGNETISM

HEAT

FIGURE 11-6   Sources of electricity. (Courtesy of Frigidaire Division, General Motors Corp.)

Friction. Electricity produced by friction is commonly called *static electricity* and is produced by rubbing different types of substances together. Trucks hauling gasoline drag a chain along the ground to prevent static electricity caused by friction from igniting the gasoline vapors and producing an explosion.

Chemical Action. Flashlight and automobile batteries are possible only because electricty can be produced by chemical action.

Magnetism. This is our greatest and most important source of electricity. Without this source, there would be no generators, motors, or many other devices in common use today.

Heat. Although the electricity produced by heat may not be as significant to us as that produced by magnetism, it does make possible accurate temperature-testing devices and safety devices on modern gas furnaces.

Pressure and Light. These sources electricity contribute to our pleasure and convenience. For example, electricity produced by pressure makes possible the recording and playing of phonograph records for homes and business establishments. Photoelectric cells and exposure meters used by photographers are possible because of electricity produced by light.

☐ Electricity by Friction ☐

As previously mentioned, electricity produced by friction is commonly called static electricity and is produced by rubbing certain types of materials together.

The types of materials used in producing static electricity are called *dielectrics*. Dielectrics are actually insulators, such as glass, rubber, paper, and plastics through which an electric current will not normally pass. They do, however, contain some free electrons that can be transferred from one substance to another to create a shortage of electrons in one and an overabundance in the other.

Due to the fact that static electricity cannot be harnessed to produce a steady flow of current, it has little practical application for the direct operation of refrigeration and air conditioning equipment. However, it is believed that a knowledge of the subject is required in order to understand the basic construction of a capacitor or condenser. A capacitor is an electrical device that is essential to the operation of numerous products such as radios, televisions, and automobiles.

Most people are familiar with many instances of static electricity. The sparks that result when a doorknob is touched after walking across a rug are static electricity developed by friction between shoes and the rug. Lightning is static electricity produced by the friction between the air and water particles. A cloud picks up electrons and then discharges them to another cloud or to the earth.

The ancient Greek philosopher Thales, who lived between 640 and 546 B.C., is said to have been the first person to have observed that amber, when rubbed, attracted straws, dried leaves, and other light objects.

There were many ancient theories as to how amber attracted objects, but none of them proved correct according to the present-day knowledge of atoms. However, they made their contributions to the subject of electricity. Not much was known about electricity until Sir William Gilbert, who was a physician to Queen Elizabeth, wrote a series of books in A.D. 1600 that summarized all that was then known about the subject and added much from his own experimentation. It was he who adopted the word *electric* from the Greek word for amber.

In 1660, Otto Van Guericke built the first static electric generator (Figure 11-7). He proved that elec-

FIGURE 11-7   First static electric generator. (Courtesy of Frigidaire Division, General Motors Corp.)

FIGURE 11-8   Uncharged atom. (Courtesy of Frigidaire Division, General Motors Corp.)

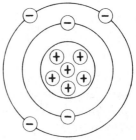

FIGURE 11-9   Charged atom. (Courtesy of Frigidaire Division, General Motors Corp.)

tricity could be generated by a machine. His generator was a globe of sulfur mounted on an axle. By rotating the globe and rubbing the dry hand on the surface, the globe became sufficiently charged to attract paper, feathers, and other light objects. Van Guericke was the first to discover that bodies were repelled as well as attracted and also that an electric charge could travel to the end of a linen thread.

For approximately the next 90 years contributions were made and theories advanced pertaining to electricity with no great significance, until Benjamin Franklin gave us the one-fluid theory of electricity. He believed that electricity existed in two states, which he called positive and negative. In 1752 he performed his famous kite experiment which proved that lightning and electricity were the same thing.

To understand lightning and other manifestations of static electricty, we must accept the fact that electrons are negatively charged and protons are positively charged.

Uncharged and Charged Atoms. In an uncharged state each atom has an equal number of electrons and protons (Figure 11-8). If one, or any number, of free electrons are removed by friction, the substance becomes positively charged (Figure 11-9). The question now arises: What happens to the free electrons that we removed? When we remove an electron from a substance it will cling to anything that is handy, thus causing that substance to become negatively charged. When this negatively charged object comes in contact with another object that has fewer electrons than protons, the electrons transfer from the negatively charged object to the positively charged object in an attempt to produce a balance or equilibrium. This electron transfer is usually accompanied by a spark or a cracking sound.

Properties. A statically charged insulator or dielectric has a field of force, and the laws of attraction and repulsion apply. Figure 11-10 illustrates a piece of fur and a hard rubber rod, both of which are neutral or uncharged. This means that the electrons and protons in each substance are equal or in equilibrium. Figure 11-11 illustrates what happens when the rod is rubbed with the fur. The friction causes some of the free electrons in the fur to cling to the rod, leaving the fur with a shortage of electrons and the rod with an oversupply. The rod is negatively charged and the fur is positively charged.

This same thing happens to silk and a glass rod, except that rubbing the glass rod with silk causes the silk to become negatively charged, leaving the glass rod positively charged. If the positively charged glass rod is suspended by a string while the end of the negatively charged rubber rod is held near the free end of the glass rod, the two will be attracted, which proves that unlike charges attract (Figure 11-12). Next, if a negatively charged rubber rod is suspended and another negatively charged rubber rod is held near it, the suspended rod will move away, proving that like charges repel (Figure 11-13).

Another characteristic of static electricity is that it has the ability to induce a charge of electricity without contact. First, a substance can be charged

INTRODUCTION TO ELECTRICITY

FIGURE 11-10 Electric charges of fur and hard rubber rod. (Courtesy of Frigidaire Division, General Motors Corp.)

FIGURE 11-11 Electric charges of fur and hard rubber rod after rubbing. (Courtesy of Frigidaire Division, General Motors Corp.)

FIGURE 11-12 Unlike charged rods.

FIGURE 11-13 Like charged rods.

FIGURE 11-14 Metal foil electroscope. (Courtesy of Frigidaire Division, General Motors Corp.)

FIGURE 11-15 Temporarily charged metal foil electroscope. (Courtesy of Frigidaire Division, General Motors Corp.)

FIGURE 11-16 Permanent charging of a metal foil electroscope. (Courtesy of Frigidaire Division, General Motors Corp.)

temporarily by induction. A device used to show a temporary charge is a metal foil electroscope (Figure 11-14).

When a negatively charged rubber rod is held near the knob, the free electrons, in an attempt to get away from the rod, travel to the foil strips (Figure 11-15). The metal strips are shown in a repelled posi-tion. Since like charges repel, the foil strips move away from each other. Removing the rod permits the repelled electrons to return to their original posi-tion. Since the strips are repelled only while the rod is held close to the knob, it is temporary induction. The same results can be obtained if the rod is posi-tively charged.

A substance can also be charged permanently with either a positive or a negative charge. Again using a metal foil electroscope, a permanent positive charge can be induced (Figure 11-16). A negatively charged rod is held near the knob. While the strips are in a repelled position, the knob is touched with the finger and the free electrons on the foil strips are attracted through the hand to the ground. This leaves the foil strips short of electrons. When the

FIGURE 11-17 Permanently charged electroscope. (Courtesy of Frigidaire Division, General Motors Corp.)

FIGURE 11-18 Pith ball electroscope. (Courtesy of Frigidaire Division, General Motors Corp.)

FIGURE 11-19 Negatively charged rod near a pith ball electroscope. (Courtesy of Frigidaire Division, General Motors Corp.)

FIGURE 11-20 Pith ball electroscope with unlike charges. (Courtesy of Frigidaire Division, General Motors Corp.)

rod and finger are removed, the strips are left with a shortage of electrons and they repel each other (Figure 11-17). The electroscope is now permanently charged positively. The electroscope can be neutralized by holding a negatively charged rod near the knob.

A substance can also be charged by contact or conduction with either a positive or negative charge (Figure 11-18). A neutral or uncharged pith will hang straight down from the stand. If the pith ball is touched with a negatively charged rod a reaction will occur (Figure 11-19). As the rod approaches the ball, just prior to contact there is an attraction of the ball to the rod. Upon contact, the ball takes a negative charge from the rod and is immediately repelled, again showing that like charges repel. This same thing would happen if a glass rod were rubbed with silk and then touched to another suspended pith ball, except that the ball would be positively charged.

The question might arise as to what happens when two pith balls with unlike charges are placed near each other? The answer is that they are attracted to each other, but as soon as they touch they both become neutral and separate (Figure 11-20).

During this discussion of static electricity, it has been demonstrated that a substance can be charged either positively or negatively, temporarily or permanently, by either induction or conduction. It has also been demonstrated that positive charges repel positive charges, negative repels negative, and that unlike charges attract each other.

☐ Electricity by Chemical Action ☐

Although electricity produced by chemical action has no association with the operation of refrigeration and air conditioning systems, it does play an important role in the operation of several types of instruments required for properly diagnosing service complaints. Also, due to the type of current produced by chemical action, and its relative simplicity, other phases of the subject of electricity can be more easily explained.

Any device that produces electricity by chemical action is commonly called a battery. To understand how such a device produces electricity, a knowledge of its ingredients and chemical behavior is required.

Electrolytes. Nature stores up much energy in many chemical compounds. Due to the fact that the amount of energy stored is not the same in the various compounds, some are more desirable than others for use in batteries.

How the stored energy of compounds can be put to a practical use is of considerable interest and is the basis of all battery operation. When certain acids or salts are dissolved in water, their molecules are broken up into positively and negatively charged particles called *ions*. An ion may be defined as an electrically charged atom or group of atoms. Any solution containing these ions is capable of conducting an electrical current and is called an *electrolyte*. Even a solution of common table salt (sodium chloride) and water may be considered an electrolyte: the sodium atoms become positively charged ions and the chloride atoms become negatively charged ions. Other solutions that contain water and acids such as hydrochloric and sulfuric are capable of conducting more current than those containing sodium chloride. They also have other more desirable characteristics for use in batteries. Regardless of the type of solution, ions are capable of moving to every part of the solution.

Electrodes. In order to take advantage of the current-carrying properties of electrolytes, a battery must have electrodes to which lights or other loads can be connected by wires or other conductors.

The chemical behavior of the electrolyte and the electrodes can best be explained by use of a simple *voltaic cell.* The term *voltaic* is applied to the cells of batteries in honor of Allessandro Volta, an Italian professor of physics who in 1796 produced the first storage battery.

Dissimilar metals can be used as electrodes; copper and zinc are excellent due to the fact that zinc has an abundance of negatively charged atoms, while copper contains an abundance of positively charged atoms. When plates or rods of these metals are submerged into an electrolyte, chemical action between the two metals begins. The zinc electrode accumulates a much larger negative charge due to the fact that it gradually dissolves into the electrolyte solution.

The atoms, which leave the zinc during the process, are positively charged. These positively charged zinc atoms are attracted by the negatively charged ions of the electrolyte. On the other hand, they repel the positively charged ions of the electrolyte, which are then forced toward the copper electrode. This causes the negative electrons to be removed from the copper, leaving it with a large excess of positive charge. If a load such as a light bulb is connected across the terminals on the electrodes, the forces of attraction and repulsion will cause the free electrons in the negative zinc electrode, connecting wires, and light-bulb filament to move toward the positively charged copper electrode. Because of this, a flow of electric current has been established.

The life of a cell is limited because the zinc electrode dissolves. When an insufficient amount of zinc remains to provide the required chemical action, the cell is no longer capable of producing electricity and the cell is dead.

Contrary to its name, a storage battery does not store electricity; it merely converts chemical energy into electrical energy in accordance with the limitations of its ingredients.

The cell just described is known as a wet cell (Figure 11–21) due to the aqueous solution used as the electrolyte. The instruments used for testing electrical components during diagnosis procedures are equipped with one or more dry cells.

A dry cell actually is not dry because the electrolyte is a relatively moist paste (Figure 11–22). From this, we can see that the outer shell, which is made of zinc, also serves as the negative electrode. The positive electrode is made of carbon. Although the electrolyte is a paste and is made of different compounds than those used in wet cells, and although carbon is used instead of copper, the chemical action within the cell is similar to the wet cell and produces the same results. Both the wet and dry type voltaic cells are called *primary cells* since they use up one of their primary ingredients.

FIGURE 11–21 **Wet-type voltaic cell. (Courtesy of Frigidaire Division, General Motors Corp.)**

FIGURE 11-22 Dry-type voltaic cell. (Courtesy of Frigidaire Division, General Motors Corp.)

FIGURE 11-23 Storage battery cell. (Courtesy of Frigidaire Division, General Motors Corp.)

Secondary Cell.

A cell that loses it chemical energy, but that can be restored by passing an electric current through it, is called a secondary cell. This is the type used in a storage battery. Most storage batteries are made up of two or more cells.

In a storage battery, the cell has electrodes or plates, one made of lead oxide, the other of lead. Both are submerged in a solution of sulfuric acid and water (Figure 11-23). The plates are kept apart by separators made of wood, rubber, or glass.

As in the dry-type voltaic cell, the ions have picked up the electrons from the lead-oxide plate, leaving it positively charged, while the others give up their electrons to the lead plate, making it negatively charged. During use, the cell slowly discharges and the lead and the sulfate from the acid combine to form lead sulfate, which is deposited on both plates (Figure 11-24). As the deposit of lead sulfate builds up on both plates, they eventually become lead sulfate and the electrolyte becomes a weak solution of sulfuric acid and water. The chemical energy has ceased.

By removing the light and connecting the wires to an outside source of current, the cell can be recharged by passing current through the cell in the opposite direction to that of the discharge. The plates again become lead and lead oxide. The solution then becomes strong with sulfuric acid. With proper care a secondary cell can be charged and discharged many times.

Regular Direct Current.

When a load, such as a lightbulb, is connected to a battery, the flow of cur-

FIGURE 11-24 Discharged storage battery cell. (Courtesy of Frigidaire Division, General Motors Corp.)

rent is always in the same direction: from the negative to the positive terminals. Because the current is flowing in only one direction and is flowing at an unfluctuating and steady rate, it is called regular dc current. The flow of regular dc current can be compared with the flow of water produced by a centrifugal pump (Figure 11-25).

The factors that affect the movement of water in a piped system also can be applied to the movement of electrons in an electrical circuit: (1) pressure, (2) quantity, (3) rate of flow, and (4) resistance.

FIGURE 11-25 Regular dc current. (Courtesy of Frigidaire Division, General Motors Corp.)

Pressure. Pressure is the force that moves water through pipes and electrons through conductors. Just as the pump produces the water pressure, the battery produces the electrical pressure.

Quantity. Quantity is the amount of water moved by the pump and is the number of electrons moved by the battery.

Rate of Flow. The speed or strength of flow of water through the pipes and electrons through the conductors is the rate of flow.

Resistance. Resistance is the opposition produced by the pipes and fittings to the flow of water and is also the opposition produced by the conductors to the flow of electrons.

Just as the pump must produce a sufficient amount of pressure to overcome the resistance in order to move a quantity of water through a pipe,

so must the battery produce a sufficient amount of pressure to move a quantity of electrons through a conductor.

Several of these factors as they pertain to electricity require further explanation.

▢ Pressure (Electromotive Force) ▢

It has been established that in a battery, chemical action produces negatively and positively charged particles that react with the negatively charged zinc electrode and the positively charged copper electrode. When a load is connected across the terminals, the imbalance in charges produces a pressure called an *electromotive force* (emf) which causes the free electrons (*current*) to flow through the circuit. Actually, this flow of current is many billions of free electrons repelling each other throughout the conductor. As stated before, the direction of force is always from the negative electrode to the positive electrode.

It should be remembered that any particular electron does not move from one end of a conductor to the other. The electron movement (Figure 11-26) may be compared to a switch engine bumping a standing string of boxcars on a track. When the engine bumps one end of the string, each car progressively bumps the next car.

If a free electron from a source of energy is forced into an atom at one end, it tends to upset the balance between the electrons and protons of the atom. This action forces another free electron from that atom to shift to an adjacent atom, thus upsetting its balance. This shifting or drifting of free electrons toward the source of energy at the other end of the

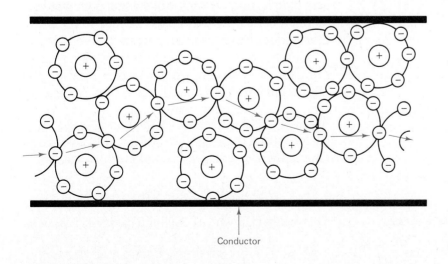

Conductor

FIGURE 11-26 Electron flow. (Courtesy of Frigidaire Division, General Motors Corp.)

conductor is called *dynamic electricity* or *current in motion*. All of this is possible because free electrons are repelled out of a substance having a high negative charge (the zinc electrode) and are attracted to another substance having fewer electrons or a positive charge (the copper electrode).

The greater the difference between the concentration of electrons, the greater the electromotive force will be. In everyday language, the term *voltage* is used to designate electromotive force.

Volt. The term *voltage* is derived from the word *volt,* which is the unit of measure for electromotive force. This unit of measure was established by Volta while developing his primary cell, and was named in his honor. How he established this unit of measure can be better explained after an explanation of the other factors pertaining to the flow of electrons in a circuit.

Potential Difference. When two substances having an equal concentration of free electrons are connected by a conductor, there is no movement of electrons from one substance to the other. The two substances are said to have the same electrical potential. However, if one substance has a higher concentration of free electrons than the other, they are said to have a potential difference between them. Again, the greater the difference between the concentration of free electrons in the electrodes, the greater the potential difference will be. According to the electron principle, free electrons always move from a point of high potential toward a point of lower potential—from a negative electrode toward a positive electrode in an electrical circuit. Using water as an analogy, the potential difference in electricity can be compared with a tank of water on a high tower. The water in the tank has a higher pressure potential than the water in a bucket placed at the end of a pipe leading from the bottom of the tank (Figure 11–27).

High Potential

Low Potential

FIGURE 11–27 Potential difference.

While the terms *electromotive force* (emf) or *voltage* and *potential difference* are closely related and are often used interchangeably, there is a technical difference between them. In brief, emf or voltage is the pressure produced throughout an electrical circuit, while potential difference indicates a ratio of electron potency or concentration between the electrodes of a battery or certain components of other current-producing devices.

Coulomb. Just as units of measure have been developed to measure the quantity of liquids, so have units been established to measure the quantity of electricity. When dealing with large quantities of water, the term *gallon* (liter) is applied. We say that a gallon (3.785 liters) is 4 quarts, 8 pints, or 128 ounces. While the electron can be considered a quantity of electricity, it is too small to be used as a standard measure. A larger unit, comprising 6.3 billion billions of electrons, is called a coulomb and is the standard electrical unit of quantity measure. The coulomb was named in honor of Charles Coulomb, a French physicist (1736–1806), who made investigations into magnetism and electricity.

Although the coulomb is a definite quantity of electrons that can be measured in a laboratory, it is of little concern to us from a practical standpoint. In other words, we do not measure coulombs directly in diagnostic procedures.

□ Rate of Flow □

Coulombs can no more be used as a measure of the strength of an electric current than the gallon (liter) alone can be used as a measure of the strength of a waterfall. Both must be considered in their relation to time. If it were said that Niagara Falls spills 5 million gallons (18.93 million liters) of water, it would mean very little unless the time required for this amount to spill was also given. If it took a year to spill that much, it wouldn't be much of a waterfall. However, it is known that it takes only an hour and, from this its strength can be determined.

Ampere. The ampere (A) is the unit of measure of current strength or the rate of flow of the electrons. One ampere is equal to 1 coulomb per second.

1 A = 1 coulomb per second

This means that when 1 coulomb passes a given point

in an electric circuit each second, the rate of flow is 1 ampere. Coulombs are quantity; amperes are the rate of flow or the strength of the current. An ordinary light bulb may require one-half A while a 36-in. searchlight may require 150 A. This shows that the searchlight is about 300 times stronger than the light bulb. This unit of measure was named in honor of André Ampère (1775–1836), a French scientist.

☐ Resistance ☐

The last, but very important factor, in the understanding of electricity is the resistance to the flow of electrons encountered in a conductor. Most metals have a low resistance and are considered good conductors because the molecules are loosely hung together and have many free electrons. In other words, the attraction between the electrons and the protons or nucleus is weak and the electrons can be readily pushed out of their orbit. Copper is one of these metals and is widely used as a conductor. Of course, all matter, including copper, presents a certain amount of resistance to the flow of current by attempting to retain its own electrons. In any circuit, the resistance of the conductor must be overcome by the force of the current. If the force is strong or the resistance small, a strong current will flow. On the other hand, if the force is weak or the resistance great, only a small amount of current will flow.

A comparison can be made between the resistance factors pertaining to water and those pertaining to electricity. Four factors determine the resistance to the flow of water through a pipe. These four factors are: (1) diameter of the pipe, (2) length of the pipe, (3) kind of pipe, and (4) velocity of flow.

The smaller the pipe, the longer the pipe, and the rougher the interior of the pipe, the more friction there is. Friction is resistance, so the greater the friction, the smaller the flow will be.

Electrical Conductors. Electrical conductors (wires) are the pipes of an electric circuit and the resistance of these wires is also dependent on four factors. These factors are: (1) diameter of wire, (2) length of wire, (3) kind of wire, and (4) temperature of the wire.

If the wire is long, or small in diameter, the flow of current will be less. Similarly, if the wire is made of iron, the flow of current will be less·than it would

be if copper were used. The resistance of most conductors increases as the temperature increases. If the resistance of a conductor is increased by any one of the four factors, the current flow is decreased.

From this, it can readily be seen that if one desires to install wires from a source of power to a load, such as a large condensing unit, the wires must be large enough to carry the load and should be of a low-resistance material such as copper.

There is one exception to the temperature-resistance factor that is of particular interest for us from a practical standpoint. When certain substances, such as oxides of manganese, nickel, and cobalt, are mixed and formed into a solid, it forms a semiconductor called a *varistor* or *thermistor*.

Thermistors are conductors whose resistance to the flow of electricity decreases as their temperature increases. In other words, at a low temperature their resistance is so high that very little current can flow through them. As the temperature of the metal increases, either by the current passing through it or by that applied externally, more current proportionately flows through it. Thermistors are being used in the control circuits of refrigeration and air conditioning equipment to provide more accurate control of the equipment.

Ohm. Just as the volt (V) is a unit of measure for electromotive force and the ampere (A) is a unit of measure for the rate of flow of electrical current, the ohm (Ω) is the unit of measure of resistance of a conductor to the flow of electrons. The unit of resistance was developed by Georg S. Ohm, a German scientist. To establish the unit of measure, he used a column of mercury 106.3 cm in height and 1 mm² in cross section. One ohm is the resistance offered by this column of mercury to the flow of 1 A of current having an electromotive force of 1 V. The temperature of the mercury column was 32 °F (0 °C). This column of mercury is comparable to 1000 ft (304 m) of number 10 wire, which is one-tenth in. (2.54 mm) in diameter or 2.4 ft (0.73 m) of number 36 wire which is 0.005 in. (0.127 mm) in diameter.

Ohm's Law. We can now define the relationship of volts, amperes, and ohms. A volt may be defined as the electromotive force required to push 1 coulomb of electrons in 1 second of time (1 ampere) through a resistance of 1 ohm. Ohm did more than experiment with resistance. He combined his own discoveries with those of Volta and Ampère and, in

1827, he formulated the all-important Ohm's law upon which all electrical measurement is based. Ohm's law states that the current in an electric circuit is equal to the pressure divided by the resistance.

$$\text{Current} = \frac{\text{pressure}}{\text{resistance}}$$

To simplify this further, letters are used to represent the words. E represents the pressure or electromotive force in volts, I represents the intensity of current in amperes, and R represents the resistance in ohms:

$$I = \frac{E}{R}$$

As practical examples, three simple circuits will be used. The proper arrangement of the formula is $I = E \div R$ (Figure 11–28). The three-cell storage battery has a voltage of 6 V; the light bulb has a resistance of 2 Ω. By simple arithmetic, the intensity of current is found to be 3 A. It is obvious that the formula $I = E \div R$ can be used to find the intensity of current only when the voltage and resistance are known. An unknown resistance can be determined if the voltage and amperage are known, however, by dividing the volts by the amperes (Figure 11–29). The formula now is $R = E \div I$. Still the battery is supplying 6 V and the intensity of the current is 3 A. Again, by simple arithmetic, the resistance is 2 Ω.

The formula can be arranged in still another manner. If the voltage is the only unknown factor, it can be readily determined by multiplying the amperes by the ohms (Figure 11–30). The formula now reads

FIGURE 11–28   Example of Ohm's law. (Courtesy of Frigidaire Division, General Motors Corp.)

FIGURE 11–29   Finding resistance using Ohm's law. (Courtesy of Frigidaire Division, General Motors Corp.)

FIGURE 11–30   Finding voltage using Ohm's law. (Courtesy of Frigidaire Division, General Motors Corp.)

$E = I \times R$. Here it is known that the intensity of the current is 3 A and the resistance is 2 Ω. By multiplying the two, it is found that the battery is supplying 6 V of pressure.

To summarize, when using Ohm's law, any one of the three factors—current, pressure, or resistance—can be computed if the other two factors have known values. Also, the current in any electric circuit increases or decreases in proportion to an increase or decrease of either of the other two factors.

□ **Electric Circuits** □

Because it has been established that electric circuits are the electron paths of an electric system, and because they can be quite complex, it is important that several fundamental facts be thoroughly under-

INTRODUCTION TO ELECTRICITY

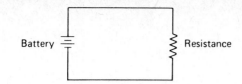

FIGURE 11-31 **Series electric circuit.**

FIGURE 11-32 **Parallel electric circuit.**

FIGURE 11-33 **Series-parallel electric circuit.**

FIGURE 11-34 **Voltage drop in a series circuit with total resistance. (Courtesy of Frigidaire Division, General Motors Corp.)**

FIGURE 11-35 **Voltage drop in a series circuit with only one-half total resistance. (Courtesy of Frigidaire Division, General Motors Corp.)**

stood. No matter how complex any particular circuit becomes, it is one of three general types: (1) series, (2) parallel, and (3) series-parallel.

The *series circuit* is a one-path circuit and can be recognized in two ways. It will never have more than one conductor connected to one terminal and there will be only one path from the source to the load and back to the source (Figure 11-31). In a *parallel circuit,* there are two or more paths between the terminals of the source of current (Figure 11-32). A *series-parallel circuit,* as the name implies, is a combination of both the series and parallel circuits (Figure 11-33).

Every electric load is designed to have a specific resistance and to operate at a certain voltage. The load resistance controls the amount of current at the rated voltage.

Voltage in Series Circuits. When a voltmeter is connected across the total circuit resistance, it will indicate the total voltage drop across the resistance. Figure 11-34 illustrates a voltage drop of 6 V. If the voltmeter is connected across half of the resistance, it will indicate one-half the total voltage, as shown in Figure 11-35. In this case the voltage drop is 3 V, indicating that half of the voltage is used for half of the resistance. If another voltage reading is taken across one-third of the resistance, the voltage for that amount of resistance would be 2 V. This experiment gives us the law of voltage drops in a series circuit.

The total voltage is the sum of all voltage drops.

A common example of this law is a string of Christmas tree lights of the series type. The rated voltage of each bulb is approximately 15 V. When they are connected in a series of eight, the combined voltage totals approximately that of the source, 115 to 120 V.

Current in Series Circuits. When two lamps and a switch are connected in series with a battery, as in Figure 11-36, the current through each part of the circuit is the same. For example, if the current strength in one bulb is 1.5 A, the other will also be drawing 1.5 A. If the switch is opened, the current through the switch becomes zero. Similarly, the current through each lamp becomes zero. Switches controlling electric loads are always in series with the load.

A broken bulb is also an open circuit and would

FIGURE 11–36   **Amperage in a series circuit. (Courtesy of Frigidaire Division, General Motors Corp.)**

FIGURE 11–37   **Resistance in a series circuit. (Courtesy of Frigidaire Division, General Motors Corp.)**

stop the current flow through the switch and the other bulbs. This is the chief reason why series connections for lighting circuits are objectionable. If a house had all the lights connected in series, turning out a light in the bedroom would cause all the lights in the house to go out.

Resistance in Series Circuits. If two resistances are connected in series in an electric circuit, the current passes through both and is, therefore, opposed by both (Figure 11–37). Since each resistance is 2 Ω, the total resistance is 4 Ω because the total resistance is equal to the sum of all resistances. That is,

$$R_t = R_1 + R_2$$

Again, the resistance of the wiring has been ignored.

Voltage in Parallel Circuits.   A parallel circuit is a multiple circuit having more than one path between the two terminals of the source of power. Figure 11–38 is a simple parallel circuit that has a conductor from each of the three lights attached to a common terminal on each side of the source of current. Three separate terminals on each side could also be used and the same result would be obtained. Regardless of the resistance offered by each light bulb, the voltage is the same across all branches.

Current in Parallel Circuits. The law for current or amperage in a parallel circuit states that the total amperage is the sum of the amperage in all branches. That is,

$$I_t = I_1 + I_2 + I_3$$

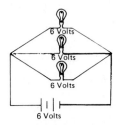

FIGURE 11–38   **Voltage in a parallel circuit.**

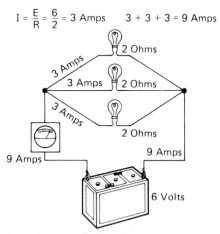

FIGURE 11–39   **Current in a parallel circuit. (Courtesy of Frigidaire Division, General Motors Corp.)**

Each of the three lights shown in Figure 11–39 has a resistance of 2 Ω. Applying the formula $I = E \div R$, we find that each branch has a current flow of 3 A (6 ÷ 2). The total current then is 3 + 3 + 3 = 9 A.

Resistance in Parallel Circuits. Because three times as much current can flow through three parallel con-

Total R = $\dfrac{\text{Total E}}{\text{Total I}}$ = $\dfrac{6}{9}$ = 0.66 Ohms

2 Ohms

2 Ohms

2 Ohms

9 Amps

6 Volts

FIGURE 11-40 Resistance in parallel circuits. (Courtesy of Frigidaire Division, General Motors Corp.)

ductors, it is plain to see that the resistance to the flow is only one-third as great. Therefore, it can be said that the total resistance is equal to the total volts divided by the total amps (Figure 11-40). For example, a total voltage of 6 and a total amperage of 9 are given. The total resistance is 6 ÷ 9 = 0.66 Ω.

Series–Parallel Circuits. Many circuits are neither simple series nor simple parallel, but a combination of both. While this may seem to present a fairly complex problem, it can be readily analyzed by distinguishing one type from the other and applying the correct laws for each type.

If a parallel–series circuit has two circuits, each with four light bulbs, and the two series circuits are connected in parallel, the laws governing series and parallel circuits must be used (Figure 11-41). The laws governing series and parallel circuits will now be applied and the volts, amperes, and the resistance determined.

As mentioned earlier, in a series circuit the total voltage is the sum of all voltage drops. Also, there is an emf of 6 V in each branch, which conforms with voltages in parallel. The amperage in a series circuit is the same in all parts. The total amperage in a parallel circuit is the sum of all the amperages in each of the branches. Therefore, the rate of flow from the batteries is 0.3 + 0.3 = 0.6 A. Now that the total voltage and the total amperage have been determined, the total resistance may be determined by applying Ohm's law.

$$R_t = \frac{E_t}{I_t} = \frac{6}{0.6} = 10 \text{ Ω}$$

5 OHMS   5 OHMS   5 OHMS   5 OHMS

1½   1½   1½   1½

(4 X 1.5 = 6V)

.3   .3

.3   .3

(4 X 1.5 = 6V)

1½   1½   1½   1½

5 OHMS   5 OHMS   5 OHMS   5 OHMS

.6 AMP.   .6 AMP.

6 VOLT

FIGURE 11-41 Series–parallel circuit. (Courtesy of Frigidaire Division, General Motors Corp.)

□ Electricity by Magnetism □

As stated previously, the production of electricity by magnetism is very important because it is the greatest and most common source of electric power. By use of magnetism, electricity can be generated, transmitted, and distributed in unlimited quantities to millions of homes, business establishments, and industrial plants.

Magnets. In order to understand how magnetism can be made to work for us in many practical ways, it is necessary to have a basic knowledge of magnets.

Natural Magnets. The power of a magnet to attract iron was the first electrical phenomenon our ancestors observed. It was discovered that nature provided a natural magnet called magnetite or lodestone (Figure 11-42). This is a species of ore common throughout the earth, but especially abundant in the province of Magnesia in Asia Minor. This type of ore has a property that enables it to attract iron or steel. According to legend, a Cretan shepherd discovered while attending his flocks that the iron nails in his sandals and the iron tip of his staff were attracted to the earth. While digging in the ground, he discovered a rock which stuck to his sandals.

BAR TYPE

HORSE SHOE TYPE

FIGURE 11-43 Artificial magnets. (Courtesy of Frigidaire Division, General Motors Corp.)

FIGURE 11-42 Magnetite or loadstone. (Courtesy of Frigidaire Division, General Motors Corp.)

The earliest practical use of the magnet was a compass used to point out a direction. It was discovered that a suspended bar of magnetite would seek a north and south position. It was for this reason that natural magnets became known as lodestones or leading stones. Magnetic compasses were developed and widely used all over the civilized world as far back as 1492 when Columbus set sail for America.

Magnetic Properties. To understand how and why a compass can determine directions, we must know some of the properties of a magnet. The most common magnetic substances are iron, steel, nickel, and cobalt. Some substances are only slightly magnetic, while others, such as bismuth, act in the opposite direction and are repelled by a magnet. These substances are called diamagnetic substances. The most important of the nonmagnetic substances are copper, brass, and aluminum.

Artificial Permanent Magnets. A magnetic substance can be artificially converted into a permanent or temporary magnet (Figure 11-43). These are known as the bar type and the horseshoe type. They can be produced by stroking pieces of steel or alloys with either a natural magnet or an artificial magnet, or by properly applying an electric current. Steel and alloys are universally used due to the fact that, although they cannot be magnetized as easily as soft iron, they retain or hold their magnetism longer after the source of magnetism has been removed.

FIGURE 11-44 Artificial temporary magnet. (Courtesy of Frigidaire Division, General Motors Corp.)

Artificial Temporary Magnets. A temporary magnet is one that passes magnetic powers only while being subjected to the magnetic force of a permanent magnet. If a piece of unmagnetized soft iron, such as a nail, is suspended from one end of a bar magnet, or bar, as shown in Figure 11-44, a second nail may be suspended from the first one, a third nail from the second, and so forth. Each nail, in turn, becomes a magnet with less force than the preceding one. When the bar magnet is removed from the first nail, however, the others will instantly fall apart. Soft iron has a high permeability, which means that it can be magnetized easily; it also has low retentivity, which means that it loses its magnetism quickly after the source has been removed.

Magnetic Induction. The reason an unmagnetized piece of iron is attached to a magnet is due to *magnetic induction* (Figure 11-45). A nail can be sufficiently magnetized to attract and hold iron filings

INTRODUCTION TO ELECTRICITY

FIGURE 11-45 Magnetic induction. (Courtesy of Frigidaire Division, General Motors Corp.)

even though it is not in direct contact with the permanent magnet. This is one of the most significant factors in the production and utilization of electricity by magnetism.

Magnetic Polarity. Every magnet has two poles, a north-seeking pole or N pole, and a south-seeking pole or S pole. This fact can be readily proven by stroking a common sewing needle with a magnet and floating the needle on a piece of straw in a pan of water. One end of the needle will turn toward the south. The end of the magnet that points toward the north is the N pole of the magnet. The fact that a magnet has poles can be readily proven by dipping a bar magnet into a quantity of iron filings. The filings will cling in tufts to both ends of the bar, but scarcely any will be seen in the center.

Repulsion and Attraction. Another property of magnets is their ability to repel and attract each other. Like poles repel and unlike poles attract. If the N poles of two bar magnets were put together, they would jump apart, but if an N pole of one magnet is placed to the S pole of the other magnet, they will be held together by magnetic attraction. This can be proven by holding one end of a bar magnet close to the end of a suspended bar magnet. If the two ends are of like polarity, the end of the suspended bar will move away from the end of the bar being held. If the polarity is different, the end

of the suspended bar magnet will move toward the other magnet.

The Earth as a Magnet. It can also be proven that the earth itself is a magnet. This is shown by the fact that a magnetized object, such as a compass needle, always points approximately north or south. It has been discovered that there is an S magnetic pole near the geographic North Pole and a magnetic N pole near the geographic South Pole. Inasmuch as the north pole of a compass points to the north and it is known that like poles repel each other, it is logical to assume that the magnetic pole in the north is S.

The magnetic S pole, which is near the north geographic pole, was discovered by Sir James Ross in 1831. Its location was 70° 30′N latitude and 95°W longitude, which is in the north central part of Canada about 1400 miles from the geographic pole. However, in 1905, Captain Roald Amundsen located it a little farther west at 70° 5′N and 96° 46′W. From these two observations, it is believed that the earth's magnetic poles slowly shift their position.

Because the magnetic and geographic poles are separated, a compass needle does not always point true north and south. Its direction also changes as the compass is moved from one place to another on the surface of the earth. The number of degrees by which the needle varies from true north and south is called the *angle of declination.*

On the *agonic line,* which is an imaginary line encircling the earth from one magnetic pole to the other (Figure 11-46), a compass will point true north and south and the angle of declination will be zero.

The magnetic lines of the earth are not parallel to its surface. The angle between the magnetic lines and the surface of the earth is known as the dip or inclination. The inclination can be readily determined in the following manner. Thrust an unmagnetized knitting needle through a cork and another needle through the cork at right angles to the first needle. Support the apparatus on two objects with one needle acting as an axis and pointing east and west. By means of a bent pin or some wax, balance the weight of the first needle until it is parallel with the table. Next, magnetize the first needle by stroking the north end extending from the cork with the N pole of a strong magnet, and in a similar manner the south end of the needle with the S pole of the magnet. It will be noted that the north end of the needle will dip or incline toward the table. The angle

FIGURE 11-46   The earth as a magnet. (Courtesy of Frigidaire Division, General Motors Corp.)

FIGURE 11-47   Magnetic field of flux. (Courtesy of Frigidaire Division, General Motors Corp.)

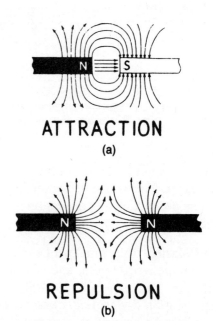

FIGURE 11-48   Flux pattern. (Courtesy of Frigidaire Division, General Motors Corp.)

of inclination will depend on how far north of the magnetic equator the experiment is being performed. At the magnetic equator, there will be no dip. At the magnetic pole, the angle will be 90°. In Chicago, the angle of inclination is 72°50′.

Magnetic Lines of Force. The fact that magnets have the properties of repulsion and attraction indicates that *lines of force* are prevalent through and about them. In 1830, Michael Faraday introduced the idea that the magnetic lines of force passed in a curved line from the N pole of a magnet around the outside to the S pole and on the inside of the magnet from S back to N.

This can be proven as follows: Place a bar magnet beneath a pan of water. Place a cork with a magnetized needle through it in the water near the N pole. The cork will move in a curved path from N to S. This is due to the fact that the forces of repulsion and attraction are continually changing as the relative distance of the moving pole changes.

Magnetic Field of Flux. The area surrounding a magnet in which its magnetic force can be detected is known as the magnetic field of flux (Figure 11-47). There are three important facts that should be noted: (1) none of the lines of force cross each other; (2) all lines are complete; and (3) all lines leave the magnet at right angles to the magnet. Using the fields of force as the basis of magnetism, the many characteristics of magnets can be readily understood.

Lines of force in a magnetic field may be seen by sprinkling iron filings on a piece of paper placed above a bar magnet.

Earlier in our discussion of magnetism we mentioned the fact that like poles of a magnet repel each other while unlike poles attract. Now, let us discuss repulsion and attraction again briefly as they relate to the field of force or the flux pattern produced by a bar magnet. Figure 11-48(a) shows the lines of force produced when two unlike poles are placed near each other, while (b) illustrates the lines of force when two like poles are placed near each other. They attempt to repel each other.

This can readily be seen by placing a piece of paper over the ends of two bar magnets, sprinkling iron filings over the area above the magnets, and observing the pattern made by the filings. One of the magnets should then be reversed and the change in the pattern observed.

UNMAGNETIZED

(a)

N   S

MAGNETIZED

(b)

FIGURE 11-49 **Molecular arrangement of iron. (Courtesy of Frigidaire Division, General Motors Corp.)**

Magnetizing a Substance. From what we have learned thus far, we can assume that every atom and molecule of a magnetic substance has a north and south pole. Each atom and molecule is a tiny magnet.

In a piece of unmagnetized iron, the molecules are jumbled together. In Figure 11–49(a), the N poles are black and the S poles are white. They cancel out each other's force. If we stroke a magnet along the piece of iron, the strong N pole of the magnet attracts the S poles in the iron, causing some of the molecules to shift so that the S poles point to the N pole of the magnet. After each stroke, more and more of the molecules are found to have shifted. After sufficient stroking, all of the S poles are pointing in the same direction, as illustrated in Figure 11–49(b).

When inducing magnetism, more strokes will produce more magnetism. However, there is a limit to which any substance may be magnetized. When all of the molecules are lined up as illustrated in Figure 11–49(b) we say that the magnet is saturated.

There are other properties and characteristics of magnetism; however, we have covered the most common and important phases that have a direct bearing on the subject of electromagnetism, which will be discussed later.

Induced Voltage and Current. Having discussed some of the more important characteristics of magnets and their effect on metals capable of being magnetized, let us now discuss the effect of a magnetic field on metals that are nonmagnetic in nature. It is the effect of magnetic fields on nonmagnetic metals that makes possible the generation of electric current and its practical application to the

FIGURE 11-50 **Induced-voltage downward motion. (Courtesy of Frigidaire Division, General Motors Corp.)**

many electrical components used on refrigeration and air conditioning equipment.

If a conductor is passed through a magnetic field, an electric voltage will be induced in the conductor (Figure 11–50). A horseshoe magnet, a metal conductor such as copper, and the wiring connecting a galvanometer to both ends of the conductor complete an electric circuit. The galvanometer is an instrument used for measuring very small electric currents. As the conductor is moved in a downward motion through the magnetic field of flux between the ends of the horseshoe magnet, an electric current is induced in the conductor and wires. The direction of the current flow is indicated by the arrows in Figure 11–50 and the amount of current is indicated on the galvanometer. Actually, it is not current that is induced. What really happens is that the transfer of some of the magnetic energy to the conductor forces the electrons in the conductor to flow.

When the conductor is moved in an upward direction, the needle of the galvanometer moves in the opposite direction, indicating that the direction of current flow has reversed. Although a flow of current is obtained while the conductor is in motion, the flow of electrons stops when the motion stops. The direction of the induced current depends on the direction in which the flux lines are cut (Figure 11–51).

Of course, the direction of the induced current is

Electricity by Magnetism

FIGURE 11-51 Induced-voltage upward motion. (Courtesy of Frigidaire Division, General Motors Corp.)

FIGURE 11-52 Induced-voltage reversed field. (Courtesy of Frigidaire Division, General Motors Corp.)

also determined by the direction of the field of flux. For example, if the conductor is being moved in a downward direction, the N pole is on the left side and the magnetic flux is to the right as shown in Figure 11-52(a). The induced voltage is to the rear as indicated by the arrow. In Figure 11-52(b), the magnetic poles and field have been reversed and, while the conductor is still cutting the field in a downward motion, the induced voltage is reversed.

An important point to remember regarding in-

duced current in a conductor is the fact that the strength of the current is determined by the position of the conductor in the magnetic field. For example, as the conductor enters the field of flux, the electron flow is weak. It increases to its maximum as the conductor reaches the center of the field. The current flow diminishes as the conductor leaves the magnetic field.

It stands to reason that the stronger the field, the stronger the current and the voltage will be. Therefore, if the poles of the magnet were increased in magnetic strength, the wire would cut more flux lines and a stronger voltage would be induced. The strength of a magnetic pole can be increased by using electromagnetism, that is, by wrapping an electricity-carrying conductor around the magnet's poles.

Since the amount of voltage induced in a wire is determined by the number of flux lines cut per second, a higher voltage can be induced in three ways: (1) speeding up the movement of the conductor, (2) increasing the magnetic field strength, and (3) coiling the conductor so that more turns of the wire will cut the magnetic field. These three factors play an important part in the construction and design of motors and generators.

The basic principle of conductors cutting a magnetic field can now be applied to generators and motors. As stated before, although generators appear to be fairly simple in construction, they are sometimes quite complex in design. Because this is an introduction to electricity, the discussion will cover only the simplest of generators.

In principle, generators and motors are identical. They are both classified as *dynamos*. If the shaft of a dynamo is connected to a prime mover, such as a turbine, and is turned, it pumps electric power out on its lines and is called a generator. If, however, the shaft is connected to a mechanical load and takes electric power in, it is called a motor. In other words, a generator *produces* electric current; a motor *utilizes* electric current.

Direct Current Generation. Earlier it was shown that a battery produces a steady unfluctuating flow of electrons in one direction only. This was called regular dc current. A dc generator also produces a flow of electrons in one direction only. Because of the method by which it is produced, however, the flow of electrons is not steady by comparison and it is called pulsating dc current.

If we use the water analogy, the dc generator may

FIGURE 11-53 Pulsating dc principle. (Courtesy of Frigidaire Division, General Motors Corp.)

FIGURE 11-54 Pulsating dc current generator parts. (Courtesy of Frigidaire Division, General Motors Corp.)

FIGURE 11-55 Dc generator action. (Courtesy of Frigidaire Division, General Motors Corp.)

be compared to a piston-type water pump connected to a closed water system (Figure 11-53). On the upstroke of the piston, the intake valve closes, the check valve opens, and water flows in the direction indicated by the arrows. On the downstroke, the check valve closes, the intake valve opens, and the flow of water momentarily stops until the piston moves upward again. Pulsating dc current can be compared also with the flow of blood in the body, which gets a push every time the heart beats.

Regardless of the complexity of any generator,

there are only three elements involved: (1) a magnetic field produced by magnets, (2) an armature for carrying a coil or wire through the magnetic field, and (3) a prime mover to turn the armature. There are two essential parts of a generator. Figure 11-54(a) shows the stationary frame, which includes the yoke and base with the two pole pieces, which in this case are permanently magnetized. The armature shown at (b) consists of a series of copper wires looped around an iron core and a *commutator*. The commutator is a series of copper segments to which the ends of the wires are attached.

DC Generator Action. To better understand what happens in an armature see Figure 11-55. One turn of the coil and two segments of the commutator have been lifted from the armature and placed in the magnetic field of the pole pieces. Two views of a coil in a magnetic field are shown. At each end of the coil is a segment of the commutator and against the segments are carbon blocks called brushes. This arrangement provides a slipping contact between the rotating armature and the stationary load. For sake of clarity, one-half of the coil and one segment are colored black and the other half is white. Note the direction of the magnetic flux and the direction of rotation.

In view A no voltage is produced because the coil is cutting no flux lines. In this position the brushes are contacting only the insulation between the segments. In view B the coil has moved at right angles. The black side of the coil is cutting downward and the white side is cutting upward. This causes the induced voltage in the coil to flow "in" the white side of the coil and "out" the black side.

In view C the armature has been turned another one-quarter turn. Also, note in view C that the same condition is apparent as in the position in view A except that the coil is upside down. Again the flow of current has stopped because the conductor is no longer cutting the lines of force.

In view D the coil is in the opposite position from view B. Now the black side of the coil is cutting upward and the current is flowing in; the white side is cutting downward and the current is flowing out. Since half of the time the current in the black portion of the coil is flowing in and the other half of the time it is flowing out, it can be said that alternating current is flowing inside the coil. In fact, a rotating coil in a magnetic field always produces alternating current. The alternating current in the loop is caused to flow in only one direction to a load (views C and D).

In views B and D the black brush is marked negative and the white brush is marked positive. The brushes are stationary, and when current is flowing the negative brush is always in contact with the positive brush. So, if it is desired to deliver dc current to a load, a commutator is always used. Since the commutator causes the alternating current in the armature to be delivered to the load in one direction, it is said to *rectify* the current. A single coil rotating in a magnetic field is like an automobile hitting on one cylinder. The output is weak and fluctuating.

Dc voltage, which is produced by a single loop rotating within a magnetic field, can be shown graphically (Figure 11–56). When the black portion

FIGURE 11–56   Graph of dc voltage and current (single loop). (Courtesy of Frigidaire Division, General Motors Corp.)

of the loop is in the upward position, no lines of force are being cut and no voltage is produced. As the loop turns in a clockwise direction, the black portion progressively cuts more lines of force and the voltage increases accordingly to the maximum. As the black portion of the loop progresses toward the downward position, the voltage decreases to zero. The loop is again in the vertical position.

At this point, the white portion of the loop begins to cut the lines of force in a downward direction as the black portion of the loop cuts them in an upward direction. This causes the voltage to increase to its maximum when the loop is horizontal and to decrease again to zero when the white portion has progressed to the downward position. Since, during this process, the black commutator has moved to the white brush and the white commutator has moved to the black brush, the current flows to and from the load in the same direction as it did before the commutator and brushes changed relationship. From this it can readily be seen that even if a one-loop coil were capable of producing sufficient current to light a lightbulb, the light would tend to go out every time the voltage dropped to zero.

The fluctuation of voltage can be reduced by increasing the number of loops and commutator segments. In Figure 11–57 an armature with two loops and a four-segment commutator is shown. When this armature is rotated, the black coil is going to be one-fourth of a revolution behind the white coil. This means that the voltage of the black coil is at zero when the white coil is at its maximum. However, the brushes riding on the commutator contact only the coil producing the greater voltage.

The solid line portion of the graph in Figure 11–57 is the voltage picked up by the brushes. Notice that the voltage is more level with two coils than it was with one coil. It is still a pulsating current, but it does not get all the way down to zero.

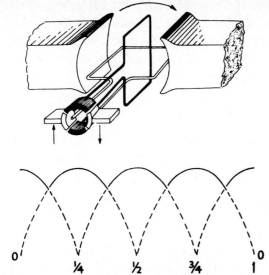

FIGURE 11-57   Two-coil voltage. (Courtesy of Frigidaire Division, General Motors Corp.)

FIGURE 11-58   Four-coil voltage. (Courtesy of Frigidaire Division, General Motors Corp.)

In Figure 11-58 a four-coil armature with eight segments in the commutator is substituted for the two-coil armature and the output is more level. As before, when the armature is rotated, the brushes pick up only the very peaks of each coil voltage. This is still pulsating dc, but only slightly pulsating. The voltage rise and fall is very short, as shown in the graph. If voltage without fluctuation is desired, all that is necessary is to add sufficient coils and increase the number of segments proportionately.

Alternating-Current Generation. While dc current, either regular or pulsating, is a one-way flow of electrons, alternating current (ac) first flows in one

FIGURE 11-59   Alternating current. (Courtesy of Frigidaire Division, General Motors Corp.)

direction, then reverses and flows in the opposite direction.

The water analogy can be used again to illustrate ac current. In Figure 11-59 a pump with a solid piston and pipes containing water are shown. If the piston is moved upward, the water flows in a counterclockwise direction (indicated by the arrows). Pulling the piston downward causes the water to flow in the opposite direction.

AC Generator Action. Since we produce ac current in the rotating coil of any generator, it is just a matter of eliminating the rectifying commutator and employing slip rings (Figure 11-60). Instead of connecting each end of the coil to a commutator segment or half a ring, each end is connected to a complete ring, and a brush is in contact with each end at all times.

In view A of Figure 11-60, the current is flowing "out" the black portion of the coil and flowing "in" through the white portion. The black ring and brush will be called negative and the white ring and brush will be called positive. In view B the coil has rotated through the neutral plane, and the black portion now cuts upward, and the current flow reverses, flowing "in" through the black ring and brush and "out" through the white ring and brush. This reverses the current through the load. The white brush becomes negative and the black brush becomes positive.

A graph of ac voltage, which is produced by a

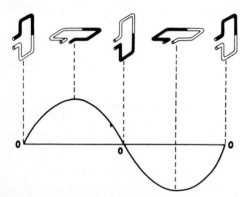

FIGURE 11-60   Ac generator action. (Courtesy of Frigidaire Division, General Motors Corp.)

FIGURE 11-61   Graph of ac voltage and current. (Courtesy of Frigidaire Division, General Motors Corp.)

single loop rotating in a magnetic field, is shown in Figure 11-61. As with dc current generation, whenever the loop is in a vertical position, no voltage or current is produced. Following the black portion of the loop through half a cycle, it can be seen that the voltage increases to its maximum positive and then decreases to zero voltage. At this point, the white portion of the loop begins its downward path through the magnetic field while the black portion begins its upward path. This causes the current to reverse its direction and, because of the slip rings, the voltage increases to the maximum negative and then back to zero. This also means that the current through the load and connecting conductors is being reversed every half turn of the loop.

As in the dc generator, a single loop would produce a very weak and fluctuating current and voltage. While the strength of ac current and voltage

can be increased by adding more loops to the slip rings, the fluctuation cannot be changed in this manner. Fluctuation can be overcome and controlled only by increasing the speed of the armature and keeping this speed constant.

☐ Electromagnetism ☐

Until now, we have established only the basic principle of producing electricity by magnetism as applied to dc and ac generators, which have a rotating armature in a stationary magnetic field.

In order to understand the production of voltage and current in a modern ac generator, as well as the basic principles of transformers and other electrical components, a knowledge of electromagnetism is required. As the name implies, electromagnetism is the production of magnetism by an electric current.

Electromagnets. An electromagnet is like a natural artificial magnet in its attraction, but differs in its control. Although an electromagnet can be powerful in its attraction, it can be turned on and off with the flick of a switch. Electromagnets are used to lift heavy pieces of metal from one place to another, to open and close contacts in relays, to open orifices in solenoid valves, and to turn rotors in dynamos or motors.

If a wire connected from the negative to the positive of a dry cell has been dipped into a pile of iron filings, some of the filings will cling to the wire as shown in Figure 11-62. This proves that any conductor carrying an electric current is surrounded by a field of flux and the current flow produces a magnetic field. If one end of the wire is disconnected from the dry cell, the filings drop off. This proves that the field exists only when the current is flowing.

While the filings are shown clinging to only a portion of the wire, the field surrounds the wire along its entire length. Although a field of flux or magnetism has been produced with a straight, single conductor, no polarity is apparent. To be able to put magnetism by electricity to a practical application, poles must be produced. To produce poles, simply form a straight conductor into a loop (Figure 11-63).

When a conductor is formed into a loop, the flux bends toward the center of the loop and produces a north pole on one side and a south pole on the other. If a small number of loops of wire are com-

246

INTRODUCTION TO ELECTRICITY

FIGURE 11-62 Magnetism by current. (Courtesy of Frigidaire Division, General Motors Corp.)

FIGURE 11-64   Basic electromagnet. (Courtesy of Frigidaire Division, General Motors Corp.)

FIGURE 11-63   Magnetic polarity of a loop. (Courtesy of Frigidaire Division, Motors Corp.)

FIGURE 11-65   Electromagnetism as applied to a dc generator. (Courtesy of Frigidaire Division, General Motors Corp.)

bined, they form a *helix coil*, which produces much stronger poles than a single loop. If a very strong magnetic coil is desired, more turns of wire are built up in layers. This produces a *solenoid coil.*

If a solenoid coil is wound around a piece of iron or other easily magnetized metal, the metal is called the core or pole piece. When the coil is energized, the core becomes temporarily magnetized and an electromagnet is produced. The core of an electromagnet has low retentivity and immediately becomes demagnetized when the flow of current stops. With-

out this characteristic, electromagnets would have no practical use.

A basic electromagnet is shown in Figure 11-64. For the sake of clarity, the illustration shows very few loops of wire as compared to those actually required. Electromagnets can be produced by using either dc or ac current.

Now that the basic principles of electromagnetism have been established, we can discuss the role electromagnetism plays in the generation of electric current. An illustration of how magnetism in the pole pieces of a dc generator can be strengthened is shown in Figure 11-65. Here we see that the pole pieces are wound with coils of wire connected to the brushes. In operation, some of the current produced by the generator is used to strengthen its own poles. An

Electromagnetism

electromagnet can always produce a stronger field than a permanent one. A permanent magnet can also be strengthened by winding an energized coil around it.

The proper polarity of the pole pieces of the generator is established and maintained by winding the poles in the opposite direction. For example, if the pole marked N were wound in the same direction as the S pole, it would also become an S pole and there would be no current generated by the rotating loop.

Electromagnetism and AC Generation. Although the simple ac generator previously described could produce voltage and current, they would both be too weak to be of practical value. By increasing the number of rotating loops and by strengthening the magnetic field through electromagnetism, stronger voltage and current can be produced. Unlike dc generators, ac generators cannot share their current to self-strengthen their pole pieces due to their inability to maintain constant polarity of the poles. They can, however, be strengthened by electromagnetism produced by current supplied from a separate dc generator. All modern power plants use dc generators to strengthen or excite a magnetic field for ac generators. When so employed, the dc generator is called an *exciter* and the ac generator is called an *alternator*.

Alternators. There are basically two types of alternators. One type uses a dc exciter to strengthen the stationary field through which the rotating armature passes as shown in Figure 11–66. This type was used in the early days of ac generation and a few may still be found in operation today in small communities. It should be remembered that both the dc exciter and alternator require a prime mover. In some instances, each has its own driving turbine, but, more often, the same turbine is used for both. This type of alternator was the first used in the generation of ac current, but it had construction limitations which made it unsuitable for the high voltages and currents required for modern transmission over long distances.

The second type is the basic design of a later development called the rotating field type which uses the principles of all modern alternators (Figure 11–67). Here the dc exciter is supplying current to the loop by means of brushes and slip rings. The pole pieces have coils wound from a continuous length of wire. The ends of the wire are connected

FIGURE 11–66 Basic rotating armature alternator. (Courtesy of Frigidaire Division, General Motors Corp.)

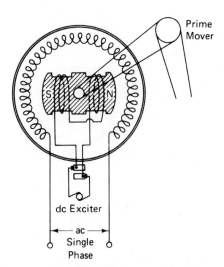

FIGURE 11–67 Ac single-phase alternator (schematic diagram). (Courtesy of Frigidaire Division, General Motors Corp.)

to conductors, which carry the ac current to the transformer and to the transmission lines. As the loop is rotated by the prime mover, the magnetic field sweeps across the pole pieces and, by magnetic induction, produces a voltage that causes the current to flow through the coils and conductors. Since the magnetic field has polarity, the induced current flows in the opposite direction every half revolution.

One of the major advantages of this type of alternator over the rotating armature type pertains to the amount of current carried by the brushes and slip rings. In the rotating armature type, the slip rings are exposed and are subject to arc-overs and short circuits when they are required to carry high voltage and current.

Because the rotating field type of alternator can generate exceptionally high voltage and current from the magnetism produced by a relatively low voltage from the dc exciter, arc-overs and short circuits at the slip rings are avoided.

Power Factor. Since the introduction of fluorescent lights and the mass production of electric-motor-powered refrigeration systems and appliances, the term *power factor* has come into fairly common use. Before entering into this discussion, we should not confuse the power factor with the efficiency of an electrical device.

The power factor of an electrical system is a very complex subject. For our purpose, power factor may be defined as the difference between the amount of power supplied by the power company for a device and that consumed by this device for which the company receives no payment. To understand how such a condition can exist, we must take into consideration that, except for capacitors, there are two classifications of current-consuming devices or loads: a "pure" resistance type and an induction type. Pure resistance loads include items such as incandescent lights and all types of heater elements.

Induction-type devices have coils of wire wound around metal cores, and electromagnetism is produced during their operation. Induction-type devices include transformers, ac motors, relays, solenoid valves, and fluorescent light equipment. For an illustration of how electric current is distributed in a total parallel circuit containing both pure resistance and induction-type loads or devices, see Figure 11–68. The induction motor can be used to drive a refrigeration compressor, the fluorescent light illuminates the interior, the incandescent light can be used for temperature warning, and the heater element can be used to defrost the evaporator.

It should be remembered that the rate of current flow returning to the source is the same as that entering a circuit from the source. The power company is quite concerned as to how the current it provides and gets back has been utilized.

Pure resistance loads have no metal cores requiring magnetism. Therefore, the electric meter records all of the energy consumed by such devices. In these instances, the power factor is said to be 100%.

With induction-type loads, however, some of the current supplied by the power company is used only

FIGURE 11–68 Distribution of current in a circuit. (Courtesy of Frigidaire Division, General Motors Corp.)

to magnetize the metal core while the remaining current overcomes resistance and performs the work. Magnetizing current does no work and, therefore, it is not recorded on the electric meter. If one-fourth of the current being supplied to an induction-type device is required to magnetize the metal and only three-fourths of it is used to produce work, the device is said to have a 75% power factor. The 25% is a direct loss to the power company due to the fact that it has spent money for fuel and equipment to supply current through the resistance of distribution lines, transformers, etc. for which no payment is received.

Realizing that induction-type devices are essential to our way of living and that some current losses are unavoidable, power companies have established a minimum 90% power factor as being acceptable for devices connected to their power lines. Power factor can also be expressed in decimal form but is usually expressed as a percentage.

▢ Electrical Unit Prefixes ▢

The simple units of electrical measure—volts, amperes, ohms, and watts—are rather clumsy when very large or very small quantities are involved. Therefore, a system of electrical unit prefixes is used to indicate large and small quantities. They include kilo, milli, mega, and micro (Table 11–2).

For example, a generator that delivers 1000 W is called a 1-kilowatt (kW) generator. Instead of say-

TABLE 11-2 Electrical Unit Prefixes

Prefix	Meaning
Kilo	Thousand [1000]
Milli	One-thousandth $\dfrac{1}{1000}$
Mega	Million [1,000,000]
Micro	One-millionth $\dfrac{1}{1,000,000}$

ing the flow of current is one thousandth (1/1000) of a volt, we say it is 1 millivolt. If we are testing insulation, we would use the term *megohms* instead of so many million ohms. The term *farad* is used as a unit of measure for electrical capacity but, due to its size, the capacity of capacitors is given in microfarads, or so many millionths of a farad.

☐ Transformers ☐

There are many applications in which transformers play a significant part in the operation of the equipment. Because of the many uses to which they can be put, transformers are manufactured in hundreds of sizes, shapes, and current characteristics. They range in size from several tons in weight for power transmission purposes to a few ounces for radio and television sets. Regardless of the size or shape, they all have the same basic components and function on the same basic principles—self-induction and mutual induction.

Self-Induction (Counter EMF). It was learned earlier that a flow of current through a solenoid coil produces a magnetic field of flux that changes polarity as the direction of current changes. It was also learned that a moving magnetic field produces a voltage and current in a conductor. There is a law of physics that states that for every force there is an opposing force. Whenever ac voltage and current flow through a solenoid coil, each cycle produces an alternating expanding and contracting magnetic field of flux over the adjacent loops of wire in the coil. This induces a voltage in the coil that opposes the applied voltage and current. This voltage is known as counter emf; its value is always less than that of the applied voltage.

In Figure 11-69 the large white arrows at each end of the coil indicate the direction of the applied voltage. The small, black arrows indicate the direc-

FIGURE 11-69 Self-induction (counter emf). (Courtesy of Frigidaire Division, General Motors Corp.)

tion of the counter emf (induced voltage). View I illustrates the relationship between the voltages during the first quarter cycle when the applied voltage is trying to build up. The counter emf opposes its buildup. View II illustrates the second quarter cycle, when the applied voltage is dropping back to zero. The counter emf now is working in the same direction as the applied voltage and is opposing the drop in voltage. Views III and IV also illustrate that the counter emf is always opposing changes in the applied voltage. The current flow resulting from the inner action of these opposing forces is smaller than it would otherwise be. Actually, the counter emf controls and regulates the amount of current through the coil, thus preventing the coil from overheating and burning. It is also interesting to note that because of the opposition between the applied voltage and induced voltage the resulting flow of current is slightly delayed so that its changes lag behind the changes in the applied voltage. This lag of current behind the voltage introduces a timing factor utilized in the operation of self-starting electric motors.

Mutual Induction. The word *mutual* means sharing. Many years ago it was discovered that, by induction, a solenoid coil could share its current and voltage with another coil of wire connected only to a load. In other words, if a coil of wire is connected to a source of current, the current also will flow through an adjacent coil of wire that is connected only to a load.

The principle of mutual induction can be readily demonstrated and explained (Figure 11–70). The coil connected to the switch and to the battery is called the primary coil. The adjacent coil, which is connected only to the galvanometer, is called the secondary coil. The galvanometer represents the load.

The iron core around which the coils are wound is not essential to the principle of mutual induction; it merely provides a stronger field. Mutual currents and voltage also can be induced across an air gap.

When the switch is closed, the primary coil and core become an electromagnet. The buildup of the magnet flux lines, which cut across the coils of the secondary winding, causes electric current to flow through the secondary coil and circuit as indicated by the galvanometer. The flow of current in the secondary circuit occurs, however, only during the instant that the magnetic field is formed. When the switch is opened, the core becomes demagnetized instantly and the collapse of the magnetic flux causes current to flow in the opposite direction. Thus it can be concluded that mutual induction depends entirely on the alternate expanding and collapsing of the magnetic field around the adjacent coils of wire. With dc current, this expanding and collapsing of the magnetic field is accomplished by rapidly making and breaking the circuit. With ac current, the same effect is produced by rapid changes in the magnetic flux as the current alternates.

When used with ac current, the device is called a transformer. A transformer is unique in that it can remain connected to an ac supply and almost no

FIGURE 11–70 Mutual induction. (Courtesy of Frigidaire Division, General Motors Corp.)

current will flow in the primary as long as its secondary circuit is opened. The reason is that the self-induction that takes place in the primary produces a counter voltage almost equal to the applied voltage. Almost no current results from these opposing voltages. When the secondary circuit is closed, however, a current is induced in the secondary coil and produces magnetism that is opposite in polarity to that produced in the primary circuit. This reduces the counter voltage, which is produced in the primary circuit by self-induction and allows a current to flow in the primary circuit. Only because of ac current is a transformer made possible, and only because of transformers is modern transmission of ac current possible.

Construction. Having discussed self- and mutual induction, let us see how they apply to transformers. A transformer core is shaped to form a continuous circular magnetic path (Figure 11–71). Instead of being made of solid metal, the core is made of laminations of thin sheets of soft iron, silicon steel, or other suitable metal. The sheets are insulated from each other by a thin coating of shellac or other nonconductive material. This insulation is done to reduce the strength of *eddy* currents, which would otherwise cause excessive heat and considerable loss of efficiency in the core. Eddy currents are induced currents that flow within the core. They are wasteful, have no purpose, and should be reduced to a minimum.

Transformers   **251**

FIGURE 11-71 Transformer construction. (Courtesy of Frigidaire Division, General Motors Corp.)

FIGURE 11-72 Step-up transformer. (Courtesy of Frigidaire Division, General Motors Corp.)

FIGURE 11-73 Step-down transformer. (Courtesy of Frigidaire Division, General Motors Corp.)

The primary and secondary coils are usually enameled copper wire. The relationship or ratio between the number of turns in the two coils is almost direct. For example, if the primary coil is connected to a 120-V source of current and it is desired to supply the load with 240 V, the secondary coil must have approximately twice the number of turns as the primary coil. This is called a step-up transformer (Figure 11-72). On the other hand, if the source is 240 V and the load requires only 120 V, the secondary coil must have approximately half the number of turns as the primary coil. This is called a step-down transformer (Figure 11-73). The number of turns in the illustration are but a few as compared to the number actually required.

While a transformer can increase or decrease the voltage and current, it cannot generate energy. For example, if the voltage is stepped up from 115 to 230 V and the load requires 2 A and consumes 47 W, the primary coil must be designed to carry 4 A so that the wattage remains the same.

FIGURE 11-74 Typical transformer. (Courtesy of Frigidaire Division, General Motors Corp.)

A typical transformer is illustrated in Figure 11-74. This type differs from the closed-core type shown in Figures 11-73 and 11-74 in that it has a shell-type core. The shell-type core further increases efficiency because it provides two parallel paths for the magnetic field.

Although some heat losses occur in any transformer, they are considered to be one of the most efficient electrical devices made. They are usually more than 90% efficient.

From the preceding explanation, it should be understood how an unlimited variety of voltage combinations can be obtained by varying the number of turns in the primary or secondary coils.*

□ Safety Procedures □

Because approximately 85% of all service calls involve electrical trouble, the service technician is constantly in danger of receiving an electric shock. Safety precautions must be observed to avoid serious personal injury or even death.

1. Always be certain that the electricity is disconnected from the unit before attempting to service electrical components.
2. If it is necessary to work on a system while the electricity is on, be certain that the service technician is not grounded.
3. Never jump a fuse or install a fuse larger than what is recommended. If a unit continues to

*See Frigidaire Home Study Course No. 3 for further reference.

INTRODUCTION TO ELECTRICITY

blow fuses, a problem exists and must be corrected.

4. Always be certain that the electrical connections are tight and are not causing a high resistance to current flow.
5. Make certain that the equipment is properly grounded.
6. Always be sure that the correct voltage is provided before connecting the electricity to the unit.
7. Replace wires whose insulation is frayed or broken.
8. Never attempt to check resistance when electric power to the unit is on.
9. In case of serious electric shock, treat the victim for physical shock.
10. Always use the proper size wire for the current being used.

▢ SUMMARY ▢

☐ The basis of the electron principle is the fact that all known matter or substance regardless of its form—solid, liquid, or gas—is fundamentally composed of very small particles called atoms.

☐ Atoms are made of smaller particles called electrons and protons.

☐ Electrons have a negative electric charge.

☐ Protons have a positive electric charge.

☐ A proton weighs almost 2000 times more than an electron.

☐ Each atom has a core or center called the nucleus.

☐ The nucleus contains protons and neutrons.

☐ A neutron is made up of one electron and one proton.

☐ Electrons rotate at high speeds in orbits or shells around the nucleus.

☐ Matter is generally defined as anything that occupies space or has mass.

☐ When matter has been reduced to its simplest form, each component is known as an element.

☐ An element may be defined as a substance containing only one kind of atom.

☐ As the number of electrons in the elements increase, the number of orbits and their spacing also increase.

☐ In a compound, two or more elements combine chemically to form a distinct substance with the same proportion of elements always present in its composition.

☐ A molecule may be defined as the smallest particle of a compound that is of the same chemical composition as the compound itself.

☐ Many compounds are composed of more than two elements. These compounds would include all the refrigerants commonly used in refrigeration and air conditioning.

☐ Positive charges repel each other, negative charges repel each other, and positive charges attract negative charges.

☐ Free electrons are those electrons that can easily move from one atom to another.

☐ Electricity is the movement of free electrons from one atom and molecule to another.

☐ Any substance with a great number of free electrons will make a good electrical conductor.

☐ Substances that have few free electrons are called insulators.

☐ Except for the splitting of atoms, energy can neither be created nor destroyed.

☐ The six sources of electricity are: (1) friction, (2) chemical action, (3) magnetism, (4) heat, (5) pressure, and (6) light.

☐ In an uncharged state, each atom has an equal number of electrons and protons.

☐ A statically charged insulator or dielectric has a field of force and the laws of attraction and repulsion apply.

☐ An ion may be defined as an electrically charged atom or a group of atoms.

☐ Any substance containing charged ions is capable of conducting an electric current and is called an electrolyte.

☐ Copper and zinc are excellent for use as electrodes. This is due to the fact that zinc has an abundance of negatively charged ions while copper contains an abundance of positively charged ions.

☐ The flow of regular dc current can be compared with the flow of water produced by a centrifugal pump.

☐ Electrical pressure is called electromotive force (emf) and causes the free electrons to flow through the circuit.

☐ The direction of electrical force is always from the negative electrode to the positive electrode.

☐ Any particular electron does not move from one end of the conductor to the other. Thus, electron movement may be compared to a switch engine bumping a standing string of boxcars on a track. When the engine bumps one end of the string, each car progressively bumps the next car.

☐ Volt is the electrical unit of measure of force.

☐ The greater the difference between the concentration of free electrons in the electrodes, the greater the potential difference will be.

☐ Emf or voltage is the pressure produced throughout an electric circuit, while potential difference indicates a ratio of electron potency or concentration between the electrodes of a battery or certain components of other current-producing devices.

☐ Coulomb is the unit of measure of the quantity of electricity.

- Ampere is the unit of measure of current strength or rate of flow of electrons.
- All substances offer resistance to the flow of electrons.
- If the force is strong or the resistance weak, a strong current will flow.
- If the force is weak or the resistance great, only a small current will flow.
- Electrical resistance of wires is dependent on four factors: (1) diameter of wire, (2) length of wire, (3) kind of wire, and (4) temperature of the wire.
- If the resistance of a conductor is increased by any one of the four factors, the current flow is decreased.
- Thermistors are conductors whose resistance to the flow of electricity decreases as their temperature increases.
- Thermistors are being used in the control circuits of refrigeration and air conditioning equipment to provide more accurate control of the equipment.
- The ohm is the unit of measure of resistance of a conductor to the flow of electrons.
- One ohm is the resistance offered by a column of mercury to the flow of one ampere of current having an electromotive force of one volt.
- Ohm's law states that the current in an electric circuit is equal to the pressure divided by the resistance.
- The three general types of circuits are: (1) series, (2) parallel, and (3) series–parallel.
- The series circuit is a one-path circuit and can be recognized in two ways: (1) it will never have more than one conductor connected to one terminal and (2) there will be only one path from the source to the load and back to the source.
- In a parallel circuit, there are two or more paths between the terminals of the source of current.
- A series–parallel circuit, as the name implies, is a combination of both the series and parallel circuits.
- The attraction of an unmagnetized piece of iron to a magnet is due to magnetic induction.
- Every magnet has two poles, a north-seeking pole and a south-seeking pole.
- In magnets, the like poles repel and unlike poles attract.
- The earth is a magnet.
- The area surrounding a magnet in which its magnetic force can be detected is known as the magnetic field of force.
- None of the lines of force cross each other.
- All of the flux lines are complete.
- All of the flux lines leave the magnet at right angles to the magnet.
- Every atom and molecule of a magnetic substance has a north pole and a south pole.
- When inducing magnetism, more strokes will produce more magnetism.

- It is the effect of magnetic fields on nonmagnetic metals that makes the generation of electric current possible.
- If a conductor is passed through a magnetic field, an electric voltage will be induced in the conductor.
- The direction of movement of the conductor and the direction of the field of flux will determine the direction of induced current.
- The strength of the current in a conductor is determined by the position of the conductor in the magnetic field and the strength of the magnetic field.
- Three ways to induce a higher voltage are: (1) speed up the movement of the conductor, (2) increase the magnetic field strength, and (3) coil the conductor so that more turns of the wire will cut the magnetic field.
- In principle, generators and motors are identical. They are both classified as dynamos.
- The three elements involved in any generator are: (1) a magnetic field, (2) an armature, and (3) a prime mover to turn the armature.
- An electromagnet is like a natural magnet, but differs in its control. It can be turned on and off with a switch.
- Electromagnets are used to lift heavy pieces of metal from one place to another, to open or close contacts in relays, to open orifices in solenoid valves, and to turn rotors in dynamos or motors.
- If a solenoid coil is wound around a piece of iron or other easily magnetized metal, the core becomes temporarily magnetized and an electromagnet has been produced.
- An electromagnet can always produce a stronger field than a permanent one.
- The proper polarity of the pole pieces of the generator is established and maintained by winding the poles in the opposite direction.
- The two types of alternators are: (1) stationary field, and (2) rotating field.
- Power factor may be defined as the difference between the amount of power supplied by the power company for a device and that consumed by this device, for which the company receives no payment.
- It should be remembered that the rate of current flow returning to the source is the same as that entering a circuit from the source.
- When ac voltage and current flow through a solenoid coil, each cycle produces an alternating expanding and contracting magnetic field of flux over the adjacent loops of wire in the coil. This induces a voltage in the coil that opposes the applied voltage and current. This voltage is known as counter emf.
- By induction, a solenoid coil will share its current and voltage with another coil of wire connected to a load.
- A transformer can increase or decrease the voltage and current, but it cannot create energy.

1. What is the basis of the electron principle of electricity?
2. How heavy is a charged proton in relation to a charged electron?
3. Of what is a neutron made?
4. Briefly define *matter*.
5. Briefly define *element*.
6. Briefly define *compound*.
7. What is the force that holds an atom together?
8. What makes a good electrical conductor?
9. What are the six sources of electricity?
10. What type of electricity is produced by use of di-electrics?
11. What electric charge does an electron have?
12. When does an atom have an equal number of electrons and protons?
13. State the laws of charges.
14. Can a substance be charged either positively or negatively?
15. Where may electricity produced by chemical action be used in refrigeration work?
16. What is an electrolyte?
17. What metals make good electrodes?
18. What limits the life of a battery cell?
19. What type of electricity flows at an unfluctuating rate?
20. What is another name for electromotive force?
21. Does a particular electron flow the complete length of the conductor?
22. What is the definition of volt?
23. When are two substances said to have an electrical potential difference?
24. Briefly define *coulomb*.
25. Briefly define *ampere*.
26. If an electric circuit has a high resistance, will the current flow be small or large?
27. On what four factors is the resistance of a circuit dependent?
28. How does temperature affect a thermistor?
29. Briefly define *ohm*.
30. State Ohm's law.
31. Name the three general types of circuits.
32. What will be the reading on a voltmeter when connected across the total resistance in a series circuit?
33. How is the amperage in a parallel circuit determined?
34. What is the greatest and most common source of electric power?
35. What was the earliest practical use of the magnet?
36. What is the reason an unmagnetized piece of iron is attracted to a magnet?
37. What are the poles on a magnet named?
38. What is the magnetic field of force?
39. How is the direction of the induced current in a conductor determined?
40. In what three ways can a higher voltage be induced in a conductor?
41. What is the name of the metal around which a solenoid coil is wound?
42. Which magnet can produce a stronger field, a natural magnet or an electromagnet?
43. What are the two basic types of alternators?
44. Briefly define *power factor*.
45. What induces the voltage and current from one coil to another in a transformer?
46. What does counter emf oppose?
47. The current in the secondary of a transformer is in what relation to the current in the primary?
48. Why is a transformer core laminated?

12 Basic Solid-State Theory

After completion of this chapter, the student should:

■ Have a working knowledge of how the various type of solid state devices are made.
■ Know how integrated circuits are manufactured and used.
■ Understand the use of printed circuit boards.
■ Be able to explain how modular circuits are used.
■ Understand how diodes work.
■ Understand the importance of heat sinks used in solid state devices.
■ Know how a thermistor operates.
■ Discuss the purpose of diacs, triacs, and unijunction transistors.
■ Be able to list the various types of power supplies used in solid state circuits.
■ Complete the Exercises with at least 80 percent accuracy.
■ Complete the Laboratory Exercises with at least 80 percent accuracy.

Electronics, or microelectronics, are being used more and more in the control of air conditioning and refrigeration systems. They are extremely more accurate and economical than conventional electromechanical control systems. With the entrance of these types of control systems and controls into the industry, the air conditioning and refrigeration technician's job is more demanding. However, those who take the time and initiative to learn about these systems and how they operate will be rewarded abundantly. A clear understanding of basic electricity is almost a necessity before a meaningful study of electronics can be undertaken. It is therefore recommended that Chapter 11 be reviewed at this time if a previous study of basic electricity has not provided the necessary confidence.

☐ Introduction to Integrated Circuits, Printed Circuit Boards, and Modular Circuits ☐

In response to the demand for smaller control components and the desired accuracy of electronic control circuits, many different manufacturing and construction techniques have been employed in the field of air conditioning and refrigeration control systems. Among the most recent developments are: integrated circuits (ICs), printed circuit (PC) boards, and modular circuits. Of these three, the most popular is probably the PC board device.

Integrated Circuits. The integrated circuit (IC) is a major development in the field of solid-state electronics. As the term "integrated" implies, several components are contained in one package. One semiconductor chip can contain two or more transistors, several resistors and capacitors, as well as many individual diodes. How this is accomplished is our next topic.

As can be seen from Figure 12–1, it is easy for one IC device to replace many separate or discrete components with a single chip by using the various properties of semiconductor elements, such as resistance, capacitance, and conductivity. The circuit shown in Figure 12–1 will in all probability be enclosed in the same-size case now associated with

FIGURE 12-1　Schematic and pictorial diagram of a simple IC device.

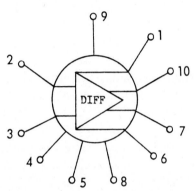

FIGURE 12-2　Schematic symbol for IC differential amplifier.

transistors. In addition to the space-saving features of ICs, the fact that they receive almost identical processing enabling them to be closely matched in characteristics. The closer such circuit elements are matched, the greater the reliability of the circuit.

ICs come in various packages but the one to be discussed at this time is the differential amplifier; a schematic symbol for this circuit is shown in Figure 12-2, while Figure 12-3 is the complete schematic. A differential amplifier is basically an amplifier that responds to the difference between two voltages or currents and effectively suppresses like voltages or currents.

FIGURE 12-3　Differential amplifier, integrated circuit device.

A closer look at Figure 12–3 shows that the signal inputs are applied to the bases of Q1 and Q2 through leads 2 and 3, respectively. These input signals are amplified by Q1 and Q2 and applied to the bases of Q3 and Q4, and the outputs of the emitter followers Q3 and Q4 are available at pins 10 and 7, respectively, of the integrated circuit. The output signals of Q1 and Q2 (the collector signals) are also available at pins 1 and 6 of the device. The constant-current source, Q5, determines the gain of the circuit by determining the emitter currents of Q1 and Q2. Control of the amplifier gain can be accomplished by applying a signal to the base of Q5 through pin 8 and R6. If it is desired to bypass the constant-current source, a signal may be applied directly to the emitters of Q1 and Q2, and pin 4. Lead 9 is the connecting lead for the positive V_{cc}, and lead 5 is the common connection for the device.

It should be noted that all components shown in Figure 12–3 are contained on the IC chip and that it might be necessary to connect external biasing or coupling components depending on the required operating conditions. For example, external resistors might be connected from pin 9 to pins 2 and 3 to provide bias for Q1 and Q2. This type of construction, with primary connections, increases the flexibility of the device; less different types are required, and thus the cost is less. The values given for Figure 12–3 are nominal and are not necessarily those that would be found in any particular IC.

Printed Circuit Boards. A PC board is an insulated surface upon which conductors and other circuit components have been "printed." To be most effective, the printed circuitry should adhere permanently to the base plate. The printed wiring should have low resistance and be able to pass the desired current without overheating. The other printed components, resistors, capacitors, and inductors should remain stable under their rated loads.

By using a mold of the desired shape it is possible to produce a uniform (as far as physical dimensions are concerned) unit. Since printed circuit boards are easily adapted to plug-in units, the elimination of terminal boards, fittings, and tie points result in a substantial reduction in the overall size of the equipment.

Printed circuit boards come in various sizes depending on the type of circuitry being used. Common sense dictates that a printed circuit board using transistors will be smaller than one using electron tubes. Regardless of the size of the printed circuit

board, you can be sure that the same circuit(s) built in the conventional manner would occupy a much larger space. The functions of a printed circuit board are to replace as many of the passive circuit components as possible.

PC Board Component Location. When servicing a piece of equipment that uses a PC board the identification of the circuitry and its components should be fairly easy. Almost all PC board manufacturers have developed a means by which each component can be identified. Most of them use either a numbering system or a lettering system or a combination of both. In most cases a pictorial guide is provided in the equipment control panel for this purpose. Be sure to leave this information in the equipment control panel for future use.

Making Electrical Measurements on a PC Board. Voltage readings can be taken from either side of the PC board without risk of damage. Usually, resistance measurements are taken from the side where the components, resistors, capacitors, etc., are connected. Most all PC boards in use today have a special coating, called a conformal coating, which is a nonconductive protective coating. Measurements taken on these boards require needlepoint meter probes to penetrate the coating. If any damage is done to this protective coating, make certain that it is repaired before placing the board back into permanent service.

When making measurements in very close quarters, avoid shorting out the components by using special probes on the test instruments. These probes reduce the possibility of shorting out the circuit and causing permanent damage to the PC board and its components.

PC Board Removal. Remember to make certain that the circuit is deenergized before removing the PC board to prevent damage to the board by surge currents and for personal protection against possible electrical shock. PC boards are designed to fit snugly into an edge card connector so that proper electrical contact is made without the need for external wiring connections. When removing the PC board from the connector, work it back and forth from side to side while pulling slightly outward. The board should come loose with a little effort. Be careful not to place any undue strains on the board to prevent breaking the circuitry or the components. In some instances, board extractors are provided to aid in removing them from the circuit. It is recommended that the PC boards be returned to the equipment manufacturer for repairs because most service

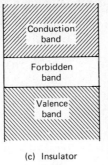

(a) Semiconductor   (b) Conductor   (c) Insulator

FIGURE 12-4 Diagrams of depletion regions for three electrical classifications of matter: (a) semiconductor; (b) conductor; (c) insulator.

shops are not equipped to make these repairs satisfactorily.

Modular Circuits. The basic building block for modular circuits is a specially constructed ceramic wafer on which are mounted or printed the various components and circuits that make a particular electronic stage. A number of these wafers are then stacked to form a module. Since there is a great similarity between parts and circuits in modern electronic equipment, and the construction of the module permits the necessary variations to conform to the various needs of circuits, the module is a very versatile subassembly unit.

Because of the difficulty in repairing the printed circuit or the subminiature portions of a module, servicing will be largely the matter of localizing the defect to a particular module. When the defective module has been found, it will be removed and a replacement module inserted into the circuit.

A semiconductor is a solid that has an area known as the depletion region (Figure 12-4). This depletion region is a great deal larger than the one in the conductor and smaller than the one in the insulator. In most instances we would think that the presence of this region would eliminate the possibility of any electrons in the conduction region of the semiconductor. However, the heat received from the room temperature provides enough energy to the electrons so that some of them can jump the depletion region and enter the conduction region of the material. Semiconductors at room temperature are therefore capable of conducting some electrical cur-

rent. This class of materials is what has made some solid-state control circuits possible.

☐ Diodes ☐

A semiconductor P-N junction is very unique in that it will allow electric current to pass easily in one direction while practically eliminating all current flow in the other direction. The semiconductor diodes have the same action as the vacuum-tube diode, but are much smaller, lighter in weight, and easier to handle. Therefore, the semiconductor diode has almost completely replaced the vacuum tube in modern system design considerations. Solid-state diodes are available in a variety of different configurations. The size and shape determine the current carrying capacity of each diode (Figure 12-5).

The most common materials used for semiconductor diodes are germanium and silicon. Neither of these elements will conduct electric current when in their pure form. They are made conductive by adding certain impurities to their basic composition. When properly doped and when a voltage of the proper polarity is applied, they will readily conduct electric current.

Diode Schematic Symbol. Semiconductor diodes are represented in unit wiring diagrams with their own schematic symbol (Figure 12-6). The bar in the symbol indicates the cathode and the point of the arrow indicates the anode. Remember that electron

Signal diode   Rectifiers   Zener diodes   Reference diodes   **FIGURE 12-5** Typical solid-state diodes.

Diodes

P-type material   N-type material

+ — ← Current flow
Cathode

+ — ← Current flow
Cathode

FIGURE 12-6 **Schematic symbol for solid-state diodes.**

flow is from the N side (cathode or bar) to the P side (anode or arrow point). Note, therefore, that electron flow is opposite to the direction that the arrowhead is pointing.

Diode Connections. All solid-state diodes, regardless of their size or shape, are equipped with two leads, or terminals. The lead that is attached to the P side is the anode lead; the lead that is attached to the N side is the cathode lead.

Diode Marking Methods. Diode manufacturers have devised several methods for determining one lead from the other (Figure 12–7). Some use a band at one end to designate the N side of the cathode. Some use the standard resistor color-code bands to identify the cathode as well as other characteristics of the diode. Still other manufacturers will print a schematic symbol on the side of the diode case. When the diode is to be used in the reverse-bias direction, the manufacturer will usually designate this with an R suffix after the type number.

Zener Diodes. The zener diode, sometimes called an avalanche diode, is unique among diodes in that it is designed to operate in the reverse-bias condition in the avalanche or breakdown region. This device is used as a voltage regulator, a clipper, a coupling device, and in many other functions. The schematic symbol for the zener diode is different in that the vertical cathode line of the standard diode symbol is slightly bent at each end or a letter at the end of the cathode line (Figure 12–8).

Zener diodes are often used as voltage sources because of the change in resistance that occurs in the P-N junction. Therefore, by connecting the zener diode in the circuit parallel to the controlled device, it will regulate the maximum voltage to the device (Figure 12–9).

Reference Diodes. Reference diodes were designed to replace zener diodes on certain applications because of the zener's temperature instability. Reference diodes provide a constant voltage over a wide range of temperatures, the important characteristic of this device. The temperature coefficient is expressed as a percentage of change of reference per degree of Celsius temperature change.

Varactor Diodes. P-N junctions exhibit capacitance properties because the depletion area represents a dielectric and the adjacent semiconductor material represents two conductive plates. Increasing the reverse bias decreases the capacitance while increasing the forward bias increases it. When the forward bias is large enough to overcome the barrier potential, the resultant high forward conduction destroys the capacitance effect, except at very high frequencies. Therefore, the effective capacitance is a

Diode marked with schematic symbol

Cathode

Cathode marked with band

Cathode

Cathode end physically larger

Cathode

Cathode

Beveled end is cathode

Cathode

Cathode end forms mounting bolt

FIGURE 12-7 **Methods of marking diodes. (Courtesy of Billy Langley,** *Solid State Electronic Controls for Air Conditioning and Refrigeration,* **© 1989, p. 60; reprinted by permission of Prentice-Hall, Inc., Englewood Cliffs, N.J.)**

BASIC SOLID-STATE THEORY

FIGURE 12-8   Zener diode schematic symbols.

FIGURE 12-9   Zener diode as a voltage regulator. (Courtesy of Billy Langley, *Solid State Electronic Controls for Air Conditioning and Refrigeration*, © 1989, p. 65; reprinted by permission of Prentice-Hall, Inc., Englewood Cliffs, N.J.)

FIGURE 12-10   Schematic symbol for varactors.

FIGURE 12-11   Tunnel diode symbol. (Courtesy of Billy Langley, *Solid State Electronic Controls for Air Conditioning and Refrigeration*, © 1989, p. 68; reprinted by permission of Prentice-Hall, Inc., Englewood Cliffs, N.J.)

function of the external applied voltage. This characteristic is undesirable in conventional diode operation, but is enhanced by special doping in the variactor or variable-capacitance (varicap) diodes. There are various schematic symbols used to designate varactor diodes (Figure 12-10).

Tunnel Diodes. The tunnel diode, sometimes called an Esaki diode, makes use of very heavily doped P and N materials in a two-terminal device (Figure 12-11). There is a heavy concentration of carriers that interact readily without passing over the potential barrier. Instead, the charge carriers pass through the potential hill with an effect known as tunneling. When the forward bias is increased in these diodes the current is also increased and its behavior is similar to that of a standard diode. This current and voltage are termed valley current and valley voltage.

During this process, the tunnel current also increases to a peak. This current is termed peak current. This process is very stable and when the peak current is flowing, the diode switches from a low-resistance to a high-resistance device.

In the reverse-bias condition the height of the potential hill is increased, which causes the minority electrons to tunnel from the P material to the N material. The current is also in the reverse-bias condition.

☐ Semiconductor Heat Sinks ☐

In semiconductors, as in any other electrical or electronic device, as the electrons flow through them a certain amount of resistance is encountered. The amount of resistance is in relation to the type of material used in the device, the type of joint at the P-N junction, and the amount of current flowing through the circuit. As the current flows through this resistance, a certain amount of heat is produced for each watt of current flow. Heat can also be generated when making soldered connections during the installation of the semiconductor into the circuit. If this heat is not removed from the device, it will eventually become overheated and will probably be destroyed, and/or its power rating will be reduced.

The usual ways of dissipating this heat from the device are from the case to the ambient air, from the case to the frame of the unit to which it is attached, and through increasing the surface area of the case of the device, such as the fins on a lawn mower engine (Figure 12-12). These fins act as a heat sink. That is, they radiate the heat from the semiconductor faster than is possible without them.

FIGURE 12-12   One type of semiconductor heat sink. (Courtesy of Billy Langley, *Solid State Electronic Controls for Air Conditioning and Refrigeration*, © 1989, p. 71; reprinted by permission of Prentice-Hall, Inc., Englewood Cliffs, N.J.)

Many times during the installation of a semiconductor, silicon grease will be applied to the surfaces where the metal meets. This silicon grease increases the thermal conductivity of the metal, which allows a faster transfer of heat from the device.

☐ Solid-State Transducers ☐

A transducer, or transfer-producer, is a device that converts one form of energy to another form. It can be used to convert magnetic, thermal, electrical, chemical, and optical forms of energy into electrical voltages and electrical currents. These converted voltages and currents may be used either directly or indirectly to operate other control circuit components. Therefore, they are very adaptable to the control circuits used in air conditioning and refrigeration systems. Solid-state transducers are now readily available and are being used to replace earlier electromechanical types of transducers.

☐ Thermistors ☐

A thermistor is a device that will have a change in resistance with a change in temperature. There are thermistors that have a nonlinear resistance change in temperature, and those that have a linear change in resistance with a change in temperature. That is, some (nonlinear) do not have an equal change in resistance for each degree of temperature change, whereas others (linear) have an equal change in resistance with each change in temperature. However, this change in resistance is the same each time for each resistor. Therefore, they are very useful as signal-producing devices for use in air conditioning and refrigeration control systems (Figure 12–13). The thermistor is a single-layer semiconductor. It is extremely temperature sensitive, which allows it to be very useful as a control circuit signaling device.

Thermistors operate on the electron–hole pair theory. On an increase in temperature of the semiconductor, the number of electron holes is increased because of the thermal agitation of the atoms. This increase in the number of electron–hole pairs causes a drop in the resistance of the material and the electrons are allowed to flow more freely. Actually, a

FIGURE 12–13 A thermistor as a transducer. [Courtesy of Billy Langley, *Solid State Electronic Controls for Air Conditioning and Refrigeration*, © 1989, p. 75; reprinted by permission of Prentice-Hall, Inc., Englewood Cliffs, N.J.]

FIGURE 12–14 Schematic symbols of (a) directly heated and (b) indirectly heated thermistors. [Courtesy of Billy Langley, *Solid State Electronic Controls for Air Conditioning and Refrigeration*, © 1989, p. 76; reprinted by permission of Prentice-Hall, Inc., Englewood Cliffs, N.J.]

thermistor can sense a change in temperature as small as one millionth of a degree Celsius.

The resistance change of the semiconductor is about five times as great as that of a comparable wire-wound resistor. Because of this, less amplification is required with thermistors than with wire-wound sensors. Also, the thermistor signal is less affected by stray voltage from neighboring circuits. Humidity can also be detected by silicon pressure transducers.

Thermistor Schematic Symbols. The schematic symbol used for transistors depends on whether or not the thermistor is directly heated or indirectly heated (Figure 12–14). Indirectly heated thermistors are used in situations where precision temperature measurements and compensation are required. Directly

FIGURE 12–15 Some popular shapes and sizes of thermistors. (Courtesy of Billy Langley, *Solid State Electronic Controls for Air Conditioning and Refrigeration*, © 1989, p. 76; reprinted by permission of Prentice-Hall, Inc., Englewood Cliffs, N.J.)

heated thermistors are used in voltage regulators, electronic time-delay circuits, and vacuum gauges.

Thermistor Types. The small size of thermistors makes them very popular for use in control circuitry. They can be installed in places that are usually inaccessible to other types of temperature-sensing devices. They can be made in the shape of discs, washers, rods, or beads, depending on where they are going to be applied (Figure 12–15). As an example, the temperature overload used in compressor motors is a thermistor embedded in the hot spot of the winding to protect it from overheating. The thermistors used for this purpose are very small and light in weight, usually no larger than a match stem and about ½ in. (12.70 mm) long.

PTC and NTC Thermistors. Thermistors are classified according to whether they have a positive temperature coefficient (PTC) or a negative temperature coefficient (NTC). PTC means that for each degree of temperature increase, the resistance of the thermistor will increase a certain amount. NTC means that for each degree of temperature increase, the resistance of the thermistor will decrease a certain amount. The reverse of this is also true. Most applications require NTC thermistors; however, some applications require PTC thermistors for proper operation.

Motor Starting Using a PTC Thermistor. In a good many single-phase compressor motors, the PTC is replacing the motor starting relay used in previous years. As the system control demands operation and the circuit is first energized, the PTC is cold and therefore has a low resistance to the flow of electricity. Most of the applied electric current is allowed to pass to the starting widing of the compressor motor. As the motor gains speed, the thermistor automatically heats up until its switch temperature is reached. At this point the thermistor rapidly changes from a low-resistance device to a high-

resistance device. While it is in the high-resistance state, virtually no electric current flows to the starting winding of the motor. The current flow through the start winding at this point is negligible and the winding is considered deenergized.

It should be noted that the PTC thermistor becomes less effective on rapid-start conditions because the temperature of the material increases with each additional start. Therefore, the resistance will be higher on each successive start.

☐ Pressure-Sensing Transducers ☐

The solid-state pressure-sensing devices are used for sensing the refrigerant pressures in the high and low sides of the system. The resistance of these devices changes with a change in pressure because of the resultant change in pressure (Figure 12–16). As these devices are used in refrigeration systems, they will open and close an electric circuit when the pressures reach predetermined points. They are designed to take the place of the electromechanical devices, which were popular on earlier models. Pressure-sensing transducers are designed to be installed directly in the refrigerant circuit and react to changes

FIGURE 12–16 Circuit of a pressure-sensitive transducer. (Courtesy of Billy Langley, *Solid State Electronic Controls for Air Conditioning and Refrigeration*, © 1989, p. 80; reprinted by permission of Prentice-Hall, Inc., Englewood Cliffs, N.J.)

FIGURE 12-17 **Refrigerant pressure sensor. [Courtesy of Billy Langley, *Solid State Electronic Controls for Air Conditioning and Refrigeration,* © 1989, p. 81; reprinted by permission of Prentice-Hall, Inc., Englewood Cliffs, N.J.]**

FIGURE 12-18 **Construction and operation of an LED. [Courtesy of Billy Langley, *Solid State Electronic Controls for Air Conditioning and Refrigeration,* © 1989, p. 87; reprinted by permission of Prentice-Hall, Inc., Englewood Cliffs, N.J.]**

in the pressure, and resulting temperature, sensed there (Figure 12-17).

▫ Light-Emitting Diodes (LEDs) ▫

The standard P-N junction diode can produce an electrical current when it absorbs light energy. However, they can also emit light. This emission of light occurs when the diode is under a forward-bias condition and is due to the recombining of the electrons and holes.

LED Operation. The free electrons moving from the N-type material scatter in the P-region when the diode is in the forward-bias condition. After moving to the P-region, these free electrons are outnumbered by the holes and they gradually recombine. In the recombining process, a free conduction electron fills a vacancy in the covalent structure of the atom and at this time becomes a valence electron. During this recombining process the electron loses some of its energy in the move from the conduction band to the valence band. When the diode is made from silicon or germanium, the energy lost by the recombining process is lost in the form of heat which is dissipated to the ambient air. However, if the diode is made from other forms of semiconductor material, such as gallium arsenide and gallium phosphide, the energy lost in the recombining process is in the form of light (Figure 12-18).

The light emitted by the diode has a frequency that is directly related to the amount of energy lost by the recombining of electrons with their holes. The greater the amount of gap in the diode, the more energy the recombining electron loses and the greater the frequency of the emitted light. That is, the light

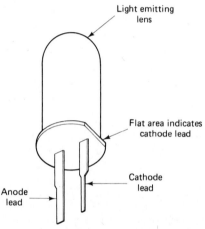

FIGURE 12-19 **Cathode lead identification. [Courtesy of Billy Langley, *Solid State Electronic Controls for Air Conditioning and Refrigeration,* © 1989, p. 88; reprinted by permission of Prentice-Hall, Inc., Englewood Cliffs, N.J.]**

has a lower wavelength. The diodes made from gallium arsenide emit energy in the infrared region of the light spectrum. Those diodes that are made from gallium phosphide emit light energy in the red region of the light spectrum. These diodes can be manufactured to emit light energy in any region of the spectrum by properly combining various materials and the use of proper doping levels. A magnifying lens is used to help in making the emitted light more usable.

In the wiring diagram, the schematic symbol used for the LED shows the arrows pointing away from the diode (Figure 12-19). In most circuits there is

FIGURE 12-20 Pictorial and schematic representation of an SCR.

☐ Silicon-Controlled Rectifiers (SCRs) ☐

The silicon-controlled rectifier is a four-layer (PNPN), three-junction (lead) semiconductor device. The three leads are the anode, cathode, and gate. The anode and cathode are very similar to the anode and cathode of the standard diode. In the SCR, the gate serves as the control point for its operation (Figure 12–20). It is normally an open circuit until a proper signal is applied to the gate element, at which time it rapidly switches to a conducting state.

Circuit Control with an SCR. The SCR can be used to control a high-power circuit (Figure 12–21). The breakover voltage with switch 1 open is higher than the voltage source (*E*); therefore, the SCR is not conducting and there is no current flow through the load. When switch 1 is closed, the breakover voltage is reduced (the forward blocking region is reduced) because of the voltage applied to the gate. The gate voltage causes the SCR to conduct, and current flows through the load. This arrangement enables a low-power line to control a high-power circuit.

☐ DIACs, TRIACs, and Unijunction Transistors ☐

In many applications, DIACs, TRIACs, and unijunction transistors (UJTs) are used in the same circuits. The use of silicon-controlled rectifiers (SCRs) in the same circuits can provide added control over a given piece of equipment. In application, TRIACs and SCRs are used as control devices. DIACs and UJTs in the circuit are used for triggering the TRIACs and SCRs.

DIACs. The DIAC is a three-layer device that can pass current in both directions. Both of the junc-

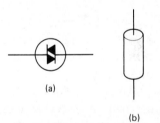

FIGURE 12-21 SCR being used to control high-power circuits.

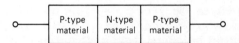

FIGURE 12-22 DIAC construction. (Courtesy of Billy Langley, *Solid State Electronic Controls for Air Conditioning and Refrigeration,* © 1989, p. 99; reprinted by permission of Prentice-Hall, Inc., Englewood Cliffs, N.J.)

FIGURE 12-23 DIAC schematic symbol (a) and casing (b). (Courtesy of Billy Langley, *Solid State Electronic Controls for Air Conditioning and Refrigeration,* © 1989, p. 99; reprinted by permission of Prentice-Hall, Inc., Englewood Cliffs, N.J.)

tions are doped heavily and evenly so that each one is just like the other (Figure 12–22). The DIAC is capable of switching voltages up to a few hundred volts. It is similar in appearance to a standard resistor used in electrical circuits (Figure 12–23).

DIAC Operation. The DIAC can pass voltage in either direction, and in this respect it acts much like two zener diodes connected in series but in opposite

FIGURE 12-24 Two zener diodes representing one DIAC. (Courtesy of Billy Langley, *Solid State Electronic Controls for Air Conditioning and Refrigeration,* © 1989, p. 99; reprinted by permission of Prentice-Hall, Inc., Englewood Cliffs, N.J.)

FIGURE 12-26 Typical trial circuit. (Courtesy of Billy Langley, *Solid State Electronic Controls for Air Conditioning and Refrigeration,* © 1989, p. 102; reprinted by permission of Prentice-Hall, Inc., Englewood Cliffs, N.J.)

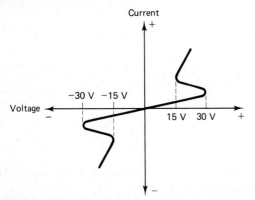

FIGURE 12-25 DIAC operation at the breakover voltages. (Courtesy of Billy Langley, *Solid State Electronic Controls for Air Conditioning and Refrigeration,* © 1989, p. 99; reprinted by permission of Prentice-Hall, Inc., Englewood Cliffs, N.J.)

FIGURE 12-27 Schematic symbol for a TRIAC. (Courtesy of Billy Langley, *Solid State Electronic Controls for Air Conditioning and Refrigeration,* © 1989, p. 102; reprinted by permission of Prentice-Hall, Inc., Englewood Cliffs, N.J.)

directions. The DIAC operates as a breakover device (Figure 12–24). In operation, when the applied voltage is increased, but still below the breakover voltage, the amount of current it will pass is reduced. Therefore, it is said that the DIAC has a negative resistance (Figure 12–25). Its primary use is for triggering another device, and when either a positive or a negative voltage reaches the breakover voltage rating of the DIAC, the resistance of the DIAC rapidly changes from high resistance to low resistance and the current is allowed to flow through the circuit.

TRIACs. The TRIAC is a semiconductor device capable of performing the same function as two SCRs connected in reverse parallel and can therefore be used for controlling an ac current load. It is a three-lead device and is often used as an ac switch. TRIACs can be triggered into the conductive state in both directions by a gate signal similar to that of an SCR. They are generally limited to relatively low current-carrying applications and cannot therefore replace an SCR in high-current applications (Figure 12–26). The schematic symbol for the TRIAC in-

cludes a gate (*G*) and two terminals (terminal 1 and terminal 2; see Figure 12–27).

TRIAC Operation. The TRIAC has high resistance and will block current in either direction until the proper triggering signal to the gate is sensed. When this signal is sensed the TRIAC will conduct in either direction, depending on the type of signal received. The TRIAC operates much like two SCRs connected in reverse parallel. The major difference between the SCR and the TRIAC is the triggering polarity of the gate signals. The TRIAC can be triggered ON when either one of the four following conditions exist:

1. T_2 positive; V_g positive
2. T_2 positive; V_g negative
3. T_2 negative; V_g positive
4. T_2 negative; V_g negative

Conditions 1 and 4 are the most popular because the least amount of gate current is required for triggering under these conditions.

After the TRIAC has been triggered ON in either direction, it is similar in action to the SCR due to the fact that the voltage drops to a very low value. The TRIAC will remain ON until the current drops below the holding current value. The gate of the TRIAC cannot be used to turn OFF the TRIAC. The TRIAC will turn off when the holding current drops below the holding value.

Since the TRIAC is either fully ON or fully OFF, there is no power being converted to heat which must be dissipated, resulting in a loss of power. Also, in the operation of the TRIAC there is no reverse breakdown of high voltages and currents because when the voltage reaches a certain value the TRIAC will turn ON and pass this high current.

TRIAC Applications. TRIACs are being used to replace the starting relay used on single-phase motor compressors used in refrigeration systems. They are popular because they operate much more rapidly and are more trouble-free than the electromechanical types of starting relays.

In operation, when the compressor is first started, the current draw to both motor windings is very high. This high current causes the TRIAC to be triggered and pass the high current. As the motor comes up to speed, the current draw is reduced because of the counter-emf generated in the motor windings. When the current draw of the compressor drops below the "holding current" of the TRIAC, it is no longer triggered and is OFF, no current is flowing, and the motor starting winding is disconnected from the circuit (Figure 12–28).

Unijunction Transistors (UJTs). The unijunction transistor is a three-terminal semiconductor device. It is sometimes referred to as a double-base diode. Its characteristics are quite different from those of the conventional, two-junction transistor. Some of its features are: (1) a stable triggering voltage, (2) a very low value of firing current, (3) a negative resistance characteristic, and (4) a high pulse current capability. UJTs are used in oscillators, timing circuits, voltage- and current-sensing circuits, SCR triggering circuits, and bistable circuits.

UJT Operation. The unijunction transistor exhibits three distinct regions of operation: (1) the cutoff region, (2) the negative resistance region, and (3) the saturation region. Figure 12–29 shows a curve depecting these regions. That portion of the emitter current graph shown at the left of the zero reference is where the emitter is either reverse biased or neutral, and this is the cutoff region, which ends

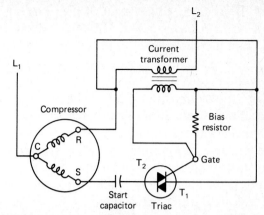

FIGURE 12–28 TRIAC used as a motor starting device. [Courtesy of Billy Langley, *Solid State Electronic Controls for Air Conditioning and Refrigeration,* © 1989, p. 104; reprinted by permission of Prentice-Hall, Inc., Englewood Cliffs, N.J.]

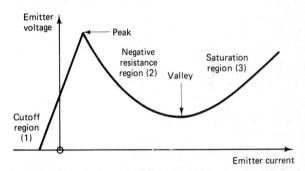

FIGURE 12–29   UJT regions of operation.

FIGURE 12–30   UJT equivalent circuit.

at the point marked "peak." At the "peak point" the current between base one and the emitter increases rapidly because now a forward-bias condition exists. That part of the resistance of the silicon bar (R_{B1} in Figure 12–30) between base 1 and the emitter is now minimal, and the emitter current is increasing while the emitter voltage is decreasing. This, in effect, is the negative resistance region shown between the peak valley points of Figure

12–29. Beyond the valley point of the resistance of the device it behaves as a positive resistance; current again increases slowly with voltage. This area of positive resistance is the saturation region of the unijunction transistor.

UJT Applications. The UJT is basically used in applications where timing and switching actions are needed. These devices are used in the air conditioning and refrigeration industry, where the precision control of temperature is desirable. When used in the precision control of temperature, the UJT is generally used as a triggering device for a TRIAC, which then turns on and off the ac electrical power.

Usually, the UJT is fired by a capacitor which charges until a high-enough voltage is reached to forward bias the emitter junction. After firing, the UJT returns to the nonconducting state. Therefore, the UJT is a voltage-sensitive device that can be used as an extremely accurate pulse generator to initiate a switching circuit.

☐ Inverters ☐

An inverter may be defined as a digital device having an output which is opposite to that of its input. For example, if it has an input of 1, an output of 0 is produced. A simple inverter circuit consists of one bipolar NPN transistor and two resistors (Figure 12–31). This circuit has four external connections, with each having a certain function, as follows: one for the input signal, one for the output signal, one for the power supply, and one for the common

FIGURE 12–31 Simple inverter circuit. (Courtesy of Billy Langley, *Solid State Electronic Controls for Air Conditioning and Refrigeration,* © 1989, p. 111; reprinted by permission of Prentice-Hall, Inc., Englewood Cliffs, N.J.)

power supply and the signal ground. When this circuit is produced in an integrated circuit configuration, there are four leads extending from the package which are used for making the necessary connections into the circuit for which it is intended to operate.

Even though this is a simple integrated circuit, it has the basic components required to amplify a analog input voltage. In addition, it can perform the basic digital operation of inversion of the input signal. For example, if a logical 1 is presented (2 or more volts) at the input terminal, the transistor will turn on completely and the output is a logical 0 (0 V), the transistor will turn off completely and the resulting output voltage will be at a logical 1 level (5 V). Therefore, because this circuit inverts a logical 1 input to a logical 0 output, and vice versa, it is referred to as an inverter.

☐ Transistor as an AC Amplifier ☐

A transistor is a device that makes use of a small change in current to produce a large change in current, voltage, or power. The transistor may therefore function as an amplifier. An amplifier is a device that increases the voltage, current, or power level of a signal applied to that device. The output of an amplifier may not be an exact replica of the input waveshape. When used in the control systems of air conditioning and refrigeration systems, the transistor ac amplifier is used to increase the small input signal currents and voltages to a large output signal to control the power supply to the equipment.

☐ Analysis of Two-Junction Transistors ☐

There are two fundamental types of junction transistors. These are the NPN and the PNP types. Their basic construction or chemical treatment is as implied by their names. That is, either a section of the P material is separated by junction barriers from two sections of N material, or a section of N material is separated by junction barriers from two sections of P material (Figure 12–32).

Note that these devices are not simply two-junction diodes. The base section is extremely thin compared to the collector and emitter sections. Also, the surface area of the emitter–base junction is less than that of the collector–base junction.

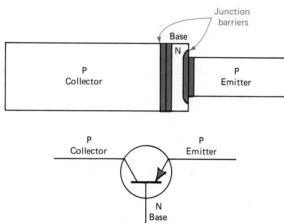

FIGURE 12-32 **Transistor cross-section diagrams and schematic symbols.**

FIGURE 12-33 **Emitter biasing a PNP transistor.**

□ Gain in Transistor Circuits □

Gain is the ratio of output voltage, power, or current with respect to an input reading in the same unit of measurement. Each transistor configuration will give a different value of gain even though the same transistor is being used. The configuration used is a matter of design consideration; however, the technician is interested in the ratio of output with respect to the input. This is important in determining whether or not the transistor is performing its circuit application function. Transistor specifications usually state the normal current gain for a given transistor, at a particular voltage.

Generally, if the output circuit has a low impedance, the transistor provides a large current or power gain. Conversely, a high output impedance is an indication of a voltage gain. Some transistor circuits have a low input impedance; others have high impedances. Some transistor circuits have low output impedances and others high output impedances.

□ Temperature Compensation □

Temperature rise is a problem in transistors because as the temperature goes up, it tends to increase the flow of electrons in both P and N materials. This allows more current to flow through the transistor for the same applied voltages. It should be mentioned that as the state of the art progresses, better transistors are being developed which are not as sensitive to rises in temperature. However, in the older types this condition causes more current to flow, hence more heat, until the transistor is entirely overheated and damaged. This effect tends to be cumulative and is commonly referred to as thermal runaway. It should be noted that a decrease in temperature would decrease the electron flow. Effects of changes in transistor parameters and temperature drifts may be greatly reduced by the insertion of a resistor in the emitter circuit (Figure 12–33).

□ Power Supplies □

The use of P-N junction rectifiers in the design of power supplies for electronic equipment is very popular. They are used because they have no requirement for filament power, the ability to begin immediate operation without the need for warm-up time, a low internal voltage drop substantially independent of the load current, low operating temperature, and generally small physical size.

Single-Phase DC Power Supplies. Electronic control circuits many times require dc voltage rather than ac voltage. This is the time when dc power supplies are used. Most electronic circuits require dc

FIGURE 12–34 Basic components of an electronic control system dc power supply.

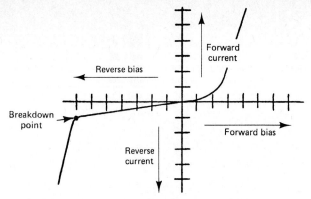

FIGURE 12–35 Diode characteristic curve.

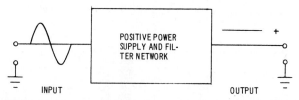

FIGURE 12–36 Positive voltage output from an ac input.

voltage for power operation. It is the purpose of the dc power supply to provide the necessary regulated dc voltage. An electronic control system has a basic set of components (Figure 12–34). In this system the transformer is used to change the ac voltage to either a higher or a lower voltage as required for system operation. The diode rectifier is the device used to change the ac voltage supply to the required dc voltage. The filter component is the section used to level out the output waveform of the diode rectifier device. The purpose of the voltage regulator is to maintain a very nearly constant voltage level over the operating range of the control circuit. The voltage divider is used to distribute the voltage at the proper level to the different loads in the circuit. It should be noted that not all control circuits make use of all the components in Figure 12–34. The particular use and requirements of the system are taken into consideration when designing the type of components required.

Diodes. Semiconductor materials treated to form P-N junctions are used extensively in electronic control circuitry. Variations in doping agent concentrations and physical size of the substrate produce diodes that are suited for different applications.

There is a wide variety of the type and size of diodes. Figure 12–5 shows a limited representation of the wide assortment in the types of case design. However, the shape of the characteristic curves of these diodes is very similar (Figure 12–35).

Rectifier Diodes. Rectifier diodes are used primarily in power supplies. These diodes are primarily of the silicon type because of this material's inherent reliability and higher overall performance compared to other materials. Silicon allows higher forward conductance, lower reverse leakage current, and operation at higher temperatures compared to other materials.

Rectification is described as the changing of an alternating current (ac) to a unidirectional or direct current (dc). The normal P-N junction diode is well

suited for this purpose, as it conducts very heavily when forward biased (low-resistance direction) and only slightly when reverse biased (high-resistance direction).

We will use a block diagram to illustrate this. Figure 12–36 shows a power supply having an ac input and a dc output from a block labeled "positive power supply and filter network." Although Figure 12–36 shows a power supply that provides a unidirectional current which causes a positive voltage output, it might well be designed to furnish a negative voltage output.

The question probably arises of why a change of ac to dc is necessary. To answer, simply, is that for proper operation many electronic circuits depend on dc voltage. As already pointed out, the P-N junction diode conducts more easily in one direction than in the other. Transistors are also unidirectional, and a constantly alternating source voltage would be undesirable.

Half-Wave Rectifier. P-N junction diodes can also be used as half-wave rectifiers (Figure 12–37). A half-wave rectifier is one that uses only half of the input cycle to produce an output.

Full-Wave Rectifier. The P-N junction diode works just as well in a full-wave rectifier circuit (Figure 12–38). The circuit shown has a negative voltage output; however, it might just as well have a positive

BASIC SOLID-STATE THEORY

FIGURE 12–37 **Positive voltage output half-wave P-N junction diode rectifier.**

FIGURE 12–38 **Unfiltered negative output full-wave rectifier.**

voltage output. This can be accomplished by either changing the reference point (ground side of R_L) or by reversing the diodes in the circuit.

Full-Wave Bridge. The full-wave bride is a combinational-type power supply (Figure 12–39). Part (A) shows a simplified schematic drawing and part (B) shows the entire schematic, including the filtering network. It is a typical arrangement of the full-wave bridge combination power supply, quite often called the "economy" power supply.

In Figure 12–39, CR1 and CR3 form the full-wave rectifier circuit and C2A, C2B, and $L2$ form the filter network. R1 is a current-limiting resistor used to protect the diodes from surge currents. $R4$ is the bleeder

resistor and also assures that the power supply has a minimum load at all times. CR1, CR2, CR3, and CR4 form the bridge rectifier circuit with $L1$, $R2$, C1A and C1B doing the filtering; $R3$ is a bleeder resistor and assures that the bridge always has a minimum load. Each circuit, itself, works in the conventional manner.

☐ Filter Circuits ☐

The filter circuits used in power supplies are of two general types, capacitor input and choke input. There are several combinations that may be used,

FIGURE 12–39 **Full-wave bridge combination power supply.**

A CAPACITOR INPUT FILTER

B CHOKE INPUT FILER

FIGURE 12-40 Filter circuits: (a) capacitor input filter; (b) choke input filter.

FIGURE 12-41 Low voltage power supply with sample capacitor filter.

FIGURE 12-42 L-section (choke input) filter showing representative waveforms.

although they are referred to by different names (pi, RC, L section, etc.). The closest element electrically to the rectifier determines the basic type of filter being used (Figure 12–40). In part (A), a capacitor shunts the load resistor, therein bypassing the majority of the ripple current which passes through the series elements. In part (B), an inductor (choke) in series with the load resistor opposes any change in current in the circuit. The capacitor input filter will keep voltage at a higher level compared to a choke input filter. The choke input will provide a steadier current under changing load conditions. From this it can be seen that a capacitor input filter would be used where voltage is the prime factor and the choke input filter is used where a steady flow of current is required.

Capacitor Input Filter. First we look at the simple capacitor input filter [Figure 12–41(A)]. The output of the rectifier without filtering is shown in Figure 12–41(B), and the output after filtering is illustrated in Figure 12–41(C). Without the capacitor, the output across R_L will be pulses. The average value of these pulses would be the E_{dc} output of the rectifier. With the addition of the capacitor the majority of

the pulse changes are bypassed through the capacitor and around R_L.

Choke Input Filter. The next filter to be presented is the choke input filter, or the L-section filter. Figure 12–42 figure shows this filter and the resultant output of the rectifier after filtering has taken place. The series inductor, L (choke), will oppose any rapid changes in current. The output voltage of this filter is less than that of the capacitor input filter since the choke is in series with the output impedance. The parallel combination of R_L and C in connection with L smooths out the peaks of the pulses and results in a steady, although reduced, output.

☐ Field-Effect Transistors ☐

Field-effect transistors are also semiconductor devices whose operation is basically that of controlling the flow of electrical current through another channel by the application of voltage (an electric field) which is perpendicular to the path of conduction. Field-effect transistors (FETs) fall into two

BASIC SOLID-STATE THEORY

FIGURE 12–43 Schematic symbols for N- and P-type JFETs.

basic major categories: the junction field-effect transistor (abbreviated JFET) and the insulated-gate field-effect transistor (IGFET), which is more commonly referred to as the metal-oxide-semiconductor field-effect transistor (MOSFET).

The junction field-effect transistor and the insulated-gate field-effect transistor operate on a similar principle (current controlled by an electric field) but differ in the method that the control element is made. This difference results in a considerable difference in device characteristics, and therefore circuit design variances become a necessity. The first FET we will discuss is the JFET (Figure 12–43).

☐ Junction Field-Effect Transistor ☐

Basically, a JFET is a bar of doped silicon, called a channel, that behaves as a resistor. The doping may be either N or P type, creating either an N-channel

or a P-channel JFET. There is a terminal at each end of the channel. One terminal is called the source, and the other is called the drain. Current flow between the source and the drain, for a given drain-source voltage, is dependent on the resistance of the channel. This resistance is controlled by a gate. The gate consists of P-type regions diffused into an N channel, or N-type regions diffused into a P channel (Figure 12–44). As with any P-N junction, a depletion region, or better, an electric field, surrounds the P-N junction when the junctions are reverse biased. As the reverse voltage is increased, the electric fields spread into the channel until they meet, creating an almost infinite resistance between the source and the drain.

☐ Insulated-Gate Field-Effect Transistor or Metal-Oxide-Semiconductor Field-Effect Transistor ☐

The details of the construction of the N-channel depletion mode (IGFET or MOSFET) are shown in Figure 12–45. As can be seen in this figure, the main body is composed of a P substrate. (In this case, the substrate is the physical material on which a circuit is fabricated. Although its primary purpose is to furnish mechanical support, it may also serve a useful thermal or electrical function.) The source and drain

FIGURE 12–44 Cutaway view of N- and P-channel junction field-effect transistors.

FIGURE 12–45 Construction of depletion-mode IGFET and schematic symbol.

material is diffused into the body. In this case it is a low-resistivity N-type material indicated by N+ in the figure. Between the source and drain materials a moderate-resistivity N channel is also diffused into the body. Covering the main body of the device is a layer of insulating oxide. Holes are cut into the oxide layer, allowing metallic contact between the source and the drain. The gate material is now overlaid in the area between the source and the drain contacts on the oxide. There is no physical contact between these metal contact areas.

E-MOSFET Symbols.   Manufacturers have adopted a system of standard symbols (Figure 12–46) that are used when the gate is separate from the remainder of the MOSFET. This indicates that the gate and channel are insulated from each other. The N-channel MOSFET has the arrow pointing in, and the P-channel MOSFET has the arrow pointing out. The broken vertical line that joins the drain and the source indicates a normally OFF device in which the channel acts like it is an open circuit.

In both symbols, however, the arrow indicates the substrate that is connected to the source. MOSFETs are also available in four-terminal devices (Figure 12–47). In these devices the substrate is brought out as a lead and is separate from the other connections.

D-MOSFET Symbols.   There are standard symbols that are used to indicate the presence of D-MOSFETs in a circuit (Figure 12–48). Note that the gate and source voltage polarity are not shown because the device can operate with either polarity. In these symbols, the arrow points inward for the N-channel device and it points outward for the P-channel device. The drain-to-source path line is not broken, indicating that the channel is normally conducting for D-MOSFET devices.

☐ Integrated Circuits ☐

Integrated circuits are becoming a very popular part of the electronic circuitry used in air conditioning and refrigeration systems. These types of systems make use of the diodes and transistors that we discussed earlier in this chapter. Here, integrated circuit construction, operation, and testing will be discussed.

FIGURE 12-49 An integrated circuit provides a complete circuit in one package.

Basic Integrated Circuit. An integrated circuit (commonly called an IC) may be defined as a complete electronic circuit made up of any or all of the many components, such as diodes, transistors, resistors, and capacitors, where the interconnecting electrical conductors are all specially processed and completely contained on a chip of silicon (Figure 12–49). The chips used for integrated circuits are available in almost any size, with the smallest being approximately $\frac{1}{16}$ in. (1.59 mm) square.

The IC chip is equipped with metal leads strong enough to be inserted into the printed circuit card without damage. The plastic package can be treated fairly rough, without damage occurring to the integrated circuit proper. The additional material included in the device merely provides protection against the ambient atmospheric conditions.

The electrical circuit used in an IC is simply a smaller version of a standard electric circuit which is made up of ordinary separate components. As a matter of fact, when ICs were first being used, the circuit was actually constructed on a breadboard using the conventional components to make certain that they would function as required and then converted into the IC form. The IC would be exactly the same as the standard circuit, with the same junctions, connections, and allied components.

Testing Integrated Circuits. After the assembly of IC chips, each individual circuit has several tests performed on it to assure that it performs in the electrical system as expected and to make certain that it performs within the specifications published in the manufacturer's data sheets. A change in temperature of the circuit components causes a change in the performance of the circuit. Because of this change, each

manufacturer includes a temperature-range specification on the data sheet.

There are two grades of IC chips, named industrial and military. The ICs are tested for satisfactory operation within 32 to 158 °F (0 to 70 °C) temperature range. This temperature range is suitable for most industrial applications. All the IC devices that pass this test are then usually tested for operation within the temperature range − 30.8 to 289 °F (− 55 to + 125 °C). Those that pass this test are suitable for use in military and space systems.

Integrated Circuit Removal. The integrated circuit chip should be removed from its socket by prying both ends loose simultaneously. If the chip is pulled from its socket by hand, the pins could be bent or broken. The best method of removing these chips is to use a commercially available extractor. The extractor holds the integrated circuit body securely to prevent putting excessive stress on any particular part of the chip. Thus the chip may be removed without damage to the pins or to the body.

Integrated Circuit Replacement. Integrated circuits are shipped with their pins bent slightly outward so that they fit securely against the socket wipers during the installation process. If the pins do not line up exactly with the socket holes, they may be slightly bent inward so that installation is possible. Be sure to bend all the pins the same amount along both sides. Do not bend them more than is necessary for proper installation. If the pins are not bent evenly, either one or more of them may bend either outward or inward and perhaps break off, ruining the chip.

☐ Optoelectronics and Fiber Optics ☐

There are two types of circuits and devices which are currently being used in highly specialized areas of air conditioning and refrigeration control systems. Optoelectronic devices are those that are used in interface situations. They use some source of external light, such as a pilot burner or a main burner flame, to generate a small current to turn on a transistor or a diode, or to operate some other device.

Fiber optics is a specialized field that makes use of a very thin flexible glass, or a plastic optical fiber for the transfer of light energy from the point of origination to the point of use. Commonly, it connects two electronic circuits. The source of light may

FIGURE 12-50 Typical fiber optic system transmission link. (Courtesy of Billy Langley, *Solid State Electronic Controls for Air Conditioning and Refrigeration,* © 1989, p. 209; reprinted by permission of Prentice-Hall, Inc., Englewood Cliffs, N.J.)

be a light-emitting diode (LED), a laser diode, or an infrared emitting diode (IRED), all of which convert electrical energy signals into light signals. A photodiode, usually located inside a receiver, changes the light signal back into electrical signals (Figure 12-50). The purpose of the receiver output circuit is to amplify the input signal that causes the desired action.

Optoelectronics. These devices are true semiconductors. They are specially made types of diodes and transistors that are designed to use light as their energy source. In fact, all diodes and transistors are, to some extent, sensitive to light, but their packaging is designed to prevent their interacting with it. However, the purpose of optoelectronics is to use the energy that is produced by light.

Optoelectronic Schematic Symbols. The photodiode and phototransistor are light sensors and have arrows indicating light going into them. Also, notice that the photodiode symbol is just an ordinary transistor symbol with the base lead omitted. The light-emitting diode (LED) emits light from it and has the arrows indicating light going from it (Figure 12-51).

Optoelectronic Device Operation. These devices all operate on the simple principle that when light strikes the semiconductor junction, the bound electrons are caused to move out of their holes. This action creates free electrons and holes in the material. When a free electron enters one of these holes a particle of light is created. This particle of light is called a "photon."

As an example, if we wanted to use light energy to operate a small dc electric motor, the circuit shown in Figure 12-52 could be used. The purpose of this circuit is to operate the motor only when a sufficient amount of light shines on the photodiode

Photodiode   NPN phototransistor   Light-emitting diode (LED)

FIGURE 12-51 Optoelectronic schematic symbols. (Courtesy of Billy Langley, *Solid State Electronic Controls for Air Conditioning and Refrigeration,* © 1989, p. 209; reprinted by permission of Prentice-Hall, Inc., Englewood Cliffs, N.J.)

FIGURE 12-52 Circuit for dc electric motor operation in response to a light source. (Courtesy of Billy Langley, *Solid State Electronic Controls for Air Conditioning and Refrigeration,* © 1989, p. 210; reprinted by permission of Prentice-Hall, Inc., Englewood Cliffs, N.J.)

junction. In this case a stronger beam of light would cause the motor to operate much faster than a weak light beam.

A *photodiode* is designed and packaged with a small lens that allows the light beam to strike the P-N junction. Our circuit makes use of a battery connected so that the electrons are trying to flow through the diode from the P region to the N region. Remember from our earlier studies that this is the

BASIC SOLID-STATE THEORY

wrong direction of flow for electrons through a diode. When the free electrons and holes are separated, a depletion layer is left around the junction. The depletion region is an area that has absolutely no free electrons or holes in it. Therefore, the only current that can flow through this area is leakage current, which is very small.

However, should a beam of light equal to one photon shine through the lens and strike the semiconductor material in the area of the depletion layer, a free electron and a hole are created in the depletion layer. As soon as the electron is freed, the battery forces it into the N region and out of the photodiode by way of the cathode terminal. At this same instant the hole is forced to go in the opposite direction. This action causes one electron to flow from the negative battery terminal to the P region, then through the N region, then through the circuit, and through the motor, thus energizing it.

Now if a strong light should strike the depletion layer of the photodiode, there would be literally millions upon millions of electrons flowing through the circuit. These electrons would in turn cause the motor to rotate. When photodiodes are used with an external power source, they are called "photoconductive." This means that they will conduct electric current when a light shines on the P-N junction and will block current flow in the absence of light on the P-N junction.

The *phototransistor* is also a "photoconductive" device (Figure 12–53). The phototransistor operates just like the photodiode discussed above with one exception: the output current is amplified. In the photodiode, for one photon of light collected, one electron of electricity was emitted by it. However, in the phototransistor for every electron that leaves the P region there are possibly a hundred that flow from the emitter to the collector and on through the circuit.

LED Operation. Light-emitting diodes operate just the opposite of photodiodes and phototransistors. A free electron moving through a piece of semiconductor material will emit one photon of light when it enters into an empty hole. The photon, however, is released in a random direction. When many photons are released together, a ray of light is produced.

In operation, an LED produces light because of a flow of free electrons through the circuit. The greater the electron flow, the greater the amount of light emitted (Figure 12–54). In this example the LED

FIGURE 12–53 NPN phototransistor construction. (Courtesy of Billy Langley, *Solid State Electronic Controls for Air Conditioning and Refrigeration*, © 1989, p. 211; reprinted by permission of Prentice-Hall, Inc., Englewood Cliffs, N.J.)

FIGURE 12–54 LED operation. (Courtesy of Billy Langley, *Solid State Electronic Controls for Air Conditioning and Refrigeration*, © 1989, p. 211; reprinted by permission of Prentice-Hall, Inc., Englewood Cliffs, N.J.)

is connected into a circuit which also has a source of power. This source of power, a battery, is forcing electrons through the circuit in a forward direction with respect to the diode. The electrons enter the semiconductor chip at the cathode and move across the N region. When crossing the N region, free electrons enter the vacated holes and the process produces photons of light. When the flow of forward current is strong enough, the P-N junction of the chip will produce enough light to be easily seen.

Light Refraction. When studying fiber optics there are four important terms with which you must become familiar. The four terms are (Figure 12–55):

1. *Normal.* This is an imaginary line that is perpendicular to the light interface of two materials.
2. *Angle of incidence.* This is the angle that is formed by an incoming light ray, or incident ray, and the normal (imaginary) line.
3. *Angle of refraction.* This is the angle that is

FIGURE 12–55   Angles of refracted light.
(Courtesy of Billy Langley, *Solid State Electronic Controls for Air Conditioning and Refrigeration,* © 1989, p. 215; reprinted by permission of Prentice-Hall, Inc., Englewood Cliffs, N.J.)

FIGURE 12–56   Critical angle of refracted light. (Courtesy of Billy Langley, *Solid State Electronic Controls for Air Conditioning and Refrigeration,* © 1989, p. 215; reprinted by permission of Prentice-Hall, Inc., Englewood Cliffs, N.J.)

formed between the refracted light ray and the normal (imaginary) line.

4. *Critical angle.* The critical angle is also the angle of incidence.

The greater the angle of incidence from normal, the fewer the chances that the incidence light ray will enter another material (Figure 12–56). The angle at which the entry of the incident light ray is prevented from entering another material is the critical angle. Each material has its own critical angle because of

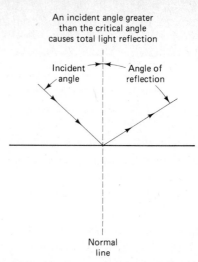

FIGURE 12–57   Angle of reflection.

the different wavelengths involved. Should the incidence light ray reach 90° refraction, the ray will not enter the second material. When the incident angle is increased to a value greater than the critical angle for the material involved, all the light is reflected back into the first material (Figure 12–57).

☐ Types of Control Systems ☐

Control systems are generally classified according to their source of power. When we consider controls from this point of view, the components that go into the makeup of a control system are either (1) electric, (2) self-actuated, or (3) pneumatic control devices.

Electric Controls. The source of power for these controls is electricity. The control units in this system are connected to the source of power by electric wires which may carry either line-voltage or low-voltage electrical current. The line voltage, when used, may either be 120 or 240 V. The low-voltage control circuit normally operates on 24 V, which is supplied by a transformer. The transformer is used to convert the 120 V or the 240 V to 24 V.

A line-voltage thermostat is used to operate a heating control motor to open a hot water or steam valve, and close it when required by the room temperature. This is an illustration of a simple application of a line-voltage electric control (Figure 12–58). A simple application of a low-voltage electric control would be a motor operating on low voltage which is supplied by a transformer (Figure 12–59). Except for the low-voltage transformer, the opera-

FIGURE 12-58   Line-voltage two-position control.

FIGURE 12-59   Low-voltage two-position control.

FIGURE 12-60   Self-contained control valve.

FIGURE 12-61   Simple pneumatic control.

tion is exactly like the line-voltage circuit presented in Figure 12–58.

Self-Contained Controls. These controls have as a source of power the sensing element, and the final control mechanism—such as a valve—is all contained in one single unit. These types of controls usually include a bellows and a bulb which are connected by a length of tubing containing a charge of liquid or vapor which changes its volume with a change in room temperature. Thus a change in the temperature of the medium surrounding the sensing element will cause the fluid inside the element to either expand or contract a sufficient amount to cause a force strong enough to operate the valve.

A simple application of the self-contained control is illustrated in Figure 12–60. The valve controlling the amount of hot water or steam that enters the heating chamber is directly operated by these changes in the pressure inside the bellows as the temperature of the surrounding medium changes. The operating principle of these controls is very similar to the operation of a pneumatic control system. The basic difference is that the variations in pressure at the bellows are produced in direct response to the variations in temperature at the sensing element.

Pneumatic Controls. When a pneumatic control system is employed, the control system components are connected together by air lines and respond to air pressure which is supplied by an air compressor.

This air is usually stored in a tank with the pressure about 100 psig (689.476 kPa). The air is then supplied to the control system at a pressure of about 15 to 25 psig (103.421 to 172.369 kPa) through a pressure regulator. These air lines may be made of either copper or polyethylene and vary in size from ⅛ to 2 in. (3.18 to 50.80 mm) in diameter. The most common sizes are from 3/16 to ½ in. (4.76 to 12.70 mm) in diameter.

A simple application of a pneumatic control system is shown in Figure 12–61. In operation, an operating pressure of 15 to 25 psig (103.421 to 172.369 kPa) is supplied to the controller through the main air line from the compressor. The thermostat is in effect a pressure regulator which is actuated by the surrounding air temperature.

The line supplying the air to the radiator valve motor maintains a pressure of about 15 psig (103.421 kPa) or less, in response to the room temperature, and is known as the branch line. The valve shown is a normally open type. That is, it will remain open as the air pressure decreases. As the thermostat responds to a fall in room temperature, the air pressure is decreased in the branch line, allowing the valve to open and pass steam to the radiator. A rise in temperature at the thermostat causes an increase in branch line air pressure, which in turn causes the pneumatic motor to close the valve and stop the flow of steam to the radiator.

The control valve motor, consisting of a bellows that is balanced against a spring, does not waste air.

However, when the valve motor is changing positions, air does not flow into or out of the motor. The valve assumes a position that is determined by the air pressure in the branch line, the tension of the spring, and the force imposed on the valve stem by the steam pressure. Therefore, the thermostat regulates the air pressure and effectively determines the position of both the motor and the valve.

□ Safety Procedures □

Electric shock is a jarring, shaking sensation resulting from contact with electric circuits or lightning. The victim usually experiences the sensation of a sudden blow, and if the voltage is sufficiently high, unconsciousness. Severe burn may appear on the skin at the place of contact; muscular spasm can occur, causing a person to clasp the apparatus or wire that caused the shock and be unable to turn loose. Electric shock can kill its victim by stopping the heart or by stopping breathing, or both. It may sometimes damage nerve tissue and result in a slow wasting away of muscles that may not become apparent until several weeks or months after the shock is received.

The following procedure is recommended for the rescue and care of electrical shock victims:

1. Remove the victim from electrical contact at once, but *do not endanger yourself*. This can be done by:
   (a) Throwing the switch, if it is nearby.
   (b) Using a dry stick, rope, leather belt, coat, blanket, or any other nonconductor of electricity.
   (c) Cutting the cable or wires to the apparatus, using insulated tools, while taking care to protect your eyes from the flash when the wires are cut.
2. Determine whether the victim is breathing. If so, keep the person lying down in a comfortable position. Loosen the clothing about his or her neck, chest, and abdomen so that the person can breathe freely. Watch the victim carefully and protect him or her from exposure to cold.
3. Keep the victim from moving about. After shock, the heart is very weak, and any sudden muscular effort or activity on the part of the patient may result in heart failure.
4. Do not give stimulants or opiates. Send at once for a medical person and do not leave until the patient has adequate medical care.
5. If the victim is not breathing, it will be necessary to apply artificial respiration without delay, even though the person may appear to be lifeless.

Insofar as practical, a technician should not undertake working on energized circuits and equipment. However, if it should become necessary, as when making operational adjustments, one should carefully observe the following precautions:

1. Ensure that you have plenty of illumination. If you are to perform the job properly and safely, you must be able to see clearly.
2. Ensure that you are insulated from ground by an approved type rubber mat, or layers of dry canvas and/or wood.
3. Where practical, use only one hand, keeping the other behind you in your pocket.
4. If the system voltage exceeds 150 V, rubber gloves should be worn.
5. An assistant should be stationed near the main switch or circuit breaker, so that the equipment may be deenergized immediately in case of an emergency.
6. A person qualified in first aid for electrical shock should be standing by during the entire operation.
7. Do *not* work alone.
8. Do *not* work on any type of electrical apparatus when wearing wet clothing or if the hands are wet.
9. Do *not* wear loose or flapping clothing.
10. The use of thin-soled shoes and shoes with metal plates or hob nails is prohibited.
11. Flammable articles, such as celluloid cap visors, should not be worn.
12. All rings, wristwatches, bracelets, and similar metal items should be removed before working on the equipment. Also, ensure that clothing does not contain exposed metal fasteners such as zippers, snaps, buttons, and pins.
13. Do not tamper with interlock switches, that is, do not defeat their purpose by shorting them or blocking them open.
14. Ensure that the equipment is properly grounded before energizing.
15. Deenergize the equipment before attaching alligator clips to any circuit.
16. Use only approved meters and other indicating devices to check for the presence of voltage.

17. Observe the following procedures when measuring voltages in excess of 300 V:
    (a) Turn off the equipment power.
    (b) Short circuit or ground the terminals of all components capable of retaining a charge.
    (c) Connect the meter leads to the points to be measured.
    (d) Remove any terminal grounds connected previously.
    (e) Turn on the power and observe the voltage reading.
    (f) Turn off the power.
    (g) Short circuit or ground all components capable of retaining a charge.
    (h) Disconnect all meter leads.
18. In all electrical circuits where the voltage is in excess of 30 V and where metal is present, the worker should be insulated from accidental ground by the use of approved insulating material. The insulating material should have the following characteristics:
    (a) It should be dry, without holes, and should not contain conducting materials.
    (b) The voltage rating for which it is made should be clearly marked on the material, and the proper material should be used so that adequate protection from the voltage can be possible.
    (c) Dry wood may be used, or as an alternative, several layers of dry canvas, sheets of phenolic insulating material, or suitable rubber mats.
    (d) Care should be exercised to ensure that moisture, dust, metal chips, and so on, which may collect on insulating materials is removed at once. Small deposits of such materials can become electrical hazards.
    (e) All insulating materials on machinery and in the area should be kept free of oil, grease, carbon, dust, and so on, since such deposits destroy insulation.

☐ SUMMARY ☐

- ☐ In response to the demand for smaller control components and the desired accuracy of electronic control circuits, many manufacturing and construction techniques have been employed in the field of air conditioning and refrigeration control systems.
- ☐ The integrated circuit (IC) is a major development in the field of solid-state electronics.
- ☐ A differential amplifier is basically an amplifier that responds to the difference between two voltages or currents and effectively suppresses like voltages or currents.
- ☐ A printed circuit (PC) board is an insulated surface upon which conductors and other circuit components have been printed.
- ☐ Voltage readings can be taken from either side of the PC board without risk of damage.
- ☐ PC boards are designed to fit snugly into an edge card connector so that proper electrical contact is made without the need for external wiring connections.
- ☐ The basic building block for modular circuits is a specially constructed ceramic wafer on which is mounted or printed the various components and circuits that make a particular electronic stage.
- ☐ A semiconductor is a solid that has an area known as the depletion region.
- ☐ The most common materials used for semiconductor diodes are germanium and silicon.
- ☐ All solid-state diodes, regardless of their size or shape, are equipped with two leads, or terminals.
- ☐ The zener diode, sometimes called an avalanche diode, is unique compared to other diodes in that it is designed to operate in the reverse-bias condition in the avalanche or breakdown region.
- ☐ Reference diodes were designed to replace zener diodes in certain applications because of the zener's temperature instability.
- ☐ P-N junctions exhibit capacitance properties because the depletion area represents a dielectric and the adjacent semiconductor material represent two conductive plates.
- ☐ The tunnel diode, sometimes called an Esaki diode, makes use of very heavily doped P and N materials in a two-terminal device.
- ☐ The usual ways of dissipating heat from a semiconductor are from the case to the ambient air, from the case to the frame of the unit to which it is attached, and through increasing the surface area of the case of the device, such as the fins on a lawn mower engine.
- ☐ A transducer, or transfer-producer, is a device that converts one form of energy to another form.
- ☐ A thermistor is a device that will have a change in resistance with a change in temperature. Therefore, they are very useful as signal-producing devices for use in air conditioning and refrigeration control systems.
- ☐ In a good many single-phase compressor motors, the PTC thermistor (positive temperature coefficient) is replacing the motor starting relay used in previous years.
- ☐ It should be noted that the PTC thermistor becomes less effective on rapid-start conditions because the temperature of the material increases with each additional start.

☐ Solid-state pressure-sensing devices are used for sensing the refrigerant pressures on both the high and low sides of the system.

☐ The standard P-N junction diode can produce an electrical current when it absorbs light energy.

☐ The silicon-controlled rectifier (SCR) is a four-layer (PNPN), three-junction (lead) semiconductor device. The three leads are the anode, cathode, and gate.

☐ The SCR can be used to control a high-power circuit.

☐ The DIAC is a three-layer device that can pass current in both directions.

☐ The TRIAC is a semiconductor device capable of performing the same function as two SCRs connected in reverse parallel and can therefore be used for controlling an ac current load.

☐ Since the TRIAC is either fully ON or fully OFF there is no power being converted to heat that must be dissipated, resulting in a loss of power.

☐ TRIACs are being used to replace the starting relays used on single-phase compressor motors used in refrigeration systems.

☐ The unijunction transistor (UJT) is a three-terminal semiconductor device. It is sometimes referred to as a double-base diode.

☐ The UJT is basically used when timing and switching actions are needed.

☐ A transistor is a device that makes use of a small change in current to produce a large change in current, voltage, or power.

☐ Gain is the ratio of output voltage, power, or current with respect to an input reading in the same unit of measurement.

☐ Temperature rise is a problem in transistors because as the temperature goes up, it tends to increase the flow of electrons in both the P and N materials. This allows more current to flow through the transistor for the same applied voltages.

☐ The use of P-N junction rectifiers in the design of power supplies for electronic equipment is very popular. They are used because they have no requirement for filament power, the ability to begin immediate operation without the need for warm-up time, a low internal voltage drop substantially independent of the load current, low operating temperature, and generally small physical size.

☐ Variations in doping agent concentrations and physical size of the substrate produce diodes that are suited for different applications.

☐ Rectifier diodes are used primarily in power supplies.

☐ The filter circuits used in power supplies are of two general types, capacitor input and choke input.

☐ Field-effect transistors are also semiconductor devices whose operation is basically that of controlling the flow of electrical current through another channel by the application of voltage (an electric field) which is perpendicular to the path of conduction.

☐ An integrated circuit (commonly called an IC) may be defined as a complete electronic circuit made up of any or all of the many components, such as diodes, transistors, resistors, and capacitors, where the interconnecting electrical conductors are all specially processed and completely contained on a chip of silicon.

☐ The IC chip is equipped with metal leads strong enough to be inserted into the printed circuit card without damage. The plastic package can be treated fairly rough without damage occurring to the integrated circuit proper.

☐ The electrical circuit used in integrated circuits is simply a smaller version of a standard electric circuit which is made up of ordinary separate components.

☐ Optoelectronics and fiber optics are two types of circuits and devices which are currently being used in highly specialized areas of air conditioning and refrigeration control systems.

☐ Optoelectronic devices are those that are used in interface situations. They use some source of external light, such as a pilot burner or a main burner flame, to generate a small current to turn on a transistor or a diode, or to operate some other device.

☐ Fiber optics is a specialized field that makes use of a very thin flexible glass, or a plastic optical fiber for the transfer of light energy from the point of origination to the point of use.

☐ Optoelectronic devices are true semiconductors.

☐ A photodiode is designed and packaged with a small lens that allows the light beam to strike the P-N junction.

☐ The photodiode is also a "photosensitive" device.

☐ Light-emitting diodes operate just the opposite of photodiodes and phototransistors.

☐ When studying fiber optics there are four important terms with which you must become familiar: (1) normal, (2) angle of incidence, (3) angle of refraction, and (4) critical angle.

☐ Control systems are generally classified according to their source of power. When we consider controls from this point of view, the components that go into the makeup of a control system are either electric, self-actuated, or pneumatic control devices.

☐ EXERCISES ☐

1. Why are electronic controls so popular in the air conditioning and refrigeration industry?
2. What constitutes an integrated circuit?
3. Define a differential amplifier.
4. Define a PC board.
5. What are the functions of PC boards?
6. How are components identified on PC boards?
7. What is the basic building block for modular circuits?

8. To what extent will servicing modular circuits be limited?
9. What is a semiconductor?
10. Can current flow through a semiconductor at room temperature?
11. What is unique about semiconductor P-N junctions?
12. What factors determine the current-carrying capacity of diodes?
13. How many leads do diodes have?
14. How will a diode that is designed to operate in the reverse-bias direction be marked?
15. What is another name for an avalanche diode?
16. Why are zener diodes often used as voltage sources?
17. Why do P-N junction diodes have capacitance properties?
18. In what type of diodes are the terms "valley current" and "valley voltage" used?
19. Why do semiconductors need heat sinks?
20. What is the purpose of a transducer?
21. Name two types of thermistors that refer to a change in temperature of them.
22. How would a thermistor be used in conjunction with a motor?
23. How are thermistors classified according to their temperature?
24. What type of thermistor is replacing the starting relay on compressor motors?
25. How can transducers be used, in regard to pressure, to control compressor operation?
26. When will a diode emit light?
27. What is a four-layer, three-junction semiconductor known as?
28. What type of semiconductor can be used to control a high-power circuit?
29. For what purpose are TRIACs and SCRs used?
30. What is the major use of DIACs and UJTs in control circuits?
31. What semiconductor device can pass voltage in either direction?
32. What semiconductor device can replace two SCRs connected in reverse parallel?
33. What causes a TRIAC to pass current?
34. What causes a TRIAC to stop passing current?
35. Name the three distinct regions of UJT operation.
36. What semiconductor device is used in the precision control of temperature in air conditioning and refrigeration applications?
37. Define *inverter*.
38. What semiconductor device uses a small change in current to make a large change in current, voltage, or power?
39. Define *gain* in a solid-state circuit.
40. What happens to the flow of electrons through a transistor on an increase in temperature?
41. Define thermal runaway in transistors.
42. In what types of circuits are rectifier diodes used?
43. Why is alternating current not desirable in electronic circuits?
44. Name two types of filter circuits used in power supplies of electronic circuits.
45. Define *integrated circuit*.
46. What is the purpose of optoelectronics?
47. Define *depletion region* of a diode.
48. What does the word "photoconductive" mean when referring to semiconductors?
49. What causes an LED to emit light?
50. In what way are control systems classified?

13 Test Equipment

Performance Objectives

After completion of this chapter, the student should:
- ■ Be able to explain the purpose of test instruments.
- ■ Know how to select instruments.
- ■ Be able to list the various types of test instruments available.
- ■ Know how to properly apply test instruments.
- ■ Complete the Exercises with at least 80 percent accuracy.
- ■ Complete the Laboratory Exercises with at least 80 percent accuracy.

Test instruments are valuable aids to service technicians. The use of test instruments seems to frighten many service technicians, however, and they will not make any effort to use these valuable tools of their trade. The use of test instruments is of great benefit to the customer as well as the service technician. Test instruments should be just as familiar to service technicians as any other tool in the toolbox.

☐ Purpose ☐

Test instruments, when used, can help the service technician to perform service operations more accurately and faster, thereby providing a savings to the customer. There are two very strong arguments for the use of test instruments: (1) prevention of the unnecessary changing of parts that usually end up in the scrap box, and (2) that time saved is money. Customer satisfaction after the sale is not promoted by the parts changer service technician, who costs the customer and his employer thousands of dollars each year in wasted parts and effort.

☐ Selection ☐

Quality tools may represent a sizable investment, but they are not expensive in the long run if care is exercised in their selection, use, and maintenance. When the task of selecting test equipment is approached properly, the errors made will be kept to a minimum. It is desirable to make a list of the features that are required, those features that would be nice but not essential, and those that are not needed. Careful shopping will probably produce a test instrument that is fairly close to what is wanted.

The cost of these instruments also should be considered. One instrument with several functions will be less expensive than an instrument for each individual purpose. It will be easier to handle, if it is not too bulky, but it will be complicated to use and will require more time in learning to use the instrument to its full capacity. The purchase of several instruments has these advantages: they are easier to use, offer lower initial cost per unit, and offer lower additional availability in that should one instrument need repairs another can take over.

☐ Repair ☐

The field repair of test instruments should not be attempted. A few instruments have components that can be replaced in the field by any competent technician. Any major, internal repairs should be left for those qualified to make them.

Most instruments have some means of adjustment or calibration that should be checked periodically to assure dependable and accurate measurement. The manufacturer's recommendations for calibration and adjustment for any instrument should be followed.

□ Care □

Some meter movements are sensitive to position. This type of instrument is more accurate when placed in a horizontal position. Any instrument should be in the position in which it is to be used when calibration is done. All instruments have a specific point at which the needle should rest when it is not in use. This point is called the *null point*.

Instrument test leads should be checked regularly for wear or deterioration of the insulation. Leads with cracked or worn insulation should be replaced. Some leads are replaceable with standard probes that can be purchased at electronic supply houses. However, some instrument leads can only be replaced with leads supplied by the instrument manufacturer.

A firm solid contact between the lead and the instrument is required. Loose-fitting leads can cause the meter to give an improper reading.

We can consider the mechanism that moves the needle across the meter face as being a motor. Improper lubrication can damage this mechanism as easily as rough handling or overloading. Use the instrument for its intended purpose and it will last forever.

Most test instruments are equipped with fuses or circuit breakers that prevent damage to the instrument. Be sure to have a few spare fuses of the proper type on hand.

Test instruments should be treated with care. When they are stored in a toolbox, they are subjected to a beating. Extremely hot or cold temperatures are not good for test instruments. They should be protected from oil, dust, and moisture. They should not be carried or stored out of their protective cases.

The batteries used in test instruments should be of good quality. Be sure to check batteries frequently for corrosion because some batteries will leak acid even before all the power is used. This acid will ruin test instruments. It is good practice to remove the batteries from instruments not used frequently and store them separately.

When storing a test instrument, be sure that the selector switch is not left in a position that calls for battery output. If a selector switch is not used, the leads should be disconnected, except when the meter is being used.

□ Application □

The proper application of test instruments will provide more accurate trouble diagnosis in the shortest possible time, providing a savings to the customer and effecting better customer relations.

Voltmeter. The voltmeter (Figure 13-1) is probably one of the most frequently used instruments. It is also the easiest instrument to read. The service technician may encounter voltages ranging from millivolts to 5000 V or more. Therefore, reasonable caution should be used.

In use, the selector switch should be set on the proper function for which the meter is to be used. When the correct voltage is not known, start checking the electrical circuit with the meter set on the highest scale possible and work down until one is found that provides a reading at approximately

FIGURE 13-1 Multimeter. (Courtesy of A.W. Sperry Instruments, Inc.)

FIGURE 13-2   Ohmmeter. (Courtesy of Robinair Manufacturing Company.)

midscale. Midscale readings are the most accurate obtained.

Ohmmeter. The ohmmeter (Figure 13-2) is used for measuring resistances, checking continuity, making quick checks on capacitors, and testing some solid-state components. Most ohmmeters are ruined because they are connected to a circuit or a component that has not been disconnected from the source voltage. Ohmmeters have their own power source and cannot be safely used with any other power source.

It is best to start with the lowest resistance scale when measuring an unknown resistance or continuity. Work up through the scales until a midscale reading is obtained.

When checking resistance, first select the proper resistance scale. Install the test probes in the proper jacks. Touch the ends of the probes together and turn the adjust knob until the meter reads zero resistance while the ends of the probes are touching. If the scale is changed, the meter must be calibrated to zero. If the meter cannot be calibrated to zero, the batteries should be replaced.

When a switch or wire is to be tested for continuity, the needle should deflect to zero, which indicates a closed circuit. The part being checked has continuity. If, on the other hand, the needle does not move from the at-rest position, the switch is open. The part being tested does not have continuity.

If the needle stops between infinity and zero, the meter is measuring the resistance of the part being tested. Infinity means that the resistance is extremely high; probably an open circuit is being measured.

Ammeter. The clamp-on type ammeter (Figure 13-3) is the most popular current-measuring instru-

FIGURE 13-3   Clamp-on ammeter. (Courtesy of Amprobe Instrument Division of SOS Consolidated Inc.)

ment used for ac circuits because the current flow can be checked without interruption of the electric circuit. All that is necessary to check the current flow is to clamp the tongs around one of the wires supplying electric power to the circuit being tested.

Clamp-on ammeters are designed primarily to measure the current flow in ac circuits only. These instruments measure the current flow in a wire by making it serve as the primary side of a transformer. The circuit to the meter movement serves as the secondary side of the transformer (Figure 13-4). These instruments are most accurate when the current-carrying wire is in the center of the tongs. For best accuracy of the instrument, the mating surfaces of the tongs should be kept clean.

When testing a circuit that has a low current flow, the wire can be wrapped around the tong of the ammeter. The sensitivity of the meter will be multiplied by the number of turns used. To obtain the correct amperage reading, divide the meter reading by the number of turns taken on the tong.

Wattmeter. The wattmeter (Figure 13-5) can be used to determine the voltage to the unit and the total wattage draw of the unit. Be sure to set the meter on the highest range possible because the starting

TEST EQUIPMENT

FIGURE 13-4 Use of clamp-on ammeter. (Courtesy of Amprobe Instrument Division of SOS Consolidated Inc.)

FIGURE 13-5 Wattmeter. (Courtesy of Robinair Manufacturing Company.)

FIGURE 13-6 Capacitor analyzer. (Courtesy of Robinair Manufacturing Company.)

FIGURE 13-7 Electronic temperature tester. (Courtesy of Mechanical Refrigeration Enterprises.)

wattage of most motors is very high. After the unit is running, the proper scale can be selected so that the needle indicates a midscale reading.

Capacitor Analyzer. The capacitor analyzer is a very important instrument to the service technician. It can be used to check for open or shorted capacitors, for the microfarad capacity, and for the power factor (Figure 13-6). Scales are provided for both the voltage at which the instrument will be used and the microfarad rating of the capacitor. It may be used to check both the running and starting capacitors.

Temperature Tester. Most electronic temperature testers (Figure 13-7) have thermistor sensing leads that are used primarily to measure temperatures below 200 °F (93 °C). Do not attempt to alter these leads in any way. If the leads do not function properly, they should be replaced. By using these leads, temperatures at several different locations may be taken without interruption of the normal operating cycle of the unit. Almost any number of leads can be used on instruments equipped with multiple connections.

Temperature testers are also available that will record both the time and the temperature on a piece of paper (Figure 13-8). These instruments are valuable for helping to locate intermittent temperature-related problems. When this type of instrument is desired be sure that the chart paper and inking supplies are available.

Application

FIGURE 13-8 Recording temperature-time meter. (Courtesy of Airserco Manufacturing Company.)

FIGURE 13-9 Micron meter. (Courtesy of Airserco Manufacturing Company.)

FIGURE 13-10 Millivolt meter. (Courtesy of Airserco Manufacturing Company.)

FIGURE 13-11 Adapter for testing thermocouple closed circuit voltage.

Micron Meter. When it is desirable to have an accurate indication of the vacuum inside a system, a micron meter is the instrument most often used (Figure 13-9). These instruments are electrically powered and use a sensing probe to measure the thermal conductivity of the gases in the system being evacuated. These instruments are designed to measure the last part of an inch of vacuum and indicate this measurement in microns. There are 25,400 microns (μm) in an inch.

The micron sensing element is not designed to measure pressures above atmospheric pressure. Therefore, some means must be provided to protect the sensor when charging the system with refrigerant.

Millivolt Meter. The millivolt meter is an instrument used to measure voltage in thousandths of a volt (Figure 13-10). In air conditioning and heating service work, these instruments are used to check the thermocouple output voltage on pilot safety devices. It is often necessary to check the closed-circuit voltage of a thermocouple. A special adapter can be obtained that will permit checking the thermocouple under loaded conditions (Figure 13-11).

Electronic Leak Detector. Electronic leak detectors (Figure 13-12) are very sensitive instruments. They operate by moving refrigerant vapor across an ionizing element. There are several ways to warn the user that a leak is present. Some instruments produce a high-pitched screech, some use a blinking light in the probe, and others use a buzzing noise. These instruments will not function when either ammonia or sulfur dioxide is the refrigerant.

Thermostat Tester. Thermostat testers are used to calibrate refrigeration thermostats. Accurate results are obtained by using this instrument. The well of

FIGURE 13-12 Electronic leak detector. (Courtesy of TIF Instruments, Inc.)

FIGURE 13-13 Cross section of thermostat tester.

the tester is partially filled with alcohol. Liquid refrigerant is bled into the surrounding shell to cool the alcohol. A thermometer and the thermostat feeler bulb are placed in the alcohol (Figure 13-13). The cut-out temperature of the thermostat can be determined from the thermometer. The cut-in temperature can be found by stopping the flow of liquid refrigerant and warming the tester. As the alcohol temperature increases, the cut-in temperature can be seen on the thermometer.

FIGURE 13-14 Hermetic analyzer. (Courtesy of Mechanical Refrigeration Enterprises.)

FIGURE 13-15 Potential relay analyzer. (Courtesy of Mechanical Refrigeration Enterprises.)

Hermetic Analyzer. Hermetic analyzers (Figure 13-14) can be used for many functions, including (1) to rock free stuck or frozen compressors by reversing motor rotation; (2) as a bank of starting capacitors; (3) as an auxiliary capacitor; and (4) to check motor windings for shorts, opens, continuity, and grounds. Some can be used as ammeter, voltmeter, capacitor analyzer, ohmmeter, and relay analyzer.

Potential Relay Analyzer. The potential relay analyzer (Figure 13-15), when used properly, can save the service technician much time and trouble. Some of the uses of these instruments are: (1) to check the relay for open, closed, or short circuit;

FIGURE 13–16 Sling psychrometer. (Courtesy of Robinair Manufacturing Company.)

(2) to indicate relay pickup and dropout voltage; (3) to operate the relay manually; (4) to test continuity; and (5) to identify defective relays.

Sling Psychrometer. The sling psychrometer (Figure 13–16) is used to determine relative humidity. It is equipped with two mercury thermometers. Both thermometers are of the ordinary dry-bulb type. One has a wick placed over the bulb end; the other is left bare. The wick is wet with water. The sling psychrometer is then twirled around until a constant reading is obtained on both thermometers. The relative humidity can then be obtained from relative humidity charts.

Heating System Analyzer. This analyzer was designed specifically for the heating and air conditioning service technician to thoroughly and rapidly check all types of heating systems and controls and to measure temperatures throughout the system. A complete analysis for a typical heating system takes only a few minutes. The instrument functions as: (1) ac-dc voltmeter, (2) millivolt meter, (3) ohmmeter, (4) device to measure milliampere output, and (5) thermocouple probe that reads the temperature of the pilot flame.

Note: It is always important to use the manufacturer's instruction sheet for the instrument being used. Always exercise caution to prevent damage to test equipment.

Electronic Sight Glass. The electronic sight glass (Figure 13–17) is a relatively new device that emits an audible signal indicating that the refrigeration system being charged is full of refrigerant. By using the principles of radar, the unit actually senses the conditions inside the refrigerant tubing. It is very useful when charging or checking the charge in a capillary tube system and on some systems used in automotive air conditioning units.

The unit can also be used on expansion valve systems. It is much more convenient and easy to hear

FIGURE 13–17 Electronic sight glass. (Courtesy of TIF Instruments, Inc.)

when the system is full of refrigerant than listening to the valve directly. The unit is simply hung on the tubing, leaving your hands free, rather than having to make constant checks of the sight glass.

These units are battery operated. The battery has a life of about 30 hr.

Electronic Refrigerant Charging Meter. These charging meters are used for charging refrigerant into the system in fractions of an ounce—without limit from a 30-lb cylinder (Figure 13–18). The charging meter measures in the refrigerant by weight but is totally different from the bathroom scale. It measures the charge dispensed into the system. It reads like a service station pump. A charging cylinder is not used with these units. The amount of charge dispensed appears on a LCD display, indicating the correct amount of refrigerant used.

Power Factor Meter. These meters (Figure 13–19) are used for determining the power factor of individual loads and systems to determine their effect on the overall plant power factor. They can be used with single, split-phase, and three-phase power sources, balanced or unbalanced. Another use is for checking motor efficiency and motor load.

FIGURE 13-18 Electronic refrigerant charging meter. (Courtesy of TIF Instruments, Inc.)

FIGURE 13-20 Digital photoelectric tachometer. (Courtesy of TIF Instruments, Inc.)

This unit uses solid-state computer electronics, but no batteries. It uses current from the power source being tested. The clamp-on jaws monitor the current so no circuits need be broken or interrupted. There are two permanently connected volt probes; a third plug-in volt probe is furnished to be used only when monitoring three-phase circuits.

Photoelectric Tachometer. These units are used for electronically measuring the rpm of a motor or shaft from 0 to 10,000 rpm. It may be used in the field or in the shop as the job demands. There is no connection between the meter and the rotating part being measured. A foolproof "electronic eye" is used (Figure 13-20). Its small size allows it to be used in tight places. They are not affected by fluorescent lighting.

Megohmmeter. The purpose of the megohmmeter (Figure 13-21) is to measure the resistance of the insulation on electric wiring. It is especially useful in refrigeration and air conditioning work for checking the condition of the insulation in the electric motors used to drive fans and compressors. Megohmmeters are available in both battery-operated and crank-operated models. We discuss only the battery-operated models because there is less possibility of making a mistake when they are used.

FIGURE 13-19 Power factor meter. (Courtesy of TIF Instruments, Inc.)

FIGURE 13-21 Megohmmeter. (Courtesy of Amprobe.)

Precautions for Personal and Instrument Protection

1. Although there is high voltage at the tips of the test leads when the megohmmeter is used, low current flow reduces any danger to relatively safe levels. However, contact with the probe tips while the instrument is energized should be avoided.

2. Do not use in an explosive atmosphere. If the equipment to be tested was in use up to the time of testing, it will probably have a capacitive charge built up. This capacitance must be discharged before making any resistance measurements.

   Equipment under test with the megohmmeter may build up a capacitive charge from testing. This charge is automatically discharged when the Model AMB-1A Amprobe megohmmeter is used and the "press-to-read" button is released. This feature may not be available on all models, so make certain how this charge is discharged with the one in use.

3. When the AMB-1A will not be used for a period of time, remove the batteries to avoid possible instrument damage due to a leaking battery.

4. Before connecting one of these instruments to any circuit as either an ohmmeter or a megohmmeter check for any voltage. The AMB-1 has a 0 to 600-V ac test band for this purpose. Do not connect any ohmmeter or megohmmeter to a live circuit.

5. Always store the instrument in its carrying case.

Operating Instructions (Model AMB-1A). Note that the lid of the AMB-1A carrying case can be detached from the case and snapped in position on the back of the carrying case.

1. Remove the panel on the back of the instrument and install eight type AA batteries observing the proper polarity. Note that the instrument is fused with a 1-A fast-blow fuse.

2. Set the range switch to the ACV/MΩ position. Use the zero adjust screw located just below the window to adjust the pointer to infinity on the M scale. *Do not press the red button.*

3. Set the range switch to the BATT (battery) position. Press the red button. If the pointer does not indicate within "BATT OK" box, replace the batteries.

4. Set the range switch to the ACV/MΩ position. Press the red button. The pointer should indicate at the "∞" scale division on the MΩ scale. Plug the test leads into the instrument and short the probes together. Press the red button. The pointer should indicate at the "0" division on the MΩ scale. If not, replace the fuse.

5. To use a megohmmeter or an ohmmeter:
   (a) Disconnect the device or circuit to be tested from the line.
   (b) Set the range switch to the ACV/MΩ position. Plug the test leads into the instrument and apply the test probes to the circuit or device to be tested. *Do not press the red button.* If the pointer moves up the red ACV band, there is voltage present in the circuit or device. Disconnect the device or circuit from any voltage supply before proceeding.
   (c) Set the range switch to the appropriate position—ACV/MΩ for megohms, Ω for low ohms.
   (d) Connect the test leads to the device or circuit to be tested.
   (e) Press the red button.
   (f) Take a reading on the scale corresponding to the range switch position.

Fenwal 24 VAC Gas Ignition Field Tester. The 05-080224-002 gas ignition field tester (Figure 13-22) is designed to conduct performance tests on Fenwal

FIGURE 13-22 24-V ac ignition field tester. (Courtesy of Fenwal Inc.)

series 05-16, 05-27, 05-29 direct spark, and series 05-21 silicon carbide ignition systems.

The following functions may be tested with this instrument:

1. Generation of high voltage for spark discharge, or power to the silicon carbide (SiC) element
2. Provides power (24 V ac) to energize the gas valve during the trial for ignition (TFI) period during the normal heating cycle
3. Proper functioning of the flame-sensing circuitry which keeps the gas valve energized during the heating cycle
4. Operation of the lockout function to remove electrical power from the gas valve and stop the sparks or removes the power from the SiC element if ignition is not accomplished
5. Reset after lockout
6. The reignition sequence that provides for a TFI period to begin if the flame sense is lost during heating cycle

The instrument must be used in accordance with the instructions accompanying it.

Fenwal 12 V dc Ignition Field Tester. The 05-080224-001 gas ignition field tester (Figure 13-23) is designed to conduct performance tests on Fenwal series 05-15 and 0-25 direct spark ignition systems (12 V dc).

FIGURE 13-23 12-V dc gas ignition field tester. (Courtesy of Fenwal Inc.)

The following functions may be tested with this instrument:

1. Generation of a high-voltage spark discharge
2. Provides power (12 V dc) to energize the gas valve during the trial for ignition (TFI) period and during the normal heating cycle
3. Proper functioning of the flame-sensing circuitry which keeps the gas valve energized during the heating cycle
4. Operation of the lockout function to remove electrical power from the gas valve and stop the generation of high voltage if ignition is not accomplished
5. Reset after lockout
6. Tests the reignition sequence that provides the TFI period to begin if the flame sense is lost during the heating cycle

This instrument must be operated according to the instructions accompanying it.

Totaltest Refrigerant Testing Device. The Totaltest device is used for checking the acid/moisture content of a system without taking an oil sample. It tests

(1) INSTRUMENT

(1) TEST TUBE PACK-
CONTAINING:

(2) TEST TUBES

(2) TEST TUBE END CAPS

(1) REPLACEMENT 'O' RING

FIGURE 13-24 **Totaltest unit. (Courtesy of Carrier Corp.)**

for corrosive acids and acid-causing moisture by using a small amount of refrigerant. (Figure 13-24).

What the Totaltest Device Does. The Totaltest device prevents a premature refrigeration or air conditioning system breakdown by:

1. Indicating when moisture levels may be considered harmful to a system. (Note: All systems contain some moisture, but small quantities are not harmful. The moisture-sensing chemical is designed to change color to indicate moisture only when the moisture content approaches a harmful level. Sightglass indicators may indicate a moisture problem even when the moisture level is acceptable).

2. Indicating when there are strong acids in a system which may corrode internal system and compressor parts. The acid-sensing chemical is designed to change color to indicate harmful acids and to disregard nonharmful acids—unlike ordinary oil sample test kits which read all acids, corrosive and noncorrosive.

The Totaltest device works with R-22, R-12, R-502, R-500, and R-11 vapor and can be used on the system whether it is operating or not. Since the Totaltest device reads refrigerant, no oil sample is needed.

WARNING: Failure to follow these instructions could result in personal injury. Wear safety goggles and gloves when handling refrigerant.

The test tubes containing the chemicals are glass; use care when handling. Break off the tips with the break-off end of the pack pointed away from the face. Do not touch the broken tip. The chemicals are not harmful, but the glass tips are extremely sharp.

Dispose of the test tubes in a proper container. Avoid exposing test tubes to extreme heat. Store in a cool, shaded area.

How to Use the Totaltest Device. Follow these steps when using the Totaltest:

1. Remove a tube from the pack (Figure 13-25).
2. Break off tips. Use the end of the pack with the small hole (Figure 13-26).
3. Take the instrument apart; insert the tube into the plastic holder. The tube must be inserted properly to work (Figure 13-27).
   IMPORTANT: Insert the chemical end of the tube in first. (Figure 13-28).
4. Couple the instrument together. "Push and twist" for proper coupling (Figure 13-29).
5. Hook the instrument to the schrader fitting on the suction side of the system. IMPORTANT: Hook the instrument to the suction side. First

1. REMOVE TUBE FROM PACK

FIGURE 13-25 **Pack of Totaltest tubes. (Courtesy of Carrier Corp.)**

2. BREAK OFF TIPS. USE END OF PACK WITH THE SMALL HOLE

FIGURE 13-26 **Breaking off tips of tubes. (Courtesy of Carrier Corp.)**

3. TAKE INSTRUMENT APART, INSERT TUBE IN PLASTIC HOLDER. THE TUBE MUST BE INSERTED PROPERLY TO WORK.

RIGHT

FIGURE 13-27 **Correct way to insert tube into holder. (Courtesy of Carrier Corp.)**

IMPORTANT-INSERT CHEMICAL END OF TUBE IN FIRST

WRONG

FIGURE 13-28 **Incorrect way to insert tube into holder. (Courtesy of Carrier Corp.)**

4. COUPLE INSTRUMENT. "PUSH AND TWIST"

"SEAT TUBE IN 'O' RING"

FIGURE 13-29 **Coupling the instrument together. (Courtesy of Carrier Corp.)**

wipe the fitting clean of moisture, grease, or dirt. *Note:* The instrument may be hooked to the discharge line between the compressor and the condenser if the pressures do not exceed 400 psig (2756 kPa). The discharge connection is better when the pressure falls below 60 psig (413.4 kPa) in the low side.

6. Leave the instrument hooked up for no longer than ten minutes! Longer readings may indicate a problem when one does not exist.

7. Disconnect the instrument from the schrader valve. Remove the test tube and compare the tube to the charts on the test tube pack (Figure 13-30). For maximum accuracy, compare the tube to the chart in the sunlight. Incandescent and fluorescent lights may show a low reading on the moisture chart.

8. Moisture Test. (*Note:* The moisture sensing chemicals will change color uniformly. The more red they are, the more moisture is present.) If the color of the chemicals in this chamber falls within the range of colors on the chart indicating low, medium, or high moisture content, follow clean-up procedure A. Procedure A is detailed on the back of the Totaltest case (Figure 13-31).

9. Acid Test. (*Note:* The stain starts from the end nearest the moisture chemical chamber and increases in length as the acid content increases.) A yellow stain will appear if there is acid present. If the yellow stain length is 0 to 1 on the scale, disregard the test. If the yellow stain length is 1

to 7, follow clean-up procedure B. For maximum accuracy, rotate the indicator tube 90° to four positions and average the readings. Procedure B is detailed on the back of the Totaltest case. If clean-up procedure B is repeated more than twice and an intense quantity of acid still persists, a more severe compressor problem may exist. Consider a compressor changeout.

Helpful Information

Timing the Test. Normally the test time is no longer than 10 minutes; however, if there is a great amount of acid in the system, the crystals in the test tube will change color in less than 10 minutes. There is no need to continue the test if this happens.

If the unit does not have sufficient clearance to allow hooking up the Totaltest instrument, use a charging hose and a union. IMPORTANT: Purge the hose and union before connecting the instrument.

Never connect the instrument to a gauge manifold because there is the possibility of contaminants in the lines and valves of the manifold, which a simple purge will not remove.

If a pink ring appears immediately at the end of the moisture crystals and the crystals do not continue to change color, there was moisture contamination in the schrader valve or not enough pressure in the system. Pressures below 60 psig (413.4 kPa) may show a pink stain at the front end of the crystals. Replace the test tube and retest.

After breaking off the tips of the test tube, use the tube immediately. Do not leave the tube open-ended any longer than necessary before taking a test because moisture or contaminants in the air may cause a false reading.

Before breaking off the tips, compare the crystals to the charts on the tube pack. If the moisture color indicates red or the acid chamber has a yellow stain, do not use the tube.

Packaged with the test tubes are two caps. These caps are used to temporarily seal the open ends of a test tube so you can show the test results to the equipment owner or operator. (Figure 13-32).

Also packaged with the test tubes is an "O" ring. The "O" ring is placed inside the aluminum sleeve of the instrument. (Figure 13-33). Replace the "O" ring if it becomes damaged or lost. It is a good idea to check to see that the "O" ring is in place before taking a test.

If the compressor is a burnout, always follow the original equipment manufacturer's recommended

FIGURE 13-30 Comparing tube reading to chart on tube package. (Courtesy of Carrier Corp.)

FIGURE 13-31 Moisture and acid test procedure. (Courtesy of Carrier Corp.)

FIGURE 13-32 Caps in place on test tube. (Courtesy of Carrier Corp.)

FIGURE 13-33 Location of "O" ring. (Courtesy of Carrier Corp.)

clean-up procedures. After completion of the clean-up procedures take another test to verify that the system is clean.

▣ Safety Procedures ▣

Test equipment is a must for efficient service procedures. The information obtained through the use of damaged test equipment is useless. Every precaution should be exercised to keep instruments functioning properly.

1. Never connect an ohmmeter to an electric circuit when the electric power is on.
2. Do not drop, kick, or throw test instruments.
3. Always start (except with an ohmmeter) with a higher than necessary scale and work downward until midscale is reached to prevent damage to the instrument.
4. Always use the correct meter for the job.
5. Never apply open flame to a charging cylinder.
6. Never subject gauges to pressures that exceed their rating.
7. Protect test instruments from adverse weather conditions.
8. Never leave a refrigerant charge in a charging cylinder.
9. Never jump a protective fuse in a test instrument.
10. Never leave thermometers in direct sunlight, as the mercury may exceed the scale and ruin the thermometer.

▣ SUMMARY ▣

☐ Test instruments can help the service technician to perform service operations faster and more accurately.

☐ The two things accomplished when test instruments are used are: (1) prevention of the unnecessary interchanging of parts and (2) savings of time.

☐ Selecting of test equipment should be approached cautiously.

☐ The field repair of test equipment should not be attempted.

☐ Some meter movements are sensitive to position. Always use them in the correct position.

☐ Never repair test leads. Replace them when worn or damaged.

☐ Test instruments stored in tool boxes are subjected to extremely rough treatment.

☐ Avoid storing test instruments in extremely hot or cold temperatures.

☐ Test instruments, when properly applied, will provide more accurate trouble diagnosis in the shortest possible time.

☐ The voltmeter is probably the most often used test instrument. It is also the easiest to read.

☐ The ohmmeter is used for measuring resistances and continuity, for making quick checks on capacitors, and for testing solid-state components.

☐ The clamp-on ammeter is the most popular current-measuring instrument used for ac circuits because the current flow can be checked without interruption of the electric circuit.

☐ The wattmeter can be used to determine the voltage to the unit and the total wattage draw of the unit.

☐ The capacitor analyzer can be used to check for open or shorted capacitors, for the microfarad capacity, and for the power factor.

☐ Temperature testers are used to sense accurate temperatures during normal operation of the equipment being tested.

☐ Micron meters are used to indicate a very high vacuum, which is not possible with ordinary service gauges.

☐ Millivolt meters are used to check the output voltage of thermocouples.

☐ Electronic leak detectors are very sensitive instruments used to detect refrigerant leaks.

☐ Thermostat testers are used to calibrate refrigeration thermostats accurately.

☐ Hermetic analyzers are very useful in hermetic compressor servicing.

☐ Potential relay analyzers are used to check the condition of potential relays.

☐ The sling psychrometer is used in determining relative humidity.

☐ Heating system analyzers are designed to aid the service technician in repairing the various components in a heating system.

☐ The electronic sight glass emits an audible signal, indicating that the refrigerant system is full of refrigerant.

☐ The electronic refrigerant charging meter is used for charging refrigerant into the system in fractions of an ounce—without limit from a 30-lb cylinder.

☐ Power factor meters are used for determining the power factor of individual loads and systems to determine their overall effect on the overall plant power factor.

☐ Photoelectric tachometers are used for electronically measuring the rpm of a motor shaft from 0 to 10,000 rpm.

☐ EXERCISES ☐

1. What is the purpose of test instruments?
2. What are two strong arguments for using test instruments?
3. What is a good method of selecting test instruments?
4. What is the fault of having all the test functions in one instrument?
5. Why should the batteries be stored separately in seldom used instruments?
6. What is the purpose of fuses and circuit breakers on test instruments?
7. In what position should the selector switch be placed when storing test instruments?
8. Which test instrument is the most often used?
9. Which test instrument is the easiest to read?
10. Which test instrument is used to check continuity?
11. Why are clamp-on ammeters popular among service technicians?
12. Name the function of a capacitor analyzer.
13. What is an advantage of using a temperature tester?
14. When is a micron meter used?
15. Under what conditions can an electronic leak detector not be used?
16. What is the fluid used in a thermostat tester well?
17. When is a hermetic analyzer especially useful?
18. What instrument is used to determine relative humidity?
19. What are the functions of a heating system analyzer?
20. On what principle does the electronic sight glass operate?
21. What does the electronic charging meter measure?
22. Where may photoelectric tachometers be used?

14 Electric Motor Theory

After completion of this chapter, the student should:

■ Understand what makes an electric motor operate.

■ Explain the use of capacitors in an electric motor.

■ Understand how the various types of electric motors operate.

■ Be able to explain how to connect polyphase motors.

■ Know what determines the amount of electric current drawn by an electric motor.

■ Be able to discuss the function of a motor overload.

■ Know how to determine the common, start, and run terminals of a compressor motor.

■ Describe how to test the windings of an electric motor.

■ Know how to determine the proper replacement capacitor.

■ Complete the Exercises with at least 80 percent accuracy.

■ Complete the Laboratory Exercises with at least 80 percent accuracy.

The laws of magnetism and magnetic induction discussed in Chapter 11 are the basis on which all electric motors operate. It is desirable that an understanding of these principles be obtained before the basic theory of electric motors is attempted.

Definition. *Electric motors* are devices used to convert electrical energy into rotating mechanical energy. They are used to drive the fans, pumps, and compressors used in refrigeration and air conditioning systems. Electric motors are designed to operate on different voltages, different directions of rotation, and at different speeds.

☐ Electric Motor Theory ☐

The most simple electric motor possible is made up of a rotor, a permanent magnet mounted on a movable shaft, and two magnetic poles mounted on the outside motor shell (Figure 14–1). These magnetic poles are actually coils of wire; they are stationary poles and are referred to as the *stator*. These coils of wire produce strong magnetic fields when electric current is passed through them.

The amount of starting torque built into the motor circuit is the basic difference among the single-phase motors used in the refrigeration and air conditioning industry. The method by which torque is obtained is the major difference in these motors.

The magnetic laws of attraction state that like magnetic poles repel and that unlike magnetic poles attract. The top pole marked N in Figure 14–1 is magnetically attracting the S pole of the rotor. Because the stator is stationary, the rotor moves toward the stator poles and rotation of the shaft begins.

The electric current in use today is 60-hertz alternating current. This means that the electric motor used must be designed to operate on this type of current. When alternating current is used, the direction of current flow is reversed alternately from a high negative potential to a high positive potential. In modern electrical systems, this change takes place 60 times each second. Therefore, the polarity of each

FIGURE 14–1   Simple electric motor.

FIGURE 14–2   Rotor repulsion and attraction.

stator coil is reversed each time the potential changes from negative to positive.

Because of this alternating polarity of the magnetic poles of the stator, there is a push–pull action on the rotor. Therefore, before the rotor can stop in line with the stator poles, the current changes direction, which also changes the magnetic polarity of the poles. The momentum of the rotor carries it past the in-line position with the pole. The south pole of the stator attracts the north pole of the rotor and repels its south pole (Figure 14–2). At the same instant, the north pole of the stator repels the north pole of the rotor and attracts its south pole. In this way, the push–pull action is kept continuous. The rotor will continue to spin as long as the poles continue to change their magnetic polarity. The rotor will theoretically adjust itself to a speed of 60 revolutions per second (rps) or 3600 revolutions per minute (rpm). This is called the *synchronous speed* of the motor. Mathematically, the synchronous speed of a motor may be determined by the following formula:

$$\text{Synchronous speed} = \frac{120 \times \text{frequency}}{\text{number of poles}}$$

FIGURE 14–3   Motor poles.

For example, the synchronous speed of a two-pole motor is

$$\frac{120 \times 60}{2} = 3600 \text{ rpm}$$

The synchronous speed of a four-pole motor is

$$\frac{120 \times 60}{4} = 1800 \text{ rpm}$$

From this, it can be determined that a six-pole motor will turn at 1200 rpm and an eight-pole motor will turn at 900 rpm. This does not take into account the amount of slipping that is inherent in the motor. Each pole will consist of two windings: one for the running, or main, winding; and one for the starting, or auxiliary, winding (Figure 14–3). One winding will be made of larger wire than the other. The winding with the larger wire is the main winding and the one made from the smaller wire is the start winding. The start winding will not stand up under the load required of the main winding and will soon overheat and burn up, ruining the motor winding, if it is connected to the main winding terminals. Caution must be observed to prevent this from happening.

From the foregoing explanation, it can be easily seen that as the stator poles continue to alternate between north and south, the rotor will continue to rotate. However, the major problem lies in starting the rotor, not running it.

Should the rotor be stopped in line with the stator poles, the rotor could not be started, regardless of the polarity of the two stator poles (Figure 14–4). In this condition, the north pole is attracting the south pole and the rotor will not move. Even when

FIGURE 14-4 Rotor in line with stator poles.

FIGURE 14-6 Equal rotor attraction between two poles.

FIGURE 14-5 Phase stator pole.

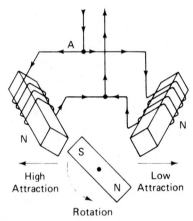

FIGURE 14-7 High and low attraction of stator poles.

the stator pole is changed to the south pole, the repelling forces will be at such an angle that the rotor will not turn.

However, if another pole is installed in the stator, the rotor can be caused to rotate (Figure 14-5). The phase north pole on the left will attract the south pole on the rotor at the same time that the north pole on the right does. The stator pole on the left will provide enough magnetic attraction to pull the motor to the left and start the rotation of the motor.

However, if the rotor should stop between these two poles, the problem of starting the rotor would still exist (Figure 14-6). As the polarity of the poles is established, the magnetic attraction of the rotor is equal in both directions.

The obvious solution would be to have a motor circuit that would build a strong magnetic strength in one stator pole before the other. In Figure 14-7 the two stator windings and the rotor have been connected to a common source of electric current. The electricity enters the motor circuit at point *A*. The current is divided at this point and part flows to the left while the rest flows to the right. If the current

can be made to flow to the left before it flows to the right, the left stator pole will obtain a strong north polarity earlier than the right stator pole. Therefore, the strong attraction to the left will overcome the weaker attraction to the right and the rotor will turn.

Finding a method of making the current in the left pole lead or flow ahead of the current in the right stator pole would solve the problem of effectively making the motor start. In other words, two-phase current must be available to produce the magnetic attraction and repulsion required to start the motor. The component used to produce this two-phase current is a capacitor.

▢ Capacitors ▢

In order to understand how a capacitor can produce the second phase of current needed to start a motor, a review of some of the basic facts about ac voltage and current is necessary.

To illustrate some of these facts a pure resistance circuit will be used. In Figure 14-8 the voltage applied from point *A* to point *B* is represented in the illustration on the right by use of the sine wave. Because the current flow through a pure resistive load will be in phase with the applied voltage (solid line) the current part of the waveform (dashed line) is shown at a peak when the voltage is at a peak and at zero when the voltage is zero.

If a capacitor is connected into an alternating current circuit, the current through the load will not remain in phase with the applied voltage. In a capacitive circuit, the load current will lead the applied voltage. Figure 14-9 shows a capacitor connected in a series with the resistive load. The voltage that is applied to points *C* and *D* has the same wave as *A* and *B* in Figure 14-8. The current waveform (the dashed line) is now shown leading the applied voltage.

From this it can be seen that installing a capacitor in series with one of the two stator windings will cause the current flowing though that winding to lead the current that flows through the winding that is connected directly to the power supply.

It has been stated that the capacitor is the component that provides the two-phase electric current necessary to start a motor. Therefore, a basic knowledge of its characteristics is required.

Capacitors are rated in millionths of a farad or in microfarads (μF). A large μF rating is the result of using either large metal plates or a small amount of insulation between the plates; a low μF rating is the result of using small metal plates or a large amount of insulation between the plates (Figure 14-10).

A large phase shift will occur between the applied voltage and the load when a capacitor with a large μF rating is connected in series with a load. This high phase shift is what provides the high starting torque for single-phase motors. A low phase shift and low starting torque is the result of using a capacitor with a low μF rating.

Also, the amount of electric current that will flow through a series load when capacitors with different μF ratings are used is important. High-μF capacitors will provide a high current flow through the series load, while low-μF capacitors will provide a low current flow through the series load.

Two types of capacitors are used to produce the second-phase current necessary to start single-phase motors (Figure 14-11). The physical dimensions have nothing to do with the μF rating of a capacitor. A

FIGURE 14-8 Pure resistive circuit.

FIGURE 14-9 Capacitive circuit.

FIGURE 14-10 Effects of plates and insulation on capacity of capacitors.

FIGURE 14-11 Motor capacitors.

running capacitor will usually have a low μF rating. It is physically larger because of the dielectric oil-type insulation. This oil dissipates the internal heat from the plates to the outside atmosphere. The running capacitor provides a low starting torque and a low current flow through the series load.

The starting capacitor usually has a higher µF rating than the running capacitor. The starting capacitor is switched out of the circuit a few seconds after the motor starts. It is necessary that the starting capacitor be switched out because it is not designed to carry the heavy current flow for an extended period of time. If the starting capacitor is left in the circuit for an extended period of time, permanent damage may be done to the capacitor.

To make a graphic illustration of this principle, the motor with the auxiliary stator pole used in Figure 14–5 is used in Figure 14–12. Here, both the pole windings have been connected across the electric power supply. A capacitor is connected in series with the pole on the right. The rotor has stopped in a position between the two poles. With the capacitor connected in series, the current leads the applied voltage.

Therefore, the current flowing through the right stator pole is leading the current through the left stator pole. Thus the north pole in the right stator pole will become strong before the pole on the left stator pole. Because of this, the south pole of the rotor will be deflected toward the pole on the right and the starting torque is created.

An instant later, the current reverses direction in the right stator pole and the polarity of the stator pole is reversed (Figure 14–13). Now the south stator pole is repelling the south pole of the rotor. This repulsion increases the amount of work that a motor can do. A more efficient motor is the result of leaving the capacitor in the circuit.

A starting capacitor found to be defective should be replaced with one of the same size in regards to the voltage and µF ratings. If the correct size is not available, it may vary from 0 to 20% of the proper µF rating for starting capacitors. When replacing starting capacitors, be sure that a 15,000- to 18,000-Ω 2-W resistor is soldered across the terminals of the capacitor. This is a precautionary measure to prevent arcing and burning of the starting relay contacts.

A running capacitor found to be defective should be replaced with one having a voltage rating equal to or greater than the one being replaced. It can be replaced with one that is within a ±10% of the µF rating of the original. Running capacitors are provided with a terminal marked with a red dot. Because of the relatively high voltage created in the starting winding, the unmarked terminal is connected

FIGURE 14–12   Use of capacitor.

FIGURE 14–13   Rotor deflection.

to the starting terminal on the motor. The red dot indicates the terminal that is most likely to short out in case of capacitor breakdown. If the terminal with the red dot is connected to the motor starting terminal, damage to the winding could result.

Also, some equipment manufacturers use the running capacitor on single-phase motor-compressors as a crankcase heater. The capacitor allows a small electric current to pass through the starting winding during the off cycle. This small current causes the winding to heat just enough to prevent the migration of refrigerant into the compressor crankcase during the off cycle, thus preventing a flooded start of the compressor. When the running capacitor on this type of unit fails it must be replaced with one of the proper size to prevent overheating the starting winding and causing possible damage to it.

There are five types of single-phase motors used on modern refrigeration and air conditioning equipment. They are: (1) split-phase, (2) capacitor-start (CSR), (3) permanent-split capacitor (PSC), (4) capacitor-start/capacitor-run (CSCR), and (5) shaded-pole.

Split-Phase Motors

These motors are popular when motor sizes ranging from $\frac{1}{20}$ to $\frac{1}{3}$ hp are required. Split-phase motors are used for such applications as oil burners, blowers, pumps, and fans. The starting torque requirements for these applications are moderate—not a great deal more than the running torque requirements.

Split-phase motors are single-phase induction motors that use a nonwound rotor and a stator whose windings are embedded in the insulated slots of a laminated steel core (Figure 14–14). The stator windings are comprised of two separate windings: the main or running winding and the starting or phase winding. These two windings are connected electrically in parallel (Figure 14–15). The phase winding is physically placed in a magnetic position to the main winding to effectively produce a two-phase field required to start a single-phase induction-type motor and help bring it up to its normal running speed. After the motor has reached approximately three-fourths of its normal running speed a centrifugal switch, actuated by a governor weight

FIGURE 14–15 **Split-phase motor connection diagram (starting).**

FIGURE 14–16 **Split-phase motor diagram in the operating position.**

mechanism, opens the electric circuit to the phase winding (Figure 14–16). After the centrifugal switch removes the phase winding from the circuit, the split-phase motor operates as a single-phase induction motor.

Capacitor-Start Motors

These motors are similar to split-phase motors with the exception that a starting capacitor is used. The starting capacitor is connected electrically in series with the phase winding as shown in Figure 14–17. The purpose of the capacitor is to increase the starting torque and reduce the starting current. The capacitor is removed from the circuit by the centri-

FIGURE 14–14 **Split-phase motor cutaway. (Courtesy of General Electric Co.)**

FIGURE 14–17 **Capacitor-start motor connection diagram.**

fugal switch when the motor reaches approximately 75% of its normal operating speed.

Since the capacitor is in the circuit only during the starting cycle, the running characteristics of split-phase motors and capacitor-start motors of the same horsepower rating are identical.

Capacitor-start motors are used on applications that require a high starting torque (150 to 350% of the running torque) or where low starting amperes are required. This group includes such components as compressors, large fans and blowers, and water pumps.

□ Motor Speeds □

The induction motors used in air conditioning and refrigeration applications, including split-phase, repulsion-start, and capacitor-type motors, are designed for use on 25-, 50-, or 60-hertz (Hz) electrical current. The motor speed is mainly dependent on two things: (1) the numbr of cycles of the electric current and (2) the number of field poles in the motor. The motor speeds are calculated from the synchronous speed. The synchronous speed has a relationship to the number of electric current cycles per second. Synchronous speed is reached when the magnetic polarity of the field poles changes with the same regularity as the alternating electric current. For the motor speed, cycle, and number of poles relationship, see Table 14–1.

Because of magnetic slippage, motors do not operate exactly at synchronous speed—especially when they are operating under loaded conditions. Therefore, motors are rated at a speed corresponding with normal load conditions rather than their synchronous speed.

Some hermetic motor-compressor manufacturers are developing these units to operate as either two-pole or four-pole motors. This design allows the unit to operate at either 1750 rpm or 3450 rpm (Figure

FIGURE 14–18 Two-speed hermetic motor-compressor that uses three-phase electric power. (Courtesy of Lennox Industries, Inc.)

14–18). The electrical diagram for a two-speed motor is shown in Figure 14–19. The motor-compressor speed is controlled by solid-state circuits.

Multispeed motors are very popular in air conditioning and refrigeration systems. They may be two-speed motors or they may have several speeds incorporated in one motor frame so that a variety of functions can be met. The technician should know how these motors operate and how to wire them into the circuit for the desired operation.

□ Two-Speed Motors □

Two-speed motors are made in both the split-phase and the capacitor-start types. Three separate windings are used in these motors: (1) the phase-winding, (2) the low-speed main winding, and (3) the high-speed main winding (Figure 14–20). The low-speed and high-speed connections are determined by an external switch or relay in the control circuit. During high-speed operation, the motor functions like a single-speed motor using only the phase and high-speed main winding. During low-speed operation, the motor starts on the phase and high-speed windings. After the starting cycle is completed, the motor switches to the low-speed main winding (Figure 14–21).

Two-speed motors are used to operate blowers in

TABLE 14–1 Synchronous and Operating Speed (rpm) for Two- and Four-Pole Motors

Number of Poles	60 Hz		50 Hz		25 Hz	
	Syn.	Op.	Syn.	Op.	Syn.	Op.
2	3600	3450	3000	2850	1500	1450
4	1800	1750	1500	1450	750	700

FIGURE 14-19 Circuit diagram of a two-speed, three-phase hermetic motor-compressor. (Courtesy of Lennox Industries, Inc.)

FIGURE 14-20 Two-speed electrical connection diagram.

FIGURE 14-21 Two-speed motor in low-speed condition.

air conditioning systems that use a modulated airflow or fans and in other applications where more than one speed is required.

□ Multispeed Motors □

Most of the motors in this category have three, four, or more speeds. They are popular in applications where one horsepower motor will handle several different loads. The speed chosen should give the desired temperature drop or rise through the equipment in use. They are equipped with an auxiliary winding and several other main windings (Figure 14–22). The figure shows a three-speed motor, but the principle is the same for any number of speeds. Some manufacturers have a color code for each speed, but few of them are consistent in their color designations.

Because of this it is almost impossible to memorize one color code for all the various types of motors. There are two ways to determine the connections for the different motor speeds: have the

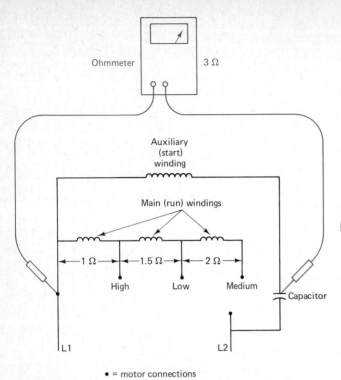

FIGURE 14–22 **Determining motor speeds.**

FIGURE 14–23 **Connections for 240-V motor (dual voltage).**

FIGURE 14–24 **Connections for 120-V motor (dual voltage).**

equipment manufacturer's wiring diagram handy, or know how to determine the speeds by taking resistance readings of the different motor winding connections.

Determining Motor Speeds by Resistance Readings. The phase winding will have the greatest amount of resistance of any single winding. Also, they are generally indicated by brown wires. When finding the phase winding, there will be only one wire or connection that will have continuity with the motor common terminal or connection. This wire will connect to one terminal of the capacitor. The other capacitor terminal will connect to the other line lead (Figure 14–22).

When finding the other motor speeds, the least resistance will be indicated between the high-speed winding lead and the common terminal or connection on the motor (Figure 14–22). The next-lower speed will be indicated by the next-higher resistance reading indicated between its lead and the common motor terminal. The next-lower speed will be indicated by the next-higher amount of resistance between its lead and the common motor terminal. This process is continued until all the motor speeds have been determined. All of the winding connections not used must be protected to prevent an electrical short or shock. The readings indicated in Figure 14–22 are for instructional purposes only. The actual resistance readings will probably be somewhat different.

☐ Dual-Voltage Motors ☐

Dual-voltage motors are popular with service technicians because they reduce the number of motors that are needed on the service truck, thus reducing the cost of the truck stock and making it more convenient for the technician and the equipment owner.

These motors are generally rated for 120 or 240 V and can be used on either voltage with only a minor change in the wiring connections. Some motor manufacturers have made this conversion very simple by installing a reversible plug on the outside of the motor frame. However, other motors do not have this convenience. When either voltage is used, the phase winding is separate from the two main windings (Figures 14–23 and 14–24). The phase winding is used only during the starting of the motor. At about 75% of the normal running speed of the motor the centrifugal switch opens and stops the current flow through the phase winding.

ELECTRIC MOTOR THEORY

240-V Motor Connections. When the motor is connected for use on 240 V, the main windings are in electrical series (Figure 14–23). The resistance when measured from line connection to line connection will be greater for a 240-V motor than that for a motor that is connected for 120-V operation.

120-V Motor Connections. When the motor is connected for 120-V operation, the two main windings are in electrical parallel (Figure 14–24). The resistance when measured from line connection to line connection will be less than half of that measured for 240-V operation.

FIGURE 14–26   Changing rotation of a 120-V motor.

☐ Reversing Motor Rotation ☐

Some motor manufacturers make motors that are designed so that the rotation can be reversed in the field. The most popular types of these motors for use in the air conditioning and refrigeration field are: capacitor-start, run (SCR); capacitor-start, capacitor-run (CSCR); and permanent-split capacitor (PSC). The rotation of CSR and CSCR dual-voltage motors may be reversed in the field by first changing the motor to the voltage desired, then wiring the motor for the desired rotation as shown in Figures 14–25 and 14–26. For PSC motors, some motor manufacturers build a reversible plug outside the motor that only need to be removed and turned around to reverse the rotation (Figure 14–27). However, when this feature is not available, most of them can be reversed by rewiring the motor (Figure 14–28). The reversing procedure changes the connection of the capacitor. In one connection, the capacitor is in electrical series with the main winding. When the connections are changed, the capaci-

FIGURE 14–27   Wiring connections for using a reversing device.

FIGURE 14–28   Wiring connections for using a reversing device.

FIGURE 14–25   Wiring connections for reversing rotation of 240-Volt motor.

tor is in electrical series with the start winding. When the capacitor is switched, the rotation of the motor is also switched from one direction to the other.

Caution: This should not be attempted on compressor motors, either hermetic or semihermetic.

There are three different types of three-phase motors in use today: (1) synchronous, (2) squirrel cage, and (3) wound-rotor motors (Figure 14–29). The power input to all three of these motors is a three-phase voltage from the power company. The squirrel cage and the wound-rotor type are very similar in operation and both use induction as their mode of starting and running. The synchronous type, however, is usually started externally with a dc power source that causes the rotor to become excited and then works on the rotating field principle.

As was learned earlier, a stator and rotor are required to make up an electric motor. Physically, the component parts of a three-phase motor resemble those of a single-phase motor: the stator is laminated, some have squirrel cage rotors, the armature is just a specific type of wound-rotor, and other such features.

The major difference is in the electric circuit. The stator winding is made up of three single-phase windings, each one located 120° from the other. Therefore, the voltage across the three windings is 120° out of phase with both the other windings. In a typical waveform diagram, if we use X voltage as the reference voltage, the Y voltage is 120° behind the X voltage and the Z voltage is 120° behind the Y voltage and 240° behind the X voltage (Figure 14–30).

There are two methods used when connecting the stator of a three-phase motor to a three-phase power source: the Y or star connection, and the delta or triangle connection (Figure 14–31). Regardless of which one is used, the windings are connected in such a manner that only three leads come from the stator, making the line connections very simple.

When the stator is connected to a source of three-phase electric power, and ac voltage is applied across the three stator windings which are 120° out of phase with each other, the motor attempts to turn (Figure 14–32).

The direction of current flow in part (A) of Figure 14–32 corresponds to the point referred to as *A* in the waveform diagram below it. Likewise, the direction of current flow in part (B) corresponds to the point referred to as point *B* in the waveform diagram. We can see that in part (A), there is a north (N) pole between one set of *X* and *Z* windings.

Referring to part (B), it can be seen that the elec-

FIGURE 14–29   Three-phase motor schematic.

FIGURE 14–30   Three-phase voltage waveform diagram.

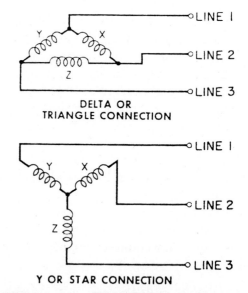

FIGURE 14–31   Methods of connecting a three-phase motor to a three-phase power source.

tric current has shifted because of an alternation of the applied voltage. This shift occurs at point *P* on the waveform diagram. The arrows in part (B) shows that this current shift has also shifted the magnetic flux around the stator. When the next alternation of the applied line voltage occurs, the shift occurs again. This shifting continues with every cycle of the applied voltage. In this manner, we have a magnetic

ELECTRIC MOTOR THEORY

FIGURE 14–32 Rotating magnetic field in a three-phase motor.

field rotating around the inner surface of the stator. As each phase of the applied voltage changes direction, the poles *X, Y,* and *Z* move across the full width of each phase.

☐ Polyphase Motor Connections ☐

The compressors used in larger systems are driven by three-phase motors. Three-phase electric current is more efficient than single-phase current because the surges of electric current are closer together than when single-phase current is used.

The sine wave curves for three-phase current are shown in Figure 14–33. These types of motors generally use 240- or 440-V electricity. These motors are equipped with nine terminals and may be wired for either 240 or 440 V (Figure 14–34).

A three-phase motor-compressor wiring circuit with its starting components and protector circuit is shown in Figure 14–35. Three-phase motors use motor starters. Notice that they do not have the starting relays used on single-phase motors.

☐ Centrifugal Switch ☐

The heart of split-phase and capacitor-start motors is the centrifugal switch. The centrifugal switch consists of four principal components: (1) the moving

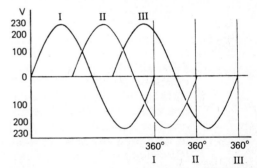

FIGURE 14–33 Three-phase power sine wave.

contact arm, (2) the stationary contact plate, (3) the governor weight, and (4) the weight spring. The contact arm pivots on the contacts, and is held in position with two pins or screws. A compression spring on the top pin holds the contacts open when the switch is in the running position. The stationary contact plate assembly includes the electric line terminal studs and the overload protection device, when used.

Operation. When the motor is not running, the switch contacts are closed with the riding edge of the governor weight against the contact arm that holds the contacts in a closed position. When the motor is energized, the rotor starts to rotate, and as it accelerates to the predetermined switching speed, the centrifugal force of the governor weight overcomes the force of the governor weight spring and snaps outward on the pin, permitting the con-

FIGURE 14-34 Schematic wiring diagram showing circuits and connections for three-phase hermetic motor: (a) circuit as connected for 220 V; (b) circuit as connected for 440 V. Note that L1, L2, and L3 are the three-phase line connections. Numbers 1-2-3, 4-5-6, 7-8-9 are the connections to the motor windings. Each motor winding coil is designed for 240 V; example: the coil between terminals 1 and 4. (Courtesy of Lennox Industries, Inc.)

FIGURE 14-35 Three-phase hermetic motor circuit wiring diagram. L1, L2, and L3 are the three-phase line connections. Overload protector is shown in L1 and L3 circuits and operates magnetic coil of starter switch (top part of drawing). Special magnetic-type starter is required for this installation. Automatic temperature control would be connected to control at top of illustration. (Courtesy of Copeland Refrigeration Corp.)

tact arm to move toward the rotor, opening the contact and deenergizing the phase winding. The contacts close, completing an external circuit such as indicator lights, heater circuits, relays, etc.

□ Permanent-Split Capacitor Motors □

These motors are used for a large number of applications in the refrigeration and air conditioning industry. They are used on fans and blowers and on some compressors. They are rated with a medium starting torque and have excellent running efficiencies. The compressors that use this motor must be installed on refrigeration systems in which the refrigerant pressures will equalize during the off cycle.

Permanent-split capacitor (PSC) motors incorporate a running capacitor connected in series with the phase winding (Figure 14–36). The running capacitor serves two important functions. First, the capacitor provides the split-phase electrical characteristics necessary to start the motor turning. Second, because the capacitor is not removed, this split-phase provides efficient motor operation. The start (phase) winding is made of much smaller wire than the run winding. It also has many more turns than the main (run) winding. Because of these differences (1) the phase winding has a higher resistance than the main winding, and (2) the start winding cannot withstand a very high current draw for a long period of time.

Because the start winding cannot withstand a con-

tinuous high current draw, the running capacitor must have a low μF rating to limit the flow of current through the start winding. When using running capacitors with a low μF rating, only a small phase angle will result, which allows the motor to have only a moderate starting torque. The running capacitor is always connected between the run and start terminals of the motor.

□ Capacitor-Start/Capacitor-Run Motors □

These motors make use of the best characteristics of both the permanent-split capacitor motor and the capacitor-start motor. The permanent-split capacitor has excellent running efficiency but only moderate starting torque. The capacitor-start motor has a high starting torque but a lower running efficiency than the PSC motor.

Therefore, the high starting torque of the CSR motor and the higher running efficiency of the PSC motor are combined to make the CSCR motor. The PSC electric circuit remains the same as with the PSC motor. The capacitor-start circuit is combined with a starting relay to take the place of the centrifugal switch that was discussed earlier. The relay is added so that the circuit can be used on hermetic and semi-hermetic compressors. The centrifugal switch is eliminated on this type of equipment because the repair of the switch would be impossible. The relay is accessible for replacement (Figure 14–37).

In operation, when the electric circuit is completed to the motor, it starts as a capacitor-start unit. As the motor reaches approximately 75% of the operating speed, the relay removes the starting capacitor from the circuit. The motor then operates as a permanent-split capacitor motor.

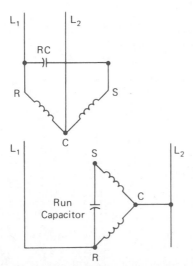

FIGURE 14–36 Permanent-split capacitor connection diagram.

FIGURE 14–37 Capacitor-start capacitor-run connection diagram.

☐ Shaded-Pole Motors ☐

These motors develop a very low torque. Therefore, they are very small in size and are usually less than ½ hp. They are used on small fans, pumps, and timer motors. They are low in cost and comparatively dependable.

The stator poles are made of laminated steel as in the other type motors. However, the pole pieces are quite different because each pole piece has a slot cut in the face (Figure 14–38). Even though it is a form of an induction motor, its windings are distributed differently than those of other induction-type motors. The slot in the face of the pole piece contains the shading coil. This shading coil may be made of a single- or a multiple-strand wire (Figure 14–39). The shading pole coil must form a closed circuit or loop. This loop is not connected to the source of electricity. The main field windings are wound around the remainder of the pole piece.

When electric current is applied the magnetic force around the wire is constantly changing direction, creating a magnetic pole with a force changing from N to S. The magnetic lines of force around the pole are built up to the maximum first in one direction and then the other. The shaded-pole banding changes the rate of speed with which the building up and collapsing of the magnetic lines occur.

Thus, a condition is created where the main pole has a magnetic field that is building up and collapsing at a normal rate while the magnetic field of the smaller shaded pole is building up and collapsing at a different rate.

The main pole and the shaded pole peak in magnetic magnitude at different rates of speed. Therefore, at some time on the way to peak, one of the poles will momentarily get ahead of the other. At that moment the magnetic strength is not the same; a torque is developed, and the rotor will turn. The rotor will turn toward the shaded pole.

These motors are, at best, only about 35% efficient. They have a shorter life span than other types of motors and they are not as good for use in high-temperature application.

☐ Motor Amperage ☐

The electric current used by any electric motor depends on the condition of the motor and the amount of load that it is carrying. Many times a motor will draw less current than what its rating in-

FIGURE 14–38 Shaded-pole motor pole piece.

FIGURE 14–39 Windings of a shaded-pole motor.

TABLE 14–2 Full Load Currents (A) for Electric Motors

| hp | Single Phase | | | | Three Phase | |
| | Shaded Pole | | PSC | | | |
	115V	230V	115V	230V	220V	460V
¹⁄₁₀₀	0.5	0.25				
¹⁄₇₀	0.7	0.35				
¹⁄₅₀	0.8	0.40				
¹⁄₄₀	1.0	0.50				
¹⁄₃₀	1.3	0.56				
¹⁄₂₀	2.0	1.00				
¹⁄₁₅	2.6	1.30				
¹⁄₁₀	3.4	1.70	1.6	0.8		
⅛	5.6	2.8	2.5	1.7		
⅙	5.8	2.9	3.0	1.9		
¼			4.4	2.2	0.9	0.45
⅓			5.0	2.5	1.1	0.57
½			6.2	3.1	2.0	1.0
¾			9.6	4.8	2.8	1.4
1			13.0	6.5	3.5	1.8
1 ½			18.4	9.2	5.0	2.5
2			24.0	12.0	6.5	3.3
3			34.0	17.0	9.0	4.5
5			56.0	28	15.0	7.5

dicates. However, all motors are designed for a certain amount of current for a given horsepower (see Table 14–2). Notice that the voltage as well as the motor horsepower determines the amount of current a motor uses. For example, a ⅙-hp 230-V motor will use only one half as much current as a ⅙-hp 115-V motor.

☐ Single-Phase Motor Protectors ☐

Most single-phase motor circuits include a switch to protect the motor against overheating, overcurrent, or both types of conditions. These overload devices may be either internal or external depending on the equipment design.

Internal Overload. These overloads may be of the line break type or the thermostat type. Both types of internal overloads are located precisely in the center of the *heat sink* portion of the motor windings and protect the motor from excessive temperatures and excessive current draw (Figure 14–40). The line break type of internal overload will interrupt the electric power line to the common connection of the motor winding if either of these conditions occurs (Figure 14–41). The internal thermostat overload protector is also mounted in the motor windings. However, the switch interrupts the control circuit to the contactor or starter (Figure 14–42). The electric connections are shown in Figure 14–43. If this overload becomes faulty, the compressor must be replaced.

External Overload. This type of overload protector is mounted externally and may be replaced when faulty. It is mounted in contact with the motor housing or the hermetic compressor housing to sense the temperature at the point of contact.

External overloads may also be of the line break or the thermostat type. These overload devices (Figure 14–44) must be placed where the manufacturer has designated. They must always be replaced with an exact replacement or proper protection will not be provided.

The line break external thermostat is operated by a bimetal disc (Figure 14–45). The switch contacts are normally closed and when an overload condition occurs the contacts open (Figure 14–46). They are electrically connected into the electric circuit to the common terminal of the motor winding (Figure

FIGURE 14–40 Internal overload location. (Courtesy of Tecumseh Products Company.)

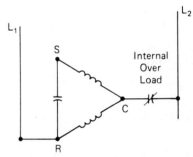

FIGURE 14–41 Line break internal overload connection.

FIGURE 14–42 Internal thermostat overload. (Courtesy of Tecumseh Products Company.)

SINGLE PHASE

To control circuit
24 V–115 V or 230 volts
(300 volts max.)

Supplementary overload

Hermetic terminal
(motor winding thermostat)

L₁ T₁

2 pole contactor

L₂ T₂

Run cap

Connector block

For PSC operation eliminate start capacitor and relay. Hook up run capacitor between start and run of connector block.

Bleeder resistor

Start cap

FIGURE 14–43 Internal overload thermostat connections. (Courtesy of Tecumseh Products Company.)

To: line

To: S cap

To: line

To: R cap

FIGURE 14–44 Location of external overload device. (Courtesy of Tecumseh Products Company.)

Wire Lead Assembly

Pin Connector

Screw and Lock Nut

Screw Terminal

Terminal No. 1

Terminal No. 3

Terminal No. 2

Contact

Heater

Bimetal Disc

Case

FIGURE 14–45 External line break overload. (Courtesy of Tecumseh Products Company.)

Open

Closed

FIGURE 14–46 External overload contacts.

ELECTRIC MOTOR THEORY

FIGURE 14–47 **External line break overload connections.**

FIGURE 14–48 **External thermostat overload. (Courtesy of Tecumseh Products Company.)**

14–47). When the resistance heater has provided sufficient heat indicating an overload condition, the contacts open and interrupt the electric circuit to the motor. The motor then stops until the overload has cooled sufficiently to allow the contacts to reset and complete the electric circuit to the motor.

The thermostat type of external overload is mounted in the same way as the line break type. However, the contacts do not open the main electric power circuit (Figure 14–48). Instead, the control or pilot circuit is interrupted, which causes the contractor or starter to interrupt the main power supply (Figure 14–49).

Time Delay Control for Short-Cycle Protection. Heat is the greatest enemy of electric motor insulation. If a motor is overheated from any cause for a number of times, failure of the winding insulation eventually occurs. Overload is a major cause of overheating and can be due to low voltage, single phasing, unbalanced voltage, or just plain mechanical overload caused by short cycling of a motor trying to start against a load.

There are models available for the protection of motors and compressors from damage due to low voltage, short cycling caused by thermostat or power failure, phase loss, and phase out of sequence.

Delay-on-Break. These are solid-state timers that prevent short cycling of compressor motors by delaying a restart due to a momentary power failure or thermostat interruption. The time delay begins when the power goes off or the thermostat opens.

Delay-on-Make. Some of these relays provide a 3-minute time delay, occurring on power resumption, providing a random restart of equipment after a control interruption or a power failure.

THREE PHASE

FIGURE 14–49 **External thermostat overload connection.**

Single-Phase Motor Protectors

Many service technicians make the false assumption tha troubleshooting motors is difficult, especially the capacitor-start/capacitor-run motors. If the service technician is familiar with the separate electric circuits that cause the motor to function, a systematic approach to troubleshooting each component will usually yield any troubles in the motor.

Also, certain safety precautions should be observed when attempting to troubleshoot any electrical system. Always be certain that the electrical system being serviced is disconnected from the electrical supply.

Capacitors can hold a high-voltage charge on their plates long after the electric supply has been disconnected. Be sure to discharge capacitors before working on them with a 20,000 Ω-resistor touched to each terminal.

The counter emf on the start winding can be much higher than the applied voltage. Therefore, all personnel and meters should be protected against injury or damage.

Testing Motor Winding Insulation. It should be remembered that the purpose of insulation on an electric conductor is to keep the current flowing through the wire, not to some place where it would be damaging or dangerous. The insulation on electric motor windings has the job of keeping the electrons flowing from one end of the winding to the other rather than bypassing some of the coils.

Even the best of insulations will allow some minute amount of leakage through it. The insulation in an electric motor is subjected to contaminants, rough handling, heat, oil, and flexing due to vibration and expansion and contraction. All of these things affect the insulation, causing it to begin to break down.

Because of this very high amount of resistance, an ordinary ohmmeter cannot detect the breakdown of the insulation until it has already happened. Therefore, the megohmmeter is used for this purpose. This instrument is very useful in detecting problems before they actually happen if used properly. A megohmmeter uses a much higher voltage than is possible with the standard ohmmeter. Usually, one in the 500-V range is the most desirable. Some megohmmeters will produce much higher voltages, but they may cause problems if not used properly.

The megohmmeter will measure from 1 to 1000 megohms (MΩ).

If there is a weakness in the insulation, this high a voltage will cause a circuit to be completed to ground. The megohmmeter can be used to detect not only weak insulation but also acid or moisture contamination in a refrigeration system before a lot of damage has been done. Impurities in the system, such as dirt, acid, or moisture, will reduce the insulation resistance to ground and cause a lower megohm reading. These conditions must be corrected or the compressor will fail very soon.

Megohmmeters are supplied with two leads that are connected between the motor winding terminal and the motor or compressor case, which then becomes the ground. Or the leads may be connected to two different motor terminals for checking shorted windings (Figure 14-50).

When the leads are properly connected, turn on the megohmmeter. The maximum resistance (megohms) has been reached when the needle stops rising. This means that this is the maximum resistance of the circuit. Each motor will have a different megohm reading for a given set of conditions. However, a general rule is that when checking hermetic compressor motors there will be approximately 1000 ohms per volt or 1 megohm per 1000 operating volts.

When taking megohm readings on a compressor motor, it is a good idea to operate the system for about 15 min before making the test so that the compressor windings will warm up to operating temperature. The following are some ideas as to what the readings mean:

100 MΩ and up indicates that the compressor is in good condition.

FIGURE 14-50 Checking resistance of a compressor or motor winding with a megohmmeter.

50 to 100 MΩ indicates that moisture is in the system. Change the drier.

20 to 50 MΩ indicates that there is an excessive amount of moisture in the system. Consider installing new driers in both the suction and liquid lines.

Below 20 MΩ indicates that the system is severely contaminated and the compressor will probably fail very soon. Consider removing the charge, tripple evacuation, new driers in both the suction and liquid lines, and recharging the system.

Locating Compressor Motor Terminals. It is a simple process to determine the motor winding terminals on a compressor. When the compressor is not marked or the manufacturer's specifications are not readily available, a good ohmmeter will provide accurate information. The first step in determining the common (C), start (S), and run (R) terminals is to draw the exact location of the terminals on a piece of paper (Figure 14–51). Use the ohmmeter to determine the resistances between each terminal and mark them on the paper in the proper place. The following rule will then apply: The greatest resistance will be between the start and run terminals; the medium resistance will be between the start and common terminals; the least resistance will be between the run and common terminals. In Figure 14–51, the greatest resistance is between terminals 1 and 2; the medium resistance is between 2 and 3, and the least resistance between 1 and 3. Therefore, terminal number 3 is common (C), terminal number 2 is start (S), and terminal 1 is run (R). The sum of the resistances between common to start and common to run should equal the resistance between run to start.

Grounded Windings. To check for grounded motor windings, the ohmmeter should be set on the highest resistance value, R X 10,000, if possible. Zero the ohmmeter and attach one lead firmly to the motor housing; then firmly touch each terminal separately (Figure 14–52). Any hermetic motor of 1 hp or less should have at least 1 million ohms resistance between the motor windings and the housing, unless the manufacturer's specifications indicate otherwise. Motors larger than 1 hp should have a minimum resistance of 1000 Ω/V. The motor should be warm. If possible, run the motor for a few minutes before checking the resistance. A warm winding will be more likely to show a ground than a cold winding.

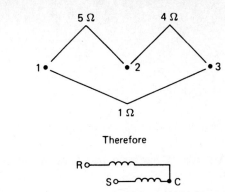

FIGURE 14–51 Locating compressor terminals.

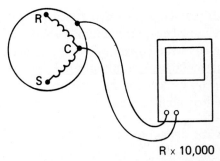

FIGURE 14–52 Checking for grounded windings.

Open or Shorted Windings. An open winding indicates that the winding has separated, leaving the motor inoperative, and will be indicated by an infinite resistance reading. A shorted winding indicates that the insulation has broken down, allowing part of the winding to be left out of the circuit. This condition is probably the most difficult condition to identify. The ohmmeter will show a lower than normal resistance on a shorted winding. To test for these conditions, the R × 1 scale on the ohmmeter should be used. Any reading different from the manufacturer's specifications indicates a bad motor winding. First, zero the ohmmeter. Be sure that the motor terminals are clean and a good connection can be made. Connect one lead of the ohmmeter to one terminal and alternately touch firmly the other lead to the other terminals (Figure 14–53).

Replacing Capacitors. Because the service technician will not always have an exact replacement capacitor, he must know how to make substitutions. There are several rules that can be applied in capacitor replacement.

FIGURE 14-53 Checking for open or shorted windings.

1. The voltage rating of any capacitor replacement must be equal to or greater than the capacitor being replaced.
2. When replacing starting capacitors, the capacitance must be equal to but not greater than 20% more than the capacitor being replaced.
3. When replacing a running capacitor, the capacitance may vary no more than ±10% of the capacitor being replaced.

Rules for capacitors in parallel:

1. All capacitors must have a voltage rating equal to or greater than the capacitor being replaced.
2. The total μF rating is the sum of the μF ratings of each capacitor.

Rules for capacitors in series:

1. The sum of the voltages must at least equal the voltage of the capacitor being replaced.
2. The total capacitance of the capacitors in series is found by using the formula

$$C_t = \frac{C_1 C_2}{C_1 + C_2}$$

☐ Safety Procedures ☐

Electric motors provide the power to drive fans, compressors, pumps, etc. on refrigeration and air conditioning equipment. They are often needlessly abused and mistreated. An electric motor will provide many years of service if properly cared for.

1. Motor compressors are sometimes very heavy and help should be obtained when moving one of these components.
2. Be careful to keep hands, feet, and clothing free of all belts.

3. Always disconnect the electric power when checking the electric circuit.
4. Be sure to discharge capacitors with a resistor to prevent personal injury or damage to the capacitor.
5. Do not tighten belts excessively.
6. Do not overload an electric motor; it may result in burnout.
7. Do not replace fuses with a higher rating than that recommended.
8. Be sure that the supplied voltage corresponds with the motor rating.
9. Be sure that an electric motor has sufficient means for cooling.
10. Never stall an electric motor because damage to the winding will usually occur.

☐ SUMMARY ☐

☐ The laws of magnetism and magnetic induction are the basis of operation for electric motors.
☐ The basic components of an electric motor are the rotor, the permanent magnet, and two magnetic poles.
☐ A phase stator pole is installed to help the motor start regardless of where the rotor stops.
☐ The phase winding allows the electric current to flow in one winding before it flows to the other, producing an out-of-phase current to start the motor.
☐ Capacitors are used to produce an out-of-phase electric current to the phase winding.
☐ The two types of capacitors are: (1) start, and (2) run.
☐ Start capacitors usually have a higher μF rating than run capacitors.
☐ High-μF-rated capacitors will permit high current flow through a series load.
☐ Running capacitors are physically larger than starting capacitors because of the oil that is used as a dielectric.
☐ Starting capacitors are to be used only for a short duration of time.
☐ When capacitors are connected in series with a load, the current leads the voltage.
☐ The five different types of single-phase motors are: (1) split-phase, (2) permanent-split capacitor (PSC), (3) capacitor-start (CSR), (4) capacitor-start/capacitor-run (CSCR), and (5) shaded-pole.
☐ Split-phase motors have moderate starting torque for use on blowers, pumps, fans, and oil burners.
☐ Capacitor-start motors have high starting torque.
☐ The starting capacitor is removed from the circuit after the motor reaches approximately 75% of its speed by the centrifugal switch.

ELECTRIC MOTOR THEORY

- [ ] Capacitor-start motors are used on compressors, large fans, blowers, and water pumps.
- [ ] Two-speed motors are made in both split-phase and capacitor-start types.
- [ ] Two-speed motors are used where a modulated air flow is required and in other applications where more than one speed is required.
- [ ] The centrifugal switch is the heart of the split-phase and capacitor-start motors.
- [ ] As the motor reaches 75% of its normal operating speed, the centrifugal switch removes the starting components from the circuit.
- [ ] Permanent-split capacitor motors are used on compressor motors, fans, and blowers.
- [ ] A running capacitor is connected in series with the phase winding to help in starting and running.
- [ ] Permanent-split capacitor motors have only a moderate starting torque.
- [ ] Capacitor-start/capacitor-run motors have both high starting torque and efficient running characteristics.
- [ ] Shaded-pole motors have a very low torque.
- [ ] Shaded-pole motors use a winding on a part of the pole to provide an out-of-phase characteristic.
- [ ] The shaded portion of the pole in a shaded-pole motor is not connected to the electric circuit.
- [ ] There are two general types of single-phase motor protectors: (1) internal type, and (2) external type.
- [ ] The internal overload is imbedded into the heat sink of the motor winding.
- [ ] The external overload is mounted securely on the outside of the housing at the hottest point.
- [ ] Overloads are also of the line break and thermostatic types.
- [ ] Always be certain that the electric power is off when servicing electric motors.
- [ ] Always discharge capacitors before handling them.
- [ ] The rule to apply when locating compressor motor terminals is: The greatest resistance will be between the start and run terminals; the medium resistance will be between the start and common terminals; the least resistance will be between the run and common terminals.
- [ ] Motor windings should have at least one million ohms resistance between the winding and the housing.
- [ ] An open winding will shown infinite resistance.
- [ ] A shorted winding will show less resistance than the specifications indicate.

◻ EXERCISES ◻

1. What are the stationary poles in an electric motor called?
2. What is the difference among single-phase motors used in refrigeration and air conditioning?
3. What causes the polarity of stator poles to change?
4. What is used in a single-phase motor to ensure that it will start each time?
5. How are capacitors rated?
6. Why is oil used as a dielectric in a running capacitor?
7. With which winding is the running capacitor connected in series?
8. Why is the starting capacitor removed from the starting circuit after the motor has started?
9. What is the purpose of a capacitor?
10. What is the purpose of a centrifugal switch?
11. At approximately what speed does the centrifugal switch open?
12. What is the purpose of the running capacitor on a PSC motor?
13. How is the capacitor electrically connected on a PSC motor?
14. Why can the run winding withstand a higher continuous current than the start winding?
15. Does the shaded-pole motor develop a high or low torque?
16. What does the term *shaded pole* mean?
17. What is the purpose of motor protectors?
18. In what location should the motor protector be installed?
19. What safety precautions should be followed when handling capacitors?
20. List the steps used in locating compressor motor terminals.
21. What should the minimum resistance between the motor windings and the housing on a 1-hp motor be?
22. What resistance will be indicated on an open winding?
23. What is the rule governing the replacement of starting capacitors?

15 Electric Motor Controls

Performance Objectives

After completion of this chapter, the student should:
- Understand the purpose of internal thermal motor protection.
- Describe why protection of the control and branch circuits is necessary.
- Be able to name the various types of starting relays and explain how they operate.
- Describe the purpose of thermostats.
- Discuss the purpose of pressure controls and explain how they operate.
- Know how oil safety switches operate.
- Understand the purpose of defrost timers.
- Know how starters and contactors are used in the control circuit.
- Complete the Exercises with at least 80 percent accuracy.
- Complete the Laboratory Exercises with at least 80 percent accuracy.

Generally, motor control equipment should provide:

1. A means of disconnecting the motor and its controller from the source of electrical supply.
2. A means of starting and stopping the motor.
3. Some type of short-circuit protection for the motor and its circuit.
4. Some means of protecting the motor against overheating.
5. Adequate protection for the equipment operator.
6. Control of the motor speed.
7. Adequate protection for the motor branch circuit wiring and the control circuit.

Definition. *Electric motor controls* and their related systems are designed to cause motors to operate and provide the desired results. They may be either safety controls or operating controls. Safety controls are designed to prevent the motor from operating when a condition is present that would cause damage to the motor. Operating controls are designed to cause the motor to operate to provide the desired results, such as temperature control, pressure control, or humidity control, just to mention a few of their uses.

☐ Internal Thermal Motor Protection ☐

The following sections describe the most popular motor protective devices. Such devices must conform to the standards set forth by the National Electrical Code® (NEC) and UL.

Thermal Protectors. The NEC defines these devices as protective devices for assembly as an integral part of a motor or motor-compressor. Their purpose is to protect the motor against the hazards of overheating from an overload and from failure to start. Thermal protectors are also used for protecting three-phase motors from overheating due to an open electrical phase in the power supply to the unit. Thermal protection may be accomplished by either a line break control or by interrupting the control circuit.

Line-Break Thermal Protectors Built in a Motor-Compressor, Sensing Both Temperature and Current. This type of protector is installed electrically in series with the motor winding. When the contact opens, the entire electrical supply to the motor is interrupted. This type of protector is most popular in single-phase and three-phase motors up to about 10 hp.

All protectors installed inside motor-compressors are hermetically sealed because arcing cannot be allowed inside the refrigerant circuit. They provide better protection than the external type for certain conditions, such as loss of refrigerant charge, restricted suction line, low ambient temperature, and stalled rotor occurrences. Externally mounted motor protectors sense only the shell temperature and the line current, but are used quite successfully. Thermal protectors mounted inside other motors can be fastened to the motor winding or they may be mounted separately, but always within the motor housing. These protectors carry the full line current; therefore, their size is based on the current-carrying capabilities of the contacts and of the stalled motor current on a continuous cycling condition.

Their application and selection are made by the motor manufacturer in cooperation with the protector manufacturer. Use of a protector other than the one supplied by the manufacturer may result in either overprotection and frequent nuisance shutdowns or underprotection and burnout of the motor windings. Any connections to the protector terminals, including lead wire sizes, should not be changed, and no additional connections should be made to the terminals. A change in any of these will change the thermal conditions and will affect the performance of the protector.

Control Circuit Thermal Protection Approved for Use with a Motor or Motor-Compressor. These devices are used with complete single-phase and three-phase motors.

The devices used for both temperature and current use a bimetal temperature sensor installed in the motor winding in conjunction with thermal overload relays. The sensors are connected in electrical series with the control circuit controlling the magnetic contactor, which in turn interrupts the power to the motor.

Thermostatic sensors of this type, depending on their size and mass, are quite capable of sensing motor winding temperature for running overloads. In cases of a locked rotor condition where the winding temperature rises very fast, the temperature lag is usually too slow for these sensors to provide the desired protection alone. When they are installed in conjunction with separate thermal overloads or magnetic relays that sense motor current; however, this combination provides the proper protection. In these instances, the thermal or magnetic relays provide protection for the initial heating cycle, and the combination of the relay and the thermostat provides protection for the remaining cycles.

Thermistor sensors may also be used with electronic current-sensing devices. Sensors that change in resistance with a change in temperature are used on temperature-sensing-only systems. This change in resistance provides a switching signal to the electronic circuitry, whose output is electrically in series with the control circuit of a magnetic contactor used to interrupt the electric power to the motor. The output of the electronic protection circuit (module) may be either an electromechanical relay or a power triac. This type of sensor may be installed directly on the stator winding end turns or it may be buried inside the windings. Because of their small size and their good thermal transfer, they are excellent for tracking the winding temperature for both locked rotor and a running overload.

There are generally three types of sensors used: (1) a ceramic material with a positive temperature coefficient of resistance, (2) a metal wire that has a linear increase in resistance with an increase in temperature, and (3) a negative temperature coefficient or resistance resistor integrated with an electronic circuit similar to the one used with the metal wire sensor.

Ceramic Material with a Positive Temperature Coefficient of Resistance. In this type of sensor, the material experiences a large rapid change in resistance at a given design temperature. This point of change is known as the anomaly point and is inherent in the sensor. The anomaly point remains set once the sensor is manufactured. Sensors with different anomaly points are available to satisfy different application requirements. A single module calibration can be supplied, however, for all anomaly temperatures of a given sensor type.

Metal Wire with Linear Increase in Resistance with Temperature. This type of sensor is manufactured with a specific value of resistance that corresponds to each desired value of response or operating temperature. It is used in conjunction with an electronic protection module calibrated for a specific resistance. Modules are available that have different calibration and are used for various values of operating temperatures.

Negative Temperature Coefficient of Resistance Sensor Integrated with an Electronic Circuit. In this type of system more than one sensor can be connected to a single electronic module either in parallel or in series according to the system design requirements. The sensors and modules, however, must be of the same design and intended for use with the particular number of sensors installed and the wiring method used. Electronic protection modules must be paired only with sensors specified by the manufacturer, unless specific alternatives are established and identified by the motor-compressor manufacturer.

FIGURE 15-1   Circuit breaker.

□ Separate Motor Protection □

The motors and motor-compressors used in air conditioning and refrigeration systems are equipped with motor protection devices by the equipment manufacturer. When added protection is desired, current-only sensing devices are generally used because the internal thermal type is not generally adaptable after the equipment has been manufactured. These types of devices consist of thermal or magnetic relays similar to those used in industrial control applications. They should provide running overload and stalled rotor protection. Quick trip devices must be used on hermetic motor windings where a stalled rotor condition might exist because the motor windings heat so rapidly due to the loss of cooling effect of refrigerant gas flow.

Note: To provide protection against abnormal running conditions that are not due to increased motor current draw or where automatic restarting is required after a trip has occurred, thermostats or thermal devices are usually used to supplement such current-sensing protectors.

□ Protection of the Control and Branch Circuit Wiring □

In addition to protection for the motor itself, the NEC requires that the wiring be protected against overload due to motor overload or failure to start. This is generally accomplished by the use of some thermal protection system that will not allow a continuous current to flow in excess of that required for proper operation of the equipment. These types of devices include current-sensing devices such as overload relays, fuses, or a circuit breaker.

FIGURE 15-2   Mounted disconnect box.

□ Circuit Breakers □

These devices (Figure 15-1) are used for disconnecting as well as circuit protection and are available in ratings for use with small household refrigerators as well as large commercial installations.

There are manual switches available for disconnecting and for holding fuses to provide short-circuit protection for the circuit (Figure 15-2). There is an attachment plug available for single-phase motors up to 3 hp that is considered to be an acceptable disconnecting device.

□ Controllers □

The type of motor control used is determined by the type and size of the motor, the power supply, and the degree of equipment automation. This control may be either manual, semiautomatic, or fully automatic.

In most cases a central air conditioning unit is located at some distance from the thermostat. Therefore, magnetic controllers must be used in these installations. All dc and large air conditioning units must be provided with in-rush current-limiting controllers.

Manual Control. The manual control for a dc or an ac motor is generally located near the motor it is to control. In these installations, an operator must be close at hand to start and stop or to change the speed of the motor by making the necessary adjustments on the control device.

Manual control is the simplest and least expensive method of controlling small ac motors for both single- and three-phase units. It is seldom used with hermetic motor-compressors. These devices usually consist of a set of main-line contacts equipped with thermal overload relays to provide motor protection.

Manual speed controllers are sometimes used for large air conditioning units using slip-ring motors. In some installations they provide reduced-current starting. They have different speed adjustments that are used to vary the amount of cooling delivered by the compressor.

☐ Across-the-Line Magnetic Controllers ☐

These devices are widely used in air conditioning and refrigeration systems. They may be applied to motors of all sizes, provided the power supply and the motor are suitable to this type of control. The across-the-line magnetic starter may be used with automatic control systems for starting and stopping the equipment. When pushbuttons are used, they may be wired for either low-voltage release or low-voltage protection (Figure 15-3).

☐ Full-Voltage Starting ☐

This method of starting motors is more desirable than the others because of its lower initial cost and the simplicity of the control system. Most ac motors are designed mechanically and electrically for full-voltage starting. However, the power companies

FIGURE 15-3   Magnetic contactor.

regulate the amount of starting in-rush current because of the voltage fluctuations that may be caused by heavy current surges. Because of this, it is sometimes necessary to reduce the starting current to a point below that obtained by across-the-line starting so that the limitations of the power company can be met. This reduction is often accomplished by placing resistors in the power to the equipment. This resistance is reduced as the motor accelerates by the use of timing relays or current relays.

In some installations the reduced starting current is reduced through the use of an autotransformer motor controller. The starting voltage is reduced, and as the motor accelerates it is disconnected from the transformer and is connected across-the-line by use of either a timing relay or current relays. These resistor starters are generally smaller and are less expensive than the autotransformer type starters when used on motors of moderate size. However, they do require more line current for a given starting torque than the autotransformer starters.

☐ Star-Delta Motor Controllers ☐

These types of controllers limit the current very efficiently, but motors designed for this type of starting are required. They are very well suited for use on centrifugal units, rotary, screw, and reciprocating compressors that start unloaded.

☐ Part-Winding Motor Controllers ☐

These devices are sometimes called incremental start controllers. They are used to limit any fluctuation in line voltage by connecting only a part of the motor winding to the line and then connecting the remaining part of the winding to the line after a time interval of from 1 to 3 sec. If there is not a heavy load on the motor, it will accelerate when the first part of the winding is connected to the line; when a heavy load is on the motor, the motor may not start until the second part of the winding is connected to the line by the controller. Regardless of which condition exists, there will be less voltage dip than would occur when a standard squirrel cage motor is used with across-the-line starting. Part-winding motors may be controlled either manually or automatically. The magnetic controller is made up of two contactors and a timing device to bring on the second contactor.

☐ Multispeed Motor Controllers ☐

Multispeed controllers provide flexibility in many types of drives requiring a variety in capacity. There are two types of multispeed motors available: (1) motors with one reconnectable winding, and (2) motors with two separate windings. Separate winding motors require a contactor for each winding and only one contactor can be closed at any time. Reconnectable winding motors are similar to motors with two windings, but the contactors and the motor circuits are different.

☐ Starting Relays ☐

Starting relays are used to remove the starting circuit from operation when the motor reaches approximately 75% of its normal running speed. Its function is basically the same as the centrifugal switch discussed earlier. There are four types of starting relays in common usage today: (1) amperage (current) relay, (2) hot-wire relay, (3) potential (voltage) relay, and (4) solid state. The horsepower size and design of the equipment regulate which type of starting relay is used.

Amperage (Current) Starting Relay. This type of starting relay (Figure 15–4) is normally used on ½-hp units and smaller. This is an electromagnetic relay. The coil of the relay is connected in series with the motor run winding. Therefore, the coil wire must be large enough to carry the amperage draw of the motor. The contacts are normally open. Amperage relays are positional-type relays. That is, they must be installed so that the weighted armature that carries the contacts will function by gravity pull.

In operation, when the motor control completes the electric circuit to the motor, the current flow through the relay coil is at its peak. The contacts are pulled together by the electromagnetic field in the relay coil (Figure 15–5). When the contacts are closed, an electric circuit is completed to the starting circuit of the motor providing the electric phase shift required to start the motor. As the motor reaches approximately 75% of its full speed, the counter emf built up in the start winding opposes the applied voltage and the amperage draw is reduced. This reduced amperage causes a reduced magnetic field in the coil of the starting relay and the contacts are pulled apart by the weight of the

GENERAL ELECTRIC
3ARR2
BRACKET TYPE

FIGURE 15–4 Amperage relay. (Courtesy of General Electric Co.)

FIGURE 15–5 Amperagle relay connections.

armature, stopping operation of the starting circuit. The motor is now in its normal operating condition. When the motor control interrupts the electric power to the motor, it will stop and the relay contacts will remain open until the electric power is again applied to the motor.

Amperage starting relays must be sized for each motor horsepower and amperage rating. A relay rated for too large a motor may not close the relay contacts, which will leave out the much-needed starting circuit. The motor will not start under these conditions. A relay that is rated for too small a compressor may keep the contacts closed at all times while electric power is applied, leaving the starting circuit engaged continuously. Damage to the starting circuit may occur under these conditions. A motor protector must be used with this type of relay.

Hot-Wire Relay. This type is a form of current relay, but it is not operated by an electromagnetic coil. It is designed to sense the heat produced by the flow of electric current. The current flowing through a resistance (hot) wire provides heat to a bimetal which operates the contacts (Figure 15–6). There are two sets of contacts in this relay, a set for starting and a set for running the motor. The contacts are normally closed.

In operation, when the motor control completes the electric circuit to the motor, the current is automatically supplied to both the running winding and the starting winding (Figure 15–7). The current flowing through the resistance wire to the main winding causes the wire to get hot. This heat causes the bimetal to warp and open the starting contacts. The main or running contacts remain closed and the relay stays in the normal operating position. Should the motor continue to draw an excessive amount of current, the hot wire will provide enough heat to cause the bimetal to warp further and open the running contacts. The electric circuit to both the running winding and starting winding is interrupted. The contacts will remain open until the bimetal has cooled sufficiently to close the contacts. Both the running and starting contacts close at the same instant to start the motor operating again.

No overload is necessary when this type of starting relay is used. However, some motor manufacturers will use an internal overload in conjunction with the hot-wire relay. Hot-wire relays are non-positional relays. They also must be sized to each motor. If the relay is too large, the starting contacts

FIGURE 15–6 Hot-wire relay.

FIGURE 15–7 Hot-wire relay connection.

FIGURE 15–8 (a) Potential starting relay (courtesy of General Electric Co.); (b) inside diagram of a potential relay.

may remain closed for too long a time or not open at all, thus possibly damaging the motor windings. If the relay is too small, the relay may stop the motor as if it were overloaded and sufficient running time will not be accomplished.

Potential (Voltage) Starting Relay. Potential starting relays (Figure 15–8) operate on the electromagnetic principle. They incorporate a coil of very fine wire wound around a core. These starting relays are used on motors of almost any size. The contacts on this relay are normally closed and are

FIGURE 15-9 Potential starting relay connections.

caused to open when a plunger is pulled into the relay coil. These relays have three connections to the inside in order for the relay to perform its function. These terminals are numbered 1, 2, and 5. Other terminals numbered 4 and 6 are sometimes used as auxiliary terminals. The relay is installed with terminal 5 connected to both the electric line to the motor and to the common terminal on the motor. Terminal 2 is connected to the motor start winding. Terminal 1 is connected to the starting capacitor.

In operation, when the motor controller completes the electric circuit to the motor, electricity is also supplied to the starting winding through the relay contacts between terminals 1 and 2 (Figure 15-9). As the motor reaches approximately 75% of its rated speed, the counter emf in the start winding has increased sufficiently to cause the relay to "pick up," opening the contacts and removing the starting components from the circuit. Remember, it is the voltage in the start winding that causes the relay to function. The relay is in the normal operating position. When the controller interrupts the electric circuit to

the motor, the motor begins slowing down. The counter emf in the start winding is reduced, and when a certain voltage is reached the relay "drops out," closing the contacts. The relay is now in the motor start position.

Potential starting relays are nonpositional. The sizing of these relays is not as critical as with the amperage and hot-wire relays. A good way to determine what relay is required is to manually start the motor and check the voltage betwen the start and common terminals while the motor is operating at full speed. Multiply the voltage obtained by 0.75 and this will be the pickup voltage of the required relay. There are sheets showing the specifications of different potential relays (Table 15-1).

Solid-State Starting Relay. These relays use a self-regulating conductive ceramic, developed by Texas Instruments, Inc., which increases in electric resistance as the compressor starts, thus quickly reducing the starting winding current flow to a milliampere level. The relay switches in approximately 0.35 sec. This allows this type of relay to be applied to refrigerator compressors without being tailored to each particular system within the specialized current limitations. These relays will start virtually all split-phase 115-V hermetic compressors up to ⅓ hp. These relays are designed for push-on installation on the compressor pin connector (Figure 15-10).

In operation, this relay is connected in the motor circuit with the ceramic material between the line and the starting terminal of the compressor (Figure 15-11). That is, it is connected in series with the starting winding. When electric power is applied to the relay, the ceramic material heats up and turns the relay off in approximately 0.35 sec. This reduces the current flow through the start winding to a milli-

TABLE 15-1 Calibration Specifications: Mars—General Electric—Potential Relays

Mars Relay No.	Size (hp)	Continuous Voltage (V)		Pickup Voltage (V)		Max. Drop-out Voltage (V)
		Max.	Min.	Min.	Max.	
Mars 63	¼-⅓-½-¾	200	115	139	153	55
Mars 64	1-1½-1¾	432	230	260	275	120
Mars 65	½-¾-1-1½	332	115	168	182	90
Mars 66	3-4-5	432	230	215	225	120
Mars 67	1¾-2-3-4-5	457	230	295	315	125
Mars 68	2-3-4-5	502	230	325	345	135
Mars 69	¾-1-1½	378	115	180	195	105
Mars 70	¾-1	253	230	285	305	177

ELECTRIC MOTOR CONTROLS

FIGURE 15–10 Solid-state motor start relay. (Courtesy of Klixon Controls Division, Texas Instruments, Inc.)

ampere level until the electric power is removed from the circuit. After the power is removed the ceramic material requires a few seconds to cool down before the next start cycle. The 4EA relay is not suitable for replacement on capacitor-start compressors, or for those systems with unusually rapid cycle rates.

From Table 15–1 it can be seen that the Mars 67 potential starting relay has a wide range of application. It can operate with a continuous voltage of 457 V. The minimum pickup voltage is 295 V with a maximum pickup voltage of 315 V. The maximum dropout voltage is 125 V.

Solid-State Hard Start Kit. This kit (Figure 15–12) is a way to provide the additional starting torque necessary to solve PSC motor starting problems. This is done through the use of PTC ceramic materials. PTC is the name given to a material mean-

ing "positive temperature coefficient" and its resistance increases as its temperature increases. This material increases its resistance with an increase in temperature. At its anomaly temperature, the resistance increase is very sharp.

One of the objectives of a PTC start-assist device is to provide a surge of current that lasts for a period of time sufficient to start the compressor. This additional current then decreases, permitting the motor to run in the normal PSC mode. When electric power is applied to the PSC motor, current flows through the start winding and through the parallel combination of the run capacitor and the low-resistance PTC (Figure 15–13).

The low resistance during starting not only increases the start winding current but also reduces its angular displacement with the main winding current. In most PSC motors, this is advantageous since the angle between these two currents is usually greater than 90%.

While the surge current increases motor starting torque, it is also flowing through the PTC and heating it to its high-resistance region. The time it takes the PTC to heat to its high-resistance state is, unlike the operation of a potential relay, independent of when the motor starts. Instead, it is a function of the PTC mass, anomaly temperature, resistance, and the voltage applied to the PTC. When a 230-V compressor having a 3 EA start assist is initially energized at nominal voltage, the 3 EA switching time is 16 electrical cycles. Starting the same compressor with 25% less voltage increases the 3 EA switching time to 32 electrical cycles, providing added assistance under these more difficult starting conditions.

The above switching times are ideal for starting PSC motors because the PTC will only be in its low-resistance state long enough to overcome the initial inertia of the motor compressor. When the PTC

FIGURE 15–11 Solid-state relay connections. (Courtesy of Klixon Controls Division, Texas Instruments, Inc.)

Starting Relays

FIGURE 15-12 Solid-state hard start kit. (Courtesy of Klixon Controls Division, Texas Instruments, Inc.)

FIGURE 15-13 Solid-state hard start kit connections. (Courtesy of Klixon Controls Division, Texas Instruments, Inc.)

switching time is longer than normal motor start times, the low-resistance PTC effectively shunts out the run capacitor. The excessive *on* time retards the motor speed while the motor is trying to overcome the increasing load.

Once the PTC heats up and reaches its anomaly temperature, its resistance increases dramatically to around 80,000 Ω, and effectively switches itself out of the circuit without requiring an electromechanical relay. While the motor is running, the 3 EA continues to draw only 6 mA of current, which is less than 1/1000 of the surge current provided by the 3 EA during the motor starting period. This low current has no effect on the start winding or on motor

running performance. When the motor is shut off, the PTC is also deenergized and starts cooling to ambient temperature. Should the motor be restarted before the PTC cools below the anomaly temperature, the motor will try starting in the standard PSC mode. Cool-down times in excess of one minute will usually provide sufficient time for the PTC to cool below its anomaly temperature and provide a start assist.

It is recommended that the 3 EA 1 be used on compressors up to 48,000 Btu or motors up to 4 hp. However, its use is not limited to this size range.

□ Thermostats □

A thermostat is a device that acts to make or break an electric circuit in response to a change in temperature. There are many types of thermostats in use today. They range from simple bimetallic switches to multiple-contact controls that operate from remote sensing bulbs. They may have fixed control points or multiple variable adjustments.

Cooling thermostats will make on a rise in temperature and break on a fall in temperature. Heating thermostats will make on a fall in temperature and break on a rise in temperature. The type used will depend on the particular installation.

In general, low-voltage thermostats should be used for the control of room temperatures for air conditioning. The low-voltage thermostats respond much faster to temperature changes than the greater-mass line-voltage devices. This has been proven on residential heating and cooling systems.

From a cost standpoint, the less expensive installation of low-voltage wiring more than offsets the extra cost of a transformer. Also, homeowners are accustomed to the safety of low-voltage thermostats.

The room thermostat is provided with a heat anticipator connected to the heating part of the control circuit (Figure 15-14). These anticipators are made of a resistance-type material that produces heat in accordance with the current drawn through them (Figure 15-15).

Heat anticipators are adjustable and are normally set to correspond with the current rating of the main gas valve or the heating part of the control circuit. The purpose of these devices is to make the room temperature more even.

In operation, when the thermostat calls for heat, the heat anticipator also starts producing heat in-

FIGURE 15–14 **Heat anticipator.**

The purpose of the cooling anticipator is to cause the cooling unit to remove more moisture from the conditioned space, thus making it more comfortable. The cooling anticipator is in the circuit when the unit is not running, causing a shorter off period. The heating anticipator is in the circuit when the unit is running, causing a shorter on period.

☐ Pressure Controls ☐

There are two types of pressure controls: (1) low pressure, and (2) high pressure. These controls are switching devices used to stop the motor compressor when the pressures reach a predetermined point.

Low-Pressure Controls. In general, low-pressure controls are connected to the low side of the compressor and are adjusted to stop the compressor if the low-side pressure drops below the desired level (Figure 15–16).

The main purposes of the low-pressure control in air conditioning are to prevent the temperature of the evaporator coil from falling below the temperature at which frost would occur and to prevent compressor overheating due to a refrigerant shortage. It may have either manual or automatic reset.

FIGURE 15–15 **Internal view of a thermostat. (Courtesy of Honeywell Inc.)**

FIGURE 15–16 **Low-pressure control. (Courtesy of Penn Controls, Inc.)**

side the thermostat. This heating action causes the thermostat to become satisfied before the room actually reaches the set point of the thermostat. Thus, the thermostat stops the heating action and the room temperature will not overshoot or go too high.

During the cooling season, a cooling anticipator is used and is located in the cooling circuit of the thermostat. The cooling anticipator is not adjustable. It provides heat in the thermostat during the off cycle of the equipment to make the unit run a bit sooner.

TABLE 15-2 Approximate Pressure Control Settings

Application	R-12		R-22		R-502		R-717	
	Out	In	Out	In	Out	In	Out	In
Ice cube maker—dry-type coil	4	17	16	37	22	45	—	—
Sweet water bath—soda fountain	21	29	43	56	52	66	33	45
Beer, water, milk cooler, wet type	19	29	40	56	48	66	—	—
Ice cream trucks, hardening rooms	2	15	13	34	18	41	5	24
Eutectic plates, ice cream truck	1	4	11	16	16	22	4	8
Walk in, defrost cycle	14	34	32	64	40	75	23	55
Reach in, defrost cycle	19	36	40	68	48	78	30	57
Vegetable display, defrost cycle	13	35	30	66	38	77	—	—
Vegetable display case—open type	16	42	35	77	44	89	—	—
Beverage cooler, blower—dry type	15	34	34	64	42	75	24	55
Retail florist—blower coil	28	42	55	77	65	89	44	67
Meat display case, defrost cycle	17	35	37	66	45	77	—	—
Meat display case—open type	11	27	27	53	35	63	—	—
Dairy case—open type	10	35	26	66	33	77	—	—
Frozen food—open type	7	5	4	17	8	24	—	—
Frozen food—open type—thermostat	2 °F	10 °F	—	—	—	—	—	—
Frozen food—closed type	1	8	11	22	16	29	—	—

Source: Sporlan Valve Company.

In refrigeration work, low-pressure controls are used as temperature controls, and sometimes as defrost controls, depending on the cut-in and cut-out settings (Table 15–2).

FIGURE 15–17 High-pressure control. (Courtesy of Penn Controls, Inc.)

High-Pressure Controls. The high-pressure control is connected into the high side of the refrigeration system so that it will stop the motor compressor when the discharge pressure exceeds the control setting (Figure 15–17). These high pressures could be caused by insufficient condenser cooling, excessive refrigerant charge, air in the refrigerant lines, or other abnormal conditions. The high-pressure control is connected to the high side of the system, preferably where the compressor discharge service valve cannot be backseated far enough to render the control inoperative (Figure 15–18).

When both the high- and low-pressure controls are incorporated into a single control housing, the

FIGURE 15–18 Pressure control connections.

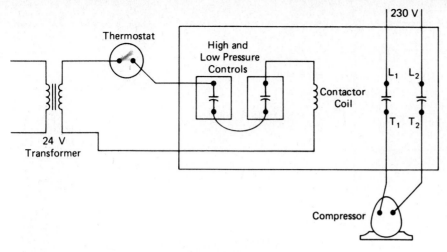

FIGURE 15-19 Typical electrical diagram of high- and low-pressure controls.

device is called a high-low pressure control or dual-pressure control. This type of construction provides a savings in space, original cost, and electrical wiring. These controls are wired into the electrical system in two ways: (1) the control circuit may be interrupted, allowing the contactor to open and stop the motor, or (2) main power to the compressor motor may be interrupted, stopping the motor. The preferred method is to interrupt the control circuit (Figure 15-19).

☐ Oil Safety Controls ☐

Oil safety controls are designed to provide compressor protection by guarding against low lubricating oil pressure (Figure 15-20). A built-in time delay switch allows for oil pressure pickup on starting and avoids nuisance shutdowns on oil pressure drops of short duration during the running cycle.

The oil safety switch is activated by a difference in pressure rather than straight system pressure. Because the lubricating system is located in the compressor crankcase, a pressure reading at the discharge of the oil pump will be the sum of the actual oil pressure plus the suction pressure. The oil safety switch measures the difference between the suction pressure and the pressure caused by the oil pump. The switch causes the compressor to be shut down if the oil pump does not maintain the oil pressure prescribed by the settings of the control. As an example, if a compressor manufacturer recommended a 20-lb net oil pressure, a compressor operating with a normal operating suction pressure of 40 lb must have at least a 60-lb oil pump discharge pressure or the control would stop the compressor.

FIGURE 15-20 Oil safety control. (Courtesy of Penn Controls, Inc.)

Because the suction pressure is present on both sides of the control diaphragm, its effect is cancelled. Operation of the control depends on the net oil pressure overcoming the predetermined spring pressure. This spring pressure must be equal to the minimum oil pressure recommended by the compressor manufacturer.

In operation, the total oil pressure is the combination of crankcase pressure and the pressure generated by the oil pump. The net oil pressure available to circulate the oil is the difference between total oil pressure and refrigerant pressure in the crankcase.

Total oil pressure − refrigerant pressure
= net oil pressure

This control measures that difference in pressure, referred to as *net oil pressure.*

When the compressor starts, a time delay switch is energized. If the net oil pressure does not increase to the control cut-in point within the required time limit, the time delay switch trips to stop the com-

FIGURE 15-21 Penn control wiring diagram used on 230-V system.

Terminal No. 1
Common

Terminal No. 2 N.C.
to Compressor

Terminal No. 3 N.O.
to Defrost Device

Ground
Terminal

Terminal No. 6
Junction Only

FIGURE 15-22 Typical defrost timer connections. (Courtesy of Gem Products, Inc.)

pressor. If the net oil pressure rises to the cut-in point within the required time after the compressor starts, the time delay switch is automatically deenergized and the compressor continues to operate normally.

If the net oil pressure should fall below the cut-out setting during the running cycle, the time delay switch is energized and, unless the net oil pressure returns to the cut-in point within the time delay period, the compressor stops.

The time delay switch is a trip-free thermal expansion device. The time delay unit is compensated to minimize the effect of ambient temperatures from 32 to 150°F (0 to 66°C). Timing is also affected by voltage variations. The manufacturer's recommendation should be followed during installation and adjustment. Normally, the oil pressure line is connected to the pressure connection labeled OIL and the crankcase line to the pressure connection labeled LOW. Oil safety controls are rated for pilot duty only. Wire as suggested for specific equipment (Figure 15-21).

☐ **Defrost Timers** ☐

Defrost timers are devices that cause a refrigeration system to go into a defrost cycle at predetermined periods of time. There are a variety of timers available and each system has its own requirements. Therefore, the manufacturer's recommendations should be followed during replacement. Defrost timers are usually driven by a synchronous motor that runs while the unit is supplied with electric power. In Figure 15-22, electric power to the motor

is connected to terminals 1 and 6. Terminal 2 is connected to the compressor and terminal 3 is connected to the defrost device.

In operation, electric power is made to the compressor from terminal 1 to terminal 2 (Figure 15-23). When the timer motor has run the required amount of time, the switch changes from terminal 2 to terminal 3, stopping the compressor and energizing the defrost device. This is known as the defrost period. After the predetermined period of time has lapsed, the switch returns to terminal 2. The defrost device, either an electric heater or a hot-gas solenoid, is

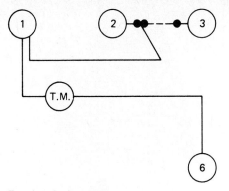

Terminal 1—Common
Terminal 2—N.C. to Compressor
Terminal 3—N.O. to Defrost Device
Terminals 4 and 5 (When Used) and Terminal 6
    are junctions only and
    have no internal connection
    to the timer

FIGURE 15-23 Internal diagram of a defrost timer. (Courtesy of Gem Products, Inc.)

removed from the circuit and the compressor is again energized and producing refrigeration.

☐ Starters and Contactors ☐

Different manufacturers build starters and contactors with a variety of armatures. Some connect the armatures to the relay frame with hinges or pivots, and others use slides to guide the armature. Except for this difference their basic operation is the same.

The armature is the moving part of the starter or contactor. A spring keeps the armature pulled away from the electromagnet when the coil is deenergized. The metallic armature is easily magnetized by the lines of flux from the energized electromagnet. The magnet consists of a coil wound around a laminated iron core; this becomes an electromagnet and pulls the armature toward it when the coil is energized.

The characteristics of the coil depend on the type of wire and the method of winding it. The potential wound type responds to some value of voltage to pick up the armature. The current-wound type responds to some value of current flowing through it. The coil terminals connect the coil to the control circuit wiring through a switching device such as a thermostat or pressure control.

The type of load is very important when selecting a contactor or a starter for a particular application. To properly match the starter or contactor to the application, the technician must consider the dif-

ferent characteristics of inductive and noninductive (resistive) loads.

In an ac circuit with an inductive load such as an electric motor, the initial in-rush of current is always much higher than the normal operating current. This situation occurs before the motor has gained sufficient speed and the back emf is very low, allowing the maximum amount of current to flow. As the motor gains speed, the back emf increases and opposes the current flow, reducing it to the normal running current. Starter and contactor contacts must, therefore, not only withstand the normal running current but also the starting current.

Contacts are made from silver cadmium alloy, which provides the maximum resistance to sticking. The contacts are bonded to a strong backing member. They operate cooly even at loads of 25% above their ratings.

Starters and contactors are available in pole configurations of one, two, three, and four poles per unit, depending on the current required for the load being started, either single phase or three phase. They may be purchased with only the poles that are required or with extra poles as well. These extra contacts are either unused or used as auxiliary contacts.

Auxiliary contacts are normally used to complete an interlock circuit such as another contactor connected to a water pump, or similar device, or to complete a circuit while in the deenergized position. These contacts are usually only for pilot duty and are not intended to withstand the heavy current used by the main load.

In refrigeration and air conditioning work, the compressor motor represents the largest switching load for the control system. The various fan motors, water pumps, and other machinery wired in parallel with the compressor motor also add to the load requirements. There are several methods of controlling the various loads but we will discuss only the contactor and starter at this time.

Contactor. A contactor (Figure 15-24) is a device for repeatedly establishing and interrupting an electric power circuit. Contactors have many common features. Each has electromagnetically operated contacts that make or break a circuit in which the current exceeds the operating circuit current of the device. Each may be used to make or break voltages that differ from the controlling voltage. A single device may be used to switch more than one circuit. A contactor is used for switching heavy current, high voltage, or both.

FIGURE 15-24
(Courtesy of Arrow-Hart, Inc.)

FIGURE 15-26 Motor starter. (Courtesy of Arrow-Hart, Inc.)

In operation, the armature is pulled in when the controller completes the control circuit (Figure 15-25). Electric current flows through the coil and causes electromagnetism which attracts the armature. The armature carries a set of movable contacts that mate with a set of stationary contacts. The electric circuit to the compressor is completed through these contacts and the motor begins to operate. When the controller is satisfied, the control circuit is interrupted. The electromagnetic force in the coil stops and the armature drops out, taking with it the movable contacts. The motor stops and awaits a signal from the controller for the next on cycle.

Starter. A motor starter (Figure 15-26) may consist of a contactor used as a means of switching electric power to a motor. A starter, however, usually has additional components, such as overload relays and holding contacts. These components may also include step resistors, disconnects, reactors, or other hardware required to make a more sophisticated starter package for large motors. The operation of a starter is the same as that of a contactor. The armature is caused to move a set of movable contacts by electromagnetism.

□ Safety Procedures □

Electric motors provide the power to drive fans, compressors, pumps, etc., on refrigeration and air conditioning equipment. They are often needlessly abused and mistreated. An electric motor will provide many years of service if properly cared for.

1. Motor compressors are sometimes very heavy and help should be obtained when moving one of these components.
2. Be careful to keep hands, feet, and clothing free of all belts.
3. Always disconnect the electrical power when checking the electric circuit.
4. Be sure to discharge capacitors with a resistor to prevent personal injury or damage to the capacitor.
5. Do not tighten belts excessively.
6. Do not overload an electric motor; it may result in burnout.
7. Do not replace fuses with a higher rating than that recommended.

FIGURE 15-25 Contactor components. (Courtesy of Arrow-Hart, Inc.)

8. Be sure that the supplied voltage corresponds with the motor rating.
9. Be sure that an electric motor has sufficient means for cooling.
10. Never stall an electric motor because damage to the winding will usually occur.

☐ SUMMARY ☐

☐ The purpose of the thermal protector is to protect the motor against hazards of overheating from an overload and from failure to start.

☐ The line break thermal protector is installed electrically in series with the motor winding.

☐ All motor protectors installed inside the motor-compressor are hermetically sealed because arcing cannot be allowed inside the refrigerant circuit.

☐ The thermal protectors used for both temperature and current sensing use a bimetal temperature sensor installed in the motor winding in conjunction with thermal overload relays.

☐ Sensors are devices that change in resistance with a change in temperature in temperature-sensing-only systems.

☐ In the ceramic material with a positive coefficient of resistance type of sensor, the material experiences a large rapid change in resistance at a given temperature.

☐ The metal wire with linear increase in resistance with temperature type of sensor is manufactured with a specific value of resistance, which corresponds to each desired value of response or operating temperature.

☐ In the negative temperature coefficient of resistance sensor, which is integrated with an electronic circuit system, more than one sensor can be connected to a single electronic module either in parallel or in series according to the system design requirements.

☐ In addition to protection for the motor itself, the National Electrical Code® requires that the wiring be protected against overload due to motor overload or failure to start.

☐ Across-the-line magnetic controllers are widely used in air conditioning and refrigeration systems.

☐ Full-voltage motor starting is more desirable than the others because of its lower initial cost and the simplicity of the control system.

☐ Star-delta motor controllers limit the starting current very efficiently, but motors designed for this type of starting are required.

☐ Part-winding motor controllers are used to limit any fluctuation in line voltage by connecting any part of the winding to the line, then connecting the remaining part of the winding to the line after a time interval of from 1 to 3 sec.

☐ Multispeed motor controllers provide flexibility in many types of drivers requiring a variety in capacity.

☐ EXERCISES ☐

1. To what standards must motor protective devices conform?
2. What is the purpose of thermal motor protectors?
3. What type of motor protector is installed electrically in series with the motor winding?
4. Why must motor protectors installed inside a motor-compressor be hermetically sealed?
5. What type of sensors are connected in electrical series with the control controlling the magnetic contactor?
6. What type of sensors may also be used with the electronic current-sensing devices?
7. In what type of system can more than one sensor be connected to a single electronic module?
8. What type of devices are generally used when added protection is desired on original manufacturer's equipment?
9. Where is the manual control generally located for dc or ac motors?
10. What type of speed controller is used for large air conditioning units using slip-ring motors?
11. What is the purpose of a starting relay?
12. Are the contacts normally open or normally closed on an amperage starting relay?
13. What causes the hot-wire starting relay to operate?
14. Between what two terminals is the coil on a potential relay?
15. Where is the voltage that operates the potential relay produced?
16. Does the solid-state starting relay need to be sized for each size of motor?
17. When is a hard start kit used?
18. What is the purpose of a thermostat?
19. Where should pressure switches be connected into the refrigeration system?
20. What are oil safety controls designed to do?
21. What is the purpose of defrost timers?
22. What is the difference between starters and contactors?

16 Domestic Refrigeration

Performance Objectives

After completion of this chapter, the student should:

■ Know how a refrigerator cabinet is made.

■ Understand the function of the various refrigeration system components used on a refrigerator.

■ Be able to explain how a domestic refrigerator works.

■ Understand sealed system repair procedures.

■ Be able to explain how to service domestic refrigeration systems.

■ Know how to read and understand domestic refrigerator wiring diagrams.

■ Complete the Exercises with at least 80 percent accuracy.

■ Complete the Laboratory Exercises with at least 80 percent accuracy.

Domestic refrigeration is usually considered as comprising three areas: (1) refrigerators, (2) freezers, and (3) room air conditioners. In most modern equipment, the refrigerator and freezer are combined into one cabinet. In early times these cabinets were fitted with one door. Today, however, it is more common to have two or three doors. These cabinets may be free-standing or built into a wall.

The freezer section of most refrigerator-freezers is small and is designed to hold frozen foods for only a short duration. The separate freezer is in a cabinet much like the refrigerator except that the inside temperature is much colder. This colder temperature allows foods to be stored for a much longer period of time.

Room air conditioning units are designed primarily to cool a single room within a structure. However, larger units have been designed to cool a complete home or several rooms where installation of a central air conditioning unit is impractical or not desired. Room air conditioners will be discussed in detail in Chapter 17.

Definition. *Domestic refrigeration units* are those that are used in residences and other domestic applications. They are used for storing fresh and frozen foods. The fresh food is usually stored for short duration. The frozen food may be kept for more than a year without spoiling. These units are generally of the self-contained type. On the larger type of domestic refrigerators and freezers, they may have a remote-type condensing unit; however, these are rarely used.

☐ Refrigerator Cabinet ☐

A refrigerator that is used for household units consists of a specially designed insulated storage box (Figure 16–1). The modern household cabinet is constructed of sheet steel. The external surfaces of these cabinets are finished with porcelain, enamel, or some other special finish. The food storage compartment is provided with wire shelves and lined with some moisture-proof material such as porcelain or enameled steel. The space between the lining and the external surface is filled with insulation.

The refrigerator cabinet has hardware fittings that enable the door to be tightly closed, minimizing leakage of heat into the box. Arrangements are made in many models for special features such as foot-controlled door openers or lights that automatically

FIGURE 16-1 Typical domestic refrigerator. (Courtesy of Frigidaire Division, General Motors Corp.)

turn on when the door is opened. There may be a number of other features along these lines, but they are all a part of the cabinet and do not directly have anything to do with the refrigeration unit.

Insulation. Insulation is considered to be anything that will retard the flow of heat. There are a large number of substances that have been used for insulation in domestic refrigerators and freezers. At present, fiberglass and foamed plastic are the most widely used insulators for this purpose.

Fiberglass is made from silicon dioxide (glass sand) and is expanded with refrigerant or carbon dioxide. This material is formed into bats approximately 3 in. (7.62 cm) thick when used in domestic refrigerators and freezers. When working with fiberglass insulation, rubber gloves and protective clothing should be worn to guard against fiberglass penetration of the skin. At one time, fiberglass was the most widely used insulating material, but in recent years foamed plastic has taken over the domestic refrigerator insulation duties.

Foamed plastic is made by expanding the plastic with refrigerant, carbon dioxide, or some other highly volatile agent. It is more expensive than the fiberglass insulation; however, only half as much is required to obtain the same insulating effect. Foamed plastic is light in weight, easily worked, and is manufactured in various forms and shapes. Therefore, it makes a useful insulating material for use in thin-walled refrigerators and freezers. The foam density ranges from approximately 1.8 to 2.2 lb/ft^3 (0.029 to 0.035 g/cm^3). The higher the density, the greater the insulating value.

Vapor Barrier. The average kitchen has a much higher relative humidity than any other place in the home except the bathroom. Condensation of moisture for the air will occur on any surface that is at a temperature below the dew-point temperature of the surrounding air. Because of this, moisture must be kept out of the insulated space between the outer shell and the inner liner. This is especially important around the freezer section. Several substances are used for this purpose. The most important substance used in domestic refrigerators with fiberglass insulation is a thin sheet of plastic placed over the outside of the insulation. The holes where the electric or the refrigerant lines go through the cabinet are filled with a water-resistant putty. Also, when fiberglass insulation is used, a tar is spread over the metal surfaces and placed in cracks to prevent the moisture from penetrating the insulation. Foamed plastic insulation does not have the number of moisture problems that fiberglass has because of its inherent vapor barrier qualities.

Door Construction. In order that the entire interior of a refrigerator cabinet be accessible to the user, the door (or doors) must cover as much as is practical of the front side of the refrigerator. The outside panel of the door is made of sheet steel. The inside of the door is usually made of high-impact plastic molded with shelves and storage areas (Figure 16-2). The inside and outside panels are separated by the proper thickness of insulation.

The four edges of the door are finished with a non-heat-conducting material called a *breaker strip,* usually made of plastic (Figure 16-3). A rubber seal or gasket is placed around the inside edges of the door so that when the door is closed the rubber gasket is pressed between the door and the cabinet, thus effecting an airtight seal (Figure 16-4). The amount of heat leakage past the door seal into the cabinet is an extremely important factor. It has been found that a single door, even though it has a total larger opening than two smaller doors, actually allows less heat to enter the cabinet. This is because

HINGE BUTT
MOUNTING SCREWS

HINGE PIN
MOUNTING SCREW

A3220

HINGE WING

HINGE WING
MOUNTING SCREWS

HINGE
BUTT

FIGURE 16-5 Adjustable door hinges. (Courtesy of Frigidaire Division, General Motors Corp.)

FIGURE 16-2 Domestic refrigerator door storage. (Courtesy of Frigidaire Division, General Motors Corp.)

Refrigerator Cabinet

Breaker Strip

FIGURE 16-3 Breaker strip.

Cabinet

Insulation

Magnet

Door

Gasket

FIGURE 16-4 Refrigerator door seal.

one large door requires less gasket than two or more doors would require. Heat leakage is more dependent on the length of the gasket than upon the area of the door.

Domestic refrigerator and freezer doors use a magnetic type of gasket. This type of gasket allows for a better seal, as well as allowing the door to be opened from the inside or the outside. They are also equipped with adjustable hinges so that the door can be adjusted to obtain the best fit possible (Figure 16-5).

Breaker Strip. Between the inner liner and the outer shell, a breaker strip must be used to reduce the flow of heat into the refrigerator by conduction. The lining of the refrigerator is the same temperature as the inside of the cabinet, while the outer shell is at ambient temperature. The breaker strip acts as an insulator between these two parts (Figure 16-6). Each manufacturer installs the breaker strip in a different manner. Some use screws, some use clamps, and others just snap them in.

These breaker strips are made of plastic and are easily broken when cold. Therefore, before attempting removal of a cold breaker strip, bathe the breaker strip with a towel that has been wetted with warm water. Before installation of a breaker strip, dip it in warm, *not hot,* water. This will make it more pliable and reduce the possibility of breakage.

Mullion Heater. The mullion heater is a strand of high-resistance wire attached to a strip of aluminum foil. These heaters are installed around the cabinet doors beneath the breaker strips (Figure 16-7). Any time the refrigerator is plugged in, electricity is sup-

Breaker Strip

FIGURE 16-6　Breaker strip location. (Courtesy of Frigidaire Division, General Motors Corp.)

Freezer Section

Mullion Heater

Fresh Food Section

Cabinet

FIGURE 16-7　Mullion heater location.

plied to the mullion heater, which produces a small amount of heat. This heat around the door area prevents the condensation of moisture. If moisture is allowed to accumulate around the freezer door, it could possibly freeze and prevent opening of the door. Also, condensation may enter the area between the inner liner and outer shell, which will deteriorate the insulation and cabinet.

Shelving. The shelves in a domestic refrigerator or freezer cabinet must be of the open grill type so that the air can circulate through them. They may be of several types, such as woven wire grill, rod or flat bar grill (formed by placing rods or bars side by side), or diamond mesh expanded metal grill. The shelves are usually supported by pegs or hooks secured to the lining, or they may rest on lugs pressed

into the lining. The various refrigerator manufacturers have their own special shelf features and arrangements, each of which has certain advantages. Aluminum, stainless steel, and heavily tinned steel are all used in the manufacture of shelves.

Installation. The refrigerator may be installed wherever practical in the home, but several factors should be considered. The floor should be firm, level, and capable of supporting the loaded cabinet's weight without vibration. If the floor is not strong enough, it should be reinforced with ¾-in. (19.05mm) plywood under the refrigerator for added support.

Refrigerators are designed to operate efficiently at room temperatures between 60 and 110°F (15.6 and 43°C). If the refrigerator is installed where the temperature will be below 60°F, the operating efficiency is affected. This may result in user complaints. Therefore, installations in ambient temperatures below 60°F are not recommended.

If possible, avoid locating the refrigerator next to a stove, radiator, or hot air register. Similarly, an installation that exposes the refrigerator to the direct rays of the sun is undesirable. The user should be advised that although the refrigerator is capable of compensating for these handicaps, it will run longer and consequently use more electricity. Level the refrigerator by adjusting either the leveling glides or the adjustable rollers after the cabinet is located (Figure 16-8).

All domestic refrigerators are equipped with a three-wire service cord with a standard three-prong plug. The electrical outlet should have a grounding-type receptacle to accommodate this type of plug.

BRAKE　　ROLLER　　PIN

FIGURE 16-8　Cabinet leveling guides. (Courtesy of Frigidaire Division, General Motors Corp.)

Refrigerator Cabinet

Do not cut off the ground terminal of the service cord plug. The wall outlet box and the receptacle itself must be properly grounded. If a properly grounded outlet is not available, one should be installed by a qualified electrician. The refrigerator should be on a separate outlet. The voltage must be checked with the compressor running to be sure that it is adequate. When the voltage is not within 10% of the 115 V required, it can cause improper operation. Extension cords should never be used. They are a source of trouble and often cause the grounding features to be lost.

With the cabinet properly located and connected to its power source, the refrigerator interior light should come on when the door is opened. With the temperature control set at mid-range, the compressor should run. Check the operation of the fan in the freezer section of the frost-free models.

A slight gurgling or bubbling sound, which may be heard when the cabinet door is open, is caused by the liquid refrigerant boiling in the evaporator. This is normal and should not be misconstrued as an indication of a malfunction.

There will be some evidence of refrigeration within 5 min after starting, but the temperature may not start to drop until after 30 min or more of operation. Explain to the customer that the first few running cycles will be quite long, but that they will shorten after the normal operating temperatures are reached. Allow 24 hr for the cycles to settle out.

□ Domestic Refrigeration System □

Before any meaningful analysis can be made, the service technician must fully understand the function of the various refrigeration system components. Figure 16–9 is an illustration of the static or gravity-type condenser system. A refrigeration system consits of an evaporator, compressor, condenser, and filter-drier all connected with tubing through which the refrigerant circulates.

Refrigeration is really the process of removing heat from within an insulated cabinet. Briefly stated, the refrigerant within the sealed system absorbs heat from inside the cabinet and through the various components of the refrigeration system and then gives up or transfers this heat to the air surrounding the refrigerator. The air surrounding the refrigerator is referred to as the ambient air. The refrigerant is cir-

FIGURE 16–9   Domestic refrigeration system—static condenser. (Courtesy of Frigidaire Division, General Motors Corp.)

culated through the system by a pressure differential created by the compressor.

The refrigeration system is divided into two parts—high-pressure and low-pressure sides. The dividing points are at the compressor and at the outlet of the capillary tube. The low-pressure side of the refrigeration system consists of the evaporator, the suction line, and part of the compressor. The high-pressure side of the system consists of the discharge line, filter-drier, capillary tube, and part of the compressor.

Evaporator. The evaporator, interchangeably called the cooling coil, freezer coil, or just plain freezer, is where the action is. The heat in the cabinet is transferred to the refrigerant circulating through the evaporator by conduction, convection currents, or forced air. As the refrigerant absorbs the heat from the foods, it begins to boil, changing its state from a low-pressure liquid to a low-pressure heat-laden vapor. This vapor is then drawn through the accumulator and suction line to the compressor. The accumulator acts as a reservoir to prevent liquid refrigerant from flooding out into the suction line during certain portions of the refrigeration cycle, long running periods, or after the cooling coil accumulates a frost buildup. In a normal operating system with a capillary tube, approximately two-thirds of the liquid refrigerant is in the evaporator and accumulator during the running cycle.

Compressor. The compressor and motor are combined in a hermetically sealed steel casing. There are no exposed moving parts. All the parts are continuously lubricated internally, eliminating the need for periodic service lubrication. The motor operates on a 115-V 60-Hz single-phase alternating current. It is a split-phase motor having two sets of windings, a start (phase) winding and a run winding. Some compressor motors run at 1725 rpm, while others run at 3450 rpm. The speed is determined by the type of stator used in the motor.

The function of the compressor is to compress the low-pressure vapor (gas) from the evaporator, changing it into a high-pressure vapor. As it is compressed, the temperature of the refrigerant vapor is increased. The temperature of the vapor must be raised so that it is warmer than the ambient air. The ambient air will then pick up the heat from the refrigerant within the condenser.

Condenser. The function of the condenser is to transfer the heat absorbed by the refrigerant to the ambient or room air. As described in the preceding paragraphs, the high-pressure gas is routed to the condenser where, as the vapor temperature is reduced, it condenses into a high-pressure liquid state. As described before, this heat transfer takes place because the discharge vapor is at a higher temperature than the air passing over the condenser. It is very important that an adequate airflow over the condenser be maintained.

Any accumulation of lint or other material covering the condenser should be removed to assure normal air flow. Remember, as the temperature of the condensing medium increases, the condensing temperature of the refrigerant increases. In other words, as the room temperature increases from about 70 to 90°F (21.1 to 32.2°C), the head pressure and temperature must also be higher to transfer the heat effectively.

Filter-Drier. A high-side filter-drier is located at the outlet of the condenser. It filters foreign matter that might restrict the refrigerant flow through the system and removes moisture that might freeze in the capillary tube, causing a restriction. When installing a filter-drier be sure that it is installed with the direc-

tion of refrigerant flow as indicated by the arrow and be certain that the outlet is always lower than the inlet.

Use of Alcohol in Refrigeration Systems. The practice of using alcohol as an antifreeze in refrigeration systems when the compressor motor is located internally in the system is not recommended. Alcohol is not compatible with aluminum or some other types of insulation used on motor windings and compressor parts in modern refrigeration systems. In addition, if alcohol is added and moisture is present in these systems, the chemical stability of the system is substantially affected. It is, however, recommended that a good filter-drier be added to the refrigeration system for the control of moisture in the system.

Capillary Line. The capillary line or *cap tube,* as it is commonly called, is a refrigerant metering device that determines the amount of high-pressure liquid refrigerant that flows from the condenser to the evaporator. This flow is dependent on the length of the cap tube as well as its diameter.

☐ Domestic Refrigerator Operation ☐

The air flow inside a refrigerator cabinet is very important. A lack of air circulation inside the cabinet will provide poor air temperatures and allow spoilage of food.

Top Freezer Frost-Free Models. The air in the freezer section is pulled through the front grill and over the entire width of the cooling coil located at the rear of the freezer compartment (Figure 16–10). The heat and moisture picked up from the food and air inside the cabinet are given up to the evaporator. This results in the discharge of cold, dry air from the air duct into the freezer section. An air deflector is used to divert the air directly over the ice service area. The blower housing is designed so that some of the air is diverted through the manual damper to the refrigerator section.

The freezer control or manual damper, located in the top rear of the refrigerator section, will determine the amount of air that flows into the refrigerator section (indicated by the open arrowheads in Figure 16–10). The air from the refrigerator section

FIGURE 16-10 **Air circulation in top freezer frost-free refrigerator. (Courtesy of Frigidaire Division, General Motors Corp.)**

returns to the cooling coil area through the slots in the divider located on each side of the control housing. It mixes with the freezer return air and passes across the evaporator and back to the blower motor housing.

Side-by-Side Frost-Free Models. These models use a single-coil evaporator mounted in a vertical position in the freezer compartment to cool both the freezer and food compartment (Figure 16-11). Only one fan is used for air distribution. Starting in the freezer compartment, air is drawn through the grill at the bottom of the compartment and through the coil where the heat and moisture are removed. The blower housing and air ducts provide direct cooling in this area. This air flows downward, cooling the lower section on its way to the return air grill. Part of the cold, dry air entering the fan housing is diverted upward through ducts to the top of the freezer compartment. A deflector, mounted on the freezer liner top, directs air over the ice service area. From here, the air filters down through the shelves to the return air grill.

The air duct from the blower housing is so designed that some of the cold, dry air is diverted to

FIGURE 16-11 **Airflow and water disposal. (Courtesy of Frigidaire Division, General Motors Corp.)**

the top of the food compartment. The amount of air entering the food compartment is controlled by a manual damper. The operation of this control is explained in the discussion of temperature controls below. As the air enters the food compartment, the cold air flows to the bottom of the compartment. In the bottom left rear corner of the food compartment an opening is provided in the foamed divider for air to return to the cooling coil in the freezer section. From here the air returns to the cooling coil and the air cycle is repeated.

Defrost Systems. Frost that forms on the cooling coil must be removed periodically to maintain cooling efficiency, airflow, and product temperatures. On the nonfrost-free models, defrosting and cleaning of the evaporator is the responsibility of the user. On frost-free refrigerators this is accomplished by a timer-controlled defrost system, usually an electric heater.

As stated before, there is a wide variety of defrost systems. In this example, we will use the Frigidaire

DOMESTIC REFRIGERATION

FIGURE 16–12 Side-by-side defrost heater location. (Courtesy of Frigidaire Division, General Motors Corp.)

FIGURE 16–13 Defrost timer. (Courtesy of Frigidaire Division, General Motors Corp.)

EVAPORATOR RAISED FOR
ILLUSTRATION ONLY

FIGURE 16–14 Defrost limiter switch location. (Courtesy of Frigidaire Division, General Motors Corp.)

8-hr defrost system. Regardless of the amount of time involved, the defrost cycle will be basically the same for all types.

Defrost Heaters. Frost-free refrigerators utilize a 500-W infrared heater enclosed in a glass tube. On top-freezer models, one of these heaters is located in a recess provided at the bottom of the evaporator. An additional drain heater prevents the drain from freezing. Some side-by-side models use two heaters wired in series (Figure 16–12). They are located in recesses provided between the fins of the coil. The lower heater directs enough heat to the drain area so that a separate drain heater is not required.

Defrost Timers. An 8-hr timer (Figure 16–13) deenergizes the compressor and freezer blower and energizes the defrost system for 20 min, three times each day. At the end of each defrost period the timer will automatically cut off the heater circuit and energize the compressor and freezer blower motor.

Defrost Limiter. The defrost timer determines when the defrost period will occur, but the duration of the defrost period, not to exceed 20 min, is normally determined by a thermostatic control called the *defrost limiter switch.* The defrost limiter is located near the evaporator and is connected electrically in series with the heater element (Figure 16–14). This switch opens at a temperature of $50 \pm 5°F$ $(10 \pm 2.78°C)$ and disconnects the defrost heater. The timer terminates the defrost cycle after 20 min and energizes the refrigeration circuit.

The defrost limiter switch resets at a temperature of $30 \pm 6°F$ $(-1.1 \pm 3.33°C)$ and is then ready for the next defrost period.

Defrost Water Disposal. On top-freezer models, the defrost water is removed from the freezer section through the drain tube and deflector mounted to the top of the refrigerator section. The water flows down the rear wall of the liner to a trough formed in the liner bottom (Figure 16–15). The defrost water then goes through a drain tap into a disposal pan located in the machine compartment where the water evaporates.

On side-by-side models, a drain trough is riveted to the evaporator housing at the bottom of the freezer compartment. The defrost water is then channeled to the freezer drain tube that extends through the liner and cabinet shell at the rear of the machine compartment (Figure 16–16). From there, a tube directs the water forward to the drain trap and disposal pan. The water in the disposal pan is then evaporated by the air that moves across the pan.

The user should be advised that periodic cleaning of the drain is necessary. The user should also be instructed on reinstalling the drain trap after cleaning. Improper placement of the drain trap can cause an air leak in the drain opening.

Domestic Refrigerator Operation _____ **343**

FIGURE 16-15 Water disposal in top-freezer frost-free models. (Courtesy of Frigidaire Division, General Motors Corp.)

FIGURE 16-16 Side-by-side water disposal. (Courtesy of Frigidaire Division, General Motors Corp.)

Temperature Controls. The temperatures within a refrigerator are controlled by a thermostatic switch that regulates the running time of the compressor and freezer blower motor, where applicable.

Keeping in mind that refrigeration of the fresh food and freezer compartments results from the cooling effect of the evaporator in the freezer compartment and the circulation of cold air by a blower, we can determine how the temperature in each compartment can be regulated.

The running time of the unit is determined by the thermostatic switch bulb, located at the top of the fresh food compartment, and by the volume of cold air that is provided through the manual damper (freezer control) from the freezer compartment (Figure 16–17).

When the fresh food compartment warms to the cut-in temperature of the thermostat, the refrigeration unit begins to run. The evaporator in the freezer gets colder and the cold air is blown, through the

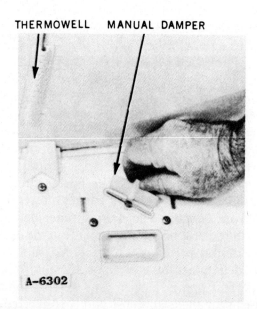

THERMOWELL   MANUAL DAMPER

A–6302

FIGURE 16-17 Thermostat bulb and manual damper location. (Courtesy of Frigidaire Division, General Motors Corp.)

manual damper, into the food compartment. When the manual damper is set at the midpoint, enough air will be diverted to the food compartment to cool it to the cut-out temperature without affecting the freezer temperature appreciably.

When a lower freezer compartment temperature is required, the manual damper is closed a little by turning the freezer control to a colder setting, restricting the volume of cold, dry air provided to the food compartment. Since more cold, dry air is circulating within the freezer compartment, the temperature will fall. The temperature in the fresh food compartment will not be greatly affected because the unit will run until the cut-out temperature in the fresh food compartment is reached. Simply stated, the longer running time compensates for the reduced volume of air that is provided for the fresh food compartment.

When a lower fresh food compartment temperature is required, the thermostat is set lower, which lowers the cut-in and cut-out temperatures of the thermostat. Setting the manual damper one increment higher will provide more air to the fresh food compartment with minimal effect on the freezer temperatures.

☐ Sealed System Repair ☐

When servicing domestic refrigerators and freezers, the malfunction is often found to be in the controls and not in the refrigeration unit. Many times the problem is solved by simply instructing the user on the proper use. At other times, it is a matter of restoring proper airflow from the cooling coil to the compartments, and the foods stored in them, and back to the cooling coil again. A lack of proper airflow is sometimes caused by fan motors not running at the proper speed. This problem also could be a result of improper setting of the controls by the user. A malfunction of the defrost system causing the evaporator to be blocked with ice could restrict the airflow. It is strongly urged that the sealed refrigeration unit be checked as a last resort. If gauges are installed, be sure that the procedures are followed properly to avoid contamination of the refrigeration system which could cause problems later.

Safety. Always of primary importance is safety: safety for yourself as you work, safety for the sur-

roundings in the area in which you are working, and safety for the customer after you leave the job.

Safety for yourself includes using the approved eye protection, using grounded tools and instruments, proper lifting and moving practices, and being aware of and prepared for any emergency that might arise as you work. You may use shortcuts or unsafe practices for years with no adverse effects, but the one time that you are caught can make all the time you have saved very expensive. *There is always time for safety.*

Safety for the surroundings means respect for the property around which you are working. Cover furniture and rugs that could be damaged. Do not track the dirt on your shoes into the house. Direct the discharge of refrigerant out of doors, being sure that shrubs, flowers, or grass are not damaged by the refrigerant and oil. Use extreme caution when moving the appliance. With the rugs and carpets used in kitchens today, moving an appliance can become troublesome. The good will derived from a successful service call can be nullified when property has been damaged or the work area left dirty.

Before leaving the job, be certain that the product is properly grounded and that the outlet into which the refrigerator is to be plugged is properly grounded. If there is any doubt about the receptacle being as it should be, advise the customer. Note on the work order that the customer has been advised and have it signed, if possible. Advise the customer to contact an electrician who can wire the receptacle in accordance with local codes.

Safety does not cost; it pays. Give electricity the respect that it demands.

Refrigeration Performance Complaints. A refrigeration performance complaint is a complaint involving food storage temperatures, percentage of running time, or both. Such complaints are much more frequent during the summer months when the refrigerator is required to work at or near full capacity. Indeed, many refrigerator performance complaints require only customer education or reassurance that the refrigerator or freezer is operating normally.

It is difficult to suggest a detailed step-by-step procedure to diagnose refrigeration performance complaints. Each service call requires a particular procedure depending on the customer's complaint, severity of the symptoms, room temperature, length of service, and other factors. In other words, there is no substitute for a knowledge of basic refrigeration principles and sound, analytical thinking when diagnosing these complaints. The following procedures should be used as a guide for every refrigeration performance complaint:

1. Measure the temperature of the frozen food packages and the food stored in the fresh food section. Always use a good reliable thermometer or thermocouple (Figure 16–18). Place the thermometer bulb or the thermocouple probe between two packages of frozen food. If the thermocouple is wet before placing it between the packages, it will freeze to the packages and result in a more accurate temperature reading. Leave the refrigerator or freezer door closed for 5 to 10 min, then read the temperature as rapidly as possible without removing the thermometer. If a temperature tester is used, the temperature can be read without disturbing the probe or opening the door. The food being measured should be located near the center of the compartment to obtain an average temperature.

   Frozen food temperatures in the freezer section ideally are between 0 and 5 °F (− 17.8 and − 15 °C) in frost-free models. Conventional models will be from 10 to 12 °F (− 12.2 to − 11.1 °C) depending on the control settings, food load, usage, and ambient temperatures. Subzero temperatures are not necessary for food preservation and one should not try to obtain lower temperatures than indicated above. Temperatures could go higher than those indicated for a brief period of time during and immediately after the defrost period on frost-free models. The temperature of food in an extremely lightly loaded freezer will fluctuate more than in a full freezer due to door opening and defrost. It is, therefore, desirable to keep the freezer section at least half full of food.

   The food temperatures in the refrigerator (fresh food) section should be between 32 and 40 °F (0 and 4.4 °C) depending on the model, running time, usage, load, and where the temperatures are taken. Cold air drops; therefore the bottom of the refrigerator section will be colder than the top except, of course, when foods are placed directly in front of the air discharge into the refrigerator section. Take the temperatures of a food such as a jar of pickles that has been in the refrigerator for at least 12 hr and is located on top of the hydrator cover as a guide. This gives a good average temperature.

2. Install a temperature recording instrument. Occasionally a temperature complaint cannot be settled to the satisfaction of the customer by a single temperature reading. Perhaps the customer agrees with the thermometer check but claims that the operation is erratic. In such cases, a temperature recording instrument is a very useful tool to settle the complaint, or perhaps prove that the operation is or is not erratic.

   Temperature recorders that record temperatures on a 24-hr chart are available from many manufacturers. They also record the number of cycles of operation, which is useful in detecting erratic operation and estimating the percentage of running time.

3. See Figure 16–19. Almost every manufacturer will have a service diagnosis section in the equipment literature, giving the problem and possible cause, along with the remedy. Recently, block diagram charts have been used (Figure 16–19). Study these charts for assistance in making a proper diagnosis.

Use of Volt-Wattmeter. One of the most valuable tools that a service technician can own is a volt-wattmeter. This meter can be used to check line voltage and line voltage drop. Most refrigerator compressors are designed to operate on 115 V ± 10%. Any voltages above 127 V or below 105 V will have an effect on the motor rpm and the overload protection. The dropout time of the starting relay (the time required for the relay to deenergize the start winding after the initial start) would also be affected.

FIGURE 16–18 Temperature tester. (Courtesy of Robinair Manufacturing Company.)

FIGURE 16-19 Diagnostic chart for an inoperative refrigeration system. (Courtesy of Frigidaire Division, General Motors Corp.)

All of this will have an adverse effect on the overall operation of the refrigerator.

Complete operating instructions accompany the volt-wattmeter. A typical procedure check is outlined below. *Caution:* Do not exceed the current ratings specified on the wattmeter dial. Always start with the instrument set on the highest scale and then switch to the scale that gives the closest reading to midscale of the dial.

1. Set the toggle switch to the 130- or 260-V position. If in doubt, use the 260-V position.

2. Set the watts selector switch to the 5000-W position. This will protect the instrument from surge overload.

3. Connect the male power cord of the instrument to a wall outlet. The voltmeter will now read the line voltage.

4. Connect the line cord of the refrigerator to the female receptacle on the volt-wattmeter front panel.

5. Turn on the temperature control switch of the refrigerator being tested. (The following

wattage readings were taken on a Frigidaire FPCI-219VN-R refrigerator.)

You will note that the instant the compressor is energized, the overswing of the needle may go above 2000 W depending on the damping of the meter movement within the wattmeter. The voltage is at 120 V and the room temperature is 70 °F (21.2 °C). The wattage then drops off to approximately 1500 W, which is the combined wattage draw of the start and run windings. It then drops off to approximately 475 W. This is the running winding wattage plus whatever other electrical components, such as condensate driers, might be incorporated within the refrigerator. All of this takes place within ½ to 1½ sec, making it difficult to establish these wattages. These wattages vary, of course, with the make and model being tested. Now switch the meter to the 500-W scale.

It should be noted that the final running wattage settles out at approximately 400 W, which includes the various condensate driers and fan motors. When the temperature control is turned off, the wattage draw is still approximately 60 W. Therefore, the net compressor wattage draw is 340 W (400 − 60 = 340).

The wattage recorded will vary slightly with the ambient temperature, make and model of the refrigerator (which determines the size of the compressor), and the other electrical components that are used, such as condensate driers, condenser fan, etc. Whether the reading is taken at the beginning or end of the running cycle (the lower wattages will be at the end of the running cycle) will make a difference. Also to be considered is the amount of airflow over the condenser and the line voltage.

Plugging the refrigerator line cord in and taking it out several times may cause the compressor overload to trip. If this should happen, wait for the compressor to cool and the overload to reset before continuing. The length of time required for it to reset will depend on the type of overload protection and whether or not it kicked off due to excessive current or because of high temperature. If a current overload is used, only a matter of minutes is required for the compressor to restart. Forced-air condenser models use temperature overloads as well as current overloads that take 1 hr or more to reset when the cause is excessive temperature. On the other hand, these forced-air units utilize starting capacitors that make it unlikely that the overload would kick out on the first few times the cord is plugged in.

Use of Compressor Test Cord with Wattmeter. Should a faulty compressor be suspected, all controls and wiring can be bypassed and the compressor checked directly. This may be done by using a compressor analyzer or a test cord in conjunction with a volt-wattmeter (Figure 16-20). When using the test cord, the three leads can be attached to the three-lead cord from the starting relay or directly to the compressor terminals. Some service technicians make an adapter from the three-lead relay to the compressor cord that attaches to the compressor terminals. The leads are properly color coded. The plug is always positioned with the leads upward.

Attach the test cord clips to the compressor terminals as follows: white to common terminal, red to start terminal, and black to run terminal (Figure 16-21). If a terminal diagram is not available, locate them as outlined in Chapter 14. Attach the ground clip to the compressor frame. If the unit has a capacitor, connect one of the same size that is known to be good to the two capacitor lead clips. If no capacitor is used, connect these two clips together.

Be sure that the fuse in the test cord is good. A *fusetron* type of fuse is preferred. Connect the male plug of the test cord into the volt-wattmeter, which has already been plugged into a wall outlet. The wattmeter should be set on the 5000-W scale. For illustration purposes, the Frigidaire model FPCI-219VN-R is used.

The stalled or locked-rotor wattage (1500 W) of the run winding will be indicated on the wattmeter when the switch on the test cord is turned from off to run. Now switch from run to start, and the combined wattage (1800 W) of the start and run windings will be indicated. Hold the switch in the start

FIGURE 16-20 Compressor test cord. (Courtesy of Robinair Manufacturing Company.)

Frigidaire RI Compressor

Necchi Compressor

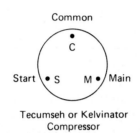

Tecumseh or Kelvinator
Compressor

FIGURE 16-21 Compressor terminal identification.

position only momentarily and then release it to run. The wattage now is 380 W.

Caution: Do not hold the switch in the start position any longer than necessary to get the compressor started or the start winding can overheat and burn out. If the compressor starts and continues to run with normal wattage, the compressor is not at fault. Normal wattages are given on the data label and wiring diagram that accompany the refrigerator. The wattage, on run settled out to 340 W, is the compressor wattage only—no fans or condensate driers. If the compressor tries to start but does not, or if it runs with continued higher than normal wattage, the compressor must be replaced.

Note: Wattages will vary depending on when they are taken. Wattages will be higher for a unit that has been running at normal operating temperatures and cycles than for a product that has been sitting idle. The compressor must run for 10 min or so to raise the oil and stator temperatures and to raise the head pressure to its normal range. Wattages will be higher if the compressor has been cycling on the overload. This could be caused by an inoperative condenser fan, for example, which will allow the compressor dome temperatures to build up.

Wattages should be used to determine if the compressor is stuck, shorted, grounded, has open windings, has a tight rotor, or has other internal malfunctions. Wattages alone should not be used to diagnose a shortage of refrigerant or an inefficient compressor.

Checking Capacitors. A faulty capacitor should be suspected if the customer states that the compressor will not run, has a loud hum or buzz at times, or tries to start and then kicks off.

If the capacitor shows visual evidence of liquid leakage or if the case is bulged or cracked, replace the capacitor even though it may seem to be good through testing.

Defective capacitors should always be replaced on a like-for-like rating. Substituting a capacitor with either a higher or lower μF rating than specified will reduce the torques available from the motor. The voltage rating of the old and new capacitors should be identical.

There are two general methods used to test capacitors:

1. *Use of an ohmmeter.* An experienced technician can readily test a capacitor by using an ohmmeter. Testing a capacitor takes a little practice because the ohmmeter setting varies with the capacitor μF rating. As a rule, the lower the rating, the higher the ohmmeter setting required. In refrigeration work, however, we usually check medium-range capacitors. First, discharge the capacitor with a 20,000-Ω resistor. Place one probe of the ohmmeter on one terminal of the capacitor to be checked. While observing the ohmmeter, place the other probe to the other capacitor terminal. The meter needle will immediately register zero resistance and then slowly increase in resistance until it approaches infinity. This, as stated earlier, requires the proper selection of the resistance scale that is compatible with the size of the capacitor being tested. If the meter reads infinity at the beginning, the capacitor is open. If the meter continues to read zero, the capacitor is shorted. In either case, it must be replaced. To check a metal-clad capacitor for a grounded condition, set the ohmmeter on the highest scale and place one probe on the metal casing of the capacitor, and alternately touch each capacitor terminal. A high resistance (250,000 Ω) or infinity should be indicated on a good capacitor.

2. *Use of a capacitor checker and a volt-wattmeter*
   (a) Connect the capacitor to the alligator clip leads of the checker.
   (b) Plug the checker into a volt-wattmeter and plug the volt-wattmeter into a power supply receptacle.
   (c) Depress the pushbutton of the checker momentarily and read both the voltage and the wattage. (The meter needles must stabilize before the readings can be taken.)
   (d) Locate the μF rating of the capacitor in Table 16-1 and select the correct wattage found in the test under the proper test-voltage column. Values within $+20\%$ to -0% are satisfactory on electrolytic starting capacitors and within $\pm 10\%$ on nonelectrolytic capacitors.

FIGURE 16–22   Capacitor check—325 to 550 μF. (Courtesy of Frigidaire Division, General Motors Corp.)

For checking capacitors with μF ratings of 325 to 550, see Figure 16–22.

1. Connect the capacitor to be checked to the alligator clipped leads of the capacitor checker.
2. Plug a volt-wattmeter into the receptacle of the capacitor checker and plug the two-wire lead of the capacitor checker into the 115-V ac power supply receptacle.
3. Depress the pushbutton momentarily and read the voltage. The correct voltages are listed in Table 16–2.

System Pressure. If a defrost cycle is not occurring, not only will the temperatures inside the cabinet be affected but also the head and suction pressures. Such conditions should be corrected before any attempt is made to check the refrigerant system pressures. We will now concentrate on only those conditions directly associated with the unit that will

TABLE 16–1 Capacitor Test Wattage [W] at Various Test Voltages

Capacitor μF Rating	Test Voltages [V]				
	100	105	110	115	120
11	4	5	5	6	6
15	8	9	10	11	12
17.5	11	12	13	14	15
20	14	15	17	18	20
28	26	29	31	34	37
32	33	37	40	44	48
35	40	44	48	52	57
40	50	55	60	66	72
42	54	60	65	72	78
45	61	67	74	81	88
50	73	80	88	96	105
54	83	91	100	110	120
60	97	107	118	128	140
75	134	148	162	177	193
85	157	172	189	207	225
105	198	218	240	263	285
115	214	236	262	187	311
125	233	257	281	308	335
130	240	265	290	318	346
145	258	285	315	345	375
155	270	298	327	361	392
167	285	313	344	375	407
190	305	336	370	404	438
210	319	352	387	422	458
240	334	369	405	442	480
270	348	382	418	459	498
315	360	398	437	479	520
Capacitor shorted	400	440	485	530	575

TABLE 16–2 Capacitor Test Voltage and μF Rating[a]

Capacitor μF Rating	Voltage Reading
350	33
400	30
450	27
500	24
550	22

[a]With line voltage higher or lower than 115 V ac, the voltage drop across the capacitor will vary with the percent of change in line voltage.

adversely affect the performance of the refrigeration system.

Previously, it was established that there is a low-pressure side and a high-pressure side within a refrigeration system. At this time, a review of this area is in order regarding the meaning of these pressure readings.

When gauges are installed it can be determined whether or not the unit is performing normally and, if not, what the problem might be.

Note: It is again emphasized that gauges should be installed only after all other possibilities have been eliminated and it has been concluded that the fault lies in the refrigeration system. The gauges should be installed on the refrigerator in the customer's home, if possible, to record the pressures under normal usage conditions. The unit should have been operating for some time before the service technician arrives. If a long or continuous running situation has existed, this is the ideal time to take pressure readings. Since the purpose of installing gauges is to detect a malfunction, do not disconnect the unit electrically to install the gauges. If the unit is already disconnected when the service technician arrives at the home, it will not be advantageous to install gauges. If the refrigerator is in the repair shop, the unit should be allowed to run for several hours prior to checking. The compressor windings must reach their normal temperature range, which will affect oil and refrigerant temperatures. The gauge readings should be recorded just prior to the end of the run cycle. The exception would be when checking idle head pressures, equalization pressures, or when a low-side leak is suspected.

In this respect common sense and judgment should prevail. Naturally, a unit should not be operated if it has a low-side leak, which would allow air and moisture to be drawn into the refrigeration system. Many times this is quite obvious. It is not necessary to have the compressor running to determine if the unit is discharged. A quick gauge check will give this indication. A thorough leak check should be made to determine where the leak is located. Running the compressor with a low-side leak not only pulls in air and moisture but can also result in damage to the various parts of the system and could result in making replacement of the entire system necessary.

Approximately 90% of sealed system entries are for repairs on the high side of the system. This includes high-side leaks, restrictions, and compressor replacement. The remaining 10% of entries are due either to moisture in the system or a low-side leak. A leak in the low side is often subjected to a vacuum in the sealed system and air and moisture are drawn in. Therefore, it must be considered a wet system.

As a general guide, the suction pressure variation has an effect on the head pressure (see Table 16–3). Head pressures and suction pressures will vary and the system performance will be adversely affected by one of the following.

Use of Different Refrigerants. When working on a refrigerator or freezer using R-12 refrigerant, different pressure readings can be expected than when working on a refrigerator using R-114, for example. The use of R-22 in air conditioners will generate higher pressures than if R-12 were used.

One Line of Products versus Another. The normal operating pressures on food freezers will be lower than those on conventional refrigerators, which in turn are lower than the pressures on air conditioners.

Heat Load. A greater heat load can result from the addition of more than a normal supply of foods, such as after doing the weekly shopping or perhaps filling all the ice cube trays at one time. The load placed in the freezer section affects the unit operation more than the load in the refrigerator section. Other items contributing to an additional heat load would

TABLE 16–3 Normal Conditions

Suction Pressure Variation	Cause	Effect on Discharge Pressure
Increase	Additional heat load primarily in the freezer compartment	Increase
Decrease	Decrease in heat load primarily in the freezer compartment; abnormal conditions	Decrease
Increase	Refrigerant overcharge	Increase
Decrease	Refrigerant undercharge [leak]	Decrease
Decrease	High-side restriction	Increase [see text]
Increase	Inefficient compressor	Near normal [see text]
Decrease	Low ambients [high ambients the reverse]	Decrease
Decrease	Low-side restriction	Normal to slight decrease
Near normal	Air in system	Increase

Source: Frigidaire Division, General Motors Corp.

be excessive door openings, poor door seals, interior lights remaining on, etc.

An increase in the amount of heat being absorbed by the refrigerant in the evaporator will affect the temperature and pressure of the refrigerant vapor returning to the compressor. Compartment temperatures, power consumption, discharge and suction pressures are all affected by heat load. The operating pressures will be higher than normal under a heavy heat load condition.

Ambient or Room Temperature. Lower ambient air temperatures reduce the condensing temperature and therefore reduce the temperature of the liquid refrigerant entering the evaporator. The increase in refrigerating effect due to operation in a lower ambient temperature results in a decrease in power consumption and running time. At lower ambient temperatures, there is a reduction in cabinet heat leakage, which is partially responsible for the lower power consumption and running time.

Note: An increase in refrigeration effect cannot be expected below a certain minimum ambient temperature. This temperature varies with the type and design of the product.

Generally speaking, ambient temperatures cannot be lower than 60°F (15.6°C) without affecting the operating efficiency. Conversely, the higher the ambient temperature, the higher the head pressure must be to raise the high-side refrigerant temperature above that of the condensing medium. Therefore, the head pressures will be higher as the ambient temperature rises.

Refrigerators installed in ambient temperatures lower than 60°F (15.6°C) will not perform as well because the pressures within the system are generally reduced and unbalanced. This means that the lower head pressure forces less liquid through the capillary line resulting in symptoms of a refrigerant shortage. The lower the ambient temperature, the more pronounced this condition becomes.

Condenser. If the efficiency of heat transfer from the condenser to the surrounding air is impaired, the condensing temperature becomes higher. A higher liquid temperature means that the liquid refrigerant will not remove as much heat during boiling in the evaporator as under normal conditions. Therefore, remove any paper sacks, lint accumulations, etc., that would restrict the normal air movement through the condenser. This condition would be indicated by a higher than normal head pressure.

Refrigerant Charge. Any deviation in the refrigerant charge will affect the suction pressure. As the suction pressure varies, the head pressure will vary in direct proportion. This is true because of the fixed size and length of the capillary tube that is the metering device. The refrigerant flow rate is also affected by the amount of refrigerant in the system and the head pressure forcing it through the capillary tube. An undercharge of refrigerant will cause a decrease in the flow rate (less heat is removed) and, therefore, a decrease in the desired refrigerating effect. This, in turn, causes a reduction in normal suction pressures. The wattage draw of the compressor will also decrease.

Incorrect Refrigerant Charge—Overcharge. An incorrect charge rarely occurs in manufacture. Various conditions could indicate an overcharge. For example, if the cooling coil is not defrosted at regular intervals due to a failure of the defrost system, the refrigerant will flood out and cause the suction line to frost or sweat (Figure 16–23). The cause of this problem should be corrected rather than purging refrigerant from the system. The indication of an overcharge of refrigerant may result from running the freezer section colder than necessary [0 to 5°F (−17.8 to −15°C) are considered normal package temperatures, except in conventional models], or from continuous running of the compressor for a variety of reasons, or even from the freezer fan motor not running.

In high-humidity conditions, sweating of the suction line may occur at various stages of the running cycle. This is usually temporary and should cause little or no problem. If need be, the suction line can be wrapped with insulation tape or a similar material to prevent moisture from dripping from the lines. The symptoms of an overcharge are:

1. Above normal freezer temperature
2. Frost on the suction line indicating refrigerant flooding into the suction line
3. High head and suction pressures
4. High running wattage

If it becomes necessary to remove refrigerant from a system, it can be accomplished by accurately removing one or two ounces at a time. Thus, the service technician knows precisely the condition with relation to the refrigerant charge.

Caution: This procedure should not be used on systems that have had previous service for leaks, compressor burnout, or other system-contaminating service. Those systems should be completely discharged and recharged with the correct amount of refrigerant.

FIGURE 16-23 **Indications of a refrigerant overcharge.**

FIGURE 16-24 **Measuring refrigerant removal. (Courtesy of Frigidaire Division, General Motors Corp.)**

1. Install the gauges to the compressor. Use a portable charging cylinder with 3 or 4 oz of liquid R-12 charge and a short charging line connected to the bottom shutoff valve. Start the threads of the charging line onto the service valve fitting.

2. Open the charging cylinder valve slightly to purge the air from the charging line. Tighten the charging line at the service valve fitting. The line is now filled with liquid R-12 (Figure 16-24).

3. Mark the refrigerant level in the charging cylinder; then with the compressor running, open the discharge service valve and condense 1 oz of R-12 from the system into the charging cylinder

by opening the bottom shutoff valve for 1-sec intervals. Observe the charging cylinder until the liquid level rises to the desired level.

Note: The compressor must be operating while the liquid is being condensed into the cylinder. The small amount of refrigerant in the cylinder should be discarded and not reused. If for some reason too much refrigerant has been removed from the system, it is best to discharge, evacuate, and recharge with the specified amount.

The refrigeration system is designed so that there is enough refrigerant for adequate heat removal in a 110°F (43°C) ambient temperature. However, there should not be an excess of refrigerant that would result in liquid refrigerant flooding the suction line and entering the compressor. An overcharge of refrigerant flooding through could cause damage to the compressor or could possibly keep the suction pressure and temperature high enough to prevent the thermostat switch from cycling the unit.

Incorrect Refrigerant Charge—Undercharging. This is a common occurrence insofar as component problems are concerned. The first thing noticed by the user is a rise in temperature of the foods. In the case of conventional refrigerators and freezers, a partial frosting of the evaporator is noticed. The refrigerant can leak out to the point that no frost is apparent anywhere on the evaporator. On some models, a slight shortage will result in very low freezer temperatures and a high food compartment temperature (Figure 16-25).

This is almost always accompanied by continuous compressor running because the thermostatic control for the unit is never satisfied. On frost-free models, the cooling coil is behind a cover panel and cannot be observed, but the customer usually complains that the fresh food compartment temperatures are too warm. Although the temperatures will rise in both the freezer sections and the food compartment, the frozen meats and vegetables will not thaw

FIGURE 16-25 **Indication of a refrigerant undercharge.**

immediately. The ice cubes will still be frozen. The customer does not associate the problem with the freezer section and will first notice that milk, vegetables, and even beer are not cold enough. Some older models use a fan cycling switch for the food compartment fan motor while others use an automatic airflow control. This section deals with a refrigerant shortage, and it is assumed that the controls are functioning properly.

Under some circumstances, such as in the case of the conventional refrigerator or freezer with a slight shortage of refrigerant, freezing in the food compartment may be experienced due to the additional running time. With a refrigerant leak, however, it always gets worse and as the refrigerant charge decreases, the temperature will continue to rise.

With a shortage of refrigerant, the capillary tube will not have a full column of liquid refrigerant available. As a result, there is a noticeable hissing sound in the evaporator. This should not be mistaken for the regular refrigerant boiling sounds that would be considered normal. The symptoms of a refrigerant shortage are:

1. A rise in the food product temperature in both compartments.
2. Longer or continuous running time.
3. Obvious traces of oil on the floor that would occur due to a broken or cracked refrigerant line. Oil dripping from the charging screw. Do not misinterpret the rust-inhibiting oil placed on the compressor charging port threads at the factory on some units as an oil leak.
4. Lower than normal wattage.
5. The compressor feeling hot to the touch because of the heat generated by the motor windings from long continuous running. It will not be as hot as it would be with a full charge accompanied by long running times for some other reason such as a dirty condenser.
6. The condenser will not be hot, but will be close to room temperature since it is not doing much work. This depends, of course, on the amount of refrigerant shortage. The capillary tube will be warmer than normal from a slight shortage.
7. If the leak is on the high side of the system, both gauges will show lower than normal readings and will show progressively lower readings as the charge decreases. The suction pressure gauge will probably be in a vacuum.
8. If the leak is on the low side of the system, the suction pressure gauge reading will be lower than

normal—probably in a vacuum—and the head pressure gauge will be higher than normal. It will probably continue to become higher because of the air being drawn in through the leak and being compressed by the compressor. This air accumulates in the high side of the system. Air and R-12 do not mix so their pressures are additive.

Restriction. Always remember that refrigeration (cooling) occurs on the low-pressure side of a partial restriction (obviously a total restriction will completely stop the circulation of refrigerant and no cooling will take place).

An experienced refrigeration service technician will physically feel the refrigeration lines when he suspects a restriction. The most common places for a restriction are at the filter-drier or at the capillary tube outlet or inlet. If the restriction is not total, there will be a temperature difference at the point of restriction; the area on the evaporator side will be cooler. In many cases, frost and/or condensation will be present (Figure 16–26). Also, a longer time will be required for the system pressures to equalize.

Any kinked line will cause a restriction, so the entire system should be visually checked. It is very difficult to diagnose a partially restricted capillary tube.

A slight restriction would give the same indication as a certain given refrigerant shortage with a lower than normal suction pressure, head pressure, and wattage, and warmer than normal product temperatures. If a total restriction is on the discharge side of the compressor, higher than normal head pressure and wattage will result. This is true only while the low side is being pumped out and if the restriction is between the compressor and the first half of the condenser.

FIGURE 16-26 Partial restriction.

To diagnose for a restriction versus a refrigerant shortage, discharge the system, replace the filter-drier, and recharge the unit with the specified amount of refrigerant. If the unit performs normally, three possibilities exist: (1) refrigerant loss, (2) partially restricted filter-drier, or (3) moisture in the system. If the unit performs as it did previously, there may be a restricted capillary tube or condenser or a kinked line. Find the point of restriction and correct it.

A restriction reduces the flow rate of the refrigerant and consequently reduces the heat removal. A complete restriction may be caused by moisture, solid contaminants in the system, or a poorly soldered joint. Moisture freezes at the evaporator inlet end of the capillary tube and solid contaminants collect in the filter drier. The compressor wattage drops because the compressor is not circulating the usual amount of refrigerant.

As far as pressure readings are concerned, if the restriction (such as a kinked line or a joint soldered shut) is anywhere on the low side, the suction pressure will probably be in a vacuum while the head pressure will be near normal. If the restriction is on the high side, the suction pressure, again, will probably be in a vacuum while the head pressure will be higher than normal during the pump-out period described earlier. In either case, it will take much longer than the normal 10 min or so for the head pressure to equalize with the low side after the compressor stops.

Air in the System. This condition can result from a low-side leak or from improper servicing. If a leak should occur on the low side, the temperature control would not be satisfied; thus, continuous running of the compressor would result. The compressor would eventually pump the low side into a vacuum, drawing air and possibly moisture into the refrigeration system. Air and R-12 do not mix so the air pressure would be added to the normal head pressure, resulting in a higher than normal head pressure.

The surest way to determine if air is in the system is to read the head pressure gauge while the compressor is idle and then take the temperature on the condenser outlet tube. This temperature should be within 2 or 3 °F (1.11 or 1.67 °C) of what the pressure temperature relation chart shows for the given idle head pressure. If the temperature of the condenser outlet is considerably lower than the idle head pressure on the gauge, it indicates that there is air in the system.

A thorough leak checking is necessary. Correct the source of the leak. Do not attempt to purge off the air because this could result in the system being undercharged. It is best to remove the refrigerant, evacuate, and recharge the system with the specified refrigerant charge.

Inefficient Compressor. This classification is used too frequently when the only sure things known are that the product temperatures are high and the compressor runs too much. A frostback, for example, or air in the system will affect the performance of the compressor in a way that could cause the service technician to suspect the compressor.

Before finally deciding to change the compressor, the technician should make one more check. Since it is necessary to cut the refrigerant lines to replace the compressor anyway, either cut the suction line and solder it shut or pinch it tightly shut just ahead of the low-side fitting and gauge. Run the compressor for a couple of minutes to see if it will pull a 26- to 28-in. (27.48- to 13.74-kPa) vacuum. If it does, it certainly is not an inefficient compressor. If it does not, try adding about 6 oz (170 g) of 525 viscosity oil and repeat the check. Any reading less than 26 in. of vacuum indicates that the compressor should be replaced.

Special Note on Refrigeration Oil. The oil within the compressor plays a vital role. An insufficient amount of oil can give the indication of an inefficient compressor because the oil coating between the parts reduces friction and results in better compression without a bypass of refrigerant vapor.

Burned Oil Check. A burned oil check is made by taking a sample of oil from the inoperative compressor and comparing it to the darkest section of an *oil sample indicator* (Figure 16–27). If the oil is darker than the darkest sample, it is an indication that the oil has undergone a chemical breakdown and has contaminated the system to the extent that the complete sealed system must be replaced. The sample should be collected and checked in a clear, clean glass container approximately 2 in. (5.08 cm) in diameter.

FIGURE 16–27 Oil sample indicator. (Courtesy of Frigidaire Division, General Motors Corp.)

Note: If water has entered the system from a low-side leak on the evaporator, the oil indicator will not show it. The oil will be gray, depending on the amount of water that has been pulled into the system. Usually the amount of water is small and is more in the form of moisture mixed with the oil.

Leak Test Procedure. Any time that a system is diagnosed as being low on refrigerant, the leak must be found and repaired. There are several types of leak detectors and each is effective and has outstanding points. The halide torch is best used where large leaks are present. Electronic leak detectors should be used when small leaks occur. Liquid leak detectors are good for locating small leaks when a particular joint or area is suspected. However, liquid leak detectors are the only type that can be used successfully around the urethane foam insulation that is used in most modern refrigerators and freezers.

The following procedure may be used when testing a refrigeration system for leaks:

1. If the systems appears to be short of refrigerant, clean the charging ports and observe for traces of oil. If oil is found, leak test the suspected port with a leak detector.
2. If no leak is found on the charging ports, install the gauges. Crack the charging screws to permit pressure to be registered on the gauges.
3. If the system has a gauge reading of 25 lb (272.74 kPa) or higher, the system has sufficient pressure to proceed with the leak test. The pressure in the low side should be as high as possible. Therefore, turn the temperature control to the off position and warm up the evaporator by placing pans of hot water directly in contact with the evaporator. On frost-free models, manually initiate a defrost cycle.
4. If the head pressure is less than 25 lb, charge refrigerant into the system until sufficient pressure is obtained so that an effective leak test can be made (Figure 16-28). The unit should be electrically disconnected for about 10 min before testing. *Note:* Avoid breathing fumes from the leak detector flame.
5. Check all lines, freezer, charging ports, etc., with a leak detector (Figure 16-29).
6. If a leak is found at one of the above-mentioned points, discharge the system, repair the leak, evacuate the system, and recharge with refrigerant using a new filter-drier.

FIGURE 16-28 Adding refrigerant for a leak test. (Courtesy of Frigidaire Division, General Motors Corp.)

Caution: On refrigerators and freezers with foam insulation, do not puncture the foam with a detector prod or remove plastic shelf pegs from areas to be leak checked. This releases trapped particles of refrigerant from the foam, which will falsely indicate a leak.

Evacuation Procedure. Evacuation is necessary before recharging the system in order to remove all traces of air and the previous refrigerant charge. The evaporator should be at room temperature, if possible, before evacuation. If not, on frost-free models, turn the defrost timer to defrost. When the defrost cycle is completed, immediately unplug the electrical cord. On conventional models, place pans of hot water in the freezer section with the electrical cord unplugged to raise the refrigerant pressure.

On units with service ports on both sides, the evacuation can be accomplished through both ports of the gauge manifold (Figure 16-30). This requires that the vacuum pump be disconnected before recharging can take place. This is not difficult, and refrigerant will not get into the system if precautions

DOMESTIC REFRIGERATION

FIGURE 16-29 Leak testing using a halide leak detector. (Courtesy of Frigidaire Division, General Motors Corp.)

FIGURE 16-30 Evacuation through both gauges. (Courtesy of Frigidaire Division, General Motors Corp.)

such as purging the charging line from the charging cylinder to the gauge set are followed. There, of course, must be no leaks on the gauge set or in any of the hose connections.

A preferred method is to evacuate the charging cylinder while evacuating the system (Figure 16-31). Notice the way the charging lines are connected. In this fashion, the system can be evacuated and recharged without disconnecting any lines.

Be sure that the evaporator is warmed before evacuating so the system can be purged of refrigerant by attaching a line from the charging port to the outdoors. Adequate ventilation should be provided so that accurate leak tests can be conducted later in the procedure. After the system is completely discharged, assemble the equipment as outlined above and as shown in Figure 16-31. Open both gauge hand valves and the charging ports. Start the vacuum pump by following the instructions that accompany the pump. Be sure that the hand valve at the base of the charging cylinder is closed during evacuation. Evacuate the system for approximately 5 min, shut off the valve at the vacuum pump, and unplug the pump. Observe the manifold gauges for a few minutes. If the gauge readings rise, there is a leak

in the system or line hookup. Recheck and correct as necessary. When this check is completed, continue evacuating for at least 20 min at 28 in. (13.74 kPa) of vacuum or lower.

Recharging Procedure. The secret to recharging is good evacuation. The compressor should not be at a higher temperature than the ambient as would be the case when it has been running for hours prior to the service call. On some compressors, the refrigerant enters through the high side of the system and vaporizes more quickly if the compressor is hot, slowing down entry of the liquid refrigerant into the system. The pressure in the charging cylinder may be increased by applying hot towels or warm air to the charging cylinder. Some charging cylinders have their own heat source. Never apply an open flame to a charging cylinder. When the refrigerant charge is measured from the charging cylinder, it is in the liquid state. Liquid refrigerant should never be charged through the low side because this tends to *scrub out* the oil from the compressor. Be safe and charge through the high side. If the occasion requires

FIGURE 16–31 Preferred method for evacuation. (Courtesy of Frigidaire Division, General Motors Corp.)

charging equipment, thoroughly check for leaks, and let the unit run. Recheck the unit the next day to determine if it is performing normally.

☐ Frigidaire Refreshment Center Refrigerator ☐

The refreshment center refrigerator (Figure 16–32) features a new method of automatically dispensing ice cubes, two beverages, and chilled water. But the refrigeration system is basically the same as that discussed before, and a discussion of that will be omitted here.

Cube Dispensing. When the user desires ice cubes, an actuator button labeled ICE CUBES is pushed. This raises the door of the cube chute. As the door

that refrigerant be added through the low side, be sure the refrigerant is in the vapor state and add it slowly while the compressor is running. This is also advisable for clearing the refrigerant from the charging hookup in order to reduce the pressure to that of the low side. Whenever the refrigerant charge is questionable, discharge and evacuate the system and recharge with the correct amount of refrigerant as shown on the model and serial number plate.

It should be noted that a reasonable amount of time should be allowed after recharging the unit for it to begin functioning properly. Since the refrigerant is charged into the high side of the system, it must flow through the small capillary tube to the evaporator. If the evaporator is warm, the refrigerant changes to a vapor instantly due to the absorption of heat. The evaporator must be reduced in temperature considerably before the system pressures begin to balance out.

If a service technician is expecting immediate cooling results or normal operating pressures on the gauges, he or she could condemn a component as being faulty when it is not. This leads to unqualified component replacements. The time required to obtain normal operating temperatures depends on many factors. The best method is to remove the

FIGURE 16–32 Refreshment center refrigerator. (Courtesy of Frigidaire Division, General Motors Corp.)

FIGURE 16-33 Ice, juice, and water-dispensing system—exploded view. (Courtesy of Frigidaire Division, General Motors Corp.)

opens to its full-open position, it engages an arm that closes the contacts of a microswitch located in the divider assembly between the fresh food and the freezer compartments (Figure 16–33). When the microswitch contacts close, a drive motor located behind the cube container in the freezer compartment is energized.

The motor is rated as a two-pole, shaded-pole type, $\frac{1}{12}$ hp, 115 V, 60 Hz, and rotates counterclockwise as viewed from the front or lead end. It is permanently lubricated and rotates at approximately 3000 rpm. It is reduced to a 30:1 ratio by a gear reduction system that drives a conveyor located in the bottom of the ice cube container (Figure 16–34). This conveyor pushes the cubes forward through an opening at the front of the container, which has a spring-loaded trap door that leads into the cube chute. The cube chute, located in the divider, leads into the food compartment door and dispenses into the user's glass. A bellows and valve arrangement is used in conjunction with the cube actuator to slow the closing action of the cube door. This prevents cubes

FIGURE 16-34   Ice service compartment. (Courtesy of Frigidaire Division, General Motors Corp.)

from being lodged in the chute when the user releases the cube actuator.

Chilled Water Dispensing. When the user depresses the WATER dispenser pushbutton, a spring is raised off of the valve seat, taking the pressure off the diaphragm. This is done so that full water pressure is not placed on the dispenser valve. Full water pressure could result in splashing in the user's glass. When the push rod is depressed to its fullest extent, a microswitch closes. When this switch closes, the electric circuit to the black solenoid of the water valve is completed, causing the water to flow by pressurizing the system. Water from the supply line flows through the water valve at a controlled rate of 334 to 366 cm³ per 10 sec to the bottom (inlet) of the storage tank, which is mounted to the rear wall of the food compartment liner. The tank is designed as a series flow but has crossovers at the ends to prevent air from being trapped within the tank. The air is forced to the top of the tank and is purged when the customer pushes the dispenser button. As a matter of fact, it is recommended that this be done by the person installing the water line to the refrigerator. The tank holds ¾ gal (2.84 liters) of water. As the WATER pushbutton is depressed and all the air is purged, water flows from the top of the storage tank through a plastic tube that connects the tank to the dispenser in the door and into the user's glass. The temperature of the chilled water will be between 40 and 48 °F (4.4 and 8.9 °C), depending on the quantity of water dispensed at one time and the temperature in the food compartment.

Beverage Dispensing. The flow of water through the system to the dispenser valve is exactly the same as it is when the user pushes the WATER button. However, when one of the BEVERAGE buttons is depressed, water leaving the dispenser valve passes through the *venturi tube* (Figure 16-35). The stream of water passing through the venturi tube creates a negative pressure in the storage container and the beverage concentrate is drawn from the storage container. The beverage concentrate and water mix in the dispenser outlet tube. The amount of water added to the concentrate is controlled by the user through a manually operated richness control valve. When this valve is set on RICH, no additional water is added to the mixture. As the valve is turned to MILD, additional water is added to the mixture. After a few days of trial period, users will determine the consistency of the beverage they desire.

Ice Maker Operation. When the water line is connected to the refrigerator and the water shutoff valve is turned on, the ice maker will begin after the unit has reached zero zone operating temperatures. At the proper temperature, the ice maker will call for a fill. The green solenoid of the two-part water valve is energized, which causes water to flow to the ice maker. The fill portion of the commutator gear allows the water to flow for 12 sec into the cubelet tray. The flow washer in this portion of the water valve meters the water flow through an orifice so that 105 to 130 cm³ of water is dispensed in this 12-sec fill time.

The polypropylene party ice tray consists of 3 rows of 7 cavities each, for a total of 21 ice cubes per harvest. The cubes are square on top and rounded on the bottom to make automatic ejection easier. It should be noted that air must be in circulation and at the desired temperatures over this tray for proper operation. Too often the ice maker is blamed for a problem when improper airflow and temperature are the cause. The best way to determine this is by taking freezer product temperatures, which should be between 0 and 5 °F (−17.8 and −15 °C) before any other analysis is attempted.

FIGURE 16-35 Beverage dispensing valve. (Courtesy of Frigidaire Division, General Motors Corp.)

Venturi Tube

Mixture

Water

Concentrate

The bulb of the ice maker thermostat senses the front, middle cube cavities and when it is satisfied—13 °F (–10.6 °C)—the ice maker mechanism starts the harvest cycle providing, of course, that the cube container is in place. The ice maker mechanism rotates the front of the cube tray to the right (clockwise) a few degrees, creating a gradual twist to break the cube loose from the molds. It returns to the horizontal position, then rotates to the left (counterclockwise). The cubes begin to fall out of the tray cavities into the container. As the tray reaches a vertical position, it engages a restrainer spring at the rear of the ice maker frame, which causes the tray to get an additional twist as the motor continues to drive the gear and slide rack that turns the tray. The twist of the tray eventually overcomes the tension of the restrainer spring and allows the tray to snap off the spring and hit the stop at the rear of the ice maker frame. This sudden jar causes any cubes still remaining in the tray to be released from the tray cavities. After this action, the tray starts to rotate back again until it reaches the horizontal position. At this time the tray is filled with water again. The overall travel time for the complete harvest cycle is approximately 3 min.

This ice maker model has no weight switch or adjustable cube level control because of the automatic dispensing feature. It does have a heavy-gauge metal wire sensing arm that moves up and down with each harvest cycle. As the rack moves, the end of it engages a gauge arm and switch shaft to which the sensing arm is attached. When the sensing arm comes back down, it senses the level of the cubes in the container. If the level of the cubes is high enough to raise the sensing arm, an actuator opens a microswitch, called the gauge switch, and cube production stops.

In addition, the automatic shutoff lever works in conjunction with the container and ice-sensing arm. If the container is removed, the automatic shutoff lever raises the ice-sensing arm, which in turn opens the gauge switch, shutting off the ice maker. The container itself holds approximately 500 cubes, but automatic operation allows somewhat less.

Freeze Cycle. Figure 16-36 shows the complete cycle diagram as well as the electrical diagram. The numbers to the left of the sequence indicate the cycle time in seconds. This cycle of operation can be followed from this diagram, which is located on the inside of the ice maker cover.

Harvest Cycle

00″ Start (#3 contact closes 16°F)

15″ Holding switch closes. Tray twists CW and untwists. Tray rotates CCW and twists against impact stop. Tray then rotates CW to untwist and return to level

145″ fill switch closes. 12 second fill time.

160″ delay switch open-motor stops till #1 contact closes. Circuit is made through stop switch and motor runs 5 seconds till stop switch opens to

172″ complete the cycle.

If ice is at max. level, gage arm will not trip switch and production stops until ice is removed. Ice maker will be reset automatic (no water in tray or thermostat not warmed up) after the refrigerator defrost cycle.

FIGURE 16–36 Ice maker cycle diagram. (Courtesy of Frigidaire Division, General Motors Corp.)

In Figure 16–37 the water is freezing into ice. This cycle requires most of the time in the cube-making process. The tray is in the horizontal position and is full of water. The thermostat has warmed up above 16°F (-8.9°C). The gauge arm is down but not touching ice because the container is not full. The container is in place, however.

The electric power comes in through the black and white leads at the ice maker terminals marked D and E, respectively. The blue lead from the water valve solenoid is attached to terminal F. The ice maker motor is not running simply because the ice maker thermostat internal contacts are in the warm position (1 and 2).

Start of Harvest Cycle. In Figure 16–38 the harvest cycle is started because the ice maker thermostat has reached its cut-out setting of 13°F (-10.6°C), causing the internal contacts to flip to positions 2 and 3. This completes the circuit from the black power lead at terminal D to the commutator, to the delay contact through the gauge switch, to the ice maker thermostat, through contacts 3 and 2 to the hold contact and the drive motor. The other side of the drive motor coil is connected to the common side of the electric power line, white. This causes the drive motor to run. As the commutator and gear advance, the hold contact comes into play by contacting the commutator. This gives direct contact to the drive motor from the commutator and bypasses the ice maker thermostat and

Freeze Cycle

1. Tray horizontal and filled with water.
2. Thermostat warm (closed 2–1).
3. Gauge arm down, not touching ice.

FIGURE 16–37 Freeze cycle. (Courtesy of Frigidaire Division, General Motors Corp.)

gauge switch. It is at this point that the tray begins its first twist (clockwise) and now cannot be interrupted by the gauge switch or ice maker thermostat.

Start Harvest Cycle

1. Thermostat contacts 2 to 3 close.
2. Tray entering pretwist position.
3. Circuit hold contact has engaged commutator.
4. Gauge arm has not yet started to rise.

FIGURE 16-38   Ice maker start of harvest cycle diagram. (Courtesy of Frigidaire Division, General Motors Corp.)

Harvest Cycle

1. Thermostat contacts 2 to 3 still closed.
2. Gauge arm has moved out of container.
3. Tray in final twist and dispensing ice.
4. Gauge switch to off position.

FIGURE 16-39   Harvest cycle continues. (Courtesy of Frigidaire Division, General Motors Corp.)

The harvest cycle continues as the drive motor turns the commutator and gear assembly. The thermostat cold contacts, across 2 and 3, are still closed although the hold contact is making on the commutator, which bypasses the thermostat and gauge switch. This is why the drive motor continues to run even though the ice-sensing arm (gauge arm in Figure 16-39) has started to rise out of the container, breaking the microswitch (gauge switch) contacts. The tray is in its final twist.

Fill Cycle. The cube tray has returned to the horizontal position, and the tray is filling with water because the commutator and gear have advanced to the portion of the commutator that energizes the fill contact and then to the water valve solenoid through the blue electric lead (Figure 16-40). The incoming water will raise the temperature—above $16°F$ $(-8.9°C)$—of the ice maker thermostat, causing the internal warm contacts 1 and 2 to be made. The ice-sensing arm enters the container but does not touch ice, so the gauge switch returns to ON—ready for the next harvest, which will occur when the water is frozen and the ice maker thermostat is satisfied at $13°F$ $(-10.6°C)$.

Full Container. When the level of the ice cubes in the container reaches a point that will cause the

Fill Cycle

1. Tray horizontal and filling with water.
2. Thermostat contact 2-3 open contacts 2-1 close.
3. Gauge arm enters container. Does not reach ice.
4. Gauge switch returns to on for next cycle.

FIGURE 16-40   Fill cycle. (Courtesy of Frigidaire Division, General Motors Corp.)

ice-sensing arm to rise, it will open the gauge switch contacts (Figure 16–41). Even though the ice in the tray is ready to be harvested, there is no electric circuit to the drive motor—thus no harvest. The ice remains in the tray until such time as the level of the ice is reduced to the point that the ice-sensing arm will complete the circuit.

Container Removed. When the container is removed for any reason, the automatic shutoff lever causes the ice-level sensing arm to move up out of the container. This causes the gauge switch contacts to open, preventing a harvest (Figure 16–42). When the container is replaced in its normal position, the system returns to a normal operating condition. As stated previously, removal of the container should be discouraged except for general cleaning purposes.

Delay Feature. The *delay feature* puts the ice maker mechanism into a hold position when there is a possibility that water might be harvested before it is frozen into cubes. A raised portion designed into the plastic drive gear protrudes through the metal commutator. This protrusion causes a momentary electrical break, sometimes referred to as a *dead spot,* that interrupts electrical contact to the delay contact or to the on contact of the gauge switch. If for any reason the ice maker thermostat fails to

Container Removed

1. Remove container.
2. Gage arm moves out of container.
3. Gage switch moves to off preventing ice harvest.
4. Replace container and system returns to a normal operating

FIGURE 16–42 Ice maker container removed. (Courtesy of Frigidaire Division, General Motors Corp.)

Full Container

1. Tray of ice is ready for harvest.
2. Ice container is full.
3. Gage arm contacting ice and will not let switch complete circuit.
4. Normal harvest will occur when level of ice is reduced to point gage arm will complete circuit.

FIGURE 16–41 Full container. (Courtesy of Frigidaire Division, General Motors Corp.)

Delay Feature

1. Delay feature prevents continuous operation of ice maker if thermostat does not reset.
2. Dead spot in commutator prevents completion of cycle and a water harvest.
3. Similar situation could occur if no water enters the tray.
4. Unless thermostat or bellows heater is faulty, reset of the cycle will automatically occur after defrost.

FIGURE 16–43 Delay feature. (Courtesy of Frigidaire Division, General Motors Corp.)

DETAIL "A"
THERMOSTAT BULB MUST BE
INSTALLED IN THIS OPENING -
NOT IN THE SPACE BELOW.

A-7761

FIGURE 16-44 **Ice maker detail part—exploded view. [Courtesy of Frigidaire Division, General Motors Corp.]**

warm up after the tray is filled and if the internal contacts do not flip back to 1 and 2, it is possible to harvest the water that just filled the tray (Figure 16–43). Should the commutator and gear assembly stop in this delay position, a reset of the cycle will automatically occur after the next defrost period.

Package temperatures do not rise appreciably during a defrost cycle, but the air temperature over the cooling coil does warm for a short time period.

This air should be warm enough to raise the temperature of the ice maker thermostat the few degrees necessary to reset the thermostat, especially if no water is in the tray. If this fails to reset the thermostat, check the electric hookup or maybe replace the thermostat to correct the condition. If you get no fill then, it is not the thermostat. Check the fill system. See Figure 16–44 for an exploded view of the ice maker.

⬜ Domestic Electric Circuits ⬜

When troubleshooting a domestic refrigeration unit, it is necessary for the service technician to be able to read properly the wiring diagrams that accompany the unit. These wiring diagrams range from very simple to very complex. However, if one has the ability to read these diagrams, he or she will realize that the more complex diagrams only have more components. Also, a basic knowledge of unit operation is very helpful.

There are basically two types of wiring diagrams in use today. They are the *pictorial* and the *schematic* diagrams. Both are very useful. The pictorial is helpful in locating components on the unit. The schematic is helpful in determining the unit's operating sequence.

Simple Freezer Diagram. When reading a wiring diagram, regardless of how simple or complex, it is less confusing to trace each circuit all the way through before starting a new one. For the following explanation, see the schematic diagram in Figure 16–45.

Beginning with the side of the electric power line labeled HOT, the first circuit that is approached is the light circuit. The light switch is shown to be a normally closed switch. The electric power progresses through the wire extending to the freezer light and to the other side of the electric power line. Therefore, any time the light switch is closed, the light will be on.

When we continue on the hot line, the next circuit approached contains only a door heater. There are no switches or other devices in this circuit. Therefore, any time the unit's electric cord is plugged in, the door heater will provide heat.

The next circuit approached is the compressor circuit. The first component encountered is the cold control (thermostat) followed by the overload protector, which is in the common line to the compressor. The electric power then goes through the compressor start winding, through the start contacts on the starting relay, which are normally open, and to the other side of the line. To complete this circuit, the electric power, after leaving the common terminal of the compressor, proceeds through the compressor run winding to the starting relay coil and to the other side of the line.

Any time the cold control and overload protector are closed, the compressor will attempt to run.

Electric power is supplied to both compressor windings through the common terminal. The high current through the run winding will cause the contacts to close on the amperage starting relay and energize the start winding. As the compressor gains speed, the amperage decreases, allowing the starting relay contacts to open and the run winding to remain in the circuit. The unit is now in its normal operating cycle. When the cold control contacts open, the electric power to the compressor is interrupted and the unit stops.

It should be evident from the foregoing explanation that any circuit can be easily traced and the operating sequence determined by properly reading and diagnosing the wiring diagram.

Complex Refrigerator Wiring Diagram. Even though this type of diagram has more components, it can be read and the operating sequence can be determined in the same manner (Figure 16–46). Each of the components is shown in the normal operating position. Simply follow through each individual circuit completely before attempting another.

⬜ Safety Procedures ⬜

Domestic refrigeration is a very large part of the refrigeration industry. It is a must that proper safety procedures be followed in order to protect the service technician, the equipment, and the products.

1. Always get help when moving a domestic refrigerator or freezer.
2. Always lift with the legs, never with the back.
3. Do not remove frost and ice from an evaporator with sharp instruments such as knives, ice picks, etc.
4. Always be certain that the unit is properly grounded electrically.
5. Always practice good housekeeping. Keep oil and water off the floor.
6. Take proper precautions to prevent electric shock because of jewelry such as watches, rings, etc.
7. When charging or purging refrigerant always wear goggles.
8. Be sure that proper ventilation is provided when repairing leaks, purging refrigerant, etc.
9. Never use carbon tetrachloride for cleaning because continued use can be fatal to the user.

Wiring Pictorial

Schematic

FIGURE 16-45 **Simple freezer wiring diagram. (Courtesy of Frigid-aire Division, General Motors Corp.)**

FIGURE 16-46 **Refreshment center refrigerator wiring diagram.**
(Courtesy of Frigidaire Division, General Motors Corp.)

10. Always remove the doors from discarded refrigerators and freezers to prevent children from possibly suffocating during play.

◻ SUMMARY ◻

◻ The areas comprising domestic refrigeration are: (1) refrigerators, (2) freezers, and (3) room air conditioners.

◻ Fiberglass and foamed plastic insulation are the types of insulation used in modern refrigerators and freezers.

◻ The density of foam insulation ranges from 1.8 to 2.2 lb (0.029 to 0.035 g/cm^2) per cubic foot.

◻ A vapor barrier is used to keep moisture out of the insulated space between the shell and the liner.

◻ Foamed plastic has an inherent vapor barrier.

◻ Magnetic door seals are used today because they provide a better seal and the door is easily opened from either side.

◻ Breaker strips are used to reduce the flow of heat into the cabinet by conduction.

◻ Breaker strips should be warmed with a warm, wet cloth to prevent breaking before attempting to remove them.

◻ A mullion heater is a strand of high-resistance wire attached to a strip of aluminum foil.

◻ Mullion heaters are installed around cabinet doors beneath the breaker strip.

◻ Mullion heaters prevent condensation of moisture around doors.

◻ The shelves in a domestic unit must be of the open grill type to allow for the circulation of air inside the cabinet.

◻ Avoid locating a domestic unit next to a stove, radiator, or hot-air register.

◻ Domestic units should be installed on a separate electric circuit.

◻ The temperature inside a domestic unit will not start to drop until after about 30 min of operation.

◻ The refrigeration system is divided into two parts—the high-pressure and the low-pressure sides.

◻ The evaporator is sometimes called the cooling coil or freezer.

◻ The evaporator is where the action is in a refrigeration system.

◻ The compressor and motor are a hermetically sealed unit.

◻ The function of the compressor is to compress the low-pressure vapor into a high-pressure vapor.

◻ The temperature of the vapor must be raised above the ambient temperature so that it can give up the heat absorbed in the evaporator.

◻ The function of the condenser is to transfer the heat absorbed by the refrigerant to the ambient air.

◻ It is very important that an adequate air flow be maintained over the condenser.

◻ The filter-drier removes foreign matter that might restrict the flow of refrigerant through the system, and it removes moisture that might freeze in the capillary tube causing a restriction.

◻ The practice of using alcohol in hermetic systems is not recommended.

◻ The capillary tube meters the flow of liquid refrigerant from the condenser to the evaporator.

◻ Poor circulation of air inside a domestic unit will cause higher air temperature and spoilage of food.

◻ An air deflector is used to divert the cold air directly over the ice service area.

◻ Side-by-side frost-free refrigerators use a single-coil evaporator mounted in a vertical position in the freezer compartment to cool both the freezer and fresh food compartments.

◻ The air that cools the fresh food section in refrigerator–freezer combination units is controlled by a damper.

◻ Most defrost systems use electric defrost heaters.

◻ The defrost timer puts the unit into defrost at the prescribed time.

◻ The duration of the defrost limiter normally determines the duration of the defrost cycle.

◻ Defrost water is disposed of by draining into a disposal pan located in the machine compartment where the water evaporates.

◻ The running time of the unit is determined by the thermostat switch bulb located at the top of the fresh food compartment and by the volume of cold air that is provided from the freezer compartment.

◻ When a lower freezer compartment temperature is required, the manual damper is closed a little by turning the freezer control to a colder setting, restricting the volume of cold, dry air that is provided to the food compartment.

◻ In servicing domestic units the malfunction, if any, is often found to be in the controls, not in the refrigeration unit.

◻ A lack of proper airflow is sometimes caused by the fan motor not running at the proper speed.

◻ Many domestic unit performance complaints require only customer education or reassurance that the unit is operating normally.

◻ The temperature in the freezer section is 0 to 5 °F (−17.8 to −15 °C) in frost-free models and from 10 to 12 °F (−12.2 to −11.1 °C) in other models, depending on the control setting.

◻ The temperature in the fresh food section should be between 32 and 40 °F (0 to 4.4 °C).

◻ Plugging the unit line cord in and taking it out several times may cause the compressor overload to trip.

- A test cord can be used to determine whether or not a compressor is bad.
- Wattages should be used to determine if the compressor is stuck, shorted, grounded, has open windings, has a tight rotor, or other internal malfunctions.
- Wattages alone should not be used to diagnose a shortage of refrigerant or an inefficient compressor.
- A faulty capacitor should be suspected if the compressor will not run, if there is a hum or buzz at times, or if it tries to start and then kicks off.
- Defective capacitors should be replaced on a like-for-like basis.
- Refrigeration gauges should be installed only after all other possibilities have been eliminated and it has been concluded that the fault is in the refrigeration system.
- Approximately 90% of the entries to a sealed system are for repairs on the high side.
- As a general guide, the suction pressure variation has an effect on the head pressure.
- Generally speaking, domestic units cannot operate efficiently in ambient temperatures lower than 60 °F (15.6 °C).
- A domestic unit is designed so that there is enough refrigerant for adequate heat removal in a 110 °F (43 °C) ambient temperature.
- The first thing noticed when a shortage of refrigerant occurs is a rise in the temperature of the foods.
- Under some circumstances, such as in the case of the conventional refrigerator or freezer with a slight shortage of refrigerant, freezing in the food compartment may be experienced due to the additional running time.
- Refrigeration occurs on the low-pressure side of a partial restriction. An experienced service technician will physically feel the refrigeration lines when he or she suspects a restriction.
- A restriction reduces the flow rate of the refrigerant and consequently reduces the heat removal.
- Since air and R-12 do not mix, the air pressure would be added to the normal head pressure, resulting in a higher than normal head pressure.
- Before changing a compressor, the service technician should seal off the suction line and pump a vacuum with the compressor to see if it will pull a 26- to 28-in. (27.48- to 13.74-kPa) vacuum.
- An insufficient amount of oil in a compressor can give an indication of an inefficient compressor because the oil coating between the parts results in better compression.
- Any time a system is low on refrigerant, the leak must be found and repaired.
- Liquid-type leak detectors are the only type that can be used successfully around urethane insulation.
- Evacuation is necessary before recharging the system in order to remove all traces of air and the previous refrigerant charge.
- The secret to recharging a system with refrigerant is good evacuation.
- Liquid refrigerant should be charged into the system through the high side to prevent washing oil out of the compressor.
- Before an ice maker can operate, the freezer must reach the zero zone operating temperature.
- Two types of electrical wiring diagrams in use today are pictorial and schematic.
- When reading a wiring diagram, it is less confusing to trace each circuit all the way through before starting a new one.

▣ EXERCISES ▣

1. What is insulation?
2. What is the density of foam insulation?
3. Is a vapor barrier necessary when polyurethane insulation is used?
4. What type of door seal is used on modern refrigerators and freezers?
5. Where is the mullion heater installed on domestic units?
6. What is the purpose of breaker strips?
7. What is the purpose of a mullion heater?
8. Why must the shelving in a domestic unit be the open grill type?
9. In what ambient temperatures are domestic units designed to operate?
10. What does a slight gurgling or bubbling sound when the door is opened indicate?
11. How much time will usually lapse after plugging in a domestic unit before a drop in temperature will be noticed?
12. What is refrigeration?
13. Where is the action in a refrigeration system?
14. What is the function of a compressor in a refrigeration system?
15. Why must the temperature of the vapor be increased?
16. What is the function of the condenser?
17. What is the purpose of the filter-drier?
18. Is it recommended to use alcohol in a refrigeration system? Why?
19. What is the purpose of the capillary tube?
20. What device is used to divert the air over the ice service area?
21. What is the purpose of the defrost system?
22. What does a defrost limiter do?
23. How is the defrost water disposed of?
24. What component limits the running time of the refrigeration unit?

25. In what components will a malfunction be most likely to occur in a domestic unit?
26. What should be the temperature range of a fresh food compartment?
27. On what voltage are most refrigerator compressors designed to operate?
28. Will the ambient temperature affect the wattage of a compressor?
29. What can be used to test a compressor to see if it or some external components are bad?
30. Why should electrical power not be applied to the start winding for long periods of time?
31. What conditions would cause a capacitor to be suspected of being faulty?
32. When should the gauges be installed on a refrigeration system?
33. What is the most common cause of seal system entry?
34. What will cause a high condensing temperature?
35. What are the symptoms of a system overcharge?
36. What is indicated by a hissing sound in the evaporator using a capillary tube?
37. Will low running wattage always be an indication of a low charge?
38. How can a restriction be located?
39. What would indicate a mixture of air and R-12?
40. What should be the vacuum reading of a good compressor?
41. Will insufficient oil indicate an inefficient compressor?
42. Why should foam insulation not be punctured when leak testing a system?
43. Why is evacuation of a system necessary?
44. What is the secret to recharging a system?
45. When will an ice maker begin operation?
46. What should be done when attempting to read an electrical diagram?

17 Room Air Conditioners and Dehumidifiers

Performance Objectives

After completion of this chapter, the student should:
- Be able to demonstrate an understanding of the basic function of room air conditioners and dehumidifiers.
- Know how the capacities of room air conditioners are measured and rated.
- Be able to describe the purpose of dehumidifiers.
- Be able to list the two basic systems in a room air conditioner.
- Complete the Exercises with at least 80 percent accuracy.
- Complete the Laboratory Exercises with at least 80 percent accuracy.

In this chapter we discuss the design and operation of window air conditioning units and dehumidifiers. These are vital sections of the air conditioning and refrigeration industry and should be understood by installation and service technicians.

☐ Room Air Conditioners ☐

A room air conditioner is an assembly designed as a unit primarily for mounting in a window or through a wall. Its purpose is to provide cool, or warm, conditioned air to the room. Some are designed to be used with ductwork, while others are not to be connected to a ductwork system. Each of these units is equipped with a refrigeration system. They provide dehumidification, a means for cir-culating and cleaning the air, and in some units a means for ventilating and/or exhausting the air.

The basic function of a room air conditioner is to provide comfort by cooling, dehumidifying, filtering or cleaning, and circulating the room air. Ventilation may also be provided by introducing outdoor air into the room and/or exhausting room air to the outside. Comfort may also be provided by controlling the temperature of the room through the use of a thermostat. Room units may provide heat by use of resistance elements, by use of a heat pump cycle, or by a combination of both these methods.

In operation, warm air passes over the cooling coil and gives up heat to the refrigerant inside the coil. This conditioned air is then circulated inside the room by either a fan or a blower (Figure 17–1).

The heat given up by the room air to the coil evaporates the liquid refrigerant inside the tubes. The evaporated vapor then carries the heat to the com-

FIGURE 17–1 Schematic view of a typical room air conditioner. (Courtesy of ASHRAE.)

pressor, which compresses the vapor and increases its temperature to a point higher than the temperature of the outdoor air. In the condenser, the hot high-pressure vapor is condensed, giving up the heat from the room air to the outdoor air. The high-pressure liquid refrigerant then passes through the flow control device, which causes a reduction in its pressure and temperature. This cool liquid refrigerant then enters the evaporator coil and the cycle is repeated.

☐ Sizes and Classifications ☐

Both the heating and cooling capacities of room units are measured and rated in terms of Btu/hr. There is a wide range of sizes available from approximately 4000 to 36,000 Btu/hr.

These units are equipped with electrical cords that may be plugged into standard or special electrical outlets. The majority of room units are designed to operate on 120-, 240-, or 208-V single-phase current. The maximum amperage of the 120-V models is generally limited to 12 A because this is the maximum allowable current for a single-outlet, 15-A circuit permitted by the National Electrical Code®.

Heat pump models are also available and are designed to operate on 208- or 240-V applications. Heat pump room units are generally designed for reverse-cycle operation when heating is required. However, some units use electric resistance heating elements for either supplementing the heat pump cycle or for providing the total heating capacity at temperatures below a given point. They are also available with these elements in a regular cooling unit to provide the total heating requirements.

☐ Room Unit Compressors ☐

There is a range of compressor capacities available from approximately 4000 to 48,000 Btu to be used in room air conditioning units. The design data are available from the various compressor manufacturers with rating conditions for both standard and high-efficiency compressors (Table 17–1). The table indicates that a standard rated room air conditioner compressor has an evaporating temperature of 45°F and a high-efficiency compressor has an evaporating temperature of 49°F. The same type of comparative

TABLE 17–1 Ratings for Standard and High-Efficiency Room Unit Compressors

	Standard	High-Efficiency
Evaporating temperature [°F [°C]]	45 [7.2]	49 [9.4]
Evaporator suction temperature [°F [°C]]	95 [35.0]	51 [10.6]
Compressor suction temperature [°F [°C]]	95 [35.0]	65 [18.3]
Condensing temperature [°F [°C]]	130 [54.4]	120 [48.9]
Liquid temperature [°F [°C]]	115 [46.1]	100 [37.8]
Ambient temperature [°F [°C]]	95 [35.0]	95 [35.0]

Source: ASHRAE.

information can be found from this table concerning room unit operation.

Manufacturers of these compressors offer complete performance curves at the various evaporating and condensing temperatures to aid in selecting the proper compressor for the application being considered.

☐ Evaporator and Condenser Coils ☐

Coils used for these purposes are generally of the tube-and-plate-fin variety or of the tube-and-spine-fin variety. Performance information on these types of coils is generally available from the manufacturers and suppliers. The design characteristics to be considered when selecting these coils are Btu/hr/ft², dry bulb and moisture content of the entering air, air-side friction loss, internal refrigerant pressure drop, coil surface temperature, air volume, and air velocity.

☐ Flow Control Application and Sizing ☐

Basically there are three types of flow control devices used on room air conditioning units: (1) the thermostatic expansion valve, which maintains a constant superheat from a point near the evaporator outlet to a point on the suction line; (2) the automatic expansion valve, which maintains a constant suction pressure; and (3) the capillary tube, also known as a restrictor tube. This is the most popular type of flow control used on window air conditioning units.

It has a low cost and a high reliability factor, even though it is not the most desirable refrigerant control over a wide range of ambient temperatures.

□ Fan-Motor and Air-Mover Selections □

There are two types of fan motors generally used on room air conditioners: (1) the shaded pole, which is a low-efficiency motor, and (2) the permanent split capacitor motor, which has a higher efficiency rating and requires the use of a run capacitor. The air movers, or blowers, are of two different types: (1) the forward-curved blower wheel, and (2) the axial, or radial, flow fan blade. In most applications the blower wheel is used to move small to moderate volumes of air through a high-resistance system, while the fan blades are used to move moderate to high volumes of air in a low-resistance system.

The applied combination of the motor and the air mover is such an important part of the system operation that manufacturers pay particular attention in their selection. The service engineer should always follow the manufacturers' recommendations when replacing either of these components.

□ Special Design Features □

The installation features for room air conditioners vary widely because they can be mounted in a variety of ways. The proper mounting for the given installation should be selected so that the user and the local codes and ordinances can be satisfied. The more common mounting methods are:

1. *Inside flush mounting.* The inside face of the unit is approximately flush with the inside wall of the building.
2. *Balance mounting.* The unit is installed approximately half inside and half outside the building.
3. *Outside flush mounting.* The outer face of the unit is either flush or slightly beyond the outside wall of the building.
4. *Special mounting.* In casement windows, horizontal windows, office windows with swing units (or swinging windows) to permit window washing, and transoms over doorways.
5. *Through-the-wall mounts or sleeves.* These devices are used for installing the window unit chassis, the complete unit, or consoles in the walls

of an apartment building, motel, hotel, and residences.

Over the years, these units have become more compact because of consumer demand for minimum loss of window light and minimum projection both inside and outside the wall. This requirement has resulted in the design of a mount that is simple to apply and has an attractive appearance. There are several types of expandable amounts now available for fast, dependable window unit installation in both single- and double-hung windows, and for windows of the horizontal sliding type.

The standard installation kit includes all the parts needed for structural mounting such as gaskets, panels, and seals for weathertight installation. The instructions and procedures provide for easy installation and should be followed carefully to assure a safe and satisfactory completion.

The NEC requires that adequate wiring and the proper-sized fuses be used for the service outlet. The necessary information is generally provided on the instruction sheets shipped with the unit or they may be stamped on the unit near the service cord or in the serial plate data. It is important that the manufacturers' recommendations be followed for the size and type of fuse. All window air conditioners are provided with a grounding-type plug cap on the service cord when shipped from the manufacturer. Receptacles with the necessary grounding contact designed to fit the unit service cord plug cap should be used when installing the unit.

One type of room unit is the integral chassis design, with the outer cabinet permanently fastened to the chassis. Most of the electrical components are accessible by partly dismantling the control area without removing the unit from the installation.

A second type is the slide-out chassis design. This design permits the outer cabinet to remain in place while the chassis is removed for service.

There are two basic systems in any air conditioner: (1) the refrigeration system and (2) the air circulating system. The sealed refrigeration system is much the same as the sealed system used in refrigerators and freezers. The refrigerant generally used in room air conditioners is R-22. The capillary tube is usually much larger in diameter than those used on refrigerators and freezers in order to operate at the higher evaporator pressure and temperature (Figure 17–2). The same service procedures used for other refrigeration systems should be applied to these units.

FIGURE 17–2 **Room air conditioner refrigeration system. (Courtesy of Frigidaire Division, General Motors Corp.)**

FIGURE 17–3 **Types of circulating air.**

FIGURE 17–4 **Indoor air system.**

The air circulation system on room air conditioning units can be divided into two categories: indoor air and outdoor air (Figure 17–3). The indoor air is the air circulated in the space to be cooled. It is circulated over the evaporator by a centrifugal blower (Figure 17–4). The centrifugal blower has high air-moving capabilities with low noise.

The outdoor fan is usually a blade-type fan. These fans are usually referred to as condenser fans and are mounted outside the conditioned space (Figure 17–5). This fan moves outdoor air over the compressor and condenser. It also has a slinger ring around the outside periphery of the blades that picks up condensate from a condensate well and distributes it over the condenser. This moisture helps cool the condenser and increase the efficiency of the unit.

The fan motor on these units may be either single speed or multiple speed. When replacing a motor, one with the same speed must be used. If the rpm of the motor is changed, the air volume will also be changed. If the air volume is changed, the Btu rating will also be affected and the comfort in the conditioning space will be changed.

☐ Capacity ☐

Room air conditioners are rated according to the results obtained from a controlled test spelled out in AHAM (Association of Home Appliance

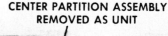

FIGURE 17–5 **Outdoor fan. (Courtesy of Frigidaire Division, General Motors Corp.)**

Manufacturers) Standard CNI. During this test the room air is controlled at 80°F (26.7°C) with a 50% relative humidity and the condenser air is 95°F (35°C) dry bulb and 38% relative humidity. The capacity of the unit will vary as these conditions vary.

Variations in ambient temperatures and relative humidity directly affect the cooling capacity and the temperature drop across the evaporator coil. Installation in restrictive or unfavorable operating locations, improper sizing of the unit for the area to be cooled, occupancy above normal, dirty filters, and improperly directed louvers are some conditions that can contribute to complaints of unsatisfactory cooling.

It is possible, however, to check the unit against established average capacity ratings, disregarding everything but the temperature of the air entering the condenser, the dry-bulb temperature drop across the evaporator, and the normal wattage draw of the unit.

Capacity Checking Procedures and Precautions. When making the capacity tests, be certain that the filter, the evaporator, and the condenser are clean and that the coil fins are not bent and obstructing the air flow. Also, be sure that the adjustable air directional louvers are directing the air up and to the right or left, not down. Air that is directed downward can circle back into the air conditioner and give a false reading.

It will be necessary to take the wet-bulb (WB) temperature, dry-bulb (DB) temperature, and the total wattage draw. The unit must be operated on maximum cooling for about 15 min before the readings are taken.

The wet- and dry-bulb temperatures, necessary for determining the relative humidity, may be determined by use of a *sling psychrometer.* The wattmeter should be connected in the circuit before the unit is started. Determine the dry-bulb temperature of the evaporator inlet and outlet airstreams.

The inlet return air temperature should be taken at the center of the return air opening. The discharge air temperatures should be taken in front of the air discharge grill. Since the temperature will vary slightly between the grills, make at least three readings across the grill area to obtain an average reading of the discharge air temperature.

The dry-bulb condenser inlet temperature can be taken with a pencil-type thermometer or with any accurate temperature tester. This temperature should be taken about ¼ to ½ in. (6.4 to 12.7 mm) away from the condenser air inlet louvers. The thermometer should not touch the louvers nor be exposed to the direct rays of the sun.

These readings can be compared to capacity check tables that can be obtained from the unit manufacturer or distributor. It is necessary to obtain a table for the specific model being checked.

☐ Thermostat Analysis ☐

Operating Range. The operating range of a thermostat is the difference between the cut-in temperature at the coldest thermostat setting and the cut-out temperature at the warmest thermostat setting. This difference usually represents approximately 20 to 21°F (11°C) temperature change in 180° rotational movement of the thermostat shaft, or a change equal to approximately 5°F (2.78°C) for every 45° of rotation.

Operating Differential. Determine the cut-in and cut-out temperature of the thermostat by sensing the return inlet air temperature with an accurate temperature tester. The temperature should be sensed in a manner similar to that described in checking the capacity of the room air conditioner. *Note:* The air conditioner must be tested in an environment where it has the capacity of cooling to the cut-out temperature of the thermostat setting.

Before and during the operating temperature check, be sure that the discharge air is not being directed or deflected directly back to the return air inlet by louvers, drapes, curtains, adjacent objects such as furniture, or by standing in front of the air conditioner.

For proper performance, it is important that the sensing bulb of the thermostat be positioned ¼ to ⅜ in. (6.4 to 9.5 mm) in front of the evaporator. If the thermostat capillary line or bulb is too close or touching the evaporator, it will sense the coil temperature directly, instead of the air temperature and cause short cycling of the compressor.

Short-Cycling Condition. Most short cycling is caused by the discharge air from the grill that hits a nearby object such as furniture, walls, or curtains and is deflected back into the air inlets. This cool air satisfies the thermostat and it shuts off the compressor before the room is really cooled. Warm air entering the inlet grill will soon cause the thermostat contacts to close and start the compressor. If the compressor motor is still hot or the unit pressures have not equalized, a hard starting condition may result, causing the motor overload protector to trip, a circuit breaker to trip, or the circuit fuse to blow.

If short cycling or premature compressor shut-off persists, it may be desirable to place a small piece of electrical tape on the back side of the clip to reduce the conductivity between it and the coil. It also may be desirable to place a small piece of sleeving over a portion of the bulb to reduce its sensitivity.

Outside air coming in the bottom of a poorly installed unit may strike the bulb and cause poor temperature control or short cycling. In a poorly sealed building or where the unit is mounted close to the floor, it may be desirable to move the bulb clips higher up on the coil so that it is better isolated from drafts.

Iced Coil Condition. Icing is usually caused by an insufficient airflow over the evaporator. This could be due to a dirt-blocked coil, dirty filter, curtains or drapes, nearby furniture, etc. Also, when the outside condensing air becomes quite cool, the air conditioner cooling capacity may increase to the point at which the evaporator coil will ice up. To guard against icing on some units, the clips holding the thermostat bulbs are configured so that they touch the evaporator. If the ice touches the clips, it will cool the bulb to the point where the thermostat will shut off the compressor, allowing the ice to melt.

Cool Ambient Temperature Operation. Occasionally an air conditioner will not run when the outside ambient temperature is less than 60 °F (15.6 °C) even though the room is warm. The metal in the air conditioner can become so cold that the liquid or vapor actuating medium in the thermostat power element condenses to the extent that its existing pressure cannot close the thermostat contacts. The volume of vapor or liquid in the power element bellows is so great as compared to the volume of vapor in the bulb that the thermostat operation will be controlled by the temperature of the thermostat body and not by the roomside thermobulb capillary line. Even heating the bulb will not expand the power element charge enough to operate the switch.

Note: If the compressor is operated at very low outside ambient temperatures to cool a room, it will probably overheat and possibly damage the compressor motor due to the lack of sufficient refrigerant passing through the capillary tube and evaporator to cool the motor windings.

Low Voltage. Low voltage is a common cause of trouble in the operation of a room air conditioning system. It therefore becomes doubly important, because of the motor size, that the service technician check the voltage when servicing room air conditioners.

Improper voltage may result in one or more of the following complaints:

1. The unit will not start.
2. Compressor motor cycling on the motor protector.
3. Premature failure of the motor protector caused by burned or pitted contact points.
4. Blown fuses.
5. Noticeable brightening of the lights when the compressor cycles off.
6. Evaporator icing. A low voltage may reduce the fan speed enough to give an inadequate airflow over the evaporator, thereby allowing it to ice up.

Low voltage can often be attributed to the use of extension cords or an inadequately wired circuit, but low voltage into the building and loose fuses or connections should not be overlooked.

High voltage. High voltage can be equally as troublesome by causing motors to overheat and cycle on their overload protectors or break down electrically. Extremely high voltages can create enough back voltage across the capacitor to cause it to fail. It will cause resistance heaters on heating models to deliver more than their rated capacity and open the safety protectors or fuse links.

Compressor. An inoperative compressor should be isolated electrically and checked before a decision is made to replace it. Winding resistance can be checked against the values on the schematic wiring diagrams. The resistances on some types of compressors will vary according to the stator manufacturer.

When testing the compressor, use a start cord and a capacitor of the proper rating. Connect the start cord as shown in Figure 17–6.

All PSC motors are designed to run with the capacitor in the circuit. Consequently, the start button must be held in or the compressor may stop when it is released.

FIGURE 17-6 Compressor test cord connections. (Courtesy of Frigidaire Division, General Motors Corp.)

FIGURE 17-7 Sleeve airflow. (Courtesy of Frigidaire Division, General Motors Corp.)

Fan and Blower Alignment. If the condenser fan and evaporator blower are not located in proper relation to the shroud orifice, reduction of airflow may result. This could give high condenser head pressures in hot weather or icing on the evaporator in cold weather.

The slinger ring on the condenser fan picks up water from the sump and "flings" it onto the condenser. This serves two purposes: (1) it disposes of the water that has condensed on the evaporator coil, and (2) it increases the capacity of the unit by removing heat from the condenser as the water evaporates. *Note:* Water should stand in the base pan; do not drain it.

Condenser Air Baffles. It is extremely important that the condenser air baffle be installed on the end of the condenser and that it be installed so that it will keep hot condenser air from recirculating (Figure 17-7). It is also important that the aluminum rear

grill be installed on the sleeve so that exhaust condenser air is not drawn back into the condenser air inlet grill.*

Dehumidifiers

The purpose of a dehumidifier is to remove moisture from the air as it is circulated through the unit. This reduced relative humidity helps in preventing rust, mildew, and rot on the surfaces inside the room or any other enclosed space where the dehumidifier is located.

The component parts of a dehumidifier are a motor-compressor unit, a refrigerant condenser, an air-circulating fan, a refrigerated surface (evaporator), some means of collecting and disposing of the condensed moisture, and a cabinet to house these various components (Figure 17-8).

In operation, the fan draws the moisture-laden air through the cooling coil where it is cooled below its dew point, causing the moisture to be removed. The moisture either drains off into a water receptacle or it may pass through a drain and into the sewer system. The cold air then passes through the hot condenser where it is reheated. The air is then heated further by other unit-radiated heat and is discharged into the room at a temperature somewhat higher and a lower relative humidity. The continuous circulation of the room air gradually reduces the humidity inside the room.

Design and Construction

Hermetic-type motor-compressors of the size that produce the rated output of the overall unit are used. The refrigerant condenser, in most cases, is a conventional finned-tube-type coil. The control of refrigerant flow is generally done with a capillary tube device. However, some units in the higher-capacity range may use an expansion valve for this purpose.

A direct-driven propeller-type fan is generally driven by a shaded-pole motor for creating the required flow of air through the unit. The airflow rate

*See Frigidaire, "Tech-Talk," Vol. 74, No. 8 for further reference.

HUMIDISTAT
MOTOR
CONDENSER (WARM)
EVAPORATOR (COLD)
HUMID AIR IN
DRY AIR OUT
COMPRESSOR
WATER RECEPTACLE

FIGURE 17–8 Typical dehumidifier unit. (Courtesy of ASHRAE.)

through these units is from 125 to 250 cfm, depending on the moisture removal capacity of the unit. Generally, the dehumidifier output is increased with an increase in the airflow. Very high rates of airflow may, however, create an objectionable noise. Residential humidifiers in most cases will maintain a satisfactory humidity level within the space when the airflow rate and unit placement permit the entire volume of air inside the space to be passed through the dehumidifier once per hour. For example, an air flow rate of 200 cfm would be sufficient to provide for one air change per hour in a room that has a volume of 12,000 ft³ (340 m³). The AHAM Dehumidification Selection Guide should be consulted for more detailed information.

The evaporator is generally of the bare-tube construction, although finned-tube coils are sometimes used if the fins are spaced to allow rapid runoff of the water droplets. Bare-tube coils mounted vertically tend to collect smaller droplets of water, allow faster runoff, and cause less water reevaporation than the finned-tube or horizontally mounted bare-tube coils. Bare-tube coils that are continuously wound in the form of a flat circular spiral (sometimes consisting of two coil layers) and mounted with the flat dimension in a vertical plane are considered a good design compromise because they have most of the advantages of the vertical-tube coil.

The dehumidifier evaporator is protected against corrosion by several methods such as painting, waxing, and anodizing (on aluminum). None of these finishes produce a considerable loss in capacity of the unit.

Most dehumidifiers are provided with removable

water receptacles that will hold approximately 16 to 24 pints of water (7.6 to 11.3 liters). Generally, they are made from plastic, which withstands corrosion better than metal. They are made so that they can be removed and emptied with a minimum amount of trouble. Most dehumidifiers are equipped with a connection for connecting a flexible hose to either the water receptacle or some other means of connecting the drain hose. The flexible hose permits direct gravity drainage to the sewer system and eliminates the need for emptying the condensate by hand.

There are several different types of cabinet design so that the interior design of the building can be matched as closely as possible with the cabinet. The more expensive the unit, the more features and higher output capacity along with a more expensive cabinet design. Some are equipped with a humidity sensing control for cycling the unit and to maintain a desired humidity inside the space. Normally, humidistats are adjustable from 30 to 80% relative humidity. Some humidistats are equipped with a continuous run setting. Most are equipped with an on and off point on the humidistat, while others will include an additional sensing and switching device that will automatically turn the unit off when the water receptacle is full and needs to be emptied. Some are equipped with a warning light indicating that the receptacle is full and that the unit is off.

These units are designed to provide the maximum performance at the standard rating conditions of 80°F (26.7°C) dry-bulb temperature and 60% relative humidity. When the room temperature drops to a point that the system is not loaded, as it would be at a condition of 65°F (18.3°C) dry-bulb temperature and 60% relative humidity, the refrigerant pressure and the corresponding evaporator temperature usually drop to a point where frost will occur on the cooling coil. This is especially noticeable on units that use a capillary tube as the refrigerant flow control device.

Some dehumidifiers are equipped with special defrost controls that will turn the compressor off when frost occurs on the evaporator. This control is generally a bimetal thermostat, strategically mounted on the evaporator tubing, that allows the dehumidifying process to continue at a reduced rate when frosting occurs. Under some frosting conditions, the humidistat can be set for a higher relative humidity setting, which will reduce the number and duration of the running cycles, permitting satisfactory operation at low-load conditions. During late

Design and Construction

fall and early spring operation of the dehumidifier, supplementary heat must be provided so that the space can be maintained within the desired conditions and to prevent frosting of the evaporator.

Dehumidifier units are generally equipped with rollers or casters so that they can be easily moved to the desired place of operation.

☐ Charging Window Units and Dehumidifiers ☐

Basically, there are three ways to charge these units properly: (1) manufacturer's charging charts, (2) superheat, and (3) subcooling.

Manufacturer's Charging Chart. This method is usually best when the charts are available. However, in most instances the service technician does not have them handy. To use this method, first connect the gauge manifold to the system as described in an earlier chapter. Then adjust the charge to meet the conditions set forth on the charts.

Charging by the Superheat. This method is sometimes very easy to use because the readings can be taken and the charge adjusted accordingly. First connect the gauge manifold to the system as described in an earlier chapter. Then attach a thermometer as close as possible to the outlet of the evaporator (Figure 17–9). Allow the unit to operate for about 10 min to stabilize the system pressures and temperatures. Take the gauge reading of the suction pressure and convert it to the saturated temperature for that pressure. Compare this temperature to that indicated by the thermometer on the evaporator outlet. There should be a 10°F (5.6°C) difference between the two. For example, on a system using R-22, the suction gauge indicates a pressure of 69 psig (475.738 kPa), which translates to a temperature of 40°F (4.44°C). The thermometer indicates a temperature of 50°F (10.00°C). Thus 40 − 50 = 10°F (5.6°C). The system is operating with a 10°F (5.6°C) superheat at the outlet of the evaporator. Therefore, the system is operating as it should—provided that everything else is as it should be.

Charging by Subcooling. This method is usually very easy to use because everything can be done at the condensing unit. However, it is sometimes a good

FIGURE 17–9 **Measuring temperature at the evaporator outlet.**

FIGURE 17–10 **Measuring condenser subcooling.**

idea to check the superheat at the evaporator outlet just to make certain that everything is operating as it should. First, install the gauge manifold to the system as described in an earlier chapter. Then connect a thermometer to the liquid line as close to the condenser outlet as possible (Figure 17–10). Take the pressure indicated by the discharge gauge and convert it to the saturation temperature of the refrigerant at that pressure. Then subtract the reading indicated by the thermometer from the saturation temperature of the discharge vapor. There should be approximately 20°F (−6.67°C) temperature difference. For example, a system operating with R-22 as the refrigerant has a discharge pressure of 300 psig (2067 kPa), which translates to a saturation temperature of 130°F (54.44°C). The thermometer indicates a liquid line temperature of 110°F (43.33°C). Thus 130 − 110 = 20°F (11.2°C) of subcooling in the condenser. The system should be operating satisfactory.

☐ Safety Procedures ☐

Domestic refrigeration is a very large part of the refrigeration industry. It is a must that proper safety procedures be followed in order to protect the service technician, the equipment, and the products.

Always get help when moving a room unit.

□ SUMMARY □

- A room air conditioning unit is an assembly designed as a unit primarily for mounting in a window or through a wall.
- The basic function of a room air conditioning unit is to provide comfort by cooling, dehumidifying, filtering or cleaning, and circulating the air.
- The majority of room units are designed to operate on 120-, 240-, or 208-V single-phase current.
- The maximum amperage of the 120-V models is generally limited to 12 A because this is the maximum allowable current for a single-outlet, 15-A circuit permitted by the National Electrical Code®.
- The coils used on room air conditioning units are generally of the tube-and-plate-fin variety or of the tube-and-spin-fin variety.
- Basically, there are three types of flow control devices used on room air conditioning units: (1) thermostatic expansion valve, (2) automatic expansion valve, and (3) the capillary tube.
- The capillary tube is the most popular flow control device used on room air conditioning units.
- There are two types of fan motors used on room air conditioning units: (1) the shaded pole, and (2) the split capacitor motor.
- Room units operate with an evaporator temperature above the freezing point.
- The two basic systems in a room air conditioner are: (1) the refrigeration system, and (2) the air circulating system.
- The outdoor fan has a slinger ring around the periphery of the blade, which picks up condensate from the evaporator and splashes it on the condenser to increase the unit's capacity.
- Most room unit short cycling is caused by air from the grill being deflected back into the air inlets.
- The National Electrical Code® requires that adequate wiring and the proper-sized fuses be used for the service outlet.
- Icing is usually caused by insufficient airflow over the evaporator.
- If a compressor is operated in low ambient temperature, it will probably overheat and possibly burn up the compressor motor due to a lack of refrigerant to cool the windings.
- Low voltage is a common cause of trouble in the operation of a room air conditioning unit.
- An inoperative compressor should be isolated electrically and checked before a decision is made to replace it.
- The purpose of a dehumidifier is to remove moisture from the air as it is circulated through the unit.
- The component parts of a dehumidifier are a motor-compressor unit, a refrigerant condenser, an air-circulating fan, a refrigerated surface (evaporator), some means of collecting and disposing of the condensed moisture, and a cabinet to house these various components.
- A direct-drive propeller-type fan is generally driven by a shaded-pole motor for creating the required flow of air through the unit.
- Residential dehumidifiers in most cases will maintain a satisfactory humidity level within the space when the airflow rate and unit placement permit the entire volume of air inside the space to be passed through the dehumidifier once per hour.
- The evaporator is generally of the bare-tube construction, although finned-tube coils are sometimes used if the fins are spaced to allow rapid runoff of the water droplets.
- Most dehumidifiers are equipped with an on and off point on the humidistat, while others will include an additional sensing and switching device that will automatically turn the unit off when the water receptacle is full and needs to be emptied.
- Some dehumidifiers are equipped with special defrost controls that will turn the compressor off when frost occurs on the evaporator.

□ EXERCISES □

1. What is the purpose of a room air conditioner?
2. Name the three methods used in room air conditioning to provide heat.
3. On what voltages are room units designed to operate?
4. What is the maximum amperage of 120-V model room units?
5. What is the design of the two coils generally used in room units?
6. Name the three types of flow control devices used on room units.
7. What are the two types of fan motors used on room units?
8. What are the five more common methods of mounting window units?
9. At what evaporator temperature does a room air conditioner operate?
10. What are the two basic systems in room air conditioners?
11. What is the purpose of the slinger ring on the condenser fan blade?
12. What is the cause of most short-cycling conditions in room units?
13. What is the purpose of a dehumidifier?
14. What type of fan and motor is used in most dehumidifiers?
15. What determines the airflow rate through a dehumidifier?

16. What airflow rate is generally required to maintain a satisfactory humidity level in a residence?
17. Of what type of construction is the evaporator used in dehumidifiers?
18. What methods are used to prevent corrosion of the dehumidifier evaporator?
19. At what conditions are dehumidifiers designed to provide maximum performance?
20. What type of control is used to prevent frost occurring on the dehumidifier evaporator?

18 Commercial Refrigeration

Performance Objectives

After completion of this chapter, the student should:
- Understand the requirements for commercial refrigeration equipment.
- Be able to name the various types of commercial cabinets.
- Understand the commercial refrigeration cycle.
- Have a better understanding of commercial refrigeration compressors.
- Be able to describe the various methods used in charging commercial refrigeration systems.
- Demonstrate a better understanding of the basic refrigerant controls used in commercial refrigeration systems.
- Describe the defrost systems used in commercial refrigeration systems.
- Have a working knowledge of heat recovery systems.
- Complete the Exercises with at least 80 percent accuracy.
- Complete the Laboratory Exercises with at least 80 percent accuracy.

There are several classifications of commercial refrigeration cabinets depending on their size and design. These classifications are dependent on the intended use of the cabinet. Grocers, butchers, and others who sell perishable food require refrigerated cabinets suitable for the preservation of these foods. In general, the size and construction of the cabinet that best meets the needs of the merchant depends on the particular line of business in which he or she is engaged and the quantity of perishable food that must be kept in stock. Because of this, commercial cabinets may be found in almost every shape, size, and description.

Definition. *Commercial refrigeration units* are those used in businesses for the preservation of perishable merchandise. Commercial refrigeration units normally use high-starting-torque compressor motors and thermostatic expansion valves. Some of the smaller units, however, may use capillary tubes as the metering device.

☐ Requirements of Commercial Refrigeration Equipment ☐

The store owner is extremely interested in the operating and maintenance cost of his or her refrigeration equipment. A quality commercial refrigeration installation must provide satisfactory service economically. Usually, this type of equipment is installed to be used as a profitable investment because every place of business is operated for a profit. Properly designed and installed refrigeration equipment will keep operating costs low and remarkably improve the preservation of stored merchandise. It can, therefore, be seen that it is necessary to estimate the requirements for the cabinet as closely as possible.

It is also necessary that the refrigeration unit be sized properly—not too small, not too large to do the work. If the capacity of the refrigeration unit is too large, the operating costs will increase. If, however, the unit is too small, it will be overloaded with longer periods of operation, more frequent starting, and more wear. The maintenance and repairs will increase and the performance probably will not be satisfactory. Therefore, it can be concluded that: (1) careful estimating of the refrigeration load is necessary, and (2) the refrigeration unit chosen should be sized according to the manufacturer's specifications.

Allowances should be made for emergencies and a proper safety factor should be applied. Properly engineered and installed refrigeration equipment is less expensive than that which is not. At the same time profits are increased due to the elimination of loss and spoilage while permitting a more effective display of the refrigerated goods.

Although there are many and varied applications of commercial refrigeration, the principles applied are the same as those that apply to domestic refrigeration. In most types of commercial refrigeration, the success or failure in performance of the system depends on the following conditions:

1. Proper sizing for the heat load.
2. Correct installation of the equipment.
3. Insulation of the proper thickness at all points where the entrance of heat is to be retarded.
4. Whether the system is installed in a refrigerator cabinet, display case, or freezer case of any type, the construction and location of the evaporator, and baffles when required, is of vital importance for successful operation.
5. The size of the openings and locations of the warm and cold air passages determine the efficiency of the cooling unit in certain types of large cabinets.
6. The general design of the interior of the cabinet should allow for free and rapid circulation of the air through the food storage and over the cooling unit to facilitate the transfer of heat from the food to the cooling unit.

☐ Types of Commercial Cabinets ☐

Commercial refrigeration cabinets are designed and constructed to fit the needs of the user. The shell and the liner are usually made of metal, although the liner may be plastic. The finishes are designed for easy cleaning and upkeep. The insulation is either polystyrene or urethane, which may be installed in pieces or, in more modern cabinets, foamed in place.

The refrigeration unit is designed to provide adequate refrigeration under the most severe usage. To prevent condensation, electric heaters are installed around the doors or other openings, like the mullion heaters used in domestic units.

Commercial refrigeration units are usually designed to be used with remote condensing units. Most commercial refrigeration units use air-cooled condensing units because they are used in freezing weather. These condensing units may be connected to several commercial cabinets.

Walk-In Coolers. Walk-in coolers generally are larger than 100 ft³ (2.83 m³) in size. These cabinets are constructed so that access to the interior is gained by entering through a door (Figure 18–1). They are usually installed in large meat markets, larger restaurants, dairies, etc.

These coolers are generally constructed of metal exterior sections and can be expanded by using additional stock sections. The entire exterior surface is also metal, thus giving protection against rats and vermin. The interior surfaces are metal as well except for the hardwood floor.

These units are also available with self-service display doors. These units allow for the storage of merchandise in the same cabinet from which it is selected, thus saving the labor of having to move the merchandise from one cabinet to another. The units may be either remote installed or self-contained.

Insulation. Refrigeration efficiency is obtained with foam-sandwich construction. With this construction technique, polyurethane foam is pumped under pressure into a form, which holds both the inner and outer case walls of all sections. As the foam expands and hardens in place, it bonds the steel outer and inner walls (Figure 18–2). The result is a wall that is exceptionally strong and solid, yet thin and lightweight. The solid wall of foam is fire-resistant and waterproof. It retains the cold much longer and requires less compressor capacity to maintain proper temperature levels than old-fashioned fiberglass insulation. There is no way for moisture to cause rotting, molding, or insulation sag.

Door Systems. Completely foamed doors are rugged but lightweight. They close automatically on heavy-duty cam left hinges. Magnetic vinyl door gasketing gives positive seal against heat leakage. When magnetic gaskets are used, the door does not require an inside safety release.

A lighting system is usually provided for the in-

FIGURE 18-1 Walk-in cooler. (Courtesy of Tyler Refrigeration Corp.)

FIGURE 18-2 Insulation of walls. (Courtesy of Tyler Refrigeration Corp.)

side of the cooler. There is a pilot light on the outside to indicate when the light is on.

Reach-In Cabinet. These cabinets usually fall between 20 and 100 ft³ (0.566 and 2.83 m³) of storage

FIGURE 18-3 Reach-in cabinet. (Courtesy of Tyler Refrigeration Corp.)

space. They may be of the single- or multiple-door design (Figure 18-3). The glass doors permit effective display of the merchandise when desirable.

A blower coil is usually installed in the top or on the rear inside of the cabinet. The merchandise should be placed so that the airflow to and from the coil will not be restricted. Reach-in cabinets are usually self-contained units that only require that drain lines be installed.

Reach-in cabinets have a variety of uses such as bottled beverage storage, dairy product storage, food storage, and dough storage in bakeries. The temperatures inside these cabinets are about the same as those for domestic units—ranging from about 32 to 40°F (0 to 4.4°C).

Florist Cabinet. The size, type of construction, and the interior arrangement of florist cabinets vary widely. They are used largely for display and are, therefore, constructed with glass doors and often have glass ends (Figure 18-4). In no case should there be less than two thicknesses of glass, and three thicknesses are commonly used, particularly for the lower temperature storage. A blower coil may be placed on the back wall or regular cooling coils may be used. Many florist cabinets are tall and have the coil installed across the top, allowing the entire lower portion to be devoted to display purposes.

FIGURE 18-4 Florist cabinet. (Courtesy of Uniflow Manufacturing Co.)

The temperatures inside a florist cabinet vary from 47 to 54°F (8.3 to 12.2°C). The humidity is kept as high as possible to prevent moisture evaporating from the flowers. This is done with a coil designed to have a large cooling surface.

Display Cases. In addition to the reach-in cabinet and the walk-in cooler, most businesses have one or more refrigerated display cases, which are also made in various sizes and designs. Display cases are used almost exclusively for the preservation and display of meats and dairy products. Although they are generally used in combination with larger cases, there are designs and sizes of display cases where the case is the sole means of refrigerated storage.

Such cases as top-display, full-vision, combination top-display, and bottom-storage or double-duty, delicatessen, candy, dairy, vegetable, multiple-deck have more than proven their worth in enticing customers to purchase more of the products so invitingly displayed. This type of display accentuates impulse purchasing, and the customer can either point to the merchandise desired or, when in self-service establishments, pick out the item personally.

Double-Duty Case. As the name suggests, the case has two compartments, an upper compartment for display and a compartment underneath for storage purposes which is also refrigerated (Figure 18-5). The storage compartment is located on the rear side of this cabinet. The refrigeration coil is located in the bottom of the cabinet where a fan is used to circulate the refrigerated air throughout the case to provide proper refrigeration for the stored foods.

Open-Type Self-Service Display Cases. The open-type refrigerated display case is the latest milestone in merchandising products that must be kept constantly under refrigeration. Foods displayed in this type of case are refrigerated by a blanket of cold air and are stored below the top level of that cold blanket. Foods stored in these cases are usually wrapped in clear, transparent material such as

$31\frac{3}{4}''$

$3''$

Optional rack

$30\frac{5}{8}''$

FIGURE 18-5 Double-duty case. (Courtesy of Tyler Refrigeration Corp.)

cellophane, which protects them from dust and germs in the air and from contamination through handling by the customer. While this type of display case has introduced some refrigeration problems, as compared to the completely enclosed case that has doors or covers, the main problem has been essentially that of providing the additional refrigeration capacity required to take care of the higher loss.

As an example, a 6-ft (1.8-m) enclosed top-display case can be adequately refrigerated by a ¼-hp condensing unit with a matching evaporator coil, while a 6-ft open-type display case will require a ½-hp unit with a correspondingly larger evaporator.

Even though the cold air is heavier than warm air and the design of the case should keep the air inside the case, the losses are so great that twice the amount of refrigeration capacity is required. While the cold air does tend to remain within the fixture, since it is in constant circulation past the evaporator and over the stored products, the top cold air comes in contact with the warm air above the case and some of it is warmed. This top layer consequently rises and leaves the case, being replaced by new warm air from outside the fixture. Since warm air contains moisture, it is necessary not only to cool the warm air down to case temperature but also to condense out the moisture it is carrying. For every pint of moisture condensed, approximately 1000 Btu (252 kcal) must be removed, and this process is occurring

continuously. The moisture condensed passes out of the case through the drain.

Even where the movement of air in the cabinet is slight, warm, moist air from outside the case will mix with the cold, dry air in the cabinet and add substantially to the refrigeration load required to cool the air and condense the moisture. Customers reaching into the fixture to remove stored products also cause air movement and mixing of cold and warm air. It is therefore most important to locate these units away from all externally induced air circulation possibilities such as fans, entrance doors which might permit blasts of air to blow past the fixture, heating duct openings, air blasts from unit heaters, and similar causes of air movement.

Modern open-type display cases usually have the evaporator located in the bottom of the cabinet (Figure 18–6). A fan is used to circulate the air over the coil and through the case. The food stored in these cases must be placed within a certain zone to prevent spoilage. This case has an optional canopy with mirrors that can be used to improve the appearance of the stored merchandise.

Frozen Food Display Case. The successful merchandising of frozen foods and packaged ice cream has proven to require the open display of such items in fixtures that attract the customer and are readily accessible for the choice of the product or flavor desired. While the open display is most suitable for

FIGURE 18-6 Top-display open meat case. (Courtesy of Tyler Refrigeration Corp.)

this purpose, the required low temperature of 0°F (−17.8°C) or lower adds problems to the use of the open-type case (Figure 18–7).

Because of the required low temperature, the difference in temperature between the room air and the cold air in the case is substantially increased, with the result that the heat absorbed by the interchange of warm air is also greatly increased. The lower temperatures induce greater infiltration of moisture-laden air with the resultant higher refrigeration load. Both the moisture-laden air and the lower temperatures required combine to increase the condensing unit size. Therefore, for a 6-ft (1.8-m) open-type frozen food storage cabinet, approximately a ¼-ton refrigeration unit would be required. To obtain efficient loading of the compressor motor for such cabinets, it is important that the unit be equipped with the proper size pulley if it is the conventional open-type compressor. If it is a hermetic compressor, the proper size cylinder is essential for low-temperature work.

Some open-type low-temperature cabinets use vertical plate-type evaporators placed about 12 in. (0.304 m) apart along the full length of the storage compartment with the packaged goods stored between the plates. The warm air penetration occurs through the top 2- or 3-in. (5- or 7.6-cm) layer of cold air just below the top edge of the refrigerated plates so that frozen foods cannot satisfactorily be stored in this area. All cabinets of this type have indicated levels beyond which the products should not

FIGURE 18–7 Tyler frozen food and ice cream cabinet. (Courtesy of Tyler Refrigeration Corp.)

FIGURE 18–8 Method of circulating air. (Courtesy of Hill Refrigeration Division, Emhart Industries, Inc.)

COMMERCIAL REFRIGERATION

be stored. Some equipment manufacturers provide a supplementary means of flowing a cold air blanket over the top of the evaporator plates to permit loading of the cabinet to the top edge of the plates. This may be done by a forced air fan, an auxiliary coil built into the structure above the case, or an extra plate placed lengthwise above the top edge of the compartment plates.

Regardless of the method used, a blanket of cold air at least 2 or 3 in. (5 or 7.6 cm) thick must separate the frozen foods and the warm moist air above the storage compartment (Figure 18–8).

□ Operating Characteristics of Self-Service Open-Type Cases □

The operating characteristics of self-service open-type refrigerated cases differ from those of the closed type. Particular attention must be given to the proper control adjustments, along with other considerations that must be taken into account if these cases are to provide the user with the desired service. Three main principles that must be kept in mind during the installation and maintenance of these units are:

1. The evaporator must be completely filled with refrigerant.
2. The cycling of the unit must be different. A longer running time is necessary to obtain the desired refrigeration.
3. Complete defrosting of the evaporator during the defrost period is necessary for the best operation of the equipment.

The one sure way to determine whether or not the evaporator is fully refrigerated is to observe the refrigerant line from the thermostatic expansion valve to the evaporator and the suction line leading from the evaporator. Both lines should be equally fed with refrigerant and fully frosted. A condition that may cause the service technician to believe the evaporator is fully refrigerated is that the suction line frosts periodically for several feet out of the case, even though the superheat setting of the thermostatic expansion valve may be too high. The "hunting" action of the thermostatic expansion valve serves to refrigerate the evaporator past the thermostatic bulb and out of the case, causing the suction line to frost several feet. The hunting action results in the evaporator being flooded and then starved of refrigerant, so that generally the evaporator is only partially refrigerated. However, during much of the running cycle the suction line will have little or no frost when compared to the evaporator coil.

The only positive check to see whether the evaporator is being fully refrigerated is to determine if the suction line and the thermostatic expansion valve outlet line to the evaporator are approximately equally frosted during the running cycle.

An evaporator that is fully refrigerated will cause the suction line to frost and sweat for several feet out of the case on humid days. To prevent this condition, a heat exchanger can be installed inside the case. The frost line should then extend through the heat exchanger but not out of the case. To obtain the best results, it is recommended that a heat exchanger be installed on cases that are maintained below 40 °F (4.4 °C).

Concerning the second principle—the long running time—it must be understood that the condensing unit is normally required to run 40 min or longer to maintain the desired temperatures.

The problem of completely defrosting the evaporator demands special attention when open-type cases are involved. Because cabinet temperatures of 28 to 36 °F (− 2.2 to 2.2 °C) are required by the store operator to properly display prepackaged meats in an open-type case, it is often necessary to operate a system using R-12 with a 4- to 6-lb (128.47- to 142.21-kPa) cut-out point, and a 26- to 28-lb (279.61- to 293.35-kPa) cut-in point. These settings result in a lower and more consistent cabinet temperature.

The question may arise as to whether or not the evaporator will build up an excess of frost and lose its efficiency at such operating pressures. This will not occur within a 24-hr period and that is the reason that the time-clock method of evaporator defrosting is in wide general use. This type of defrost system uses an electric time clock wired into the control circuit.

□ Methods of Heat Transfer □

Commercial refrigeration systems make use of four primary methods of heat transfer, as follows:

1. *Direct-contact cooling* with the cooling coils hung from the ceiling or walls of the cooler, thus cooling the air directly. The air in turn circulates

naturally, carrying the heat from the products to the evaporator.

2. *Indirect-contact cooling* by means of a brine tank located at the ceiling or on the walls of the cooler.
3. *Combination direct and indirect cooling* where half of the evaporator is exposed while the other half is immersed in the tank. Thus, the heat storage capacity of the brine tank minimizes temperature variation.
4. *Forced-draft cooling* by means of a blast of air blown over the evaporator into the cooler.

□ Commercial Refrigeration Cycle □

The commercial refrigeration cycle is basically the same as the refrigeration system used in domestic refrigerators. The major differences are in the types of controls used. These controls will vary with the design and requirements of the particular system. The refrigerant pressures will also vary with the type of system. On low-temperature systems, extreme caution must be exercised to prevent the entrance of moisture into the refrigeration system. Moisture in these systems will probably be noticed first with freezing at the flow control device. When freezing at the flow control device is experienced, refrigeration will be at least partially reduced, depending on the severity of freeze-up. Refrigeration can be completely stopped because of moisture freezing at the flow control device.

□ Commercial Refrigeration Compressors □

It was shown earlier in the discussion of the refrigeration cycle that the compressor has one purpose. That is to raise the refrigerant vapor pressure from evaporator pressure to condenser pressure. The compressor must deliver the compressed refrigerant vapor to the condenser at a pressure and temperature at which the condensing process can be accomplished.

Single-Stage Low-Temperature Systems. Low-temperature single-stage refrigeration systems have become increasingly critical from a design and application standpoint as the desired evaporating temperature is decreased. The combination of high compression ratios, low operating temperatures, and

thin return vapor can cause lubrication and overheating problems, thus making the compressor more vulnerable to permanent damage due to moisture and contaminants in the refrigeration system.

The selection of the compressor, the suction vapor temperature, and the application must be made so that the discharge line temperature taken within 1 to 6 in. (2.54 to 15.2 cm) of the compressor discharge service valve does not exceed 230°F (110°C) for refrigerants R-12, R-22, and R-502. Under these conditions, the estimated average temperature at the discharge port—measured at the discharge valve retainer on the valve plate—will be approximately 310°F (154°C) for refrigerants R-12 and R-502, and 320°F (160°C) for refrigerant R-22.

The compressor displacement, pressure-limiting devices, and quantity of cooling air or water must be selected to prevent the compressor motor winding temperature from exceeding the following limits:

1. 210°F (99°C) when protected by inherent protectors affected by line current and motor temperature
2. 190°F (88°C) when protected by motor starters

The compressor motor temperature should be determined by the resistance method and should be measured when the motor is tested in the highest ambient temperature in which it is expected to operate, at 90% of the rated voltage, with 90°F (32.2°C) suction gas temperature. For longer motor life, the maximum operating temperatures of 170 to 190°F (77 to 88°C) are recommended.

To prevent the discharge vapor temperature and motor temperature from exceeding the recommended limits, it is desirable, and in some cases absolutely necessary, to insulate the suction lines, which will allow the suction vapor to return to the compressor at a lower than normal temperature. This is especially important when R-22 is used with suction-cooled compressors.

Most suction-cooled compressors require some auxiliary cooling by an air blast on the compressor when the compressor is used for operating temperatures below 0°F (−17.8°C).

To prevent compressor motor overload during pull-down periods or after a defrost period, either the evaporator must be properly designed or a pressure-limiting device such as an expansion valve or a crankcase valve must be used.

Some compressor manufacturers recommended that R-502 be used in all single-stage low-temperature applications where the evaporator temperature is −20°F (−28.9°C) or below. Now that R-502 is readily available, the use of R-22 should be avoided in single-stage low-temperature compressors of 5 hp and larger. The lower refrigerant vapor discharge temperatures encountered when using R-502 have resulted in much more trouble-free operation.

An adequate supply of oil must be maintained in the compressor crankcase at all times to ensure continuous lubrication of the parts. If the refrigerant velocity in the system is so low that rapid return of the oil is not assured, an adequate oil separator must be used. The normal oil level should be maintained at or slightly above the center of the compressor oil sight glass (Figure 18–9). An excess of refrigerant or oil in the system must be avoided because it may result in liquid slugging of the compressor and damage to the compressor valves, pistons, or cylinders.

The refrigerant lines must be designed and installed to prevent oil trapping. The highest refrigerant velocity possible without excessive pressure drop is recommended.

Care must be exercised to prevent the evaporating temperature from dipping so far below the normal system operating temperature that the refrigerant velocity becomes too low to return the oil to the compressor. The low-pressure control cut-out setting should not be below the lowest published rating for the compressor.

The smallest practical size tubing should be used

in evaporators and condensers in order to hold the system refrigerant charge to a minimum. When large refrigerant charges are unavoidable, recycling pump-down control methods should be used.

When air-cooled condensing units are required for operation in low ambient temperatures, some means of head pressure control to prevent the condensing pressure from falling too low is highly recommended to maintain the required refrigerant velocities.

A filter-drier of generous size must be installed in the liquid line, preferably in the cold zone. The desiccant used must be able to remove moisture to a low end point and be capable of removing a reasonable quantity of acid. It is most important that the filter-drier be equipped with an excellent filter to prevent the circulation of carbon and foreign particles inside the system. A permanent suction line filter is highly recommended to protect the compressor from any contaminants that may have been left in the system during installation.

After installation has been completed, all systems should be thoroughly evacuated with a high-grade vacuum pump and dehydrated to ensure that no air or moisture is left in the system. The triple-evacuation method should be used, breaking the vacuum each time with dry refrigerant charges into the system through a drier. The compressor motor must not be operated while the high-vacuum pump is in operation. If the motor is run, permanent damage to the windings could occur.

The system should be charged with clean, dry refrigerant through a drier (Figure 18–10). Other substances such as liquid driers or alcohol must not be used.

Two-Stage Low-Temperature Systems. Because of their basic design and operation, two-stage systems are inherently more efficient and encounter fewer operating difficulties at the lower operating temper-

FIGURE 18–9 Oil level in sight glass.

FIGURE 18–10 Gauge manifold hookup using a drier.

atures than do single-stage units. The two-stage compressor, however, has its limitations. When evaporator temperatures below − 80 °F (− 62.2 °C) are encountered, they lose their efficiency. Compressor motor overheating also becomes an increasing problem. At evaporating temperatures below − 80 °F, a cascade system is recommended. However, for applications with evaporating temperatures in the − 20 to − 80 °F (− 28.9 to − 62.2 °C) range, the two-stage compressor efficiency is high, the refrigerant discharge temperatures are low, and field experience with properly applied two-stage compressors has been excellent.

Two-stage refrigeration systems are somewhat more complex and sophisticated than simple single-stage systems. Many of the operating problems encountered on two-stage systems stem from the fact that too often their application has been made without sufficient appreciation of the safeguards that must be taken in proper system design.

Volumetric Efficiency. Three definitions given previously are of importance in analyzing two-stage systems.

1. The compression ratio is the ratio of the absolute discharge pressure (psia) to the absolute suction pressure (psia).
2. The absolute pressure is gauge pressure plus atmospheric pressure, which at sea level is standardized at 14.7 psi (101.32 kPa).
3. Volumetric efficiency is defined as the ratio of the actual volume of the refrigerant vapor pumped by the compressor to the volume displaced by the compressor pistons.

A typical single-stage volumetric efficiency curve is shown in Figure 18–11. Note that as the compression ratio increases, the volumetric efficiency decreases.

Two factors that cause a loss of efficiency with an increase in compression ratio are:

1. The density of the residual vapor remaining in the cylinder clearance space after the compression stroke is determined by the discharge pressure—the greater the discharge pressure, the greater the density. Since this vapor does not leave the cylinder on the discharge stroke, it reexpands on the suction stroke, thus preventing the intake of a full cylinder of vapor from the suction line. As the compression ratio increases,

FIGURE 18–11 Typical single-stage low-temperature compressor efficiency curve. (Courtesy of Copeland Refrigeration Corp.)

the added space in the cylinder on the intake stroke is filled by the residual vapor.
2. The high temperature of the cylinder walls resulting from the heat of compression is a factor in the loss of efficiency. As the compression ratio increases, the heat of compression increases, and the cylinders and compressor head become very hot. The suction vapor which enters the cylinder on the suction stroke is heated by the cylinder walls and expands, resulting in a reduced weight of refrigerant vapor entering the compressor.

Therefore, a single-stage compressor has its limitations as the compression ratio increases. The effective low limit of even the most efficient single-stage system is approximately − 40 °F (− 40 °C) evaporating temperature. At lower evaporating temperatures, the compression ratio becomes so high that the compressor capacity falls rapidly. The compressor may no longer be handling a sufficient weight of return refrigerant vapor for proper motor cooling. Also, because of the decreased vapor density, oil may no longer be properly circulated throughout the system.

Two-Stage Compression and Compression Efficiency. In order to increase the compressor's operating efficiency at low evaporating temperatures, the compression ratio can be done in two steps or stages. For two-stage operation, the total compression ratio is the product of the compression ratio of each stage. In other words, for a total compression ratio of 10:1, the ratio of each stage might be 4:1; or compres-

FIGURE 18–12 Typical six-cylinder two-stage compressor. (Courtesy of Copeland Refrigeration Corp.)

FIGURE 18–13 Typical three-cylinder two-stage compressor. (Courtesy of Copeland Refrigeration Corp.)

sion ratios of 4:1 and 5:1 in separate stages will result in a total compression ratio of 20:1.

Two-stage compression may be accomplished with the use of two compressors with the discharge of one pumping into the suction of the second. However, because of the difficulty of maintaining the proper oil levels in the two crankcases, it is more satisfactory to use one compressor with multiple cylinders. A greater volume of the low-stage cylinders is necessary because of the difference in specific volume of the refrigerant vapor at low and interstage pressures. While the compression ratios of the two stages are seldom exactly equal, they will be approximately the same. Two-stage compressors generally use an external manifold and a de-superheating expansion valve (Figure 18–12). A typical three-cylinder two-stage compressor with its external manifold is shown in Figure 18–13.

The volumetric efficiency of compressors can be entered on graphs or curves to indicate their performances. The three straight lines in Figure 18–14 are typical single-stage curves—one for an air conditioning compressor, one for a typical multipurpose compressor, and one for a low-temperature compressor. There are some variations in the compressor design involved, but the primary difference in characteristics is due to the clearance volume.

The two vertical curved lines represent the comparative efficiency of a two-stage compressor. Actually, each separate stage would have a straight

FIGURE 18–14 Typical compressor volumetric efficiency curves. (Courtesy of Copeland Refrigeration Corp.)

393

line characteristic similar to the single-stage curves to enable a comparison with single-stage compressors. The overall volumetric efficiency has been computed on the basis of the total displacement of the compressor, not just the low-stage displacement.

The solid black curve represents the efficiency of a two-stage compressor without a liquid subcooler. Note that the efficiency is relatively constant over a wide range of total compression ratios, and that the crossover in efficiency with the best low-temperature single-stage compressor is at a compression ratio of approximately 13:1. In other words, at compression ratios lower than 13:1, a single-stage compressor will have more capacity than a two-stage compressor of equal displacement without liquid subcooling.

The dashed curve represents the efficiency of the same two-stage compressor with a liquid subcooler. In the subcooler, the liquid refrigerant being fed to the evaporator is first subcooled by liquid refrigerant fed through the interstage de-superheating expansion valve, and a much greater share of the refrigeration load has been transferred to the high-stage cylinders. Since the high-stage cylinders operate at a much higher suction pressure, the refrigeration capacity there is far greater per cubic foot (0.0283 m³) of displacement than in the low-stage cylinders.

In effect, the capacity of the compressor has been greatly increased without having to handle any additional suction vapor returning from the evaporator. Note that with the liquid subcooler, the crossover point in efficiency as compared with a single-stage compressor is at a compression ratio of approximately 7:1. In other words, at compression ratios lower than 7:1, the single-stage compressor will have more capacity for an equal displacement, but at compression ratios higher than 7:1, the two-stage compressor will have more capacity.

Table 18–1 lists comparative operating data at varying evaporating temperatures for a Copeland compressor, which is available either as a single-stage or a two-stage compressor. Although the displacement, refrigerant, and motor are the same, the rapidly increasing advantage of two-stage operation as the evaporating temperature decreases is plainly shown.

Compressor Overheating at Excessive Compression Ratios. In addition to efficiency, the extremely high temperatures created by operation at abnormally high compression ratios make the use of single-stage compressors impractical for ultralow temperature

TABLE 18–1 Efficiency Comparison of Single-Stage versus Two-Stage Compression[a]

	Evaporating Temperature (°F)		
	−30	−40	−50
Condensing temperature (°F)	120	120	120
Condensing pressure (psig)	280.3	280.3	280.3
Condensing pressure (psia)	295	295	295
Evaporating pressure (psig)	9.40	4.28	0.04
Evaporating pressure (psia)	24.1	18.97	14.74
Single stage			
Compression ratio	12.5:1	15.6:1	20:1
Capacity (Btu/hr)	46,000	32,000	23,000
Btu/watt	3.42	2.86	2.3
Two stage with subcooler			
Compression ratio—low stage	3.74	3.68	4.05
Compression ratio—high stage	3.26	4.23	4.95
Capacity (Btu/hr)	61,000	50,000	38,500
Btu/watt	4.15	3.84	3.35
Increase in capacity Two stage vs. single stage (%)	32	56	67

Source: Copeland Refrigeration Corp.
[a]Capacity data based on equal displacement.

applications. A valve plate can be ruined with carbon formation due to lubricating oil breakdown from excessive heat. Excessive cylinder temperatures also can cause rapid piston and cylinder wear, cylinder scoring, and early failure of the compressor. With two-stage compressors, the interstage expansion valve maintains safe operating temperatures and this type of damage is prevented.

Low-Pressure Control Settings. The low-pressure control used on commercial applications must never be set at a pressure lower than the design limits of the compressor. The control must be set to stop the compressor at this temperature regardless of the temperature inside the case. In some cases it may take several hours and several operating cycles before the desired case temperature is actually reached. The operating differential of the pressure control does not need to be exact as long as the compressor is not allowed to short cycle, causing a temperature buildup in the compressor motor windings.

Medium-Temperature Applications. In this as well as with any other type of system, the compressor must operate within the intended design limits. In these applications the operating differential of the low-pressure control must be set so that the coil will not frost, causing refrigerant floodback to the compressor. If the application requires a temperature that will cause the coil to ice over, a defrost cycle must be used to protect the compressor.

Suction Line Insulation. The suction line must be insulated on any application, but it is especially important on medium- and low-temperature applications. The insulation helps to maintain the suction gas at a lower temperature, which also results in a lower discharge gas temperature. This is especially true when the refrigerant lines are routed through high-temperature areas.

High Discharge Gas Temperatures. In medium- and low-temperature applications, the discharge gas temperature is the most frequent cause of compressor failure on these systems. The discharge gas temperature in medium- and low-temperature applications should not be allowed to rise above 230 °F (110 °C). A reliable gauge manifold and an electronic thermometer is needed to determine these temperatures. The gauges must be accurate and the thermometer sensors must be securely fastened to the line and insulated so that the ambient temperature will not affect them. The line temperatures must be taken within 6 in. (152.40 mm) of the compressor service valve.

To determine the discharge temperature, use the following procedure: Connect the gauges to the system as described in an earlier chapter. Securely fasten the thermometer to the discharge line within 6 in. (15.2 cm) of the compressor service valve and insulate the sensor from the ambient temperature (Figure 18–15). Take the discharge pressure gauge reading and convert it to the saturated refrigerant temperature. Then take the temperature indicated on the thermometer. Subtract the saturated refrigerant temperature from the thermometer reading. The results should be a difference of 230 °F (110 °C) or less: for example, a system operating with R-502 refrigerant and a discharge pressure of 230 psig (1584.7 kPa), which relates to a saturated refrigerant

temperature of 105 °F (40.56 °C). The thermometer indicates a temperature of 300 °F (168 °C). Thus 300 − 105 = 195 °F (90.56 °C). This system is operating within the desired limits of the discharge gas temperature.

Some of the more common causes of a high discharge gas temperature are:

1. Insufficient air over the compressor top
2. System low on refrigerant charge
3. Dirty air-cooled condenser
4. Scaled water-cooled condenser
5. Defective condenser fan motor or capacitor
6. Too low a cutout setting on the low-pressure control
7. Faulty low-pressure control
8. Lack of or inadequate suction line insulation
9. Superheat setting too low on expansion valve
10. Too much oil in the system

Basic Two-Stage System. The basic flow of refrigerant in a six-cylinder two-stage compressor is different from the single-stage system (Figure 18–16). The suction gas returning from the evaporator enters the four low-stage cylinders directly from the suction line. Since the discharge vapor from the first-stage cylinders is heated from compression, it must be cooled by the de-superheating expansion valve before entering the compressor motor chamber. The de-superheated refrigerant vapor, now at the interstage pressure, enters the high-stage cylinders, is compressed, and then discharged to the condenser.

Figure 18–17 is a schematic view of a typical two-stage system showing the various components that are necessary for operation.

De-superheating Expansion Valve. Expansion valves currently supplied as original equipment with two-stage compressors are of the nonadjustable superheat type. In the event of failure, standard replacement valves with adjustable superheat that have been approved by the manufacturer may be used. Improper valve selection can result in the compressor overheating and possible damage due to improper liquid refrigerant control.

Liquid Refrigerant Subcooler. Two-stage systems may be operated either with or without liquid subcoolers. The function of the subcooler is to cool the liquid refrigerant being fed to the evaporator; this is accomplished by the evaporation of refrigerant fed through the de-superheating expansion valve. This transfers a greater portion of the refrigeration load to the high-stage cylinders, and because of the

FIGURE 18–15 Checking discharge gas temperature.

FIGURE 18-16 Schematic two-stage system. (Courtesy of Copeland Refrigeration Corp.)

Typical installation
6 cyl 6RBI and 6T two-stage series compressors
with liquid subcooler

FIGURE 18-17 Two-stage system with six-cylinder compressor (with liquid subcooler). (Courtesy of Copeland Refrigeration Corp.)

greater compressor capacity at higher suction pressures, the system capacity is greatly increased.

The temperature of the liquid refrigerant being fed to the evaporator is reduced in the subcooler to within approximately 10°F (5.56°C) of the interstage saturated evaporating temperature. The increase in system capacity can only be realized if the subcooled liquid is maintained at this low temperature and heat transfer into the liquid line is prevented. Normally this requires insulation of the liquid line.

When selecting expansion valves for two-stage systems with liquid subcoolers, the designer must keep in mind that the expansion valve will have greatly increased capacity due to the low temperature of the refrigerant entering the valve. Unless this is taken into consideration, the increased refrigerating effect per pound of refrigerant may result in an oversized expansion valve with resulting erratic operation.

☐ Charging Commercial Refrigeration Systems ☐

The proper performance of any refrigeration or air conditioning system is greatly dependent on the proper charge of refrigerant. An undercharged system will result in a starved evaporator, resulting in excessively low compressor suction pressures, loss of capacity, and possible compressor motor overheating. On the other hand, overcharging can flood the condenser, resulting in high discharge pressures and in potential compressor damage caused by liquid refrigerant flooding the compressor. Most refrigeration systems have a reasonable tolerance for some variation in charge. However, some small systems will actually have a critical refrigerant charge that is essential for proper operation.

Each refrigeration system must be considered separately. Systems with the same capacity or horsepower rating may not necessarily require the same refrigerant charge. Therefore, it is important first to determine the type of refrigerant required for the system. The unit nameplate normally identifies both the type and weight of refrigerant required.

Liquid Charging. Charging a system with liquid refrigerant is much faster than charging with vapor. Because of this factor, liquid charging is almost always used on large field-installed systems. Liquid charging requires either a charging valve in the liquid line, a process fitting in the high-pressure side of the system, or a receiver outlet valve with a charging port. It is recommended that liquid charging be done through a filter-drier to prevent any contaminants from entering the system. Never charge liquid into the compressor suction or discharge service valve ports because this can damage the compressor valves.

On original installations, the entire refrigeration system should be thoroughly evacuated. Weigh the refrigerant drum and attach the charging line from the refrigerant drum to the charging valve. If the approximate weight of refrigerant required is known, or if the charge must be limited, the refrigerant drum should be placed on a scale so that the weight of the

refrigerant remaining in the drum can be checked frequently.

Purge the charging line of any air and open the cylinder liquid valve and the charging valve. The vacuum in the system will cause the liquid to flow through the charging connection until the system pressure is equalized with the pressure in the refrigerant cylinder.

Close the receiver outlet valve and start the compressor. Liquid refrigerant will now flow from the refrigerant cylinder to the liquid line through the evaporator and will be collected in the condenser and receiver.

To determine if the charge is approaching the amount required by the system, open the receiver outlet valve, close the charging valve, and observe the operation of the system. Continue charging until the proper charge has been introduced into the system. Again weigh the refrigerant drum, and make a record of the weight of refrigerant charged into the system.

Watch the discharge pressure gauge closely. A rapid rise in discharge pressure indicates that the condenser is filling with liquid and the system pumpdown capacity has been exceeded. If this occurs, stop charging from the cylinder immediately and open the receiver outlet valve.

On factory assembled package units using welded compressors, charging is normally accomplished by drawing a deep vacuum on the refrigeration system, and introducing the proper refrigerant charge by weight into the high side of the system by means of a process connection that is later sealed and brazed closed. To field charge such systems, it may be necessary to install a special process fitting or charging valve, and weigh in the exact charge required for that unit.

Vapor Charging. Vapor charging is normally used when only small amounts of refrigerant are to be added to a system, possibly up to 25 lb (11.34 kg), although it can be more precisely controlled than liquid charging. Vapor charging is usually accomplished by use of a charging manifold connected to the compressor suction service valve port. If no valve port is available—for example, on welded compressors—it may be necessary to install a piercing valve or fitting in the suction line.

Service gauges are connected to read both the suction and discharge pressure. When adding refrigerant to a system, the discharge pressure should be observed to be sure the system is not being overcharged and that the refrigerant is not being added too rapidly. A higher than normal discharge pressure indicates that either the condenser is filling with liquid or the compressor is being overloaded by too rapid charging. The charging manifold permits throttling of the vapor from the cylinder. The cylinder is mounted on a scale to measure the amount of refrigerant being charged into the system. An approved valve wrench is used to operate the cylinder valve.

The refrigerant cylinder must remain in the upright position with the refrigerant being withdrawn only through the vapor valve to ensure that only vapor is reaching the compressor. The vaporizing of the liquid refrigerant in the cylinder will chill the remaining liquid and reduce the cylinder pressure. To maintain cylinder pressure and expedite charging, warm the cylinder by placing it in warm water or by using a heat lamp. *Do not apply heat with a torch.*

To determine if sufficient charge has been introduced, close the refrigerant cylinder valve and observe system operation. Continue charging until the proper charge has been added. Watch the compressor discharge pressure closely during the charging operation to be certain that the system is not overcharged.

How to Determine the Proper Charge. There are several methods that can be used to determine whether or not a system contains enough refrigerant. The method used will depend to a great extent on the system design and the personal preference of the service technician. The following is a list of these methods.

Weighing the Charge. The most accurate charging procedure is to actually weigh the refrigerant charged into the system. This method can only be used when the system requires a complete charge and the amount of that charge is known. Normally, the data are available on packaged unitary equipment. If the charge is small, it is common practice to remove the charge to the atmosphere when repairs are required. The system is completely recharged after the repairs have been made.

Using a Sight Glass. The most common method of determining whether or not the system is properly charged is by use of a liquid indicator (sight glass) in the liquid line. Because a solid head of liquid refrigerant is required for proper operation of the expansion valve, the system can be considered prop-erly charged when a clear stream of liquid refrigerant is visible in the sight glass. Bubbles or flashing usually indicate a shortage of refrigerant.

Keep in mind that if there is only vapor and no liquid in the sight glass, it will also appear clear. However, the service technician should be aware of the fact that at times the sight glass may show bubbles as flash gas even when the system is fully charged. A restriction in the liquid line ahead of the sight glass may cause sufficient pressure drop to cause flashing of the refrigerant. If the expansion valve feed is erratic or surging, the increased flow when the expansion valve is wide open can create enough pressure drop to create flashing at the receiver outlet. Rapid fluctuations in the condensing pressure can be a source of flashing. For example, in a temperature-controlled room, the sudden opening of shutters or the cycling of a fan can easily cause a change in the condensing temperature of 10 to 15 °F (5.56 to 8.34 °C). Any liquid refrigerant in the receiver may then be at a temperature higher than the saturated temperature equivalent to the changed condensing pressure and flashing will occur until the liquid temperature is again below the saturation temperature.

Some systems may have different refrigerant charge requirements under different operating conditions. Low ambient head pressure control systems for air-cooled applications normally depend on partial flooding of the condenser to reduce the effective surface area. Under these conditions, a system operating with a clear sight glass may possibly require twice the amount of refrigerant for proper operation under low ambient conditions.

While the sight glass can be a valuable aid in determining the proper charge, the system performance must be carefully analyzed before placing full reliance on it as a positive indicator of the system charge.

Using a Liquid Level Indicator. On some systems, a liquid level test port may be provided on the receiver. The proper charge can then be determined by charging until liquid refrigerant is found when the test port is cracked (Figure 18–18). With less than a full charge, only vapor will be available at the test port. Larger receiver tanks may be equipped with a float indicator to show the level of liquid in the receiver in much the same manner as a gasoline gauge on an automobile.

Checking Liquid Subcooling. On smaller systems, if no other means of checking the refrigerant charge

FIGURE 18-18   Liquid level indicator.

FIGURE 18-20   Checking superheat.

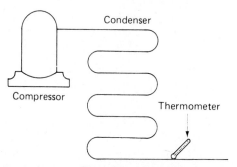

FIGURE 18-19   Checking liquid subcooling.

is available, a test of the liquid subcooling at the condenser outlet can be used. With the unit running under stabilized conditions, compare the temperature of the liquid line leaving the condenser with the saturation temperature equivalent to the condensing pressure (Figure 18-19). This provides an approximate comparison of the condensing temperature and the liquid temperature leaving the condenser. Continue charging the unit until the liquid line temperature is approximately 5 °F (2.78 °C) below the condensing temperature under maximum load conditions. This charging procedure is normally used only on factory-packaged systems. However, it does provide a means of emergency field checking that should indicate proper system operation.

Charging by Superheat. On small self-contained systems equipped with capillary tubes, the operating superheat may be used to determine whether or not the system contains the proper charge.

If a service port is available so that the suction pressure can be determined, the superheat may be calculated by determining the difference between the temperature of the suction line approximately 6 in. (15.24 cm) from the compressor and the saturation temperature equivalent of the suction pressure. If no means of determining the pressure is available, the superheat can be taken as the difference between the suction line temperature 6 in. from the com-

pressor and a temperature reading on the evaporator tube (not a fin) at the midpoint of the evaporator (Figure 18-20).

With the unit running at its normal operating condition, continue charging the unit with refrigerant until the superheat is found to be approximately 20 to 30 °F (11.2 to 16.68 °C). A superheat approaching 10 °F (5.56 °C) indicates an overcharged condition. A superheat approaching 40 °F (22.24 °C) indicates an undercharge.

Charging by Manufacturer's Charging Charts. Some manufacturers of self-contained equipment provide charging charts so that the proper charge can be determined by observing the system operating pressures. Follow the manufacturer's directions for determining the proper charge if the unit is to be charged in this fashion.

☐ Crankcase Pressure-Regulating Valves ☐

Crankcase pressure-regulating (CPR) valves are installed in the suction line ahead of the compressor. The valve controls the maximum pressure at the compressor suction (Figure 18-21). This provides overload protection for the compressor motor. The valve is adjusted for the maximum pressure specified by the condensing unit manufacturer.

At conditions of overload, the valve modulates to prevent suction gas pressures at the compressor suction greater than the pressure for which the valve is adjusted. When the overload has passed and the pressure drops below the valve setting, the valve assumes a wide open position. The valve setting is determined by a pressure spring, as the valve modulates from fully open to fully closed in response to the outlet pressure, closing on a rise in outlet pressure.

FIGURE 18-21 **Crankcase pressure regulator installation.**

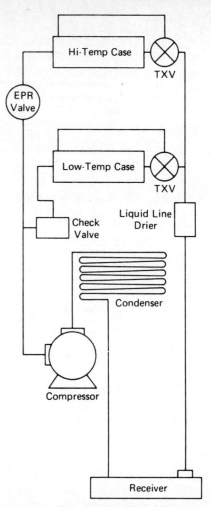

FIGURE 18-22 **Cross section of a crankcase pressure-regulating valve. (Courtesy of Singer Controls Division.)**

These valves assure positive protection against motor overload on systems using hot-gas defrosting. For example, high coil temperatures following defrosting plus an accumulation of liquid in the evaporator results in high pressures at the start of the normal refrigerating cycle. Also, high system heat loads may result in low side pressures above that for which the compressor motor is designed.

☐ Evaporator Pressure Regulator ☐

Evaporator pressure regulators (EPRs) are installed in the suction line between the evaporator and compressor (Figure 18-22). They maintain the desired minimum evaporator pressures within close limits.

On systems with multiple evaporators operating at different temperatures or on systems where the evaporating temperature cannot be allowed to fall below a given temperature, an evaporator pressure regulator valve is frequently used to control the evaporating temperature. This valve, often called an EPR valve, acts like the crankcase pressure regulator, except that it is responsive to the inlet pressure (Figure 18-23).

Since the valve responds to the inlet pressure, it opens when the inlet pressure equals or exceeds the opening point for which the valve is set. As the evaporator continues to rise, the valve stroke increases, permitting an increased flow of refrigerant. Close-off is caused when the inlet pressure drops below the valve setting, preventing the evaporator temperature from dropping below this point.

FIGURE 18-23 Two temperature system with EPR valve.

□ CDA (Close on Drop in Air Temperature) Valve □

CDA valves are usually factory installed in the suction branches or on the header defrost assembly by the manufacturer (Figure 18-24). They may also be field installed. These valves modulate by a change in the magnetic pull of the dc-operated solenoid coil.

There are three sizes: the CDA-12, the CDA-15, and the CDA-20. They have nominal 1-ton and 3-ton capacity ratings, respectively. These valves are designed and manufactured by Sporlan for precise and energy efficient control of the temperature in commercial refrigerators. The proper temperature is maintained by regulating the flow of refrigerant through the coil assembly.

The CDA temperature control is designed to replace evaporator pressure regulators and the conventional thermostats used on most commercial refrigeration equipment. The CDA valve temperature control system is composed of the CDA valve and the electronic components that operate it. We will now discuss these valves and their operation.

Valve Construction. The CDA-12 valve differs slightly in construction from the CDA-15 and the CDA-20, but the operation is identical for all sizes (Figures 18-25 and 18-26).

FIGURE 18-25 CDA-12 valve. (Courtesy of Hussmann Corp.)

FIGURE 18–26 CDA-15 and CDA-20 valves. (Courtesy of Hussmann Corp.)

Valve Operation. The CDA is a pilot-operated valve, requiring discharge gas pressure to close the valve. The valve is serviceable without being removed from the suction line. It consists of a solenoid, pilot assembly, piston, piston return spring, and a valve body. An exploded view of the CDA-15 and CDA-20 is shown in Figure 18–34.

The main port of the CDA-20 is modulated by bleeding a controlled amount of discharge gas pressure to the chamber above the valve piston. The discharge pressure enters the piston chamber through an orifice and then drains through the modulating pilot valve to the suction line downstream of the valve; thus as the pilot valve opens, the pressure decreases in the piston chamber and the piston return spring pushes the piston upward, opening the valve.

CDA Control Components. The components which control the CDA valve consists of (1) a plug-in thermostat, (2) a panelboard, (3) a refrigerator air temperature sensor, and (4) a diagnostic board.

Plug-In Thermostat. The plug-in thermostat is the brain of the CDA system. It has a temperature adjustment dial, operational amplifier (op-amp), full-wave rectifier, and a Wheatstone bridge circuit for modulating the electrical output to the CDA valve solenoid. Plug-in thermostats can easily be checked for proper operation. Thermostat ranges are shown on the temperature adjustment dial (Figure 18–27). Operation of the plug-in thermostat is as follows (Figure 18–28):

FIGURE 18–27 Plug-in thermostat. (Courtesy of Hussmann Corp.)

1. The 24-V ac power supply from the transformer is rectified to a variable dc voltage (0 to 34 V dc).
2. When the resistance of the temperature is increased by a drop in temperature, a voltage difference occurs across the bridge circuit.
3. The voltage difference across the bridge is measured and amplified 50 times by the operational amplifier.
4. The operational amplifier transmits the amplified signal to the base of the transistor.
5. The current flow through the current transistor increases to a value equal to 100 times the base current. This current flows through the CDA solenoid coil to reposition the CDA valve and thereby regulate the air temperature.

FIGURE 18-28 **CDA plug-in thermostat diagram. (Courtesy of Hussmann Corp.)**

Panelboard. The thermostat plugs into the panelboard. Each section of the panelboard has slots to accept four thermostats. The panelboard has four wiring connections for each thermostat: two wires to the CDA valve solenoid coil, and two wires to the temperature sensor in the refrigerator. There are also terminals for wiring the defrost clock and for the 24-V power supply from the transformer. The panelboard is equipped with a solid-state relay that will cause the CDA valve to close when the defrost clock applies 208 to 240 V to the CDA panelboard connections (Figure 18-29).

Temperature Sensor. The temperature sensor consists of a thermistor with a protective shield and a neoprene-covered lead wire (Figure 18-30). It has a nominal resistance at 77 °F (25.00 °C) of 1000 Ω. One sensor is required for each walk-in refrigerator or preparation room. It is installed in the discharge airstream of one unit cooler. One sensor is sometimes factory installed in the master refrigerator of each case lineup.

Figure 18-31 shows the change in resistance of the temperature sensor as compared to changes in temperature. The sensor consists of a carbon disc which increases in resistance with a decrease in temperature.

Diagnostic Board. The diagnostic board plugs into the panelboard slot and is used to check operation during the startup and service operations. There are four lights, which check:

1. 24-V ac supply

2. Defrost clock wiring and operation
3. CDA solenoid wiring for opens or shorts
4. Temperature sensor wiring for opens or shorts

Figure 18-32 shows a front view of the diagnostic board. See ''Service Tips'' for possible causes and remedies for malfunctions.

Installation and Setting Procedure. Use the following information for completing the installation of the sensor wiring and procedures for setting the CDA valve.

Installation of the Sensor Wiring. The only field wiring required for the CDA system is connecting the temperature sensor to the panelboard. Temperature sensors are readily mounted in display refrigerators but wires must be run from the sensor to the panelboard. Use 14-gauge wire with 600-V insulation. Route the sensor wiring in conduit to prevent damage. Each system (case or lineup) has been assigned a number on the store legend. Corresponding system numbers are located on the panelboard terminal strip (Figure 18-29).

The temperature sensors for unit coolers and preparation areas must also be field installed. Mount the temperature sensor in the discharge air according to Figure 18-33.

CDA Valve Setting Procedure. Use the following steps when setting the CDA valve:

1. Insert the diagnostic board to make sure that all systems are operational and that no systems are

FIGURE 18-29 Wiring Terminals on the CDA panelboard. (Courtesy of Hussmann Corp.)

FIGURE 18-30 Temperature sensor. (Courtesy of Hussmann Corp.)

Lights when the 24-V ac circuit is complete from the transformer to the plug-in slot

24 V

For the Koolgas system only; lights when the timer is in the defrost mode; fails to light if the timer output is less than 80 V

DEFROST

Lights when the circuit through the temperature sensor is complete

SENSOR

Lights when the circuit through the CDA solenoid is complete

COIL

FIGURE 18–32 **Diagnostic board. (Courtesy of Hussmann Corp.)**

FIGURE 18–33 **Temperature sensor installation. (Courtesy of Hussman Corp.)**

on defrost (Figure 18–32). Then remove the diagnostic board and install the thermostat.

2. Set the CDA thermostat to the required temperature for the refrigerator in use. (See the manufacturer's data for the correct settings.)

3. Start the compressor unit and allow the fixtures to pull down to their normal operating temperature. *Note:* This procedure assumes that the refrigerators are too warm at the time of startup.

4. Place a thermometer in the airstream of the control fixtures. For unit coolers, place the thermometer in the return airstream. *Note:* Further reference of control fixture and slave fixture will be as follows:

(a) *Control fixture:* that refrigerated fixture that contains the CDA temperature sensor.

(b) *Slave fixture:* the other refrigerated fixtures that are on the same suction line controlled by the CDA valve but do not contain the temperature sensor.

5. Allow a minimum of 1-hr operation at this setting. If the air temperature is more than 3 °F (1.68 °C) off the desired setting, adjust the CDA thermostat to bring it within this tolerance. Check those systems with the highest operating temperature first.

6. With the control fixtures adjusted to the correct temperature tolerance, the thermostatic expansion valve must be adjusted. To adjust the expansion valve properly, the CDA valve must not be allowed to control the system at this time. Mark the adjusted position of the CDA thermostat (Figure 18–27). Turn the thermostat adjustment to the coldest temperature (counterclockwise). This will make the CDA valve open up and stay open until the TEV adjustments are completed. Adjust all expansion valves that are controlled by the CDA valve at this time by use of the following procedure.

(a) The expansion valve must be adjusted to feed the evaporator fully. Before attempting to adjust the valve, make certain that the evaporators are either clear or only lightly covered with frost, and that the fixtures are within 10 °F (5.60 °C) of their expected operating temperature. Adjust the valve as follows:

CDA (Close on Drop in Air Temperature) Valve

(b) Attach two sensing probes (either thermocouple or thermistor) to the evaporator: one under the clamp holding the expansion valve bulb, and the other securely taped to one of the return bends two-thirds of the way through the evaporator circuit.

(c) Some "hunting" of the expansion valve is normal. The valve should be adjusted so that during the hunting action the greatest difference between the two probes is 3 to 5 °F (1.68 to 2.80 °C). With this adjustment, during a portion of the hunting action, the temperature difference between the probes will be less than 3 °F (1.68 °C) [at times as low as 0 °F (0 °C)]. Make adjustments of no more than one-half turn of the valve stem at a time and wait for at least 15 min before rechecking the probe temperature and making further adjustments.

Upon completion of setting the TXVs, reset the CDA thermostat to the reference setting. This should return the fixtures to the correct temperature tolerance.

7. Check the air temperature in the control fixture and make any necessary adjustments on the CDA thermostat to bring it to the desired operating temperature. When making final adjustments, a minimum of 30 min must be allowed for the system to balance to the new setting before checking the results. This step should be repeated as many times as necesary to get the correct control temperature.

8. Check the air temperature in all the fixtures and compare them to the control fixture temperature. A tolerance of 3 °F (1.68 °C) is an acceptable deviation from the control fixture. If one or more slave fixtures are out of this tolerance, the cause must be found and corrected.

Service Tips. The following tips are suggested for use when a problem occurs with the system.

Plug-In Thermostat. With a dc voltmeter (0 to 30 V dc) connected across the valve terminals, jumper the sensor terminals. The voltage should drop to 5 V dc or less signifying that the valve is open. Remove the jumper and disconnect one of the sensor leads. The valve should go to the full-voltage mode, signifying that the valve is closed. If either test indicates a wrong voltage, replace the plug-in thermostat with another of the proper temperature

range. If the problem is still present, check the wiring to the sensor and CDA solenoid coil according to the following procedure.

Solenoid Coil. To check the coil, measure its resistance. The reading should be approximately 52 Ω at 70 °F (21.11 °C).

Temperature Sensor. To check the temperature sensor, measure its resistance. The reading should be approximately 1000 Ω at 77 °F (25 °C). See Figure 18–31 for resistance at other temperatures. See the equipment manufacturer's specifications for the proper location of the temperature sensor for the cabinet in use.

Valve Replacement. The following steps should be followed when installing the CDA valve:

1. Wrap the valve in a wet cloth.
2. Solder or braze the body into the suction line. Heat should be directed away from the valve body. Low-melting-point, high-strength solders such as "Staybright" 95% tin and 5% silver are recommended to keep the valve body at as low a temperature as possible.
3. Reconnect the solenoid coil wires to the panelboard.
4. Test the system by plugging in the diagnostic board.
5. Adjust the CDA valve (see the setting procedures).

Disassembly of the CDA Valves. Use the equipment manufacturer's recommended procedures if disassembly of the CDA valve is necessary (Figure 18–34). Each manufacturer may have specific instructions, depending on the piping of the system.

Introduction to the Diagnostic Chart. Poor refrigeration performance can be the result of causes other than the CDA valve. To eliminate guesswork, carefully observe the problem and screen out other possible causes before concluding that the CDA valve is malfunctioning. For example:

1. Be sure that all refrigerators controlled by the valve are performing poorly. If only some are, the CDA valve is probably not at fault; the fault is probably within the refrigerators.
2. If all refrigerators are performing poorly, install pressure gauges at the nearest taps on each side of the CDA valve.

FIGURE 18–34   CDA-15 and 20 exploded view. (Courtesy of Hussmann Corp.)

3. Note the present setting of the thermostat and turn the dial to the coldest possible setting. The pressure difference across the valve should be 2 psig (13.78 kPa) or less, which indicates that the valve is wide open.
4. Next, turn the thermostat to the highest setting. The pressure difference across the valve should be significantly higher.
5. If both gauge readings are correct, the problem is not with the CDA valve or its control (Table 18–2).

☐ Air-Cooled Condenser Pressure Regulator ☐

The function of this valve is to maintain adequate high-side (receiver) pressure during low ambient temperatures. Sufficient high-side pressure must be

FIGURE 18–35   Typical low ambient control system for air-cooled condensers.

available at all times to provide an adequate pressure drop across the system expansion valve.

The recommended location for the condenser pressure regulator valve is at the condenser outlet between the condenser and liquid receiver. In this location the valve handles liquid and will perform as a high-capacity device (Figure 18–35).

It is also very important to note that these valves are not recommended for installation at the condenser inlet or other locations where compressor discharge pulsations may reach the valve. This regulator is responsive to the valve inlet pressure, opening on an inlet pressure increase and closing on an inlet pressure decrease. If the inlet pressure changes in rapid sequence (compressor discharge pulsations, etc.), the valve will be adversely affected. When a condition develops to cause the valve to open and close with each revolution of the compressor, the life of the valve bellows will be extremely short.

TABLE 18-2 Troubleshooting

Malfunction	Diagnostic Board Indicator Lights	Possible Cause	Action
Valve stays closed	1. Defrost [lit] 2. Sensor [unlit] 3. Beyond scope of diagnostic board	1. Defrost in "on" mode 2a. Open circuit in temperature sensor wiring b. Temperature sensor damaged 3. Plug-in thermostat circuit defective or wrong range	1. Readjust or repair defrost timer 2a. Locate and correct b. Replace sensor 3. Replace circuit
Valve stays open	1. 24 V [unlit] 2. Sensor [unlit] 3. Coil [unlit] 4. Defrost [unlit] 5. Beyond scope of diagnostic board	1. No 24 V ac to panel board 2a. Short circuit in temperature sensor wire b. Temperature sensor damaged c. Low temp case; frost on temperature sensor in flue 3a. Open circuit in coil wiring b. Coil has open winding 4. No voltage or low voltage defrost signal 5a. Discharge to suction pressure difference below 45 psi b. Plug-in thermostat circuit defective or wrong range c. Blocked orifice or restricted strainer d. Valve held open by solder or dirt e. Panel board damaged f. Panel board relay defective	1. Verify 24 V ac supply 2a. Locate and correct b. Replace sensor c. Relocate sensor away from evaporator 3a. Locate and correct b. Replace coil 4. Readjust or repair defrost timer 5a. Raise discharge pressure b. Replace circuit c. Repair or replace Restrictor Assembly d. Repair or replace valve e. Replace panel board f. Replace panel board
Valve does not control temperature	1. Sensor [unlit] 2. Beyond scope of diagnostic board	1a. Temperature sensor damaged b. Open or short circuit in temperature sensor 2a. Plug-in thermostat circuit defective b. Valve damaged	1a. Replace sensor b. Locate and correct 2a. Replace circuit b. Repair or replace valve

The addition of a dampener weight assembly in this case will not totally protect the valve, but it will assist in prolonging valve life.

☐ High- to Low-Side Bypass Valve ☐

These valves are installed between the system high and low sides and respond to the valve outlet (system low side) pressure. The valve opens when the low-side pressure is reduced to the valve setting. The machine capacity is thus reduced and the evaporator temperature ceases to drop as the suction pressures are maintained (Figure 18–36).

The bypass valve is an effective way to control system capacity as well as to limit a minimum evaporator temperature. Continuous machine operation is made possible without a reduction of evaporator temperature below predetermined limits.

The bypass gas may be piped through the valve to either the suction line ahead of the compressor or to the inlet of the evaporator downstream from the expansion valve. When the bypass gas is circu-

FIGURE 18-36 High- to low-side bypass valve location. (Courtesy of Singer Controls Division.)

FIGURE 18-37 Bypass control valve for air-cooled condensers.

lated through the evaporator, compressor protection against high superheat is provided by an expansion valve de-superheating effect.

Bypass Control Valve for Air-Cooled Condensers

These valves function as condenser bypass devices on air-cooled condensers with a winter start control. The valves are installed in the condenser bypass line between the compressor discharge and the liquid receiver (Figure 18–37).

Operating in response to receiver (valve outlet) pressure, the valve is open when the receiver pressure is below the valve set point. At the start of each machine on cycle, high-pressure discharge vapor is admitted directly to the receiver and quickly builds up adequate receiver pressure. When the valve opening pressure is reached, the valve closes and the compressor discharges directly into the condenser.

Were it not for the bypass control, some delay in reaching adequate receiver pressure would occur due to the time required for the larger condenser mass to reach operating temperatures.

Capacity Control Devices and Methods

On many refrigeration and air conditioning systems, the refrigeration load will vary over a wide range. This may be due to differences in product load, ambient temperatures, usage, occupancy, or other factors. In such cases, some type of compressor capacity control is necessary for satisfactory system performance.

On–Off Operation. The simplest form of capacity control is one-off operation of the compressor. This method is acceptable for small compressors, but it is seldom satisfactory for larger compressors because of the fluctuations in the controlled temperature. When light-load conditions are experienced, the compressor may short cycle. On refrigeration applications where the formation of ice on the coil is not a problem, users frequently lower the low-pressure cut-out setting to a point below the design limits of the system in order to prevent compressor short cycling. As a result, the compressor may operate for long periods at extremely low evaporating temperatures. Since the compressor capacity decreases rapidly with a reduction in suction pressure, the reduced refrigerant density and velocity is frequently inadequate to return the oil to the compressor. Operation of the system at temperatures below those for which it was designed may also lead to overheating of the motor-compressor. Both of these conditions can cause compressor damage and ultimate failure.

Compressor Cylinder Unloaders. To provide a means of changing compressor capacity under fluctuating load conditions, larger compressors are frequently equipped with cylinder unloaders.

Unloaders used on reciprocating compressors are of two general types. On some compressors, the suction valves on one or more of the cylinders are held open by some mechanical means in response to a pressure control device. With the suction valves open, refrigerant vapor is forced back into the suction chamber rather than the discharge chamber during the compression stroke, and the cylinder performs no pumping action.

Copelametic® compressors with unloaders have a bypass valve arranged so that the unloaded cylinder, or cylinders, are isolated from the discharge pressure created by the loaded cylinders. A three-way valve connects the discharge ports of the cylinder either to the normal discharge line when loaded or to the compressor suction chamber when unloaded (Figure 18–38). Since the piston and cylinder do no work when in the unloaded cycle other than pumping vapor through the bypass circuit (and only suction vapor), the problem of cylinder overheating while unloaded is practically eliminated. At the same time the electric power consumption of the compressor motor is greatly reduced because of the reduction in the work performed. The

reduced electric power consumption and better temperature characteristics of this type of compressor unloading, when properly applied, are major advantages over external hot-gas bypass unloading where all the cylinders of the compressor are working against the condensing pressure.

Because of the decreased volume of available suction vapor returning to the compressor from the system for compressor motor cooling, the operating range of unloaded compressors must be restricted. In general, compressors with capacity control are recommended only for high-temperature applications, but in some instances they can be satisfactorily applied in the medium temperature range. Because of the danger of overheating the compressor motor on low-temperature systems, either cycling the compressor or the hot-gas bypass method of capacity control is recommended.

Hot-Gas Bypass. Compressor capacity control by means of using hot-gas bypass is recommended where normal compressor cycling or the use of unloaders may not be satisfactory. Basically, this is a system of bypassing the condenser with the compressor discharge gas to prevent the compressor suction pressure from falling below a desired setting.

All hot-gas bypass valves operate in a similar manner. They open in response to a decrease in downstream (low side) pressure and modulate from fully open to fully closed over a given range of pressures. The introduction of the hot, high-pressure vapor into the low-pressure side of the system at a metered rate prevents the compressor from lowering the suction pressure further.

The control setting of the valve can be varied over a wide range by means of an adjusting screw. Because of the reduced power consumption at lower suction pressure, the hot-gas bypass valve should be adjusted to bypass vapor at the minimum suction pressure within the operating limits of the compressor, which will result in acceptable system performance.

If a refrigeration system is properly designed and installed, field experience indicates that maintenance may be greatly reduced if the compressor operates continuously within the design limitations of the system as opposed to frequent cycling of the compressor. Electrical problems are minimized, compressor lubrication is improved, and liquid refrigerant migration is avoided.

Therefore, on multiple-evaporator systems where

DE-ENERGIZED

Open to
Suction
Pressure →

Discharge from
Compressor Cylinder

To Condenser

ENERGIZED

To Compressor
Suction
Passage ←

Discharge from
Compressor Cylinder

Open to
Condenser
Pressure

FIGURE 18-38 Compressor-unloading valve. (Courtesy of Copeland Refrigeration Corp.)

the refrigeration load is continuous but may vary over a wide range, hot-gas bypass may not only provide a convenient method of capacity control, but may also result in more satisfactory and more economical system operation.

Bypass into Evaporator Inlet. On close connected systems using a single evaporator, it is sometimes possible to introduce the hot vapor into the evaporator immediately after the expansion valve (Figure 18-39). Refrigerant distributors are available with side openings for this purpose. An artificial cooling load is created by the bypass vapor entering at this point. Because the regular system thermostatic expansion valve will control the flow of refrigerant

required to maintain its superheat setting, the refrigerant returns to the compressor at normal operating temperatures. Thus no compressor motor heating problem is encountered. Because high refrigerant velocities are maintained in the evaporator, the oil return is also aided. Because of these advantages, this type of control is the simplest, least costly, and most satisfactory bypass system.

Bypass into Suction Line. Where multiple evaporators are connected to a single compressor, or where the condensing unit is located some distance from the evaporator, it is sometimes necessary to bypass the hot vapor into the refrigerant suction line (Figure 18-40). The suction pressures can be satisfactorily

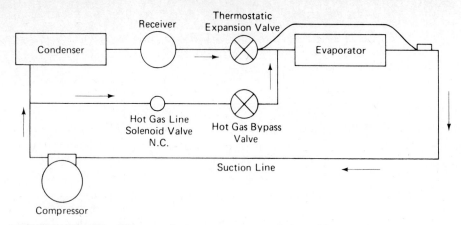

FIGURE 18-39 **Bypass into evaporator inlet. (Courtesy of Copeland Refrigeration Corp.)**

FIGURE 18-40 **Typical hot-gas bypass into suction line. (Courtesy of Copeland Refrigeration Corp.)**

controlled by using this method. However, a desuperheating expansion valve is required to meter the liquid refrigerant into the suction line in order to keep the temperature of the refrigerant vapor returning to the compressor within the allowable temperature limits. It is necessary to thoroughly mix the bypassed hot vapor, the liquid refrigerant, and the suction vapor from the evaporator so that the refrigerant mixture entering the compressor is at the correct temperature. A mixing chamber is recommended for this process. A suction line accumulator can serve as an excellent mixing chamber and at the same time protect the compressor from liquid floodback.

Another commonly used method of mixing the refrigerant is to arrange the piping so that a mix-

ture of discharge vapor and liquid refrigerant is introduced into the suction line at some distance from the compressor, in a suction header if possible (Figure 18-41).

□ **Pumpdown Control** □

Refrigerant vapor will always migrate to the coldest part of the system. If the compressor crankcase can become colder than the other parts of the system, the refrigerant in the condenser, receiver, and evaporator will vaporize and travel through the system. This vaporized refrigerant will condense in the compressor crankcase.

FIGURE 18–41 Alternative method of bypass into suction line. (Courtesy of Copeland Refrigeration Corp.)

Because of the difference in vapor pressures of the lubricating oil, the refrigerant vapor is attracted to the lubricating oil. Even though no pressure or temperature difference exists to cause the flow, the refrigerant vapor will migrate through the system and condense in the oil until the oil is saturated. During any equipment off cycles that extend over several hours, it is possible for liquid refrigerant to almost fill the compressor crankcase completely. Because of the attraction to equalize at an ambient temperature of 70°F (21.1°C), the oil–refrigerant mixture in the crankcase will be about 70% refrigerant before pressure equilibrium is reached.

The most positive and dependable method of preventing refrigerant migration is to use a system pumpdown cycle. By closing a liquid line solenoid valve, the refrigerant can be pumped into the condenser and receiver. The compressor operation can be controlled by use of a low-pressure control (Figure 18–42). The refrigerant can then be isolated during periods when the compressor is not in operation. Therefore, refrigerant migration to the compressor crankcase is prevented.

System pumpdown control can be used on all systems equipped with a thermostatic expansion valve with the addition of a liquid line solenoid valve, provided, of course, that adequate receiver capacity is available. A slight refrigerant leakage may occur through the solenoid valve during the off cycle, which will cause the suction pressure to rise gradually. Because of this rise in suction pressure, a recycling-type low-pressure control is recommended

FIGURE 18–42 Typical pumpdown system.

to repeat the pumpdown cycle as needed. The occasional short cycle usually is not objectionable.

A pumpdown cycle is highly recommended whenever it can be used. If a nonrecycling pumpdown cycle is required, then consideration should be given to the use of a crankcase heater in addition to the pumpdown for more dependable compressor protection.

□ Commercial Refrigeration Defrost Systems □

Refrigerated cases are equipped with evaporator coils in which liquid refrigerant is evaporated. This evaporating liquid absorbs heat from the case. Relatively warm, moist air that passes through and over the coil gives up its heat and as a result loses its moisture in the form of frost being deposited on the fins and tubes of the evaporator. This frost gradually builds up until the air movement is reduced and eventually completely stopped if the evaporator is not defrosted. Since proper refrigeration of the case is dependent on air circulation, it is necesary to defrost the evaporator periodically so that refrigeration may continue.

Evaporator defrosting is accomplished by the following basic methods:

Off-Cycle Defrost. This method uses a low-pressure control to control the compressor operation (Figure 18–43). By setting the control cut-in point at a pressure and hence a temperature above freezing, the evaporator will be completely clear of frost before the compressor starts again. This method of defrost is used for closed meat cases, produce cases, and coolers.

Timed Defrosts. This type uses a defrost control, usually a time clock, to provide the desired time and duration of the defrost cycle.

No Supplemental Heat. This method is used on open and multishelf meat and dairy cases, multishelf produce cases, and coolers. Since the evaporator does not completely defrost between compressor cycles or when the compressor does not cycle at all, an enforced off period allows a natural warming to clear the evaporator of any accumulated frost.

With Electric Heaters. This method is sometimes used with meat and dairy cases. It is usually a standard method used with frozen food and ice cream cases.

FIGURE 18–43 Low-pressure control. (Courtesy of Penn Controls, Inc.)

FIGURE 18–44 Typical hot-gas defrost system.

With Hot Gas. This method is used on some unit cooler evaporators but almost never on display cases. This method uses valves operated by a defrost control to circuit the compressor discharge vapor directly to the evaporator (Figure 18–44). Usually some method must be used to avoid flushing liquid refrigerant from the evaporator to the compressor. Extra heat is often used to supplement the available sensible heat of the hot vapor since only the heat of friction and some motor heat is added after the first few minutes of defrosting.

Latent Heat Defrosting. The latent heat defrost method is used sometimes where there are multiple evaporators. In its simplest form, there would be three evaporators with one defrosting while the other two refrigerate (Figure 18–45). The coil being defrosted actually acts as the condenser for the refrigerant that is being evaporated in the two refrigerating evaporators. This procedure is sequenced usually by a programmed time clock so that each of the three evaporators is defrosted as often as needed in rotation.

Direct Electric Defrost. When supplemental heat is required for defrost, some manufacturers equip their cases with supplemental electric heat strips and a defrost limit switch. The advantages of electric defrost are:

Latent Heat Defrost as Used on a Combined System
(Do not confuse this with reverse cycle defrost with all cases being defrosted at once.)
Only one-third to one-fourth of the total Btuh load can be defrosted at any one time.

Refrigeration cycle — Liquid refrigerant is expanded through the expansion valve, absorbs heat in the coil and is drawn as a vapor back to the condensing unit.

The multi-circuit time clock control operates solenoid-pilot-operated three-way valves when a defrost is scheduled.

Defrost cycle — three-way valve (at condensing unit or near case line-up) sends hot gas from compressor thru suction line. Drain pan is heated first and frost is melted from the coil as the hot gas cools and condenses. Condensed liquid is returned to liquid line through the by-pass and check valve around the expansion valve. EPR's (when used) are located either downstream from the three-way valves or upstream. An accumulator prevents condensed liquid from harming the compressor.

Case line-up on defrost cycle

Case line-up on refrigeration cycle

Bypass and check valve

Case line-up on refrigeration cycle

EPR

Liquid line

Three-way valves

Valve

Normal position

Suction line

Locations

Defrost position

(When used)

Normal position

Receiver

Compressor

Condenser

Liquid line solenoid, N.O., closes during a defrost of any part of the system. Pressure control on liquid line overrides solenoid, opening it when pressure drops too low.

Accumulator

INSTALLATION NOTICE — Since suction lines (used as hot gas lines during defrost) may lengthen up to 2" in 100 feet when the hot gas heats the line, vibration absorbers are recommended to prevent line damage. The lines must also be allowed to "float free" in the trenches to allow full freedom of movement.

Note: These lines must be insulated.

Vibration eliminators

FIGURE 18–45 Latent heat defrosting cycle. (Courtesy of Tyler Refrigeration Corp.)

1. Low initial cost
2. Simple installation
3. Efficient operation
4. Easy maintenance and service

All these factors in the lowest total operating cost on a year-to-year basis. Additionally, the defrost limit switch affords a built-in safety device used also to terminate reliably the defrost cycle by using electric current sensing relays.

Methods of Terminating Defrost. All time-clock defrost periods are initiated on a time basis. Termination is accomplished mainly by three methods.

Straight Time Termination. The system is controlled entirely by the time clock. The length of the defrost cycle, a constant preset value, must be determined by past experience with similar systems or by adjustment after installation. While this is a reliable method, it has the disadvantage of making no allowance for changing conditions of the service load, humidity, or other variables. To ensure proper operation at all times, the length of the defrost period must be set for the most severe conditions, which could result in excessively long defrost periods much of the time.

Pressure Termination. A pressure switch incorporated in the control is connected to the low side of the refrigeration system (Figure 18–46). At a predetermined and preset pressure—most often around 45 psig (410.14 kPa) for low-pressure systems and 40 psig (375.79 kPa) for nonelectric medium temperature systems using R-12 as the refrigerant—the defrost is terminated. The controls are usually provided with a fail-safe time termination in case of failure of the pressure switch. The pressure-terminated control has the advantage of adjusting automatically to the changing conditions of service load and humidity. This results in defrost periods of only such duration as are needed to clear the evaporator of frost.

Note: Cold suction lines or cold machines—outdoor condensing units—will prevent this type of termination from being fully reliable. It works best when the pressure rise is consistent throughout the year.

Solenoid Reset Termination. This type of defrost clock (Figure 18–47) usually has a fail-safe dial and can be used in a number of ways.

1. Close on rise of thermostats on evaporators in the cases. A series pilot circuit must be installed

FIGURE 18–46 Pressure-terminating defrost control clock. (Courtesy of Paragon Electric Corp.)

FIGURE 18–47 Solenoid defrost control clock. (Courtesy of Paragon Electric Corp.)

Current Relay
(RBM Controls)

Limit Switch
(Texas Instrument)

Portion of control panel wiring diagram showing use of a current relay. Defrost limit switches are sometimes mounted to return bends of case coils.

FIGURE 18–48 Tyler TG control defrost. (Courtesy of Tyler Refrigeration Corp.)

to all cases. The solenoid resets when all the thermostats are satisfied completing this pilot circuit. It can be used on either electric or hot-gas defrost systems.

2. Open on rise of thermostats on the evaporators in series with electric heater elements. The electric supply to the heaters from the defrost contactor or time clock is wired through a current relay. As long as the current flows through the defrost heaters, the relay holds open an electric circuit to the solenoid (Figure 18–48). When all of the case thermostats in the cases have opened, zero current draw to the electric heaters allows the relay contacts to close and reset the solenoid. The advantages here are that each case is protected individually from excessive defrost heat, and no extra wiring is required to the cases.

3. Close on rise—pressurestat activates the solenoid. This is just a variation of the pressure termination method of defrost.

Demand Defrost. Solid-state components have made accurate monitoring of simple devices possible. One method monitors the case air velocity, initiating a defrost cycle. A lowered air velocity indicates that the evaporator has accumulated all the frost it can hold and still maintain the desired case temperature.

Another method employs the *fluidic principle* by monitoring the air pressure in the front and the back of the evaporator. The fluidic principle amplifies small pressure changes so that a defrost can be initiated when necessary. Termination for either method is usually controlled by a thermostat.

There are basically two types of controls used for defrost systems. These are the low-pressure switch and the time clock. Most variations will have a combination of these two types of controls.

Pressure Control. Since a pressure–temperature relationship is known for each type of refrigerant (Table 18–3), regulating the suction pressure of the compressor will also regulate the operating temperature of the evaporator in the case.

The dual pressure control used by most equipment manufacturers has three adjustable settings (Figure 18–49). One setting is the high-pressure cut-out, a safety device which is set according to the type of refrigerant used and the manufacturer's recommendation. The other two settings are the cut-in pressure—the pressure at which the compressor starts—and the differential—the number of pounds of pressure that the compressor suction will pull the system down to before stopping. The cut-out is the value determined by subtracting the differential setting from the cut-in setting. Do not confuse this cut-out setting with the high-pressure cut-out setting.

Defrost Time Clocks. The 24-hr time-clock-operated devices have both normally open (NO) and normally closed (NC) contacts. Up to six defrost periods can be scheduled per 24 hr with the pins provided. The defrost controls used by most manufacturers initiate a defrost period when a pin on the main dial trips the switches (Figure 18–50). The normally closed contacts allow electric current to pass through the defrost control to the dual pressure control, which in turn starts and stops the condensing unit. During a defrost period the condensing unit is shut off and the normally open contacts close, energizing the defrost heater circuit, if one is used (Figure 18–51).

Defrost methods are named by their mode of defrost initiation and termination. The following are the basic types.

1. TC (straight time clock defrost—time clock termination). The length of the defrost period is set on the inner dial and can be set from 2 to 110 min.

2. TP (time pressure termination). The length of the defrost period varies with the time required for the suction pressure to raise the cut-in setting

TABLE 18-3 Pressure–Temperature Chart[a]

°F	R-12	R-22	R-502	°F	°C
−60	19.0	12.0	6.9	−60	−51.1
−55	17.3	9.1	3.6	−55	−48.3
−50	15.4	6.0	0	−50	−45.0
−45	13.2	2.6	2.0	−45	−42.7
−40	11.0	0.6	4.3	−40	−40.0
−35	8.3	2.6	6.7	−35	−37.2
−30	5.5	5.0	9.4	−30	−34.4
−28	4.2	6.0	10.5	−28	−33.3
−26	3.0	7.0	11.7	−26	−32.2
−24	1.6	8.0	12.9	−24	−31.1
−22	0.2	9.1	14.2	−22	−30.0
−20	0.6	10.2	15.5	−20	−28.9
−18	1.3	11.3	16.9	−18	−27.8
−16	2.1	12.6	18.3	−16	−26.7
−14	2.8	13.9	19.7	−14	−25.6
−12	3.7	15.2	20.2	−12	−24.4
−10	4.5	16.6	22.8	−10	−23.3
−8	5.4	18.0	24.4	−8	−22.2
−6	6.3	19.3	26.0	−6	−21.1
−4	7.2	21.0	27.7	−4	−20.0
−2	8.2	22.5	29.4	−2	−18.9
0	9.2	24.1	31.2	0	−17.8
2	10.2	25.7	33.1	2	−16.7
4	11.2	27.5	35.0	4	−15.6
6	12.3	29.2	37.0	6	−14.4
8	13.5	31.1	39.0	8	−13.3
10	14.7	33.0	41.1	10	−12.2
12	15.9	34.9	43.2	12	−11.1
14	17.1	36.9	45.4	14	−10.0
16	18.4	39.0	47.7	16	−8.9
18	19.7	41.0	50.1	18	−7.8
20	21.1	43.3	52.4	20	−6.7
22	22.5	45.5	54.9	22	−5.6
24	24.0	47.9	57.4	24	−4.4
26	25.4	50.1	60.0	26	−3.3
28	27.0	52.7	62.7	28	−2.2
30	28.5	55.2	65.4	30	−1.1
32	30.1	57.8	68.2	32	0
34	31.8	60.5	71.1	34	1.1
36	33.5	63.3	74.1	36	2.2
38	35.2	66.1	77.1	38	3.3
40	37.0	69.0	80.2	40	4.4
45	41.7	76.6	87.7	45	7.2
50	46.8	84.7	96.9	50	10.0
55	52.1	93.3	107.9	55	12.8
60	57.8	102.4	115.6	60	15.6
65	63.9	112.1	125.8	65	18.3
70	70.3	122.5	136.6	70	21.1
75	77.1	133.4	147.9	75	23.9
80	84.3	145.0	159.9	80	26.7
85	92.0	157.2	172.5	85	29.4
90	100.0	170.0	185.8	90	32.2
95	108.4	183.6	199.7	95	35.0
100	117.4	197.9	214.4	100	37.8
105	126.8	212.9	229.7	105	40.6
110	136.7	228.6	245.8	110	43.3
115	147.1	245.2	266.1	115	46.1
120	158.0	262.5	280.3	120	48.9
125	169.5	280.7	298.7	125	51.7
130	181.5	299.7	318.0	130	54.4
135	194.1	319.6	338.2	135	57.2
140	207.2	340.3	359.2	140	60.0
145	221.0	362.0	381.1	145	62.8
150	235.4	384.6	404.0	150	65.6
155	250.4	406.3	427.8	155	68.3
160	266.1	443.3	452.6	160	71.1

Source: Tyler Refrigeration Corp.

[a]Vacuum in inches of mercury, pressure in psig.

FIGURE 18-49   Dual pressure control. (Courtesy of Penn Controls, Inc.)

Defrost Initiating Pin

Fail Safe Dial

Fail-Safe Setting

FIGURE 18-50   Defrost clock dial. (Courtesy of Tyler Refrigeration Corp.)

(Table 18-4). TP controls can be used successfully only when pressure-limiting expansion valves are not used. These valves prevent a normal pressure rise during the defrost period.

3. TG (temperature guard temperature termination). This is a Tyler Refrigeration Corp. innovation that utilizes a current relay that holds the time clock in defrost until all of the case limit switches are open, or the fail-safe dial setting takes effect.

Note: All cases equipped for electric defrost systems have defrost limit switches attached to the evaporator tubing to cut off the electric current to the defrost heater(s) when the evaporator temperature reaches 50 °F (10 °C). Thus each case is individually protected from overheating during the defrost period. When all the limit switches have opened, the electric current ceases to flow, allowing the current relay to close. This, in turn, activates a solenoid, tripping the time clock back to normal operation. As in the first two controls, the inner dial is set for maximum or fail-safe time (Figure 18-52).

Setting Defrost Controls. When the defrost period is dependent on the off-cycle defrost method, the pressure switch cut-in setting must be at 36 or 37 psig (348.31 or 355.18 kPa) when R-12 refrigerant is used. This is most accurately determined with a pressure gauge because the indicator on the pressure

On a TC Control — No Extra Devices
On a TP Control — Tubing to Low Side
On a TG Control — Current Relays Would
          be in the Control Box.

Tubing to Low Side of Compressor

Tubing to High Side

Circuit Breaker
(or Pull-Fuses)

Motor Contactor

Dual Pressure Control

Defrost Wiring to Cases

To Compressor

Defrost Control

FIGURE 18-51 Typical control box with controls. (Courtesy of Tyler Refrigeration Corp.)

TABLE 18-4 Reset Pressure Settings (psi) for TP Controls

Refrigerant	Low-Temperature Cases	Normal Temperature Cases	
		Without Extra Heat	With Electric Heaters
R-12	45	40	52
R-502	95	84	100

Source: Tyler Refrigeration Corp.

FIGURE 18-52 TP defrost control. (Courtesy of Paragon Electric Corp.)

control is not that accurate. The use of a pressure gauge when setting this control is also recommended when a timed defrost control is used.

Setting the timed defrost control involves setting the length, in minutes, of each defrost period and setting the number of defrost periods each day. Most manufacturers publish their recommendations as follows:

1. On a case setting label inside each case
2. In a case setting card packed with the condensing unit
3. In the instructions applicable to the case

These recommendations may not always serve the need of a particular installation. However, they do serve as a starting point. They are based on laboratory and field experience.

Variable Factors. Factors that affect the defrost cycles are:

1. Store humidity and temperatures vary throughout the year. The amount of frost to be removed, therefore, varies throughout the year.
2. Voltage to the defrost heaters may vary causing the amount of heat available also to vary.
3. The use of doors (coolers and reach-ins) varies.
4. The operating temperature of the case (particularly with multishelf cases, low and normal temperatures) may be lower than necessary. This will cause an extra heavy frost load on the evaporator.
5. Drafts across the case permit too much warm, moist air to enter the case, thereby loading the evaporator excessively.

Principles of Defrost. The duration of defrost must be long enough to clear the evaporator completely each defrost period, while the defrost frequency must be often enough to keep the refrigerated air circulating freely throughout the case.

A recording temperature gauge is the most efficient method of determining when a case evaporator is becoming choked with frost. When the temperature "sags" before a defrost period, another defrost period per day is required. Make sure, however, that the factors causing the temperature sag (case or ambient temperature, humidity, usage, or drafts) have been controlled as much as possible.

□ Multiple (Combined) Systems □

Not too long ago almost everyone subscribed to the theory that a single refrigeration unit was required for each case. Some still think that satisfactory operation will be obtained only if not more than two display cases per condensing unit are installed. In general, however, this type of thinking has advanced to the point that more people are combining more systems. Most store planners and operators are finding it entirely satisfactory and desirable to combine not only similar cases but cases of differing temperature requirements as well.

Combining systems is certainly not new. The combination of refrigeration systems on very large, continuously operating condensing units has been in use since the 1920s. However, the advantages of these large systems are still somewhat nullified by the complexities caused by involved combining.

The choices the design engineer must make when designing a system are many and varied. He (or she) can use from 1 to 20 compressors, depending on compressor size. If 1 or 2 compressors are chosen, the designer must combine systems extensively. If 20 compressors are chosen, they are small and less efficient compressors, along with 20 small refrigeration systems. Most experienced designers agree that the best route lies between these extremes.

Basically, combining is the grouping of more than one refrigeration function on one compressor. This procedure usually requires the use of larger condensing units. Some of the reasons for combining are:

1. Good, relatively inexpensive, larger condensing units are now available.
2. Improved and varied controls are now on the market.
3. Refrigeration technology has improved.
4. The prepackaging of machinery can now take more of the installation burden from the local level.

However, the most important pros are economic, saving the user money in:

1. Original costs and installation costs
2. Operating costs
3. Maintenance costs

Motor efficiency is one of the attributes of a larger compressor. Above 5 hp, however, there is not too

FIGURE 18–53 Motor efficiency comparison. (Courtesy of Tyler Refrigeration Corp.)

FIGURE 18–55 Simple multiple system. (Courtesy of Tyler Refrigeration Corp.)

FIGURE 18–54 Simple system. (Courtesy of Tyler Refrigeration Corp.)

Ice Cream Cases

Frozen Food Cases with Ice Cream Defrost Heaters and an Evaporator Pressure Regulator on the Suction Line.

FIGURE 18–56 Simple combined system. (Courtesy of Tyler Refrigeration Corp.)

much efficiency to be gained (Figure 18–53). It also must be considered that the larger the compressor, the more complicated the total system will tend to be.

Simple Systems versus Combined Systems. A simple system is one case connected to one condensing unit as shown in Figure 18–54. The electrical wiring for this system is also simplified.

A simple multiple system is two or more cases of the same suction pressure—evaporator or suction temperature—and defrost requirements connected to one condensing unit (Figure 18–55). The electrical wiring and the refrigerant system are more complicated than in the simple system.

A combined system involves cases of different suction pressures and/or defrost requirements on one condensing unit (Figure 18–56).

One thing that must be considered when combining cases and/or coolers of different suction pressures is the operating premium that must be paid. A rule of combining is that the entire system must be sized to match the Btu, which are figured at the lowest suction pressure required in the system

(Table 18–5). The operating costs rise in this situation.

Complex combined systems are designed to handle all low-temperature (or all normal temperature) cases and coolers on one condensing unit.

Ultracomplex combined systems include all of the refrigeration equipment in the store on one condensing unit (usually with an air conditioning or other unit available as an emergency spare).

TABLE 18–5 Total Heat (1000 Btu) Rejected from Condensing Units

	3 hp	5 hp	5 hp	7½ hp	7½ hp	10 hp	15 hp	20 hp
High-Back-Pressure Compressors, R-12								
Designation	300H		500H	750H		1000H		2000H
Displacement CFH	698		1080	1620		2380		3580
+35°	39.2		58.4	90.0		134.0		202.0
+20°	28.2		43.4	68.0		103.0		156.0
+10°	25.4		34.0	52.5		85.0		128.0
Medium-Back-Pressure Compressors, R-12								
Designation	300S		500S		760S	1000S	1500S	2000S
Displacement CFH	815		1375		2120	2380	3020	3580
+20°	34.0		55.2		88.0	103.5	188.0	155.0
+10°	27.8		45.0		72.3	85.0	145.0	128.0
Low-Back-Pressure Compressors, R-502 Only								
Designation	300L	450L	500L	750L	760L	1000L	1500L	2000L
Displacement CFH		1080	1375	1620	2120	2380	3020	3580
−25°	Marginal	31.2	39.0	45.0	58.4	70.8	88.4	94.5
−35°		24.7	30.8	36.8	48.0	56.0	69.5	79.4

Source: Tyler Refrigeration Corp.

All of these systems are mechanically feasible. The equipment and controls to make them work are all practical and available. The equipment, which will serve the customer best through the years that follow the installation, should be given first priority in making the selection.

The maintenance of commercial refrigeration equipment is also of prime importance and is usually dependent on competent local service technicians. The equipment to be used should be within their ability to service properly.

In general, the following statements serve well for most supermarket installations:

1. The use of 5-, 7½-, 10-, and sometimes 20-hp units will reduce the number of condensing units in an average supermarket to 6 or 8. A saving of space will be realized. The condensing units can even be put in a prepackaged room outside the store. In addition, these units are fairly common, with parts or entire replacements available on an over-the-counter basis (Figure 18–57).
2. The resultant refrigeration systems will range between the simple multiple type and the simple combined type.
3. Service on these sizes of compressors and on these types of systems is generally available locally.

It can be seen that every mechanical selection is made by weighing positive and negative factors.

When the positive factors outweigh the negative, then it is a good selection and a good compromise. Negative factors do, nevertheless, exist and have to be recognized for what they are.

Maintaining Case Temperature Differences. To separate cases and coolers of different temperatures, there are two commonly used methods. One is the use of liquid line solenoid valves controlled by thermostats (Figure 18–58). This method is acceptable in situations where ups and downs of temperature (due to the thermostat differential) inside the case will not be critical. However, this method introduces a secondary responsibility for the performance of the whole system. Extra electric circuits must be installed (an extra installation cost) and these circuits must be maintained.

The second method is to use evaporator pressure-regulating (EPR) valves. These are nonelectric, pressure-actuated devices that control the refrigerant pressure in the evaporator (Figure 18–59). This method has proven to be highly reliable. EPR valves are installed in the suction line of a refrigeration system. Initially, they must be adjusted to the desired pressure (hence the desired temperature). Once they are adjusted, they will maintain a constant evaporator temperature. With a constant refrigeration load, as is the case with open supermarket equipment, constant evaporator temperatures will main-

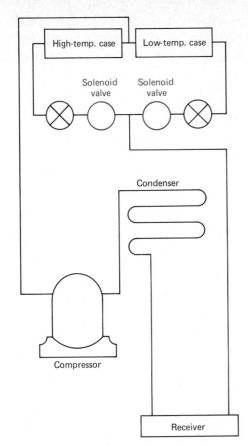

FIGURE 18-57 Solenoid valve location.

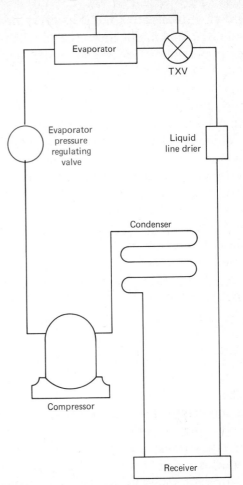

FIGURE 18-59 Evaporator pressure-regulating installation.

FIGURE 18-58 Composite chart—typical thermostat and solenoid valve action. (Courtesy of Tyler Refrigeration Corp.)

FIGURE 18-60 Composite chart—EPR operation. (Courtesy of Tyler Refrigeration Corp.)

tain almost perfectly constant case temperatures (Figure 18-60).

Maintaining Defrost Requirement Differences. Sometimes these defrost differences can be resolved by making the defrosts alike. For instance, a top-display produce case can tolerate the four defrosts per day required for a multishelf produce case. However, this is not generally done. Another equalizing method is that of equipping frozen food cases

with an extra defrost heater so that it can be defrosted in the same way as ice cream cases.

An ordinary defrost time clock is not always suitable for defrosting the combined system. A multicircuit time clock has been developed that is capable of handling six different functions with its six pairs of normally open and normally closed switches. Each pair of switches has its own 24-hr clock dial with initiating tab slots every hour and a terminating dial timing up to 110 min. This clock makes it possible to delay the cooler fans after defrost. It can also be used to synchronize defrosts permanently for case lineups on more than one condensing unit.

Compressor unloaders could be required when most of the load on a system is isolated during defrost while the compressor is still running. Actually, all compressors have a built-in unloading feature. As case requirements are satisfied and the suction pressure drops, the amount of refrigerant vapor pumped is reduced. The cut-out on the pressure control should be set low enough to permit this. Avoid the unnecessary use of unloaders because they make a compressor inefficient. A half-unloaded compressor still draws 70% of full-load amperes.

□ **Heat Recovery Systems** □

People in most businesses today must get the maximum out of every dollar of operating cost. The food store operator is certainly no exception. This is the reason that so many people have been so interested in the technique of heat recovery when considering new stores or planning an extensive remodeling program.

The major purpose of all heat reclaim equipment is to reduce the energy consumption and costs of operating a building or a process by transferring energy between the supply and exhaust airstreams. A second purpose is the reduction in size and capital costs of the supporting utility equipment such as boilers, chillers, and burners used in the building or process. The proper application of heat reclaim equipment, in many cases, can result in very little if any additional first cost for the building or process, while at the same time providing long-term benefits through reduced energy consumption.

Heat recovery is a most complex subject and, if it is discussed in its entirety, would leave the uninitiated completely baffled. To understand it fully, one must have more than a nodding acquaintance with basic electricity, refrigeration, heat transfer, and cost accounting.

Let us suppose that a food store operator or a chain store group has decided to use heat recovery systems. Possible yearly projected savings indicate that their investment can be paid for in five years, which is acceptable.

The savings that justify the expense of added equipment come from the simple fact that a refrigeration system is made to move heat. It removes heat from display cases and discharges the heat elsewhere—either directly to air or water. The refrigeration compressor is the pump, and the refrigerant is the fluid that acts as a vehicle for the heat.

The heat extracted from the display cases comes from the store in the beginning, so if the heat is discharged to the outside air during the heating season, it will have to be replaced by conventional heating sources. By discharging the heat into the store, some or most of the normal heating costs can be saved. There is an added bonus in that the heat generated by normally operating electric motors, which are powering the refrigeration compressors, is also recovered at the condenser. With well-engineered and installed heat recovery systems, there have been amply documented reports that a large portion of the winter heating cost is eliminated. In some areas, particularly the southern states, no makeup heat is necessary in an average year.

The amount of heat available for use in a heat recovery system is tremendous. For instance, a 15-hp unit operating dairy cases at 10°F (−12.2°C) suction temperature has 145,000 Btu (36,540 kcal) of heat to reject, all of which is available to heat the store when properly channeled (Table 18–5). This single unit rejects more heat than is required to heat a large home! The heat from this and other units found in a typical supermarket can quickly add up enough heat to maintain the store at comfortable levels on all but the coldest days of a normal heating season in the midwestern United States.

Care must be taken in the design of heat recovery systems to eliminate every unnecessary cost. These costs, though often hidden, must come out of the budget. A primary example is unnecessary electrical consumption for moving air or water in transferring the heat.

Basic Approaches to Heat Recovery. No one can state with any certainty the origin of recovering heat in food stores, but it is easy for the "old timers" to recall homemade arrangements that did the job. Some of these date back close to 20 years.

Direct-Air Method. In the summertime, the air from an air-cooled condenser was discharged to the atmosphere, but in the winter that air was utilized to supplement store heating. These systems worked. Even so, they were cumbersome, moved great quantities of unfiltered air, and often required blower fans with high initial and operating costs. In these arrangements, even in a store that used no natural fuel for heating, the electrical costs for moving the air around the store could not be ignored. They frequently involved 10-hp motors or larger, whose monthly operating costs would be approximately $100 at 0.02 cent per kWh.

Recognizing the shortcomings of the first approach, various other schemes have been used or tried. One variation of the direct-air approach used a single multiple-circuit condenser with face and bypass dampers on both sides. Some even used two-speed fans so that in the wintertime the large volume of air was not required. The basic flaw in that particular scheme lay in the fact that the building had to be specifically prepared to utilize this type of approach (Figure 18–61).

Closed-Circuit Water Method. The northwest area of the country has for years utilized closed-circuit water-cooling for the condensing side of refrigeration units. This scheme led some, quite normally, to simply circuit the hot water to a coil in an air handler, specially put there to discharge the heat from the water into the store air during the wintertime. In the summertime, a three-way valve would isolate that coil and allow the water to be cooled in the evaporative condenser where the heat was discharged to the atmosphere.

On first examination, this scheme has merit, but further study discloses the fact that once again pumping horsepower must be considered. With these arrangements frequently using 5-, 7½-, or 10-hp

water pumps, the cost of operating the water pump must be taken into account on the debit side. Furthermore, the closed-circuit arrangement must utilize closed-circuit water running at 100°F (37.8°C) or cooler. A 100°F heat source is not particularly good for the store because it is far too cool. The temptation has been to raise the head pressure in order to raise the water temperature so that more comfortable and reasonable conditions may be maintained in the store. However, this leads to other concealed costs that come from higher maintenance because of the higher pressures and the higher cost of getting the refrigeration job done using higher discharge pressures.

Dual-Condenser Method. Any mechanical alternative has faults; it is, in its own way, a compromise. However, the dual-condenser or alternate condenser technique seems to have the fewest shortcomings.

This is an arrangement whereby air or remote air condensers may be used for summertime cooling. Thus, maximum effectiveness of the refrigeration equipment is achieved during the most trying season. A refrigerant vapor valve bypasses the primary condenser and causes the vapor to be circulated in an appropriate coil located in the store air-handling system.

Figure 18–62 illustrates a typical arrangement showing the relationship of the air conditioning cooling coil, fans, filter, the heat recovery coil, and the makeup furnace. The heat recovery coil can be single or multiple circuited depending upon the number of condensing units from which heat is recovered. Ob-

FIGURE 18–61 Direct air recovery system. (Courtesy of Tyler Refrigeration Corp.)

Heat Recovery Systems

425

Pilot operated
hot gas diverting
valve

This condenser used when heat
is not required in the store

When heat is required, the hot gas goes
directly to the H.R. coil to heat store air

Furnace bypass (as required)

Other H.R.
circuits

Check valves

To cases   Make-up
air

Blower
isolating
section

Air duct to store

Return
air

Air conditioning
coil section

Multi-
circuit
heat
recovery
coil

Blower section (blower
should operate continually
to maintain proper pressure
balance in the buliding)

Optional gas
(or oil) make-up
heat section

Filter
section

Optional
electric make-up
heat section

**FIGURE 18–62   Dual condenser heat recovery system. (Courtesy of
Tyler Refrigeration Corp.)**

viously, the piping of six condensing units to such
a coil is comparatively simple. The piping of 8 or
10 might be quite acceptable, but the use of more
than a dozen condensing units with a heat recovering technique might be deemed impractical for an
installation.

Nevertheless, regardless of the number of units
employed, the heat is recovered for the least possible cost. Head pressures are still maintained at an
optimum level—very low, but not so low as to cause
refrigeration problems. The temperature of the
vapor being circulated through the recovery coil is
high—most of it around 150 °F (66 °C) or higher.
Furthermore, there is no cost for pumping the heat
in any fashion except that which would normally be
required because the fan in the air conditioning section would be required to distribute heat even
without heat recovery equipment. Even better, while
the heat recovery system is operating, condenser fans
and/or water pumps are turned off on the alternate
outdoor condensers.

Illustrated in Figure 18–61 is the whole scheme as
it relates to a food store, showing the relationship
of makeup air to the return ducting as well as the
entire air handler package to the store. In this con-

nection, it is important to mention two significant
points. In most situations, it can be flatly stated that
the best heating and air conditioning and distribution of these throughout the store will be achieved
with constant running of the fan in the air handler.

The second point is one being adopted by one of
the companies promoting doorless openings for food
stores. Their contention that "You can't heat a va-
cant lot" seems to make sense. This leads to the
second significant statement. Allowing the air from
a building to escape through various uncontrolled
openings can lead to loss or control of air distribution throughout the store. It can allow cold winter
air to find its own entrances into the building and
let shoppers be subjected to it before the air is
warmed to a comfortable level. Tightening the
building by eliminating unnecessary openings seems
a reasonable requirement with the idea that controlled makeup air will come into the building as it
is required. The above thoughts would seem to apply
equally to a store with conventional doors as to one
with a doorless arrangement.

Process-to-Process. In this type of heat reclaim
system, heat is captured from the process exhaust

FIGURE 18-63 Process-to-process sensible heat device. (Courtesy of ASHRAE.)

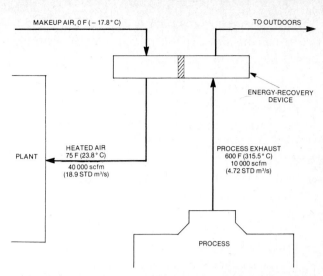

FIGURE 18-64 Process-to-comfort sensible heat device. (Courtesy of ASHRAE.)

airstream and transferred to the process supply airstream. Higher-temperature processes are generally preferred because more heat is available for reclaiming. Equipment that can handle process exhaust temperatures as high as 1600°F (870°C) is available from different manufacturers.

Some of the more popular applications of this type of equipment are driers, kilns, and ovens. An application for a typical oven for an air-to-air heat reclaim unit for 70% sensible heat reclaim, and operating under typical winter design conditions, is shown in Figure 18-63.

When applications of the air-to-air heat reclaim systems from process exhaust are being considered, the following conditions should be carefully evaluated:

1. *Effects of corrosive materials.* Process exhaust air frequently contains compounds that require compatible construction materials for the heat reclaim unit.
2. *Effects of condensables.* The process exhaust may contain a compound with a concentration high enough to cause it to condense upon being cooled in the heat reclaim unit. The condensed compound can, in some cases, be recovered.
3. *Effects of contaminants.* When the process exhaust contains particulate contaminants of condensables, the heat reclaim unit should be easily cleaned. Air purification and the selection of heat reclaim units of the open type to minimize the cleaning procedure should be considered.
4. *Effects on other equipment.* Removing the heat from the process exhaust air may reduce the cost

of pollution control because less expensive bags may be used in the baghouses or by improving the efficiency of electronic precipitators. As a result, heat reclaim and pollution control can often be coupled with beneficial effects.

Generally, process-to-process heat reclaim units will recover only sensible heat and do not transfer latent heat (humidity). In most cases, the transfer of moisture is detrimental to the process. Because maximum heat recovery is desired, modulation of the recovery process is not desired in a process-to-process heat reclaim unit. In some cases that involve condensables, modulation may be necessary to prevent overheating the process exhaust airstream.

Process-to-Comfort. In these types of applications, waste heat is taken from the process exhaust and used to heat the building makeup air during the winter months (Figure 18-64). The more typical applications include foundries, strip coating plants, can plants, plating operations, pulp and paper plants, and other processing areas having a heated process exhaust and large makeup air volume requirements.

Full heat reclaim is generally desirable at all times in a process-to-process application; heat reclaim modulation is necessary in process-to-comfort applications to prevent overheating the makeup air during warm weather. No heat reclaim is used during the summer months. Due to the fact that heat reclaim is used only during the winter months and the process is modulated during periods of moderate weather, the annual energy savings is less than that using the process-to-process applications.

Heat Recovery Systems

When the process-to-comfort heat reclaim systems are being considered, the effects listed below should be taken into consideration:

1. *Effects of corrosives.* The exhaust from the process sometimes contain compounds that require special construction materials for the heat reclaim unit.
2. *Effects of condensables.* There may be compound concentration in the process exhaust high enough to cause condensation on cooling in the heat reclaim unit.
3. *Effects of contaminants.* In cases when the process exhaust has heavy particulate contamination, the heat reclaim unit must be equipped with facilities for cleaning. Also, some consideration may be given to air prefiltration and to the selection of a heat reclaim unit that has an open-type construction to reduce the amount of cleaning required.

Process-to-comfort heat reclaim units generally reclaim only sensible heat and do not transfer moisture between the different airstreams.

Comfort-to-Comfort. When comfort-to-comfort applications are used, the heat is transferred from the exhaust air to the supply air of the building. In this type of application, the total heat of the building supply air is lowered during the warmer weather and is increased during the cooler weather.

In general, there are two categories of air-to-air heat reclaim units for comfort-to-comfort applications. They are sensible heat reclaim units and total heat reclaim units.

Sensible Heat Reclaim Units. These types of units transfer only sensible heat (dry-bulb temperature) from the exhaust to the supply airstreams. The one exception is in applications where the exhaust airstream may be cooled down below the dew point. At this point, condensation will occur and the system's performance will depend on the type of heat reclaim unit being used.

Total Heat Reclaim Units. This type of unit transfers both sensible and latent heat (humidity) between the exhaust air and the supply airstreams (Figure 18–65). The transfer of latent heat is generally desirable in comfort-to-comfort type systems.

A typical comfort-to-comfort heat reclaim system will have a sensible heat reclaim of approximately 70%. However, systems may also have an efficiency of 70% on both sensible and latent heat recovery when operating under the same conditions (Figure 18–66).

When these two examples are compared, we can see that when they are operating under typical summer design conditions, the total heat reclaim unit will recover almost three times as much energy as the sensible heat reclaim unit. When operating under typical winter design conditions, the total heat reclaim unit will recover more than 25% more energy than the sensible heat reclaim unit. The supply air process for both of these example systems for both summer and winter operation can be plotted on a psychometric chart, which permits a direct comparison of the amount of energy transferred (Figure 18–67).

When comfort-to-comfort heat reclaim units are being considered, the following items should be carefully evaluated:

1. *Effect of particulate contaminants.* If the exhaust air from the building contains particulate matter, such as dust, lint, and animal hair, special provisions should be made for an adequate prefiltering of the building exhaust air to prevent plugging of the heat reclaim unit. Prefiltration of the same quality should be provided for the building supply air as would be provided for any other type of air conditioning unit.

FIGURE 18–65   Comfort-to-comfort sensible heat device. (Courtesy of ASHRAE.)

FIGURE 18–66   Comfort-to-comfort total heat device. (Courtesy of ASHRAE.)

FIGURE 18-67 Comfort-to-comfort sensible heat versus total heat recovery. (Courtesy of ASHRAE.)

2. *Effect of gaseous or vaporous contaminants.* When gaseous or vaporous contaminants such as hydrocarbons, sulfur compounds, and water-soluble chemicals are in the exhaust air from the building, the effects of these contaminants should be carefully investigated.

3. *Effect of indoor humidity level.* The winter performance of the heat reclaim unit should be investigated thoroughly when the building requires humidity control conditions during the winter months or the climate is cold. In these types of applications special controls, preheating of the supply airstream, or auxiliary heating of the heat reclaim unit may be required to prevent frosting of the heat reclaim unit or to prevent other operational problems during operation in the winter months.

4. *Comparison of heat reclaim units.* All types of heat reclaim units should be given careful consideration; even when the design engineer is considering an application having total heat reclaim, both sensible and latent heat, is possible. Even sensible heat reclaim units should be considered because of their individual features, reclaim efficiency, and cost.

☐ Commercial Refrigeration Piping ☐

Refrigeration piping design procedures are a series of compromises, at best. A good piping system will have a maximum capacity, be economical, provide proper oil return, provide minimum power consumption, require a minimum amount of refrigerant, have a low noise level, provide proper refrigerant control,

and allow perfect flexibility in system performance from 0 to 100% of unit capacity without lubrication problems. Obviously, it is impossible to obtain all of these goals because some of them are in direct conflict with others. To make an intelligent decision as to what type of compromise is most desirable, it is essential that the piping designer have a thorough understanding of the basic effects on the system performance of the piping design in the different points of the refrigerant system.

Pressure Drop. Pressure drop, in general, tends to decrease system capacity and increase the amount of electric power required by the compressor. Therefore, excessive pressure drops should be avoided. The amount of pressure drop allowable will vary depending on the particular segment of the system involved. Each part of the system must be considered separately. There are probably more charts and tables available that cover refrigeration line pressure drop and capacities at a given pressure, temperature, and pressure drop than any other single subject in the field of refrigeration and air conditioning.

The piping designer must realize that there are several factors that govern the sizing of refrigerant lines and that pressure drop is not the only criterion to be used in designing a system. It is often required that refrigerant velocity, rather than pressure drop, be the determining factor in system design. Also, the critical nature of oil return can produce many system difficulties. A reasonable pressure drop is far more preferable than oversized lines, which may hold refrigerant far in excess of that required by the system. An overcharge of refrigerant can result in serious problems of liquid refrigerant control, and the flywheel effect of large quantities of liquid refrigerant in the low-pressure side of the system can result in erratic operation of the refrigerant flow control device.

The size of the refrigerant line connection on a service valve supplied with a compressor, or the size of the connection on an evaporator, condenser, or some other system accessory does not determine the correct size of the refrigerant line to be used. Equipment manufacturers select a valve size or connection fitting on the basis of its application to an average system, and other factors such as the application, length of connecting lines, type of system control, variation in load, and a multitude of other factors. It is entirely possible for the required refrigerant line size to be either smaller or larger than the fittings on various system components. In such cases, reducing fittings must be used.

Oil Return. Since oil must pass through the compressor cylinders to provide the proper lubrication, a small amount of oil is always circulating through the system with the refrigerant. The oils used in refrigeration systems are soluble in liquid refrigerant, and at normal room temperatures they will mix completely. Oil and refrigerant vapor, however, do not mix readily, and the oil can be properly circulated through the system only if the velocity of the refrigerant vapor is great enough to carry the oil along with it. To assure proper oil circulation, adequate refrigerant velocities must be maintained not only in the suction and discharge lines, but in the evaporator circuits as well.

Several factors combine to make oil return most critical at low evaporating temperatures. As the suction pressure decreases and the refrigerant vapor becomes less dense, it becomes more difficult for the refrigerant to carry the oil along. At the same time, as the suction pressure decreases, the compression ratio increases and the compressor capacity is reduced. Therefore, the weight of the refrigerant circulated decreases. Refrigeration oil alone becomes as thick as molasses at temperatures below 0°F (−17.8°C). As long as it is mixed with a sufficient amount of liquid refrigerant, however, it flows freely. As the percentage of oil in the mixture increases, the viscosity increases.

At low-temperature conditions several factors start to converge and can create a critical condition. The density of the refrigerant vapor decreases, the velocity decreases, and as a result more oil starts to accumulate in the evaporator. As the oil and refrigerant mixture becomes thicker, the oil may start logging in the evaporator rather than returning to the compressor. This results in wide variations in the compressor crankcase oil level in poorly designed systems.

Oil logging in the evaporator can be minimized with adequate refrigerant velocities and properly designed evaporators even at extremely low evaporating temperatures. Normally, oil separators are necessary for operation at evaporating temperatures below −50°F (−45.6°C), in order to minimize the amount of oil in circulation.

Equivalent Length of Pipe. Each valve, fitting, and bend in a refrigerant line contributes to the friction pressure drop because of its interruption or restric-

TABLE 18–6 Equivalent Lengths (ft) of Straight Pipe for Valves and Fittings

OD Line Size (in.)	Globe Valve	Angle Valve	90° Elbow	45° Elbow	Tee Line	Tee Branch
½	9	5	0.9	0.4	0.6	2.0
⅝	12	6	1.0	0.5	0.8	2.5
⅞	15	8	1.5	0.7	1.0	3.5
1⅛	22	12	1.8	0.9	1.5	4.5
1⅜	28	15	2.4	1.2	1.8	6.0
1⅝	35	17	2.8	1.4	2.0	7.0
2⅛	45	22	3.9	1.8	3.0	10.0
2⅝	51	26	4.6	2.2	3.5	12.0
3⅛	65	34	5.5	2.7	4.5	15.0
3⅝	80	40	6.5	3.0	5.0	17.0

tion of smooth flow. Because of the detail and complexity of computing the pressure drop of each individual fitting, normal practice is to establish an equivalent length of straight tubing for each fitting. This allows the consideration of the entire length of line, including the fittings, as an equivalent length of straight pipe. Pressure drop and line sizing tables and charts are normally set up on a basis of a pressure drop per 100 ft (30.4 m) of straight pipe, so the use of equivalent lengths allows the data to be used directly (see Table 18–6).

For accurate calculations of pressure drop, the equivalent length for each fitting should be calculated. As a practical matter, an experienced piping designer may be capable of making an accurate overall percentage allowance unless the piping system is extremely complicated. For long runs of piping of 100 ft (30.4 m) or greater, an allowance of 20 to 30% of the actual length may be adequate. For short runs of piping, an allowance as high as 50 to 75% or more of the lineal length may be necessary. Judgment and experience are necessary in making a good estimate, and estimates should be checked frequently with actual calculations to ensure reasonable accuracy.

For items such as solenoid valves and pressure-regulating valves, where the pressure drop through the valve is relatively large, data are normally available from the manufacturer's catalog so that items of this nature can be considered independently of lineal length calculations.

Pressure Drop Tables. There are pressure drop tables available that show the combined pressure drop for all refrigerants. Figures 18–68, 18–69, and

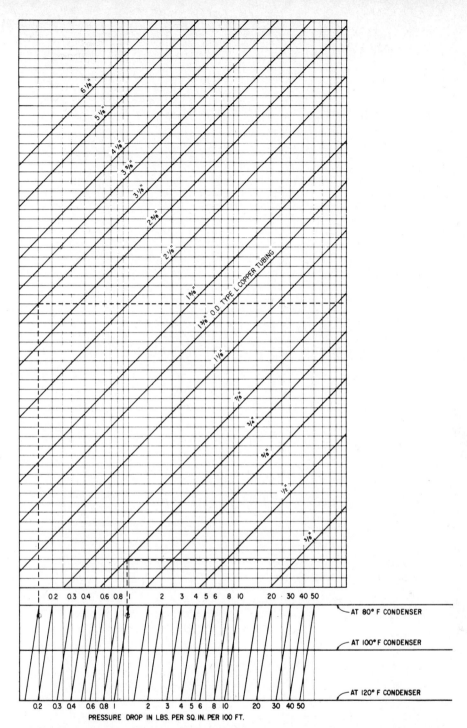

FIGURE 18-68 Freon-12 refrigerant pressure drop in lines (65° evaporator outlet). (Courtesy of Copeland Refrigeration Corp.)

18–70 are for refrigerants R-12, R-22, and R-502. Pressure drops in the discharge line, suction line, and the liquid line can be determined from these charts for condensing temperatures ranging from 80 to 120°F (26.7 to 49°C).

To use the chart, start in the upper right-hand corner with the design capacity. Drop vertically downward on the line representing the desired capacity to the intersection with the diagonal line representing the operating condition desired. Then move horizontally to the left. A vertical line dropped from the intersection point with each size of copper tubing to the design condensing temperature line allows the pressure drop in psi (kPa) per 100 ft (30.4 m) of tubing to be read directly from the chart. The diagonal pressure drop lines at the bottom of

FIGURE 18-68 *(Continued)*

the chart represent the change in pressure drop due to a change in condensing temperature.

For example, in Figure 18–70 for R-502, the dotted line represents a pressure drop determination for a suction line in a system having a design capacity of 5.5 tons or 66,000 Btu/hr (16,632 kcal) operating with an evaporating temperature of −40 °F (−40 °C). The 2⅜-in. (66.7-mm) suction line illustrated has a pressure drop of 0.22 psi (1.51 kPa) per 100 ft (30.4 m) at 85 °F (29.4 °C) condensing temperature. The same line with the same capacity would have a pressure drop of 0.26 psi (1.79 kPa) per 100 ft at 100 °F (37.8 °C) condensing temperature, and 0.32 psi (2.19 kPa) per 100 ft at 120 °F (49 °C) condensing temperature.

In the same manner, the corresponding pressure

COMMERCIAL REFRIGERATION

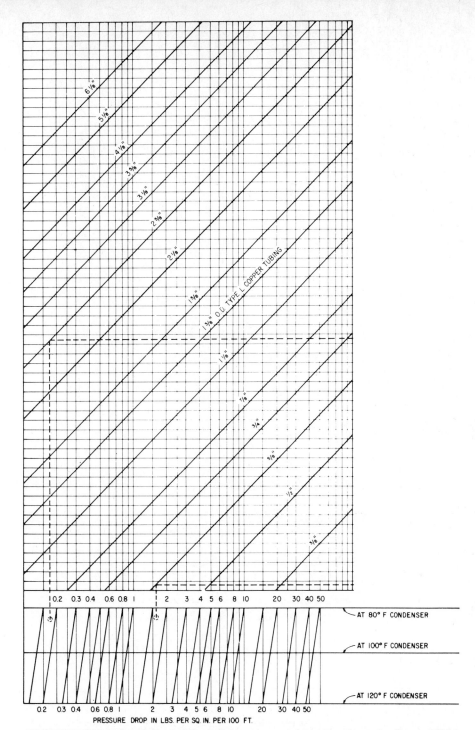

PRESSURE DROP IN LBS. PER SQ. IN. PER 100 FT.

**FIGURE 18-69 Freon-22 refrigerant pressure drop in lines (65°
evaporator outlet). (Courtesy of Copeland Refrigeration Corp.)**

drop for any line size and any set of operating conditions within the range of the chart can be determined.

Sizing Hot-Gas Vapor Discharge Lines. Pressure drop in discharge lines is probably less critical than in any other part of the system. Frequently, the effect on capacity of the discharge line pressure drop is overestimated since it is assumed that the compressor discharge pressure and the condensing pressure are the same. In fact, these are two different pressures. The compressor discharge pressure is greater than the condensing pressure by the amount of the discharge line pressure drop. An increase in pressure drop in the discharge line might increase the compressor discharge pressure materially, but

TONS OF REFRIGERATION

NOTE: Pressure drops do not allow for pulsating flow. If flow is pulsating, use next larger pipe size.
Liquid line determined at 0°F evap. & 80°F cond.
Discharge lines at 0°F evap. Other conditions do not appreciably change result. Vapor at evap. outlet assumed to be at 65°F.

EXAMPLE
5.5 Tons at −40°F Evap. and 85°F Cond.
2 ⅝" Suction Line Pressure Drop = 0.18 p.s.i /100 Ft.
⅝" Liquid Line Pressure Drop = 1.7 p.s.i /100 Ft.

FIGURE 18–69 *(Continued)*

have little effect on the condensing pressure. Although there is a slight increase in the heat of compression for an increase in discharge pressure, the volume of vapor pumped is decreased slightly due to a decrease in volumetric efficiency of the compressor. Therefore, the total heat to be dissipated through the condenser may be relatively unchanged, and the condensing temperature and pressure may be quite stable, even though the discharge line pressure drop and, because of the pressure drop, the compressor discharge pressure might vary considerably.

The performance of a typical compressor, operating at air conditioning conditions with R-22

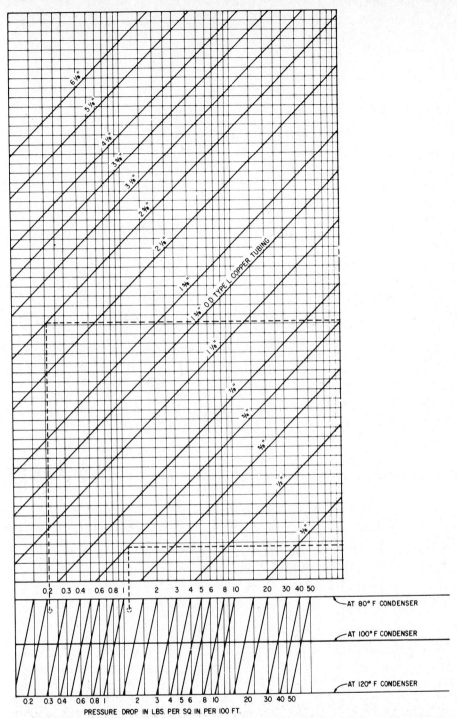

FIGURE 18-70 **Freon-502 refrigerant pressure drop in lines (65 evaporator outlet). (Courtesy of Copeland Refrigeration Corp.)**

and an air-cooled condenser, indicates that for each 5 psi (34.35 kPa) pressure drop in the discharge line, the compressor capacity is reduced less than ½ of 1%, while the electric power required is increased about 1%. On a typical low-temperature compressor operating with R-502 and an air-cooled condenser, approximately 1% of the compressor capacity will be lost for each 5 psi pressure drop, but there will be little or no change in electric power consumption.

As a general guide, for discharge line pressure drops up to 5 psi, the effect on the system performance would be so small it would be difficult to measure. Pressure drops up to 10 psi (68.7 kPa) would not be greatly detrimental to system perform-

Commercial Refrigeration Piping

TONS OF REFRIGERATION

NOTE: Pressure drops do not allow for pulsating flow. If flow is pulsating, use next larger pipe size.
Liquid line determined at 0°F evap. & 80°F. cond.
Discharge lines at 0°F evap. Other conditions do not appreciably change result. Vapor at evap. outlet assumed to be at 65°F.

EXAMPLE:

5.5 Tons at -40°F Evap. and 85°F Cond.

2 ⅜" Suction Line Pressure Drop = 0.22 p.s.i./100 Ft.

¾" Liquid Line Pressure Drop = 1.2 p.s.i./100 Ft.

FIGURE 18–70 *(Continued)*

ance provided the condenser is sized to maintain reasonable condensing pressures.

Actually, a reasonable pressure drop in the discharge line is often desirable to dampen compressor discharge pulsation, and thereby reduce noise and vibration. Some discharge line mufflers actually derive much of their efficiency from a pressure drop through the muffler.

Discharge lines on a factory built condensing unit usually are not a field problem, but on systems installed in the field with remote condensers, line sizes must be selected to provide proper system performance.

Because of the high temperatures that exist in the discharge line, the oil circulation through both horizontal and vertical lines can be maintained

COMPRESSOR
CAPACITY
(BTU/Hr.)

Line Size
O.D.

Example: At 40° F. evaporating temperature and a system capacity of 100,000 BTU/hr, the maximum size tubing that can be used to insure oil return is 1 ⅝″ O.D.

EVAPORATING TEMPERATURE (°F.)

FIGURE 18–71 Maximum recommended vertical discharge line sizes for proper oil return R-12. [Courtesy of Copeland Refrigeration Corp.]

satisfactorily with reasonably low refrigerant velocities. Since oil traveling up a riser usually creeps up the inner surface of the pipe, oil travel in vertical risers is dependent on the velocity of the gas at the tubing wall. The larger the pipe diameter, the greater will be the required velocity at the center of the pipe to maintain a given velocity at the wall surface. Figures 18–71 and 18–72 list the maximum recommended discharge line riser sizes for proper oil return for varying capacities. The variation at dif-

ferent condensing temperatures is not great, so the line sizes shown are acceptable on both water-cooled and air-cooled applications.

If horizontal lines are run with a pitch in the direction of refrigerant flow of at least ½ in. (12.7 mm) in 10 ft (3.04 m), there is normally little problem with oil circulation at lower velocities in horizontal lines. However, because of the relatively low velocities required in vertical discharge lines, it is recommended wherever possible that both horizontal and

Commercial Refrigeration Piping

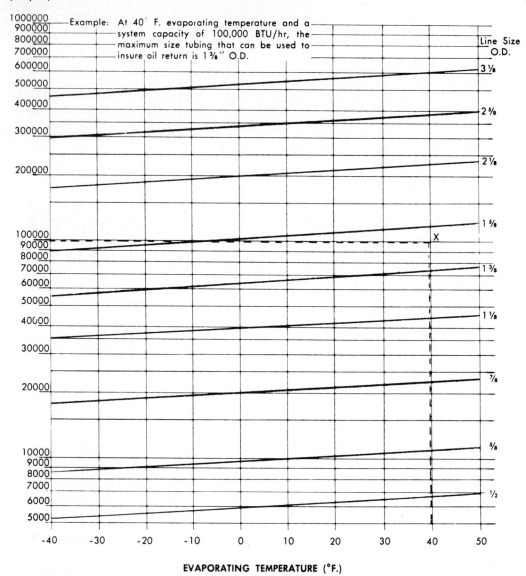

COMPRESSOR
CAPACITY
(BTU/Hr.)

Example: At 40° F. evaporating temperature and a
system capacity of 100,000 BTU/hr, the
maximum size tubing that can be used to
insure oil return is 1 ⅜″ O.D.

Line Size
O.D.

EVAPORATING TEMPERATURE (°F.)

**FIGURE 18-72 Maximum recommended vertical discharge line sizes
for proper oil return R-502. (Courtesy of Copeland Refrigeration
Corp.)**

vertical discharge lines be sized on the same basis.

To illustrate the use of the chart we will assume a system operating with R-12 at 40°F (4.4°C) evaporating temperature has a capacity of 100,000 Btu/hr (25,200 kcal). The intersection of the capacity and evaporating temperature lines at point X on Figure 18–73 indicates the design condition. Since this is below the 2⅛-in. (53.97-mm)-OD line, the maximum size used to ensure oil return up a vertical riser is a 1⅝-in. (41.27-mm)-OD line.

Oil circulation in discharge lines is normally a problem only on systems where large variations in system capacity are encountered. For example, an air conditioning system may have steps of capacity control allowing it to operate during periods of light load at capacities possibly as low as 25 or 33% of the design capacity. The same situation may exist on commercial refrigeration systems where compressors connected in parallel are cycled for capacity control. In such cases, vertical discharge lines must be sized to maintain velocities above the minimum velocity necessary to properly circulate oil at the minimum load condition.

For example, consider an air conditioning system using R-12 having a maximum design capacity of 300,000 Btu/hr (75,600 kcal) with steps of capacity

COMPRESSOR CAPACITY (BTU/Hr.)

Line Size O.D.

To Avoid Excessive Noise, Discharge Line Velocities Should Not Exceed 3000 FPM

DISCHARGE LINE VELOCITY (Feet Per Minute)

FIGURE 18–73 Discharge line velocities for various Btu/hr capacities, R-22 and R-502. (Courtesy of Copeland Refrigeration Corp.)

reduction up to 66%. Although the 300,000-Btu/hr (75,600-kcal) condition could return oil up to a 2⅝-in. (66.67-mm)-OD riser at light-load conditions, the system would have only 100,000 Btu/hr (25,200 kcal) capacity, so a 1⅝-in. (41.27-mm)-OD riser must be used. In checking the pressure drop chart (Figure 18–72) at maximum load conditions a 1⅝-in. (41.27-mm)-OD pipe will have a pressure drop of approximately 4 psi (27.48 kPa) per 100 ft (30.4 m) at a condensing temperature of 120°F (49°C). If the total equivalent length of pipe exceeds 150 ft (45.6 m), in order to keep the total pressure drop within

reasonable limits, the horizontal lines should be the next larger size or 2⅝ in. (53.97 mm) OD, which would result in a pressure drop of only slightly over 1 psi (6.87 kPa) per 100 ft.

Because of the flexibility in line sizing that the allowable pressure drop makes possible, discharge lines can almost always be sized satisfactorily without the necessity of double risers. If modifications are made to an existing system that result in the existing discharge line being oversized at light-load conditions, the addition of an oil separator to minimize oil circulation will normally solve the problem.

Commercial Refrigeration Piping

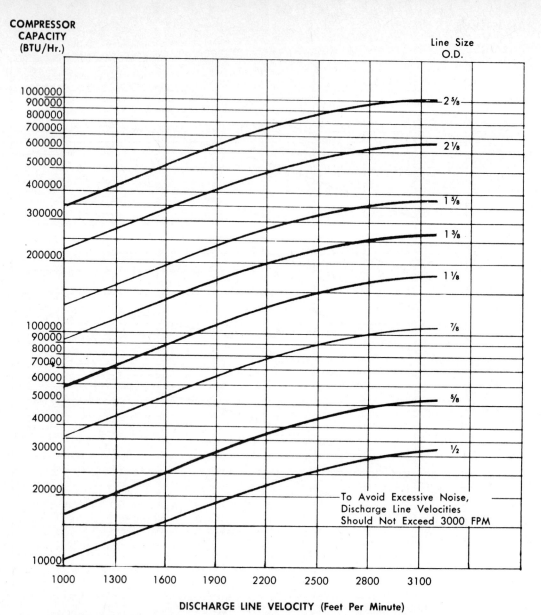

COMPRESSOR
CAPACITY
(BTU/Hr.)

Line Size
O.D.

DISCHARGE LINE VELOCITY (Feet Per Minute)

FIGURE 18-74 Discharge line velocities for various Btu/hr capacities, R-12. (Courtesy of Copeland Refrigeration Corp.)

One other limiting factor in discharge line sizing is that excessive velocity can cause noise problems. Velocities of 3000 ft per minute (fpm) (912 m/min) or more may result in high noise levels, and it is recommended that maximum refrigerant velocities be kept well below this level. Figures 18–73 and 18–74 give equivalent discharge line gas velocities for varying capacities and line sizes over the normal refrigeration and air conditioning range.

Sizing Liquid Lines. Since liquid refrigerant and oil mix completely, velocity is not essential for oil circulation in the liquid line. The primary concern in liquid line sizing is to ensure a solid liquid column

of refrigerant at the expansion valve. If the pressure of the liquid refrigerant falls below its saturation temperature, a portion of the liquid will flash into vapor to cool the remaining liquid refrigerant to the new saturation temperature. This can occur in a liquid line if the pressure drops sufficiently due to line friction or vertical lift.

Flash gas in the liquid line has a detrimental effect on system performances in several ways.

1. It increases the pressure drop due to friction.
2. It reduces the capacity of the flow control device.
3. It can erode the expansion valve pin and seat.
4. It can cause excessive noise.

5. It can cause erratic feeding of the liquid refrigerant to the evaporator.

For proper system performance, it is essential that liquid refrigerant reaching the flow control device be subcooled slightly below its saturation temperature. On most systems, the liquid refrigerant is sufficiently subcooled as it leaves the condenser to provide for normal system pressure drops. The amount of subcooling necessary, however, is dependent on the individual system design.

On air-cooled and most water-cooled applications, the temperature of the liquid refrigerant is normally higher than the surrounding ambient temperature, so no heat is transferred into the liquid, and the only concern is the pressure drop in the liquid line. Besides the friction loss caused by the refrigerant flow through the piping, a pressure drop equivalent to the liquid head is involved in forcing liquid refrigerant to flow up a vertical riser. A vertical head of 2 ft (0.6 m) of liquid refrigerant is approximately equivalent to 1 psi (6.87 kPa). For example, if a condenser or receiver in the basement of a building is to supply liquid refrigerant to an evaporator three floors above, or approximately 30 ft (9.12 m), a pressure drop of approximately 15 psi (103.05 kPa) for the liquid head alone must be provided for in system design.

On evaporative or water-cooled condensers where the condensing temperature is below the ambient air temperature, or on any application where the liquid lines must pass through hot areas such as boiler or furnace rooms, an additional complication may arise because of heat transfer into the liquid. Any subcooling in the condenser may be lost in the receiver or liquid line due to temperature rise alone unless the system is properly designed. On evaporative condensers when a receiver and subcooling coils are used, it is recommended that the refrigerant flow be piped from the condenser to the receiver and then to the subcooling coil. In critical applications, it may be necessary to insulate both the receiver and the liquid line.

On the typical air-cooled condensing unit with a conventional receiver, it is probable that very little subcooling of the liquid is possible unless the receiver is almost completely filled with liquid refrigerant. Vapor in the receiver in contact with the subcooled liquid will condense, and this effect will tend toward a saturated condition. At normal condensing temperatures, a relation between each 1 °F (0.56 °C) of

TABLE 18-7 Relationship of Subcooling and Saturation Pressure

Refrigerant	Subcooling [°F (°C)]	Equivalent Change in Saturation Pressure [psi (kPa)]
R-12	1 [0.56]	1.75 [12.02]
R-22	1 [0.56]	2.75 [18.89]
R-502	1 [0.56]	2.85 [19.58]

Source: Copeland Refrigeration Corp.

subcooling and the corresponding change in saturation pressure applies (Table 18–7).

To illustrate, 5 °F (2.78 °C) subcooling will allow a pressure drop of 8.75 psi (60.11 kPa) with R-12, 13.75 psi (94.46 kPa) with R-22, and 14.25 psi (97.90 kPa) with R-502 without flashing in the liquid line. For the previous example of a condensing unit in a basement requiring a vertical lift of 30 ft (9.12 m), or approximately 15 psi (103.05 kPa), the necessary subcooling for the liquid head alone would be 8.5 °F (4.72 °C) with R-12, 5.5 °F (3.06 °C) with R-22, and 5.25 °F (2.92 °C) with R-502.

The necessary subcooling may be provided by the condenser used, but for systems with abnormally high vertical risers a suction to liquid heat exchanger may be required. Where long refrigerant lines are involved, and the temperature of the suction vapor at the condensing unit is approaching room temperatures, a heat exchanger located near the condenser may not have sufficient temperature differential to adequately cool the liquid. Individual heat exchangers at each evaporator may be necessary.

In extreme cases, where a great deal of subcooling is required, there are several alternatives. A special heat exchanger with a separate subcooling expansion valve can provide maximum cooling with no penalty on system performance. It also is possible to reduce the capacity of the condenser so that a higher operating condensing temperature will make greater subcooling possible. Liquid refrigerant pumps also may be used to overcome large pressure drops.

Liquid line pressure drop causes no direct penalty in electric power consumption, and the decrease in system capacity due to friction losses in the liquid line is negligible. Because of this, the only real restric-

FIGURE 18-75 Liquid line velocities for various pressure drops, R-12, R-22, and R-502. [Courtesy of Copeland Refrigeration Corp.]

tion on the amount of liquid line pressure drop is the amount of subcooling available. Most references on pipe sizing recommend a conservative approach with friction pressure drops in the 3 to 5 psi (20.61 to 34.35 kPa) range. Where adequate subcooling is available, however, many applications have successfully used much higher design pressure drops. The total friction includes line losses through such accessories as solenoid valves, filter-driers, and hand valves.

To minimize the refrigerant charge, liquid lines should be kept as small as practical, and excessively low pressure drops should be avoided. On most systems, a reasonable design criterion is to size liquid lines on the basis of a pressure drop equivalent to 2 °F (1.11 °C) subcooling.

A limitation on liquid velocity is possible damage to the piping from pressure surges or liquid ham-

mer caused by the rapid closing of liquid line solenoid valves, and velocities above 300 fpm (91.2 m/min) should be avoided when they are used. If liquid line solenoid valves are not used, higher refrigerant velocities can be employed. Figure 18-75 gives liquid line velocities corresponding to various pressure drops and line sizes.

Sizing Suction Lines. Suction line sizing is more important than that of the other lines from a design and system standpoint. Any pressure drop occurring due to frictional resistance to flow results in a decrease in the refrigerant pressure at the compressor suction valve, compared with the pressure at the evaporator outlet. As the suction pressure is decreased, each pound of refrigerant that returns to the compressor occupies a greater volume, and the weight of the refrigerant being pumped by the com-

TABLE 18-8 Pressure Drop Equivalent for 2°F Change in Saturation Temperature at Various Evaporating Temperatures

Evaporating Temperature [°F (°C)]	Pressure Drop [psi (kPa)]		
	R-12	R-22	R-502
45 [7.2]	2.0 [13.74]	3.0 [20.61]	3.3 [22.67]
20 [−6.7]	1.35 [9.18]	2.2 [15.11]	2.4 [16.49]
0 [−17.8]	1.0 [6.87]	1.65 [11.33]	1.85 [33.32]
−20 [−28.9]	0.75 [5.15]	1.15 [7.9]	1.35 [9.27]
−40 [−40]	0.5 [3.44]	0.8 [5.50]	1.0 [6.87]

Source: Copeland Refrigeration Corp.

pressor decreases. For example, a typical low-temperature R-502 compressor operating at a −40°F (−40°C) evaporating temperature will lose almost 6% of its rated capacity for each 1 psi (6.87 kPa) suction line pressure drop.

The normally accepted design practice is to use a design criterion of a suction line pressure drop equal to a 2°F (1.11°C) change in saturation temperature. Equivalent pressure drops for various operating conditions are developed and placed in tables (Table 18-8).

The maintenance of adequate velocities to properly return the lubricating oil to the compressor is also of great importance when sizing suction lines. Studies have shown that oil is most viscous in a system after the suction gas has warmed up a few degrees higher than the evaporating temperature so that the oil is no longer saturated with refrigerant. This condition occurs in the suction line after the refrigerant vapor has left the evaporator. The movement of oil through the suction line is dependent on both the mass and the velocity of the suction vapor. As the mass or density decreases, higher refrigerant velocities are required to force the oil along.

Nominal minimum velocities of 700 fpm (212.8 m/min) in horizontal suction lines and 1500 fpm (456 m/min) in vertical suction lines have been recommended and used successfully for many years as suction line sizing design standards. The use of one nominal refrigerant velocity provided a simple and convenient means of checking velocities. However, tests have shown that in vertical risers the oil tends to crawl up the inner surface of the tubing, and the larger the tubing, the greater the velocity required in the center of the tubing to maintain tube surface velocities that will carry the oil. The exact velocity required in vertical suction lines is dependent on both the evaporating temperature and the size of the line.

Under varying conditions, the specific velocity required might be either greater or less than 1500 fpm (456 m/min).

For better accuracy in line sizing, revised maximum recommended vertical suction line sizes based on the minimum vapor velocities have been calculated and are plotted in chart form for easy use (Figures 18-76 and 18-77). These revised recommendations supersede previous vertical suction riser recommendations. No change has been made in the 700 fpm (212.8 m/min) minimum velocity recommendation for horizontal lines (Figures 18-78 and 18-79).

To illustrate, again assume that a system operating with R-12 at a 40°F (4.4°C) evaporating temperature has a capacity of 100,000 Btu/hr (25,200 kcal). On Figure 18-76 the intersection of the evaporating temperature and capacity lines indicate that a 2⅛-in. (53.97-mm)-OD line will be required for proper oil return in the vertical risers.

Even though the system might have a much greater design capacity, the suction line sizing must be based on the minimum capacity anticipated in operation under light-load conditions after allowing for the maximum reduction in capacity from the capacity control, if used.

Since the dual goals of low pressure drop and high velocities are in direct conflict with each other, obviously some compromises must be made in both areas. As a general approach, in suction line design, velocities should be kept as high as possible by sizing lines on the basis of the maximum pressure drop that can be tolerated. In no case, however, should the gas velocity be allowed to fall below the minimum levels necessary to return the oil to the compressor. It is recommended that a tentative selection of suction line sizes be made on the basis of a total pressure drop equivalent to a 2°F (1.11°C)

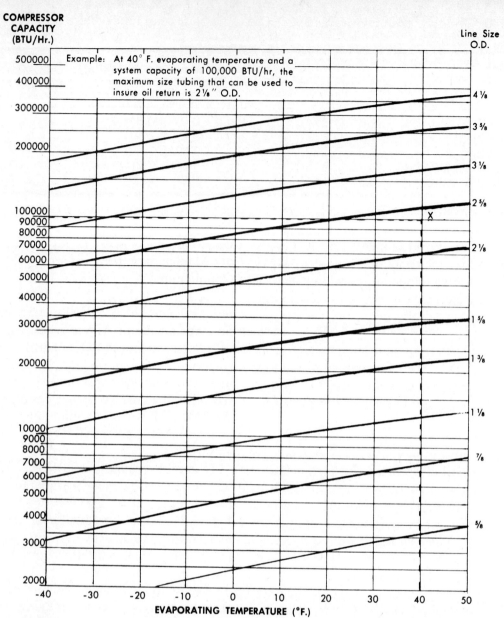

COMPRESSOR
CAPACITY
(BTU/Hr.)

Line Size
O.D.

Example: At 40° F. evaporating temperature and a system capacity of 100,000 BTU/hr, the maximum size tubing that can be used to insure oil return is 2⅛″ O.D.

FIGURE 18-76 **Maximum recommended suction line sizes for proper oil return, vertical risers, R-12. (Courtesy of Copeland Refrigeration Corp.)**

change in the saturated evaporating temperature. The final consideration always must be to maintain velocities adequate to return the lubricating oil to the compressor, even if this results in a higher-pressure drop than is normally desirable.

Double Risers. On systems equipped with capacity control compressors or where tandem or multiple compressors are used with one or more compressors cycled off for capacity control, single-suction line risers may result in either unacceptably high or low vapor velocities. A line sized properly for light-load

conditions may have too high a pressure drop at maximum load. If the line is sized on the basis of full-load conditions, the velocities may not be adequate at light-load conditions to move the oil through the tubing. On air conditioning applications where somewhat higher pressure drops at maximum-load conditions can be tolerated without any major penalty in overall system performance, it is usually preferable to accept the additional pressure drop imposed by a single vertical riser. But on medium- or low-temperature applications where pressure drop is more critical, and where separate risers from in-

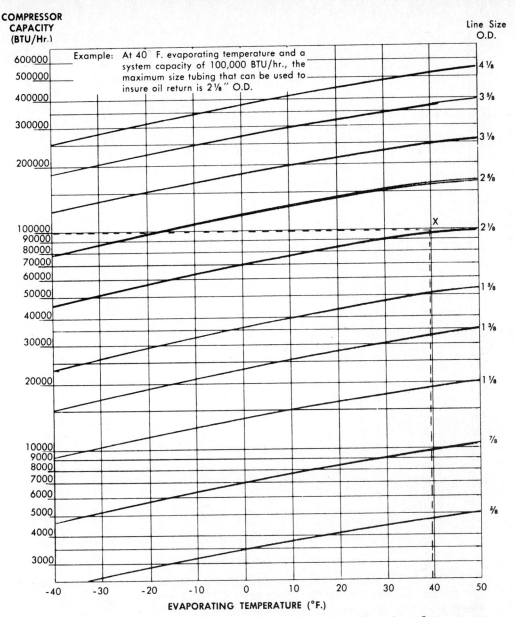

FIGURE 18-77 Maximum recommended suction line sizes for proper oil return, vertical visers, R-22 and R-502. (Courtesy of Copeland Refrigeration Corp.)

dividual evaporators are not desirable or possible, a double riser may be necessary to avoid an excessive loss of capacity.

The two lines of a double riser should be sized so that the total cross-sectional area is equivalent to the cross-sectional area of a single riser that would have both satisfactory gas velocity and an acceptable pressure drop at maximum-load conditions (Figure 18–80). The two lines are normally of different sizes, with the larger line being trapped as shown. The smaller line must be sized to provide adequate

velocities and acceptable pressure drop when the entire minimum load is carried in the smaller riser.

In operation, at maximum-load conditions, gas and entrained oil will be flowing through both risers. At minimum-load conditions, the vapor velocity will not be high enough to carry the oil up both risers. The entrained oil will drop out of the refrigerant vapor flow, and accumulate in the P trap forming a liquid seal. This will force all the flow up the smaller riser, thereby raising the velocity and assuring oil circulation through the system.

Commercial Refrigeration Piping

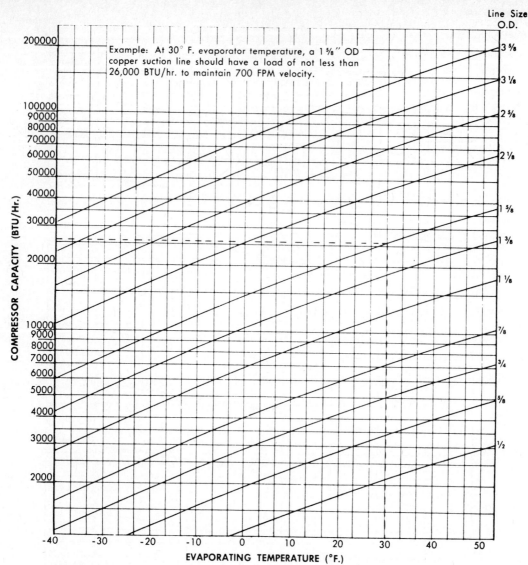

Example: At 30° F. evaporator temperature, a 1 5/8" OD copper suction line should have a load of not less than 26,000 BTU/hr. to maintain 700 FPM velocity.

Line Size O.D.

3 5/8
3 1/8
2 5/8
2 1/8
1 5/8
1 3/8
1 1/8
7/8
3/4
5/8
1/2

COMPRESSOR CAPACITY (BTU/Hr.)

EVAPORATING TEMPERATURE (°F.)

FIGURE 18-78 Maximum recommended horizontal suction line sizes for proper oil return, R-12. (Courtesy of Copeland Refrigeration Corp.)

As an example, assume a low-temperature system as follows:

Maximum capacity	150,000 Btu [37,800 kcal]
Minimum capacity	50,000 Btu [12,600 kcal]
Refrigerant	R-502
Evaporating temperature	−40°F [−40°C]
Equivalent length of horizontal piping	125 ft [38 m]
Vertical riser	25 ft [7.6 m]
Desired design pressure drop [equivalent to 2°F [1.11°C]]	1 psi [0.87 kPa]

A preliminary check of the R-502 pressure drop chart indicates for a 150-ft (45.6-m) run with 150,000 Btu/hr (37,800 kcal) capacity and a total pressure drop of approximately 1 psi (6.87 kPa), a 3⅛-in. (79.37-mm)-OD line is indicated (Figure 18–70). At the minimum capacity of 50,000 Btu/hr (12,600 kcal) a 3⅝-in. (92-mm)-OD horizontal suction line is acceptable (see Figure 18–81). However, Figure 18–77 indicates that a maximum vertical riser size is 2⅛ in. (53.97 mm) OD. Referring again to the pressure drop chart of Figure 18–70, the pressure drop for 150,000 Btu/hr (37,800 kcal) through a 2⅛-in. (53.97-mm)-OD tubing is 4 psi (27.48 kPa) per 100 ft (30.4 m),

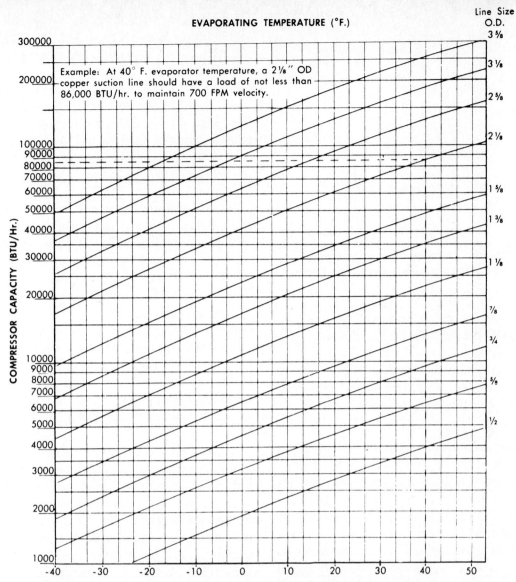

EVAPORATING TEMPERATURE (°F.)

Line Size O.D.

Example: At 40° F. evaporator temperature, a 2⅛″ OD copper suction line should have a load of not less than 86,000 BTU/hr. to maintain 700 FPM velocity.

FIGURE 18–79 Maximum recommended horizontal suction line sizes for proper oil risers, R-22 and R-502. (Courtesy of Copeland Refrigeration Corp.)

or 1 psi (6.87 kPa) for the 25-ft (7.6-m) suction riser. Obviously, either a compromise must be made in accepting a greater pressure drop at maximum-load conditions, or a double riser must be used.

If the pressure drop must be kept to a minimum, then the size of the double riser must be determined. At maximum-load conditions, a 3⅛-in. (79.37-mm)-OD riser would maintain adequate velocities, so a combination of line sizes approximating the 3⅛-in. (79.37-mm)-OD line can be selected for the double riser. The cross-sectional area of the line sizes to be considered are:

Line Size [in. (mm)]	Cross-Sectional Area [in.² (cm²)]
3⅛ [79.37]	6.64 [42.82]
2⅝ [66.67]	4.77 [30.76]
2⅛ [53.97]	3.10 [19.99]
1⅝ [41.27]	1.78 [11.48]

At the minimum-load conditions of 50,000 Btu/hr (12,600 kcal), the 1⅝-in. (41.27-mm)-OD line will have a pressure drop of approximately 0.5 psi (3.43 kPa), and will have acceptable velocities, so a com-

FIGURE 18–80 Suction line double riser. (Courtesy of Copeland Refrigeration Corp.)

bination of 2⅝-in. (66.67-mm)-OD and 1⅝-in. (41.27-mm)-OD tubing should be used for the double riser.

In a similar manner, double risers can be calculated for any set of maximum and minimum capacities where single risers may not be satisfactory.

Suction Piping for Multiplex Systems. It is common practice in supermarket applications to operate several cases, each with a liquid line solenoid valve and expansion valve control, from a single compressor. Temperature control of individual cases is normally achieved by means of a thermostat that opens and closes the liquid line solenoid valve as necessary. This type of system, commonly called multiplexing, requires careful attention in design to avoid oil return problems and compressor overheating.

Since the cases fed by each liquid line solenoid valve may be controlled individually, and because the load on each case is relatively constant during operation, individual suction lines and risers are normally run from each case, or group of cases, which are controlled by a liquid line solenoid valve for a minimum pressure drop and maximum efficiency in oil return. This provides excellent control as long as the compressor is operating at its design suction pressure. However, there may be periods of light load when most or all of the liquid line solenoid valves are closed. Unless some means of controlling compressor capacity is provided, this can result in the compressor short cycling or operating at excessively low suction pressures. Either condition can result not only in overheating the compressor but

in reducing the suction pressure to a level where the vapor becomes so rarified it can no longer return oil properly in lines sized for much greater vapor density.

Because of the fluctuations in the refrigeration load caused by the closing of the individual liquid line solenoid valves, some means of compressor capacity control must be provided. In addition, the means of capacity control must be such that it will not allow extreme variations in the compressor suction pressure.

Where multiple compressors are used, cycling of the individual compressors provides satisfactory control. Where multiplexing is done with a single compressor, a hot-gas bypass system has proven to be the most satisfactory means of capacity reduction. This system allows the compressor to operate continuously at a reasonably constant suction pressure while compressor cooling can be safely controlled by means of a de-superheating expansion valve. In all cases, the operation of the system under all possible combinations of heavy load, light load, defrost, and compressor capacity must be studied carefully to be certain that operating conditions will be satisfactory.

Close attention must be given to piping design on multiplex systems to avoid oil return problems. The lines must be properly sized so that the minimum velocities necessary to return the oil are maintained in both horizontal and vertical suction lines under minimum-load conditions. Bear in mind that although a hot-gas bypass maintains the suction pressure at a proper level, the refrigerant vapor being bypassed is not available in the system to aid in returning the oil.

Piping Design for Horizontal and Vertical Lines. Horizontal suction and discharge lines should be pitched downward in the direction of flow to aid in oil drainage, with a downward pitch of at least ½ in. (12.7 mm) in 10 ft (3.04 m). Refrigerant lines should always be as short and should run as directly as possible.

Piping should be located so that access to system components is not hindered and so that any components that could possibly require future maintenance are easily accessible. If piping must be run through boiler rooms or other areas where they will be exposed to abnormally high temperatures, it may be necessary to insulate both the suction and liquid lines to prevent excessive heat transfer into the lines.

FIGURE 18-81 Suction line riser. (Courtesy of Copeland Refrigeration Corp.)

FIGURE 18-82 Discharge line riser. (Courtesy of Copeland Refrigeration Corp.)

Every vertical suction riser greater than 3 to 4 ft (0.9 to 1.2 m) in height, should have a *P* trap at the base to facilitate oil return up the riser (Figure 18-81). To avoid the accumulation of large quantities of oil, the trap should be of minimum depth and the horizontal section should be as short as possible.

Prefabricated wrought copper traps are available, or a trap can be made by using two street ells and one regular ell. Traps at the foot of hot-vapor risers are normally not required because of the easier movement of oil at higher temperatures. However, it is recommended that the discharge line from the compressor be looped to the floor prior to being run vertically upward to prevent the drainage of oil back to the compressor head during shutdown periods (Figure 18-82).

For long vertical risers in both suction and discharge lines, additional traps are recommended

for each full length of pipe [approximately 20 ft (6.08 m)] to ensure proper oil movement.

In general, trapped sections of the suction line should be avoided except where necessary for oil return. Oil or liquid refrigerant accumulating in the suction line during the off cycle can return to the compressor at high velocity as a liquid slug on start-up, and can break the compressor valves or cause other damage.

Suction Line Piping Design at the Evaporator. If a pumpdown control system is not used, each evaporator must be trapped to prevent liquid refrigerant from draining back to the compressor by gravity during the off cycle. Where multiple evaporators are connected to a common suction line, the connections to the common suction line must be made with inverted traps to prevent drainage from one evaporator from affecting the expansion valve bulb control of another evaporator.

Where a suction riser is taken directly upward from an evaporator, a short horizontal section of tubing and a trap should be provided ahead of the riser so that a suitable mounting for the thermal expansion valve bulb is available. The trap serves as a drain area and helps to prevent the accumulation of liquid under the bulb that could cause erratic expansion valve operation. If the suction line leaving the evaporator is free-draining, or if a reasonable length of horizontal piping precedes the vertical riser, no trap is required unless necessary for oil return (Figure 18-83).

☐ Water Coolers ☐

Water coolers are used to provide drinking water for the persons who occupy or visit a building. They are different from chilled water systems which provide cool water for cooling the air inside the building. The water used in these systems is not suitable for drinking. Basically, there are three different types of water coolers: (1) the bottled water cooler, (2) the pressure-type water cooler, and (3) the remote-type water cooler (Figure 18-84). They are also available in the freestanding models or the wall-mounted type.

The capacity of these units is expressed in gallons (liters) per hour and refers to the amount of water that can be cooled in 1 hr from a given inlet water temperature, usually 90 °F (32.22 °C), to a specified

FIGURE 18-83 Typical suction line piping with multiple evaporators. (Courtesy of Copeland Refrigeration Corp.)

(a)

(b)

(c)

FIGURE 18-84 Types of water coolers: (a) bottled-type water cooler; (b) pressure-type water cooler; (c) wall-mounted pressure-type water cooler. (Courtesy of EBCO Manufacturing Co.)

outlet water temperature, usually 50°F (10.00°C). Their capacity is generally from about ½ gal (1.89 liters) to 30 gal (114 liters) per hour, depending on the need for the specific application.

The refrigeration system is of the hermetically sealed type. The air-cooled compressor pumps the refrigerant through the condenser. A capillary tube flow control device is used to meter the refrigerant to the cooling tank and the compartment evaporators. The evaporator is wrapped around the cooling tank to increase the unit efficiency.

Bottled Water Coolers. This type uses a bottle for the water reservoir, which will be cooled inside the cabinet when needed. These units are equipped with a faucet or some other means of filling glasses, cups, or other types of containers. They are also equipped with a wastewater receptacle. These units are available for cold water only or for hot or cold water.

The heating tank, when used, is vented to the atmosphere through the two-way action of the hot water faucet assembly to prevent it from rupturing if it should get too hot (Figure 18–85). The electri-

FIGURE 18-85 **(a) Bottled-water cooler: (a) parts breakdown; (b) parts list. (Courtesy of EBCO Manufacturing Co.)**

451

FIGURE 18–85(b)

Item No.	Description	DBA1R-001 DBA1R-001-D	DBA1RH-001 DBA1RH-001-D	DBA1RH-D-50/60
*1	Plastic bottle gasket	020864-002	020864-002	020864-002
	Water guard bottle gasket	026757-001	026757-001	026757-001
2	Air filter	026972	026972	026972
3	Top—tan	017317-005	017317-005	017317-005
4	Cold faucet	025386-001	025386-001	025386-001
5	Hot faucet	—	025386-003	025386-003
6	Cooling tank fitting	020485	020485	020485
7	Gasket (water outlet tube)	027139	027139	027139
8	Upper front panel—tan	017411-002	017410-002	017410-002
9	Water outlet tube	027310-002	—	
10	Shelf insulation	027301-001	027301-001	027301-001
11	Faucet gasket	025568-001	025568-001	025568-001
12	Hot tank assembly	—	027666-001	027666-002
13	Cover	017396-002	017396-002	017396-002
14	Cooling tank assembly	027319-003	027319-003	027319-003
15	Cooling tank insulation	027891-001	027891-001	027891-001
16	Service cord	026540-003	026540-003	026540-003
17	Lower front panel—tan	017409	017409	017409
18	Side panel—tan	017408	017408	017408
19	Rear panel—tan	017405	019119	019119
20	Service door plate	—	019121	019121
21	Hot control	—	027813-001	027813-001
22	Hinge retainer	017706	017706	017706
23	Top hinge	023460	023460	023460
24	Bottom hinge	023459	023459	023459
25	Cold control	027040-002	027040-002	027040-002
26	Door assembly (includes item 27)	028105-001	028105-001	028105-001
27	Door gasket	022535	022535	022535
28	Antisplash tray	017371	017371	017371
29	Receptor assembly	017795-004	017795-004	017795-004
30	Bracket	017394	017394	017394
31	Relay	016941	016941	—
	Relay (Danfoss)	026867	026867	026309
32	Overload	017560	017560	
33	Control knob	019838	019838	019838
34	Compressor assembly (includes items 31 and 32)	016939	016939	—
	Compressor assembly (Danfoss) (includes item 31)	026868-002	026868-002	026883-002
35	Ice cube tray	018159-002	018159-002	018159-002
36	Evaporator	017379	017379	017379
37	Compartment	019803	019803	019803
38	Base assembly	018379-001	018379-001	018379-001
39	Condenser	027300-001	027300-001	027300-001
40	Fan thermostat	017480	017480	017480
41	Fan motor	017079	017079	024938
42	Fan blade	017067	017067	017067
43	Baffle	025430	024398	024398
44	Receptor bracket	017605	017605	017605
45	Deflector	—	024402	024402
Not shown:				
	Heat limiter	—	024373-002	024373-002
	Hot tank insulation	—	025792	025792
	Heat exchanger assembly	027656-001	027656-001	027656-001
	*Bottle gasket kit (includes both types of gaskets)	027553-001	027553-001	027553-001

COMMERCIAL REFRIGERATION

FIGURE 18–86 Typical bottled water cooler wiring diagram. (Courtesy of EBCO Manufacturing Co.)

TABLE 18–9 Service Check for Bottled Water Coolers

Trouble	Probable Cause	Remedy
Unit not running	No electrical power	Check power source. Check that cold control is not in off position.
	Defective cold control	Replace.
	Defective relay	Replace.
Unit running but not cooling	Refrigerant leak or low refrigerant	Add refrigerant.
Unit running continuously	Defective cold control	Replace.
	Low refrigerant	Add refrigerant.

Source: EBCO Manufacturing Co.

cal circuit includes overload protection, hot and cold water temperature controls, and a thermostatically controlled fan for additional condenser cooling when required (Figure 18–86). Table 18–9 is a service check chart for bottled water coolers.

Pressure-Type Water Coolers. These types of units provide drinking water at the rate of 7 to 16 gal (26.49 to 60.56 liters) per hour. They are free-standing and are self-supporting. Connections to an electrical, water supply, and wastewater drain are required (Figure 18–87). Some of these types of units have stainless steel components to provide resistance to corrosive atmospheres.

The refrigeration system is hermetically sealed. The air-cooled compressor pumps the refrigerant through the fan-cooled condenser. A capillary tube is used as the refrigerant flow control device. The evaporator is wrapped around the cooling tank to improve the unit efficiency. Also, a precooler routes the supply water around the cold drain water line to cool the incoming water (Figure 18–88). The electrical circuit includes automatic overload protection and an adjustable cold control for water temperature regulation (Figure 18–89). Table 18–10 is a service check chart for pressure-type water coolers.

Wall-Mounted Pressure-Type Water Coolers. These types of units provide drinking water cooled to 50°F (10.00°C) at 8, 14, and 16 gal (30.28, 52.99, and 60.56 liters) per hour and up to 45 cups (10646.47 ml) per hour of hot water (Figure 18–90). These units mount on the wall and require connections to the electricity, water supply, and a wastewater drain.

The refrigeration system on these units is hermetically sealed. The air-cooled compressor pumps the refrigerant through the fan-cooled condenser. A

FIGURE 18–87 Pressure-type water cooler. (Courtesy of EBCO Manufacturing Co.)

capillary tube is used for the refrigerant flow control device. The evaporator is wrapped around the cooling tank to increase the unit efficiency. Some models have a precooler included to route the supply water around the cold drain line to help cool the incoming water (Figure 18–91).

TABLE 18–10 Service Check for Pressure-Type Water Coolers

Trouble	Probable Cause	Remedy
Unit not running	No electrical power	Check power source. Check that cold control is not in off position.
	Defective cold control	Replace.
	Defective relay	Replace.
Unit running but not cooling	Refrigerant leak or low refrigerant	Add refrigerant.
Unit running continuously	Defective cold control	Replace
	Low refrigerant	Add refrigerant.

Source: EBCO Manufacturing Co.

FIGURE 18-88 Pressure-type water cooler; (a) parts breawkdown; (b) parts list. (Courtesy of EBCO Manufacturing Co.)

FIGURE 18–88(b)

Item No.	Description	7P-2 7P-D	13P-2	13PL-2
1	Bubbler valve	026551-002	026551-002	026551-002
2	Valve body gasket	008983	008983	008983
3	Friction washer	009281	009281	009281
4	Drain screw	026655-002	026655-002	026655-002
5	Spud screw	026680-005	026680-005	026680-005
6	Hole cover	016994	016994	016994
7	New style top	015444	015444	015444
8	Supply spud gasket	010545	010545	010545
9	Waste gasket	027389-001	027389-001	027389-001
10	Top supply line	027411-002	027411-002	027411-002
11	Upper cabinet bracket	027407-001	027407-001	027407-001
12	Insulated precooler assembly	027412-001	027412-001	027412-001
13	Cooling tank insulation [2 required]	018797	—	—
	Cooling tank insulation—right	—	025114	025114
	Cooling tank insulation—left	—	025115	025115
14	Insulated cooling tank	027443-100	027474-100	027474-100
15	Cold control	027040-007	027040-007	027040-007
16	Condenser	018252	018252	018252
17	Front panel—tan	024442-005	024442-005	024442-009
	Front panel—stainless steel	024442-003	024442-003	024442-011
18	Cabinet—tan	015570-002	015570-002	015569-002
	Cabinet—stainless steel	015570-005	015570-005	015569-005
19	Compressor [includes items 20 and 21]	027245-001	027256-002	027255-002
	Compressor—Danfoss [includes item 21]	026868-003	—	—
20	Overload	027268-002	017977	017977
21	Relay	027267-003	017978	017978
	Relay—Danfoss	026867	—	—
22	Base—tan	018560-001	018560-001	018560-001
	Base—stainless steel	018560-004	018560-004	018560-004
23	Service cord	026540-003	026540-005	026540-005
24	Fan blade	019199	022977	022977
25	Fan motor	027354-020	027354-008	027354-008
26	Fan bracket	017530	017530	017530
27	Fan shroud	017821	017821	017821
Not shown:				
	Insulated heat exchanger	027442-001	027470-001	027470-002

The hot tank is vented to the atmosphere through the two-way action of the hot faucet, eliminating the need for pressure–temperature devices allowing outgassing of water additives. The electrical circuit includes automatic overload protection and adjustable hot and cold controls for temperature regulation of the water (Figure 18–92). Table 18–11 is a service check chart for wall-mounted pressure-type water coolers.

Remote-Type Water Coolers. These units may be either field-built-up or factory-assembled units which are connected to a piping system for water distribution. They use a typical type of refrigeration system which is installed at a remote location to provide cool drinking water to several points of use (Figure 18–93). The components used on these types of systems are described below.

Compressor. A standard hermetic or semi-hermetic compressor is used to pump the refrigerant through the system. They may use either R-12 or R-22 refrigerant.

Condenser. The condenser may be either water-cooled or air-cooled, depending on the requirements of the system. The larger air-cooled condensers are usually located at a remote location so that a sufficient quantity of air can be available.

Evaporator. A direct-expansion evaporator is generally used. It may be of the shell-and-tube design with an additional section providing an external

FIGURE 18-89 Typical pressure-type water cooler wiring diagram. (Courtesy of EBCO Manufacturing Co.)

storage tank, or it may be an immersion type placed directly in the storage tank. When a separate storage tank is used, a pump is required to pump the water from the evaporator to the storage tank. In some localities special requirements, such as two thicknesses of metal between potable water and a refrigerant circuit is required. Check the local regulations for special requirements.

Water Storage. The storage tank should have the capacity of about one-half the rated chiller capacity. However, some situations may require that a different capacity be used. The type of material used must also be given careful consideration. Check the local codes for special requirements.

Water Pumps. The pumps used on drinking water systems are generally of the close-coupled, single-stage type with bronze fittings and mechanical seals. The pump is sized according to the amount of water in the system. The pump should move the water at a velocity to prevent a temperature rise of more than 5 °F (2.8 °C) when it flows through the complete system.

FIGURE 18-90 Wall-mounted pressure-type water cooler. (Courtesy of EBCO Manufacturing Co.)

FIGURE 18–91 Wall-mounted pressure-type water cooler: (a) parts breakdown; (b) parts list. (Courtesy of EBCO Manufacturing Co.)

Item No.	Description	DM WELCO8	DP8MH WELH08	DP14M WELC14	DP16M-1 WELC16-001
1	Gooseneck	—	017590	—	—
2	Handle	—	017925	—	—
3	Tee nut	—	017926	—	—
4	Cover	—	017927	—	—
5	Nut	—	017583	—	—
6	Top screw—stainless steel	026630-005	026630-005	026630-005	026630-005
7	Drain screw—stainless steel	026655-002	026655-002	026655-002	026655-002
8	Valve assembly	026551-002	026551-002	026551-002	026551-002
9	Gasket	008983	008983	008983	008983
10	Friction washer	009281	009281	009281	009281
11	Screw	026680-005	026680-005	026680-005	026680-005
12	Top	017930	017930	017930	017930
13	Gasket (under top)	010545	010545	010545	010545
14	Waste gasket	018050	018050	018050	018050
15	Fan motor	018909	018909	018909	018909
16	Fan blade	022977	022977	022977	022977
17	Fan motor bracket	018559	018559	018559	018559
18	Valve spud	025136	025136	025136	025136
19	Hanger	016698-002	016698-002	016698-002	016698-002
20	Hot valve gasket	—	017582	—	—
21	Hot valve body	—	021516	—	—
22	Check valve	—	019690	—	—
23	Top pads	011641	011641	011641	011641
24	Cold control	027040-007	027040-007	027040-007	027040-007
25	Hot control	—	027040-013	—	—
26	Panel clip	018156-001	018156-001	018156-001	018156-001
27	Right-side panel—painted tan	021866R	021866R	021866R	021866R
	Right-side panel—chestnut	018937-002	018937-002	018937-002	018937-002
	Right-side panel—charcoal	018938R	018938R	018938R	018938R
	Right-side panel—stainless steel	018939R	018939R	018939R	018939R
	Right-side panel—walnut	018940R	018940R	018940R	018940R
	Right-side panel—sandstone	018937-004	018937-004	018937-004	018937-004
	Right-side panel—painted lt. beige *W/W	024644	024644	024644	024644
	Right-side panel—beige *W/W	024577	024577	024577	024577
28	Condenser assembly	024076-001	024076-001	024076-001	024076-001
29	Terminal base	—	023265	—	—
30	Control mounting bracket	—	018933	—	—
31	Compressor (includes items 32 and 33)	027256-002	027256-002	027256-002	027256-002
32	Overload	017977	017977	017977	017977
33	Relay	017978	017978	017978	017978
34	Hot tank		026279-003	—	—
35	Hot tank insulation	—	025792	—	—
36	Precooler or waste assembly	019010	025356	025198	025198
37	Precooler insulation	018073	018073	018059	018059
38	Cooling tank	025139-001	025305-001	025239-001	025239-001
39	Cooling tank insulation ½	025114	025114	025114	025114
40	Cooling tank insulation ½	025115	025115	025115	025115
41	Drain plug	016899	016899	016899	016899
42	Service cord	026540-004	026540-013	026540-004	026540-004
43	Front panel—painted tan	021871	021871	021871	021871
	Front panel—chestnut	019644-001	019644-001	019644-001	019644-001
	Front panel—charcoal	019645	019645	019645	019645
	Front panel—stainless steel	019646	019646	019646	019646
	Front panel—walnut	019647	019647	019647	019647
	Front panel—sandstone	019644-002	019644-002	019644-002	019644-002
	Front panel—painted lt. beige *W/W	024638	024638	024638	024638
	Front panel—beige *W/W	024479	024479	024479	024479

FIGURE 18-91(b)

Item No.	Description	DM WELC08	DP8MH WELH08	DP14M WELC14	DP16M-1 WELC16-001
44	Left-side panel—painted tan	021866L	021866L	021866L	021866L
	Left-side panel—chestnut	018937-001	018937-001	018937-001	018937-001
	Left-side panel—charcoal	018938L	018938L	018938L	018938L
	Left-side panel—stainless steel	018939L	018939L	018939L	018939L
	Left-side panel—walnut	018940L	018940L	018940L	018940L
	Left-side panel—sandstone	018937-003	018937-003	018937-003	018937-003
	Left-side panel—painted lt. beige *W/W	024643	024643	024643	024643
	Left-side panel—beige *W/W	024556	024556	024556	024556
45	Frame assembly	025145	025145	025145	025145
46	Partition plate	025148	025148	025148	025148

*W/W, White Westinghouse.
[Courtesy of EBCO Manufacturing Co.]

FIGURE 18-92 **Typical wall-mounted pressure-type water cooler wiring diagram. (Courtesy of EBCO Manufacturing Co.)**

TABLE 18-11 Service Check Wall-Mounted Pressure-Type Water Coolers

Trouble	Probable Cause	Remedy
Unit not running	No electrical power	Check power source. Check that cold control is not in off position.
	Defective cold control	Replace.
	Defective relay	Replace.
Unit running but not cooling	Refrigerant leak or low refrigerant	Add refrigerant.
Unit running continuously	Defective cold control	Replace.
	Low refrigerant	Add refrigerant.

Source: EBCO Manufacturing Co.

FIGURE 18–93 Typical remote-type water cooling system.

System Controls. These types of systems typically have the following controls for operation and safety: (1) high- and low-pressure controls, (2) freeze control, (3) water temperature control, and (4) a water flow sensing switch to make certain that there is a sufficient amount of water flowing through the system to prevent freezing before the compressor is allowed to start.

System Options. The following are some of the more common options that should be considered for a clean, economical, and safe drinking water system: (1) split refrigeration system using timers, (2) tank anodes, (3) duplex system pumps using timers, and (4) a water filtration system used in combination with an activated carbon treatment program.

Description and Diagnosis of Electrical Components. The primary components of a typical refrigeration electrical circuit include the compressor, starting relay, compressor overload, and the cold control (Figure 18–94). In this type of circuit the electricity must first flow through the switching devices. The switching devices are the cold control (thermostat) and the overload protector. By following this

pattern, when either switch opens, the electrical power will be stopped.

These systems also include a thermostatically controlled fan motor (Figure 18–95). The fan thermostat measures the condenser temperature. When this temperature reaches a predetermined level the thermostat closes, turning on the fan to cool the condenser.

On some units there may also be a hot tank added in addition to the fan and thermostat (Figure 18–96). At the terminal board there is a wire on the return (white lead) side which runs to the hot tank heating element; this is the last leg of the circuit. The hot tank receives its power from a "piggyback" terminal on the cold control. This power must flow through the switching devices just as in the basic circuit. The power must first flow through the heat limiter. This heat limiter acts as a fuse in the circuit, and should the hot tank overheat it will blow. After the heat limiter, the power must flow to the hot thermostat. This device measures the temperature of the water in the hot tank. If the water is hot enough, the control will remain open and the power will not pass on to the heating element. When the water temperature drops below the designed temperature, the switch closes, allowing the power to flow into the heating element. The power in the heating element flows through a small wire coil, creating heat as it passes through, increasing the water temperature.

External Overload. The overload is a switching device used to protect the compressor from overheating (Figure 18–97). This device operates by use of a bimetal strip. When heated, the bimetal strip bends, opening the electrical contacts inside the overload. When the overload cools, the bimetal strip inside straightens back out, again closing the electrical contacts. The overload is mounted inside the box on the side of the compressor.

Internal Overload. On some water coolers the overload may not be mounted on the side of the compressor. In this case the compressor has an internal overload. This overload is embedded in the motor windings inside the compressor housing. It senses the temperature of the motor windings and a bimetal plate will open the electrical circuit should the temperature become too high. These types of overloads, since they are internal, cannot be replaced.

To test this type of overload the compressor must be cool. (It may require waiting for an hour.) Using an ohmmeter, check for continuity between the

FIGURE 18-94 Typical refrigeration electrical circuit. (Courtesy of EBCO Manufacturing Co.)

COLD CONTROL

OVERLOAD

POWER (BLACK OR RED)

NEUTRAL (BLACK OR WHITE)

RELAY

common and the run winding terminals on the compressor. If there is continuity between these terminals the overload is good. If there is no continuity and the compressor is cool, the overload is defective. In this case the compressor must be replaced.

Starting Relay. The starting relay is the component that starts the compressor (Figure 18–98). Each compressor has two sets of windings, the start winding and the run winding. The compressor in the refrigeration system is under pressure and requires

- – - – NEUTRAL (BLACK OR WHITE)
– +⊢ POWER (BLACK OR RED)

COLD CONTROL

FAN MOTOR

FAN
THERMOSTAT

OVERLOAD

RELAY

TERMINAL
BOARD

FIGURE 18-95 System with thermostatically controlled fan motor. (Courtesy of EBCO Manufacturing Co.)

a high starting torque. Therefore, tremendous electric power is required to start it pumping. This is the reason for the start winding. When the start winding is energized, it provides extra power to start the compressor running; once the compressor is running, the start winding is deenergized, and the run winding keeps the motor running. Two types of starting relays are used on water coolers: (1) the current relay, and (2) the PTC temperature-activated (ceramic) relay.

Water Coolers

POWER (BLACK OR RED)

NEUTRAL (BLACK OR WHITE)

HEAT LIMITER

HOT CONTROL

COLD CONTROL

FAN MOTOR

OVERLOAD

RELAY

TERMINAL BOARD

FAN THERMOSTAT

FIGURE 18-96 System with thermostatically controlled fan and hot tank. (Courtesy of EBCO Manufacturing Co.)

Use the following steps to test the starting relay:

1. *Continuity to the main (run) winding.* With the relay in the upright position, check for continuity between the line terminal and the main (run) ter-

minal. The ohmmeter must show continuity for the part to be good.

2. *Continuity to the start winding.* With the relay upside down, check for continuity between the line terminal and the start terminal. The ohm-

FIGURE 18-97   External overload. (Courtesy of EBCO Manufacturing Co.)

FIGURE 18-98   Starting relay. (Courtesy of EBCO Manufacturing Co.)

meter must show continuity for the part to be good. Do not shake this part trying to get continuity; the armature inside must move freely without interference. This does not apply to temperature-activated (ceramic) relays.

Ceramic PTC Relay.   It is difficult to test these relays, due to their heat sensitivity; therefore, it is best to have a known good replacement that can be installed to check the unit operation.

Cold Control.   The cold control is a switching device used to control the cold water temperature in the cooling tank (Figure 18-99). The control operates by using a temperature-sensitive gas sealed inside the capillary, or thermobulb. This bulb is attached to a bellows assembly inside the control housing. When the temperature increases, it causes the gas to expand; this in turn forces the bellows to expand, causing the contacts inside the housing to energize the compressor motor.

Once the water temperature begins to decrease, the sensitive gas will contract until the bellows shrinks and separates the two contacts, deenergizing the compressor motor.

There are two primary failures common to this component. One is that the control is open. This means that the control does not allow the electricity to pass through to the compressor motor. The most common cause for this failure is because the sensitive gas inside the thermobulb has escaped through a hole in the thermobulb or capillary tube. This can be checked by following the continuity test procedure. The second most common cause of failure is that the control will not start or stop the compressor motor when the temperature of the water demands it. This problem is the result of a miscalibration of the control. If this condition exists, the control must be replaced. It is not advisable to adjust the calibration screws on the control. The slightest turn of these screws makes a large change

FIGURE 18-99 Hot control. (Courtesy of EBCO Manufacturing Co.)

in the temperature at which the control will cycle. Doing this could also cause a cooling tank to freeze solid and burst creating a great deal of damage. This results in a great cost when compared to the price of replacing the control. One other defect that may be encountered would be a grounded or shorted control. This condition would be caused by one of the terminals making contact with the outside of the housing. This can be determined by using the grounded test procedure.

Use the following steps when testing the cold control.

1. *Continuity test.* With the control bulb warm and the control in the on position, continuity must be found between the terminals (Figure 18-99).
2. *Grounded.* Make sure that there is no continuity between the terminals and the outside housing.
3. *Miscalibration.* Install a new control. *Do not try to recalibrate.*

Hot Control. The hot control is used to regulate the water temperature in the hot tank (Figure 18-100). This control uses a bimetal disc to sense the water temperature and turn the heater off. Use the following steps to test the hot control.

1. *Continuity test.* When the control is cool, continuity must be present between the terminals.
2. *Miscalibration.* If the control is suspected of being out of calibration, install a new control. *Note:* If the tank continues to overheat after the new control is installed, a limiting condition may exist.

FIGURE 18-100 Cold control. (Courtesy of EBCO Manufacturing Co.)

FIGURE 18-101 Heat limiter. (Courtesy of EBCO Manufacturing Co.)

Heat Limiter. The heat limiter is a device used to protect the hot tank from overheating (Figure 18-101). There are two types of this control: (1) disposable, and (2) resettable.

The disposable heat limiters are made of a small wire that runs between the two electrical terminals. This type works in the same way as a fuse. When the hot tank reaches a predetermined temperature, the wire opens, shutting off the power to the tank

COMMERCIAL REFRIGERATION

heater. When this heat limiter opens, it cannot be repaired; it must be replaced.

The resettable heat limiter has a bimetal disc, which when heated above 195 °F (91 °C) bends, opening the electrical contacts. When this heat limiter cools down to approximately 110 °F (43 °C), the reset button can be pushed, and the device will work again.

Due to the possible adverse effect of an overheating hot tank, the heat limiter should not be altered or bypassed under any circumstances. Use the continuity test to test the heat limiter. Check for continuity between the terminals. The ohmmeter must show continuity for the part to be good.

◻ Safety Procedures ◻

A very good safety precaution is to check the unit before working on it to be certain that it is installed in accordance with local and national fire, electrical, plumbing, and building codes. The board of health has regulations that apply to all equipment used around food and beverages.

1. Check floor load limits before installing heavy commercial equipment.
2. Be sure not to trap liquid refrigerant inside a system without gas space; hydraulic pressure may build up and cause a rupture.
3. Obtain help when moving or lifting heavy cases and compressors.
4. Never purge burned refrigerant into a building where food is stored since the food might absorb the burned refrigerant and spoil.
5. Never use oxygen to test for leaks because oxygen and oil form an explosive mixture.
6. Be sure to disconnect the electrical service before attempting to work on the electrical components.
7. Always provide plenty of ventilation when welding, soldering, or brazing on refrigerant lines.
8. Wear goggles when charging or purging refrigerant from a system.
9. Wear rubber gloves when working with acidic materials.
10. Keep hands, feet, and clothing away from belts and moving parts.

◻ SUMMARY ◻

◻ Commercial refrigeration units are those used in business for the preservation of perishable merchandise.
◻ A quality commercial refrigeration installation must provide satisfactory service economically.
◻ It is necessary that the refrigeration unit be sized properly.
◻ Commercial refrigeration cabinets are designed and constructed to fit the needs of the user.
◻ Walk-in coolers generally are larger than 100 ft³ (2.83 m³) in size.
◻ Foam-in-place insulation is used to improve refrigeration by a foam-sandwich type construction.
◻ Reach-in cabinets generally have between 20 and 100 ft³ (0.566 and 2.83 m³) storage space.
◻ Reach-in cabinets are generally designed with glass doors to permit effective display of merchandise when desirable.
◻ Reach-in cabinets have a variety of uses such as bottled beverage storage, dairy product storage, food storage, and dough storage in bakeries.
◻ Florist's cabinets are designed for display. They have glass doors and often have glass ends.
◻ Florist's cabinet temperatures range from 47 to 54 °F (8.3 to 12.2 °C).
◻ Display cases are used almost exclusively for the preservation and display of meats and dairy products.
◻ Double-duty cases have two compartments: an upper compartment for display and a lower compartment for storage.
◻ Foods displayed in the open-type self-service display case are stored below a blanket of cold air.
◻ The main problem with open self-service cases is that of providing the additional refrigeration capacity required to take care of the greater loss of cold air.
◻ One sure way to determine whether or not the evaporator is fully refrigerated is to observe the refrigerant line from the expansion valve to the evaporator and the suction line leading from the evaporator. Both must be frosted.
◻ It is recommended that a heat exchanger be installed on cases maintained below 40 °F (4.4 °C) to obtain the best results.
◻ The commercial refrigeration cycle is basically the same as the refrigeration system used in domestic refrigerators.
◻ Extreme caution must be exercised to prevent the entrance of moisture into the refrigeration system.
◻ Low-temperature single-stage refrigeration systems become increasingly critical from a design and application standpoint as the desired evaporating temperature is decreased.

- [ ] In an effort to prevent the discharge gas and motor temperatures from exceeding the recommended limits, it is desirable and in some cases absolutely necessary to insulate the suction lines, which will allow the suction gas to return to the compressor at a lower than normal temperature.
- [ ] Some compressor manufacturers recommend that R-502 refrigerant be used in all single-stage low-temperature applications where the evaporator temperature is $-20\,°F$ ($-28.9\,°C$) or below.
- [ ] The refrigerant lines must be designed and installed to prevent oil trapping.
- [ ] When large refrigerant charges are unavoidable, a recycling pumpdown control should be used.
- [ ] To prevent possible permanent damage to the motor windings, the compressor motor must not be operated while the high-vacuum pump is in operation.
- [ ] Two-stage low-temperature units are more efficient and encounter fewer operating difficulties at lower operating temperatures than single-stage units.
- [ ] As the compression ratio increases, the volumetric efficiency decreases.
- [ ] Two-stage compressors generally use an external manifold and a de-superheating expansion valve.
- [ ] Below compression ratios of 7:1, the single-stage compressor will have greater capacity; however, two-stage compressors are more efficient above these ratios.
- [ ] De-superheating expansion valves are used to keep low-temperature compressors cool.
- [ ] The function of the subcooler is to cool the liquid refrigerant being fed to the evaporator by the evaporation of refrigerant fed through the de-superheating expansion valve.
- [ ] The proper performance of any refrigeration or air conditioning system is greatly dependent on the proper charge of refrigerant.
- [ ] Never change liquid refrigerant into the compressor suction or discharge valve ports because the compressor valves may be damaged.
- [ ] Vapor charging is normally used when only small amounts of refrigerant are to be added to a system.
- [ ] The refrigerant charge can be more precisely controlled by using vapor charging rather than liquid charging.
- [ ] Crankcase pressure-regulating valves (CPR) are installed in the suction line ahead of the compressor suction valve and control the maximum pressure at the compressor suction.
- [ ] The function of the condenser pressure-regulating (CPR) valve is to maintain adequate high-side pressure during low ambient temperature.
- [ ] The CPR valve is located between the condenser and receiver.
- [ ] High- to low-side bypass valves are installed between the system high and low sides and respond to the valve outlet pressure. The valve opens when the low-side pressure is reduced to the valve setting.
- [ ] Bypass control valves function as condenser bypass devices on air-cooled condensers with a winter start control.
- [ ] The simplest form of compressor capacity control is on-off operation of the compressor.
- [ ] Cylinder unloaders are used on large equipment operating under a fluctuating load condition.
- [ ] Hot-gas bypass capacity control is recommended where normal compressor cycling or the use of unloaders may not be satisfactory.
- [ ] The most positive and dependable method of preventing refrigerant migration is to use a system pumpdown cycle.
- [ ] Since proper refrigeration of a case is dependent on air circulation, it is necessary to defrost the evaporator periodically so that refrigeration may continue.
- [ ] Off-cycle defrost uses a low-pressure control to control the compressor operation.
- [ ] Timed defrost systems use a defrost control, usually a time clock, to provide the desired time and duration of the defrost cycle.
- [ ] When supplemental heat is required for defrost, some cases are equipped with supplemental electric heat strips and a defrost limit switch.
- [ ] The methods used in defrost termination are: (1) straight-time termination, (2) pressure termination, and (3) solenoid reset termination.
- [ ] Demand defrost is accomplished by use of solid-state components in two ways: (1) air velocity, and (2) the fluidic principle.
- [ ] Pressure controls are used to control the compressor in response to the suction or discharge.
- [ ] Defrost methods are named by their mode of defrost initiation and termination.
- [ ] The duration of defrost must be long enough to clear the evaporator completely each defrost period, while defrost must occur often enough to keep the refrigerated air circulating freely throughout the case.
- [ ] Basically, combining is the grouping of more than one refrigeration function on one compressor.
- [ ] A simple system is two or more cases of the same suction pressure and defrost requirements connected to one condensing unit.
- [ ] A combined system involves cases of different suction pressures and/or defrost requirements on one condensing unit.
- [ ] Complex combined systems are designed to handle all low-temperature (or all normal temperature) cases and coolers on one condensing unit.
- [ ] Ultracomplex combined systems include all of the refrigeration equipment in the store on one condensing unit.

- To separate cases and coolers of different temperatures there are two commonly used methods: (1) liquid line solenoid valves, and (2) evaporator pressure-regulating (EPR) valves.
- Heat recovery systems take the heat extracted from the refrigerated cases and return it to the store.
- The three approaches to heat recovery are: (1) direct-air method, (2) closed-circuit water method, and (3) dual-condenser method.
- Refrigerant line pressure drop tends to decrease system capacity and increase the amount of electric power required by the compressor.
- Compressor lubricating oil can be circulated through the system only if the velocity of the refrigerant vapor is great enough to carry the oil along with it.
- Normally, oil separators are necessary for operation at evaporating temperatures below $-50\,°F$ ($-45.6\,°C$) in order to minimize the amount of oil in circulation.
- Each valve, fitting, and bend in a refrigerant line contributes to the friction pressure drop because of its interruption or restriction of smooth flow and must be figured as part of the equivalent length of pipe.
- Pressure drop in discharge lines is probably less critical than in any other part of the system.
- The primary concern in liquid line sizing is to ensure a solid column of refrigerant at the expansion valve.
- Liquid line pressure drop causes no direct penalty in electric power consumption, and the decrease in system capacity due to friction losses in the liquid line is negligible.
- The normally accepted design practice in sizing suction lines is to use a design criterion of a suction line pressure drop equal to $2\,°F$ ($1.11\,°C$) change in saturation temperature.
- Nominal minimum velocities of 700 fpm (212.8 m/min) in horizontal suction lines and 1500 fpm (456 m/min) in vertical suction lines have been recommended and used for years as suction line sizing design standards.
- The two lines of a double riser should be sized so that the total cross-sectional area is equivalent to the cross-sectional area of a single riser that would have both satisfactory vapor velocity and an acceptable pressure drop at maximum-load conditions.
- Close attention must be given to piping design on multiplex systems to avoid oil return problems.
- Horizontal suction and discharge lines should be pitched downward in the direction of flow to aid in oil drainage.
- If a pumpdown control system is not used, each evaporator must be trapped to prevent liquid refrigerant from draining back to the compressor by gravity during the off cycle.

▣ EXERCISES ▣

1. Define commercial refrigeration units.
2. Why is it necessary that the refrigeration unit be sized properly?
3. What size cabinet is considered to be a walk-in cooler?
4. For what are the glass doors used on reach-in boxes?
5. For what are reach-in cabinets used?
6. What should the temperature range be in a florist's cabinet?
7. What type of case has a storage area and a display area?
8. How are foods stored in open-type self-service cases?
9. What has proven to be the main problem in open-type self-service cases?
10. What should the temperature be in a frozen food display case?
11. What should be the thickness of the cold air blanket that separates the frozen food from the ambient air?
12. What is the one sure way to determine whether or not the evaporator is fully refrigerated?
13. How will moisture probably be noticed in low-temperature refrigeration systems?
14. How is a decreased evaporating temperature hazardous to single-stage low-temperature compressors?
15. When should compressor motor windings be tested?
16. Below what temperature do most compressor manufacturers recommend that R-502 be used in single-stage low-temperature applications?
17. Where should the filter-drier be installed in the liquid line on commercial cases?
18. Why should the compressor motor not be operated when it is under a high vacuum?
19. Why are two-stage low-temperature systems more efficient than single-stage low-temperature systems?
20. What are the two factors that cause a loss of efficiency with an increase in compression ratio?
21. Why are two-stage compressors used in low-temperature applications?
22. What is the function of a liquid subcooler?
23. What does a starved evaporator indicate?
24. Why should liquid refrigerant not be fed into the compressor suction?
25. Is liquid charging or vapor charging more accurate?
26. What will a higher than normal discharge pressure indicate?
27. What is the most common method of determining whether or not a system is properly charged?
28. What is the purpose of the crankcase pressure-regulating valve?
29. What is the purpose of the evaporator pressure-regulating valve?

30. Where is the condenser pressure-regulating valve located?

31. Can the high- to low-side bypass valve be used for capacity reduction?

32. When are bypass control valves for air-cooled condensers used?

33. When are compressor cylinder unloaders most likely to be used?

34. Name the methods used in hot-gas bypass capacity control.

35. Under what conditions is a pumpdown control recommended?

36. Why do refrigerated cases need a defrost period?

37. Name the basic methods used to defrost evaporators in commercial cases.

38. What are the three settings on a dual pressure control?

39. How are defrost methods named?

40. Define a combined system.

41. What are the two methods used in maintaining case temperature differences?

42. Where is the heat obtained for a heat recovery system?

43. How can oil logging of an evaporator be prevented on a low-temperature system?

44. What are the two main considerations given to refrigerant piping design?

45. What are the maximum desirable refrigerant velocities in discharge lines?

46. What is the main consideration when sizing liquid lines?

47. Why is the sizing of the suction line more important than the sizing of the other lines?

48. How should horizontal suction and discharge lines be installed?

19 Commercial Ice Machines

Performance Objectives

After completion of this chapter, the student should:

■ Be able to demonstrate a better understanding of the basic operation of ice machines.

■ Understand the basic operating principles of the Frigidaire ice machine.

■ Understand the basic operating principles of the Ice-O-Matic Flaked ice machine.

■ Understand the basic operating principles of the Kold-Draft Automatic ice cuber (electric).

■ Complete the Exercises with at least 80 percent accuracy.

■ Complete the Laboratory Exercises with at least 80 percent accuracy.

There are many brands of commercial ice machines on the market. They are generally referred to as *flakers and cubers*. This reference is to the style of ice the machine produces. It is impossible to cover every brand of ice machine in this book. Every manufacturer of ice machines will have a different operating sequence and sometimes more than one operating sequence is found on different models of the same brand. Therefore, the following examples of ice machine operation will deal only with the Frigidaire SCK and MCK models, the Ice-O-Matic flaked ice machine, and the Kold-Draft automatic ice cuber.

Definition. *Commercial ice machines* are designed to produce the ice used in commercial applications, such as restaurants, cafeterias, and other applications requiring large amounts of ice on a daily basis. They are usually self-contained units which are connected to the city water supply and electricity so that the ice can be made automatically.

☐ Frigidaire Ice Machines ☐

Frigidaire ice machines are designed primarily for back bar installation, but can be used almost anywhere. The storage bin opening is designed for easy ice cube removal. The storage bin door also is placed on an inclined plane, which makes possible the full opening and closing of the door without the use of a latch (Figure 19–1). The ice bin will store approximately 70 lb (31.75 kg) of ice cubes.

The unit is completely serviceable from the front. Both the front and rear machine compartment openings are the same size, permitting the use of an additional panel for freestanding installations.

Water Reservoir. The stainless steel water reservoir is mounted on the cabinet walls of the rear right-hand corner of the ice bin storage compartment (Figure 19–2).

Thermostats. The ice thickness thermostat and the bin thermostat are located behind the access plate just above the grid on the right-hand side of the bin storage compartment. The ice level in the bin can be varied by removing the two mounting screws in the ice bin thermostat well-mounting flange and rotating the well one-half turn.

Expansion Valve. The expansion valve is located on the upper left-hand wall of the cabinet liner at the side of the evaporator. The expansion valve bulb is

FIGURE 19-1 Ice bin. (Courtesy of Manitowoc Equipment Company.)

FIGURE 19-2 Rear view of ice cuber with panel removed. (Courtesy of Manitowoc Equipment Company.)

attached to the suction line approximately in the center of the cabinet, under the evaporator.

Machine Compartment. The condensing unit consists of the following components: compressor, air-cooled condenser, upright receiver, fan and motor, connecting refrigerant lines, hand shutoff valves, and a hot-vapor solenoid valve (Figure 19-3). Each of these components is individually replaceable.

The entire condensing unit is easily removed through the front of the machine compartment. The unit is fastened to the cabinet base by a hold-down clip in the rear and two hold-down bolts in the front.

The ¾-in. (19.05-mm) female drain and ¼-in. (6.4-mm) female pipe water supply connections are located in the upper left-hand side of the cabinet shell to accommodate the lines to these connections.

Electrical Connections and Control Box. A choice of two openings is provided in the control box for the entrance of electrical supply lines—one in the left-hand side of the box and one in the back. The two screw terminals on the terminal board are the line connections. The control box contains a terminal

FIGURE 19-3 Front view of compressor compartment. (Courtesy of Manitowoc Equipment Company.)

FIGURE 19-4 Frigidaire SCK-11A wiring diagram (high voltage). (Courtesy of Frigidaire Division, General Motors Corp.)

board, transformer, grid circuit fustat, starting relay, starting capacitor, and service switch. The wiring diagram is located on the inside of the control box cover (Figures 19–4 and 19–5).

Cycle of Operation. Water from the circulating pump in the reservoir cascades over the freezer plate or plates and freezes out until it builds up into a slab of the desired thickness. The ice thickness control then opens the electric circuit to the water pump and energizes the hot-vapor solenoid valve. The hot vapor from the receiver passes into the freezer plate and releases the ice slab, which slides down onto the

cutting grids. As soon as the water pump is shut off by the ice thickness control, all the water in the system drains back into the reservoir, resulting in a siphoning action that drains almost all the water from the reservoir. The reservoir then fills with fresh water for the next freezing cycle. After the ice slabs are released, a hinged mercury heater switch is tripped to add heat to the thickness control. The thickness control then opens the circuit to the hot-vapor solenoid and closes the circuit to the water pump to complete the cycle. The compressor operates continuously until the storage bin thermostat is satisfied, at which time all electric circuits

FIGURE 19-5 Frigidaire SCK-11A wiring diagram (low voltage). (Courtesy of Frigidaire Division, General Motors Corp.)

are opened except for the low-voltage circuit to the cutting grids, which are energized as long as the ice machine is connected to the external electric power supply.

☐ Ice-O-Matic Flaked Ice Machine ☐

These basically consist of a hermetically sealed refrigeration system designed to freeze ice in a cylinder. The water is fed by a float valve into the cylinder where it is frozen against the outside wall. The ice is then chipped from the wall by a rotating auger. A thermostat installed in the ice bin shuts the refrigeration unit off when the bin is full of ice (Figure 19-6).

Refrigeration System. The sealed refrigeration system consists of a hermetic motor-compressor, a condenser (which can be either air or water cooled), a heat exchanger, a refrigerant drier, an automatic expansion valve (to meter the refrigerant to the evaporator maintaining a constant pressure and temperature in the freezing chamber), a freezing chamber, the refrigerant lines, and the refrigerant (Figure 19-7).

1.	Bin Control—Safety	9. Crank	17. Bin Rod—Safety
2.	Compressor Electric Box	10. Drive Plate	18. Water Purge Line
3.	Condenser	11. Grid Adjuster	19. Grid Water Return Trough
4.	Electric Inlet Box	12. External Equalizer Line	20. Electrical Junction Box
5.	Heat Exchanger	13. Refrigerant Distributor	21. Freezing Plate
6.	Hot Gas Defrost Line	14. T.X. Valve	22. High Temperature Safety Control Bulb
7.	Pivot Arm	15. Drains	23. Water Distribution Tube
8.	Crosshead	16. Water Inlet	24. T.X. Valve Bulb

FIGURE 19–6 Component location. (Courtesy of Mile High Equipment Co.)

Water System. The water system on these flakers consists of a float-operated valve that maintains a constant water level in the evaporator. This water level is just below the ice discharge opening and above the top of the freezing chamber. Units equipped with water-cooled condensers use a water-regulating valve to meter the flow of water through the condenser to maintain a predetermined head pressure on the high side of the refrigeration system. To adjust the level of the freezing water, loosen the two screws on the adjustable float tank bracket and move the float up or down as needed (Figure 19–8).

Automatic Expansion Valve. The automatic expansion valve is designed to meter the refrigerant into

FIGURE 19-7 Refrigeration cycle. (Courtesy of Mile High Equipment Co.)

FIGURE 19-8 Water system description. (Courtesy of Mile High Equipment Co.)

FIGURE 19-9 Automatic expansion valve. (Courtesy of Mile High Equipment Co.)

TABLE 19-1 Expansion Valve Settings for Different Altitudes

Below 3000 ft Altitude		Over 3000 ft Altitude	
Model No.	Setting	Model No.	Setting
225	15	225	16
350	15	350	16
600	13	600	14
750	12	750	13
1000	11	1000	12

Source: Mile High Equipment Co.

the evaporator and maintain a constant evaporator pressure.

The diaphragm has a pressure exerted on it from the top by the diaphragm spring and also by atmospheric pressure (Figure 19-9). These pressures are opposed by the valve needle spring and the evaporator pressure. The stainless steel push rod transfers the diaphragm movements to the valve needle.

The valve needle is a stainless steel ball that eliminates sticking and ensures smooth action. Because the diaphragm uses atmospheric pressure to operate, some additional adjustment may be necessary at the installation location by means of the adjusting screw. The automatic expansion valve should be adjusted to provide a certain suction pressure on specific unit models at certain altitudes (Table 19-1). This suction pressure should allow no live frost on the suction line at the compressor.

Adjusting Bin and Safety Controls. Sometimes after a unit has been installed, it becomes necessary to make internal adjustments at the range screw. However, in most cases the external adjustment will be adequate and no internal adjustment is necessary.

Caution: Before any internal adjustments are made, adjust the control externally. If conditions then require internal adjustments use the settings in Table 19-2 for internal adjustment of the range screw.

After the altitude correction has been completed, fine adjustment of the control can be obtained by turning the screwdriver slotted shaft (Figure 19-10). The bin control should allow the bin to fill without the bin overflowing or the evaporator being obstructed by packing of the ice in the chute.

To set the safety control, shut off the water supply to the float and leave the machine running. When the water in the float chamber and the evaporator

TABLE 19-2 Bin and Safety Control Temperature Settings and Altitude Correction Adjustments

Setting							
Bin Control—300035				**Safety Control—300036**			
Cold Out	Norm Out	Norm In	Warm In	Cold Out	Norm Out	Norm In	Warm In
31°±2	36°±1	42°±1	47°±2	25°±2	30°±1	38.5°±1	43.5°±2
Altitude Correction (CW Turns of Range Screw)							
Feet	Turns	Feet	Turns	Feet	Turns	Feet	Turns
2000	$5/32$	8000	$27/32$	2000	$3/16$	8000	$1\,1/16$
4000	$13/32$	9000	$31/32$	4000	$1/2$	9000	$1\,3/16$
6000	$5/8$	10000	$1\,1/16$	6000	$25/32$	10000	$1\,5/16$

Source: Mile High Equipment Co.

Bin Control
Part Number 300035

Safety Control
Part Number 300036

Screw driver SLOTTED shaft for turning the cam

FIGURE 19-10 Bin and safety control. (Courtesy of Mile High Equipment Co.)

line has been used, the control should stop the unit. Turn the water supply back on. The safety control should restart the unit after a period of 3 to 5 min.

Caution: The safety control setting must allow a continuous flow of water to the evaporator.

Removing the Chassis. To remove the ice maker chassis from the cabinet, first remove the side and front panels (Figure 19–11).

Caution: Switch off the electric power before removing the panels.

Disconnect the water inlet line from the float. Disconnect the electrical wiring at the junction box at the rear of the cabinet. Disconnect all drain lines. Loosen the ice chute at the evaporator and push it out of the way toward the ice bin. Remove the bin rod tube. Remove the hold-down bolt at the front of the lower shelf of the chassis. The chassis can now be lifted over the angle and slid out the front of the cabinet.

SHIPPING BRACKET
(To be removed at installation)

FIGURE 19-11 Chassis removal. (Courtesy of Mile High Equipment Co.)

Note: Disconnect the high-pressure water line from the Procon Pump.

Replacing a Seal. Use the following steps when replacing a seal:

1. Switch off all electric power to the unit.
2. Remove the rear and left-hand-side panels.
3. Remove the water line to the evaporator.
4. Remove the auger drive motor.
5. Unscrew the upper evaporator nut.
6. Loosen the setscrew in the top half of the drive coupling. Pry up the auger with a screwdriver under the spider in the coupling until it can be removed from the top of the evaporator.
7. Remove the evaporator condensation boot.

Ice-O-Matic Flaked Ice Machine

8. Unscrew the bottom evaporator nut.

9. Remove the lower bearing assembly by driving lightly against the bearing. *Caution:* Do not drive against the seal. Use a 1⅛ in. × 12 in. (28.575 mm × 304.8 mm) wooden driving rod for best results.

10. Use a screwdriver to pry out the seal. The seal is pressed in at the factory.

11. Clean the housing for installation of the new seal.

12. Install the new seal. Care should be taken when installing the new seal; be certain that it goes into the bearing housing straight. Do not try to tap it in with a hammer. Use a steady pressure. Lubricate the seal freely with standard cup grease for better installation results. Part of the seal (a bronze and hard rubber assembly) is assembled on the auger shaft. Replace this part by using hand pressure.

13. Reassemble the unit. Reverse the procedure outlined above.

Leak Testing the Unit. The insulation used on the evaporator is expanded by use of a Freon agent. Caution must be exercised with any leak detector because breaking of a bubble of the insulation will cause a simulated leak. Therefore, when a refrigerant leak exists in the system and cannot be located on any of the exposed tubing, it will be necessary to proceed slowly in removing the insulation from the evaporator. Allow enough time for the vapor to dissipate before completing the leak test. Sometimes it may even be necessary to expose the entire tubing and freezing cylinder and use a soap-bubble solution to make a satisfactory test.

Replacing an Evaporator Shell Assembly. Use the following steps when replacing an evaporator shell assembly:

1. Switch off the electric power before removing the panels.

2. Remove the chassis from the cabinet. (See the section on chassis removal.)

3. Disconnect the water inlet tube from the evaporator.

4. Disconnect the drain tube from the evaporator.

5. Remove the safety control cap tube from the well on the lower section of the evaporator.

6. Remove the upper evaporator nut.

7. Loosen the setscrews in the upper part of the drive coupling located at the bottom of the auger. Insert a screwdriver in the coupling under the fiber spider and pry up until the auger can be lifted from the top of the evaporator assembly.

8. Purge all refrigerant from the system, disconnect the suction line at the service valve and cut the liquid line at the compressor end of the heat exchanger.

9. Remove the two bolts holding the evaporator to the frame and lift the evaporator assembly out.

10. Remove the lower nut from the evaporator assembly.

11. Remove the drain boot by pressing down on one side.

12. Remove the lower bearing housing assembly by driving lightly against the bearing. *Caution:* Do not drive against the seal. Use a 1⅛ in. × 12 in. (28.575 mm × 304.8 mm) wooden driving rod for best results.

13. Reassemble the unit by reversing the outlined procedure.

14. Evacuate the refrigerant system.

15. Install a new refrigerant drier.

16. Charge the unit with the proper amount of R-12 as indicated on the nameplate.

Evaporator-Auger Cleaning Procedure. It should be noted that cleaning the auger regularly cannot be overemphasized. An auger that has deposits on it can cause excessive pressure on the auger assembly as well as reduce the unit capacity. The auger should be cleaned at least every three months and more often in areas that have severe water conditions. It is best to observe the auger during the first few months of operation to determine the required frequency of cleaning. Inspection of the clear plastic float chamber will indicate the condition of the auger; a large amount of scale here will indicate a large amount of scale on the auger.

Before doing any cleaning, remove all ice from the storage bin or place a sanitary container in the bin to catch all the ice produced during the cleaning procedure and for 30 min afterward to prevent contamination of the ice already in the bin. Any ice produced during and immediately after cleaning will have an acid taste and must be thrown away.

Use the following steps in cleaning the auger in an air-cooled unit.

1. Turn off the electric power supply to the unit.

2. Remove the front and side panels.

3. Turn off the water supply to the unit.
4. Drain the evaporator and float reservoir by removing the water feedline from the float and draining the water into a container.
5. Replace the water feedline to the float reservoir.
6. Mix the ice machine cleaning solution. Use 1 gal of water to 2 oz of ice machine cleaner. *Caution:* Do not mix the solution stronger than recommended.
7. Fill the float reservoir with the solution.
8. Replace the panels.
9. Turn on the electric power supply to the unit.
10. Use the following chart to determine the length of running time of the machine before adding additional cleaning solution.

    B-225   6 min
    B-350   5 min
    B-600   4 min
    B or L-750   3 min
    B or L-1000 2 min

11. Repeat steps 1, 2, 7, 8, and 9 until the 1 gal of cleaning solution has been used.
12. Turn off the electric power supply to the unit and repeat steps 2, 4, and 5.
13. Replace all panels.
14. Turn on the electric power supply to the unit.
15. Turn on the water supply to the unit.
16. Allow the unit to operate and throw away all ice produced in the first 30 min of operation.

When cleaning the auger on a water-cooled unit, use the same procedure as that used on air-cooled units with one exception: Do not turn off the water supply to the unit. Instead, block up the float ball in the water reservoir to prevent water from entering the reservoir during the cleaning operation.

Water Treatment. The quality of water varies from location to location, from season to season, and from day to day. Most city water supplies contain solids and chemicals that tend to decrease the performance and life expectancy of an ice machine.

The end product—the ice—can only be as good as the raw materials fed into the unit. Therefore, it is recommended that a local water treatment company be contacted for specific filtering recommendations at your location. Some of the most common conditions and causes are listed in Table 19-3.

Magnetic Water Conditioning. Awareness of water conditioning alternatives is essential for providing

TABLE 19-3 Water Condition Diagnosis

Condition	Probable Cause
Metallic taste in water	High iron or acid content
Water smells like rotten eggs	High sulfur content
Soft ice	High solid mineral content
Stainless steel bin liner appears rusted or stained	High chlorine content

FIGURE 19-12   Magnetic water conditioning units. (Courtesy of Aqua-Flo Inc., Baltimore, Md.)

good customer service. In many situations, hard water problems cause the need to replace piping and water heaters as well as other heat transfer surfaces or to use a salt ion-exchange water softener. Many times the salt ion-exchange process is impractical for homeowners and businesses that serve customers who have heart problems or hypertension because it increases the sodium level of the drinking water to unacceptable levels.

This problem may be solved through the use of a magnetic water conditioning unit (Figure 19-12). This system works this way. Natural water is made up of chaotic randomly positioned molecules. The magnetic water conditioner polarizes the water, causing all the water molecules and the scale-forming minerals to be attracted to each other.

This causes two sets of permanently high charged magnetic systems. As the water flows through the magnetic field, the hard-water minerals are polarized and remain in liquid suspension, eliminating the problem of these particles clinging to the water pipes and the walls and surfaces of heat transfer equipment.

Essentially, the scale-forming minerals end up suspended in the water and all prior existing scale is magnetically attracted to the water. This descaling period, known as the stabilization period, lasts from two to six months. During this stabilization period, the scale that is removed from the heat transfer surfaces turns into a fine sediment that settles in the bottom of the system. It is recommended that during the stabilization period that water is drained from the lower tap once a month until it is clear. Manufacturers state that while scale is being removed from the system, green stains also are removed from the fixtures. This is a sign that the pipes have stopped leaching (pitting internally).

After the polarizing process, the water is still very safe to drink, reports R.E. Meyers, Inc. "Those scale producing minerals are exactly what people pay extra for when drinking quality mineral water. The medical community says that these minerals are important to our health. Because the magnetic condition also causes de-gasification of the water which naturally releases such gases as sulfur and chlorine, magnetic water conditioning often improves the taste of drinking water." This water may be used in residential systems for water treatment. Ice makers, coffee makers, and vending machines are some of the commercial uses for these units (Figure 19–13).

When installing one of these units on a commercial ice machine for the treatment of lime scale and chlorine, use the following instructions (Figure 19–14).

1. Install the model 10 clear housing with a granular activated carbon filter (A), Model GAC-10 (5 μm nominal) and (optional) polyphosphate cartridge (B), Model PCC-1.
2. Install an Aqua-Flo conditioner (C) on the circulating line after the circulating pump. (Size Aqua-Flo according to the circulating line size.)
3. Install a grounding wire (D) on the inlet of the Aqua-Flo conditioner.

When the system is operating, the following results: The carbon filter will remove chlorine. If the chlorine is not removed, it will create a mild hydrochloric acid condition that will etch away at the heat transfer surface. The Aqua-Flo conditioner will dissolve the existing lime scale and prevent further lime scale buildup on the surfaces.

Another fact is the shortening of the bonding between the hydrogen and oxygen atoms. This makes the water wetter, thereby increasing its solubility index.

The National Bureau of Standards (NBS) reports that ⅜ in. (9.53 mm) of scale in the hot water heater uses 48% more fuel to heat. When the scale is removed, a more efficient water heating system is created (Figure 19–15). Also, the efficiency of all heat transfer equipment will be increased by the elimination of the scale buildup on the transfer surfaces.

Preparing the Unit for a Period of Storage. Use the following steps in preparing a unit for a period of storage or inoperation:

FIGURE 19–13 Beverage Equipment installation. (Courtesy of Aqua-Flo Inc., Baltimore, Md.)

FIGURE 19–14 Magnetic water conditioner installed on a commercial ice machine. (Courtesy of Aqua-Flo Inc., Baltimore, Md.)

COMMERCIAL ICE MACHINES

FIGURE 19–15 Typical residential conditioner installation. (Courtesy of Aqua-Flo Inc., Baltimore, Md.)

1. Turn off the water supply to the unit.
2. Turn the toggle switch to the off position.
3. Let the unit stand idle for ½ hr and let the ice in the auger assembly melt.
4. Disconnect the evaporator-reservoir tube from the evaporator assembly.
5. Drain the complete system. Do not replace the tube removed in step 4.
6. Wipe out the storage bin.

Wiring Diagrams. Figures 19–16 to 19–19 show the wiring diagrams for the different Ice-O-Matic flaked ice machines. They should be followed when troubleshooting or making wiring repairs to the unit.

Troubleshooting. See Table 19–4.

☐ Kold-Draft Automatic Ice Cuber (Electronic) ☐

The following is a description of the Kold-Draft Automatic Ice Cube GB models. It should be remembered that the inspection panel must always be in place in order to get the most satisfactory performance from the machine. When it is removed, it will affect the operation of the expansion valve and the length of the freezing time for the unit.

The entire cuber is contained in one unit and is mounted on an ice bin at the time of installation. Should extra capacity be desired, another cuber may be installed on top by merely removing the top cover

FIGURE 19–16 Wiring diagram for 120-V air-cooled unit. (Courtesy of Mile High Equipment Co.)

FIGURE 19–17 Wiring diagram for 120-V water-cooled unit. (Courtesy of Mile High Equipment Co.)

FIGURE 19–18 Wiring diagram for 208/240-V air-cooled unit. (Courtesy of Mile High Equipment Co.)

FIGURE 19–19 Wiring diagram for 208/240-V water cooled unit. (Courtesy of Mile High Equipment Co.)

TABLE 19-4   Flaker Service Diagnosis

Condition	Possible Cause or Remedy
Low ice production	Check for obstructions in flow of water to the evaporator.
	Check float adjustment.
	Corroded auger, clean as per instructions.
	Stopped up water float valve.
	Restricted filter-drier.
	Dirty condenser.
	Valves in compressor not functioning properly.
	System under- or overcharged.
	Condenser fan not working.
	Low line voltage.
	Water-regulating valve set too low (water-cooled units only).
Unit runs, but no ice production	Water supply shut off.
	Water float valve plugged.
	Inlet water tube frozen at evaporator due to safety control set too cold.
	Combination of no water to evaporator and defective safety control.
	Unit out of gas.
	Motor compressor not pumping.
	Defective T.X. valve.
Vibration in water reservoir assembly	Too high water pressure or defective water pressure regulator.
	Partially stopped up float valve.
Excessive noise in evaporator	Auger needs cleaning.
	Defective bearing.

TABLE 19–4 [Continued]

Condition	Possible Cause or Remedy
Low suction pressure	Water restriction to evaporator. Restricted liquid flow through filter-drier. System low on refrigerant. Moisture in refrigeration system.
Water leaking at bottom of evaporator assembly	Retaining nut loose on lower housing assembly. Lower water seal leaking.
Wet ice [water being carried out with ice]	Water level set too high.
Wet ice [soft]	System low on refrigerant. Back pressure set too high. Valve in compressor not pumping properly or reeds in compressor coked [burned deposits].
Noisy motor compressor	High head pressure. Defective compressor. Compressor low on oil.
Compressor cuts out on overload	High head pressure. Low line voltage [it should be within 10% of rated voltage]. Defective compressor unit, starting capacitor, relay, or overload device. Loose electrical connection, probably in compressor junction box.
High head pressure	Dirty condenser. System overcharged. Moisture in refrigeration system.
Machine frozen up	Dirty auger. Defective bearing. Loose V-belt or pulley. Defective auger motor.
Brass in ice	Defective bearing.
Gear reducer input shaft turns, but output shaft does not	Gear stripped.
Motor-compressor and auger motor will not run	Check 120- or 240-V supply. Check on–off switch. Activate bin control by placing hand on bulb for 15 sec. Readjust if necessary. Check safety control. Check for loose connections. Defective auger motor. Defective wiring harness.
Auger motor runs, but motor compressor does not	Loose connections at compressor junction box. Defective wiring harness. Defective overload, relay, starting capacitor, or motor compressor. Low voltage causing motor compressor to short circuit on overload.
Motor-compressor runs, but condenser fan does not	Loose electrical connection. Fan blade cannot turn due to obstruction. Fan motor burned out.
Auger motor does not work	Loose connection. Defective auger motor.
No voltage at auger motor or compressor junction box	Check voltage at on/off switch. Check bin control and/or safety control by placing a jumper across their two terminals.
Machine fails to shut off when bin is full	Bin control out of adjustment or it is defective.

Source: Mile High Equipment Co.

FIGURE 19-20 Components layout.

and attaching the additional cuber, using a multiplex kit which provides the parts required to stack cubers, or you may order a multiplex cuber which comes with a multiplex kit. See Figure 19-20 for the components layout.

Skin. The skin of the ice cuber consists of the top, the left end, right end, back panel of the ice making section, and inspection panel. To remove the inspection panel, loosen the fastener in the center and lift up. With the inspection panel off, the top may be removed by pulling forward. The ends are removed by pulling to the front and lifting. They are held in place by clips on the inside back and bottom, which clip over the frame.

Condensing Unit. The condensing unit consists of a ¾-hp, 10,000- or 14,000-Btu compressor with enough power to handle the defrost load, carrying a normal charge of 3½ lb R-12, except for the R and X models, which require 10 lb. The divider between the condensing unit compartment and the ice-making compartment is removable for easy service on the condensing unit. To slide out the divider, remove the deflectors or chute, pull to the left to clear the

frame, and pull out. Both the metal sheet and the insulation will come out at the same time. When replacing, make sure that the top corners, and the front and rear are properly sealed with packing so that warm air from the condenser fan does not pass through to the freezing compartment.

Evaporator. The Full Cube Evaporator is made from 108 cells, 1¼ in. each way. The Half Cube Evaporator (KK) is made up of 216 cells 1¼ in. × 1¼ in. × ⅝ in. deep. The Cube-Let Evaporator is made up of 216 cells, 1¼ in. × ⅝ in. × ⅝ in. deep. They are made from copper and the entire assembly is tinned, thus preventing corrosion, and making its use acceptable to any sanitary board. A seal is not required between the evaporator and the water plate. Normally, there is about 1/32 in. of clearance between them.

Expansion Valve. A thermostatic expansion valve is used on all current GB models. Replacement valves for field use are the 1-ton size (½-ton GB4), which is a suitable replacement for any valve used in any GB model cuber. It has been regulated properly on tests before being shipped to give maximum superheat and maximum flooding of the evaporator and should not require any adjustment.

Sometimes after shipping or storage, the expansion valve sticks and allows more refrigerant to pass than necessary, increasing the low-side pressure and temperature, thus creating an excessive frostback and a long on-cycle. If this condition does not correct itself during the second freezing cycle, it will be necessary to adjust the superheat on the expansion valve, closing the valve clockwise $\frac{1}{8}$ to $\frac{1}{4}$ turn to increase the superheat, reducing the suction pressure and preventing frostback. If the valve hunts (varying suction pressure up and down) more than 2 or 3 lb when the suction pressure is below 13 lb, this is an indication that the valve should be opened more. (Also see discussion on cubes in evaporator under startup.)

Water Plate and Tank Assembly. The water plate is made of approved plastics and is used to distribute the water through jet holes to the freezing cells, and to return the water through the holes in the water plate to the circulation tank, which is fastened to the bottom of the water plate.

Circulation Strainer. This is a large screen inserted in the tank outlet to prevent dirt or particles of precipitated mineral from clogging the jet holes or the control stream. It also protects the pump impeller. If the screen becomes clogged with precipitated minerals, it is advisable to clean the whole circulation system with an ice machine cleaner.

Water Pump. This unit uses a direct-drive, centrifugal-type water pump. The inlet tube is at the bottom of the water circulation tank, and the outlet is at the top of the pump and is connected to the header of the water plate. It is made entirely of molded plastic.

Water-Level Probe Assembly. This device replaces the control tank and switch and is connected to the main water circulation tank by means of a flexible tubing. The water height in the probe assembly indicates the height of water in the circulation tank. Thermistor probes determine water valve on–off levers.

Pump and Defrost Switch. A spring-loaded switch located in the control box controls the defrost circuit and water pump and is operated by an adjustable screw in the left plate, which is attached to the water plate. As the water plate closes after the defrost period, the adjustable screw in the lift plate pushes up the pump switch to cut one connection to the defrost circuit and start the circulation pump.

Water Inlet Valve. The water inlet valve is mounted in the left front of the freezing compartment and controls the rate of flow of water into the water tank when the cuber is filling. It is of a constant-flow type and will allow about 0.6 to 1 gpm (depending on model) of water to flow regardless of any variation in water pressure, the minimum of which must be 15 psi. Cubers produced in and after 1974 are equipped with plastic-bodied water valves mounted in the wiring channel. An external Y-type strainer is used with these valves. For pressures over 100 psi or if there is a water hammer, use a pressure regulator.

Actuator Motor and Cam Assembly. This assembly is mounted just to the right and front of the evaporator. The actuator motor drives the cam shaft directly and is reversible. A cam on each end of the shaft forces the water plate down, separating the water plate from the ice in the evaporator. The two springs on the cams with reverse direction of the actuator motor, pull the water plate up at the end of the defrost cycle and hold the water plate against the bottom of the cams during the freezing cycle. To prevent the water plate from opening after the electric current has stopped flowing to the actuator motor, a drift stop spring with a plastic end presses against the actuator motor shaft on the front of the motor. If the drift stop spring is not aligned with the motor shaft, it can be removed easily and bent into shape.

Actuator Control. The actuator is controlled by signals from the evaporator thermistor probe mounted on the evaporator [adjustable on the printed circuit (PC) card in the control box] and the toggle switch mounted in the control box and is operated by the actuator motor.

Cold Water Control. The cold water control receives signals from the evaporator thermistor probe, which is mounted on the evaporator (adjustable on the PC card in the control box). Its function is to prevent the water plate from opening prematurely due to cold water, causing a resetting of the actuator control before the water level control shuts the water valve off. Also, during the fill cycle, it keeps circulating water that is warm enough to thaw out the jet holes in the water plate that may have been frozen shut at the end of the preceding freezing cycle. When the incoming water is below 50 °F (10 °C) and the circulating water drops to 40 °F (4.4 °C), the action of the cold water control opens the defrost valve, allowing hot gas to warm up the evaporator, thus also warming the water and preventing the resetting of the actuator control until the water fill cycle is completed. If the circulating water warms up to 50 °F, the cold water control cuts off the electric current to the defrost valve and the water valve.

Bin Control. The bin control receives signals from the bin thermistor probe mounted in the ice bin (adjustable on the PC card in the control box). Its function is to cut off the entire cuber when the bin fills up with ice. The bin thermistor probe hangs about 4 in. into the bin and clips into one of the four plastic brackets molded into the side of the plastic chute, or into a bracket in the bin.

Sequence of Operation. After the cuber has been installed, the water turned on, and the electricity properly connected, the water flows through the water valve (the water solenoid valve being energized) and the water plate in the up position. The pump and the defrost switch is held up by the lift bolt on the water plate, causing the pump to operate and the defrost valve to be closed. Water will gradually come into the water level probe assembly and touch the upper thermistor probe, causing the valve to close. The water is being pumped from the bottom of the circulation tank into the water header at the left of the water plate, through the laterals and through the small openings in the laterals, and in a jet stream up in the center of each evaporator cell. As the evaporator cools, the water is also cooled, and in about 5 or 10 min the water will be sufficiently cooled so that it will form minute layers of ice on the inside of the evaporator cells. As this process continues, the water level will decrease in the circulation tank and water level probe assembly.

When the circulating water reaches a temperature of 32 °F (0 °C), it *may* be supercooled and it *may* partially crystalize in the water tank. If this occurs, the flow of water in the control stream nozzle will stop or fluctuate considerably and most of the circulation will stop for about 30 sec. This is strictly a normal operation and the control stream should not be adjusted at this time.

The control stream, on the front left corner of the water plate, should flow so that it does not go over the dam and, under normal freezing conditions, should not be too high or too low. The screw at the inlet of the control stream may be adjusted until the stream is set for proper operation. The water continues to freeze until each cell is almost completely full of ice. As the ice comes closer to the jet streams, in the center of each cell, the head pressure in the water circulation system increases to a point where the control stream will rise and flow over the dam in the control stream box. This water is directed to the drain pan, causing the level of the water to decrease, and the low water level will leave the bottom thermistor probe exposed to the air.

When this condition occurs, the water control completes the circuit through the actuator control and through the actuator toggle switch to the actuator motor, causing the actuator shaft to turn counterclockwise (looking from the front) separating the water plate from the evaporator. (Also the water inlet and hot-gas circuits are complete.) The cams may be seen rotating down the left of the actuator motor. The actuator shaft continues to turn for one-half revolution, at which time the trip lever on the actuator motor gear shaft snaps the toggle switch to the stop down position and opens the actuator motor electric circuit, stopping the actuator motor.

As the water plate is lowered to its harvesting position, the pump and defrost switch lever drops causing the pump to stop. This switch also completes a second circuit to the defrost valve to keep it open during the harvest cycle. The machine is now in the defrost cycle, and the hot gas from the compressor is passing through the evaporator coil, warming up the evaporator so that the cubes will drop down to the water plate as a sheet. There is a small fin connecting the bottom of the cubes together so that they will come down in unison and clear the water plate without any difficulty and slide into the bin where they will break apart.

As long as ice is in the evaporator, it remains cool (32 to 35 °F) (0 to 1.7 °C), keeping the actuator control cool and keeping the cuber in defrost.

After the ice is out of the evaporator, the evaporator and the evaporator thermistor probe warm up rapidly [40 to 45 °F (4.4 to 7.2 °C)]. The actuator control moves to the warm position, completing the electric circuit from the defrost circuit to the "reversing" side of the actuator toggle switch and the actuator motor. (*Note:* The actuator control does not start the defrost cycle; it only ends it after all the ice falls out.) The motor revolves clockwise (cams may be seen rotating up on the left of the actuator motor) until the water plate is in its full up position where the actuator toggle arm will snap the actuator toggle over to the left, causing the actuator to stop. When the water plate rises to the up position, the lift plate and lift bolt push up the pump and defrost spring-loaded switch, starting the pump to begin another freezing cycle. When the water level is high enough to touch the upper thermistor in the water probe assembly, it breaks the electric circuit to the water valve.

Should some of the cubes be left on the water plate keeping it partially open when it moves to the up position, the actuator motor will continue to operate and the springs will stretch to allow the cams to take their vertical position and snap the actuator toggle switch. Since the lift plate cannot push up the defrost switch, the electric circuit is completed through this switch to the stop up side of the actuator toggle switch. When the actuator toggle arm snaps the actuator toggle switch to the stop up position, the actuator motor will immediately reverse itself and open the plate. The captive ice then falls off. When the plate is in the lowest position, the actuator toggle arm will again reverse the actuator toggle switch and the actuator motor causing the plate to close. This action will continue until the water plate is clear and the lift plate can push up the defrost switch, breaking the electric circuit through this switch to the stop up side of the actuator toggle switch. The motor will stop with the cams up when the actuator toggle arm pushes the actuator toggle switch to the stop up position.

The cuber has now completed its full cycle and started another freezing cycle. This will be regularly repeated until the bin is full of ice and the bin control shuts off the cuber automatically. When some ice is removed from the bin, the cuber will start up and refill the bin.

Electronic Cuber Startup Operation Checklist. The following information provides details on the startup and operating procedures for the electronic ice cuber.

Water Fill. Note the flow of water from all the small inlet holes. The tank should be completely filled in about 3 to 4 min.

Water Level. The water fills to the tip of the high-water probe. To adjust the water fill, sight the tip of the high-water probe carefully across the control tube.

FIGURE 19–21 Control stream and adjustment. (Courtesy of Kold-Draft Division, Uniflow Manufacturing Co.)

After one water fill cycle, check the water level and adjust the probe, if necessary, for a more accurate level by raising or lowering the probe by the same amount that the water level is off. A single readjustment is all that is needed after the initial eyesight adjustment of the probe tip.

Cold Stream. After the circulating water is cold, note the control stream and compare with Figure 19–21. The adjustment screw A should be adjusted so that the control stream strikes as shown in the illustration. (*Note:* After the water is cold, it may subcool and form crystals in the circulation tank for a minute or two, partially stopping circulation and the control stream temporarily. Do not adjust the control stream at this time.) If the control stream adjusting screw is changed at the end of the freeze cycle to make the stream go over the dam, recheck it during the beginning of the next freeze cycle to be sure that it does not go over the dam until all the cubes are ready for harvest.

Water Level at End of the Freeze Cycle. The water level after the cubes are frozen will normally be 4½ in. from the top of the white circulation tank for all Kold-Draft machines. If it is above this point when the control stream rises and if the control stream goes over the dam for more than a minute, this is an indication that the water level may be lowered to shorten the cycle, if desired. Lower the high-water level probe to shorten the cycle, or raise the lower water level probe to cut down on the amount of time the water stream must go over the dam.

Control Stream at the End of the Freeze Cycle. When the cubes are fully frozen, they freeze over some of the jet holes, increasing the water pressure in the circulation system. This makes the control stream rise and go over the dam, dumping water down the drain and lowering the water level rapidly so that the control low-water probe will send a signal to the solid-state controller to start the defrost without delay. The circulating water pressure should increase, and the control stream should go over the dam for a period of time from a few seconds to one minute before the low-water probe initiates the defrost cycle.

If the control stream does go over the dam and the control tube initiates a defrost, check the cubes. If many have large holes, there was insufficient water taken in at the beginning of the cycle or there is a leak in the system. Eliminate the leak or adjust the high-water level probe upward to increase the water level on the next fill cycle. If the cubes are full with only small dimples, the exact *minimum* amount of water was taken in. If the control stream goes over the dam for more than one minute before the water clears the low-water probe, check the cubes. If many cubes have large holes, the control stream was set too high. Reset it slightly lower during the beginning of the next freeze cycle. If the cubes are full with only small dimples, more water was in the machine than necessary. This is lengthening the cycle time a few minutes. If maximum ice production is necessary, lower the high-water level probe to decrease the water fill level.

Expansion Valve Setting. During the freeze cycle, note the suction line near the compressor. There should be no frostback on the suction line. Frostback indicates that the valve is open too wide (or the valve is slow in responding due to the machine being shut off for an extended period of time and the valve may take one or two cycles to come back to normal). Frostback will lengthen the running cycle time and make a lag in the control stream operation. However, if there is frostback while the pressure is low, check to see if something is preventing a defrost cycle. (See discussion on cuber that will not harvest.)

Defrost Cycle. The various steps of the defrost cycle are explained in full detail under the sequence of electric circuits. Learn these steps, so you can follow them as the machine goes through a defrost cycle.

Cubes in Evaporator. Check the bottoms of the cubes in the evaporator. If the cubes in the front row and the right-hand rows have excessively large holes while the rest of the cubes are solid, the evaporator has been starved. Either the expansion valve is too far shut (high superheat) or the system is short of refrigerant. If there is an occasional cube

that did not form at all or is quite cloudy and only partly formed, note which cell it is in and clean out the corresponding jet hole in the water plate with a $1\frac{1}{16}$-in. drill. (Take care not to drop the drill through one of the larger return holes in the water plate; it may be difficult to fish out of the water tank, although a small magnet will come in handy if it happens.) After cleaning out the jet holes, remove the plug at the end of the water distribution tube and allow it to flush out at the beginning of the next fill cycle to remove the debris.

Ice Fall Out. If the cubes are full without large holes, they should drop out substantially all at the same time. If some cubes hang up in the evaporator, twist and become distorted after others have fallen out, the fin between the cubes can be increased by lowering the hinges (front and rear) at the left of the water plate.

Actuator Control Adjustment. After the cubes fall out, the cams should start rotating clockwise to close the water plate within 10 to 30 sec. If the water plate stays down too long, adjust the actuator potentiometer slightly clockwise until the cams start to turn. If the water plate starts to close before the ice falls out, turn the actuator potentiometer all the way counterclockwise to stop the cams. After the ice falls out, turn the potentiometer slowly clockwise until the cams start to close the water plate. If the actuator is adjusted, it is advisable to check it again, after another cycle. An orange LED is mounted under the actuator control adjustment potentiometer as a service aid. The LED is on when the actuator control is tripped to the cold side.

Quality of Cubes. It is normal for the cubes to have a small dimple on one side (bottom when they are in the evaporator). This dimple will vary from $\frac{3}{8}$ in. deep in some cubes to $\frac{1}{8}$ in. deep in a large number of cubes. It would take an excessively long freeze cycle to freeze the dimples shut in all of the cubes. If there are large holes in the cubes see discussion on control stream and adjust the water level up or down as indicated. If the water level and control stream are correct, there may be a leak, and water is being lost during the freeze cycle. Check to see if any water is dripping into the drain pan under the water tank during the freeze cycle. An occasional drip will not matter, but a steady drip should be traced and repaired.

Bin Control Setting. Hold two or more ice cubes against the bin thermistor probe. The cuber should shut off in about 1 min. If it does not, while holding one cube on the probe, turn the control adjustment slowly counterclockwise.

After the cuber stops, take the cubes away from the bulb. The cuber should start in about 1 min unless the ambient temperature is quite low.

Caution: Do not make the above check or try to shut the cuber off when it is filling with water. The control card has a circuit to fill the water tank before shutting off. During the filling procedure, the bin control is ignored and any attempt to adjust it is invalid.

Maintenance. The cuber is self-flushing and, since the water circulation system is entirely plastic, there should be a minimum buildup of mineral deposits, although there may be discoloration, and most cubers require cleaning only once a year. However, since water in each area has different characteristics, some cubers may require more frequent cleaning if, after a period of operation, cubes come out cloudy due to clogging of the tank outlet strainer, reducing the circulation of water. (For details, see the discussion on cleaning instructions.) If the water is exceptionally hard, requiring more frequent cleaning than is practical, it can be treated with scale inhibitor at a nominal cost and a minimum amount of trouble for the owner, or in extreme cases a water softener may be used.

Should the water supply be heavily laden with suspended matter to such an extent that the water inlet screen in the water solenoid clogs frequently causing a slow water fill at the beginning of the cycle, it may be necessary to put a special filter on the inlet water line.

The stainless steel bin interior may discolor, particularly above the ice storage line if not cleaned periodically. Follow the cleaning instructions given below and on the decal affixed to the outside of the bin.

Bin Cleaning. The bin should be cleaned periodically. If the bin drain has any horizontal run, remove the ice from the left side of the bin and flush with two quarts of hot water monthly. (Long drains should be flushed weekly.) The inner door of the GBN-2 bin may be removed by first removing the triangular guards on each side and then by holding the two sides of the inner door so that they will slide past the stops until the door is horizontal. The hinge pins on the bottom of the inner door can be lifted

and pushed back out of their slots at the bottom front of the bin, the inner door will then slide forward and free the outer door.

Lubricating the Water Pump and Actuator Motor. The front and rear bearing on the GA-208 water pump should be oiled every 3 months with GB-1655 Kold-Draft oil. The actuator motor operates only about one or two revolutions per hour, and it should require little lubrication (one drop on the armature shaft). The GB-208 water pump has sealed bearings requiring no lubrication.

Cleaning an Air-Cooled Condenser. The air-cooled condenser should be cleaned weekly with a stiff brush and a vacuum cleaner to remove dust and dirt for more efficient operation of the unit. To determine that the condenser is clean, a light held at one side of the condenser will be clearly visible from the other side.

Cleaning a Water-Cooled Condenser. The Halstead condensers used on some units have end plates that can be removed for mechanical cleaning if necessary. Water-cooled condensers without end plates can be cleaned by flushing with a condenser cleaning solution. Acid recirculating pumps and solution are available at refrigeration supply houses. If the head pressure is excessive, and water usage is higher than normal, the condenser must be cleaned.

Water Plate Silicone Treatment. In certain areas of the country where the water is unusually pure with practically no dissolved solids, the water plates must have fairly frequent applications of silicone in order to prevent ice adherence.

Application of Silicone. This treatment is recommended every three months, or whenever the cuber is serviced, where water conditions cause the ice to stick to the water plate.

1. Defrost the unit by lifting the control tank or pulling down on the water plate. With the wash switch in the WASH position, allow the unit to refill with water to warm up the plate and melt off any accumulated ice. Open the plate to dump out the tank.
2. Turn off the electric power with the plate down. Wipe water drops off the evaporator, and with several rags wipe the water plate as dry as possible.
3. Apply Kold-Draft water plate spray to the water plate, being sure to avoid the last half inch on

the right-hand side. If the silicone coating is on the edge, the water will run off too readily into the bin during the defrost period.

4. Turn the electric power on and place the WASH switch in the on position.

Ice Machine Cleaning Instructions. The following instructions are recommended when cleaning a Kold-Draft Automatic Ice Cuber.

1. Mix one bag of CSCO ice machine cleaner into 1 quart of water at approximately 130 °F (54 °C).
2. Push the washing switch to the WASH position and pull down on the water plate by hand to start the cleaning operation with an empty water tank. As soon as the water plate closes, pour the mixed solution into the machine through the control stream box while the machine is filling with water. If the solution is not pre-mixed, the machine may be turned off after the defrost cycle, but before the water plate closes. The cleaning powder is then poured onto the water plate and the machine is restarted. Be sure that the washing switch is in the WASH position before starting the machine.
3. After approximately 15 min of washing, pull the water plate down by hand to dump the solution. After the tank fills with water, pull the water plate down by hand to dump the rinse water. Repeat this filling and dumping two more times to thoroughly rinse the machine.
4. After completing the wash–rinse cycle, allow the water plate to close and the water tank to fill until the weight control tank drops with the cycle switch still in the WASH position.
5. Mix a sanitizing solution of 2 oz of 5¼% sodium hypochlorate (household-strength laundry bleach) and about 1 quart of clean water.
6. Slowly add about ½ pint (8 oz) of this solution into the weight control tank.
7. Add the remaining solution into the control stream box until the water level is at the top of the weight control tank.
8. Allow 15 min of circulation time, then dump this solution and rinse two times as described above.
9. Push the washing switch into the on position and the machine is ready to make ice.

Note: Do not use ammonia solutions in cleaning any part of the ice maker.

Warning: Be sure to rinse all ice machine cleaner from the cuber before adding the sanitizing solution (bleach). Mixing of ice machine cleaner with bleach can cause a chemical reaction that will release poisonous fumes.

Status Indicator Operation Checklist. The following steps indicate the procedure for checking the operation of the electronic ice machine (Figure 19–22).

Ice Harvest Switch. This switch is mounted on the end of the water plate and resets a lockout time on the status indicator PC card. Falling ice during the harvest cycle operates the switch. To test the ice harvest switch operation, trip the switch and observe the yellow flickering light on the front panel. This test will not work on the PC cards manufactured before 1980.

Green Light. This light blinks when the cuber is making ice.

Yellow Light. This light is on when the cuber is off because the bin probe shuts the cuber down.

Green and Yellow Lights. These lights are on when the cuber has shut itself off due to a hot evaporator (140 °F) (60 °C). The cuber will return itself to service when the evaporator cools down (120 °F) (49 °C).

Red and Yellow Lights. These lights are on when the cuber has shut down due to a timer "timeout." This occurs when the ice has not triggered the ice harvest switch and is a "service required" condition. A cuber left in the wash position or a defect will cause this condition. A cuber can be put back into operation by tripping the ice harvest switch, removing power temporarily or by unplugging the status indicator.

Timeout Potentiometer. This control is mounted on the status indicator PC card and is adjustable from 15 min to 1½ hr. This timeout period can be accurately measured by timing 17 pulses of the green light while the cuber is operating. The time in seconds is equal to the time in minutes that the timeout timer is set for. Adjustment is made from the back of the inspection panel through the small hole in the plastic status indicator PC card cover.

The time setting should exceed the normal cycle by 50% to allow for a longer cycle on startup. Doubling this time period for cublet machines will eliminate possible "no harvest trips," due to the small size of the cubes not tripping the switch the first time. Doubling the time gives the switch a second try to sense the harvest.

Test Procedure with Cuber Analyzer. The following test procedure will thoroughly check all opera-

Without the use of special tools or testers, these tests can quickly determine any major faults with either the control module card or probes. An ordinary pocket screwdriver can be ground down to fit slots on potentiometers which have been designed so they cannot be adjusted with a standard screwdriver.

BEFORE STARTING TEST, BE SURE THERE IS POWER TO CUBER AND CHECK FUSES IN CONTROL MODULE BOX.

Kold-Draft Electronic Cuber

Speedy Test

(A Screwdriver May Be Used To Short Pins)

	B	E	W	L	H	
1	Open	Open	Open	Open	Open	Cuber Stops, Orange & Red L.E.D. ON
2	Short	Open	Open	Open	Open	Cuber Runs, Orange & Red L.E.D. ON
3	Short	Short	Open	Open	Open	Cuber Stops, Orange & Red L.E.D. OFF
4	Short	Open	Open	Short	Open	Harvest Begins (Allow plate to open fully)
5	Short	Short	Open	Open	Open	Press Bin Relay Plunger, Water Plate Closes

— STATUS INDICATOR TESTS —

	B	E	W	L	H	
6	Short	Open	Open	Open	Short	Yellow L.E.D. on Status Indicator ON
7	Open	Open	Open	Open	Open	Yellow L.E.D. on StatusIndicator ON
8	Short	Open	Open	Open	OPen	Green L.E.D. on Status Indicator BLINKING
9	Short	Short	Open	OPen	OPen	Yellow & Green L.E.D. on Status Indicator ON

Above Tests confirm a good Control Module.

See Note "D" for Probe Tests.

NOTES

A. "Open" terminals produce a "Cold" signal to the P.C. Card.

B. "Short" terminals produce a "Hot" signal to the P.C. Card.
C. Turn wash switch to"wash" to prevent compressor short cycling during tests.
D. To test probes, connect each in turn to the "E" pins with all other pins left open. Turn Cold water pot full warm (CCW) to increase Red L.E.D. sensitivity. A warm probe will turn the L.E.D. Off. Submerging the probe in ice water will turn the Red L.E.D. On. Reaction time, 5 to 20 seconds.

E Be sure to return cold water pot to original position after probe tests.

F. To test Status Indicator "time out", short "B" pins (or insert good, warm probe), and short the test pins (TP1 and TP2) at the Status indicator socket on the P.C. Card. Green L.E.D. will blink 2 to 4 times, then cuber will shut off. Status Indicator Red and Yellow L.E.D.'s will turn On.

FIGURE 19–22 Cuber functional tests. (Courtesy of Kold-Draft Division, Uniflow Manufacturing Co.)

tions of the control module PC card by simulating correctly operating bin, evaporator, and water level probe signals. If a particular probe is suspected of causing trouble, the proper operation of that probe may be simulated by using the cuber analyzer. If a particular part of the cuber does not respond to one or more of these tests, either the part itself or the PC card may be defective. The suspected part should always be checked before a replacement PC card is installed. If a new probe is required, it should be installed and the cuber watched through a minimum of one cycle to assure proper operation.

To use the Cuber Analyzer (Figure 19–23), follow the instructions below:

1. Remove all the probe plugs from the upper left-hand plug. Once they are removed, install the analyzer plug. There is an interlocking lip that will mate with another lip on the PC plug. At this time the thermistor probes must be connected to the back of the analyzer plug. The correct order is, from left to right: bin, evaporator, water level, and harvest switch.
2. Now that the analyzer is correctly hooked up, you may send the correct logical signal to the associated circuit by moving a switch or group of switches.
3. On the analyzer plug, a small slide switch is incorporated to allow testing of all PC cards. The new PC cards have the new part number (GBB-03135-02-P or GBB-03130-P) inked on the transformer bracket and/or screened on the circuit card. This switch interchanges the water level probe connections to the analyzer. For the old cards, this switch is moved to GBB-03130, and on the new cards to GBB-03135.

Bin Thermostat Switch. When switched to the "full" side, the cuber will shut down. If a status indicator is attached, the yellow LED will come on. If the switch is pushed to the "full" side during the fill cycle, the cuber may not shut down until the water reservoir is full.

Water-Level Switch. When the machine is taking on water, you can move the switch to the high position. This will shut off the water valve. To continue filling, move the switch to the low position and then release it. If you move the switch to the low position, the water valve will open. However, if the red LED is on, the water inlet valve and the defrost valve will open, and if the orange and red LEDs are both

on, the water valve, defrost valve, and the actuator motor will all be energized.

Actuator Thermostat Switch. Moving this switch to the "cold" position will simulate that the actuator and cold water thermistors are cold. This will be evident by the orange and red LEDs being lighted. Once the switch is released, the LEDs will go out, unless the evaporator is also cold [below 40 °F (4.4 °C)], in which case the red or both of the LEDs will stay on; pushing the switch up (warm) will cause the LEDs to go out. If you would like to initiate a harvest cycle, hold the actuator switch in the cold position and move the water level switch to the low position, which will energize the water inlet valve, defrost valve, and the actuator motor. Moving this switch to the warm position will raise the water plate to the full position after the harvest cycle is completed.

Cold Water Thermostat Switch. Moving this switch to the cold position will turn the red LED light on and close the circuit for the defrost valve. If you energize this circuit during the period when the water is coming in, the defrost valve will open. You can also perform this function by moving the cold water switch to the cold position and moving the water level switch to the low position. If the evaporator probe is good and the evaporator is near 32 °F (0 °C), the orange LED may also come on.

Safety Thermostat Switch. Move this switch to the warm position and it will shut down the cuber on an overheated evaporator. If a status indicator is attached, the yellow and green LEDs will come on.

Timeout. Timeout of the cuber may be simulated by connecting the two alligator clip leads to the two test points located adjacent to the status indicator terminal block. On PC cards manufactured before July 1980, these test points will be labeled TP1 and TP2. The status indicator PC card must be connected to simulate this feature. (On old circuit boards without test points, connect the alligator clips to left and right pins of the status indicator terminal block.) With the clips connected, move the switch to the timeout position. Within a few blinks of the green LED, the cuber will timeout and the red and yellow LEDs on the status indicator will turn on. Releasing the switch should start the cuber, turn off the red and yellow LEDs, and start the green LED blinking. Moving the same switch to the trip position should turn the green LED off and the yellow LED

CUBER ANALYZER (GBB-03130-03)

As an aid to trouble shooting, Kold Draft has available, a tester which can be carried in the serviceman's tool kit. The tester consists of a small metal box which contains manual switches and electronic devices used to provide test signals to the printed circuit card. To use, the serviceman simply removes the thermistor probe leads from the card, connects the tester's plugs into the same socket and connects the probes to the tester plug. The cuber then can be run through it's complete cycle in a matter of minutes by activating the manual switches on the tester in the proper sequence. If the cuber operates properly by using the tester, the serviceman knows that one of the probes may be faulty. If the cuber will not operate properly with the tester connected to the card, the indication would be that the printed circuit card may require replacement. The tester is used to trouble shoot the electronic system only; if the cuber has failed due to refrigeration or mechanical malfunctions, these problems must be corrected before testing the solid state control system.

1. 1979-1980 SOME P.C. CARDS DO NOT HAVE VOLTAGE SELECT AND ARE WIRED FOR 208-230 VOLTS.

1a. PRIOR TO AUGUST 1980 115 VOLT CUBERS REQUIRED P.C. CARDS WITH VOLTAGE SELECT JUMPER.

2. P.C. CARDS GBB-03135-02 AFTER AUGUST 1980 WILL HAVE VOLTAGE CONNECTIONS IN TB-1 AND CUBERS WILL HAVE HARNESS TO SELECT CORRECT TERMINAL AUTOMATICALLY.

MAT'L. NOTE:

.004" WHITE PLIAPRINT PRESSURE SENSITIVE STOCK WITH PERMANENT ADHESIVE-PRINTED BLACK - OVERVARNISHED - BACK SPLIT

FIGURE 19–23 Cuber Analyzer. (Courtesy of Kold-Draft Division, Uniflow Manufacturing Co.)

on (on new cards). For more information refer to Table 19–5.

Wiring Diagrams. See Figures 19–24 to 19–27.

Troubleshooting. See Table 19–5.

A very good safety precaution is to check the unit before working on it to be certain that it is installed in accordance with local and national fire, electrical,

TABLE 19–5 Cuber Service Diagnosis

Trouble	Cause/Symptom	Remedy
Cuber will not start	1. Line fuse blown	1. Check circuit for short or ground. Replace fuse.
	2. Bin full of ice	2. Use some ice.
	3. Open circuit in cord or feed wires	3. Repair or replace.
	4. No money in meter if meter is used	4. Feed meter.
	5. Room too cold (below 45°F)	5. Warm room. Consult factory for cold room adaptation.
	6. Blown fuse on control module card	6. Replace fuse.
	7. Bin probe disconnected or loose; set too warm counterclockwise	7. Install bin probe properly. Set slightly clockwise.
	8. Defective bin probe	8. Jumper bin probe pins (B) on PC card, if cuber starts, replace probe.
	9. Defective control module PC card	9. Check with cuber analyzer. Replace control module PC card.
	10. Status indicator (if used) timed out	10. See status indicator operation page to reset.
	11. Overheated evaporator	11. If evaporator is hot, allow to cool. Check defrost circuit.
	12. Shorted evaporator probe	12. Replace.
Condensing fan operates but not the compressor	1. Compressor stuck	1. Jar with mallet.
	2. Inoperative capacitors or relay	2. Replace capacitors or relay.
	3. Overload switch defective	3. Replace overload switch or compressor with internal overload.
	4. Open wash switch	4. Switch to on or replace.
	5. Open high- or low-pressure cut-out	5. Check charge and condenser.
	6. Defective compressor	6. Replace compressor.
Compressor operating but fan off	1. Circuit not complete	1. Check circuit.
	2. Fan motor burned out	2. Replace motor.
Condenser fan operating, but condensing unit operating intermittently during freezing cycle, wait until the end of defrost to see if unit returns to normal operation	1. Dirty condenser coil	1. Clean coil.
	2. Low voltage	2. Correct to proper voltage—not less than 5% below that stated on nameplate. Install automatic brownout voltage booster number 5-1320 for 120-V cuber.
	3. Excessive refrigerant	3. Bleed off some refrigerant.
	4. Fuse blown one leg of three-wire system	4. Replace fuse.
Compressor cuts out	1. Defective run capacitor	1. Run capacitor should draw 1·to 3 A, GB2 and GB4, 4 A.
	2. Open high- or low-pressure cut-out	2. Check refrigeration system pressure.

TABLE 19-5 Continued

Trouble	Cause/Symptom	Remedy
Water plate closes and opens constantly; water plate closes all the way when cams are up but defrost valve stays open and pump does not run	1. Maladjusted pump and defrost switch lift bolt 2. Water plate does not close all the way	1. Adjust lift bolt on water plate to push switch lever up, closing hot gas valve and starting pump when water plate is up. Tighten jam nut. 2. Remove obstruction. Adjust hinge for clearance between evaporator and water plate. Make sure Teflon brackets on water plate are tight against cams. Check springs.
Water plate opens before water probe assembly tube is full	1. Spring missing or springs weak allowing water plate to lower slightly [as water fills tank] until pump switch drops and plate opens under power 2. Drift stop not adjusted; cams drift counterclockwise until water plate lowers slightly and pump switch drops 3. Slow fill-cold incoming water. Orange and red LEDs on control module PC card go on	1. Replace springs. 2. Drift stop on front of actuator motor. Remove drift stop and bend spring for more tension on motor shaft. 3. Adjust cold water control on control module PC card or replace control module PC card. Improve water supply. Clean strainer.
Water plate will not completely close	1. Obstruction between evaporator and water plate 2. Lift bolt for pump toggle on water plate too high, holding plate away from cams	1. Remove obstruction. Check clearance between water plate and evaporator. 2. Adjust lift bolt so water plate comes up against cams and lift bolt holds pump and defrost toggle switch up without binding and holding water plate down.
Water plate closes before cubes dropped	1. Actuator pot on control module PC card adjusted too cold 2. Faulty evaporator probe 3. Faulty control module PC card	1. Adjust to warmer position [counterclockwise]. Water plate should remain in down position 10 to 30 sec after ice drops. 2. Replace probe. 3. Check with cuber analyzer. Replace control module PC card.
Water plate stays wide open after defrost and all ice is out of evaporator	1. Orange LED stays lit on control module PC card 2. Orange LED is off, but no voltage to the yellow actuator motor lead.	1. Adjust actuator control slightly clockwise. Check and replace the evaporator probe or control module PC card if adjustment has no effect. 2. Check wiring. Replace actuator toggle switch GB-897.
Water plate open—evaporator will not defrost	1. Red LED on control module PC card is on 2. Red LED is off	1. Check voltage at defrost valve coil, if not 120-V change control module card. 2. Check evaporator probe [5600 Ω at 32°F]. If probe is functional, replace control module PC card.

Safety Procedures

TABLE 19-5 Continued

Trouble	Cause/Symptom	Remedy
	3. Refrigerant charge low	3. Check for leaks and recharge.
	4. Inadequate hot gas volume	4. Check for tube obstruction or cold condenser.
Water pump does not operate	1. Fuse blown in transformer box, or in control module box	1. Replace fuse.
	2. Pump bearings defective	2. Replace pump motor.
	3. Pump windings burned out or off on thermal overload	3. Allow to cool, or replace motor, check for 120 V ± 10%.
	4. Circuit incomplete between water pump and pump defrost switch	4. Check circuit and switch.
Water pump motor running but not pumping water	1. Impeller loose	1. Replace impeller.
	2. Strainer in tank outlet to pump clogged	2. Clean or replace screen.
	3. Impeller broken	3. Replace impeller, replace screen in tank outlet.
Most cubes not fully formed	1. Not enough pressure from water pump	1. Check bearings. Check voltage. Replace pump.
	2. Clogged strainer in tank outlet to pump	2. Clean strainer.
	3. Leak in water circulation system	3. Fix leak or replace water plate.
	4. Water plate not aligned	4. Check alignment with evaporator.
A few cloudy cubes, others okay	1. Some holes in water plate clogged	1. Unplug with $\frac{1}{16}$-in. drill. Flush laterals by removing plugs.
Holes in left-hand cubes (evaporator inlet)	1. Expansion valve too far open	1. Close $\frac{1}{8}$ turn at a time (clockwise).
Holes in right-hand cubes (evaporator outlet)	1. Shortage of refrigerant	1. Check for leak and recharge.
	2. Expansion valve too far closed	2. Open $\frac{1}{8}$ turn at a time (counterclockwise).
Holes in all cubes sometimes and solid cubes most of the time	1. Power shut off while water is filling tank or temporary shut off near end of freeze cycle	1. Correct power source if possible.
	2. Bin control shuts the cuber off during water fill	2. Interlock between water fill control and bin control not operating. Replace control module PC card.
Holes in cubes all the time. Control stream does not go over the dam at end of freeze cycle	1. Water level too low	1. Measure from the top of the circulation tank down to the water level in the water level control tube. Sight carefully across water in the control tube.
	2. Low water level probe too high	2. Adjust low water level probe to remain immersed in water in control tube at least 10 sec after control stream starts "going over the dam."
Holes in cubes all the time. Control stream does go over the dam	1. Control stream too high, allowing water to splash over the dam during freeze cycle (should only go over the dam after cubes are fully formed)	1. Lower control stream, turn adjusting screw clockwise.

TABLE 19-5 Continued

Trouble	Cause/Symptom	Remedy
Cuber will not harvest, water plate will not come down	1. Control stream obstructed	1. Loosen adjusting screw to flush out foreign matter.
	2. Orange LED does not come on	2. Check or replace evaporator probe [5600 Ω at 32°F]
	3. Orange LED will not come on but probe is okay	3. Replace control module PC card.
	4. Inoperative lower probe	4. Check water level probe connections or replace probe assembly. Replace control module PC card.
	5. Actuator motor problem	5. Check motor and circuit.
	6. Warm air infiltration from compressor compartment or room	6. A. Secure all skin panels. B. Skin gaskets must seal. C. All panels must seal to prevent air from compressor compartment getting into ice making compartment. Check especially, top cover over partition.
Cuber stops when bin is not full	1. Bin control adjusted too warm	1. Readjust bin control slightly clockwise.
	2. Bin probe connector loose or dirty	2. Clean connector and install properly.
	3. Defective bin probe	3. Replace if considerably more than 5600 Ω at 32°F.
	4. Intermittent evaporator probe	4. Replace.
Cubes do not harvest in a slab but some cubes hang up in the evaporator and become distorted after others fall out	1. Fin too thin	1. Adjust water plate hinges to $\frac{1}{32}$-in. fin thickness.
	2. Deformed evaporator cells	2. Straighten cells with smooth jaw pliers or tool 5-1028.
Slab does not break up into individual cubes	1. Fin too thick	1. Adjust hinges up or evaporator down. Leave $\frac{1}{32}$-in. space between water plate and evaporator.
Unusually long cycles	1. Voltage below required potential at the cuber	1. Check power source for full voltage. Run at least No. 12 wire directly to cuber to prevent line loss.
	2. Dirty condenser	2. Clean.
	3. Hot air leaks between condensing unit compartment and freezing compartment	3. Check for leaks and close with permagum or Presstite tape. All skin parts must be tight.
	4. Expansion valve too far open	4. Close valve $\frac{1}{8}$ turn at a time so that there will be no frost back to compressor and pressures are according to "Chart of Water Levels, etc."
	5. Expansion valve too far shut and large holes in right-hand rows of evaporator	5. Open expansion valve $\frac{1}{8}$ turn, but recheck to see that there is no frost-back to compressor at end of freeze cycle [see 2 and 3 above].
	6. Water level too high after water fill	6. Adjust water level according to "Chart of Water Levels, etc."

TABLE 19-5   Continued

Trouble	Cause/Symptom	Remedy
	7. Refrigerant low	7. Check for leak and add R-12.
	8. Compressor defective	8. Close refrigerant valve on receiver, pump down; low side should go to 15 in. of vacuum, with LP cutout jumpered. If not, replace compressor.
	9. Control stream too low	9. Adjust control stream up, but not so high that it goes over dam at beginning of cycle.
	10. Fan not operating	10. Check fan wires, replace motor if necessary.
Some cubes do not form in right-hand corners of evaporator	1. Jet holes in ends of laterals frozen shut and will not thaw because of very low incoming water temperature	1. Thaw out by shutting off unit, and adjust cold water control warmer ccw. Adjust expansion valve ⅛ turn closed.
Water plates out of synchronization on GB4 cubers		1. See GB4 section.
Water valve stays closed	1. Water level probe connector loose or dirty	1. Clean connector and install properly, low probe lead to right (close). Use NC123 or any electrical contact cleaner.
	2. Defective water probe assembly	2. Check with probes standing in 32°F ice–water mixture. Resistance should be 5600 Ω ± 20%.
	3. Circuits okay, 120 V to water valve; coil open; flow control jammed cockeyed	3. Replace coil. Clear valve passages or replace valve.
	4. Defective PC card	4. Replace PC card.
Water valve stays open after upper probe covered, will not shut off	1. Water level probe connector loose or dirty	1. Clean connector and install properly, low probe lead to right (close). Use NC123 or any electrical contact cleaner.
	2. Defective water probe assembly	2. Check with probes standing in 32°F ice mixture. Resistance should be 5600 Ω ± 20%.
	3. Defective PC card	3. Replace PC card.
	4. Water pressure below 15 psi	
	5. Defective water valve	
Water valve stays open more than 5 sec after upper probe covered, then shuts off	1. Upper probe covered with scale	1. Clean cuber with ice machine cleaner CSCO. If necessary, remove and clean probe carefully.
Status indicator shuts cuber off on "service required" but cuber operates normally most of the time and timer resets when ice harvest switch triggered manually (cubers built after June 1980—yellow light flashes when ice drops)	1. Cubelet ice does not trip harvest switch	1. Trip wire on switch too high. Bend trip wire until it is only ⅜ in. higher than water plate surface.

TABLE 19-5 Continued

Trouble	Cause/Symptom	Remedy
	2. Maladjusted impact switch	2. Turn adjusting screw in to increase sensitivity
	3. Time out period too short	3. Lengthen time out period by adjusting status indicator pot slightly clockwise.
Status indicator shutdown sometimes occurs after being off on full bin with status indicators built before March 1980	1. Circuit voltage borderline; timer does not reset upon startup after bin shutdown	1. Replace status indicator PC card [new style revised].

plumbing, and building codes. The board of health has regulations that apply to all equipment used around food and beverages.

1. Check floor load limits before installing heavy commercial equipment.
2. Be sure not to trap liquid refrigerant inside a system without gas space; hydraulic pressure may build up and cause a rupture.
3. Obtain help when moving or lifting heavy cases and compressors.
4. Never purge burned refrigerant into a building where food is stored since the food might absorb the burned refrigerant and spoil.
5. Never use oxygen to test for leaks because oxygen and oil form an explosive mixture.
6. Be sure to disconnect the electrical service before attempting to work on the electrical components.
7. Always provide plenty of ventilation when welding, soldering, or brazing on refrigerant lines.
8. Wear goggles when charging or purging refrigerant from a system.
9. Wear rubber gloves when working with acidic materials.
10. Keep hands, feet, and clothing away from belts and moving parts.

☐ SUMMARY ☐

☐ Commercial ice machines are generally referred to as flakers and cubers.
☐ Periodic maintenance on ice machines is necessary for satisfactory performance.
☐ Water from the water reservoir is circulated over the evaporator where it freezes to the desired thickness.
☐ The ice machine compressor operates continuously until the storage bin thermostat is satisfied.
☐ Frigidaire ice machines are designed primarily for back bar installation but can be used almost anywhere.
☐ The ice thickness control opens the electric circuit to the water pump and energizes the hot-vapor solenoid.
☐ A hinged mercury switch is tripped to add heat to the ice thickness control after the ice slabs are released.
☐ The compressor operates continously until the storage bin thermostat is satisfied, at which time all electric circuits are opened except for the low-voltage circuit to the cutting grids, which is energized as long as the ice machine is connected to the external electric power supply.
☐ The Ice-O-Matic flaked ice machine basically consists of a hermetically sealed refrigeration system designed to freeze ice in a cylinder.
☐ The water system of the Ice-O-Matic flakers consists of a float-operated valve that maintains a constant water level in the evaporator.
☐ The automatic expansion valve is designed to meter the refrigerant into the Ice-O-Matic evaporator and maintain a constant evaporator pressure.
☐ The bin and safety control sometimes needs to be adjusted after the unit is installed.
☐ The insulation used on Ice-O-Matic evaporators is expanded by use of Freon agents. Breaking of a bubble of the insulation will cause a simulated leak.
☐ Regular cleaning of the evaporator-auger on a flaker cannot be overemphasized.
☐ The inspection panel must always be in place during operation on the Kold-Draft automatic cuber in order to get the most satisfactory performance from the machine.
☐ The condensing unit of the Kold-Draft unit consists of a ¾-hp, 10,000- or 14,000-Btu compressor with enough power to handle the defrost load carrying a normal charge of 3½ lb of R-12, except for the R and X models, which require 10 lb.
☐ All current Kold-Draft GB models use a thermostatic expansion valve. On the Kold-Draft unit, the circula-

Chart of Water Levels, Pressures, Cycles and Charge [Higher Than Average Temperatures Increase Pressures and Cycle Times]

	GBN1-GT1	GBN1- Cube-Let-K Half Cube-KK	GBN5- GBN5-J	GBN5- Cube-Let-K Half Cube-KK	GY3-	GY3- Cube-Let-K Half Cube-KK	GY7-GTN7	GY7-GTN7 Cube-Let-K Half Cube-KK
Full water level distance below top of main tank	2½"	3⅜"	2½"	3⅜"	2¼"	3⅛"	2¼"	3⅛"
Tank insert used, or hanger wire extention loop	Tank insert	Extension loop	Tank insert	Extension loop	Tank insert	Extension loop	Tank insert	Extension loop
Approximate trip up level above tank bottom	½" ±	½" ±	½" ±	½" ±	½" ±	½" ±	½" ±	½" ±
Suction pressure after defrost	25#	25#	30#	30#	27#	27#	25#	25#
Suction pressure before defrost	1#-2" Vac	4#-1#	7#-3#	9#-5#	6#-3#	7#-4#	4#-1#	6#-3#
Defrost pressure	40#-60#	40#-60#	40#-60#	40#-60#	40#-60#	40#-60#	40#-60#	40#-60#
Defrost pressure with control valve	35# Set 25-40# OK	35# Set 25-40# OK	—	—	—	—	—	—
Cycle time, approximate	30 min	16 min	60 min / 40 min	20 min / 40 min	60 min	40 min	30 min	20 min
GB5N-J								
					18 oz GY3A 11 oz GY3W	18 oz GY3A 11 oz GY3W	15 oz GY7A & 9 oz GY7W &	15 oz GTN7A 9 oz GTN7W
F-12 charge	3½-5#	3½-5#	3½-5#	3½-5#				
Approx. # ice per batch	8#	4¾#-K 4¾#-KK	8#	4#-K 4¾#-KK	4#	2#-K 2⅜#-KK	4#	2#-K 2⅜#-KK

FIGURE 19-24 Wiring diagram. (Courtesy of Kold-Draft Division, Uniflow Manufacturing Co.)

FIGURE 19-25 Wiring diagram. [Courtesy of Kold-Draft Division, Uniflow Manufacturing Co.]

FIGURE 19-26 Wiring diagram. (Courtesy of Kold-Draft Division, Uniflow Manufacturing Co.)

503

FIGURE 19-27 Wiring diagram. [Courtesy of Kold-Draft Division, Uniflow Manufacturing Co.]

tion strainer is a large screen inserted in the tank outlet to prevent dirt or particles of precipitated minerals from clogging the jet holes or the control stream.

☐ A spring-loaded switch located in the control box controls the defrost circuit and water pump and is operated by an adjustable screw in the left plate, which is attached to the water plate.

☐ The cold water control receives signals from the evaporator thermistor probe, which is mounted on the evaporator (adjustable on the PC card in the control box).

☐ The bin control receives signals from the bin thermistor probe mounted in the ice bin (adjustable on the PC card in the control box).

☐ The water fills to the tip of the high-water probe on the Kold-Draft unit.

☐ The water level after the cubes are all frozen will normally be 4½ in. from the top of the white circulation tank for all Kold-Draft machines.

☐ When the cubes are fully frozen, they freeze over some of the jet holes, increasing the water pressure in the circulation system.

☐ After the ice cubes fall out, the cams should start rotating clockwise to close the water plate within 10 to 30 sec.

☐ The ice bin should be cleaned periodically.

☐ The ice harvest switch is mounted on the end of the water plate and resets a lockout time on the status indicator PC card.

☐ EXERCISES ☐

1. How are commercial ice machines referenced?
2. When do ice machine compressors stop running?
3. For what purpose are Frigidaire ice machines primarily designed?
4. On the Frigidaire ice machine, where is the expansion valve bulb attached?
5. What is the purpose of the ice thickness control on the Frigidaire ice machine?
6. What controls the flow of water into the cylinder of the Ice-O-Matic flaked ice machine?
7. What type of flow control device is used on the Ice-O-Matic machine?
8. How often should the auger be cleaned on the Ice-O-Matic ice flaked ice machine?
9. What should be remembered about the inspection plate on Kold-Draft ice machines?
10. What sometimes happens to the expansion valve after shipping or storing?
11. What is the purpose of the circulation strainer?
12. What is the purpose of the water inlet valve on the Kold-Draft unit?
13. From where does the cold water control receive its signal on the Kold-Draft unit?
14. What may happen when the circulating water reaches a temperature of 32 °F in the Kold-Draft ice machine?
15. What is the proper water level in a Kold-Draft ice machine?
16. What happens to the pressure in the water circulating system when all of the ice cubes are frozen in the Kold-Draft unit?
17. On a Kold-Draft ice machine, what does frostback on the suction line indicate?
18. What should be done to the Kold-Draft ice machine if the water plate stays down too long?
19. Can the Kold-Draft ice machine be shut off when it is filling with water?
20. When must a water-cooled condenser on a Kold-Draft ice machine be cleaned?

20 Air Conditioning (Heating)

Performance Objectives

After completion of this chapter, the student should:

- Be able to name the various types of heat sources.
- Understand the process of combustion.
- Define the purpose of orifices and burners.
- Describe what causes the different types of flames.
- Better understand the purpose of heat exchangers.
- Describe the purpose of humidity control.
- Understand the basic heating controls and control systems.
- Explain the operation of the Lennox Pulse furnace.
- Better understand the operation of the Heil Condensing Gas furnace.
- Be able to explain the operating principles of solid state electronic ignition systems.
- Complete the Exercises with at least 80 percent accuracy.
- Complete the Laboratory Exercises with at least 80 percent accuracy.

The term *air conditioning* has for many years been misunderstood. Air conditioning involves treating all of the properties of air. Total air conditioning involves circulating, cooling, heating, humidifying, dehumidifying, and cleaning the air. Therefore, heating is only a portion of total air conditioning.

Definition. The *heating process* involves air circulation, heating, humidification, and cleaning. The other processes are accomplished during the cooling process.

☐ Heat Sources ☐

A heat source is any substance that produces heat while going through a given process. The main sources of heat are in the form of solids, liquids, gases, electricity, and solar energy. The value of a source of heat is derived from the amount of heat released when it is processed. A good heat source must be readily obtainable in large quantities at relatively low prices.

Solids. The main solid used for heating purposes for the general public is coal. Coal is obtained by one of two methods: open-pit and deep mining. Coal is divided into four classifications: anthracite, bituminous, subbituminous, and lignite. The anthracite and bituminous types are classified according to their carbon content when they are dry. Subbituminous, and lignite are classified according to their calorific value.

Anthracite and bituminous coals are the most popular types used for heating purposes. Lignite coal is used in electric power generating plants in some areas of the country. Anthracite coal will release from 13,000 to 14,000 Btu/lb (7150 to 7700 kcal/kg) when burned with 9.6 lb (4354.56 g) of air. Bituminous coal will release from 12,000 to 15,000 Btu/lb (6600 to 8250 kcal/kg) when burned with 10.3 lb (4672 g) of air. Dry air at about 70°F (21.1°C) has a volume of 13.3 ft³/lb (0.376 m³/g).

Coal is burned by use of several methods. Hand firing is the oldest method. The coal is thrown onto

Green coal
Distillation zone
Ashes
Hot zone
Grates
Primary air

FIGURE 20-1 Coal bed.

the grates through the opening through which the primary air is drawn into the firebox. The fuel bed consists of a layer of ashes lying directly on the grates (a hot zone where the combustion process occurs), a cooler (distillation) zone where the gases are driven from the coal, and then a layer of green coal (Figure 20-1). There are two reasons why it is necessary to supply secondary air over the fire to obtain complete combustion: (1) hot carbon and carbon dioxide react to form combustible carbon monoxide, and (2) there is a distillation of gases from the distillation zone.

During the hand-firing process, one method was to put the green coal in the front of the firebox and push the hot coals to the back. The secondary air entering the firebox would carry the gases over the hot coals, causing them to burn more completely than if the green coal were placed on the top.

The next method used in firing coal as a fuel was the mechanical firing process. During this process, coal and air were fed simultaneously into the firebox. The coal was pushed into the firebox by means of a rotating worm or screw. The hotter part of the fire was either pushed to the rear or to the top of the firebox, depending on whether front or bottom feed was used.

Liquids. Fuel oil is the most common liquid being used as a heat source. Fuel oil is a petroleum product and is made up of a mixture of liquid hydrocarbons; it is produced as a by-product of the petroleum refining process. Petroleum has been known for thousands of years. Seepages of oil and gas around the Caspian and Black Seas were known and used for heating and cooking before the birth of Christ. The Chinese drilled for oil long before the Christian era.

Fuel oil is a strong competitor of natural gas in the heating industry. It is graded and classified according to the range of distillation. The grades range from 1 to 6, omitting number 3. Grades 1, 2, and 4 are used for heating, with grades 1 and 2 being the most popular for domestic use. Number 4 is

generally used in light industrial furnaces. The lower-graded numbers are more expensive than the higher-numbered fuel oils because of fewer impurities, such as asphalt. Grades 5 and 6 are too thick for use in domestic oil-burning equipment and require preheating to ensure a steady flow to the burner. The specifications that govern fuel oils are set forth by the U.S. Department of Commerce and conform to American Society for Testing and Materials (ASTM) specifications.

The proper combustion of fuel oils can be accomplished only when the oil is properly atomized and mixed with air. The heat given off by burning fuel oil ranges from 137,000 to 151,000 Btu/gal (34,524 to 38,052 kcal/3.7 liters) depending on the grade. The heat content of number 1 fuel oil is 137,000 Btu/gal (34,524 kcal/3.7 liters), and number 2, the most popular domestic fuel oil, is 140,000 Btu/gal (35,280 kcal/3.7 liters) (Table 20-1). The flash points of fuel oils will vary considerably because of the refining methods.

The storage of fuel oil is a contributing factor in the efficient operation of the heating system. In cold climates, the oil should be stored indoors or in some type of heating device. Steam or hot water pipes around the tank or electric emersion heaters can be used to preheat the oil before entering the combustion area. The fuel oil may be stored outdoors in milder climates without the problem of having it become excessively thick.

The use of fuel oil for domestic purposes is a result of the convenience and cleanliness of oil as compared to coal. Despite the competition of natural gas, fuel

TABLE 20-1 Heat Ratings of Fuel Oil

Grade	Btu	kcal
No. 1	132,900–137,000	33,490.8–34,524
No. 2	135,800–141,800	34,221.6–35,733.6
No. 4	143,100–148,100	36,061.2–37,321.2
No. 5	146,800–150,000	36,993.6–37,800
No. 6	151,300–155,900	38,127.6–39,286.8

oil is still a leader as a source of heat in heating equipment, especially in the northeastern sections of the United States.

Gases. There are two major classifications of gaseous fuels used in heating equipment today. They are natural gas and liquefied petroleum (LP) gas. Natural gas is distributed by means of a pipeline system, while LP gas is usually transported by truck and stored where it is to be used.

Natural Gas. This is the lightest of all petroleum products. It is generally found where oil is found; however, in some gases oil is not present. Theorists have long argued about the exact origin of natural gas. However, most agree that natural gas was formed during the decomposition of plant and animal remains buried in prehistoric times. Because these plants and animals lived during the same period as those that are presently found as fossils, natural gas is sometimes called a fossil fuel.

The gas industry may be broken down into the following areas: exploration, production, transmission, and distribution.

The exploration section of the industry actually finds the areas where the gas is located. The people who work in exploration examine new areas and make the necessary reports, purchases, and perform other essential duties prior to the actual drilling.

When all the exploration functions are completed, the production department does the actual drilling of the well. The gas, or crude oil, is brought to the earth's surface and is blocked at this point. The well remains in this state until the gas or oil is needed.

When the need arises, the transmission department receives the gas from the well at pressures from 500 to 3000 psig (3535.99 to 20,710.99 kPa). Even with these high pressures, the resistance given by the pipe and the distance to be covered necessitate the use of booster pumps to transfer the gas from the well to the refinery. At the refinery, the gas passes through a drying process that removes moisture, propane, and butane. During this process, most of the odor is also removed from the raw gas. An odorant is therefore added to aid in leak detection.

Natural gas has a specific gravity of 0.65, an ignition temperature of 1100 °F (593 °C), and a burning temperature of 3500 °F (1909 °C). One cubic foot (0.0283 m³) of natural gas will emit from 900 to 1400 Btu/ft³ (226.8 kcal/m³ to 352.9 kcal/m³) with most of the gas emitting approximately 1100 Btu/ft³ (277.2 kcal/m³). The Btu (calorie) content of natural gas will vary from area to area. The local gas company should be consulted when the exact Btu content is desired.

Natural gas is made up of 55 to 90% methane (CH_4), 1 to 14% ethane (C_2H_6), and 0.5% carbon dioxide (CO_2). It requires 15 ft³ (0.4245 m³) of air per ft³ (0.0283 m³) of gas for proper combustion. It is lighter than air. Because methane and ethane have such low boiling points—methane −258.7 °F (−160 °C) and ethane −127 °F (−89 °C)—natural gas remains a gas under the pressures and temperatures encountered during distribution. Because of the varying amounts of methane and ethane, the boiling point of natural gas will vary according to the mixture.

Liquefied Petroleum (LP). Liquefied petroleum is either butane or propane, and in some cases a mixture of the two. These two fuels are refined natural gases and were developed for use in rural areas.

When these fuels are extracted from raw gas at the refinery, they are in the liquid state, under pressure, and remain in this state during storage and transportation. Only after vapor is drawn from above the liquid does the liquefied petroleum become a gaseous fuel.

LP gas has at least one definite advantage in that it can be stored in the liquid state and, thus the heat content is concentrated. This concentration of heat makes it economically feasible to provide service anywhere that portable cylinders may be used. The fact that 1 gal (3.785 liters) of liquid propane turns to 36.31 ft³ (1.02 m³) of gas when it is evaporated illustrates its feasibility.

When LP gas is stored in a tank, it is in both the liquid and gaseous state (Figure 20–2). To make the action of LP gas more easily understood, let's review briefly the boiling point and pressures of water.

When a pressure cooker is first filled with water with no heat applied and the top remaining off, there is no pressure on the surface of the water [Figure 20–3(a)]. However, when the top is in place and heat is applied, as in Figure 20–3(b), the pressure will

FIGURE 20–2 LP gas storage tank.

FIGURE 20-3 Pressure cooker.

begin to rise after the boiling point of the water is reached. As more heat is applied, more pressure will be created above the water by the evaporating liquid. When a constant temperature is maintained, a corresponding pressure will also be maintained. Similarly, when the pressure is reduced, the boiling point is reduced.

If we apply this principle to LP gas stored in a tank, it can readily be seen that as vapor is withdrawn from the tank, more liquid will evaporate to replace that which was withdrawn. It must be remembered that each liquid has its own boiling point and pressure.

Butane (C_4H_{10}), like propane (C_3H_8) and natural gas, is placed in the hydrocarbon series of gaseous fuels, because it is comprised of hydrogen and carbon. Butane has a boiling point of 31.1 °F (− 0.6 °C), a specific gravity of 2, a heating value of 3267 Btu/ft^3 (823.3 kcal/m^3) of vapor. It requires 30.97 ft^3 (0.87 m^3) of air per ft^3 (m^3) of vapor for proper combustion. At sea level, it has a gauge pressure of 36.9 lb (354.49 kPa) at 100 °F (37.8 °C). Butane expands to 31.75 ft^3 (0.898 m^3) of vapor per gal (3.785 liters) of liquid. It is heavier than air. The ignition temperature is approximately 1100 °F (593 °C) and the burning temperature is 3300 °F (1798 °C).

Propane has a boiling point of − 43.8 °F (6.6 °C), a specific gravity of 1.52, a heating value of 2521 Btu/ft^3 (635.29 kcal/m^3) of vapor. It requires 23.82 ft^3 (0.67 m^3) of air per ft^3 (m^3) of vapor for proper combustion. At sea level, it has a gauge pressure of 175.3 lb (1305.3 kPa) at 100 °F (37.8 °C). Propane expands to 36.35 ft^3 (1.02 m^3) of vapor per gal (3.785 liters) of liquid. It is also heavier than air. It has an ignition temperature of 1100 °F (593 °C) and a burning temperature of 2975 °F (1616.9 °C).

When we study the physical properties of LP gases, we can see that each has both good and bad

qualities. These qualities should be given a great deal of consideration when determining which fuel to use for any given application. The two characteristics deserving the most consideration are the Btu (cal) content and vapor pressure. See Table 20-2 for a comparison of vapor pressures.

When considering these fuels, the pressure is the major limiting factor, especially in colder climates. As we study the table, we can see that when the temperature of liquid butane reaches 30 °F (− 1.1 °C) or lower, there is no pressure in the tank. Therefore, butane would not be a suitable fuel at lower temperatures without some source of heat for the storage tank. This source of heat may be steam or hot water pipes around the tank, electrical heaters around the tank, or even burial of the tank in the ground. However, these all add to the initial cost of the equipment.

If we look at the pressures of propane, we see that they are suitable throughout a wide range of temperatures. Therefore, from the pressure standpoint, propane would be the ideal fuel. On the other hand, its low Btu (cal) rating makes it less desirable than butane. To overcome this dilemma, the two fuels may be mixed to obtain some of the desirable characteristics of each gas. An example of this may be a mixture of 60% butane and 40% propane. At a temperature of 30 °F (− 1.1 °C), it will have a vapor pressure of approximately 24 psig (265.87 kPa) and a heat content of 2950 Btu/ft^3 (734.4 kcal/m^3). Since these fuels are usually mixed before delivery to the local distributor it is difficult to determine exactly the tank pressure and Btu/ft^3 (cal/m^3). As long as there is enough pressure in the storage tank to allow 11 in. (279.4 cm) water column of pressure to enter the house piping, there is little or nothing the service technician can do.

Before getting too involved in working with LP

TABLE 20-2 LP Gas Vapor Pressures [psi [kPa]]

Temperature [°F (°C)]		Propane		Butane	
0	[−17.8]	38.2	[363.42]		
10	[−12.2]	46	[417.0]		
20	[−6.7]	55.5	[482.27]		
30	[−1.1]	66.3	[533.78]		
40	[4.4]	78	[636.85]	3	[121.60]
50	[10]	91.8	[731.66]	6.9	[148.39]
60	[15.6]	107.1	[836.77]	11.6	[180.68]
70	[21.1]	124	[592.87]	16.9	[217.09]
80	[26.7]	142.8	[1082.01]	22.9	[258.31]
90	[32.2]	164	[1227.67]	29.8	[305.71]
100	[43]	187	[1385.68]	37.5	[358.61]
110	[43]	212	[1557.43]	46.1	[417.70]
120	[49]	240	[1749.79]	56.1	[486.40]
130	[54]	272	[1969.63]	66.1	[555.10]

gas, the state and local authorities should be consulted. Some states maintain strict control over the personnel working with these fuels.

Electricity. Electricity is becoming more popular as a source of heat for heating applications. The early applications of the heat-producing ability of electricity were limited. It was used only in a few specialized areas, mainly in industrial processes and in portable heaters to help supplement inefficient heating systems. Today, electric heating is adaptable to almost any type of construction and any climate.

The major methods of using electricity for heating applications are resistance heating, heat pump, and a combination of the two. There are three types of resistance elements used for electric heating: (1) the open wire, (2) the open ribbon, and (3) the tubular-cased wire (see Figure 20-4). An electric resistance heating element may be defined as an assembly consisting of a resistance wire, insulated supports, and terminals for connecting the electrical supply wire to the resistance wire. Resistance heating will convert electric energy to heat energy at the rate of 3412 Btu/kW (859.8 kcal/kW). Theoretically, electric heating is 100% efficient, that is, for each Btu (cal) input to the heating equipment, 1 Btu (cal) in usable heat is recovered.

Open Wire Heating Elements. Open wire heating elements are usually made of *nichrome* wire, which is wire made from nickel and chromium, without iron, wound in a springlike shape and mounted in ceramic insulators to prevent electric shorting to the metal frame. The open wire elements have a longer life span than the others because they operate cooler

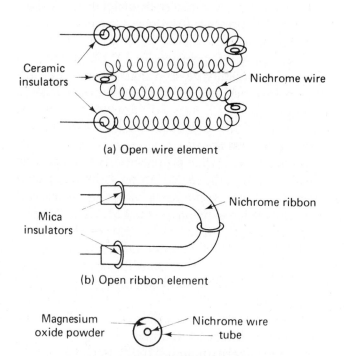

(a) Open wire element

(b) Open ribbon element

(c) Tubular cased element

FIGURE 20-4   Resistance elements.

since they release all the heat directly into the airstream. They also have a lower air pressure drop as compared to the other types of elements.

Open Ribbon Heating Elements. The open ribbon elements are also made of nichrome wire and are insulated in the same manner as the open wire elements. The flat design allows more intimate contact between the wire and the air because it has more surface area. Its efficiency is thereby increased over the open wire element. However, due to the increase

AIR CONDITIONING [HEATING]

in manufacturing costs, the open wire element is more popular.

Tubular-Cased Heating Element. The tubular-cased heating element is the same type as that used in cooking stoves. The nichrome wire is placed inside a tube and insulated from it by magnesium oxide power. Thus, tubular-cased heating elements do not require external insulation as do the other two resistance elements. They are also less efficient because of the energy loss caused by the extra material that the heat must pass through before reaching the moving air. They are, however, safer to use because of the interior insulation used in the manufacturing process. The tubular-cased elements have a shorter life than either of the other two elements because of higher operating temperatures. The control of these elements is more difficult than with the others because of the extra material involved. Also, the air must be circulated sufficiently long to ensure the proper cooling of these elements.

Heat Pump. The heat pump is the most efficient method of heating in use today. A heat pump is a refrigeration unit that reverses the flow of refrigerant according to the different seasons of the year. By use of a series of valves, it cools the space in the summer and provides heat for it in the winter. Electricity is used to drive the compressor and if the system design conditions—usually 72 °F (22.2 °C) and 50% relative humidity (RH) indoor conditions—are maintained, a heat pump will release up to eight times as much heat as could be obtained from resistance elements. Therefore, with an input of 1 kW (3412 Btu or 859.2 kcal) approximately eight times the input of 27.280 Btu (6.8745 kcal) are released by the heat pump.

A heat pump is less efficient at lower outdoor temperatures because the evaporator temperature must be lower than the ambient temperature in order to absorb enough heat to evaporate the refrigerant. This lower temperature is accompanied by a lower suction pressure that reduces the compressor capacity. Also, at the lower temperatures, there is not enough heat absorbed to replace the heat loss through the walls of the building.

When these units are installed in cold climates, additional heat is usually required. Because of the reduced efficiency, resistance elements are used in conjunction with these systems (Figure 20–5).

If the cost of equipment per hour of operation is considered, a heat pump is more economical to own than a cooling system with a separate heating unit. The reason for this is that when one part (either

FIGURE 20–5 Supplementary heating element and heat pump.

heating or cooling) is used, the other part is not operating. Therefore, the cost per hour of operation is more.

☐ Combustion ☐

The available energy contained in a fuel is converted to heat energy by the combustion process. Combustion may be defined as the chemical action of a substance with oxygen resulting in the evolution of heat and some light. There are three basic requirements for combustion: (1) sufficiently high temperature, (2) oxygen, and (3) fuel (Figure 20–6).

Temperature. When the air–fuel mixture is admitted to the combustion chamber, some means must be provided to bring the mixture to its flash point. This usually is done by a pilot light or an electric arc. If for any reason the temperature of the mixture is reduced below its flash point, the flame will automatically be extinguished. For example, if the temperature of a mixture of natural gas and air is reduced below its flash point of 1100 °F (593 °C), there will be no flame.

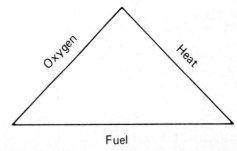

FIGURE 20–6 Combustion triangle.

Oxygen. Also, an ample supply of properly distributed oxygen must be supplied. The oxygen requirements governing the combustion process vary with each fuel. They also depend on whether or not the fuel and air are properly mixed in the correct proportions.

Fuel. The third requirement for combustion is the fuel. Properties of fuel were dealt with earlier in this chapter. The physical properties of each fuel must be considered when determining its requirements for combustion. All of the basic requirements for combustion must be met or there will be no combustion.

An important factor to keep in mind when making adjustments involving gaseous fuels is the limits of flammability, which are stated in percentages of gas in the air of the mixture that would allow combustion to take place. To simplify this, if there is too much gas in the air, the mixture will be too rich to burn. If there is too little gas in the air, the mixture will be too lean to burn. The upper and lower limits are shown for the more common gaseous fuels in Table 20–3.

Complete combustion can be obtained only when all of the combustible elements are oxidized by all the oxygen with which they will combine. The products of combustion are harmless when all of the fuel is completely burned. These products are carbon dioxide (CO_2) and water vapor (H_2O).

The rate of combustion, or burning, depends on three factors:

1. The rate of reaction of the substance with oxygen
2. The rate at which the oxygen is supplied
3. The temperature due to the surrounding conditions

All the oxygen supplied to the flame is not generally used. The excess is commonly called excess oxygen or excess air. This excess oxygen is expressed as a percentage, usually 50% of the air required for the complete combustion of a fuel. An example of this is that natural gas requires 10 ft³ (0.28 m³) of air for each ft³ (m³) of gas. When 50% excess air is added to this figure, the quantity of air supplied is calculated to be 15 ft³ (0.42 m³) of air to each ft³ (m³) of natural gas.

There are several factors governing the excess air requirements: (1) the uniformity of air distribution and mixing, (2) the direction of gas flow from the burner, and (3) the height and temperature of the combustion area. Excess air constitutes a loss and should be kept to a minimum. However, it cannot usually be less than 25 to 35% of the air required for complete combustion.

Excess air has both good and bad effects in the combustion area operation. It is added as a safety factor in case the 10 ft³ (0.28 m³) of air required is reduced for some reason. Dirty burners, improper primary air adjustments, or a lack in the supply of primary air can cause such a reduction. The adverse effect is that the nitrogen in the air does not change chemically and tends to reduce the burning temperature and the flue gas temperature, thereby reducing the efficiency of the heating equipment. The air supplied for combustion contains about 79% nitrogen and 21% oxygen.

The products of combustion produced when 1 ft³ (m³) of natural gas is completely burned are: 8 ft³ (0.23 m³) of nitrogen, 1 ft³ (0.0283 m³) of carbon dioxide, and 2 ft³ (0.057 m³) of water vapor (Figure 20–7). These products are harmless to human beings. In fact, carbon dioxide is the ingredient added to water that makes soft drinks fizz.

The by-products of incomplete combustion are: carbon monoxide—a deadly product; aldehydes—colorless, inflammable, volatile liquids with a strong pungent odor that are irritating to the eyes, nose, and throat; ketones—used as paint removers; oxygen acids; glycols; and phenols. These by-products are harmful and must be guarded against by the proper cleaning and adjustment of heating equipment.

When considering the combustion requirement for any given installation, the air supply to the equipment must be calculated. The air supply is governed by the size of the equipment in Btu (cal) rating, type of fuel, size of equipment room, building construction tightness, exhaust operation, and the local code requirements.

Heating equipment that is regulated by experienced personnel will produce clean, economical, efficient combustion. However, when the design of

TABLE 20–3 Fuel Gas Limits of Flammability

Gas	Upper Limit	Lower Limit
Methane	14	5.3
Ethane	12.5	3.2
Natural	14	3
Propane	9.5	2.4
Butane	8.5	1.9
Manufactured	29	4

1 ft³ natural gas + 10 ft³ air + Flame = 8 ft³ nitrogen +

1 ft³ carbon dioxide + 2 ft³ water vapor

FIGURE 20-7   Elements of combustion.

the equipment and the fundamentals of combustion are ignored, a potential hazard exists.

☐ Orifices ☐

The orifice may be defined as the opening through which gas is admitted to the main burner (Figure 20-8). It is mounted on the gas manifold by means of pipe threads and projects into the burner (Figure 20-9). It is normally referred to as the *orifice spud*.

For a burner to operate satisfactorily, it must be supplied with proper Btu (cal) input. This is the function of the orifice. The size of the orifice, the gas manifold pressure, and the gas density are the factors that determine the rate of gas flow to the burner. The orifice and burner must be matched in Btu ratings. If the type of gas is changed, say from natural to LP gas, the orifice size must also be changed. When the orifice is oversized, an insufficient amount of primary air is drawn into the burner. Thus incomplete combustion results and a yellow, inefficient flame is produced.

An orifice that is too small will also have an adverse effect on the burner. There will be delayed ignition accompanied by a loud boom when the burner is ignited. Also, the burner will not operate to its full capacity, resulting in poor heating.

The orifice must direct the gas stream exactly down the center of the burner (Figure 20-10). The velocity of gas down the burner tube causes the primary air to be drawn in through the burner face and mixed with the gas. If the gas velocity is reduced, insufficient primary air will be the result. Therefore, the orifice must be drilled straight and in the exact center of the spud. This may be accomplished easi-

FIGURE 20-8   Orifice location.

FIGURE 20-9   Orifice spuds.

Primary air passage

Gas stream

FIGURE 20-10   Orifice in relation to burner.

Orifices   **513**

FIGURE 20-11 **Cutaway diagram of orifice.**

ly by drilling the hole from the rear (Figure 20-11). The V-shape will ensure the desired hole position.

The foregoing information should point up the fact that the orifice is a precision piece of equipment, and it should be treated as such when service is required. To clean an orifice a soft instrument, such as a broom straw, wire brush bristle, etc., must be used. Care must be taken not to change the shape or size of the orifice. Table 20-4 can be used to ensure the proper orifice size for the job.

▢ Main Gas Burners ▢

A gas burner is defined as a device that provides for the mixing of gas and air in the proper ratio to ensure satisfactory combustion. The first burners were used for lighting purposes. The carbon in the flame was the result of incomplete combustion that produced the light. This flame was not good for heating because of the lower temperature produced.

The blow pipe was the first burner to produce a blue flame and, therefore, any great heat intensity. The blue flame was produced by blowing primary air into the base of the flame (Figure 20-12). The modern blow pipe accomplishes the same thing by blowing compressed air in the same end that the gas enters. The gas and air are thus mixed in the pipe as in the burners in modern furnaces.

Modern furnaces make use of four types of main gas burners: (1) drilled port, (2) slotted port, (3) ribbon, and (4) inshot (Figure 20-13). They are made of either cast iron or stamped steel, depending on the type and purpose. Usually, the drilled port burner is made of cast iron. The others are made by forming steel into the desired shape. The type of burner used depends on the particular equipment design. Each manufacturer will use the burner that lends itself most effectively to the requirements of his equipment.

Primary Air Openings. The control of primary air is accomplished on the face of the main burner. There are several methods used for this purpose

TABLE 20-4 Orifice Capacities for Natural Gas at 1000 Btu/ft³ [252 kcal/m³] Manifold Pressure, 3½ in. [8.89 cm] Water Column

Wire Gauge Drill Size	Rate/hr			
	ft³ [m³]		Btu [kcal]	
70	1.34	[0.037]	1,340	[337.7]
68	1.65	[0.046]	1,650	[415.8]
66	1.80	[0.050]	1,870	[471.2]
64	2.22	[0.062]	2,250	[567]
62	2.45	[0.069]	2,540	[640]
60	2.75	[0.077]	2,750	[693]
58	3.50	[0.099]	3,050	[768.6]
56	3.69	[0.104]	3,695	[931.1]
54	5.13	[0.145]	5,125	[1,291.5]
52	6.92	[0.195]	6,925	[1,745.1]
50	8.35	[0.236]	8,350	[2,104.2]
48	9.87	[0.279]	9,875	[2,488.5]
46	11.25	[0.318]	11,250	[2,835]
44	12.62	[0.357]	12,625	[3,181.5]
42	15.00	[0.424]	15,000	[3,780]
40	16.55	[0.468]	16,550	[4,170.6]
38	17.70	[0.501]	17,700	[4,460.4]
36	19.50	[0.552]	19,500	[4,914]
34	21.05	[0.596]	21,050	[5,304.6]
32	23.70	[0.671]	23,075	[5,814.9]
30	28.50	[0.807]	28,500	[7,182]
28	34.12	[0.966]	34,125	[8,599.5]
26	37.25	[1.054]	37,250	[9,387]
24	38.75	[1.097]	39,750	[10,017]
22	42.50	[1.203]	42,500	[10,710]
20	44.75	[1.267]	44,750	[11,277]

FIGURE 20-12 **Basic blow pipe.**

(Figure 20-14). The adjustable shutter, however, is the most common method used. The adjustment is made by enlarging or reducing the size of the opening through which the primary air is admitted.

The threaded shutter is less common and is used only on cast-iron burners. The deflector type, or ad-

FIGURE 20-13 Types of burners.

(a) Adjustable shutter   (b) Threaded shutter

(c) Nonadjustable shutter   (d) Deflector shutter

FIGURE 20-14 Primary air adjustment methods.

FIGURE 20-15 Forced-draft burner.

FIGURE 20-16 Power gas burner. (Courtesy of Ray Burner Co.)

justable baffle, is gaining acceptance in the industry. This burner has a larger air opening than necessary. This is an advantage because it will not clog up as fast as the other types. The primary air adjustment is made by pushing the deflector into the gas stream. This added restriction reduces the gas velocity and reduces the intake of primary air. Some burners have fixed openings and no adjustment can be made. These openings are sized according to the burner Btu (cal) rating and the proper flame is automatically obtained when the correct orifice size is used.

Forced-Draft Burners. The burners discussed previously were atmospheric types. Another type of burner that is gaining popularity, especially with the advent of roof-mount units, is the forced-draft burner. The equipment using this type of burner makes use of a small blower to provide the combustion air (Figure 20-15). The combustion area is sealed from atmospheric conditions, and the air is forced into the combustion zone in which there is a positive pressure. The amount of air supplied and the inside pressure are controlled by an adjustment on the flue outlet. By forcing a fixed amount of air

into the combustion chamber, preset combustion conditions are maintained.

Forced-draft burners are ideal for outdoor heating equipment where unusual conditions prevail and where vent pipe heights are restricted. The main gas valve cannot open on these units until a definite pressure is reached inside the combustion zone. When making adjustments involving this pressure, the manufacturer's specifications must be consulted. This pressure will vary from manufacturer to manufacturer and sometimes on different models of the same manufacturer.

Power Gas Burners. Another type of burner that is now being used is the power gas burner (Figure 20-16). These burners are small, highly efficient, and suited for use with natural gas. Power burners have a very wide range of capacities. The maximum capacity is 500,000 Btu (126,000 kcal) and the minimum is 120,000 Btu (30,240 kcal). They are appropriate for any heating application where oil or an equivalent capacity is presently being used or as an integral component of a new boiler–burner system. They also can be directly installed and fired through existing doors of cast iron or steel firebox boilers without requiring further brickwork or modification of the boiler base.

Power burners furnish 100% of the combustion air through the burner for firing rates up to their rated capacities. These burners have locked air shutters, with a reduced input start by means of a slow opening gas valve. Power burners are easy to install and to adjust due to the use of an orifice that meters the gas and a dial that adjusts the air shutter to match the gas orifice used. An automatic electric pilot ignition is used that incorporates a 6000-V transformer to provide the necessary spark to ignite the pilot gas.

In operation, these burners are noisier than the atmospheric-type burner. Power burners are applicable with small boilers, residential, and industrial furnaces. They are also used as antipollution afterburners. These burners are automatic and are equipped with the latest types of safety and operating controls.

☐ Pilot Burners ☐

A pilot burner is defined as a small burner used to ignite the gas at the main burner ports and as a source of heat for the pilot safety device. Approximately 50% of all heating equipment failures can be attributed to this burner.

The automatic pilot is probably the most important safety device used on modern gas heating equipment. During normal operation, the pilot flame provides the heat necessary to actuate the pilot safety device. Also, the pilot is located in the combustion chamber in proper relation to the main burners so that the flame will ignite the gas admitted when the thermostat demands heating. Should something happen to cause the pilot to become unsafe, the pilot safety device will prevent any unburned gas from escaping into the combustion chamber.

There are basically three types of pilot burners: (1) the millivolt, (2) the bimetal, and (3) the liquid filled.

Millivolt Pilot Burner. The millivolt pilot burner can be divided into two types: the *thermocouple* and the *thermopile*. The thermocouple is an electricity-

FIGURE 20–17

FIGURE 20–18   Thermopile principle.

producing device made of dissimilar metals. One end of the metal is heated and is called the hot junction, while the other end remains relatively cool and is called the cold junction (Figure 20–17). The output of a thermocouple is approximately 30 mV when heated to a temperature of about 3200 °F (1742 °C). The thermopile is a series of thermocouples wired so that the output voltage will be approximately 750 mV (Figure 20–18). The greater the number of thermocouples in series, the higher will be the output voltage.

Bimetal Pilot Burner. The bimetal pilot is so named because of the sensing element. The sensing element is made of two dissimilar metals welded together so that they become one piece (Figure 20–19). The two metals have different expansion rates, and this causes them to bend when heated. As the bimetal element is heated by the pilot flame, a movable contact is pushed toward a stationary contact. When these two contacts touch, an electric circuit is completed through them (Figure 20–20). The bimetal pilot is usually employed to operate in the control or low-voltage circuit.

Liquid-Filled Pilot Burner. The liquid-filled pilot functions from the force of pressure exerted when a liquid is heated and a vapor is formed. This liquid-vapor is contained in a closed system so that more accurate control can be obtained. The bulb is located in the pilot flame (Figure 20–21), and when the

FIGURE 20-19 Bimetal element.

FIGURE 20-20 Operation of bimetal element.

FIGURE 20-23 Yellow flame.

designated temperature and pressure are reached, a circuit is completed and the heating unit can function.

☐ Flame Types ☐

There are basically two types of flames: yellow and blue flame. There are, however, variations of these two flames that are the result of changes in the primary air adjustment. Primary air is the air that is mixed with the fuel before ignition. Secondary air is the air that is admitted to the flame after ignition (Figure 20–22). It is also called excess air.

Yellow Flame. This flame has a small blue-colored area at the bottom of the flame (Figure 20–23). The outer portion (outer envelope) is completely yellow

and is usually smoking. The smoke is unburned carbon from the fuel. The yellow or luminous portion of the flame is caused by the slow burning of the carbon. Soot is the result of the yellow flame, which also provides a lower temperature than the blue flame. The yellow flame is an indication of insufficient primary air. *Incomplete combustion* is also indicated by a yellow flame—a hazardous condition that cannot be allowed.

Yellow-Tipped Flame. This condition occurs when not quite enough primary air is admitted to the burner. The yellow tips will be on the upper portion of the outer mantle (Figure 20–24). This flame is also undesirable because the unburned carbon will be deposited in the heat exchanger, thus eventually restricting the passage of the products of combustion.

Orange-Colored Flame. Some orange or red color in a flame, usually in the form of streaks, should not cause any concern. These streaks are caused by dust particles in the air and cannot be completely

FIGURE 20-21 Liquid-filled element.

FIGURE 20-22 Primary and secondary air supporting combustion.

FIGURE 20-24   Yellow-tipped flame.

FIGURE 20-25   Orange or red streaks in flame.

FIGURE 20-26   Soft, lazy flame.

FIGURE 20-27   Sharp, blue flame.

FIGURE 20-28   Lifting flame.

eliminated (Figure 20-25). When there is a great amount of red or orange in the flame, however, the combustion area location should be changed or some means provided for filtering the combustion air before it enters the equipment.

Soft, Lazy Flame. This type of flame appears when just enough primary air is admitted to the burner to cause the yellow tips to disappear. The inner cone and the outer envelope will not be as clearly defined as they are in the correct, blue flame (Figure 20-26). This type of flame is best for high–low fire operation. It is not suitable, however, whenever there may be a shortage of secondary air. This flame will burn blue in the open air. When it touches a cooler surface, however, soot will be deposited on that surface.

Sharp, Blue Flame. When the proper ratio of gas to air is obtained, there will be a sharp, blue flame (Figure 20-27). Both the outer envelope and the in-

ner cone will be pointed and the sides will be straight. The flame will be resting on the burner ports, and there will not be any noticeable blowing noise. The flame will ignite smoothly on demand from the thermostat. Also, it will burn with a nonluminous flame. This is the most desirable flame for heating purposes on standard gas heating units.

Lifting Flame. When too much primary air is admitted to the burner, the flame will actually lift off the burner ports. This flame is undesirable for many reasons. When the flame is raised from the ports, there is a possibility that the intermediate products of combustion will escape into the atmosphere (Figure 20-28). The flame will be small and will be accompanied by a blowing noise, much like that made by a blow torch. The ignition will be rough, and if enough secondary air is not available, ignition may be impossible. Intermediate products of combustion are hazardous and must be avoided.

Floating Flame. A floating flame is an indication that there is a lack of secondary air to the flame. In severe cases, the flame will leave the burner and have a cloudlike appearance (Figure 20-29). Again the intermediate products of combustion are apt to escape from between the burner and the flame. The flame will actually be floating from one place to another wherever sufficient secondary air for combustion may be found.

This flame is sometimes hard to detect because as the equipment room door is opened sufficient air will be admitted to allow the flame to rest on the burner properly. However, the obnoxious odor of aldehydes will still be present.

Another possible cause of this flame is an improperly operating vent system. If the products of combustion cannot escape, fresh air cannot reach the flame. This is a hazardous situation that must be eliminated.

□ Heat Exchangers □

A heat exchanger may be defined as a device used to transfer heat from one medium to another. There are two types of heat exchangers: (1) primary heat exchanger, and (2) secondary heat exchanger. For a heat exchanger to be efficient, the transfer of heat must be made with as little loss as possible.

Intermediate products
of combustion

Flame has cloud-like
appearance

FIGURE 20–29 Floating flame.

FIGURE 20–30   Forced-air heat exchanger.

Restriction

(a)

Baffles

(b)

FIGURE 20–31   Gas vent restrictions.

Primary Heat Exchangers. The heat exchanger is the heart of the heating plant (Figure 20–30). It is designed to transfer heat from the combustion gases to the heating medium flowing through the passages. On direct-fired heating systems, it also serves as the combustion area. Openings in the lower section permit installation of the main burners and allow secondary air to reach the flame. As the fuel is burned, the flue gases rise through the vent passages. Restrictions are incorporated in the heat exchanger to control the flow of combustion air and reduce the amount of excess air to within the 35 to 50% limits. These restrictions are made in several different ways: (1) the top section is reduced in cross-sectional width, (2) baffles are placed in the exhaust openings at the top of the heat exchanger, and (3) a combination of the two (Figure 20–31). These restrictions also permit maximum heat transfer by reducing the vent gas velocity, which allows the maximum heat to be extracted from it. The vent gases then leave the heat exchanger and enter the venting system.

There are two general classifications of primary heat exchangers: (1) the barrel, and (2) the sectionalized (Figure 20–32). They may be constructed of either cast iron or steel. Due to the weight, expense, and slow response to temperature change of cast iron, steel has dominated the forced-air furnace heating industry in recent years.

The major problem with steel is the noise that sometimes accompanies the expansion and contraction associated with the heating and cooling of the heat exchanger. This noise can, however, be reduced

by placing dimples and ribs in the metal during the stamping operation. Ceramic coating is also used to help eliminate noise as well as providing corrosion protection. This ceramic coating does not hinder the transfer of heat from the heat exchanger.

The application of the heat exchanger determines its physical shape. The heat exchanger used in a horizontal forced-air furnace (Figure 20–33) would need to be shaped differently from one used in an upflow furnace [Figure 20–32(b)]. In most instances, there is very little difference between the upflow heat exchanger to the downflow heat exchanger.

There are two flow paths through a heat exchanger. Both the heated air and the combustion gases pass through the unit without any mixing of the two. The heated medium flows around the outside of the heat exchanger, while the flue gases pass through the inside (Figure 20–34).

The heat exchanger is normally a trouble-free apparatus, especially when the burners are installed and adjusted properly. From time to time, however, due to the continuous expansion and contraction, a hole

(a) Barrel

(b) Sectionalized

FIGURE 20-32 **Types of primary heat exchangers: (a) barrel; (b) sectionalized.**

FIGURE 20-33 **Horizontal heat exchanger.**

Vent gas flow

Air flow

FIGURE 20-34 **Flow path through a heat exchanger.**

will develop in the metal. This is a dangerous situation. When the flame is burning before the blower starts, carbon monoxide could be admitted to the airstream. After the blower starts, the flame will be agitated due to the greater pressure in the duct system. The amount of agitation will, naturally, depend on the size of the hole. A large hole could possibly cause the flame to be blown from the front of the furnace and set fire to anything in the immediate area, along with depositing the products of combustion in the building. The only difference between a large hole and a small hole is that the small hole is more difficult to detect. In any case, a leaking heat exchanger must be replaced.

Secondary Heat Exchangers. Hydronic heating (the use of circulated hot water or steam) enjoys a fair share of the heating industry. The boilers used for hydronic heating use both primary and secondary heat exchangers. The primary heat exchangers used with hydronic heating heat water rather than air (Figure 20-35). When hydronic heating is used, a secondary heat exchanger must be used. When hydronic heating (hot water or steam) is used, the heating medium is passed from the boiler to the secondary heat exchanger, located in the area to be heated, through a series of pipes and returned to the boiler where it is reheated and recirculated. In some installations, the secondary heat exchanger is attached to a duct system and used as a central heating unit. The secondary heat exchanger is usually in the form of a finned coil (Figure 20-36).

Another type of secondary heat exchanger is the embedded coil. This type of exchanger consists of a pipe embedded in the floor, wall, or ceiling (Figure 20-37). This system of piping is connected to the boiler in the conventional manner. The coil has no fins and uses the surrounding structure to radiate the heat. It is ideal in playrooms or dens where a warm floor is desired. These exchangers, naturally, should be installed during the construction of the room.

The efficient operation of any hydronic heating unit requires that all air be kept from collecting in the piping. Most installations have air vents on top of the coil as well as in the highest point of the entire system. Should these exchangers become air locked, the heating will be reduced and sometimes completely stopped. In order to maintain maximum output from these units, the air must be continuously removed from the piping, coils, and boilers.

FIGURE 20-35 **Water and vent gas flow through a boiler. (Courtesy of Weil-McLain Co., Inc., Hydronic Division.)**

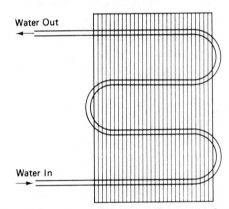

FIGURE 20-36 **Secondary heat exchanger.**

FIGURE 20-37 **Embedded hydronic heat exchanger.**

Electric heating is beneficial from the standpoint of ecology because the fuel being burned in the production of electricity is consumed in one remote location. At the consumer level, there is no waste to be removed. There are no vent pipes to be installed as with gas, oil, or coal.

However, when considering electric heating, there are other factors to be weighed. The user must be willing to pay higher electric bills. There is also an increased initial cost of construction. This cost is accrued by the added electrical service and wiring and the extra insulation required when electric heating is used.

Resistance heating involves a large number of operations, all of which are merely variations but are considered different in use and application. Resistance heating includes resistance elements, baseboard elements, heating cable, portable and unit type heaters, and duct heaters. However, regardless of the use or application, resistance heating wire is selected for high resistance per unit or length, and for the stability of this resistance at high temperatures. The type of wire used for this purpose is a high-nickel chrome alloy resistance wire. Resistance heating does not use a centrally heated medium. The electric energy is expended directly into the airstream.

Durability must also be considered when determining the type of wire to be used for resistance heating. The high nickel content in this type of wire retards oxidation and separation of the wire. If, however, the resistance wire does become broken, it may be repaired by cleaning the two ends, overlapping them, and applying borax liberally to the two pieces. Apply heat to the cleaned area until the two ends fuse together. After the wire has cooled to room temperature, remove any excess borax with a wet cloth. If this borax is allowed to remain on the wire, an immediate burnout will occur because the borax will cause excessive spot heating.

Resistance Elements. These elements are more commonly used in forced-air control heating systems. They are so termed in order to differentiate them from duct heater strips, and are mounted directly in the airstream inside the furnace (Figure 20-38). The elements are staggered for more uniform heat transfer, to eliminate hotspots, and to ensure maximum heat transfer. Each element is protected by

FIGURE 20-38 Resistance element installed in a furnace flame. [Courtesy of Electric Products Manufacturing Co.]

FIGURE 20-39 Element compartment. [Courtesy of Electric Products Manufacturing Co.]

a temperature limit switch and a fuse link. These elements are installed inside the furnace proper in an element compartment (Figure 20-39). These elements will emit approximately 3412 Btu/kW (859.8 kcal/kW) input.

Baseboard Elements. Baseboard elements may be used for heating the whole building or for supplementary heating. These elements are usually enclosed in a housing that provides safety as well as efficient use of the available heat.

These units are available in a wide variety of sizes and may be joined together, usually end to end, to provide the Btu (cal) required for any given installation. The ideal location is at the point of greatest heat loss. The element in these units is normally of the enclosed type. That is, the electric resistance element has fins added to increase the surface area of the heating element, thereby increasing its efficiency. Baseboard units have a direct Btu (cal) output per watt input.

Duct Heaters. Electric duct heaters are factory-assembled units consisting of a steel frame, open coil heating elements, and an integral control compartment. These matched combinations are fabricated to order in a wide range of standard and custom sizes, heating capacities, and control modes. They are prewired at the factory, inspected, and ready for installation at the job site.

The duct heater's frame is made in two basic configurations—slip-in (Figure 20-40) or flanged (Figure 20-41). The slip-in model is normally used in ducts up to 72 in. wide by 36 in. high (182.8 cm wide by 91.4 cm high); the flanged model is used in larger ducts or where duct layout would make it impossible or impractical to use the slip-in type.

The slip-in heater, which is standard, is designed so that the heater frame is slightly smaller than the duct dimensions. The frame is inserted through a rectangular opening cut in the side of the duct, with the face of the heater at a right angle to the airstream. It is secured in place by sheet metal screws running through the control compartment.

The flanged heater is designed so that the frame matches the duct dimensions. The frame is then attached directly to the external flanges of the duct.

Airflow. Electric heaters differ from steam or hot water coils in that the Btu/hr (cal/hr) output is constant as long as the heater is energized. It is, there-

AIR CONDITIONING [HEATING]

FIGURE 20-40　Slip-in duct heater. (Courtesy of Gould, Inc., Heating Element Division.)

FIGURE 20-41　Flanged duct heater. (Courtesy of Gould, Inc., Heating Element Division.)

fore, necessary that sufficient and uniform airflow be provided to carry away this heat and to prevent overheating and nuisance trapping of the thermal cutouts. The minimum air velocity required is determined on the basis of entering air temperature and watts/ft^2 of the cross-sectional duct area.

☐ Furnaces ☐

Furnaces are generally considered as being the heating part of an air conditioning system. They also serve as the air handling and filtering section of a complete system. Furnaces may be designed to use the various types of gases, fuel oil, or electric resistance elements as the heat source. The type of heat source will also govern the type of heat exchanger construction.

Gas Furnaces. Gas furnaces are manufactured in three designs: upflow (Figure 20–42), downflow, and horizontal. This terminology refers to the direction of the airflow. Each furnace is specifically designed for a particular application. Notice that the burners and flue pipe connection are in a fixed location and cannot be changed.

In the downflow furnace, the fan is located on top and the air is blown down through the heat exchanger (Figure 20–43). The ductwork, which distributes the air to the separate areas, is connected to the bottom of the furnace. The horizontal furnace is generally installed in the attic or a crawl space under the building. The air is blown in a horizontal direction (Figure 20–44).

Generally, these furnaces may be operated with either natural or LP gases. Minor adjustments may be needed along with changing the orifice and in some cases the burner assembly to the proper size. The manufacturer's recommendations should be followed regarding these items. Natural gas furnaces require a gas pressure regulator. LP furnaces do not require a gas pressure regulator. LP gas is regulated at the storage tank.

Condensing Gas Furnaces. These type of furnaces, also known as pulse combustion units, have an efficiency from 92 to 96%. This is a tremendous improvement over the standing pilot furnaces, which have efficiencies ranging from 55 to 65%. This increased efficiency is possible because most of the moisture in the flue gases are condensed, thus recovering the latent heat of condensation as well as the sensible heat of the flue gases, which are both added to the air used for heating the structure. This condensation occurs in a secondary heat exchanger that cools the flue gases below the condensing temperature. The secondary heat exchanger is located below the primary heat exchanger and directly above the indoor blower. Thus the cooler air blows first over the secondary heat exchanger condensing the moisture and adding heat to the cooler air before it enters the primary heat exchanger.

Because there are two heat exchangers used in these furnaces, the vent gas temperature is much less than in the standard-type furnace. The sizing of condensing gas furnace vents is much more critical than

PARTS ARRANGEMENT

FLUE PIPE

BURNER ACCESS PANEL

FAN/LIMIT CONTROL

UNIT RATING PLATE

GAS VALVE

PILOT

BURNERS

WIRING MAKE-UP BOX

HAMMOCK FILTER
(Slab Filter On Some Models)

BLOWER ACCESS PANEL

LINERS

HEAT EXCHANGER

BLOWER MOTOR

BLOWER

FIGURE 20-42 Cutaway of upflow furnace. (Courtesy of Lennox Industries, Inc.)

1. Burners
2. Heat Exchanger
3. Blower
4. Controls
5. Draft Diverter
6. Limit Control
7. Fan Control
8. Cabinet

FIGURE 20-43 Cutaway of counterflow furnace. (Courtesy of Southwest Manufacturing Company.)

       AIR CONDITIONING (HEATING)

1. Burners
2. Heat Exchanger
3. Blower
4. Controls
5. Draft Diverter
6. Limit Control
7. Fan Control
8. Cabinet

FIGURE 20–44 Horizontal furnace. (Courtesy of Southwest Manufacturing Company.)

with standard gas furnaces. The manufacturer's literature must be adhered to strictly during the installation procedures.

Lennox Pulse Combustion Furnace. The unit has a regular valve, but from there several different components are used (Figure 20–45). These different components are a gas expansion tank with a flapper valve, an air intake mixer valve and blower assembly, a finned cast combustion chamber, a stainless steel "tailpipe," and a spark plug for initial ignition. Outside air is used in the sealed combustion system, so there is no loss of conditioned air from within the space as with other types of units. The vent gases are cooled to 100 to 120 °F (37.8 to 49 °C). Because of these low vent gas temperatures a small PVC plastic pipe can be used for the vent pipe. A drain is required to remove the condensate created in the secondary heat exchanger. The secondary heat exchanger is similar to a standard air conditioning type coil.

The Lennox Pulse Combustion Furnace operates on a pulse combustion principle. This principle ignites small quantities of gas–air mixture at a rate of 60 to 70 times per second. Each combustion produces from one-fourth to one-half of a Btu. The process begins as air and gas are introduced into the chamber to mix near an ignitor [Figure 20–46(a)]. A spark creates the initial combustion, which in turn causes a positive pressure buildup that closes off the gas and air inlets [Figure 20–46(b)]. This pressure then relieves itself by forcing the products of combustion down a tailpipe [Figure 20–46(c)].

As the combustion chamber empties, its pressure becomes negative, drawing in air and gas for the next ignition [Figure 20–46(d)]. At the same instant, part of the pressure pulse is reflected back from the end of the tailpipe. It reenters the combustion chamber, causing the new gas–air mixture in the chamber to ignite and continue the cycle [Figure 20–46(e)]. Once combustion is started, it feeds upon itself, allowing the combustion blower and spark ignitor to be turned off.

Heil Condensing Gas Furnace. The Heil furnace makes use of a more standard type of heat exchanger and combustion principles than the Lennox Pulse Combustion Gas Furnace.

Sequence of Operation. This type of furnace makes use of an induced-draft combustion blower and motor to control the flow of air through the furnace combustion zone (Figure 20–47). To aid in reaching the higher energy ratings, a spark ignition system is used. The standing pilot is not used on these models of furnace. They are equipped with a negative pressure switch and a blower relay to prove that there is sufficient air for combustion before the gas valve is energized to admit gas to the main burner.

In operation, when the thermostat calls for heat, the following sequence of events occurs:

1. When the thermostat contacts close an electric

Air Intake Pipe

Tailpipe

Insulation

Combustion Chamber

Decoupler

Air Intake Mixing Valve and Blower Assembly

Gas Valve

Expansion Tank

Flue Vent Pipe

Secondary Heat Exchanger Coil

Control Box

Blower

FIGURE 20–45 Component location on condensing gas furnace. (Courtesy of Lennox Industries, Inc.)

circuit is completed to the induced-draft blower relay.

2. The induced-draft relay is energized, completing the line voltage circuit to the induced-draft blower motor.

3. The induced-draft blower then builds up a negative pressure inside the blower housing.

4. The negative pressure switch senses this airflow. A set of contacts located in the blower housing is made when the proper pressure is detected through a tube which is connected to the induced-draft blower housing and the ignition system is

energized (Figure 20–48). (The operation of this switch will be discussed later.)

5. With the completion of this circuit the gas control and the ignition circuits are energized. The gas is ignited.

6. When the thermostat is satisfied, the induced-draft blower relay is deenergized. The relay also deenergizes the spark ignition circuit and the main gas valve.

When a replacement part is needed, be sure to use an exact replacement. One part may look exactly like

(a)

(b)

(c)

(d)

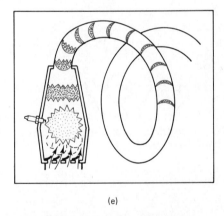

(e)

FIGURE 20-46 (a) Introduction of gas and air into combustion chamber; (b) Combustion created by initial spark; (c) Pressure forcing products of combustion down a tailpipe; (d) Air and gas drawn in for next ignition; (e) Ignition of new gas and air mixture. (Courtesy of Lennox Industries, Inc.)

another, but they may operate in a different manner and cause a dangerous situation. If a substitute part must be used, be sure to check with the equipment manufacturer to make certain that it is an exact replacement.

Operation of the Power Vent System Single Switch Assembly (Figure 20-48). In normal operation, as the fan starts to turn, a positive (+) pressure starts to build in the chamber of the power vent. The sensor tube picks up the pressure where it is felt in the pressure side of the air switch. The pressure moves the diaphragm, closing the normally open switch, completing the 24-V ac path to the DSI module. As the positive (+) pressure builds in the chamber of

Fan
Limit
Control

Combustion Air
Proving Switch

Redundant
Gas Valve

Induced
Combustion
Blower

Low Voltage
Terminal Strip

Blower Door
Safety Switch

Stainless Steel
Primary Heat
Exchanger

Stainless Steel
Secondary Heat
Exchanger

Ignition
Module

Multi-Speed
Blower

FIGURE 1

FIGURE 20–47   Heil condensing gas furnace. (Courtesy of Heil-Quaker Corp.)

AREA OF
NEGATIVE
PRESSURE

CHAMBER

NORMALLY
OPEN SWITCH
NOW CLOSED

FIGURE 20–48   Negative pressure switch. (Courtesy of Heil-Quaker Corp.)

the power vent, notice the venturi in the center of the chamber with "pitot" tube or J tube extending into the center of the venturi. As the positive (+) or pressurized air passes the tube, a small area of negative (−) pressure is formed above the tube. The negative pressure is felt in the blockage side of the air switch and helps hold the switch closed.

Oil Furnaces. These furnaces are designed to use domestic fuel oil as the heat source. They also are available in upflow, horizontal, and counterflow. The horizontal units are designed to be installed on combustible flooring in an attic space, on a slab in a crawl space, or they can be suspended in a basement, utility room, furnace room, or crawl space

RETURN AIR ►   SUPPLY AIR ►

Crawl Space Horizontal Installation

RETURN AIR ►   SUPPLY AIR ►

Suspended Horizontal Installation

FIGURE 20–49   Horizontal furnace installations. (Courtesy of Lennox Industries, Inc.)

(Figure 20–49). The counterflow oil furnace is designed for installation in either the counterflow or horizontal position (Figure 20–50).

Electric Furnaces. Electric furnaces can be installed in the upflow, downflow, and horizontal positions (Figure 20–51). Note that these furnaces do not have burners or a combustion-type heat exchanger as do the gas- and oil-fired furnaces. Instead, electric heating elements are used as the heat source. There are no vent pipe or gas pipe connections—only electrical connections. It is mainly for this reason that these furnaces are multipositional.

All furnaces should be installed with the proper clearances on the rear, sides, and front. These clearances are specified in the correct installation instructions for the particular furnaces being installed.

Heat Rise. The heat rise through a furnace can be defined as the temperature increase of the air as it passes through the heating device. That is, the temperature of the supply air minus the return air temperature. For example, air is entering the heating unit at 70 °F (21.1 °C) and leaving at 145 °F (63 °C). The heat rise is 145 °F − 70 °F = 75 °F (41.7 °C).

The heat rise through a unit is determined by the amount of heat input to the furnace and the volume of air flowing through the furnace. Gas and oil furnaces are rated from 40 °F (22.4 °C) to 100 °F (56 °C) heat rise. Electric furnaces are generally rated from 40 °F (22.4 °C) to 60 °F (33.6 °C) heat rise.

FIGURE 20-50 Counterflow installation. (Courtesy of Lennox Industries, Inc.)

☐ Oil Burners ☐

Fuel oils, like natural and LP gases, are excellent heating fuels. However, the lighting and burning of fuel oil requires special equipment. Fuel oil in its liquid form will not burn; it must be either atomized or vaporized. From this, we can define a fuel oil burner as a mechanical device that prepares oil for combustion.

The actual burning of the fuel takes place in the firebox, with the atomizing method being the most popular. The atomizing type of burner prepares the oil for burning by breaking it up into a foglike vapor. This is accomplished in three ways: (1) by forcing the oil under pressure through a nozzle (air or steam may be forced through with the oil, or the oil may be alone), (2) by allowing the oil to flow out the end of small tubes that are rapidly rotated on a distributor, or (3) by forcing the oil off the edge of a rapidly rotating cup that may be mounted either vertically or horizontally.

Fuel oil burners are divided into three types depending on the operating principle: (1) high-pressure, (2) low-pressure, and (3) rotary-type burners (Figure 20-52).

High-Pressure Oil Burner. A high-pressure oil burner is made up of an electric motor, blower, fuel unit, and an ignition transformer. The motor is mounted on the motor shaft and the fuel unit is directly connected to the end of the motor shaft. The fuel unit consists of a strainer, a pump, and a pressure-regulating valve.

High-pressure oil burners do not use primary and secondary air as such. They are designed to atomize

Horizontal installation with cooling coil and electronic air cleaner

Up-Flo installation with cooling coil.

Down-Flo installation with cooling coil.

Horizontal installation in closet.

FIGURE 20-51 Electric furnace installations. (Courtesy of Lennox Industries, Inc.)

Oil Burners

FIGURE 20-52 Low-pressure oil burner. (Courtesy of Carlin Company.)

FIGURE 20-53 High-pressure oil burner operation.

the fuel oil and mix this atomized oil with the air before ignition (Figure 20-53).

The fuel oil is forced through the burner nozzle at approximately 100 psig (787.99 kPa). The atomized oil has enough velocity at the nozzle to cause a low-pressure area into which the combustion air is drawn. The combustion air is also under force from the fan, which creates a turbulence in the air. All of these together are used to create a complete mixing of the fuel oil and the air. This combustible mixture is then ignited at the end of the burner.

In operation, when the thermostat demands heat for the building, the fuel oil is drawn from the storage tank (Figure 20-54). It passes through a strainer, where the solid particles are removed, to the pump which forces it to the pressure-regulating valve. At this point, the oil hesitates until the pressure has reached 100 psig (787.99 kPa). Then the pressure-regulating valve opens and allows the oil to proceed to the burner nozzle where it is atomized. After the fuel leaves the nozzle, it is mixed with the proper amount of air and ignited by a high-voltage transformer—approximately 10,000 V and 25 mA.

When sufficient heat has been supplied to the structure, the thermostat will signal the burner motor to stop. As it reduces its speed, the oil pressure from the fuel pump is also reduced. As the oil pressure is reduced below 90 psig (719.29 kPa), the pressure-regulating valve stops the flow of oil to the nozzle and the flame is extinguished because of the lack of fuel. The oil burner will remain at rest until the building requires additional heat, at which time the sequence is started again.

Low-Pressure Oil Burner. The low-pressure oil burner is similar in appearance to the high-pressure burner. It also operates much the same as the high-pressure burner except for the method used to atomize the oil. In the low-pressure burner, the oil is atomized by compressed air, similar to a paint spray gun. The primary air and fuel are forced through the nozzle at the same time (Figure 20-55) with 1 to 15 psig (107.86 to 204.04 kPa) pressure. The secondary air is supplied in the same manner as the air used by the high-pressure burner.

Low-pressure oil burners operate quite satisfactorily, but each manufacturer uses different mechanical means of mixing the air and the oil. This requires special service knowledge and techniques for each burner. Because simplicity of service is almost essential, low-pressure oil burners are not a common item.

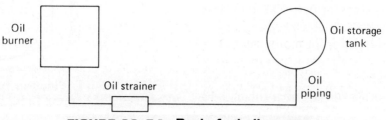

FIGURE 20-54 Basic fuel oil system.

FIGURE 20-55 **Low-pressure oil burner operation.**

Rotary Oil Burners. Rotary oil burners incorporate a completely different principle from either the high-pressure or the low-pressure burners. The rotary burner employs a rotating cup to atomize the oil. The fuel oil is forced to the edge of the cup in small droplets because of the high speed of the cup. As the droplets leave the edge of the cup, they are dispersed into a current of rapidly moving air which is provided by a centrifugal blower (Figure 20-56). These droplets are further atomized by this air and the foglike fuel is projected parallel to the axis of the cup. The atomized oil is then ignited in much the same manner as any other type of oil burner. In some rotary burners the cup may be replaced with small tubes extending to the face of the rotating member. The oil is forced to the ends of the tubes where centrifugal force distributes the oil into the airstream.

□ Boilers □

Hydronic heating maintains comfort in the home by circulating hot water to the secondary heat exchanger. This system basically consists of a boiler, a pump, a secondary heat exchanger, and a series of piping (Figure 20-57). The principle of hydronics dates back to ancient times when the Romans used it by circulating warm water through the walls and floors. Today, however, hydronics bears no resemblance to the ancient methods. Modern systems incorporate many refinements, such as zoning, which allows the temperature to vary in different areas of the building.

Hot Water Boilers. A hot water boiler maintains water temperatures between 120 and 210°F (49 and 99°C). This water is pumped through piping to the secondary heat exchanger. The term boiler may bring to mind large heating plants such as those found in schools or other large buildings. There are, however, commercial-type boilers. Actually, a modern cast-iron boiler used in residential heating is very com-

FIGURE 20-56 **Rotary oil burner operation. (Courtesy of Ray Burner Co.)**

FIGURE 20-57 **Basic hot water heating system.**

pact. Most units are the size of an automatic washing machine, and some are as small as a suitcase and even can be hung on the wall (Figure 20-58).

Steam Boilers. The widespread use of steam for space heating today points up a long recognized fact that steam as a heating medium has numerous basic characteristics that can be advantageously employed. Some of the most important advantages are as follows.

Steam Has the Ability to Give Off Heat. Properties of saturated steam may be found in steam tables, which give much information regarding the temperature and heat contained in 1 lb (453.6 g) of steam for any pressure (Table 20-5). For example, to change 1 lb of water to steam at 212°F (100°C) at atmospheric pressure requires a heat content of 1150.4 Btu (289.9 kcal). This is made up of 180.1

FIGURE 20-58   Water boiler. (Courtesy of Weil-McLain Co., Inc., Hydronic Division.)

FIGURE 20-59   Steam and condensate flow-in lines. (Courtesy of ITT Hoffman Specialty.)

Btu (45.4 kcal) of sensible heat (the heat required to raise 1 lb of water from 32 to 212 °F (0 to 100 °C) and 970.3 Btu (244.5 kcal) of latent heat. Latent heat is the heat added to change the 1 lb of water at 212 °F into steam. This stored-up heat is required to transform the water into steam, and it reappears as heat when the process is reversed to condense the steam to water.

Because of this fact, the high latent heat of vaporization of 1 lb (453.6 g) of steam permits a large quantity of heat to be transmitted efficiently from the boiler to the heating unit with little change in temperature.

Steam Promotes Its Own Circulation. For example, steam will flow naturally from a higher pressure (as generated in the boiler) to a lower pressure (existing in the steam lines). Circulation or flow is caused by the lowering of the steam pressure along the steam supply mains and in the heating units due to the pipe friction. Also, the condensation process of the steam as it gives up heat to the space being heated aids in circulation (Figure 20–59). Because of this fact, the natural flow of steam does not require a pump such as that needed for hot water heating, or a fan as used in warm air heating.

Steam Heats More Readily. Steam circulates through a heating system faster than other fluid media. This can be important where fast pickup of the space temperature is desired. It will also cool down more readily when circulation is stopped. This is an important consideration in spring and fall when comfort conditions can be adversely affected by long heating-up or slow cooling-down periods (Figure 20–60).

Piping Arrangements. Steam heating systems fall into two basic classifications—one pipe and two pipe. These names are descriptive of the piping arrangement used to supply steam to the heating unit and to return condensate from the unit. A one-pipe system consists of a heating unit with a single pipe connection through which steam flows to it and condensate returns from it using the same line at the same time. In a like manner, a two-pipe system consists of a heating unit with two separate pipe connections—one used for the steam supply and the other for condensate return.

One-Pipe System. Modern, automatically fired boilers promote rapid steaming and assure quick pickup of the space temperature from a cold start. The natural circulation of steam in the system, in combination with the simplicity of piping and air venting, makes this type of system a desirable method of heating as well as the most economical.

TABLE 20-5 Properties of Saturated Steam (approx.)

Absolute Pressure	Gauge Reading at Sea Level	Temperature (°F)	Heat in Water (Btu/lb)	Latent Heat in Steam (Vaporization) (Btu/lb)	Volume of 1 lb Steam (ft³)	Weight of Water (lb/ft²)
0.18	29.7	32	0.0	1076	3306	62.4
0.50	28.4	59	27.0	1061	1248	62.3
1.0	28.9	79	47.0	1049	653	62.2
2.0	28	101	69	1037	341	62.0
4.0	26	125	93	1023	179	61.7
6.0	24	141	109	1014	120	61.4
8.0	22	152	120	1007	93	61.1
10.0	20	161	129	1002	75	60.9
12.0	18	169	137	997	63	60.8
14.0	16	176	144	993	55	60.6
16.0	14	182	150	969	48	60.5
18.0	12	187	155	986	43	60.4
20.0	10	192	160	983	39	60.3
22.0	8	197	165	980	36	60.2
24.0	6	201	169	977	33	60.1
26.0	4	205	173	975	31	60.0
28.0	2	209	177	972	29	59.9
29.0	1	210	178	971	28	59.9
30.0	0	212	180	970	27	59.8
14.7	0	212	180	970	27	59.8
15.7	1	216	184	968	25	59.8
16.7	2	219	187	966	24	59.7
17.7	3	222	190	964	22	59.6
18.7	4	225	193	962	21	59.5
19.7	5	227	195	960	20	59.4
20.7	6	230	196	958	19	59.4
21.7	7	232	200	957	19	59.3
22.7	8	235	203	955	18	59.2
23.7	9	237	205	954	17	59.2
25	10	240	208	952	16	59.2
30	15	250	219	945	14	58.8
35	20	259	228	939	12	58.5
40	25	267	236	934	10	58.3
45	30	274	243	929	9	58.1
50	35	281	250	924	8	57.9
55	40	287	256	920	8	57.7
60	45	293	262	915	7	57.5
65	50	298	268	912	7	57.4
70	55	303	273	908	6	57.2
75	60	308	277	905	6	57.0
85	70	316	286	898	5	56.8
95	80	324	294	892	5	56.5
105	90	332	302	886	4	56.3
115	100	338	309	881	4	56.0
140	125	353	325	868	3	55.5

Source: ITT Hoffman Specialty.

Boiler
Water
Level

(a) Steam Boiler
Water at Boiler Water Level Only

(b) Hot Water Boiler
System Completely Filled With Water.

FIGURE 20-60 Boiler water comparison. (Courtesy of ITT Hoffman Specialty.)

Supply Valve

Vent Valve

Supply Main

Header

Main Vent

Dry Return

A

Drip Connection

Boiler

2 in.

Water Line

Hartford Loop

Wet Return

FIGURE 20-61 Basic one-pipe upfeed system. (Courtesy of ITT Hoffman Specialty.)

A one-pipe system, properly designed for gravity return of the condensate to the boiler with open air vents on the ends of the mains on each heating unit, requires a minimum of mechanical equipment (Figure 20–61). The result is a low initial cost for a very dependable system.

Two-Pipe System. Although a two-pipe system has fundamental differences from a one-pipe system, many components and piping installation practices are common to both systems. Also, the two-pipe system employs many of the advantages of using steam as a heating medium. They are applicable in a variety of structures from small residences to large commercial buildings, office buildings, apartment buildings, and industrial complexes.

Two-pipe systems are designed to operate at pressures ranging from subatmospheric (vacuum) to high pressure. Although they may use many practical piping arrangements to provide upflow or downflow systems, they are conveniently classified by the method of condensate return to the boiler—by gravity or by use of any one of several mechanical means.

▢ Humidification ▢

One of the most important aspects of air conditioning, humidification, is unfortunately one of the least understood. This is undoubtedly because it is intangible; you cannot see it, you touch it, it has no odor, no color, and no sound.

To be sure that we are starting out on common ground, let's define some of the words that we are going to use in the discussion of humidity and humidification.

Humidity: the water vapor within a given space.
Absolute humidity: the weight of water vapor per unit volume.

Percentage humidity: the ratio of the weight of water vapor per pound of dry air to the weight of water vapor per pound of dry air saturated at the same temperature.

Relative humidity: the amount of water vapor (percent) actually in the air compared to the maximum amount that the air could hold under the same conditions. The warmer the air, the more moisture it can hold. Air in a building heated to 70 °F (21.1 °C) can hold about 8 grains (0.52 g) of moisture per cubic foot (0.0283 m³). This is 100% relative humidity. If there are only 2 grains (0.13 g) per cubic foot (0.0283 m³) in the building, this is one-fourth of the amount of moisture that the air can hold. Therefore, the relative humidity is one-fourth or 25%; the air could hold four times as much water.

The important thing to remember is that when air is heated it can hold more moisture. This reduction of relative humidity is taking place in every unhumidified building where heating is used.

To solve this problem, we add moisture so there is more water available for the air to absorb. We humidify because there are benefits that are as important as heating to overall indoor comfort and well-being during the heating season.

These benefits can be grouped into three general classifications: (1) comfort, (2) preservation, and (3) health.

Comfort. Have you ever stepped out of the shower and noticed how warm the bathroom is? It is probably about 75 °F (23.9 °C) in the bathroom, and the relative humidity is probably about 70 to 80% because of the water vapor that was added to the air while showering. Now, if the phone rings, and you step out into the hall to answer it, you nearly freeze. Yet the temperature there is probably about 70 °F (21.2 °C)—just 5 °F (2.78 °C) cooler than in the bathroom, and you are shivering. This shivering is because you just became an evaporative cooler. The air out in the hall is dry, the relative humidity is possibly about 10 to 15%. You are wet and this thirsty air absorbs moisture from your skin. The water evaporates, and as it does, your skin is cooled. This same type of thing happens day after day, every winter, in millions of buildings. People are turning thermostats up to 75 °F and higher in order to feel warm because it feels drafty and chilly when the evaporative cooling process is going on. The prop-

er relative humidity level makes 70 °F feel more like 75 °F.

The chilling effect is not the only discomfort caused by too dry air. Static electricity is usually an indication of low relative humidity and a condition that is consistently annoying. Proper relative humidity will alleviate this discomfort.

Preservation. The addition or reduction of moisture drastically affects the qualities, the dimensions, and the weight of hygroscopic materials. Wood, leather, paper, and cloth, even though they feel dry to the touch, contain water—not a fixed amount, but an amount that will vary greatly with the relative humidity level of the surrounding air. For example, 1 ft³ (0.0283 m³) of wood with a bone-dry weight of 30 lb (13.6 kg) at 60% relative humidity will hold over 3 pints (1.41 liters) of water. If the relative humidity is lowered to 10%, the water held by the wood will not even fill a 1-pint (0.473-liter) bottle. We have, in effect, withdrawn 2½ pints (1.2 liters) of water from the wood by lowering the relative humidity from 60 to 10%.

This type of action goes on, not only with wood, but with every single material in the building that has the capability of absorbing and releasing moisture. Paper, plaster, fibers, leather, glue, hair, skin, and so on will shrink as they lose water and swell as they take on water. If the water loss is rapid, warping or cracking takes place. As the relative humidity changes, the condition and dimensions of the materials change as constantly as the weather. This is why proper relative humidity is important.

Too much moisture also can be damaging. Everyone has seen windows fog during the winter, maybe a little fog on the lower corners, maybe a whole window fogging or completely frosting over. This latter condition is an indication that the indoor relative humidity is too high.

The formation of condensation is due to the effect of vapor pressure. Dalton's law explains vapor pressure. It states, "In a gaseous mixture, the molecules of each gas are evenly dispersed throughout the entire volume of the mixture." Taking the building as the volume involved, water vapor molecules move throughout the entire building. Because of the tendency of these molecules to disperse evenly, or to mix, the moisture in the humidified air moves toward drier air. In other words, in a building the moist indoor air attempts to reach the drier outside air. It moves toward the windows, where there is a

FIGURE 20–62 Outside wall construction. (Courtesy of Research Products Corp.)

lower temperature, and therefore, an increase in relative humidity to a point at which the water vapor will condense out of the air and onto the cold surface of the window. This is the dew point and it occurs at various temperatures, depending on the type of windows in the building.

Usually, condensation on inside windows is a type of measurement of the allowable relative humidity inside a building. We can further assume that if this condensation activity is taking place on windows, it also is taking place within the walls if there is no vapor barrier (Figure 20–62).

The typical outside wall has drywall (or plaster), a vapor barrier (on the warm side of the insulation), the insulation, air space, sheathing, building paper, and siding. Given an indoor temperature of 70°F (21.2°C) and a relative humidity of 35%, and an outside temperature of 0°F (−17.8°C), the temperature through the wall drops to about 63°F (17.2°C) at the vapor barrier, down to 17°F (−8.3°C) at the sheathing, and on down to 0°F (−17.8°C) outside. If a psychometric chart was checked, we would discover that 70°F indoor air at 35% relative humidity has a dew point of 41°F (5°C). That temperature occurs right in the middle of the insulation. This is where condensation occurs—and where the trouble is—without a vapor barrier and a humidifier that can be controlled.

Properly controlled relative humidity is the important factor in avoiding damaging effects of too dry air and too high relative humidity.

Health. In the struggle between the nose and the air conditioning equipment, sometimes the heater wins and sometimes the cooler, but seldom does the nose win. The nasal mucus contains some 96% water. It is more viscous than mucus elsewhere in the body, and even slight drying increases the viscosity enough to interfere with the work of the cilia. Demands on the nasal glands are great even under excellent conditions and they cannot cope with extreme dryness indoors in winter.

Experience has shown that with approaching winter, the first wave of dry nose patients appears in the doctor's office when the relative humidity indoors falls to 25%. It would seem, therefore, that 35% would be regarded as a passing grade, but 40% is something to try for. It boils down to this: a pint of water (0.473 liter) is a lot of water for a small nose to turn out. In disease or old age, it simply does not deliver enough moisture; drainage stops and germs take over.

Dr. Joseph Lubart, an expert on the common cold, states in the New York State *Journal of Medicine:* "Prevention of the common cold at the present is our nearest approach to a cure. The most important prevention measure would appear to be proper regulation of the humidity especially during the heating season with its distressing drying of the indoor air and the creation of an environment favorable to the cold bug."

Types of Humidifiers. There are many types of humidifiers available. They vary in price, capacity, and in principle of operation. For classification purposes, it is simpler, and more logical, to consider humidifiers in three general types:

1. Pan type
2. Atomizing type
3. Wetted element type

The pan type is the simplest type of humidifier. It has a low capacity. On a hot radiator, it may evaporate 0.07 lb (0.317 kg) of water per hour. In the warm air plenum of a heating furnace, it would evaporate approximately 1.5 lb (0.680 kg) of water per hour.* To increase the capacity, the air-to-water surfaces can be increased by placing water wicking plates in the pan. The capacity goes up as the air temperature in the furnace plenum increases. Greater capacity is also possible through the use of a steam, hot water, or electric heating element immersed in the water (Figure 20–63). A 1200-W heating element,

---

*A plenum is a pressurized chamber with one or more distributing ducts connected to it.

FIGURE 20-63   Pan-type humidifier.
(Courtesy of Research Products Corp.)

FIGURE 20-64   Atomizing-type humidifier.
(Courtesy of Research Products Corp.)

FIGURE 20-65   Wetted-element-type
humidifier. (Courtesy of Research Products
Corp.)

FIGURE 20-66   Humidifer installation.
(Courtesy of Research Products Corp.)

FIGURE 20-67   Bypass-type humidifier.
(Courtesy of Research Products Corp.)

for example, in a container with water supplied by a float valve could produce 4 lb (1.8 kg) of moisture per hour.

The second type of humidifier is the atomizing type. This device atomizes the water by throwing it from the surface of a rapidly revolving disc. It is generally a portable or console unit. However, it also can be installed so that the water particles will be directed into a ducted central system (Figure 20–64).

The third type is the wetted element type. In its simplest form, it operates in the manner of an evaporative cooler. Here air is either pushed or pulled through a wetted element or filter and evaporative cooling takes place. By increasing the airflow or by supplying additional heat, the evaporation rate of the humidifier can be increased. The heat source for evaporation can be from an increase in water temperature or an increase in air temperature (Figure 20–65).

The air for evaporation can be taken from the heated air of the furnace plenum and directed through the humidifier by the humidifier fan (Figure 20–66). It also can be drawn through the wetted element by the air pressure differential of the furnace blower system (Figure 20–67).

Furnace-mounted humidifiers, usually of the wetted element type, can be constructed so that they produce 10 lb (4.536 kg) or more of moisture per hour. Because of their high capacity, this type usually has a humidistat or control that will actuate a relay or water valve and start a fan that operates until the control is satisfied. Normally, more water is supplied to the unit than is evaporated, and this flushing action washes a large portion of the hardness salts from the evaporative element to the floor drain, eliminating them from the humidifying system.

Humidification

FIGURE 20-68 **Wiring diagram of a humidifier.**

Sequence of Operation. When the furnace is in operation, infiltration air is drawn in through the cracks around the doors, windows, and other cracks in the building construction. As this air passes through the furnace and is heated, it expands. When it expands, the relative humidity is reduced, drying out the air. The purpose of the humidifier is to replace this moisture and bring the relative humidity back within the desired comfort, or required process, level.

The heat required to evaporate the moisture in the humidifier is generally taken from the discharge air from the furnace when it is the warmest. This moisture vapor is then distributed into the leaving air, which then carries it to the conditioned space, where the relative humidity is raised.

Most of these units are powered by both 120- and 24-V electrical circuits. The 120-V circuit is commonly referred to as the power circuit, and the 24-V circuit is commonly referred to as the control circuit (Figure 20-68). The voltage for the 24-V circuit is supplied by a transformer which receives its power from the furnace blower circuit. Therefore, the humidifier will operate only when the furnace blower is in operation.

The components in the control circuit are the humidistat and the electromagnetic relay coil. The components in the power circuit are the transformer primary coil, the water valve coil, and the humidifier fan motor. The purpose of the fan motor is to pull the heated air into the humidifier, where it will evaporate the water and push the moisture-laden air into the furnace discharge air plenum. In some installations the humidifier fan motor and the water valve coil are powered from a separate 120-V circuit, as shown in the diagram.

When the furnace fan motor is energized, 120 V is supplied to the primary of the humidifier transformer. Then 24 V is supplied to the control circuit. If the humidistat is demanding humidification, the

FIGURE 20-69 **Airflow through a humidifier.**

electromagnetic relay coil is energized, closing the relay contacts which complete the 120-V electric circuit to the water valve coil and the humidifier fan motor. The water flows into the humidifier and is evaporated into the air passing through the furnace (Figure 20-69). When either the heating thermostat or the humidistat is satisfied, the humidifier will also stop operating. The humidifier is ready for another operating cycle on demand from either the humidistat or the thermostat.

☐ Basic Heating Controls ☐

Certain controls are required in order for a heating system to provide safe, economical, and automatic heating for a building. Other more sophisticated controls of the system may be added to these basic controls.

Thermostat. The purpose of a thermostat is to signal the equipment when to operate. During the heating season, when the temperature inside the conditioned space drops below the thermostat set point, an electric circuit will be completed and the furnace

FIGURE 20-70 Single-stage thermostat. (Courtesy of Honeywell Inc.)

FIGURE 20-71 Double-stage heating, double-stage cooling thermostat. (Courtesy of Honeywell Inc.)

will provide the required amount of heat. When sufficient heat has been added to the building, the thermostat will interrupt the electric circuit and the heating equipment will stop the heating process.

Thermostats are available to provide almost any operation desired. The majority of thermostats are single-stage heating and single-stage cooling (Figure 20-70). However, some equipment operation will demand that other staging be used, such as double-stage heating and double-stage cooling (Figure 20-71). These thermostats are capable of controlling heating equipment in two stages, or inputs. The cooling can also be controlled in two parts. When less conditioning is used, only that much of the equipment is operated. However, if the lower capacity is not capable of maintaining the desired temperature the thermostat will cause the equipment to operate at full capacity.

Heat Anticipator. Thermostats are provided with heat anticipators that cause the thermostat to stop the heating equipment just a little bit before the space is actually up to temperature. This is to prevent over-

FIGURE 20-72 Main gas valve. (Courtesy of White-Rodgers Division, Emerson Electric Co.)

heating of the space. The heat anticipator is a resistor wired in series with the temperature control circuit and provides heat inside the thermostat. This heat is what causes the thermostat to signal the heating equipment early. The heat anticipator is usually adjustable and should be set to match the current draw of the temperature control circuit.

Main Gas Valve. The purpose of the main gas valve is to allow gas to flow to the main burners or stop this flow on demand from the thermostat. The valve is mounted physically in the gas line to the heating equipment. Electrically it is placed in the temperature control circuit (Figure 20-72). These valves usually contain all the elements of a gas manifold. These components are solenoid gas stop valve, pilot safety control, gas pressure regulator, pilot burner control, and main line gas shutoff. It is generally referred to as the combination gas control.

The main parts are the valve body, valve seat, valve disc, and the valve stem. Gas is allowed to flow to the burners when the valve disc is lifted off the valve seat. Gas valves are classified by the type of power that is used to open and close them. They are classified as follows: solenoid valve, magnetic diaphragm valve, bimetal diaphragm valve, heat motor valve, and automatic spark ignition valve.

Solenoid Valve. The solenoid valve is generally considered an electrically operated stop valve. It is normally closed and is the most popular type used on heating systems. When the electrical circuit is completed to the solenoid coil, the valve opens. When the circuit is interrupted, the valve closes. The noise made by the closing of the solenoid valve is lessened by the addition of a soft seat made of rubber or plastic. This, in turn, is given the name of soft-seated solenoid. A spring is added above the valve to slow down the opening and to help stop the clacking noise made when the valve opens. An improvement on this valve is the oil-filled solenoid valve. The oil reduces friction and the operating noise is almost completely eliminated.

Magnetic Diaphragm Valve. The magnetic diaphragm valve makes use of the available gas pressure as the major opening and closing force. On demand from the thermostat, the control circuit energizes the valve coil. When the valve is energized a lever is magnetically pulled, closing the gas supply passage. At the same time the gas pressure above the diaphragm is vented out through the bleed orifice and to the pilot burner, where it is immediately burned. The gas pressure below the diaphragm is now greater than that above the diaphragm, and the weight and disc are lifted off their seat, opening the gas valve. When the control circuit is deenergized, the bleed gas is stopped, allowing the gas pressure above and below the diaphragm to equalize. The weight of the valve disc closes the valve.

Bimetal Diaphragm Valve. The bimetal diaphragm valve and the magnetic diaphragm valve are identical in operation except that a bimetal is used to open and close the bleed orifice. The bimetal diaphragm valve uses the principle of dissimilar metals, which warp when heated to open and close the bleed orifice. An electric heater is placed either on or near the bimetal element. When the control circuit is completed to the heating coil and sufficient heat is added to the bimetal, the bimetal will warp. The bimetal opens the bleed orifice and the gas pressure is allowed to vent to the pilot burner. This reduces the pressure above the diaphragm and allows the greater pressure below the diaphragm to force the valve open. When the control circuit is deenergized, the bimetal cools, returns to its original shape, and closes the bleed orifice. This allows gas pressure to build up above the diaphragm and the seat. The weight of the diaphragm and seat closes the valve.

Heat Motor Valve. The heat motor valve uses the heat given off by an electrical heating coil to move an expandable rod to open the valve. The movement of this heated rod provides the force necessary to operate the valve mechanism. When the thermostat demands heat, the control circuit to the heating coil is energized and the coil is heated. The heat given off by the coil causes the expandable rod to expand and the valve to snap open. When the thermostat is satisfied, the control circuit is opened and the coil is allowed to cool. When the rod has cooled sufficiently, it shrinks in length and the valve snaps closed.

Note: Most of these valves have a certain amount of time delay before they open or close. This is to provide a smoother ignition and extinguishing of the flame.

Automatic Spark Ignition Valve. The automatic spark ignition gas valve is equipped with a redundant pilot solenoid, main gas regulator, and the necessary wiring connections for the make and model valve being used. They use electrically operated solenoid valves controlled by a room thermostat or other temperature control (Figure 20–73). The main gas valve is controlled by the proper flame sensor, which plugs into the valve. Both the pilot and the main gas valve are controlled through the proper relight-timer control, which deenergizes both valve coils if a pilot flame is not established within the lockout time period. These valves should be installed only on systems that are equipped with the proper ignition systems as recommended by the manufacturer for proper operation.

Main Valve Sequence of Operation. On demand from the thermostat, the solenoid coil is energized from the 24-V circuit, and the valve lever opens the cycling valve (Figure 20–74). The inlet gas now flows through the control orifice past the cycling valve. At this point, gas flows in two directions, as follows:

1. Part of the flow is to the back of the diaphragm by means of internal passageways. The resulting increase in pressure pushes the main valve to the open position, compressing the diaphragm springs lightly.
2. Part of the flow is through the seat of the master regulator into the valve outlet by means of internal passageways. This causes the master regulator to begin its function.

The gas valve remains in this position and the master regulator continues to regulate until the relay coil is deenergized, at which time the cycling valve seals off. As the cycling valve closes, the regulator spring causes the seat of the master regulator to close

AIR CONDITIONING (HEATING)

off. The function of the bypass orifice is to permit gas in the passageways to escape into the outlet of the valve, thereby causing the main gas valve to close.

Redundant Gas Valve. The purpose of these valves is safety. They contain two independently operated valves to the main gas burner. If one valve should fail to close, the other will close and shut off the gas flow to the main burner (Figure 20–75). This type of valve is used on gas-fired heaters and boilers, with or without intermittent pilot ignition, in place of the regular gas valve.

The majority of problems with these valves are electrical rather than mechanical. The redundant gas valve has two electrically controlled internal shut-off valves in series with each other. Before gas can flow to the main burner, both of these valves must be open. However, only one needs to close to stop the flow of gas.

Safety, therefore, is accomplished. It is almost impossible for both of these individual valves to stick in the open position when either the thermostat or the limit control opens. The standard main gas valve does not provide this safety.

One make of redundant gas valve uses an instant-acting solenoid for controlling the flow of gas at the valve outlet and a time delay valve on the inlet to the valve. The time delay on the bimetal valve is

FIGURE 20–74 Main gas valve cutaway. (Courtesy of White-Rodgers Division, Emerson Electric Co.)

FIGURE 20-75   Redundant gas valve.

FIGURE 20-76   Two 24-V coils in one valve.

about 10 sec, thus providing a time delay on start-up of the main burner.

The redundant gas valve used on furnaces with standing pilots is similar to the standard type of gas valve, with the exception that an internal heat motor valve is included. There are two 24-V coils used in these valves (Figure 20-76). One coil is shown as a solenoid valve and the other is shown as a resistor. They are in parallel electrically, and both are in series with the limit switch and the thermostat. The resistor is for the heat motor that operates the second or redundant gas valve.

These valves have a third internal valve that is the 100% safety shutoff valve. It is energized with the electricity (millivolts) produced by the thermocouple.

Should the pilot flame not heat the thermocouple sufficiently, the 100% safety shutoff will close, stopping the flow of gas to the main burner.

When the contacts in either the thermostat or the limit switch open, the solenoid instantly closes. The heat motor valve, however, takes a few seconds to cool down and close the valve.

When the standing pilot valve is to be treated, use the following steps:

1. With the pilot burning, the open-circuit thermocouple test should show a minimum of 21 mV.
2. When the thermostat is calling for heat and the limit switch is closed, 24 V of electricity should be at both the solenoid coil and the heating coil of the heat motor.

If proper voltage is indicated in both of these tests, the problem is not in the electric circuit. At this point check for gas at the valve inlet. If gas is present, the valve has a mechanical problem and should be replaced.

When an intermittent pilot ignition system is used, each time the thermostat demands heat, pilot gas will be supplied and lighted electrically. The redundant gas valve will then open and admit gas to the main burner. If the pilot gas should not be lighted within the specified amount of time, the system will shut down and will lock out on safety.

The valve used on this type of system has two solenoid valves wound on the same core (Figure 20-77). When the thermostat demands heat, an electric circuit is completed to both of the coils and to the pilot gas ignitor. The pilot gas ignitor will direct a high-voltage spark across the pilot, igniting the pilot gas.

The coil in the pilot safety circuit, which consists of a set of normally open and a set of normally closed contacts, is then energized by the millivolts produced by the heated thermocouple.

When the pilot flame has been established, the millivoltage generated by the thermocouple will cause the contacts in the pilot safety to change position. At this point, the solenoid 1 coil and the pilot gas ignitor is deenergized. The coil in the solenoid 2 coil will remain energized until the thermostat is satisfied. In this type of valve, both solenoids must be energized for the valve to open, but only one is required to keep the valve open.

As the contacts in the pilot safety change positions, an electric circuit to the main gas valve is completed. The main gas valve is operated by the heat

FIGURE 20-77 Two solenoid coils on one core.

motor valve. It will be approximately 30 sec before gas is admitted to the main burner.

For gas to flow to the main burner, both solenoids must be energized when the thermostat demands heat. The pilot flame must be established before the main gas valve will open.

The two solenoids being wound on the same core prevents cycling of the main burner because of sudden opening or closing of the thermostat contacts or a momentary electric power failure. Should either of these conditions occur, solenoid 2 will close and will not be energized again until the thermocouple is cooled down and resets the circuit.

When troubleshooting this valve for electrical troubles, use the following steps:

1. With the pilot lit, there should be 24 V to the main gas valve heater coil.
2. There should be 24 V to the solenoid 2 coil.
3. When these conditions are met and still no gas is admitted to the main burner, the valve is defective and must be replaced.

Pilot Safety. The purpose of a pilot safety device is to interrupt the electric circuit, or interrupt both the gas supply and the electric control circuit when the pilot burner will not safely light the main burner. These are safety devices and should not be bypassed to allow the furnace to operate. They are termed 90% safety and 100% safety. A 90% safe system will allow pilot gas to flow at all times unless it is manually shut off, while a 100% safe system will not allow any gas to flow when the pilot is not safe

FIGURE 20-78 One-hundred percent pilot safety shutoff. (Courtesy of Honeywell Inc.)

(Figure 20–78). Most pilot safety devices are incorporated in the main gas valve and are 100% safe systems. When LP gas is being used in heating equipment, the 100% safe system must be used. However, 90% safe systems may be used on natural gas units.

Thermocouple. The thermocouple provides the electric power necessary to operate the pilot safety device. These devices are made of dissimilar metals so that when heated one metal will repel electrons while the other will attract them. Thermocouples are mounted in the pilot flame on one end and connected to the pilot safety device on the other end. When the pilot gas is lit and the flame encases the thermo-

Basic Heating Controls

FIGURE 20–79 **Thermocouple principle.**

couple, a small electric current of approximately 25 to 35 mV is generated (Figure 20–79).

Thermocouple Testing. There are three tests that can be made on a thermocouple to determine if it is producing enough current: (1) open-circuit voltage, (2) closed-circuit voltage, and (3) dropout voltage.

Open-Circuit Voltage. The pilot gas must be burning while making this test. Remove the end of the thermocouple connected to the pilot safety device. Check the voltage output at this end by touching one lead of the millivolt meter to one ter-

minal while touching the other meter lead to the other terminal (Figure 20–80).

The output of the thermocouple is dc voltage, and the meter leads may need to be reversed in order to obtain a reading. Remember that the pilot must be burning during this test. The output voltage should be between 30 and 35 mV. If it is lower than this, replace the thermocouple. The thermocouple must be reconnected to the pilot safety device for normal operation.

Closed-Circuit Voltage. To make this test, an adapter is required between the thermocouple and the pilot safety device. This adapter has points for checking the voltage on the inner contact of the thermocouple (Figure 20–81).

The output voltage is checked while the pilot gas is lit and the thermocouple is hot. The output should be a minimum of 17 mV. If the output is lower than this, replace the thermocouple. When this test is completed, remove the adapter and reconnect the thermocouple to the pilot safety device.

FIGURE 20–80 **Open-circuit thermocouple test.**

FIGURE 20–81 **Closed-circuit thermocouple test.**

Dropout Voltage. The dropout voltage is measured with an adapter between the thermocouple and the pilot safety device (Figure 20–81). Turn off the pilot gas and observe the voltage. When the voltage has dropped enough for the pilot safety device to close off the gas to the main burner, a low dull thud can be heard. The voltage at this point is the dropout voltage and is usually about 5 mV. Remove the adapter and reconnect the thermocouple to the pilot safety device, relight the pilot gas and put the unit back in operation. There is no recommended point for the dropout voltage, but it should occur within approximately 3 min, or preferably less.

Limit Control. The limit control is a safety device that interrupts the temperature control circuit if temperatures inside the furnace heat exchanger should reach 200 °F (93 °C). It has one set of normally closed contacts that are operated by a bimetal sensing element. The element is placed inside the furnace heat exchanger. This element may be of the helical type, blade type (Figure 20–82) or disc type (Figure 20–83). These controls are not usually adjustable and are set to open the contacts at approximately 200 °F and to close the contacts at approximately 170 °F (77 °C). These controls are sometimes manufactured in the same housing as the fan control, but their function is completely different.

Fan Control. The purpose of the fan control is to stop and start the fan in response to the temperature inside the furnace heat exchanger. In operation, when the heat exchanger temperature reaches approximately 150 °F (66 °C), the normally open contacts close and start the fan motor. When the temperature drops to approximately 100 °F (37.8 °C), the contacts open and stop the fan motor. This is done to prevent the circulation of cold air within the conditioned space. These controls also are operated by a bimetal sensing element (Figure 20–84). In some instances it is desirable to have a fixed time before the fan will come on. This is accomplished by use of an electrically operated fan control (Figure 20–85). The heater element in this control receives electric power from the control circuit at the same time the gas valve is energized. After a preset time, the fan is started by the contacts closing (Figure 20–86).

Fire-Stat. The fire-stat is a safety device mounted in the fan section of the furnace that senses the return air temperature. It has a set of normally closed contacts that open and stop fan operation when the re-

FIGURE 20–82   Blade-type limit control. (Courtesy of White-Rodgers Division, Emerson Electric Co.)

FIGURE 20–83   Disc-type limit control. (Courtesy of White-Rodgers Division, Emerson Electric Co.)

FIGURE 20–84   Fan control. (Courtesy of Honeywell Inc.)

FIGURE 20-85 Fan timer switch. (Courtesy of Honeywell Inc.)

FIGURE 20-86 Fan timer switch wiring diagram.

FIGURE 20-88 24-V: top, foot mounted; bottom, plate mounted. (Courtesy of Honeywell Inc.)

FIGURE 20-87 Fire-stat wiring diagram.

turn air temperature reaches approximately 160 °F (71 °C) (Figure 20-87). This is done to prevent the fan from agitating a fire in the building.

Transformer. The transformer is a device used to provide the low voltage for the temperature control system. Line voltage is connected to the primary side and 24 V is provided by the secondary side of the transformer (Figure 20-88). Transformers are rated by their voltampere (VA) capacity. Usually, a 40-VA transformer is used on heating and cooling applications.

Boiler Pressure Control. This device is used to control the pressure inside a steam-heating boiler. A bellows is connected to the inside of the boiler. When the steam pressure reaches the set point, the main

FIGURE 20–89 **Protectorelay control. (Courtesy of Honeywell Inc.)**

gas valve is closed and steam is no longer generated. These controls also may be used as boiler pressure limit controls. The control setting is merely set at a higher pressure.

Immersion Control. This is a temperature control used inside a hot water boiler. The sensing element is inserted directly into the hot water. The action of the main gas valve is controlled by this device in response to water temperature. This control may also be used as a high limit control.

Protectorelay Control. This control is used for the automatic recycling of an intermittent ignition oil burner heating system (Figure 20–89). The control senses the flue gas temperature of an oil burner system. If the flue gas temperature has not reached the designated temperature within a predetermined period of time, the control will stop the oil burner. After a given period of time the control will recycle the oil burner in an attempt to provide heat to the building. A schematic of a protectorelay control is shown in Figure 20–90.

Basic Warm Air System Wiring Diagram. There are three different circuits and voltages in a modern 24-V temperature control system (Figure 20–91): (1) fan or circulator circuit, (2) pilot safety circuit, and (3) temperature control circuit. The fan or circulator circuit is line voltage. The pilot safety circuit is millivolts. The temperature control circuit is 24 V (low voltage).

FIGURE 20–90 **Wiring diagram for protectorelay. (Courtesy of Honeywell Inc.)**

FIGURE 20-91 **Basic warm air system wiring diagram.**

FIGURE 20-92 **Glow coil and spark ignition.**

FIGURE 20-93 **Direct spark ignition system burner mounting.**

Warm Air Furnace Sequence of Operation. In operation, when the thermostat demands heat, the main gas valve opens and allows gas to flow to the main burners. When the temperature inside the furnace heat exchanger reaches approximately 150 °F (66 °C), the fan control starts the fan motor, which circulates the air inside the conditioned space. The furnace is now in the heating cycle. When the

thermostat is satisfied, the electric circuit to the main gas valve is interrupted and the valve closes. The flame inside the furnace is extinguished and the temperature inside the furnace starts to fall. When this temperature falls to approximately 100 °F (37.8 °C), the fan control stops the fan motor and the air circulation stops. The furnace now awaits a command from the thermostat to start a heating cycle.

Retrofitting for Energy Conservation. Almost all gas-burning equipment in use today is equipped with standing (constant-burning) pilots and with standard (nondamper) venting systems. It has been estimated that turning the pilot off during the off cycles can save from 3 to 5% of the fuel used. The installation of a vent damper can save from 15 to 20%, while a temperature setback thermostat can save approximately 7 to 8%. Thus a 25 to 33% savings in operation costs can be achieved when all three systems are installed on a gas furnace.

Pilot. The standing pilot is replaced with one of several devices in use to convert them to intermittent-type pilots. The two most popular pilot relight systems are the glow coil and the spark ignition unit (Figure 20-92). An alternate to these is the direct spark ignition system. These systems light the main burner directly without using a pilot (Figure 20-93).

Caution: Some of these retrofit units are not to be used on LP gas. Be sure to check the manufacturers' suggestions before making the changes.

The retrofit spark ignition system requires that some specially designed components be used to replace those in the existing control system. These control systems light the pilot flame by providing an electric spark to the pilot burner. The pilot gas is ignited and burns during each running cycle (in-

FIGURE 20-96 **Thermopilot relay in deenergized position.**

FIGURE 20-97 **Thermopilot relay in energized position.**

FIGURE 20-94 **Wiring diagram for retrofit spark ignition system using automatic vant damper. (Courtesy of Johnson Controls, Inc.)**

FIGURE 20-95 **Glow coil pilot ignitor.**

termittent pilot). When the pilot flame is proven, the spark ignition is discontinued.

The main gas valve is permitted to open only when the pilot flame is proven. If a loss of pilot flame occurs, the main gas valve is deenergized and the spark ignition is reactivated almost immediately. Both the main burner and the pilot burner are off during the off cycle. These units must be correctly wired to provide the desired functions (Figure 20-94).

There are models available that provide a slow opening pressure regulator to gradually increase the flow of main burner gas after the pilot flame has been proven.

The glow coil pilot ignitors use a coil of resistance wire placed by the pilot gas stream. When electricity is applied to the glow coil, it gets red hot and ignites the pilot gas (Figure 20-95). A single-pole double-throw thermopilot relay is used in conjunction with the glow coil (Figure 20-96). When the pilot is not burning, the relay contacts are in a position to supply 2.5 V electric current to the glow coil. After the pilot is ignited and the thermocouple heated sufficiently, the movable contact moves to complete the 24 V electric current circuit to the temperature control circuit (Figure 20-97). The complete control system must be properly wired to provide the desired results (Figure 20-98).

The direct spark ignition system uses an electric spark to ignite the main burner flame directly without the use of a pilot (Figure 20-99). The electric spark is directed across the burner port and the gas is ignited when it leaves the burner port. See Figure 20-100 for a suggested wiring diagram.

Automatic Vent Damper. The purpose of an automatic vent damper is to keep heat from going up the chimney during the heating unit off cycle and being wasted (Figure 20-101). In operation, upon a call from the thermostat for heat, the damper operator rotates the damper blade to the open position (Figure 20-102). It takes about 15 sec to completely open. When the damper has reached the fully open position, the ignition control is energized through the damper operator and lights the pilot

FIGURE 20-98 24-V temperature control circuit with automatic relight.

FIGURE 20-99 Direct spark ignition system burner mounting.

flame. The pilot then lights the main burner. The products of combustion are then vented in the traditional manner.

When the thermostat is satisfied, the ignition control is shut off. The damper operator then rotates the damper blade to the closed position in about 15 sec. This is a sufficient amount of time for the products of combustion to be properly vented (Figure 20-103). For a typical wiring diagram see Figure 20-104. Be sure to follow the vent damper manufacturer's recommendations during installation.

Temperature Setback Thermostat. These thermostats have two temperature settings: one for use when the heated space is normally occupied and one for use during periods when occupant activity is at a minimum or when there are no occupants in the building (Figure 20-105). These two settings are set

FIGURE 20-100 Wiring diagram for DSI system.

FIGURE 20-101   Automatic vent damper. (Courtesy of Johnson Controls, Inc.)

FIGURE 20-102   Automatic vent damper in open position. (Courtesy of Johnson Controls, Inc.)

FIGURE 20-103   Automatic vent damper in closed position. (Courtesy of Johnson Controls, Inc.)

FIGURE 20-104   Wiring diagram for automatic vent damper and a retrofit spark ignition system. (Courtesy of Johnson Controls, Inc.)

FIGURE 20-105   Night setback thermostat (Chronotherm). (Courtesy of Honeywell Inc.)

to be used during the desired time period. For example, the lower setting can be set to take over the system at 11:00 P.M. or bedtime. The higher setting can be set to take over about one hour before the occupants wish to get out of bed. This temperature will be maintained until bedtime. However, if all the occupants are gone every day, the lower setting can be adjusted to operate the equipment during the unoccupied period. This provides more savings.

☐ Solid-State Electronic Ignitors ☐

The purpose of electronic ignitors are to light or relight, automatically and instantly, either the pilot gas or the main burner gas on demand from the room thermostat. The manufacturer of the equipment will choose the type that is best suited for their particular equipment.

Basically, there are two types of electronic ignition systems: (1) spark, and (2) hot surface ignition (HSI). The spark-type ignition systems can be further divided into: (1) those that light the pilot gas with a spark across the pilot burner gas opening, (2) those that light the main burner gas with a spark across an orifice on the main burner, and (3) those that use a spark plug to light the main burner gas. These sparks jump from one electrode to another across the gas opening. The high temperature of the spark causes the gas to ignite and perform the desired function. The spark ignition types are generally referred to as either direct spark ignition devices (DSI) or indirect spark ignition devices (IID). The hot surface ignition systems use a surface that is heated with a resistance element to a temperature hot enough to light the gas. HSI systems are also sometimes referred to as silicon carbide ignition systems.

When the intermittent ignition device is used on pilot light and relight installations, there are certain changes that must be met before the system will function as required. In these types of systems, solid-state circuitry and flame sensors are used to replace the thermocouple and pilot safety device. The solid-state device reacts so much faster than the thermocouple system that it is a much more desirable type of installation, at least from the safety standpoint.

Principles of Operation. In the thermocouple-type safety devices, heat is used to increase the temperature of the thermocouple, which then generates a small amount of dc electric current. However, when the solid-state electronic ignition systems are used, heat is not used except in the HSI types of systems. The DSI and the IID systems use a principle known as *flame conduction* or *flame rectification*.

Flame Conduction. How a flame can conduct electricity must be known before the principle of flame rectification can be understood. When a gas is burning with the proper amount of oxygen there are three flames present (Figure 20–106). There are three zones or cones in a flame:

1. *Inner cone:* a gas-rich area that will not burn because there is not enough oxygen to support combustion.
2. *Intermediate cone:* the blue envelope that surrounds the inner cone. In this area where the

FIGURE 20–106 Structure of a gas flame.

flame is burning properly and is ionized. This type of flame will conduct electricity.
3. *Outer cone:* the blue envelope that surrounds the other two cones. It is generally referred to as the outer cone. Because this flame is exposed to the atmosphere, it contains an excessive amount of oxygen and does not burn properly.

Cone 2, the intermediate cone, is the most important in flame rectification because this is where proper combustion occurs. Therefore, this is the best area for sensing whether or not the flame is burning.

In actuality a flame is merely a series of very small explosions occurring in a controlled atmosphere. Because of the ionization that occurs during complete combustion, electricity can be conducted through this area with little difficulty. Therefore, this is where the flame sensor of the electronic ignition system is located (Figure 20–107).

The flame in this instance can be thought of as a switch or a diode that is located between the tip of the pilot burner and the intermediate cone of the flame being sensed. In operation, the flame must be in contact with both the pilot burner tip and the flame sensor. If the flame loses contact with either of them for only a second or two, the system will shut down and it must be reset before operation can resume. In Figure 20–100 the flame is used to conduct an ac signal. Both of the flame probes have approximately the same amount of area exposed to the flame, and an electric current flows through the flame. This type of system is not good enough for use as a safety signal because it does not identify the current that is conducted by the flame. Because of this, a shorted condition could cause the system to indicate that a flame is present when it is not. The difference between these two systems is what is known as flame rectification (Figure 20–108).

Flame Rectification. In the flame rectification process, the flame probes are used much the same as with the previous system but with one major difference. The area of one of the probes is larger than the other.

FIGURE 20-107 Principle of flame rectification.

If we refer to Figure 20-108, it can be seen that the flame is still used to conduct the ac current. Both of the probes are still in contact with the flame, but the probe having the greater area will conduct more electrons than the other one. The larger probe therefore takes on a negative electrical charge. It should also be noticed that the ac sine wave has not changed, but the negative part of the sine wave is not shown, making it a pulsating dc voltage. This is what is known as flame rectification.

When this principle is used, a pilot and a flame sensor have replaced the two probes previously used (Figure 20-109). After the flame has ignited, a dc microampere electric current flows through the flame in a direction from the flame to the sensor (the positive probe) and then to the pilot tip (the negative probe). The pilot tip now becomes the negative probe and completes the electrical circuit to ground. In this manner, the sensing circuit passes a dc current to energize a relay, which in turn energizes the gas valve.

Sequence of Operation. Following is the sequence of operation of electronic ignition systems.

1. On a call for heat, a high-voltage transformer in the control circuit and the pilot gas valve are energized at the same time.
2. A spark is generated across the gap over the gas opening.
3. When the gas is ignited, the flame is sensed by the sensor and the ignition control is deenergized stopping the spark. When the spark is stopped, the main gas valve is automatically energized.

FIGURE 20-108 (a) Normal ac circuit; (b) flame out-switch is open, no current is flowing ; (c) pilot gas lit flame is sensed, current is flowing.

FIGURE 20–109 Electrical rectification by use of a pilot flame.

pressure should never fall below the recommended manifold gas pressure.

When natural gas is used the inlet pressure should never go above 10.5 in. (266.7 mm) wc. When LP gas is used, the supply gas pressure should never be higher than 14 in. (355.6 mm) wc.

3. *Temperature.* The electronic ignition control module should never be exposed to temperatures higher than 150 °F (66 °C) or lower than −40 °F (−40 °C).

4. *Pilot applications.* The two most important parts of the IID system are the pilot and the flame sensor. The pilot flame must at all times be in contact with both the pilot burner and the flame sensor. A microammeter is used to determine the amount of current flowing through the flame. The amount of current that should be flowing depends on the type and the manufacturer of the control module. If the current flow falls below the minimum required for proper operation, the system will automatically shut down. In some installations this may be as little as 0.8 sec from flame failure. If the pilot flame wavers or blows off the pilot burner, the flame could be lost or the burner caused to cycle rapidly. In most cases the burner will not even ignite. The position of the flame sensor is very important. It should be placed so that the maximum microampere current will flow through the system.

When 100% lockout models are used, the complete system is shut down if the flame is not sensed when the trial-for-flame period is over.

4. The pilot lights the main burner gas and the system warms the building.
5. When the room thermostat is satisfied it opens the control circuit and deenergizes the main gas valve, stopping the flame.

Application Guidelines. Every IID system is affected by the following factors:

1. *Voltage.* There are certain ranges within which the voltage should be supplied to electronic ignition systems. For 120-V systems the range is from 102 to 132 V ac. For 24-V systems the range should be from 21 to 26.5 V ac.
2. *Gas pressure.* The pressure of the gas supplied to the equipment should be at least 1 in. (25.4 mm) water column (wc) more than the recommended manifold gas pressure. The supply gas

DSI System Operation. The following is a description of the Fenwal Series 05-14 Proof of Flame Direct Spark Ignition System (Figure 20–110).

Warning: This ignition system complies with the Z21.20 ANSI Standard for gas burner automatic ignition systems when installed with Fenwal-specified electrodes and high-voltage lead wires. The replacement of components or any alterations to this control board will invalidate the Fenwal warranty, void agency certifications, and can create hazardous conditions that may cause property damage, personal injury, or even death. This is why only qualified personnel should work on these systems.

Principle of Operation. On a call for heat, input power is applied to the control board, sparking is then initiated, and the gas valve is energized: The sparking continues with the gas valve electrically energized for a specified "trial-for-ignition" period. The timing period may vary from model to model. Be sure to check the specifications for the model being used.

AIR CONDITIONING (HEATING)

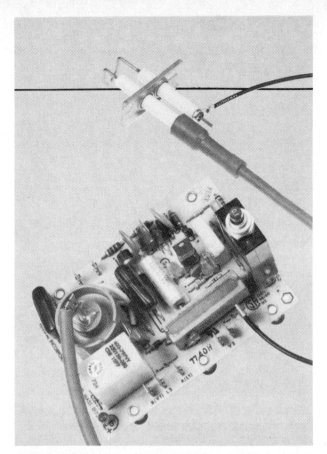

FIGURE 20-110 Fenwal Series 05-14 proof of flame direct spark ignition system. (Courtesy of Fenwal Inc.)

In normal operation, as soon as the flame is established and proven by the flame-sensing circuitry, sparking will cease immediately and the system will remain on, monitoring the flame until the end of the heating cycle.

Should a flame-out occur during the heating cycle, the control module will reactivate the spark within 0.8 sec, to provide for reignition. The flame will either be reestablished within the trial-for-ignition time, or the system will lock out in the normal manner. Should a lockout occur, the control module must be reset (manual or thermostatic) before the unit can make another cycle. This reset action may be accomplished at either the thermostat or at the control unit circuit breaker, depending on the model number of the unit being used. If the ignitor is a manual reset model, the circuit breaker will trip within a few seconds after the sparking stops and the gas valve is deenergized. If the ignitor is designed for thermostatic reset, the circuit breaker will trip only after several attempts at reignition.

Thermostatic Reset (Line Interruption). Should the Series 05-14 with the thermostatic reset go into a lockout due to a loss of flame, the thermostat should momentarily be turned below room temperature and then returned to a setting above room temperature, or the system switch on the thermostat turned off and then back on to start recycling the system. Lowering the thermostat below room temperature and then back above room temperature will normally recycle the system. If several attempts at using the thermostatic reset do not result in ignition, a manual reset button on the ignitor module will actuate, preventing further recycling of the unit. The manual reset must then be reset to resume furnace operation, or checkout procedures.

Manual Reset. The manual reset feature, available on models designated 05-140X01-11X, requires manual resetting of the system in the event of a lockout. This manual reset button can be located either on the ignitor board or at some remote location in a system control panel. (Consult the control manufacturer for remote mounting agency approvals.) Should the flame not be established during the ''trial for ignition'' period, the system will lock out. Should flame lockout occur at any time during the running cycle, the system will retry for ignition, one time, and then lock out. The manual reset button will then trip within several seconds. After a minimum of 5 sec, depress the reset button to reset the control module.

Flame Sensing. The Series 05-14 ignitor is designed to operate with a single electrode assembly, which provides the spark and senses the presence of flame (Figures 20-111 and 20-112). Remote sensing models, however, allow for the flame-sensing function to be separated from the spark function and located remotely (Figures 20-113 and 20-114). Using the remote sensing model ensures flame propagation across an entire burner rack.

Electrode Application. The application of the electrode is designed by the equipment manufacturer to meet the requirements of their particular piece of equipment. The following steps are only to provide you with the basic procedure used by the original equipment manufacturer's burner design engineer in designing a properly operating ignition system and to help you in understanding how the system is designed and how it functions. These procedures should not be used by field service and installation technicians because to do so may cause a hazardous situation. These applications should be changed only at the equipment manufacturer's instructions.

Solid-State Electronic Ignitors

FIGURE 20-111 Fenwal local sensing 120-V ignitor wiring diagram. (Courtesy of Fenwal Inc.)

FIGURE 20-112 Fenwall local sensing isolated contact ignitor wiring diagram. (Courtesy of Fenwal Inc.)

FIGURE 20-113 Fenwal remote sensing 120-V ignitor wiring system. (Courtesy of Fenwal Inc.)

1. Manually ignite the burner flame to observe the flame characteristics.
2. Temporarily mount the electrode directly over the burner ports in the blue part of the flame pattern. The electrode tip should be ¼ to ½ in. (6.35 to 12.70 mm) into the flame with the ⅛-in. (3.18-mm) spark gap directly over a burner port (Figure 20-115).

FIGURE 20-114 Fenwal remote sensing isolated contact wiring diagram. (Courtesy of Fenwal Inc.)

FIGURE 20-115 Proper location of ignitor electrodes. (Courtesy of Fenwal Inc.)

3. Connect the electrode lead wires to the control module. See the wiring diagrams in Figures 20-111 to 20-115. Make certain that the remaining system wiring is complete.
4. With the gas supply manually shut off, set the thermostat to demand heat to operate the control module. Make certain that steady sparking occurs at the electrode tips. Sparking should not occur elsewhere. The system should shut down after the prescribed numbers of trial for ignition have been attempted.
5. Set the thermostat below room temperature to deenergize the system, or depress the module reset button. Manually turn the gas supply back on.
6. Set the thermostat to demand heat. Check to see that sparking occurs immediately and that the flame is established almost instantly. Once the flame is established, sparking will cease.
7. If ignition is achieved almost instantly and sparking terminates immediately after the flame is established, the system operation should be functionally sound. There are many such tests as these done at the factory in designing a properly operating ignition system. However, should these requirements not be met, consult the "System Checks" section later in the chapter for repair recommendations.

AIR CONDITIONING (HEATING)

Note: Electrode assemblies are precision components and should not be adjusted or disassembled. Electrodes should have a gap spacing of 0.125 ± 0.031 in. (3.18 ± 0.79 mm). If this spacing is not correct, return the electrode assembly to the factory for replacement. Electrodes are not field adjustable.

System Checks. The following is a description of the system checks that should be completed after installation of the furnace and during each seasonal startup procedure:

1. *Preliminary system checks.* It is desirable to check the system after the original installation of the furnace in the building and before the gas supply is turned on. Be sure that the input electrical power is polarized as shown on the wiring diagram and the complete installation is electrically grounded. The burner chassis, electrode bracket, and the burner must have a common ground. Connect as shown in the proper wiring diagram in Figures 20–111 to 20–114.
2. *Polarity check*
   (a) With a voltmeter or indicator light, measure from terminal L_1 to the burner chassis, electrode bracket, or the burner ground. The voltage should be a nominal 120 V.
   (b) Measured from L_2 to ground should be 0 V.
   (c) If 1 and 2 above are opposite, reverse the electrical input leads to ensure correct polarity.

Initial Operation. Use the following steps for the initial operation of the 05-14 Fenwal ignition system.

1. Check the installation, mounting, and the electrode gap to make sure that they conform to the equipment manufacturer's specifications. These specifications are listed in the brochure shipped with the furnace.
2. With the gas supply manually shut off, apply electrical power to the system by either closing the electrical power switch to the furnace or by setting the thermostat to call for heat.
3. Make sure that sparking occurs at the electrode tips across the ⅛-in. (3.18-mm) gap during the trial-for-ignition period specified for the control module being used. Check the equipment manufacturer's specifications to be sure that the ignition system is properly installed and adjusted.
4. The Fenwal 05-142 thermostatic reset models will lock out after the trial-for-ignition period. The 05-140 manual reset models will lock out after the trial-for-ignition period and a few seconds later will trip the circuit breaker on the control module.
5. Manually open the gas supply line and set the room thermostat above the room temperature or depress the manual reset button to recycle the unit.
6. Check to see that ignition has been accomplished within the trial-for-ignition period. Sparking will cease immediately after ignition.
7. If the system ignites but fails to hold-in, check the input voltage and polarity and check to see that the complete piece of equipment is properly grounded through the equipment ground (green wire). If the system ignites but drops out and the input voltage is correct, check the amount of flame current (5 μA minimum). Check the burner and other points of ground for proper ground.
8. For a multiple-furnace installation, make certain that all of the furnaces are powered by a common supply voltage and all are correctly polarized.

Safety Checks. An annual inspection and test, conducted prior to the start of the heating season *by qualified gas service personnel,* is required to verify that the automatic gas ignition system, all associated components (i.e., fan and limit switches), and the appliance combustion chamber components operate according to the furnace manufacturer's specifications. Any components that have been physically damaged or that otherwise fail to perform their function according to specifications must be replaced immediately to avoid creating a fire. Explosive and/or toxic hazards can result in property damage, personal injury, or even death.

Test the control module as follows:

1. Manually shut off the gas supply and apply electric power to the control module. After the trial-for-ignition period has lapsed, check to see that there is no voltage between terminals $B(V_1)$ and V_2, using a suitable voltmeter.
2. Manually open the gas supply line and reactivate the control module. Sparking should occur and then immediately cease when the flame is established. While the unit is running, manually close the gas supply line. Sparking should promptly reappear when the flame is extinguished. The spark should remain on for the trial-for-ignition period and then lock out. Check to see that there is no voltage between terminals $B(V_1)$ and V_2 as described in the section above.

Repairs and Replacements. The following warnings must be adhered to:

1. The Fenwal 05-14 ignition system is not field repairable. Any alteration or modification may result in fire, explosion, toxic hazards, property damage, personal injury, or even death. Consult your local Fenwal authorized heating wholesaler for replacement systems.
2. Use caution to avoid miswiring a replacement controller. It is recommended that the technician remove one wire at a time from the defective unit and reconnect it to the corresponding terminal of the new control. Misconnection can result in conditions that are hazardous to life and property.
3. The Fenwal 05-14 Direct Spark Ignition System is designed for use only on new products by manufacturers of gas-fired equipment. It may be used only as such, or as a replacement for an existing Fenwal Spark Ignition System with the same catalog number. Any substitution or applications must be approved by Fenwal or the manufacturer of the equipment. Improper substitutions or applications may result in malfunction of the equipment, such as loss of a flame-sensing safety circuit, causing an explosive situation.

FIGURE 20–116   Fenwal Series 05-24 silicon carbide (HSI) ignition system. (Courtesy of Fenwal Inc.)

HSI System Operation. The Fenwal Series 05-24 Silicon Carbide Ignition System (Multiple Trial) ignition system combines an interrupted means of ignition, using a high-energy heat source, with the flame rectification method of monitoring the presence of flame. Specifically designed for use in residential gas heating applications, the 05-24 can provide a positive ignition over a wide range of air–gas mixtures (Figure 20–116).

Principle of Operation. The system uses a silicon carbide element that performs a dual function of ignition and flame detection. The ignitor is an electrically heated resistance element which thermally ignites the gas. The flame detector circuit uses flame rectification for monitoring the gas flame.

Upon a call for heat, the element is powered from the 120-V ac line and allowed to heat for 45 sec (typical). Then the main burner valve is energized, permitting gas to flow to the burner for the trial-for-ignition period, which is typically 7 sec long (other timings are available). At the completion of this time period, the ignitor is switched from its heating function to that of being a flame probe which checks for the presence of flame. If a flame is present, the system will monitor it and hold the main gas valve open. If a flame is not established within the trial-for-ignition period, the system will automatically recycle the specified number of times before locking out, deenergizing the main gas valve, and shutting off the electric power to the ignitor. If during the heating cycle a loss of flame occurs, the system will recycle through the ignition sequence.

Mounting. The series 05-24 is not position sensitive and can be mounted vertically or horizontally. The case, or printed circuit board, may be mounted with the proper hardware. Stand-offs made on the bottom of the PC board provide proper electrical and thermal isolation between the board and the mounting surface.

Wiring. Caution: Note that both 24 and 120 V ac are required for system operation. Interrupt both power supplies before proceeding. Do not apply electric power to the control module until the wiring is completed and the system is properly grounded.

FIGURE 20-117 Connecting plastic connectors to edge of PC board. (Courtesy of Fenwal Inc.)

FIGURE 20-118 Fenwal 05-24 system schematic. (Courtesy of Fenwal Inc.)

Cable Assembly. Use Fenwal assembly P/N 05-128070-0XX (last two digits indicate the length of the wire in inches).* Insert the plastic connector into the edge of the PC board as shown in Figure 20-117. The connector will not fit if it is reversed. Connect the leads as follows (Figures 20-118 and 20-119). (*Caution:* This is 120 V.)

Black wire to the hot side of the 120-V ac line
Gray wire to the silicon carbide ignitor; not polarized
Blue wire to the silicon carbide ignitor; not polarized
White wire to the neutral side of the 120-V ac line
Brown wire to the gas valve (24 V ac)
Red wire to the 24-V ac supply
Green wire to the ground (the control module, 24-V ac supply, and the gas valve must have a common ground)

Silicon Carbide Ignitor/Sensor. The proper location of the silicon carbide ignitor–sensor is important for proper system performance. It is recommended that the ignitor–sensor be mounted according to the furnace manufacturer's specifications. The ignitor–sensor should be mounted so that its tip extends ¾ in. (19.05 mm) through the center of the flame and about ½ in. (12.70 mm) above the base of the flame.

Preliminary System Checks. The system must be checked after the furnace installation and before the

*If the cable assembly is not ordered from Fenwal, component parts may be obtained, or made, using either of the following nine-pin edge connectors: AMP 640136-9 with terminals 61668-1 or Molex 09-01-1091 with terminals 08-01-0101.

FIGURE 20-119 Fenwal 05-24 system schematic with test adaptor in place. (Courtesy of Fenwal Inc.)

NO   NO   YES

FIGURE 20-120 Position of the ignitor in the pilot flame. (Courtesy of Fenwal Inc.)

gas supply is turned on. Be sure that the system is electrically grounded.

Initial Operation. Use the following steps for initial operation:

1. Check the installation, mounting, and position of the ignitor in the flame envelope (Figure 20-120).
2. Apply 120 V ac electric power to the system. Manually shut off the gas supply and apply 24 V ac power to the system by setting the thermostat to call for heat.
3. Check to make sure that the ignitor–sensor "glows" during the trial-for-ignition period specified for the control module being used. The system will automatically recycle the specified number of times, then lock out. Set the thermostat to the lowest setting.
4. Manually open the gas supply line and set the thermostat to call for heat to recycle the system.
5. Check to see that ignition has been accomplished within the trial-for-ignition period.
6. If the system ignites but fails to hold in, check for 5 μA minimum flame-sensing current using the Fenwal Test Adaptor, P/N 05-080223-001, and a 0–50 microammeter (Figure 20-119). Also, be sure that the system is properly grounded (Figure 20-118).

Safety Checks. Complete the following safety checks before placing the furnace in operation:

1. Manually shut off the gas supply and apply 120 and 24 V ac electrical power to the control module by setting the thermostat to call for heat. After the system has locked out, check to see that there is no voltage across the gas valve terminals with a suitable voltmeter. Set the thermostat to the lowest setting.
2. Wait 15 sec, then manually open the gas supply and reactivate the system by moving the thermostat to a setting above room temperature. The

ignitor should glow brightly and then diminish after the flame has been established. While the system is operating, manually shut off the gas supply. The ignitor should begin to glow brightly as soon as the flame is extinguished. The ignitor should continue to glow brightly for the specified number of trial-for-ignition periods. The system will then lock out. Check to see that there is no voltage across the gas valve terminals.

Service Checks. Flame current is the current that passes through the flame from the ignitor–sensor to the ground to complete the primary safety circuit. The minimum flame current for proper operation is 5 μA dc. To measure the flame current, first disconnect both sources of electric power (120 and 24 V ac). Then connect the Fenwal Test Adaptor, P/N 05-080223-001, as shown in Figures 20-118 and 20-119. The meter reading should be 5 μA or higher after reconnecting the power and establishing the flame.

If the meter reads below "0" on the scale, the meter leads are reversed. Disconnect the electric power and reconnect the leads for proper polarity.

Repairs and Replacements. Warning: The Fenwal Series 05-24 Silicon Carbide (HSI) Ignition Module is not repairable. Any modifications or repairs to Series 05-24's will invalidate Fenwal's standard warranty as well as agency certifications *and may create hazardous conditions that could result in property damage, personal injury, or even death from fire, explosions, and/or toxic gases.* Faulty units should be returned to the factory for repair or replacement. Wire the 05-24 Fenwal system according to Figure 20-121.

FIGURE 20-121 Fenwal Series 05-24 system wiring diagram. (Courtesy of Fenwal Inc.)

☐ Safety Procedures ☐

Heating equipment is probably the most dangerous of all the equipment used in the industry. This is because of the combustion process that provides heat for the building. Therefore, a service technician must practice safety procedures to prevent loss of life, property, and equipment.

1. Do not tamper with the safety controls. If a problem arises with one, replace it.
2. Avoid lighting the pilot until the firebox has been cleared of any unburned fuel.
3. Follow the manufacturer's directions carefully. Much time and money have been spent developing these directions and they should be heeded.
4. Always follow the building and safety codes effective in your locality.
5. Be sure the unit is properly grounded electrically.
6. Do not make adjustments without knowing what the results will be.
7. Do not alter the venting system in any way.
8. Be sure that proper ventilation air is supplied in the equipment room.
9. Never allow a unit to be operated when a bad heat exchanger is suspected.
10. Never leave a furnace in operation that is not functioning properly.

☐ SUMMARY ☐

☐ A heat source is any substance that produces heat while going through a given process.
☐ The main sources of heat are solids, liquids, gases, and electricity.
☐ Coal is the main solid used for heating purposes.
☐ Fuel oil is the most common liquid used for heating purposes.
☐ Natural and LP gases are the most common gases used for heating purposes.
☐ LP gas is a mixture of butane and propane.
☐ Butane has a boiling point of 31.1 °F (−0.6 °C) and a heat content of 3267 Btu/ft³ (23.3 kcal/m³).
☐ Propane has a boiling point of 43.8 °F (−42 °C) and a heat content of 2521 Btu/ft³ (635.29 kcal/m³).
☐ Electricity has a heating value of 3412 Btu/kW (859 kcal/kW).
☐ The types of electrical heating elements are: (1) open wire, (2) open ribbon, and (3) tubular cased.

☐ The major methods of electric heating are resistance, heat pump, and a combination of the two.
☐ The heat pump is the most efficient electric heating method.
☐ Auxiliary heat is usually required when heat pumps are used in cold climates.
☐ The three basic requirements for combustion are: (1) fuel, (2) oxygen, and (3) heat.
☐ The limits of flammability should be observed when making adjustments on gas burners.
☐ Ten cubic feet (0.28 m³) of air are required to burn just 1 ft³ (m³) of natural gas.
☐ Excess air is supplied at the rate of 50% of the required air for combustion.
☐ The products of combustion produced when 1 ft³ (m³) of natural gas is completely burned are: 8 ft³ (0.23 m³) of nitrogen, 1 ft³ (0.0283 m³) of carbon dioxide, and 2 ft³ (0.057 m³) of water vapor.
☐ The orifice is the opening through which gas is admitted to the main burner.
☐ The orifice and the burner must be matched in Btu (cal) ratings.
☐ A gas burner is defined as a device that provides for the mixing of gas and air in the proper ratio to ensure satisfactory combustion.
☐ Primary air is admitted to the main burner through the face of the burner.
☐ Forced-draft burners are used on rooftop units.
☐ Power burners have a very wide range of capacities.
☐ The automatic pilot is probably the most important safety device used on modern gas heating equipment.
☐ The millivolt-type pilot burner can be divided into two types: thermocouple and thermopile.
☐ The bimetal pilot uses a sensing element made of dissimilar metals.
☐ The liquid-filled pilot functions from the force exerted when a liquid is heated and a vapor is formed.
☐ There are basically two types of flames: yellow and blue.
☐ The heat exchanger is the heart of the heating system.
☐ The two general classifications of primary heat exchangers are: (1) barrel, and (2) sectionalized.
☐ Hydronic heating uses both primary and secondary heat exchangers.
☐ The building requires additional insulation when electric heating is used.
☐ Electric resistance heating involves resistance elements, baseboard elements, and duct heaters.
☐ The minimum air velocity required for electric resistance heating is determined on the basis of entering air temperature and watts/ft² of the cross-sectional duct area.
☐ Furnaces are manufactured in upflow, downflow, and horizontal designs.

- The heat rise through a furnace can be defined as the temperature increase of the air as it passes through the heating device.
- There are basically three types of oil burners: (1) high pressure, (2) low pressure, and (3) rotary.
- In the high-pressure oil burner the oil is forced through the burner nozzle at approximately 100 psig (787.99 kPa).
- The low-pressure oil burner forces the oil through the nozzle at from 1 to 15 psig (107.86 to 204.04 kPa).
- The rotary oil burner uses a cup rotating at high speed to atomize the oil.
- A hot water boiler maintains water temperature between 120 and 210°F (49 and 99°C).
- Steam heating has numerous basic characteristics that can be advantageously employed.
- Steam-heating systems can be divided into two classifications: one pipe and two pipe.
- Humidification is one of the most important aspects of air conditioning.
- Relative humidity is the amount of moisture actually in the air as compared to the amount that it can hold.
- When air is heated, it can hold more moisture.
- The three reasons for controlling humidity are: (1) comfort, (2) preservation, and (3) health.
- The three types of humidifiers are: (1) pan type, (2) atomizing type, and (3) wetted element type.
- The purpose of the thermostat is to signal the equipment when to operate.
- The purpose of the main gas valve is to allow gas to flow to the main burners or stop this flow on demand from the thermostat.
- The purpose of the pilot safety device is to interrupt the electric circuit, or interrupt both the gas supply and the electric circuit when the pilot burner will not safely ignite the main burner.
- The thermocouple provides the electric power necessary to operate the pilot safety device.
- The limit control is a safety device that interrupts the temperature control when the temperature inside the furnace heat exchanger reaches approximately 200°F (93°C).
- The purpose of the fan control is to stop and start the fan in response to the temperature inside the furnace heat exchanger.
- The fire-stat is a safety device installed in the fan compartment that stops the fan when the return air temperature reaches approximately 160°F (71°C).
- A transformer is used to reduce line voltage to 24 V for use in the temperature control circuit.
- Boiler pressure controls are used to control the steam pressure inside a steam-heating boiler.
- Immersion controls are used to control the water temperature inside a hot water boiler.
- Protectorelay controls are used for the automatic recycling of intermittent ignition oil burner heating systems.

◻ EXERCISES ◻

1. What are the four basic sources of heat used in modern heating units?
2. What is LP gas made of?
3. What is the boiling point of propane?
4. What is the ignition point of natural gas?
5. Name the components of natural gas and their percentages.
6. What is a definite advantage of LP gas?
7. Name the major methods of using electricity for heating purposes.
8. Why is a heat pump less efficient at low outdoor temperatures?
9. Define the combustion process.
10. What are the basic requirements for combustion?
11. What is excess oxygen?
12. Define a gas orifice.
13. Define a main gas burner.
14. Name the four different types of main gas burners.
15. Where is the primary air introduced into the main burner?
16. Where are forced-draft burners popular?
17. What is probably the most important safety device used on modern gas heating equipment?
18. Name the types of pilot burners.
19. What are the two basic types of flames?
20. What controls the flow of combustion and excess air through a heat exchanger?
21. In what types of heating equipment are secondary heat exchangers used?
22. Why is electric heating beneficial from an ecological standpoint?
23. How is the airflow determined in electric heaters?
24. Define heat rise through a furnace.
25. Name the different types of oil burners.
26. What is the temperature range desired on a hot water boiler?
27. Name the piping arrangements used in steam-heating systems.
28. Which can hold the most moisture, cold air or warm air?
29. Name the three classifications of benefits of controlling humidity.
30. Name the three types of humidifiers.
31. What is the purpose of the thermostat?
32. What is the purpose of the pilot safety?
33. What is the limit control?

34. What provides the low voltage for the temperature control system on a modern heating furnace?

35. What device provides automatic recycling of intermittent ignition oil burners?

36. What percentage of energy can be saved by using an intermittent pilot on a gas furnace?

37. How much energy can the installation of a vent damper save?

38. Are all intermittent pilot ignition systems usable on LP gas?

39. On the Lennox pulsating furnace, when is the spark ignitor turned off?

40. What is the approximate vent gas temperature from the Lennox pulsating furnace?

21 Heat Pump Systems

Performance Objectives

After completion of this chapter, the student should:

■ Be able to name the various operating cycles of a heat pump system.
■ Be able to list the various types of defrost systems used on heat pump systems.
■ Be able to explain the purpose of the supplemental heat use on heat pump systems.
■ Know how a reversing valve operates.
■ Describe the different methods of installing a ground-coupled water-source heat pump system.
■ Be able to discuss the basic operating principle of a ground-coupled water-source heat pump system.
■ Complete the Exercises with at least 80 percent accuracy.
■ Complete the Laboratory Exercises with at least 80 percent accuracy.

FIGURE 21-1 Heat pump classifications.

Definition. A *heat pump* can be defined as an air conditioning system that moves heat both to and from the conditioned area. Through the proper use of controls, these units may be used during all seasons of the year. In the cooling season, the unit functions as a regular cooling unit. During the heating season however, the refrigerant flow is reversed and it becomes a heating unit.

Heat pumps are classified according to their heat source and the medium to which the heat is transferred during the heating season. The classifications are: (1) air-to-air, (2) water-to-air, (3) water-to-water, and (4) ground-to-air (Figure 21–1). Because of the expense of installation and equipment, the air-to-air heat pump is the most popular.

☐ Cooling Cycle ☐

During the cooling season, the heat pump operates as a normal cooling system. The refrigerant is compressed by the compressor and discharged to the outdoor coil (condenser), where it is condensed to a liquid (Figure 21–2). It then travels to the flow control device and into the indoor coil (evaporator)

FIGURE 21-2   Heat pump cooling cycle. (Courtesy of Carrier Corp.)

FIGURE 21-3   Heat pump heating cycle. (Courtesy of Carrier Corp.)

where it is evaporated and absorbs heat; then back to the compressor and the cycle is repeated. Notice that the names of the coils have been changed. This is to eliminate confusion when discussing heat pumps. The evaporator is termed the *indoor coil* and the condenser is termed the *outdoor coil*. As the season changes, the functions of the coils change.

☐ Heating Cycle ☐

During the heating season, the heat pump operates because of the reversed refrigerant flow. The refrigerant is discharged by the compressor into the indoor coil where it is condensed into a liquid refrigerant. The heat removed from the refrigerant is discharged

into the conditioned area. The liquid refrigerant is then directed to the outdoor coil where it is evaporated before returning to the compressor (Figure 21-3).

It should be noted that the reversed flow is accomplished by use of a four-way valve and check valves. Some manufacturers will use different variations of this setup, but all accomplish the same thing.

The indoor coils used on heat pump units are usually designed especially for heat pumps and many times a coil is designed for a particular model unit. In most instances, the indoor coil used on a regular cooling unit will not function properly on heat pump installations. This is because of the reduced airflow over the indoor coil. The proper refrigerant-to-air ratio must be maintained for economical performance.

For these units to be accepted by the public, they must operate as economically as possible. Every possible Btu (cal) must be squeezed from every 1 kW of input. For each 1 kW put into a heat pump, more than 3412 Btu (859.8 kcal) is released. A properly designed heat pump can produce three to four times the amount of heat released by resistance heating equipment. Some of the newer, more efficient units can produce as much as eight times the input Btu (cal). This is known as the coefficient of performance (COP). The COP is determined by dividing the Btu output by the Btu paid for.

For example, if we have a 10-ton heat pump which has been found to be delivering 86,000 Btu (21,672 kcal), the COP would be:

$$COP = \frac{86,000 \text{ Btu (21,672 kcal)}}{10 \text{ BHP (brake horsepower)}}$$
$$\times \text{ 2545 Btu/hp (641.3 kcal/hp)}$$
$$= 3.3 \text{ total heat available}$$

That is, the COP of this particular unit is 3.3 or the Btu (cal) output is 3.3 times greater than the Btu (cal) input. It should be noted here that the higher the COP, the greater the amount of heat available. In comparison, direct resistance heating has a COP of only 1.

When properly designed, installed, and maintained, a heat pump can be superior to a gas heating unit. This can be attributed to a lower discharge air temperature and a higher relative humidity, which provides a more even temperature during the heating cycle.

However, these units do not operate as satisfactorily in extremely cold climates. Some manufacturers introduce an automatic means of stopping the compressor completely when the outdoor temperature drops to around 20°F (−6.7°C). In ambient temperatures below this, the COP will fall below the economical point. Also, the outdoor coil must operate at a temperature below the ambient temperature in order to absorb heat for the evaporation process. At these low temperatures, the suction pressure is low enough to cause damage to the compressor. Some units are equipped with an electric heater on the air inlet side of the outdoor coil to provide for the evaporation process at low ambient temperatures (Figure 21-4). Because of the added expense in manufacturing and operation, this is done only in special cases.

FIGURE 21-4   Electric outdoor heat for a heat pump.

FIGURE 21-5   Solar-assisted heat pump.

☐ Solar-Assisted Heat Pump ☐

Solar-assisted heat pumps use the same principle as supplemental electrical heat. The major difference is that the electric heater is replaced by a water coil (Figure 21-5). The solution inside the coil is heated by use of a solar panel. This added heat keeps the operating efficiency as high as possible without the added operating expense of using electricity.

☐ Defrost Cycle ☐

During the heating season, it is quite possible that the outdoor coil will operate at a temperature below freezing. At this temperature, any moisture which may be removed from the air will immediately freeze on the surface of the coil. This frost on the coil will continue to build up and cause the outdoor coil to lose its efficiency. This frost will also act as an insulator, which will reduce the heat transfer and reduce the coil efficiency even more. When the coil efficiency has been reduced enough to affect the performance of the system, the frost must be removed.

There are three basic steps involved in the defrost cycle:

1. Detection of the frost buildup (defrost initiation).
2. Frost removal.
3. Detection of sufficient frost removal to allow the unit to return to the heating cycle (defrost termination).

Defrost Initiation. The methods used for frost detection are many and varied. The following are some of the more common methods.

Air Pressure. This method uses the air pressure differential across the outdoor coil (Figure 21-6). This type of control has a time delay to prevent false initiation. When the air pressure reaches a given drop across the coil, the defrost cycle is initiated.

Time-Temperature. This control uses a temperature bulb attached to the outdoor coil to initiate the defrost cycle at the prescribed temperature (Figure 21-7). Incorporated in this control is a time clock that will only allow the defrost period to be initiated at given intervals. This method requires that both a low temperature be reached and a sufficient amount of time elapse before defrost initiation can occur.

Solid State. Some units are equipped with a solid-state circuit board that initiates the defrost cycle. Thermistors within the solid-state system sense the difference between the ambient air temperature and the refrigerant temperature. When the temperature difference exceeds the differential band, the defrost cycle is activated and continues until the defrost pressure switch terminates it.

In operation, when the ice on the coil has reached a predetermined point, the defrost cycle is initiated. The unit is returned to the cooling cycle so that the hot discharge gas now passes through the condenser. The outdoor fan is stopped so that the cold outdoor air will not flow through the coil and adversely affect the defrost efficiency. The indoor fan is kept running to evaporate the refrigerant before it returns to the compressor. Also, the auxiliary heat is turned on to prevent cold air from blowing into the conditioned area. When the ice has all melted, the defrost termination control will return the unit back to the normal heating cycle.

Defrost Termination. Many of the controls that were used for defrost initiation are also used to terminate the defrost cycle.

Air Pressure. After the frost has been cleared from the outdoor coil, the air pressure drop across the coil is reduced. The air pressure on both sides

FIGURE 21-6   Air pressure defrost initiation.

FIGURE 21-7   Time-temperature defrost control method.

of the coil is almost equal. The unit is then automatically returned to the heating cycle.

Time. This type of control is usually used in conjunction with any of the previously mentioned initiation methods. In operation, after a predetermined amount of time has elapsed, the unit will return to the heating cycle if the defrost initiation control has reached the cut-in setting.

Refrigerant Pressure. This type of defrost termination control functions because the refrigerant pressure increases when the frost has been melted. As the pressure inside the system increases, the con-

FIGURE 21-8 Electric defrost heater.

FIGURE 21-9 Location of supplementary duct heaters.

trol puts the unit back into the heating cycle at a predetermined setting.

Some manufacturers incorporate electric strip heaters attached to the outdoor coil to speed up the defrosting process. These heaters are energized by the defrost initiation control (Figure 21-8).

Reheating

A heat pump should not be used for reheating purposes, especially in the summer. The compressors used in these units have a higher compression ratio than normal air conditioning compressors. It is not difficult to obtain a 300°F (149°C) discharge gas temperature. This is the maximum allowable temperature and many manufacturers will include controls to stop the compressor before it is reached. This higher temperature is caused by too low an airflow over the indoor coil. A high indoor air temperature [90°F (32.2°C) or higher] will require more air to be moved across the indoor coil. Caution must be exercised to ensure that any supplemental heating equipment is installed in the airstream after the indoor coil.

Supplementary Heat

These units are sized to satisfy the cooling load, and supplementary heat is added to bring the unit to the desired heating capacity. Each manufacturer will provide charts indicating the cooling and heating capacity in Btu (cal) for any given unit. From these charts and the calculated heat load, the amount of supplementary heat may be determined. This supplementary heat is usually in the form of duct heaters installed in the supply duct of the equipment (Figure 21-9).

Supplementary heat is used only when the heat pump cannot supply the desired amount of heat. It is controlled from two points: (1) the second stage

on the indoor thermostat, and (2) the outdoor thermostats, which are mounted under the eaves of the building. When both of these thermostats demand heat, the supplementary heat will be energized. The design engineer will determine the temperature setting of the outdoor thermostats. If these thermostats are set too high, the electricity bill may be excessive. On the other hand, if they are set too low, an insufficient amount of heat may be supplied to the conditioned area. There must be a point where comfort and economy are maintained.

Reversing Valve

The reversing valve is the device that changes the direction of the refrigerant flow in a heat pump system (Figure 21-10). It is installed in the discharge

FIGURE 21-10 Reversing valve. (Courtesy of Ranco, Inc.)

SHOWN IN COOLING POSITION

FIGURE 21–11 **Cutaway of reversing valve. (Courtesy of Ranco, Inc.)**

line of the compressor and switches the flow of refrigerant on demand from the thermostat or the defrost control.

The refrigerant gas path is schematically diagrammed through the main valve showing the sliding port at a position of rest over two tube openings as it transfers both coils from the opening phases of cooling, defrosting, and heating (Figure 21–11). The solenoid coil is not energized in a normal cooling cycle.

When adding refrigerant to these units or making any type of adjustment, the manufacturer's specifications must be followed. The trial and error method is too time consuming and the desired conditions may never be reached because of the numerous overlaps and interlocks incorporated into these units. Therefore, for comfort and economy's sake follow the proper specifications.

☐ Water Source Heat Pump Systems ☐

Through research, development, and manufacture of materials that are capable of better heat transfer, water source heat pumps are gaining popularity in the industry. These systems are generally referred to as closed-loop earth-coupled and open-loop (water well or groundwater) systems. When the closed-loop earth-coupled system is used, the water is circulated through special piping that is buried in the ground. This solution may also be a mixture of water and antifreeze. Therefore, there is no wastewater to be disposed of. The open-loop systems take water from the source and discharge it to an appropriate area. Sometimes, disposing of the used water is a problem and is more expensive than the source of the water.

The basic operating principle of this system is that the heat is rejected to the water after being removed from the air being cooled by the refrigerant circuit during the cooling cycle, and heat is extracted from the water by the refrigeration circuit and delivered to the air to be heated during the heating cycle. Dehumidification is achieved in the same manner as with any cooling unit, by its removal in the form of condensate. The heat transfer medium is the refrigerant inside the refrigeration system. The refrigeration system components, with the exception of the water to the refrigerant heat exchanger, are the same as those used on any other type of heat pump system. This type of system is most popular for heating and cooling residential structures where an adequate source of water or ground area is available.

☐ Closed-Loop Earth-Coupled Systems ☐

In the closed-loop earth-coupled heat pump system a solution is circulated through special pipes buried in the ground. This solution may be water or a water–antifreeze mixture. There are three methods of installation for these types of heat pump systems: (1) horizontal, (2) vertical, and (3) lake loops.

Typical Closed-Loop Earth-Coupled Methods. These types of systems may use either vertical or horizontal tubing buried in the ground. The vertical system is the most popular when the site area is too small to accommodate the amount of piping needed for the horizontal-type system. In these types of systems a number of boreholes are drilled into the ground of sufficient size and depth to allow the proper amount of piping to be buried for the size of system being installed (Figure 21–12). Note that the letters shown on Figures 21–12 and 21–13 match the corresponding "Feet per Nominal Ton" category in Table 21–1. The specifications and installation instructions for all methods shown are available from the manufacturer of the equipment being used. The earth coupling length information given here applies only to and has been specifically designed for Command-Aire Earth Energy Heat Pump systems to be installed in residential and light commercial applications in the Waco, Texas area.

The horizontal system is used in applications where the building site is large enough to allow the proper amount of pipe to be buried horizontally (Figure 21–13). This type of system is especially advantageous when drilling is not desirable.

The choice of vertical, horizontal, or lake loop earth coupling should be based on the characteristics of the particular application. Horizontal and vertical systems are designed to provide the same fluid temperature under a given set of conditions. The lake loop system will provide slightly lower fluid temperatures, but the reduced installation cost should compensate for any minor reduction in performance. The three earth coupling methods should be considered at each application, with the most cost-effective method chosen after all methods have been evaluated.

Earth-Coupled System Design. The design of an earth-coupled system is divided up into the following steps: (1) determine the structure design heating load in Btu/hr loss and design cooling load in Btu/hr gain, (2) select a water source heat pump, and (3) select the earth-type coil and materials to be used.

Determine the Structure Design Heating and Cooling Loads. An accurate heat load calculation is a

FIGURE 21–12 Vertical piping application. (Courtesy of Command-Aire Corporation.)

TABLE 21–1 Command-Aire Earth Coupling Design Information

Zone	Feet Per Nominal Ton[a]				Anti-freeze[b]	Pumping Unit		
	A	B	C	D		LPK1	LPK2	LPK3
1	200	550	650	875	0	060–130	190–350	410–610
2	210	575	675	900	0	060–130	190–350	410–610
3	220	600	725	975	0	060–130	190–350	410–610
4	240	675	800	1075	0	060–130	190–350	410–610
5	260	750	900	1200	0	060–130	190–350	410–610
6	280	850	1025	1350	0	060–130	190–350	410–610
7								
8								
9								

Source: Command-Aire Corporation.

[a]Horizontal length is lineal feet of pipe per ton; vertical length is lineal feet of borehole per ton.

[b]Percentage of volume.

STYLE B

STYLE C

STYLE D

FIGURE 21–13 Horizontal piping applications. (Courtesy of Command-Aire Corporation.)

necessity if the system is to perform satisfactorily. Therefore, it is recommended that some nationally accepted method be used for these calculations.

Select a Water Source Heat Pump. The water source heat pump selected for use on an earth-coupled system may be required to operate at entering water temperatures that vary between 30 and 100 °F (– 1.67 and 37.78 °C) Because of this, it is necessary that the minimum and maximum temperatures of the water source selected remain within that range through all seasons. There are several models on the market which have a much smaller operating range of entering water temperature. Some of these units are not satisfactory for use on earth-coupled heat pump installations.

The heating and cooling capacities of water source heat pumps should be determined from the manufacturer's specifications for the local groundwater temperatures and design conditions. Following is a suggested method of sizing water source heat pump systems.

1. *Heat pumps sized for cooling.* The sensible heat output capacity of the cooling equipment selected should in no case be less than the calculated total sensible heat load. Also, it should never be more than 25% greater than the calculated sensible heat load. The latent heat capacity should never be less than the calculated total latent heat load. The sensible and latent heat capacities of the equipment should be obtained from the performance data listed in the manufacturer's catalog. The capacities of the equipment should be verified for the local indoor design conditions and the local groundwater temperature.

2. *Heat pumps sized for heating.* Systems that are designed for heating-only applications should provide a minimum of 75% and a maximum of 115% of the total Btuh heat loss. In applications where the system is undersized, auxiliary heat strips must be installed to bring the system up to the calculated heat loss of the structure. In some localities emergency heat that will provide either a large part or the total heat loss must be provided, depending on the local requirements.

3. *Heat pumps designed for both heating and cooling.* Heat pump systems that are designed to provide both heating and cooling must be sized to satisfy the cooling requirements of the structure. When designed in this manner, the heating thermal balance point will be determined by the cooling requirements. In applications when a

lower balance point is desired, the system may be oversized by a maximum of 25% of the system capacity. Auxiliary heat strips are then used to provide the amount of heat that is lacking between the balance point and the output of the heat pump system. Some localities may require that emergency heat be provided that will produce the total or some percentage of the heat loss of the structure.

Select the Earth-Type Coil and Materials. In earth-coupled systems, the earth coupling is a particular method which uses water or another type of solution that is circulated through special pipes buried in the ground. The heat is transferred to and from the water through the walls of the pipe. These types of systems are used in areas where there is an insufficient amount of available groundwater, or where the drilling of a well would be impractical. The required piping may be buried by either the vertical or the horizontal method.

Of particular importance when designing an earth-coupled heat pump system is the balance between the heat pump unit and the earth-coupled loop. When this balance is achieved, the earth-coupled loop will remove all the heat that is transferred to the water by the heat pump during the cooling cycle, and will provide all the heat that is required to heat the water during the heating cycle. The final result of a perfectly balanced system is that the change in water temperature through the heat pump system is offset by an equal and opposite change in temperature through the earth-coupled loop. As an example, if when operating in the cooling cycle, the heat pump unit causes a temperature rise in the circulating water of 15 °F (8.4 °C), the loop must respond with a corresponding but opposite temperature drop of 15 °F (8.4 °C).

It should be remembered that even though the earth-coupled loop is designed for balanced rise and fall of the water temperature, which suggests that the net average loop water temperature remains constant, because the ground temperature may vary ±15 °F (8.4 °C) from season to season, the loop water temperature may vary ±20 °F (11.2 °C) from the balance point temperature. This is because the ground is able to overcool the loop water during the heating season, and may undercool it during the cooling season. For this reason, the water temperature entering the water source heat pump system may drop below 30 °F (– 1.11 °C) during the heating season or rise above 100 °F (37.78 °C) during the cooling season. This range in the temperature of the entering water is extremely important because water source heat pump systems are designed to operate within specific operating temperature ranges. The manufacturer's specifications for the particular unit being considered must be checked to make certain that the unit is designed to operate within the entering water temperature range at the point of application. The temperature ranges are established to protect both the heat pump unit and the water loop piping. Also, these temperature ranges are generally based on water only passing through the system. The low temperature limit of 40 °F (4.44 °C) in a water source heat pump unit is established to protect the loop water from freezing. Again, this low limit is assumed for water only circulating through the system. However, if the circulating water is mixed with a nontoxic antifreeze solution, the entering water temperature can be allowed to fall to 30 °F (−1.11 °C).

1. *Heat pump.* Use only a water source heat pump that can be operated on loop water temperatures that fall below 40 °F (4.44 °C) and perhaps down to 25 °F (−3.89 °C). Information on when to use an antifreeze solution in a ground-coupled water-source heat pump system is given in the following discussions on vertical and horizontal installation methods.

2. *Pipe.* It is generally preferred that polybutylene (PB) or polyethylene (PE) pipe be used for horizontal coils, vertical U-bend wells, and for service lines to the wells and lake heat exchangers. IPS (iron pipe size) PB pipe is used with insert fittings and clamps. CTS (copper tube size) PB pipe is fused together with the appropriate fittings using a fusion tool. PE pipe is heat butt fused together with the appropriate fittings using a fusion tool.

3. *Cleanliness.* During the installation procedure it is very important that all trash, soil, and small animals be kept out of the piping system. Leave the tape on the ends of the pipe until they are to be connected to the service lines or the equipment room piping.

4. *Pressure testing.* The plastic pipe assemblies should be pressure tested at twice the anticipated system operating pressures before the backfilling procedure is started. Normal static equipment room pressure is 50 psig (344.738 kPa).

5. *Backfill.* The narrow trenches that are made with a chain trenching machine may be backfilled with the tailings provided that no sharp rocks are present.

The wider trenches made with a backhoe may be backfilled with the dirt that was removed from the trench if it is a loose, granular type of material. If there are clumps of clay or rocks in the material, the plastic pipe must be covered with sand before filling the trench with the clumps and rocks.

Boreholes drilled to 4 to 6 in. (101.60 to 152.40 mm) in diameter are commonly used for vertical geothermal wells. The backfill may be of any granular material that does not contain sharp rocks or other sharp objects, including drilling tailings, sand, pea gravel, or bentonite mud.

6. *Location markers.* The locations of important points such as well heads must be marked for subsequent recovery. A steel rod driven to a point just below the surface of the earth is a good way to identify these features or to mark the outline of an entire serpentine earth coil.

7. *As-built plans.* The components of the earth coupling unit should, if at all possible, be drawn on a site plan as it will be installed. This step will aid in the location of the key components. A simple way to locate key components is to make two measurements (sides of a triangle) from two corners of a building to the specific component being identified. These measurements should be recorded in the table on the plans.

Reasons for Using an Earth-Coupled System. Some of the major reasons for using earth-coupled systems are as follows:

1. Unlike a standard solar system, the loop operates day or night, rain or shine, all year long, delivering heat to and from the heat pump unit.

2. It is cost-effective in either northern or southern climates.

3. Because the water circulates through a sealed closed loop of high-strength plastic pipe, it eliminates scaling, corrosion, water shortage, pollution, waste, and disposal problems which are possible in some open-well water systems.

☐ Vertical Earth-Coupled System ☐

A vertical earth-coupled system consists of one or more vertical boreholes through which water flows in a plastic pipe. A distinct advantage of a vertical

PIPE: High strength polyethylene, fusion joined

or

polybutylene, joined with type 300 stainless steel bands and stainless screws or heat fusion fittings

FLOW PATH: SERIES
PIPE SIZES: 3/4, 1, 1-1/4, 1-1/2 & 2 INCH
BORE LENGTH: 100 to 175 FEET/TON
PIPE LENGTH: 200 to 350 FEET/TON

Approximately 140 ft. per ton depending on location and type soil. A single borehole should not exceed 300 feet depth.

TRANSFER FLUID–WATER NOTE:

Antifreeze solutions are needed in northern zones.

U-Bend Fitting (fusion joined) or (clamped stainless steel)

Min. 10 to 15 ft.

FIGURE 21–14 **Vertical (series) system. (Courtesy of Bard Manufacturing Company.)**

system over a horizontal system is that the vertical system requires less surface area (acreage). In areas where the ambient groundwater (average well water) temperature is less than 60°F (15.56°C), the use of an antifreeze solution such as propylene glycol to avoid freezing the loop is recommended.

Boreholes are drilled 5 to 6 in. (127.00 to 152.40 mm) in diameter for 1½-in. (38.10-mm)-diameter pipe. For ¾-in. (19.05-mm)-diameter pipe loop systems, the vertical loops are connected in parallel to a 1½-in. (38.10-mm)-diameter pipe header. A borehole of 3 to 4 in. (76.20 to 101.60 mm) in diameter is used for ¾-in. (19.05-mm)-diameter loops; this procedure lowers the drilling costs. The ¾-in. (19.05-mm)-diameter pipe also costs less per ton of heat pump capacity. The smaller pipe is easier to handle, yet there is no sacrifice in the pressure rating. Also, having two loops in one hole reduces borehole length. The depth for these systems is usually between 80 and 180 ft (24.384 and 54.864 m).

The basic components of a vertical earth-coupled system are shown in Figure 21–14. Each borehole

contains a double length of pipe with a U-bend fitting at the bottom. Several boreholes may be joined in series or in parallel (Figure 21–15). Sand or gravel packing is required around the piping to assure heat transfer. In addition, the bore around the pipes immediately below the service (connecting) lines must be cemented closed to prevent surface water contamination of an aquifer, in accordance with local health department regulations (Figure 21–16).

Series U-Bend. A series U-bend earth coupling is one in which all the water flows through all the pipe, progressively traveling down and then up each well bore. Series wells need not be of equal length.

Pipe. The 1½-in. (38.10-mm) CTS (copper tube size) or IPS (iron pipe size) polybutylene or polyethylene pipe is commonly used in 5- and 6-in. (127.00- and 152.40-mm) boreholes. IPS PB pipe is used with insert fittings and clamps. Turn the clamps so that they face inward and will not be chafed by the well bore. Tape the clamped section of the U-bend with duct tape to provide added protection

PIPE - Header, 1-1/2" O.D. Vertical, 3/4" O.D. or 1" O.D. polyethylene high strength fusion joined or high strength polybutylene joined with Type 300 stainless steel bands and screws or heat fusion fittings

HEADERS

Detail B

FLOW PATH: PARALLEL
PIPE SIZES: 3/4 & 1 INCHES
BORE LENGTHS: 100 to 200 FEET/TON
PIPE LENGTHS: 200 TO 400 FEET/TON

Approximately 170 ft. + per ton depending on location and type soil

Min. 10 to 15 ft. Between Loops

TRANSFER FLUID-WATER
NOTE:
   Antifreeze solutions are needed in northern zones.

1 1/2" PIPE

3/4" X 1 1/2" SERVICE SADDLE

DETAIL B

U-Bend Fitting (fusion joined)

FIGURE 21–15 Vertical (parallel) system. (Courtesy of Bard Manufacturing Company.)

Grade level

Recommended 4 ft depth

To be backfilled with material having no sharp objects

Clay grout or concrete plug

Plastic pipe

Depth of pipe varies with system

Granular packing

Fuse or clamp close to return bend fitting

FIGURE 21–16 Vertical borehole and piping detail for earth-couples system.

to the clamps while the pipe is being installed into the well. CTS PB pipe is heat fused together with fittings. PE pipe is heat fused together with butt joints.

Stiffener. Tape the last 10 to 15 ft (3.048 to 4.572 m) of pipe above the U-bend together to a rigid piece of pipe or conduit. This will make it easier to install the pipe in the well.

Fill and Pressure Test. Fill the system with water and pressure test before lowering the U-bend into a well bore. When drilling with air, a bore can be completed that contains no water. If unfilled plastic pipe is lowered into the bore, it will be crushed as the hole slowly fills.

Multiple Wells. Multiple 100-ft (30.48-m) wells connected in series are the easiest to drill and install in most areas. It will be difficult to sink water-filled plastic U-bends into mud-filled holes over 150 ft (45.72 m) deep without weights. Wells are generally spaced 10 ft (3.048 m) apart in residential systems.

TABLE 21-2 Minimum Diameters (in.) for Boreholes

Nom Pipe Size	Single U-Bend	Double U-Bend
¾	3¼	4½
1	3½	5½
1¼	4	5¾
1½	4¾	6
2	6	7

Source: Bard Manufacturing Company.

Service Lines. Follow the guidelines for the horizontal earth coil when installing service lines to and from a U-bend well.

Parallel U-Bend. A parallel U-bend earth coupling is one in which the water flows out through one header, is divided equally, and flows simultaneously down two or more U-bends. It then returns to the other header. Headers are reverse-return plumbed so that equal-length U-bends have equal flow rates.

Lengths of individual parallel U-bends must be within 10% of each other to ensure equal flow in each well. Either 1½-in. (38.10-mm) polybutylene or polyethylene pipe is used for the headers with 1- or ¾-in. (25.40- or 19.05-mm) pipe used for the U-bends. Four-inch (101.60-mm) boreholes are sufficient for placement of the 1-in. (25.40-mm) U-bends. Follow the instructions given previously under "Series U-Bend Well" for the stiffener, fill and pressure test, multiple wells, and service lines. For the minimum diameters for boreholes, see Table 21-2.

□ Horizontal Earth-Coupled System □

A horizontal earth-coupled system is similar to a vertical system in that water circulates through underground piping. However, the piping in this type of system is buried in a trench (Figures 21-17 to

PIPE: High strength polyethylene, fusion joined

or

polybutylene, joined with Type 300 stainless steel bands and screws or heat fusion fittings

FLOW PIPE: SERIES
TYPICAL PIPE SIZE: 1-1/4 TO 2 INCHES
NOMINAL LENGTH: 350 to 500 FEET/TON
BURIAL DEPTH: 3.5 to 6 FEET
MAXIMUM HEAT PUMP SIZE: 5 TONS

TRANSFER FLUID-WATER
NOTE:
Antifreeze solution needed in northern zones.

FIGURE 21-17 Horizontal (series) system—one pipe in trench. (Courtesy of Bard Manufacturing Company.)

HEAT PUMP SYSTEMS

Total Piping Length: Per ton of capacity depends on location and type of soil. First pipe installed at base of trench then partially backfilled with 2' of soil for second return pipe. A single loop should not exceed 2000 feet length.

PIPE - High strength polyethylene, fusion joined.

or

polybutylene, joined with Type 300 stainless steel bands and screws or heat fusion fittings

TRANSFER FLUID-WATER
NOTE:
Antifreeze solutions are needed in northern zones.

BURIAL DEPTH:
4 FEET AND 6 FEET - SOUTHERN
3 FEET AND 5 FEET - NORTHERN

Min. Space
4 to 6 feet

FLOW PATH: SERIES

TYPICAL PIPE SIZE: 1-1/4 TO 2 INCHES

PRACTICAL LENGTH: 210 TO 300 FEET OF TRENCH/TON

420 TO 600 FEET OF PIPE/TON

FIGURE 21-18 Horizontal (series) system—two pipes in same trench. (Courtesy of Bard Manufacturing Company.)

FIGURE 21-19 Horizontal multilevel (parallel) system. (Courtesy of Bard Manufacturing Company.)

Be sure the buried pipe system is properly designed for the heat pump load, soil type, climate, the pipe used, and the operating cycle pattern.

System uses 3/4" or 1" pipe, installed four ft. deep, spaced one foot apart vertically. Return bends as shown. Trenches spaced four feet apart. In Northern Zone, pipes are installed at 6 ft., 5 ft., 4 ft., and 3 ft. depths. In Southern Zones, pipes are spaced at 7 ft., 6 ft., 5 ft., and 4 ft. depths.

The use of smaller diameter pipes results in a thinner pipe wall and thus better heat transfer without sacrificing the pipe pressure rating. Parallel hookups are usually required in order to keep water pressure drops from being too high.

Partially backfill between pipe layers

Dig and backfill header trenches last

Attach pipe ends to headers above ground

The 3/4" parallel system shown with four pipes in a 6 ft. trench, separated by one foot of soil, results in a minimum site area. A 3 ton parallel 3/4" buried pipe system can be installed in an area of approximately 16 ft. by 150 to 200 ft. The small diameter pipe can be bent around 90° intersecting trench corners by shaving off the inside corner of the trench to the required radius with a shovel to prevent kinking.

FLOW PATH: PARALLEL

TYPICAL PIPE SIZE: PARALLEL PATHS 3/4 TO 1 INCHES
HEADERS 1-1/2 TO 2 INCHES

PARALLEL PIPE LENGTH: 500 FT. MAX. PIPE LENGTH (3/4 INCH)
750 FT. MAX. PIPE LENGTH (1 INCH)

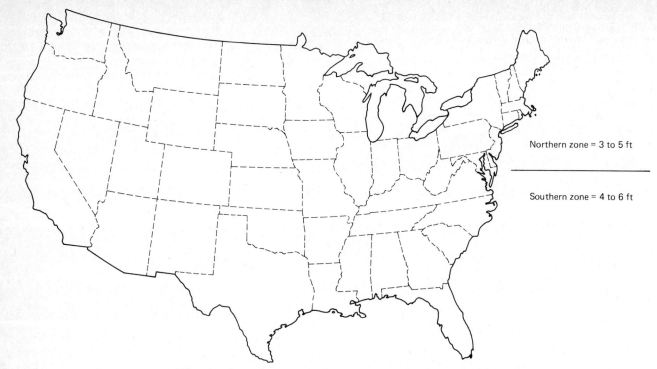

Northern zone = 3 to 5 ft

Southern zone = 4 to 6 ft

FIGURE 21-20 Pipe depths for different regions.

21–19). The pipe depths in the northern zone of the United States should be 3 to 5 ft (0.9144 to 1.524 m) (Figure 21–20). Burying the pipes at excessive depths will reduce the ability of the sun to recharge the heat used during the heating cycle. Pipe depths in the southern zone of the United States should be 4 to 6 ft (1.2192 to 1.8288 m), so that the high temperature of the soil in late summer time will not seriously affect the performance of the system.

Antifreeze will be necessary in the northern zone to prevent freezing of the circulated water and to allow the system to gain capacity and efficiency by using the large amount of heat released when the water contained in the soil is frozen. The antifreeze solution used in these systems is nontoxic propylene glycol or calcium chloride.

The use of multiple pipes in a single trench substantially reduces the total trench length. If a double layer of pipe is laid in the trench, the two layers should be set 2 ft (0.6096 m) apart to minimize thermal interference (Figure 21–21). For example, consider a 1½-in. (38.10-mm) series horizontal system with pipes at 5- and 3- (1.524- and 0.9144-m) depths. After installing the first pipe at 5 ft, partially backfill to the 3-ft depth using a depth gauge stick before installing the second pipe in the trench. Install with the return line running closest to the surface and the supply line running below it. This arrangement will

(a)

(b)

FIGURE 21-21 Separation of the horizontal earth-coupled pipes in the trench for (a) single- and (b) double-layer designs.

maximize the overall system efficiency by providing warmer water in the heating mode and cooler water for the cooling mode. Connect the pipe ends to the heat pump after the pipe temperature has stabilized, so that shrinkage will not pull the pipe loose.

Two pipes in the same trench, one above the other, separated by 2 ft (0.6096 m) of earth require a trench 60% as long as that of a single-pipe system. The total length of pipe would be 120% as long as a single-pipe system, due to the heat transfer effect between the pipes. In addition, when laying a double layer of pipe, be careful to avoid kinks when making the return bend (Figure 21–22).

Backfill the trench by hand when changing direction. If it is necessary to join two pipes together in the trench, use the fusion technique for IPS 304 stainless steel or brass fittings for greater strength and durability, then mark the fitting locations for future reference by inserting a steel rod just below the grade (Figure 21–23). The steel rod will aid later in locating the fittings with a metal detector.

Trenches can be located closer together if the pipe in the previous trench can be tested and covered before the next trench is started. This also makes backfilling easier. Spacing of 4 to 5 ft (1.2192 to 1.524 m) between the pipes is good.

In areas with dry climates and heavy clay soil, any heat that is dissipated into the soil may significantly reduce the thermal conductivity of the soil. In such cases the designer may specify additional feet of pipe per ton of heat pump capacity. A few inches (millimeters) of sand may also be put in with the pipe, or a drip irrigation pipe buried with the top pipe to add occasional amounts of water to the soil.

In series horizontal earth couplings, all the water flows through all the pipe. These may be made of 1-, 1½-, and 2-in. (25.40- 38.10- and 50.80-mm) pipe either insert coupled or fused.

Narrow Trenches. Narrow trenches are dug with trenching machines. The trenches are usually 6 in. (152.40 mm) wide. Generally speaking, the trencher will require about 5 ft (1.524 m) between trenches. This is sufficient spacing for horizontal earth coils.

The pipe can be coiled into an adjoining trench. Since the trencher spaces the trenches about 5 ft (1.524 m) apart, looping the coil from one trench to another will give a return of large enough diameter. The end trench should be backhoed to give enough room for the large-diameter bend.

If the pipe is brought back in the same trench, bend the pipe over carefully to avoid kinking the pipe, and handfill the area around the return bend (Figure 21–22). To reduce the bend radius, elbows may be used. However, keeping the number of fit-

FIGURE 21-22 Detail of narrow trench return bend used on double layer horizontal earth-coupling system.

(a)

(b)

(c)

FIGURE 21-23 Hose fitting and location detail: (a) socket-fused coupling; (b) clamp installation; (c) fitting location indicator.

tings underground to a minimum may be preferable since the potential for leaks is reduced.

If a double layer of pipe is used, the incoming water to the heat pump unit should be from the deepest pipe (Figure 21–24). This provides the heat pump with the coolest water during the cooling

Water in — Water out — Existing grade

3 ft minimum

2 ft

FIGURE 21-24 Double-layer horizontal earth coil water flow connection schematic.

2 ft minimum

Existing grade

Earth backfill

Sand backfill

Earth coupling lines

3 ft minimum

FIGURE 21-25 Backhoed trench.

Bend pipe with caution to avoid kinking

Trench

2 ft wide

Pit excavated to at least 4 ft deep with a diameter large enough to place a full-diameter loop of pipe

FIGURE 21-26 Detail of wide trench return for either a single- or double-layer horizontal earth coil.

season and the warmest water during the heating season.

Backhoe Trenches. If a backhoe is used, the trench will probably be about 2 ft (0.6096 m) wide. In a wide backhoed trench, two pipe may be placed side by side, one on each side of the trench. The pipes in the trench must be at least 2 ft (0.6096 m) apart.

Backfill carefully around the pipe with fine soil or sand. Do not drop clumps of clay or rock onto the pipe (Figure 21-25). A pit may be dug at the end of the trench to accommodate a 4-in. (101.60-mm)-diameter return bend (Figure 21-26).

Service Lines. The recommendations for the horizontal earth coils also apply for the installation of the service lines to and from the U-bend wells and pond or lake heat exchanger. Bury the service lines a minimum of 3 ft (0.9144 m) for single-layer pipe and 3 and 5 ft (0.9144 and 1.520 m) for double-layer installations. If two pipes are buried in the same trench, keep them 2 ft (0.6096 m) apart.

A parallel horizontal earth coupling is one in which the water flows out through a supply header, is then divided equally, and then flows simultaneously into two or more earth coils. The water then returns to the other header. Headers are reverse return plumbed so that equal-length earth coils have equal flow rates. The length of individual parallel earth coils must be within 10% of each other to ensure equal flow in each coil.

Freeze Protection. The antifreeze solutions that are used in earth loop systems must be nontoxic and noncorrosive: nontoxic because should a leak develop in the loop system, the groundwater will not be contaminated and noncorrosive for the protection of the metal components of the system that come into contact with the circulating solution.

In situations when the local well water temperature is below 60 °F (15.56 °C), the water in the earth loop should be protected from freezing at temperatures down to 18 °F (−7.78 °C). The recommended antifreeze solution is propylene glycol. The amount of antifreeze needed for the earth loop is determined as follows: Calculate the approximate volume of water in the loop system by using Table 21-3. This table gives the gallons (liters) of water per 100 ft (30.48 m) of pipe. Add 2 gal (7.57 liters) for the equipment room devices and heat pump.

Propylene Glycol. When the groundwater at a depth of 100 ft (30.48 m) is 66 °F (18.89 °C) or less, a 20% solution by volume of propylene glycol is required to provide the necessary protection. The percentage of antifreeze solution depends on the geographical location. A solution 20% by volume of propylene glycol will provide freeze protection down to 18 °F (−7.78 °C). For example, for a system that holds 100 gal (378.54 liters) of water in the loop, 20 gal (75.708 liters) of propylene glycol is needed to provide the desired protection.

To add the antifreeze, two small pieces of hose, a bucket, and a small submersible pump are needed. Block the system by closing a ball valve. Blocking the flow prevents the antifreeze from being pumped into one boiler drain and out the other.

HEAT PUMP SYSTEMS

TABLE 21-3 Water Capacity of Pipe

Pipe Material	Nominal Pipe Size (in.)	Gallons per 100 ft of Pipe
Polyethylene		
SDR-11	¾	3.02
SDR-11	1	4.73
SDR-11	1¼	7.52
SDR-11	1½	9.85
SDR-11	2	15.40
SCH 40	¾	2.77
SCH 40	1	4.49
SCH 40	1¼	7.77
SCH 40	1½	10.58
SCH 40	2	17.43
Polybutylene		
SDR-17 IPS	1½	11.46
SDR-17 IPS	2	17.91
SDR-13.5 CTS	1	3.74
SDR-13.5 CTS	1¼	5.59
SDR-13.5 CTS	1½	7.83
SDR-13.5 CTS	2	13.38
Copper	1	4.3

Source: Bard Manufacturing Company.

Attach the hoses to the boiler drains. Run the uppermost hose to a drain. Connect the other hose to the submersible pump in the bucket. Put full-strength propylene glycol in the bucket and pump the desired amount into the system. When the required amount has been pumped in, turn off the pump, close the boiler drains, disconnect the hoses, and open the isolation flange or gate valve.

Calcium Chloride. A 20% by weight solution of calcium chloride and water may also be used as an antifreeze in the earth-coupled heat pump system. It is also nontoxic, a better heat conductor, and less expensive than propylene glycol. However, it is mildly corrosive. To determine the amount of calcium chloride needed for the system, multiply the gallons (liters) of water in the loop system by 1.4841 to find the pounds (kilograms) of 94 to 98% pure calcium chloride required to provide the desired 18 °F (−7.78 °C) freeze protection.

□ Closed-Loop Lake Heat Exchanger □

The closed-loop earth-coupled heat pump systems that use a pond or a lake for their heat source are by far the most economical of all closed-loop earth-coupled systems. The use of pond loops for residen-tial applications are restricted to bodies of water having at least 1 acre of surface area and a depth of 10 ft (3.048 m) or more at the location of the loop during worst-case low-water-level conditions. The acceptability of any body of water varies depending on the size, average depth, and water source (springs, river or creek, runoff, etc.). Most manufacturers recommend that they be consulted to discuss any application larger than 5 tons, or any commercial installation.

Also, there are special considerations that must be given to pond loop applications. *Never,* under any circumstances, place a loop in a river or stream, because the drifting objects at flood conditions will in all probability cause damage to the loop. The pond loop gets colder than either a horizontal or a vertical earth coupling, so regardless of the location, the system must use a 20% by volume antifreeze solution. Evaluate the location of the pond loop in relation to the building. There is no need to install a pond loop when the trench length to the pond and back to the building is equal to or greater than that required for a horizontal earth coil.

Heat Exchanger. The length of the loop is determined to a great extent by the temperature of the water in the pond. For example, if the well water temperature is between 50 and 68 °F (10 and 20 °C), use 60 ft (18.288 m) of ¾-in. (19.05-mm) copper pipe per nominal ton of refrigeration. The water temperature information is available from the local water well association of the National Water Well Association. In areas where the well water temperature is indicated to be either above 68 °F (20 °C) or below 50 °F (10 °C), use 80 ft (24.384 m) of ¾-in. (19.05-mm) copper pipe per nominal ton of refrigeration (Figure 21-27).

Location. The pond loop will not perform correctly if it is allowed to settle down into the mud. It should be supported at least 1 ft (0.3048 m) off the bottom, and at least 9 ft (2.7432 m) below the surface of the water. Suspending the pond loop under a pier is an easy application method. Another is to secure an old tire to each of the corners and allow the pond loop to settle to the top of the tires. The tires act as spacers to keep the loop about 1 ft (0.3048 m) or so off the bottom of the pond.

Installation. The service lines for the pond loop should be buried 4 ft (1.2192 m) deep or below the frostline, whichever is deeper. The lines should be

Male adapter

Header

3/4" Copper

20'

1'

Male adapter

FIGURE 21-27 **Typical lake heat exchanger. (Courtesy of Command-Aire Corporation.)**

about 2 ft (0.6096 m) apart in a wide trench. If a narrow trench is used, the pipes should be at depths of 4 and 6 ft (1.2192 and 1.8288 m). Do not attempt to use the pond loop if the service lines cannot be buried across the shore of the lake and out into the water. *Never* bring service lines up from the loop and run them underneath the pier—always leave the lines in the water and as deep as possible.

□ Open-Loop (Groundwater) Application □

In an open-loop system the water is taken into the system, passed through the heat pump, and then discharged. It is not recirculated as in the closed-loop groundwater systems.

Since the water is a crucial element in the operation of these types of systems, many manufacturers recommend that the services of an experienced local well driller be obtained. His expertise is valuable in assessing an existing well and equipment or in making recommendations for a new installation. If no local driller is available, contact your state well drilling association or the National Water Well Associa-

tion at 6375 Riverside Drive, Dublin, Ohio 43017 for referral. Their telephone number is (614) 761-1711.

The water used in these systems does not need to be suitable for human consumption; however, it should meet certain standards (Table 21-4). The local or state health department can run water tests to see if it will meet the manufacturer's recommendations.

Most heat pumps require a water flow rate of approximately 2 gpm (7.5708 liters/m) per nominal ton of refrigeration. The required flow rate may need to be somewhat higher in northern climates during winter operation. Under extreme weather conditions the heat pump(s) may run continuously for extended periods. Therefore, the water supply must be able to deliver the appropriate continuous water flow for 24 hr or more. A thorough "drawdown" testing of the well, according to the methods described by the National Water Well Association, will indicate if it has the capacity to handle the extreme conditions demand. If the well is intended to supply both the heat pump and the household water supply, be sure to conduct the "drawdown" test using the combined

TABLE 21-4. Water Standards for Open-Loop (Groundwater) Heat Pump Application

Scaling	
Calcium and magnesium salts (hardness)	Less than 350 ppm
Iron oxide	Low
Corrosion	
pH	5–10
Hydrogen	Less than 50 ppm
Carbon dioxide	Less than 75 ppm
Chloride	Less than 600 ppm
Total dissolved solids	Less than 1500 ppm
Biological growth	
Iron bacteria	Low
Suspended solids	Low

Source: Command-Aire Corporation.

DETAIL A

FIGURE 21-28 Heat exchanger valve location. (Courtesy of Command-Aire Corporation.)

water requirements, not just the heat pump water flow. It is the responsibility of the driller to be knowledgeable as to all applicable water well codes and regulations and to perform the installation accordingly.

A final consideration for the water well is its pump sizing and depth. Careful evaluation of the pumping power requirements may prevent an installation where all the energy saved by the heat pump is consumed by the well pump.

After an adequate water supply is assured, the water must be delivered to the heat pump. When tapping into an existing pressure line, the tap should be made immediately after the pressure tank prior to down-line taps. The tank should be sufficiently large to prevent rapid cycling of the well pump. If the tank is not of sufficient size, a new tank must be installed, or it might be desirable to bypass the tank and cycle the pump with the heat pump unit. Remember, the electrical power consumed by the pump is an important consideration. These are decisions that need to be made for each installation, based on the existing equipment and water availability.

Control Valve Installation. Because of the open-loop nature of all natural well applications, the installer should take precautions against scaling and fouling of the water-to-refrigerant heat exchanger. By keeping the well water under pressure and not introducing air into the system, some of the more severe problems can be avoided. Installation of water controls on the discharge side of the heat pump will

keep the water under pressure in the heat exchanger.

Pete's Plugs, boiler drains, and ball or gate valves should be installed in the "water in" and "water out" heat pump piping runs (Figure 21-28). Pete's Plugs are self-sealing orifices that allow a probe-type thermometer or pressure gauge to be inserted directly into the system water. The temperature differential and the pressure drop across the heat exchanger are easily measured using one thermometer or pressure gauge. Measuring and recording the pressure drop across the heat exchanger provides a reference point to use when checking for scaling problems. Gate or ball valves and boiler drains will permit isolation and cleaning of the heat exchanger should scaling occur.

The water flow to the heat pump can be controlled by a slow-closing solenoid valve, a motorized slow-closing valve, or a pressure-regulating valve assembly. Some manufacturers recommend the size and type of valve to be used on their equipment. Their recommendations should be followed.

The water flow control valve should be installed in the proper location (Figure 21-29). Based on normally closed valves, the wiring for the solenoid valve is shown in Figure 21-30. The alternative motorized valve should be wired as shown in Figure 21-31.

Some units make use of a pressure regulating valve to control the flow of water through the unit (Figure 21-32). Once the valve assembly is installed, con-

FIGURE 21-29 Control valve location. (Courtesy of Command-Aire Corporation.)

FIGURE 21-30 Solenoid valve wiring diagram. (Courtesy of Command-Aire Corporation.)

FIGURE 21-31 Motorized valve wiring diagram. (Courtesy of Command-Aire Corporation.)

FIGURE 21-32 Pressure regulating valve assembly. (Courtesy of Command-Aire Corporation.)

nect the sensing line tee to the refrigerant port provided on most residential units. If there is no port, a port must be added to the refrigerant line between the reversing valve and the water-to-refrigerant heat exchanger. The assembly contains two valves, one that opens in response to an increase in discharge pressure during the cooling cycle and the other that opens in response to a decrease in suction pressure during the heating cycle. The correct flow may be established by setting the opening point to its lowest setting if the ambient temperature is below the set point. If the ambient temperature is too low to allow the valve to open, the system may be manually flushed by lifting the range spring follower with a screwdriver at two sides of the lower spring cap to open the valve. This will not affect the valve adjustment. To optimize performance, with the refrigerant gauges attached, operate the heat pump in the heating mode. Adjust the heating-cycle water-regulating valve for an increase in the suction pressure. Adjust the valve to the point where there is little or no increase in the pressure reading by further opening. Any increase beyond this setting will be wasting water. The object is to gain a higher pressure reading for optimal heating capacity, and this is gained by opening the valve for heating.

Now operate the unit in the cooling mode and adjust the cooling cycle valve according to Table 21-5. Be sure to use the recommended pressure settings for the unit under consideration. Each manufacturer will have their own settings to follow.

After completing these adjustments, remove the refrigerant gauges and tightly replace the caps on the Schrader service ports to avoid refrigerant leakage. It is the responsibility of the installer to be knowledgeable as to all applicable plumbing and

TABLE 21-5 Water-Regulating Valve Operating Pressures (psi)

Water Temperature (°F)	Cooling: 80°F Return Air Discharge Pressure	Heating: 70°F Return Air Suction Pressure
45	160 min	46–54
50	160 min	52–60
55	160 min	56–66
60	166–190	60–72
65	178–204	62–74

Source: Command-Aire Corporation.

HEAT PUMP SYSTEMS

electrical codes and regulations and to perform the installation accordingly.

Water Discharge. Three common methods of discharge disposal are:

1. *Recharge well.* In most cases, the use of a recharge well is recommended for water from the heat pump. There should be at least 50 ft (15.24 m) between the supply and the recharge well to prevent thermal interference between the wells.
2. *Drain fields.* In most areas, the use of a drain field is not recommended. Drain fields require soil conditions that allow very rapid percolation. Remember: If a 5-ton heat pump runs continuously for 24 hr, it will dischage 14,400 gal (54,509.9 liters) of water.
3. *Surface Discharge.* Trouble-free, with no clogged screen, percolation, or service problems, the surface discharge application is very simple. Regardless of dry land or surface water discharge, the requirements are as follows:
   (a) Discharge a minimum of 1 ft (0.3048 m) above flood level at the discharge point. This prevents organic growth or animal entry into the system. Siphoning due to a loss of pressure is also prevented.
   (b) Protect the discharge pipe end with a screen or flap valve to keep foreign objects out of the pipe.
   Note: Surface discharge with a secondary discharge to allow use of the water is an excellent idea. Uses such as irrigation or livestock watering allow optimum utilization of the dollars spent for pumping power. Be sure that the secondary circuit is properly designed and does not restrict the required flow of water to the heat pump.

Acceptance of these methods of disposal varies nationwide, so it is the responsibility of the installer to be knowledgeable as to all applicable codes and regulations and to perform the installation accordingly.

Special Applications. The majority of special applications are the result of attempting to use surface water to supply a heat pump or having insufficient water supply from the well. Surface water should not be used! Installations are attempted quite often with what initially seems to be great success, only to fail later. The following problems make long-term success almost impossible:

1. Suspended solids damage or destroy water controls or heat exchangers.
2. Plant life, small shellfish, and small fish always appear to be able to bypass protective devices to reach and restrict or clog a heat exchanger.
3. Temperature! Heated water rises and cooled water sinks. Any body of water, regardless of current or wave motion, is continually moving due to the effects of air and ground temperatures. These currents continually change the water temperature throughout the body of water. For this reason, it is difficult to rely on surface water continually to supply water no colder than 42 °F (5.56 °C). Entering water less than 42 °F (5.56 °C) will trip the freeze-stat and shut down the heat pump.

When an insufficient water supply is the problem, solutions can vary. Some manufacturers might recommend a closed-loop earth-coupled application. If you do not select that type of system, the following are basic descriptions of alternative methods for utilizing low-flow wells. The most common solution is to supply from and return to the same well. You can utilize one of the following methods to supply the heat pump(s).

1. *Use the well as a heat exchanger.* If the well is deep enough to have a large volume of water, you may be able to supply from and return to the same well without wasting any water. This method requires at least 100 ft (30.48 m) of standing water per ton and the owner's willingness to accept fluctuations in household water temperature if it comes from the same well.
2. *Waste some—save some.* If the well volume is too small to allow its use as a heat exchanger, you can return to the well while discharging some portion elsewhere. Typical examples are:
   (a) Any time that the heat pump is running, the primary discharge circuit will return most of the water to the well, while a secondary circuit wastes the remainder. This prevents depletion of the well by allowing its limited capacity to replenish what is wasted.
   (b) By using temperature sensors, the secondary circuit mentioned previously will waste water only if the supply temperatures become colder or hotter than desired. As the well replaces the wasted water, it also moderates the overall well water temperature.

In all cases, the well return discharge point should be as far from the pump intake as possible. Never discharge above the well's water level since aeration can cause clouding and increase the possibility of scaling and bacteria problems. Other solutions can be used with normal discharge methods or with the ''supply from and return to'' methods mentioned previously. Depending on water temperature and system requirements, you can utilize one of the following methods to conserve water.

1. Where the supply water temperature at the first heat pump is 50°F (10.00°C) or warmer, the water can be run through two heat pumps in series prior to disposal. The water flow provided should be the amount required by the largest heat pump if they are of different sizes. Figure 21-33 is a plumbing schematic for this type of application. Figure 21-34 is a wiring schematic using a 24-V solenoid valve.

2. When a mixing tank is used, a portion of the water already circulated through the heat pump can be mixed with the well supply water to provide a larger volume than the well can deliver. Figure 21-35 is a plumbing schematic for this type of application. Figure 21-36 is a wiring schematic using a Grundfos $\frac{1}{12}$-hp pump.

Heat Exchanger Cleaning. If scaling is a problem, the coil can be cleaned with a solution of an inhibited acid to minimize corrosion to copper or cupronickel heat exchanger tubing. Use the gate or ball valves shown in Figure 21-28 to isolate the heat pump from the remainder of the water system. Then, using a bucket and pump, circulate the acid solution through the heat exchanger by utilizing both boiler drains. Allow several hours for cleaning.

FIGURE 21-34 Wiring schematic for multiple heat pumps in series using a solenoid valve. (Courtesy of Command-Aire Corporation.)

FIGURE 21-35 Heat pumps using a mixing tank. (Courtesy of Command-Aire Corporation.)

FIGURE 21-36 Pump wiring schematic. (Courtesy of Command-Aire Corporation.)

☐ Safety Procedures ☐

The old staying that ''safety doesn't cost, it pays'' is certainly true. It is most important that proper safety procedures be followed, especially by workers in the air conditioning and refrigeration industry. Since these people generally work alone, help may be difficult to obtain should a serious injury occur. Safety may be divided into three general categories: safety of the worker, safety of the equipment, and safety of the contents.

Safety of the worker is, of course, of prime importance. Machinery and materials can always be

FIGURE 21-33 Plumbing schematic for multiple heat pumps in series. (Courtesy of Command-Aire Corporation.)

replaced, but a human life cannot. When the equipment and tools are used properly, there is very little danger to the worker. Always use the leg muscles when lifting heavy objects; never use the back muscles. Ask for help when lifting objects weighing more than 30 to 35 lbs (13.6 to 15.9 kg). Keep the floor and area clear of obstructions and debris. A slip or a trip while carrying heavy objects usually causes a serious injury.

Air conditioning and refrigeration equipment use electricity as their source of power. Be sure that all electrical circuits are disconnected from the power source before working on them. The disconnect switch or the circuit breaker will usually break the circuit. Never work a "hot" circuit.

Always wear safety goggles when working with refrigerants. If refrigerants should get in the eye, a frozen eye will result. When this happens, gently wash the eye with tap water and see a physician as soon as possible. Protect the skin from frostbite when working with refrigerants.

Safety of the equipment is also of great concern. Many components on refrigeration and air conditioning equipment are easily broken. Some of these components are very expensive and all of them take time to replace. Use the proper torque and sequence when tightening bolts and nuts. Before operating the equipment be sure that the fan blades and belts are clear of all objects.

Safety of the contents is a basic requirement and will depend to a great extent on the care given to the equipment during installation and servicing. The contents of a given installation must be kept at the desired temperature; this is the responsibility of the service technician. Technicians must know what the desired conditions are and maintain the equipment to produce these conditions. Most manufacturers provide tables and charts showing the desired operating conditions and the required control settings.

☐ SUMMARY ☐

☐ A heat pump may be defined as an air conditioning system that moves heat both to and from the conditioned area.

☐ Heat pumps are classified according to their heat source and the medium to which the heat is transferred during the heating season. The classifications are (1) air-to-air, (2) water-to-air, (3) water-to-water, and (4) ground-to-air.

☐ During the cooling season the heat pump operates as a normal cooling system.

☐ To avoid confusion on heat pump systems, the evaporator is termed the indoor coil and the condenser is termed the outdoor coil.

☐ During the heating season, the heat pump operates because of the reversed refrigerant flow.

☐ It should be noted that the reversed flow is accomplished by the use of a four-way valve and check valves.

☐ The indoor coils used on heat pump units are designed especially for heat pumps, and many times a coil is designed for a particular model unit.

☐ When properly designed, installed, and maintained, a heat pump can be superior to a gas heating unit.

☐ Solar-assisted heat pumps use a hot water coil to add heat to the building or to the unit for defrosting purposes.

☐ There are three basic steps involved in the defrost cycle: (1) detection of the frost buildup (defrost initiation), (2) frost removal, and (3) detection of sufficient frost removal to allow the unit to return to the heating cycle (defrost termination).

☐ Some of the more common methods of frost detection are: (1) air pressure, (2) time–temperature, and (3) solid state.

☐ Some of the more common methods of defrost termination are: air pressure, time termination, and pressure termination.

☐ A heat pump should not be used for reheating purposes.

☐ These units are designed to satisfy the cooling load, and supplementary heat is added to bring the unit to the desired heating capacity.

☐ Supplementary heat is used only when the heat pump cannot supply the desired amount of heat. It is controlled from two points: (1) the second stage on the indoor thermostat, and (2) the outdoor thermostats.

☐ A reversing valve is a device that changes the direction of the refrigerant flow in a heat pump system.

☐ When adding refrigerant to these systems or making any kind of adjustment, the manufacturer's specifications must be followed.

☐ Water source heat pumps are generally referred to as closed-loop earth-coupled and open-loop (water well or groundwater) systems.

☐ The basic operating principle of these systems is that the heat rejected to the water after being removed from the air being cooled by the refrigerant circuit during the cooling cycle and heat is extracted from the water by the refrigerant circuit and delivered to the air to be heated during the heating cycle.

☐ In a closed-loop earth-coupled heat pump system a solution is circulated through special pipes buried in the ground.

☐ There are three methods of installing these types of heat pump systems: (1) horizontal, (2) vertical, and (3) lake loops.

☐ The choice of vertical, horizontal, or lake loop earth coupling should be based on the characteristics of each application.

☐ The design of an earth-coupled system is divided into the following steps: (1) determine the structure design heating load in Btu/hr loss and design cooling load in Btu/hr gain, (2) select a water source heat pump, and (3) select the earth-type coil and materials to be used.

☐ Of particular importance when designing an earth-coupled heat pump system is the balance between the heat pump unit and the earth-coupled loop. When this balance is achieved, the earth-coupled loop will remove all the heat that is transferred to the water by the heat pump during the cooling cycle, and will provide all the heat that is required to heat the water during the heating cycle.

☐ Use only a water source heat pump that can be operated on loop water temperatures that fall below 40°F (4.44°C) and perhaps down to 25°F (−3.89°C).

☐ A vertical earth-coupled system consists of one or more vertical boreholes through which water flows in a plastic pipe.

☐ A series U-bend earth coupling is one in which the water flows through all the pipe, progressively traveling down and then up each well bore.

☐ CTS PB pipe is heat fused together with fittings. PE pipe is heat fused together with butt joints.

☐ A parallel U-bend earth coupling is one in which the water flows out through one header, is divided equally, and flows simultaneously down two or more U-bends.

☐ A horizontal earth-coupled system is similar to a vertical system in that water circulates through underground piping. However, the piping in this type of system is buried in a trench.

☐ Antifreeze will be necessary in the northern zone to prevent freezing of the circulated water and to allow the system to gain capacity and efficiency, by using the large amount of heat released when the water contained in the soil is frozen.

☐ If a double layer of pipe is used, the incoming water to the heat pump unit should be from the deepest pipe.

☐ The antifreeze solutions used in earth loop systems must be nontoxic and noncorrosive.

☐ A 20% by weight solution of calcium chloride and water may be used as an antifreeze in an earth-coupled heat pump system.

☐ Closed-loop earth-coupled heat pump systems that use a pond or a lake for their heat source are by far the most economical of all closed-loop earth-coupled systems.

☐ The length of the loop is determined to a great extent by the temperature of the water in the pond.

☐ The pond loop will not perform satisfactorily if the loop is allowed to settle down into the mud.

☐ In the open-loop type of system the water is taken into the system, passed through the heat pump, and then discharged. It is not recirculated as in the closed-loop groundwater systems.

☐ The water used in these systems does not need to be fit for human consumption; however, it should meet certain standards.

☐ A final consideration for the water well is its pump sizing and depth. Careful consideration of the pumping power requirements may prevent an installation where all the energy saved by the heat pump is consumed by the well pump.

☐ Three common methods of discharge disposal are: (1) recharge well, (2) drain fields, and (3) surface discharge.

☐ In all cases, the well return discharge point should be as far from the pump intake as possible.

☐ If scaling is a problem, the coil can be cleaned with a solution of an inhibited acid to minimize corrosion to copper or cupronickel heat exchanger tubing.

☐ EXERCISES ☐

1. How are heat pumps classified?
2. Name the classifications of heat pumps.
3. Name the coils on heat pump systems.
4. During the heating cycle, how does the refrigerant flow through the system?
5. When properly designed, installed, and maintained, why are heat pumps superior to gas heating units?
6. Why are defrost periods necessary on heat pump systems?
7. Name the basic steps in a defrost cycle.
8. Name the methods of frost detection on heat pump systems.
9. Name the types of defrost termination controls.
10. Why should a heat pump system not be used for reheating purposes?
11. How are heat pump units sized?
12. What is the purpose of supplementary heat on heat pump systems?
13. What is the function of the reversing valve used on heat pump systems?
14. What should be done when adding refrigerant to a heat pump system?
15. How are water source heat pumps generally referred to?
16. Name three ways of installing the closed-loop earth-coupled heat pump system.

17. What are the steps used in the design of earth-coupled heat pump systems?
18. What is the maximum oversize capacity recomended for heat pump systems?
19. What is of particular importance when designing an earth-coupled heat pump system?
20. What types of pipe are recommended for earth-coupled heat pump systems?
21. What is the purpose of location markers?
22. For 1½-in. (38.10-mm) pipe, what size borehole is used?
23. What must be done to the bore around the pipes immediately below the service lines?
24. Define a series U-bend earth coupling.
25. Define a parallel U-bend earth coupling.
26. How is the piping in a horizontal earth-coupled heat pump system installed?
27. What types of antifreeze solutions are used in earth-coupled heat pump systems?
28. When two pipes are laid in the same trench, how much longer must the total pipe be?
29. If a double-layer pipe system is used, which pipe should be connected to the unit?
30. How much distance should there be between two pipes buried in the same trench?
31. What determines the amount of antifreeze solution to be used in a system?
32. Which is the better heat conductor, propylene glycol or calcium chloride?
33. What type is the most economical of all the earth-coupled heat pump systems?
34. Why should a loop never be installed in a river or stream?
35. What determines the length of the loop for a closed-loop lake heat exchanger?
36. How much water is recirculated through an open-loop heat pump system?
37. What precautions should the installer take when installing a control valve on an open-loop system?
38. Describe the adjusting procedure for the heating-cycle water-regulating valve.
39. Name three common methods of discharge water disposal.
40. How can the coil on a water source heat pump system be cleaned?

22 Air Conditioning (Cooling)

Performance Objectives

After completion of this chapter, the student should:
- ■ Be able to define the term air conditioning.
- ■ Be able to list the primary reasons for using air conditioning.
- ■ Know the conditions that affect body comfort.
- ■ Understand the air distribution system used in air conditioning systems.
- ■ Be familiar with the terminology used in air conditioning.
- ■ Have a working knowledge of psychrometrics.
- ■ Have a basic understanding of heat load calculation.
- ■ Know more about the different types of equipment used as air conditioning systems.
- ■ Complete the Exercises with at least 80 percent accuracy.
- ■ Complete the Laboratory Exercises with at least 80 percent accuracy.

Most people have little appreciation of the basic principles of air conditioning, probably because the public in general only became conscious of air conditioning about 1920. It was at this time that the large-scale use of air conditioning on trains and in theaters began. It was these systems that first exposed large numbers of people to the comfort and advantages of summer cooling and at the same time caused the erroneous impression that air conditioning is synonymous with cooling.

The cooling process involves air circulation, cooling, dehumidifying, and cleaning. The other processes are accomplished during the heating process.

Definition. The accepted definition of the term *air conditioning* is the simultaneous mechanical control of temperature, humidity, air purity, and air motion. Unless all of these conditions are controlled, the term *air conditioning* cannot be properly applied to any system or equipment. It should be noted that the control of temperature can mean either cooling or heating. The control of humidity can mean either humidifying or dehumidifying. Thus an industrial system that provides an indoor condition of 150 °F (66 °C) dry bulb (DB) at 75% relative humidity (RH) can just as well be called air conditioning as a system designed to provide indoor conditions of 80 °F (26.7 °C) DB and 50% RH. Similarly a system that only cools a space without regard to the relative humidity or air purity or motion cannot be properly called a true air conditioning system. An air conditioning system can maintain any atmospheric condition regardless of variations in the outdoor atmosphere.

□ Human Comfort □

The two primary reasons for using air conditioning are to improve the control of an industrial process and to maintain human comfort. The conditions to be maintained in an industrial process are dictated by the very nature of the process or the materials being handled. In a comfort system, however, the conditions to be maintained are determined by the requirements of the human body. Therefore, an understanding of the essential body functions is basic to an understanding of air conditioning.

Human comfort is dependent on how fast the body loses heat. The human body might be compared to a heating unit that uses food as its fuel. Food is composed of carbon and hydrogen. The energy contained in the fuel, food in our case, is released by oxidation. The oxygen used in the process comes from the air, and the principal products of combustion are carbon dioxide and water vapor. Doctors call this process metabolism.

The human body is basically a constant-temperature machine. The internal temperature of the human body is 98.6°F (37°C), which is maintained by a delicate temperature-regulating mechanism. Because the body always produces more heat than it needs, heat rejection is a constant process. The main object of air conditioning is to assist the body in controlling the cooling rate. This is true for both the heating and cooling seasons. In summer, the job is to increase the cooling rate; in winter, it is to decrease the cooling rate.

In the air conditioning process, there are three ways that the body gives off heat: (1) convection, (2) radiation, and (3) evaporation. In most cases the body uses all three methods at the same time.

Convection. In the convection process, the air close to the body becomes warmer than the air farther away from the skin. Because the warm air is lighter than the cool air, the warm air rises. This warm air is replaced by the cooler air, and the cooling by convection is a continuous process. As this air becomes warm, it also floats upward. Even though the deep body temperature remains at 98.6°F (37°C), the human skin temperature will vary. The skin temperature may vary from 40 to 105°F (4.4 to 40°C) in relation to the temperature, humidity, and velocity of the surrounding air. If the temperature of the surrounding air drops, the temperature of the skin also will drop.

Radiation. Heat radiates directly from the body to any cooler surface just as the rays of the sun travel through space and warm the surface of the earth. Heat may flow from the skin to any surface or object that is cooler than the body. This process is independent of the convection process. The temperature of the air between the person and the cooler surface has no effect on the radiation process. The same principle applies when a person is warmed by a camp fire. The side next to the fire gets warm while the other side is cool. The air temperature between the person and the fire is the same as the air temperature on the person's other side.

Evaporation. Evaporation heat regulation is the body process that maintains life outside of an air-conditioned space. In this process moisture, or perspiration, is given off through the pores of the skin. When this moisture evaporates, it absorbs heat from the body and cools it. The effect of evaporation can be felt more easily when alcohol is put on the skin because the alcohol vaporizes more readily and absorbs heat faster. This evaporation turns the moisture into low-pressure steam or vapor and is a continuous process. When drops of sweat appear on the skin, it means that the body is producing more heat than it can reject at the normal rate.

◻ Conditions that Affect Body Comfort ◻

There appears to be no set rule as to the best conditions for all people. In the same atmospheric conditions, the young, healthy person may be slightly warm while an elderly one may be cool.

The three conditions that affect the ability of the body to give off heat are: (1) temperature, (2) relative humidity, and (3) air motion. A change in any one of these conditions will either speed up or slow down the cooling process.

Air Temperature. Air at a temperature lower than the skin will speed up the convection process. The cooler the air, the more heat the body will lose through convection. Heat always flows from a warm place to a cooler place. The greater the temperature difference, the faster the heat will flow. If this difference in temperature is too great, the body will lose heat more rapidly than it should, and it will become uncomfortable.

If the air temperature is higher than the skin temperature, the convection process will be reversed. The body heat will be increased. It can be seen that the air temperature has a very important effect on human comfort. Experience shows that air temperatures ranging from 72 to 80°F (22.2 to 26.7°C) feel comfortable to most people.

The temperature of any surrounding surfaces is also important because this temperature affects the rate of radiation from the body. The lower the surface temperature, the more heat is given off by the body through radiation. As the temperature difference between the surface and the body is decreased, the rate of radiation is decreased. The radiation process will be reversed if the surrounding surface temperature is higher than the body temperature. When this happens, the body must give off more heat through the convection and evaporation processes.

Relative Humidity. Relative humidity regulates the amount of heat that the body can reject through evaporation. Relative humidity is a measure of the amount of moisture in the air. It is an indication of the ability of the air to absorb moisture. Relative humidity is basic to the air conditioning process.

As an example, let us consider 1 ft³ of air at 70 °F (21.1 °C) that contains 4 grains (0.26 g) of water vapor (Figure 22–1). A grain is a very small amount of water. Actually, it takes 7000 grains (453.6 g) to make 1 lb.

The relative humidity can be determined in the following manner. There are only 4 grains (0.26 g) of water vapor in the cubic foot of air. If that cubic foot of air held all the moisture it could possibly hold at that temperature, it would actually hold 8 grains (0.52 g) of water vapor (Figure 22–2), and would be said to be saturated.

To find the relative humidity, divide the moisture actually present in the air by the amount that the air could hold at a saturated condition at the same temperature (Figure 22–3). This process tells us that the relative humidity is 50%. Relative humidity is an indication of the amount of moisture in the air compared to the amount of moisture that could be present at that same temperature. The relative humidity will change with a change in temperature.

For example, the temperature of the air is raised to 92 °F (33.3 °C) without adding any moisture. When the humidity tables are checked, we can see that 1 ft³ of air at 92 °F will hold 16 grains (1.04 g) of water vapor when saturated. The relative humidity in this example is 4 grains (0.26 g) divided by 16 grains or 25% (Figure 22–4).

If the air that surrounds the body has a low relative humidity, the body will give off more heat through evaporation. If the air surrounding the body has a high relative humidity, the body will give off less heat through evaporation. A conditioned air temperature at 80 °F (26.7 °C) and 50% RH will be reasonably comfortable.

Air Movement. An increase in the rate of evaporation of perspiration from the body is largely a result

FIGURE 22–1 One cubic foot of air at 70 °F (21.1 °C) with 4 grains (0.26 g) of water vapor.

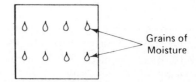

FIGURE 22–2 Cubic foot of saturated air.

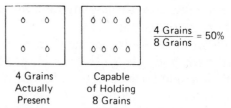

FIGURE 22–3 Determination of relative humidity.

FIGURE 22–4 Relative humidity at a higher temperature.

of air movement. Evaporation is dependent on the ability of air to absorb moisture. As the air moves across the skin, the moisture-laden air is replaced by drier air that allows more moisture to evaporate from the skin.

If the air is allowed to remain still (static), the air next to the skin will absorb moisture until the saturation point is reached. As the saturation point is reached, the evaporation process is slowed down. The moisture will evaporate more slowly and will eventually stop when the saturation point is reached and the person will feel uncomfortable.

The movement of air also speeds up the convection process. This is possible because the warm air next to the skin is replaced by cooler air and heat is given up from the body to the air.

Air movement also removes heat from other

substances such as walls, ceilings, and other objects surrounding the body, thus, tending to speed up the radiation process. It must be remembered that air motion is one of the conditions that affects the comfort of human beings.

Air
Flow

Air Flow

Fan
Motor

FIGURE 22–5 Axial flow fan.

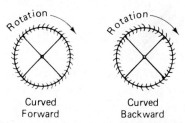

Rotation

Rotation

Curved
Forward

Curved
Backward

FIGURE 22–6 Centrifugal fans.

□ Air Distribution □

The air introduced into a conditioned space should be distributed so that there will be only minor temperature differences between the floor and ceiling—from floor level to 6 ft (1.8 m) above the floor—and between the inside and outside walls. The proper quantity of air can be delivered to the different areas of the room by a careful consideration of the cooling and heating requirements. However, these requirements must be met without drafts. Generally, an air velocity of 15 to 25 fpm (4.6 to 7.6 m/min) is considered still air, while air moving with a velocity of 65 fpm (19.8 m/min) is considered drafty to most people.

Air Distribution Requirements. A good air distribution system must do the following:

1. Mix the conditioned air with enough room air so that upon reaching the occupied zone the airstream will not be so cold as to be objectionable.
2. Allow the air velocity to be reduced sufficiently before reaching the occupied zone so as to prevent drafts.
3. Provide a turbulent, underflowing air motion within the occupied zone.
4. Keep any air noise from the supply and return grills to an unobjectionable level.

The requirements may seem difficult to meet, but it should be remembered that an air conditioning unit may have the very best of components and still be, from the owner's viewpoint, unsatisfactory if the air distribution does not provide comfort for the occupants.

Fan Requirements. The fans used in modern air conditioning systems fall into two general classes depending on the direction of the airflow through them. The first class is the axial flow type, through which the air flows in the direction of and parallel to the fan shaft. This type of fan is commonly known as a *propeller fan* (Figure 22–5). It consists essen-

tially of two or more blades which extend from the shaft. Each blade is bent or twisted to give the desired pitch. The blades are often made from formed sheet metal but are sometimes made from castings and molded from plastic. This type of fan is best suited for applications where it is necessary to move large volumes of air against low resistances, such as air-cooled condensers, humidifiers, and cooling tower fans.

The second class of fan is the radial flow type, through which the airflow is outward from the fan shaft. This outward airflow is produced by centrifugal force. Therefore, this type is commonly known as a *centrifugal fan* or blower. It consists of a series of blades mounted around the circumference of a circle with the shaft in the center. The blades themselves are parallel to the shaft and may be either forward curved or backward curved (Figure 22–6).

The fan using forward-curved blades will generally run at a lower speed than a backward-curved fan for the same air delivery conditions. However, the fan that uses backward-curved blades will tend to be nonoverloading while a forward-curved fan will tend to overload as the pressure is reduced and the air volume is increased. Therefore, unless a very careful analysis is made as to the exact operating pressure, there is considerably more danger of overloading and burning out the motor with forward-curved fans.

The fan wheel of a centrifugal blower is not sufficient by itself to provide a usable flow of air. A casing or housing is always necessary to collect and direct the airflow as it leaves the fan blades. This housing, or scroll, usually starts with a very small area just after the discharge outlet and progressive-

FIGURE 22-7 Centrifugal fan scroll.

ly increases in size around the circumference of the shell until the maximum size is reached at the discharge opening (Figure 22-7). This shape is known as a *volute*. The air inlet to the wheel usually takes the form of a circular opening in the casing in one or both sides. The diameter of this opening is as large as possible without exceeding the inside diameter of the blower wheel.

Centrifugal fans are capable of delivering large volumes of air against considerable resistance. It is for this reason that they are the first choice in air conditioning and ventilating systems where it is necessary to overcome the resistance of the air filters, cooling coils, heat exchangers, ducts, and outlets.

Fan Noises. Fans and blowers can be designed and operated at such speeds as to provide almost any desired velocity and air volume. However, the difficulty in air conditioning work is the noise produced by the fans and blowers. The noise is caused mainly by the successive air waves caused by the fan blades and by the turbulence caused by the resistance of the air being forced through the system. A small amount of noise may be caused by the vibrating fan blades. This noise is relatively minor, however, when compared with the noise that may be created by a high-velocity fan. The humming noise produced by a fan is in effect a miniature siren. The loudness and the pitch of the sound depend on the tip speed of the fan, the general shape of the blades, and the number and angle of the blades. Thus, it can be seen that there is a definite relation between the amount of noise and the quantity of air being delivered. To be suitable, an air conditioning system must deliver the proper quantity of air with a minimum of fan noise.

Heating Chamber. In most air conditioning systems the next device the air encounters is the heating chamber. This device can be either a gas-fired furnace heat exchanger, an electric heating element, or a hot water or steam coil, depending on the system design. It is here that the air is heated during the heating season. The heat gained by the air is termed

FIGURE 22-8 Cooling coil. (Courtesy of Southwest Manufacturing Company.)

the *temperature rise*. During the cooling season this apparatus does not generally supply heat.

Cooling Chamber. Upon leaving the heating chamber, the air flows through the cooling chamber. The cooling process is generally done by a refrigeration evaporator (Figure 22-8). However, the cooling process may also be accomplished by a chilled-water coil. During the cooling process the temperature of the air is lowered. When there is excessive moisture in the air, some of it is removed to lower the humidity. There must be a drain line connected to drain away this condensate. The moisture or humidity level must be kept low to produce effective cooling. The cooling chamber is not generally used during the heating process even though air will still flow through it.

Humidifier. As the air leaves the cooling chamber, it enters the area where a humidifier may be located (Figure 22-9). Not all systems include a humidifier. The humidifier is used to add moisture to the air during the heating season. As the air leaves the heating chamber the relative humidity is lower. This warm, dry air should be humidified for comfort and health reasons. The desired moisture is added to the air by evaporation, or by spraying moisture directly into the airstream, depending on the type of humidifier used. These units are not generally used during the cooling season because the removal of moisture is part of the cooling process. To add moisture at this time would be expensive and undesirable.

AIR CONDITIONING (COOLING)

Supply Ducts. The air is then forced through a series of pipes known as air ducts that direct the air to the desired area. This duct system must be properly designed if satisfactory operation is to be obtained. The air outlet location must also receive a great deal of consideration in order to maintain proper air distribution within the conditioned space. These air ducts are insulated to prevent the escape of heat during the heating season and the absorption of heat during the cooling season. This insulation must include a vapor barrier to prevent the condensation of moisture on the colder surfaces during the cooling process (Figure 22-10). The duct system is used during both the heating and cooling processes.

Supply Outlets. Supply outlets are devices placed on the end of the supply duct in the conditioned space (Figure 22-11). They help to distribute the air properly in the room. Some outlets are used to fan the air, while others are used to direct the air in a jet stream. Other types are designed to do both. They are used to control the direction of airflow. The directional control, along with the location and number of outlets in the room, contribute tremendously to the satisfactory operation and comforting effect of the air pattern.

Conditioned Space. The conditioned space is one of the most important parts of the air distribution system. If an enclosed space did not exist, the air could not be reclaimed and the circulation of conditioned air would not be possible. It is because of this that the conditioned space is so important. The material and workmanship used to build the space are also important because they help to control the heat loss or heat gain inside the space (Figure 22-12).

Return Outlets. Return outlets are the openings through which the air from the conditioned space is allowed to enter the return duct system. They are generally located opposite the supply outlet (Figure 22-13). This is not always possible nor feasible but should be done when possible. They should be properly sized to allow air to leave the room with as little resistance as possible.

Return Ducts. Return ducts are used to connect the conditioned space and the air-handling equipment together (Figure 22-14). The return ducts are generally designed for less resistance to airflow than the supply ducts. This is done to ensure sufficient

FIGURE 22-9 Humidifier location.

FIGURE 22-10 Air duct with insulation and vapor barrier.

FIGURE 22-11 Supply outlet.

FIGURE 22-12 Conditioned space.

FIGURE 22-13 Location of return outlets.

FIGURE 22-14   Return duct.

FIGURE 22-15   Location of filter.

air return to the fan. Return ducts are sometimes, but not always, insulated. If the duct is to be run through a hot or cold space, it should be insulated to prevent heat loss or heat gain into the air inside the duct.

Filters. Filters are located at the inlet of the fan to prevent dust particles from entering the equipment (Figure 22–15). Their only purpose is to clean the air. They should always be properly located to protect the fan, heating chamber, and cooling chamber from dust particles. If this dust is allowed to enter the equipment, it will gradually collect on the surfaces and will reduce the airflow and adversely affect system operation. Filters are made from many materials such as spun glass and composition plastic. More efficient types of filters operate on the electronic principle.

▢ Air Conditioning Terminology ▢

Before a study of air conditioning can be undertaken, an understanding of the terms used is necessary. The following definitions explain the most common terms.

Dry Air. Dry air is air that contains no water vapor. In nature, air contains some water vapor.

Absolute Humidity. Absolute humidity is the actual quantity or weight of moisture present in the air in the form of water vapor. It is expressed as the weight of water vapor in grains per pound of air. There are 7000 grains of moisture to one pound of moisture. A rise or fall in temperature without a change in the quantity of moisture in the air does not affect the absolute humidity.

Relative Humidity. Relative humidity (RH) is the ratio between the moisture content of a given quantity of air at any temperature and pressure compared to the maximum amount of moisture which that same quantity of air is capable of holding at the same temperature and pressure. Relative humidity is expressed as a percentage. Note particularly the difference between relative humidity and absolute humidity. The relative humidity can be found by dividing the amount of moisture actually present by the maximum moisture-holding capacity of the air at the same temperature and pressure.

EXAMPLE
If 1 ft³ (0.0283 m³) of air contains 4 grains (0.26 g) of moisture and is capable of holding 8 grains (0.52 g) of moisture, what is the relative humidity?

Solution

$$RH = \frac{\text{absolute humidity}}{\text{maximum amount it can hold}} = \frac{4}{8} = 50\%$$

A rise in air temperature will cause a decrease in the ratio or relative humidity. A fall in air temperature will cause a rise in the ratio or relative humidity, unless the saturation point (100% RH) has been reached. Beyond this point, condensation will take place as the temperature falls. When air contains all the moisture it can hold at any given temperature and pressure, it is said to be saturated. When air is saturated, the relative humidity is 100%.

Humidification. The process of adding moisture to the air is called humidification.

Dehumidification. The process of removing moisture from the air is called dehumidification.

Saturated Air. Saturated air is air that contains all the water vapor it is capable of holding at that particular pressure and temperature. When air is saturated the dry-bulb, the wet-bulb, and the dew-point temperatures are the same.

Dry-Bulb Temperature. The temperature of the air as indicated by an ordinary thermometer is known as the dry-bulb (DB) temperature and is a measure of the sensible heat content of the air.

Wet-Bulb Temperature. The wet-bulb (WB) temperature is the temperature of the air as measured by an ordinary thermometer with the bulb covered with wet cloth or gauze (Figure 22–16). The temperature is taken after the thermometer has been exposed to a rapidly moving airstream. The temperature indicated on a wet-bulb thermometer will be depressed lower than a dry-bulb thermometer reading because of the evaporation of the moisture in the wick. The heat necessary for the evaporation of the water from the wick is removed from the air while the air is passing over the thermometer. The thermometer has been exposed to the airstream for a sufficient length of time when the temperature falls to a point of equilibrium, which is determined by the rate of moisture evaporation from the wick and the quantity of sensible heat in the air.

The point of equilibrium is the temperature at which the rate of sensible heat transfer from the air to the water on the wet bulb is equal to the rate at which heat is transferred from the wet bulb to the air by evaporation of the water. For any given dry-bulb temperature and any given relative humidity, the point of equilibrium will be the same. Since the rate of evaporation from the wick is determined by the quantity of moisture in the air, as well as the sensible heat, it can be seen that the wet-bulb temperature is indicative of the total heat contained in the air.

Web-Bulb Depression. The difference between the dry-bulb temperature and the wet-bulb temperature for any given condition is called the wet-bulb depression, except at the saturation point when the two temperatures are the same.

Dew-Point Temperature. The dew-point temperature is the temperature at which water vapor will start to condense out of the air. The quantity of water vapor in the air is always the same at any given dew-point temperature. Therefore, the quantity of moisture in the air can be measured by the dew-point temperature. When air is at the dew-point temperature, it is holding all the moisture it can hold at that temperature. As long as there is no removal or addition of moisture, the dew-point temperature will remain constant. No latent heat is given to or

FIGURE 22–16 Wet-bulb thermometer.

released from a mixture of air and water vapor unless some moisture is added to or removed from the air.

Saturation Temperature. When air is at the dew-point temperature, it is also at the saturation temperature and is considered to be saturated with moisture. The relative humidity is 100% at the saturation temperature. That is, it contains all the moisture it can hold at that temperature.

Total Heat. When the sensible heat and the latent heat are added together, the sum is known as the total heat. In air conditioning, 0°F (−17.8°C) is usually taken as the point from which heat content is measured. Total heat is also indicated by a wet-bulb thermometer.

Saturated Gas (Vapor). When a gas or water vapor is at the temperature of the boiling point which corresponds to its pressure, it is also at its saturation temperature. This saturated state is the condition under which a gas exists above its liquid in a closed container, as in a refrigerant drum or an evaporator. A saturated gas contains no superheat.

Pound of Air. When this term is used it means a pound of *dry* air. The following terms are examples of the use of the term *pound of air:* total heat per pound of air, moisture per pound of air, or latent heat per pound of air. In each case this means a pound of dry air and does not refer to a pound of the mixture.

Cubic Foot of Air. This term merely indicates a cubic foot of air and is used in expressing the quantity of moisture present in one cubic foot of dry air when it is saturated.

Ventilation. The process of supplying or removing air from a space is known as ventilation whether it is conditioned air, used air, or fresh air.

Effective Temperature. Effective temperature is a temperature determined by experiment. It is not a temperature measured on a thermometer, but is a measure of comfort as felt by a person. Effective temperature is produced by the correct combination of dry-bulb temperature, relative humidity, and air motion.

Comfort Zone. This is the range of temperature and humidity in which most people feel comfortable. The outer limits of the comfort zone are not clearly defined due to dependence on outside conditions.

Overall Coefficient of Heat Transmission. The quantity of Btu (cal) transmitted each hour through 1 ft² (0.09 m²) of any material, or combination of materials, for each degree of temperature difference between the two sides of a partition is known as the overall coefficient of heat transmission. Note that the difference between the two sides of the material is the difference in temperature of the air on both sides, and not the difference in temperature between both surfaces of the material.

Static Pressure. Static pressure is the pressure a gas or air exerts at right angles against the walls of an enclosure/duct. Static head or static pressure of air is usually measured in inches (mm) of water column by use of manometers and pitot tubes. The frictional loss in air ducts is known as pressure drop or loss of static head.

Velocity Pressure. This is the pressure due to the air movement. Velocity pressure is created by the energy of motion or kinetic energy in the moving air. It is measured in inches (mm) of water column.

Total Pressure. Total pressure is the sum of the static pressure and velocity pressure and is a measure of the total energy of the air.

Stack Effect. This is an upward flow or draft created by the tendency of heated air to rise in a vertical or inclined duct. Warm air rises because it expands and becomes lighter when heated. This principle is of special importance in the heating of tall buildings, where the entire building acts as a chimney (stack) due to the warmer air inside of it. The warm air rises and escapes out of openings or cracks on the upper floors while cold air is drawn in through doors, windows, cracks, etc. on the lower floors to replace the warm, rising air. As a result, extra radia-tion must be used on the lower floors to offset the chilling effect of the incoming cold air.

Plenum Chamber. The plenum chamber is an equalizing chamber or air supply compartment to which the various distributing ducts of an air conditioning or ventilating system are connected. The plenum chamber is maintained under pressure, which causes the air to flow into the different distribution ducts.

☐ Psychrometrics ☐

The science that deals with the relationships that exist within a mixture of air and water vapor is known as *psychrometrics*. In air conditioning, psychrometrics involves the measuring and determining of the different properties of air both inside and outside of the conditioned space. Psychrometrics also can be used to establish the conditions of air that will provide the most comfortable conditions in a given air conditioning application.

☐ Psychrometric Chart ☐

A psychrometric chart is a diagram that represents the various relationships that exist between the heat and moisture content of air and water vapor. Different air conditioning manufacturers have slightly different versions of the psychrometric chart, locating the various properties in different places on the chart. The factors shown on a complete psychrometric chart are a dry-bulb temperature, wet-bulb temperature, dew-point temperature, relative humidity, total heat, vapor pressure, and the actual moisture content of the air.

Identification of Lines and Scales on a Psychrometric Chart. The following illustrations will help to locate the different lines and scales on a psychrometric chart. Consider the psychrometric chart as being a shoe with the toe pointing to the left and the heel on the right (Figure 22–17).

Dry-Bulb Temperature Lines. The dry-bulb temperature scale is located on the "sole" of the psychrometric chart (Figure 22–18). The dry-bulb lines extend vertically upward from the sole. There is one line for each degree of temperature.

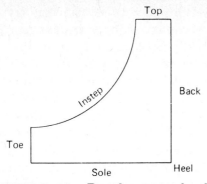

FIGURE 22-17 Psychrometeric chart outline.

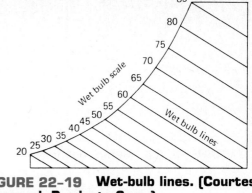

FIGURE 22-19 Wet-bulb lines. (Courtesy of Research Products Corp.)

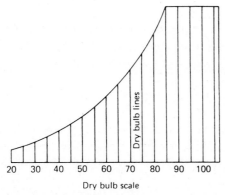

FIGURE 22-18 Dry-bulb lines. (Courtesy of Research Products Corp.)

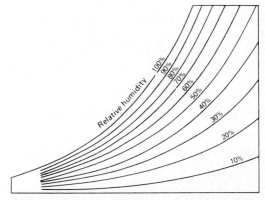

FIGURE 22-20 Relative humidity lines. (Courtesy of Research Products Corp.)

Wet-Bulb Temperature Lines. The wet-bulb temperature scale is found along the "instep" of the chart, extending from the toe to the top (Figure 22-19). The wet-bulb lines extend from the instep diagonally downward to the right. There is one line for each degree of temperature.

Relative Humidity Lines. On a complete psychrometric chart, the relative humidity lines are the only curved lines on it (Figure 22-20). The various relative humidities are indicated on the lines themselves. There is no coordinate scale as with the other air properties.

Absolute Humidity Lines. The scale for absolute humidity is a vertical scale on the right side of the psychrometric chart (Figure 22-21). The absolute humidity lines run horizontally to the left from this scale.

Dew-Point Temperature Lines. The scale for the dew-point temperature is identical to the scale for the wet-bulb temperature (Figure 22-22). The dew-point temperature lines run horizontally to the right, however, not diagonally as is the case with the wet-bulb temperature lines.

FIGURE 22-21 Absolute humidity lines. (Courtesy of Research Products Corp.)

FIGURE 22-22 Dew-point lines. (Courtesy of Research Products Corp.)

FIGURE 22–23   Cubic feet lines.

FIGURE 22–24   Enthalpy lines.

Specific Volume Lines. The specific volume scale is located along the sole of the chart from 12.5 to 14.5 ft³ (Figure 22–23). The cubic feet lines extend diagonally upward to the left from the sole to the instep of the chart. The specific volume lines on the chart are identified in terms of cubic feet per pound of air.

Enthalpy Lines. The enthalpy scale is located along the instep of the chart (Figure 22–24). The enthalpy lines are the same as the wet-bulb lines on the psychrometric chart. Enthalpy is total heat content. It can be used with the psychrometric chart to measure any heat change that takes place in any given psychrometric process. Both sensible and latent heat can be measured by use of enthalpy. It is a quick means of finding either of these factors.

When we put together the seven charts we have just covered, we will have a complete psychrometric chart (Figure 22–25).

Use of the Psychrometric Chart. Any two of the foregoing properties located on the psychrometric chart will determine definitely the condition of the air. If we choose any dry-bulb temperature and any wet-bulb temperature, the point at which these two lines intersect on the chart is the point that indicates the condition of the air at those given temperatures. The condition of the air at this point is definitely fixed. Similarly, the condition of the air at any other point on the psychrometric chart is fixed by the par-

ticular dry-bulb and wet-bulb temperatures. Because the possible combinations of any two temperatures are unlimited, there are an infinite number of possible conditions of air, and an equally infinite number of possible points that may be plotted on the chart.

When a fixed air condition has been located at a point on the chart, all of the other properties of that sample of air can be determined from the chart. Similarly, by use of the psychrometric chart, any two properties of the air and water vapor combination are sufficient to determine a condition of the air and all its other properties.

EXAMPLE 1

Given a dry-bulb temperature of 95 °F (35 °C) and a dew-point temperature of 54 °F (12.2 °C), find the wet-bulb temperature (Figure 22–26).

Solution

1. Locate 95 °F (35 °C) on the dry-bulb scale.
2. Draw a line straight upward to the instep.
3. Follow the instep down until 54 °F (12.2 °C) is found.
4. Extend this point horizontally to the right until the 95 °F (35 °C) dry-bulb line is crossed.
5. Extend the point upward to the left to the wet-bulb scale on the instep.
6. Read the wet-bulb temperature of 68.5 °F (20.3 °C).

EXAMPLE 2

Given a wet-bulb temperature of 74 °F (23.3 °C) and a dew-point temperature of 70 °F (21.1 °C), find the dry-bulb temperature (Figure 22–27).

Solution

1. Locate 74 °F (23.3 °C) on the wet-bulb scale.
2. Draw a line diagonally down the right to the back of the chart.
3. Locate 70 °F (21.1 °C) on the dew-point scale.
4. Extend this point horizontally to the right until the wet-bulb line is crossed.
5. Extend this point vertically downward to the dry-bulb scale.
6. Read the dry-bulb temperature of 84 °F (28.9 °C).

EXAMPLE 3

Given a wet-bulb temperature of 73 °F (22.8 °C) and a dry-bulb temperature of 81 °F (27.2 °C), find the dew-point temperature (Figure 22–28).

ENTHALPY–Btu per lb. of dry air and associated moisture

FIGURE 22–25 **Psychrometric chart. (Courtesy of the Trane Co.)**

FIGURE 22-26 Finding wet-bulb temperature.

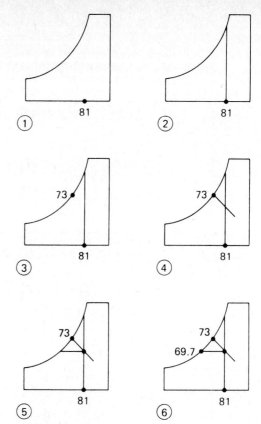

FIGURE 22-28 Finding dew-point temperature.

Solution

1. Locate 81 °F (27.2 °C) on the dry-bulb scale.
2. Draw a line vertically upward to the instep of the chart.
3. Locate 73 °F (22.8 °C) on the wet-bulb scale.
4. Extend this point diagonally downward until the dry-bulb line is crossed.
5. Extend this point horizontally to the left to the instep.
6. Read a dew-point temperature of 69.7 °F (20.9 °C).

EXAMPLE 4

Given a dry-bulb temperature of 95 °F (35 °C) and a dew-point temperature of 54 °F (12.2 °C), find the relative humidity (Figure 22-29).

Solution

1. Locate 95 °F (35 °C) on the dry-bulb scale.
2. Draw a line vertically upward to the instep of the chart.
3. Locate 54 °F (12.2 °C) on the dew-point scale.
4. Extend this point horizontally to the right until the dry-bulb line is crossed.

FIGURE 22-27 Finding dry-bulb temperature.

AIR CONDITIONING (COOLING)

FIGURE 22–30 Finding total heat content using wet-bulb temperature.

FIGURE 22–29 Finding relative humidity.

5. Read the relative humidity of 25% at this point.

EXAMPLE 5
Given a wet-bulb temperature of 74°F (23.3°C) and a dew-point temperature of 70°F (21.1°C), find the total heat content of the air (Figure 22–30).

Solution

1. Locate 74°F (23.3°C) on the wet-bulb scale.
2. Draw a line diagonally upward to the left to the total heat scale.
3. Read the total heat to be 37.8 Btu (9.46 kcal) per pound (0.4536 kg) of air.

EXAMPLE 6
Given a dry-bulb temperature of 95°F (35°C) and a dew-point temperature of 54°F (12.2°C), find the total heat content of the air (Figure 22–31).

Solution

1. Locate 95°F (35°C) on the dry-bulb scale.
2. Draw a line vertically upward to the instep of the scale.
3. Locate 54°F (12.2°C) on the dew-point scale.

FIGURE 22–31 Finding total heat content using dry-bulb temperature.

4. Extend this point horizontally to the right until the dry-bulb line is crossed.
5. Extend this point diagonally upward to the left to the total heat scale.
6. Find the total heat to be 32.32 Btu (8.14 kcal) per pound (0.4536 kg) of air. This total heat is found by interpolation.

Total heat at 69 °F (20.6 °C) wet bulb
$$= 32.71 \text{ Btu (8.24 kcal)}$$
Total heat at 68 °F (20 °C) wet bulb
$$= 31.92 \text{ Btu (8.04 kcal)}$$
Difference = 0.79 Btu (0.199 kcal)
$$68.5° - 68° = 0.5°$$
$$0.5° \times 0.79 = 0.395$$
Total heat at 68.5 °F (20.3 °C) wet bulb
$$= 31.92 + 0.395$$
$$= 32.315 \text{ or } 32.32 \text{ Btu (8.14 kcal)}$$

Change in Total Heat Content. During the cooling season, we are primarily interested in the amount of heat that must be removed so that the air is cooled enough to fulfill the inside design conditions. In the heating season, heat is added to the air to meet the inside design conditions. Suppose, for example, that the outside design conditions is 75 °F (23.9 °C) wet bulb, and it is desired to maintain a 67 °F (19.4 °C) wet-bulb temperature inside the conditioned space. The amount of total heat that must be removed per pound (453.6 g) of dry air is found by the following method:

Total heat at 75 °F (23.9 °C) wet bulb
$$= 38.7 \text{ Btu/lb or } 21.5 \text{ kcal/kg of air}$$
Total heat at 67 °F (19.4 °C) wet bulb
$$= 31.7 \text{ Btu/lb or } 17.6 \text{ kcal/kg of air}$$
Difference = 7.0 Btu/lb or 3.89 kcal/kg of air

Therefore, the total heat that must be removed in cooling air from 75 °F (23.9 °C) wet bulb down to 67 °F (19.4 °C) wet bulb is 7.0 Btu (1.76 kcal) per pound (453.6 g) of dry air (Figure 22–32).

Sensible Heat. Sensible heat is heat that can be added to or removed from a substance without a change of state. If that substance is air, changing the sensible heat will result only in a change in temperature. The sensible heat content is indicated by the dry-bulb temperature. A change in dry-bulb temperature will result in a change in sensible heat only because there is no change of state.

EXAMPLE 7
During the heating season, air is to be heated from a 65 °F (18.3 °C) dry-bulb temperature and a 50 °F (10 °C) wet-bulb temperature to an 88 °F (31.1 °C) dry-bulb temperature and a 60 °F (15.6 °C) wet-bulb temperature. Find the sensible heat that must be added per pound of dry air (Figure 22–33).

FIGURE 22–32   Change in total heat.

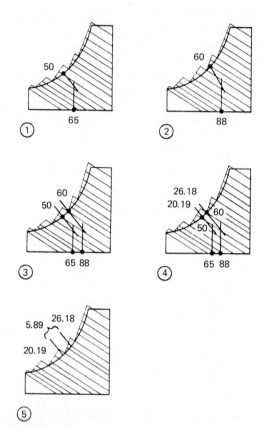

FIGURE 22–33   Sensible heat change on heating air.

Solution

1. Locate 65 °F (18.3 °C) dry bulb and 50 °F (10 °C) wet bulb.
2. Locate 88 °F (31.1 °C) dry bulb and 60 °F (15.6 °C) wet bulb.
3. Extend these points diagonally upward to the left to the total heat scale.
4. Read 20.19 and 26.5 Btu (11.1 and 14.4 kcal/kg).
5. Obtain the difference:

Total heat at 60°F (15.6°C) wet bulb
= 26.5 Btu/lb (14.57 kcal/kg)

Total heat at 50°F (10°C) wet bulb
= 20.19 Btu/lb (11.1 kcal/kg)

Sensible heat added
= 6.3 Btu/lb (3.47 kcal/kg of dry air)

This heat change is sensible heat only because there was no change in the moisture content of the air.

Latent Heat. The dew-point temperature is always an indication of the moisture content of the air. Any change in the dew-point temperature will result in a change in moisture content. The moisture content can only be changed by changing dew-point temperature. Therefore, it should be noted that as long as the dew-point temperature remains constant there will be no change in the moisture content or water vapor present in the air.

The moisture content scale used on most charts indicates grains of moisture per pound (453.6 g) of dry air. The dew-point temperature is read on the same scale as the wet-bulb temperature.

The latent heat of vaporization is the amount of heat in Btu (cal) required to change a liquid to a gas at a constant temperature. If we now have a given number of grains of moisture per pound (453.6 g) of dry air, there must have been a certain amount of heat required to vaporize this moisture into the air. This is the latent heat content of the air and water vapor mixture.

EXAMPLE 8

Air at 75°F (23.9°C) dry bulb and 57°F (13.9°C) wet bulb is to be conditioned to obtain a 75°F (23.9°C) dry-bulb temperature and a 70°F (21.1°C) wet-bulb temperature. Find the amount of latent heat added and the grains of moisture added (Figure 22–34).

Solution

1. Locate 75°F (23.9°C) dry bulb and 57°F (13.9°C) wet bulb.
2. Extend this point to the total heat scale and read 24.5 Btu (6.2 kcal).
3. Locate 75°F (23.9°C) dry bulb and 70°F (21.1°C) wet bulb.
4. Extend this point to the total heat scale and read 34 Btu (8.6 kcal).

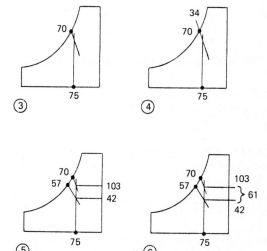

FIGURE 22–34 Latent heat change and grains of moisture added.

5. The heat added to this air is 34 − 24.5 = 9.5 Btu (2.4 kcal) per pound (453.6 g) of dry air.
6. Extend these points to the right to the grains of moisture scale and read 42 grains (2.85 g) at the beginning of the process and 102 grains (6.61 g) of moisture per pound (453.6 g) of dry air at the end of the process.
7. The grains of moisture added per pound (453.6 g) of dry air is 102 − 44 = 58 grains of moisture added.

This heat change is latent heat only because there was no change in the dry-bulb temperature of the air.

Sensible Heat Ratio. Sensible heat ratio is the sensible heat removed in Btu (cal) divided by the total heat removed in Btu (cal). The sensible heat ratio indicates the sensible heat percentage of total heat removed. The sensible heat ratio is above 50% in most comfort air conditioning. This is because most comfort air conditioning systems remove more sensible heat than latent heat.

The ratio will vary with the type of installation. A residence may have an 0.7 or 0.8 sensible heat fac-

FIGURE 22-35 Sensible heat ratio line.

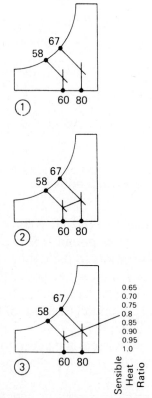

FIGURE 22-36 Finding sensible heat ratio (line method).

indicates that half of the heat removed is sensible and half is latent.

EXAMPLE 9

The desired room conditions are 80 °F (26.7 °C) dry bulb and 67 °F (19.4 °C) wet bulb. The supply air conditions are 60 °F (15.6 °C) dry bulb and 58 °F (14.4 °C) wet bulb. Find the sensible heat factor (Figure 22-36).

Solution

1. Plot the given conditions on the psychrometric chart.
2. Draw a straight line between the two points.
3. Extend the line to the sensible heat ratio scale.
4. Read 0.73 on the sensible heat ratio scale.

Another method of finding the sensible heat ratio is to use a ratio of the total heat removed to the sensible heat removed from the air. If we use the same conditions as stated in Example 9, we can figure the sensible heat ratio as follows (Figure 22-37).

1. Plot the given conditions on the psychrometric chart.

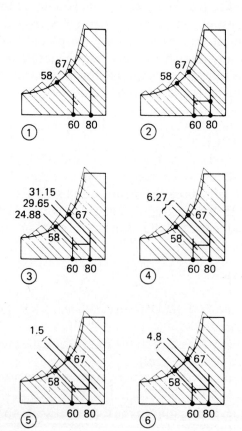

FIGURE 22-37 Calculated sensible heat ratio.

tor, while a restaurant may have an 0.5 or 0.6 sensible heat ratio. This indicates that 70 or 80% of the total change in heat removed from the air is sensible heat.

The sensible heat ratio scale is located on the right-hand side of the psychrometric chart. When the conditions of air are plotted on the chart and a line is drawn through them and extended to the sensible heat ratio scale, the ratio is read directly (Figure 22-35). When the line is at approximately a 45° angle, the sensible heat ratio is 50% or 0.50. This

AIR CONDITIONING (COOLING)

2. Extend the final air condition [60°F (15.6°C) dry bulb and 58°F (14.4°C) wet bulb] point horizontally to the right until the 80°F (26.7°C) dry-bulb line is crossed.
3. Extend the 80°F (26.7°C) dry-bulb line vertically downward to the sole of the chart. This forms a reverse "L."
4. Extend the room air point, the supply air point, and the intersection of the two sides of the L to the total heat scale.
5. Read the total heat for each line extended; room conditions 31.15 Btu (7.85 kcal); supply conditions 24.88 Btu (6.3 kcal); and intersection of L 29.65 Btu (7.5 kcal).
6. Calculate total heat removed: 31.15 − 24.88 = 6.27 Btu (1.5 kcal).
7. Calculate latent heat removed: 31.15 − 29.65 = 1.5 Btu (0.38 kcal)
8. Calculate sensible heat removed: 29.65 − 24.88 = 4.8 Btu (1.2 kcal).
9. Calculate sensible heat ratio 4.8/6.27 = 0.76%.

This indicates that 76% of the heat removed is sensible heat.

Mixing Two Quantities of Air at Different Conditions. The science of air conditioning involves the process of mixing air, such as when the conditioned air supply is mixed with the room air or when it is mixed with bypass air. Another example is when fresh air is supplied in some proportion for ventilation purposes. When the initial condition and the quantities of these two air sources are known, the condition of the final mixture can easily be found by use of the psychrometric chart.

EXAMPLE 10
Outdoor air at 95°F (35°C) dry bulb and 75°F (23.9°C) wet bulb, point *A*, is to be mixed with return air at 70°F (21.1°C) dry bulb and 10% relative humidity, point *B*. The mixture is to consist of 25% outdoor air and 75% return air. Find the resulting dry-bulb and wet-bulb temperatures of the mixture (Figure 22–38).

Solution

1. Plot the two points, *A* and *B*, on the chart.
2. Draw a line between the two points.
3. Locate the dry-bulb temperature by adding the percentages of each dry-bulb temperature, that is

FIGURE 22–38 Mixing two quantities of air.

25% of 95° = 23.75 (−4°C)
75% of 70° = 52.5 (−11.4°C)
Resulting dry-bulb temperature of mixture
= 76.25 (24.4°C)

4. Locate 76.25°F (24.4°C) dry bulb on the scale and extend this point to the mixture line, point *C*.
5. Extend point *C* to the wet-bulb scale and read 55.7°F (13°C).

It is very important to note that it is permissible to find the resulting dry-bulb temperature of a mixture of air by the percentage method, but it is not permissible to use this method for determining the resulting wet-bulb temperature. For example, if the percentage method was used in the preceding example the results would be as follows:

25% of 75° = 18.75
75% of 47° = 32.25
51.00 *incorrect* resulting wet-bulb temperature

The percentage method can be used to find the resulting wet-bulb temperature by indirect application. In order to do this it is necessary first to find the total heat for each condition of the air and then to apply the percentage method to find the resulting total heat of the air mixture. From the total heat of the mixture it is then possible to locate the corresponding correct wet-bulb temperature. The preceding example would then be:

EXAMPLE 11
The total heat at 75 °F (23.9 °C) is 37.81 Btu (9.53 kcal). The total heat at 47 °F (8.3 °C) is 18.60 Btu (4.7 kcal). Find the corresponding wet-bulb temperature.

Solution

25% of 37.81 = 9.45 Btu (2.38 kcal)
75% of 18.60 = 13.95 Btu (3.52 kcal)
Total heat of moisture = 23.40 Btu (5.89 kcal)

The corresponding wet-bulb temperature is 55.6 °F (13.1 °C).

☐ Heat Load Calculation ☐

Everything in nature reacts against the air conditioning process. During the summer, the outdoor elements tend to make the cooling equipment work harder to remove the unwanted heat from within the space. During the winter, the outdoor elements tend to make the heating equipment work hard to supply warmth to the conditioned space.

The process of estimating the size of air conditioning equipment required to serve the conditioned space is much like the process used in refrigeration calculations. The chief difference is that there are more sources of heat involved in air conditioning estimating.

An accurate load estimate will indicate four things:

1. The true cooling or heating load requirements
2. The most economical equipment selection
3. The possibilities for least cost with the greatest load reduction
4. The most efficient air distribution system

Heat Sources. Heat sources are listed under two general classifications: (1) those that result in an internal heat load on the conditioned air space, and (2) those that result in an external heat load on the conditioned space. The external heat load requires extra evaporator capacity, but it does not affect the conditioned air after it has been delivered to the conditioned space. The following list is a classification of these sources.

1. Heat sources that result in an internal heat load:
   (a) Heat conduction through the walls, roof, windows, etc.
   (b) Solar heat gain
   (c) Duct heat gain (when outside the conditioned space)
   (d) Occupant heat load
   (e) Lighting load
   (f) Electric equipment and appliances
   (g) Outside air infiltration
2. Heat sources that result in an external heat load:
   (a) Ventilation air
   (b) Any heat added to the air after it leaves the conditioned space

The heat that enters a conditioned space may come from any or all of the above listed sources. An accurate survey must be made on every estimate so that the heat load can be calculated correctly. The purpose of cooling equipment is to remove unwanted heat and maintain the space at the desired conditions. Therefore, caution must be exercised to make sure that the total heat load used is the load for the peak period.

Many of the heat gains are at their peak at different times. For example, a restaurant may have a peak sun load at noon, while the peak occupant load may not be until 7:00 P.M. These two peak loads should not be added together. Another example would be when the lighting heat load is larger than the solar load from the sun and would not occur at the same time.

It can be seen that the heat load on an air conditioning system is subject to great variation. Because of this, the heat load is calculated on the basis of 1 hr, not 24 hr. The cooling system selected must be able to handle this *peak load hour*. It must be remembered that the peak load hour of the day is when the sum of the heat loads is at a peak and not a sum of the heat loads.

Conducted Heat Load. Heat always flows from a warm temperature to a cooler temperature. Therefore, when the temperature inside a conditioned space is lowered, heat will be conducted through the

Wall

T_1 = Warmer Outside Temperature

T_2 = Cooler Inside Temperature

U = Heat Transfer

FIGURE 22–39 Heat conducted through a wall.

walls, doors, windows, etc., from the outside to the inside (Figure 22–39). The amount of heat that will flow through a wall whose temperatures differ on each side will depend on three factors:

1. The area of the wall
2. The heat-conducting characteristics of the wall, called the *overall coefficient of heat transmission,* commonly designated as the U factor
3. The difference in the temperature between the spaces separated by the wall

These three factors also are applied to all forms of construction such as doors, windows, and roofs. The formula used to calculate heat flow is

$$Q = A \times U \times \Delta t$$

where

Q = heat transferred per hour [Btu (cal)]

A = net area of the wall, window, door, etc.

U = number of Btu (cal) which will flow through 1 ft² (0.09 m²) of the material, in 1 hr, with a temperature difference of 1 °F (0.56 °C) between the two sides

The U factors for most materials have been tabulated and placed in tables for convenience (Table 22–1). A more complete listing of U factors can be found in the American Society of Heating, Ventilating, and Refrigerating Engineers' Society *Guide and Data Book.* The guide provides the most accurate information available on this subject. When a heat load is to be calculated, each type of construction must be considered and calculated separately.

EXAMPLE 12

Consider a wall 80 ft (24.32 m) long and 8 ft (2.43 m) high and constructed of 8-in. (20.32-cm)-thick

TABLE 22–1 Partial List of Heat Transfer Coefficient *U* Factors

Exterior Walls	
Wood siding, 2 in. by 4 in. studs, plastered	0.25
Wood siding, 2 in. by 4 in. studs, plastered ½ in. insulation	0.17
Brick veneer, 2 in. by 4 in. studs, plastered	0.27
Brick veneer, 2 in. by 4 in. studs, plastered ½ in. insulation	0.18
Stucco, 2 in. by 4 in. studs, plastered	0.30
Brick wall, 8 in. thick, no inside finish	0.50
Brick wall, 8 in. thick, furred, plastered on wood lath	0.30
Brick wall, 12 in. thick, furred and plastered on wood lath	0.24
Hollow tile, 10 in. thick	0.39
4-in. brick hollow tile, 12 in. thick, furred, plastered wood lath	0.20
4-in. brick-faced concrete, 10 in. thick	0.48

Interior Walls or Partitions	
Studding, wood lath and plaster both sides	0.34
Studding, wood lath and plaster both sides, ½ in. insulation	0.21
4-in. hollow tile, plastered both sides	0.40
4-in common brick, plastered both sides	0.43

Floors, Ceilings, and Roofs	
Plastered ceiling without flooring above	0.62
Plastered ceiling with 1 in. of flooring	0.28
Plastered ceiling on 4 in. of concrete	0.59
Suspended plastered ceiling under 4 in. of concrete	0.37
Concrete 4 in. [floor or ceiling]	0.65
Concrete 8 in. [floor or ceiling]	0.53
Average wood floor	0.35
Average concrete roof	0.25
Average wood roof	0.30

Glass and Doors	
Window glass, single	1.13
Double glass windows	0.45
Doors, thin panel	1.13
Doors, heavy panel [1¼ in.]	0.59
Doors, heavy panel [1¼ in.] with glass storm door	0.38

brick, furred and plastered with wood lath. There is one thin panel door, 6 ft by 2½ ft (1.8 by 0.76 m). There will be a temperature difference (Δt) of 15 °F (8.4 °C). The U factors found in Table 22–1 for this type of construction are 0.30 for the wall and 1.13 for the door.

Heat Load Calculation

Solution

Gross area of wall = 640 ft² (59.10 m²)
Area of door = 15 ft²   (1.37 m²)
$\overline{}$ 625 ft² (57.73 m²)

Substitute values in the formula $Q = A \times U \times \Delta t$:

Heat gain through wall
= 625 × 0.30 × 15 = 2812.5 Btu (708.7 kcal)

Heat gain through door
= 15 × 1.13 × 15 = 254.3 Btu   (64 kcal)
$\overline{}$
Total heat gain = 3066.8 Btu (772.8 kcal)

EXAMPLE 13 (Part 1)

This example is somewhat more complicated than Example 12 and will use Figure 22–40 to complete the computations, which will be completed in three separate steps. The exterior walls will be considered first. The wall is 10 ft long and 8 ft high (3.04 m × 2.43 m) and constructed of 8-in. (20.32-cm)-thick brick, furred and plastered on wood lath.

Solution

Gross area of exterior wall =
3(10 ft × 8 ft) = 240 ft²
Area of exterior glass = 2(4 ft × 5 ft) =   40 ft²
$\overline{}$
Net exterior wall = 200 ft³

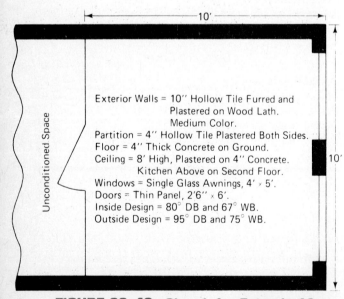

FIGURE 22–40   Sketch for Example 13.

The sketch shows:
Exterior Walls = 10″ Hollow Tile Furred and Plastered on Wood Lath. Medium Color.
Partition = 4″ Hollow Tile Plastered Both Sides.
Floor = 4″ Thick Concrete on Ground.
Ceiling = 8′ High, Plastered on 4″ Concrete. Kitchen Above on Second Floor.
Windows = Single Glass Awnings, 4′ × 5′.
Doors = Thin Panel, 2′6″ × 6′.
Inside Design = 80° DB and 67° WB.
Outside Design = 95° DB and 75° WB.
Unconditioned Space

Table 22–1 indicates:

U factor for wall = 0.30
U factor for glass = 1.13
Δt = 95°F − 80°F = 15°F

Heat gain through wall
= 200 × 0.30 × 15 = 900 Btu/hr

Heat gain through glass
= 40 × 1.13 × 15 = 678 Btu/hr (170.85 kcal)

This heat loss is also sensible heat because it was calculated on a difference in dry-bulb temperatures.

Temperature Differences (Δt). In practice, the 15°F (8.34°C) does not remain constant for all surfaces when calculating the conduction heat load. For example, if the conditioned space is located on the opposite side of a partition from a space which has a high temperature, like a boiler room, it is obvious that a 15°F temperature differential is not adequate. In cases where a 15°F Δt for the exterior walls would be normal, the Δt across the partition would be higher than 15°F. Table 22–2 gives the Δt in degrees that should be used if the normal differential for the exterior walls is 15°F. If the normal Δt to be maintained is 20°F (11.12°C) for the exterior walls, then 5°F (2.8°C) must be added to the Δt given in the table. For example, for "ceilings with laundries or kitchens located above" the table shows an indicated Δt of 25°F (14°C) when the normal Δt is 15°F. If the normal Δt is 20°F, then 5°F must be added to the 25°F which indicates a Δt of 30°F (16.8°C) for this particular ceiling.

EXAMPLE 13 (Part 2)

To calculate the heat gain through the partition, floor, and ceiling we will use Table 22–2.

Solution

Gross area of partition = 10 ft × 8 ft = 80 ft²
Gross area of door =   6 ft × 2½ ft = 15 ft²
$\overline{}$
Net area of partition = 65 ft²

Table 22–1 indicates:

U factor for partition = 0.40
U factor for thin-panel door = 1.13

Table 22–2 indicates that partitions located next to unconditioned spaces have a Δt of 10°F (5.6°C)

TABLE 22-2 Temperature Differential for Abnormal Conditions (Based on 15°F normal differential for exterior walls)[a]

Structure	Δt
Walls—exterior	15
Glass in exterior walls	15
Glass in partitions	10
Partitions next to unconditioned rooms	10
Partitions next to laundries, kitchens, boiler rooms, or store show windows having a large lighting load	30
Floors above unconditioned rooms	10
Floors on ground or over unvented spaces or basements	0
Floors above laundries, kitchens, or boiler rooms	40
Floors above vented spaces	15
Ceilings with unconditioned rooms above	10
Ceilings with laundries, kitchens, etc., above	25
Ceilings with roof directly above (no attic)	15
Ceilings with totally enclosed attic and roof above	20
Ceilings with cross-vented attic and roof above	15

[a]If normal differential is 20°F, add 5°F to above values.

when the normal Δt of 15°F (8.4°C) is used. This Δt also applies to the door.

Heat gain through the partition
$$= 65 \times 0.40 \times 10 = 260 \text{ Btu/hr}$$
Heat gain through the door
$$= 15 \times 1.13 \times 10 = 169.5 \text{ Btu/hr}$$
Area of the floor = 10 ft × 10 ft = 100 ft²
Area of the ceiling = 10 ft × 10 ft = 100 ft²

Table 22-1 indicates:

U factor for the floor = 0.65
U factor for the ceiling = 0.59

Table 22-2 indicates:

Δt for the floor = 0°F
Δt for the ceiling = 25°F

Heat gain through the ceiling
$$= 100 \times 0.59 \times 25 = 1475 \text{ Btu/hr}$$
Heat gain through the floor
$$= 100 \times 0.65 \times 0 = 0 \text{ Btu/hr}$$

The total heat gain by conduction can now be found by adding together all the calculated values:

Exterior walls	=	780.0 Btu/hr
Exterior glass	=	678.0 Btu/hr
Partition	=	260.0 Btu/hr
Door in partition	=	169.5 Btu/hr
Ceiling	=	1475.0 Btu/hr
Floor	=	0.0 Btu/hr
Total heat gain	=	3362.5 Btu/hr

The air conditioning unit designed for this building must remove 3362.5 Btu (847.35 kcal) each hour for the heat gained by conduction. This is the amount of heat that will be conducted through the exterior walls and surfaces. The other sources of heat that also must be removed by the air conditioning unit will be taken up in the following examples.

Design Temperatures. The various parts of the country undergo a considerable variation in temperatures daily. For example, a city like Minneapolis, located in the northern part of the country, may have very high temperatures during parts of the summer season. However, on an average the temperatures in Minneapolis are lower than the average temperatures in Dallas. When calculating the conducted heat load for an air conditioning system, it is necessary to choose the proper design temperature. The Weather Bureau records indicate that extremely high temperatures occur during less than 10% of the time. Therefore, the design temperature for air conditioning is not based on the highest temperature occurring in that locality, but on its average maximum temperature. It would not be economically feasible to install an air conditioning unit that would provide the cooling required for the highest temperatures, since they would occur for only a small percentage of the time.

The design dry-bulb temperatures for the various cities throughout the United States have been placed in tables for convenience (Table 22-3). The wet-bulb temperature and the summer wind velocity in miles per hour are also shown. For example, the correct outside design conditions for Detroit would be 95°F (35°C) dry bulb, 75°F (23.9°C) wet bulb, and a wind velocity of 10.3 mph (16.58 km/hr).

Solar Heat Load. When a roof, ceiling, wall, or window is exposed to the direct rays of the sun, the surface will warm up very quickly, thus giving the effect of a greater Δt between the inside and outside surfaces. The factors that affect the quantity of solar heat gained are:

TABLE 22-3 Design Wet-Bulb and Dry-Bulb Summer Temperatures and Wind Velocities

City and state	DB	WB	mph	City and state	DB	WB	mph
Alabama				**Nebraska**			
Birmingham	95	78	5.2	Lincoln	95	75	9.3
Mobile	95	80	8.6	**Nevada**			
Arizona				Reno	95	65	7.4
Phoenix	105	76	6.0	**New Hampshire**			
Arkansas				Manchester	90	73	5.6
Little Rock	95	78	7.0	**New Jersey**			
California				Trenton	95	78	10.0
Los Angeles	90	70	6.0	**New Mexico**			
San Francisco	90	65	11.0	Santa Fe	90	65	6.5
Colorado				**New York**			
Denver	95	64	6.8	Buffalo	93	75	12.2
Connecticut				New York	95	75	12.9
New Haven	95	75	7.3	**North Carolina**			
Delaware				Asheville	90	75	5.6
Wilmington	95	78	9.7	Wilmington	95	79	7.8
Dist. Columbia				**North Dakota**			
Washington	95	78	6.2	Bismarck	95	73	8.8
Florida				**Ohio**			
Jacksonville	95	78	8.7	Cincinnati	95	70	6.6
Georgia				Cleveland	95	75	9.9
Atlanta	95	76	7.3	**Oklahoma**			
Idaho				Oklahoma City	101	76	10.1
Boise	95	65	5.8	**Oregon**			
Illinois				Portland	90	65	6.6
Chicago	95	75	10.2	**Pennsylvania**			
Indiana				Philadelphia	95	78	9.7
Indianapolis	95	76	9.0	**Rhode Island**			
Iowa				Providence	93	75	10.0
Des Moines	95	77	6.6	**South Carolina**			
Kansas				Charleston	95	80	9.9
Wichita	100	75	11.0	**South Dakota**			
Kentucky				Sioux Falls	95	75	7.6
Louisville	95	76	8.0	**Tennessee**			
Louisiana				Memphis	95	78	7.5
New Orleans	95	79	7.0	**Texas**			
Maine				Dallas	100	78	9.4
Portland	90	73	7.3	El Paso	100	69	6.9
Maryland				Houston	95	78	7.7
Baltimore	95	78	6.9	**Utah**			
Massachusetts				Salt Lake City	92	63	8.2
Boston	92	75	9.2	**Vermont**			
Michigan				Burlington	90	73	8.9
Detroit	95	75	10.3	**Virginia**			
Minnesota				Richmond	95	78	6.2
Minneapolis	95	75	8.4	**Washington**			
Mississippi				Seattle	85	65	7.9
Vicksburg	95	78	6.2	**West Virginia**			
Missouri				Parkersburg	95	75	5.3
Kansas City	100	76	9.5	**Wisconsin**			
St. Louis	95	78	9.4	Milwaukee	95	75	10.4
Montana				**Wyoming**			
Helena	95	67	7.3	Cheyenne	95	65	9.2

AIR CONDITIONING (COOLING)

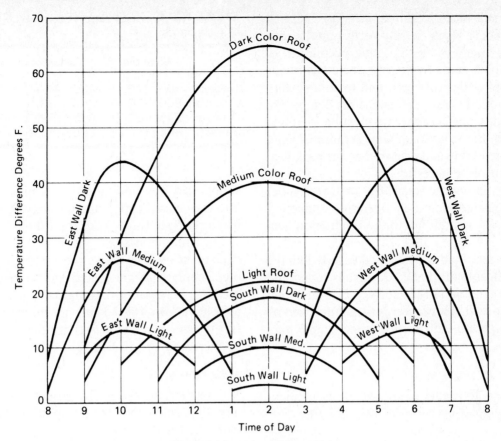

FIGURE 22-41 Solar chart.

1. The time of day
2. The direction in which the exposed surfaces face
3. The color of the exposed surface
4. The type of surface, whether it is smooth or rough
5. The latitude

The time of day is very important because the solar effect will vary greatly from one hour to another. There are tables and charts designed to provide the excess differential, due to the solar effect, for any wall or roof for any hour of the day (Figure 22-41). Notice that the solar effect on the east wall is the greatest at 10:00 A.M. It is the greatest on the south walls and the roof at 2:00 P.M., and on the west walls at 6:00 P.M. When the rays of the sun are perpendicular to the surface, the rate of heat absorption is the greatest. However, due to the time lag in the heat passing through the wall, it is not noticed in the conditioned space until sometime later.

This time lag will vary depending on the construction of the surface. For instance, a light frame construction may have a lag of less than one hour while a heavy masonry type construction may have a lag

of several hours. The average type construction will usually have a time lag of two hours. For example, at noon the rays of the sun are perpendicular to the roof, but the peak effect is not felt until 2:00 P.M.

There is no time lag through glass because the sun's rays pass directly through the glass and release heat when they strike objects in the room. The solar heat gain through glass is, therefore, maximum during the time when the sun shines through the window with the greatest intensity. When a building is to be air conditioned, the windows and glass are usually protected from the direct rays of the sun.

Walls and roofs constructed with dark surfaces absorb a great deal more heat than light-colored surfaces. Because of this, roofs are usually coated with a light color or a reflective type paint. The solar heat may be omitted when there is a ventilated air space, attic, or room above the conditioned space, or when the roof is shaded by adjacent buildings, etc., except for a very short period at noon. When roofs are partly shaded, calculate the heat gain for the exposed part only. When the roof is sprayed with water to cool the surface, the solar heat load need not be calculated for the roof.

The peak solar heat load in a one-story building will usually occur when the sun's effect on the roof is at a peak, except when steps have been taken to reduce the solar load. Ordinarily, the areas of the east and west walls and glass, and to some extent the south walls and glass, are the factors that govern the heat load. Considerable care must be exercised in calculating the solar load. Good common sense used when inspecting and determining the peak heat load is an absolute necessity.

When calculating the solar heat load, remember that it is calculated separately from the conduction heat load and is in addition to the conduction heat load.

The solar heat load is calculated exactly like the conduction heat load; that is, the area is found and multiplied by the same U factor which is a characteristic of the wall. This factor is then multiplied by the solar temperature difference found from Figure 22-41.

EXAMPLE 13 (Part 3)

Referring back to Figure 22-40, we can now calculate the solar load on the east wall and glass.

Solution

Gross area of the wall = 10 ft × 8 ft = 80 ft²

Area of the glass = 2 × 4 ft × 5 ft = 40 ft²

As before,

U factor for the wall = 0.3

U factor for the glass = 1.13

Figure 22-41 indicates that the peak solar load on a medium-colored east wall would occur at 10:00 A.M. The combined solar heat load will be as calculated in the table on page 615.

Notice especially that the combination heat gain may not come at a time when either of the separate loads is at a peak. This calculation shows that the peak solar load on this east wall and glass will be 2371.2 Btu (597.54 kcal) per hour at 8:00 A.M. Remember that this is in addition to the conducted heat load and would occur only while the sun is shining. The conducted heat load would be effective even when the sun is not shining.

Occupant Load. The heat gain for people has also been calculated and placed in tables (Table 22-4). Notice that this is the first item that has been divided into sensible and latent heat. The use of this table

TABLE 22-4 Body Heat [Btu/hr [kcal]] of Occupants

	Sensible	Latent	Total
Theater	195 [49.1]	155 [39]	350 [88.2]
Office	200 [50.4]	250 [63]	450 [113.4]
Dancing	245 [61.7]	605 [152.5]	850 [214.2]
Bowling	465 [117.2]	985 [248.2]	1450 [365.4]

simply involves finding out how many people fall into each classification and multiplying this number by the proper value from the table.

The amount of heat given off by a person is dependent on how active the person is. For example, people who are seated and at rest, as in the home, in restaurants, or in theaters, give off approximately 350 Btu (88.2 kcal) per hour of total heat. In the case of medium activity the total heat may be increased to 850 Btu (214.2 kcal) per hour.

EXAMPLE 14

In a restaurant having five employees and a capacity of 30 customers, the occupant heat load can be calculated as follows:

Solution

Sensible heat customers = 30 × 195 = 5850 Btu/hr

Sensible heat employees = 5 × 245 = 1225 Btu/hr

Latent heat customers = 30 × 155 = 4650 Btu/hr

Latent heat employees = 5 × 605 = 3025 Btu/hr

Total body heat
= 5850 + 1225 + 4650 + 3025 = 14750 Btu/hr

Infiltration. Infiltration is a source of external heat load for the air conditioning apparatus. Infiltration will occur in almost every building because it is virtually impossible to build an airtight room. Air will gain entrance into the conditioned space through the cracks around doors, windows, etc., unless the space is kept under a positive pressure. In a residence or small office where there are no excessive tobacco smoke or other objectionable odors, natural infiltration will supply enough air to satisfy the ventilation air requirements for the space. To satisfy the ventilation air requirements the opening must be in excess of, or at least equal to, 5% of the floor space. Also, the space must be large enough to allow 50 ft² (15.2 m²) of space and a minimum of 500 ft³ (14.15 m³) of volume for each occupant. When these

A	8:00 A.M. $U \times \Delta t$	9:00 A.M. $U \times \Delta t$	10:00 A.M. $U \times \Delta t$
East wall	40 ft² × 0.30 × 2°F = 24	40 ft² × 0.30 × 18°F = 216	40 ft² × 0.30 × 26°F = 312
East glass	40 ft² × 1.13 × 52°F = 2350.4	40 ft² × 1.13 × 46°F = 2079.2	40 ft² × 1.13 × 32°F = 1446.4
Total Btu	2374.4	2295.2	1758.4

A	8:00 A.M. $U \times \Delta t$	9:00 A.M. $U \times \Delta t$	10:00 A.M. $U \times \Delta t$
East wall	3.7 m² × 1.46 × 1.12°C = 6.05	3.7 m² × 1.46 × 10°C = 54.02	3.7 m² × 1.46 × 14.56°C = 78.6
East glass	3.7 m² × 5.51 × 29.12°C = 593.6	3.7 m² × 5.51 × 25.76°C = 525.17	3.7 m² × 5.51 × 17.9°C = 364.9
Total kcal	599.65	579.19	443.5

conditions are fulfilled and sufficient infiltration is obtained for the fresh air requirements, forced ventilation will not be necessary unless it is required by local codes and ordinances (Table 22–5). The better designed installations do not depend on air infiltration to satisfy the ventilation air requirements.

Tables have been constructed so that air infiltration through doors and windows can be more easily determined by the designer (Table 22–6). When there are two exposed walls, use the wall that has

the greater air leakage. If more than two walls are exposed, use either the one wall that has the greatest amount of leakage or one half of the total of all the walls, whichever has the larger quantity of infiltration air.

EXAMPLE 15

Determine the air infiltration into a room that has three exposed walls. One wall has three average fit weather-stripped double-hung windows measuring 2 ft × 5 ft (0.6 m × 1.5 m). The opposite wall has one industrial-type steel sash window measuring 3 × 3 ft (0.9 × 0.9 m). The third wall has a poorly fitted weather-stripped door measuring 3 × 6 ft (0.9 × 1.8 m). The average wind velocity is 5 mi (8 km) per hour during the cooling season. Determine the total crack length around the windows.

Solution

The cracks at the top, center, and bottom of

$$\text{Windows} = 2 \text{ ft} \times 3 \text{ ft} = 6 \text{ ft}$$
$$\text{Vertical cracks} = 5 \text{ ft} \times 2 \text{ ft} = 10 \text{ ft}$$
$$\text{Total crack length for 1 window} = 16 \text{ ft}$$
$$\text{Total crack length for 3 windows} = 3 \times 16 = 48 \text{ ft}$$

A wind velocity of 5 mi (8 km) per hour for an average fit weather-stripped double-hung window allows an infiltration of 5 ft³ (0.14 m³) per hour per foot (0.304 m) of crack.

Infiltration per hour in first wall
$$= 48 \text{ ft} \times 5 = 240 \text{ ft}^3$$

TABLE 22–5 Minimum Outdoor Air Required for Ventilation [Subject to local code regulations]

Application	cfm per Person
Apartment or residence	10–15
Auditorium	5–7½
Barber shop	10–15
Bank or beauty parlor	7½–10
Broker's board room	25–40
Church	5–7½
Cocktail lounge	20–30
Department store	5–7½
Drugstore	7½–10
Funeral parlor	7½–10
General office space	10–15
Hospital rooms [private]	15–25
Hospital rooms [wards]	10–15
Hotel room	20–30
Nightclubs and taverns	15–20
Private office	15–25
Restaurant	12–15
Retail shop	7½–10
Theater [smoking permitted]	10–15
Theater [smoking not permitted]	5–7½

TABLE 22–6 Infiltration [ft³] per Foot of Crack per Hour[a]

Type of Opening	Condition	Wind Velocity (mph) 5	10	15	20	25	30
Double-hung wood window	Average fit, not weather-stripped	6	20	40	60	80	100
Double-hung wood window	Average fit, weather-stripped	5	15	25	35	50	65
Double-hung wood window	Poor fit, weather-stripped	6	20	35	50	70	90
Double-hung metal window	Not weather-stripped	20	45	70	100	135	170
Double-hung metal window	Weather-stripped	6	18	30	44	58	75
Steel sash	Casement, good fit	6	18	30	44	58	75
Steel sash	Casement, average fit	12	30	50	75	100	125
Steel sash	Industrial type	50	110	175	240	300	375
Doors	Good fit, not weather-stripped	30	70	110	155	200	250
Doors	Good fit, weather-stripped	15	35	55	75	100	125
Doors	Poor fit, not weather-stripped	55	140	225	310	400	500
Doors	Poor fit, weather-stripped	30	70	110	155	200	250

[a]Use of storm sash permits a 50% infiltration reduction on poorly fitting windows only.

Total crack length of steel sash window
$$= 3 \text{ ft} \times 4 \text{ ft} = 12 \text{ ft}$$

Infiltration air for industrial steel sash at 5 mph wind velocity is 50 ft³ per hr.

Infiltration per hour in second wall
$$= 12 \text{ ft} \times 50 = 600 \text{ ft}^3$$

Infiltration through door in third wall
$$= 2(6 \text{ ft} + 3 \text{ ft}) = 18 \text{ ft}$$

A poorly fitted weather-stripped door admits 30 ft³ (0.85 m³) of infiltration air per hour at 5 mi (8 km) per hour wind velocity.

Infiltration per hour in third wall
$$= 18 \times 30 = 540 \text{ ft}^3$$

Total infiltration through the three walls
$$= 240 \text{ ft}^3 \ (\ 6.8 \text{ m}^3) \quad \text{first wall}$$
$$600 \text{ ft}^3 \ (16.98 \text{ m}^3) \quad \text{second wall}$$
$$\underline{540 \text{ ft}^3 \ (15.28 \text{ m}^3)} \quad \text{third wall}$$
Total infiltration $= 1380 \text{ ft}^3$ (39 m³)

Half of total infiltration $= 1380 \div 2$
$$= 690 \text{ ft}^3 \ (27.16 \text{ m}^3)$$

The quantity is greater than the infiltration through any individual wall. This should be the quantity used in determining whether or not there is sufficient fresh air being admitted to the space by infiltration. Because the air conditioning unit must cool all the air that enters by infiltration, it must be added to the load on the unit. However, the quantity of air entering by infiltration is not sufficient to provide the minimum fresh air requirements. Fresh air must be taken in by the cooling unit to provide the requirements (Table 22–5). This extra quantity of fresh air will prevent natural infiltration. This extra amount of air being introduced into the space will build up an internal pressure that will cause the air to leak out rather than in. This is the superior arrangement because all the air that enters the conditioned space must first pass through the air conditioning equipment.

Infiltration Heat Load. From the foregoing explanation and examples, it should be evident that infiltration air will be an internal heat load on the space. Therefore, it must be divided into its sensible and latent heat components. This is not the same for ventilation air, which is a heat load on the cooling unit but not on the room. When ventilation air is supplied to the conditioned space, it is only necessary to figure the total heat load of the air.

EXAMPLE 16

The total infiltration air figured in Example 15 was 690 ft³ (19.53 m³) per hour. Since the basis of the psychrometric chart which we use to find the properties of air is 1 lb (0.4536 g) of dry air, the cubic feet (m³) per hour must be changed to pounds (grams) per hour. If there is an outside air temperature of 95 °F (35 °C) dry bulb and 75 °F (23.9 °C) wet bulb, which are the design conditions for New York City (Table 22–3) and if the inside design conditions are 80 °F (26.7 °C) dry bulb and 67 °F (19.4 °C) wet bulb, the psychrometric chart indicates that dry air at 95 °F dry bulb weighs 0.0715 lb/ft³ (0.0011 g/cm³).

Total air per hour $= 690 \times 0.0715$
$$= 49.335 \text{ lb (22.38 kg) of dry air per hour}$$

Total heat at 75 °F (23.9 °C) wet bulb
$$= 37.81 \text{ Btu/lb (20.79 kcal/kg) of dry air}$$

Total heat at 67 °F (19.4 °C) wet bulb
$$= \underline{31.15 \text{ Btu/lb (17.13 kcal/kg)} } \text{ of dry air}$$

Total heat removed
$$= 6.66 \text{ Btu/lb (3.66 kcal/kg) of dry air}$$

Total heat from infiltration
$$= 49.335 \text{ lb air (22.38 kg)} \times 6.66 \text{ Btu/lb}$$
$$(3.66 \text{ kcal/kg}) = 328.57 \text{ Btu/hr}$$
$$(82.80 \text{ kcal/kg) of dry air}$$

Now divide the total heat load into the latent and sensible heat loads.

Sensible heat content at 95 °F (35 °C) dry bulb
$$= 22.95 \text{ Btu/lb (12.62 kcal/kg) of dry air}$$

Sensible heat content at 80 °F (26.7 °C) dry bulb
$$= \underline{19.32 \text{ Btu/lb (10.6 \ kcal/kg)} } \text{ of dry air}$$

Total sensible heat removed
$$= 3.63 \text{ Btu/lb (1.99 kcal/kg) of dry air}$$

Total sensible heat from infiltration
$$= 49.335 \text{ lb air (22.38 kg)} \times 3.63 \text{ Btu/lb}$$
$$(1.99 \text{ kcal/kg}) = 179.09 \text{ Btu/lb}$$
$$(98.5 \text{ kcal/kg) of dry air}$$

The latent heat to be removed is total heat less sensible heat. Therefore, the latent heat removed = 328.57 Btu (180.7 kcal/kg) × 179.09 Btu (98.5 kcal/kg) = 149.48 Btu (82.2 kcal/kg).

Ventilation Heat Load. The amount of infiltration air should be compared with the minimum fresh air requirements and if it is found to be insufficient, ventilation air must be brought in through the air conditioning equipment. The ventilation heat load and the infiltration heat load should not be used on the same job. In Example 15, it was found that 690 ft³ (19.53 m³) of air entered the room by infiltration. If the room is an office that has two occupants, Table 22-5 indicates that 1200 ft³ (33.96 m³) of air per hour are required because each occupant requires 10 ft³ (0.28 m³)/min, i.e., 2 occupants × 10 cfm (0.28 m³) × 60 min/hr = 1200 ft³ (33.96 m³) of air per hour.

Therefore, the amount of infiltration air would not be sufficient and the heat load due to the introduction of 1200 ft³ (33.96 m³) of air would be added to the conduction heat load. It would not, however, be necesary to divide the total ventilation heat load into the sensible and latent heat loads. The ventilation heat load would be calculated exactly like the infiltration heat load.

Electric Appliance Load. Only 5% or less of the electricity used by an electric light is converted into light. The other 95% is given off in the form of heat. Because of this, the lighting load on a conditioned space must be calculated on the basis of the total amount of electricity consumed by the light. The figure used in calculating this head load is 3.41 Btu (859.3 cal) per watt. Only the lights used during the peak load period should be included in the load calculation. A business located on a lower floor of a large building probably uses lights all the time, while it would receive very little, if any, solar effect. These conditions require that the lighting load be included in the estimate. If a unit is installed where the sun will shine into the windows and additional lighting is also required, both loads must be included in the calculation. When fluorescent lighting is used, there must be an addition of 25% of the lighting load included in the estimate because of the heat generated by the ballast.

Appliances such as toasters, hair driers, etc., are figured on the same basis as electric lighting; that is, 3.41 Btu (859.3 cal)/watt. If all the heat generated by these appliances is dissipated into the conditioned space, the total input wattage is multiplied by 3.41. Consideration should be given to the amount of time and the period that these appliances are operated. A toaster in a restaurant, for example, may operate continuously during the breakfast period, but not be used at all during the peak load period. It, therefore, should not be included in the estimate.

Varying amounts of heat are given off by electric motors. Larger motors are more efficient than the smaller horsepower motors and, therefore, generate less heat to be dissipated into the conditioned space. Equivalent values for motors have been placed in tables for convenience (Table 22-7). It should be noted that the values for motors are given on the basis of Btu (cal) per horsepower per hour.

Gas and Steam Appliances. These appliances include good cooking and warming devices, such as steam tables, coffee urns, etc. (Table 22-8). Each

TABLE 22-7 Heat from Electric Appliances

Item	Amount of Heat
Lights	3.41 Btu per watt [0.86 kcal]
Appliances [toasters, hair driers, etc.]	3.41 Btu per watt [0.86 kcal]
Motors	
⅛ to ½ hp	4250 Btu per hp per hr [1071 kcal]
¾ to 3 hp	3700 Btu per hp per hr [932.4 kcal]
5 to 20 hp	3000 Btu per hp per hr [756 kcal]

TABLE 22-8 Heat [Btu/hr [kcal/hr]] from Gas and Steam Appliances

Item[a]	Sensible	Latent	Total
Steam tables, ft² top surface	1,000 [252]	1,000 [252]	2,000 [504]
Restaurant coffee urns[b]	5,000 [1260]	5,000 [1260]	10,000 [2520]
Natural gas, ft³ [m³]	500 [126]	500 [126]	1,000 [252]
Manufactured gas, ft³ [m³]	275 [69.3]	275 [69.3]	550 [138.6]
Steam condensed in warming coils, lb [0.4536 kg]	480 [120.9]	480 [120.9]	960 [241.92]

[a]If there are exhaust hoods over steam tables and coffee urns, include only 50% of the heat load from these sources.

[b]Coffee urns may also be figured on the basis of consuming approximately 1 ft³ of natural gas or approximately 2 ft³ of manufactured gas per hour per gallon rated capacity.

TABLE 22-9 Heat Transfer Factors for Ducts

Duct	Btu per ft² per hr per 1°F Difference
Sheet metal, not insulated	1.13 [0.28 kcal]
Average insulation, ½ in. thick [12.7 mm]	0.41 [0.10 kcal]

pound of steam condensed by the coil in a steam table must be added to the heat load. This is equal to 960 Btu/lb (528 kcal/kg) of steam condensed. Because of the moisture given off by the heated food, this heat must be considered as half sensible heat and half latent heat. Coffee urns heated by gas emit the total amount of heat generated by the burning gas. This heat also is considered as being half sensible and half latent. The amount of heat given off for coffee urns is calculated by figuring one cubic foot (0.0283 m³) of natural gas or two cubic feet (0.057 m³) of manufactured gas per gallon of rated capacity of the urn. If exhaust fans are placed over the urns, the sensible and latent heat loads may be reduced by 50%.

Duct Heat Gain. When ducts are to be installed in hot places such as attics or boiler rooms, there is an additional load placed on the cooling unit. Any heat added to the air from the time it leaves the cooling coil until it gets back to the coil must be removed by the air conditioning unit. Heat transfer factors have been placed in tables for convenience (Table 22-9). It should be noted that when the entire supply duct is located inside the conditioned space, no duct heat gain needs to be calculated. Any heat gain will come from the conditioned space. This becomes a credit on the amount of cooling required in the exact amount of heat gain to the ducts.

Since the amount of air that must be supplied to the conditioned space is determined from the amount of sensible heat that must be removed, it follows that the duct sizes are indirectly related to the sensible heat gain of the space. Therefore, it is reasonable that the duct heat gain be calculated as a percentage of the sensible heat gain. This percentage will vary from 0% to a maximum of 5%. The length of the duct run through an unconditioned space, as well as the surrounding air temperature, must be given consideration. The above-mentioned percentages are based on the assumption that all the ducts installed in unconditioned spaces will be insulated.

The procedure for figuring duct heat gain is the same used for calculating the heat gain for any other heat flow (Table 22-9). The area of the exposed duct is multiplied by the U factor found in the table. This total is then multiplied by the dry-bulb temperature difference between the air inside the duct and the air outside the duct.

Fan Heat Gain. To maintain the proper air circulation, a fan is used to push the air through the system. The fan depends on the amount of air circulated, which is also dependent on the total sensible heat calculated. The fan power allowance also may be taken as a percentage of the sensible heat. This percentage will usually be between 3 and 4% of the total sensible heat load. It is, however, customary to allow 5% as a combined factor for the duct heat gain and the fan heat gain.

Safety Factor. Because of the many variables involved, heat load calculation is not an exact science like mathematics, chemistry, physics, etc. For example, heat transfer coefficients are calculated with great accuracy for most types of construction, but the design engineer has no positive assurance that the construction under consideration is identical in all respects to the test panels from which the coefficients were established. Also, internal loads such as people, lighting, and equipment, as well as shading factors, color of the exterior surfaces, etc. may vary from those upon which the design was based. Because of this, it is advisable to apply a factor of safety to the calculated heat load. This safety factor should vary from a maximum—on very small installations where any variation may seriously affect the performance of the system—to a minimum—on very large installations where there is a greater chance that the possible variations may cancel out one another. For the average installation, a safety factor of 10% is usually adequate. It must be noted that the safety factor must be applied to both the sensible and latent heat loads so that the relation between the two is not upset.

Equipment Selection. The Btu (cal) capacity of the equipment must be as close as possible to the calculated value. However, it must always be equal to or greater than the calculated heat gain. The equipment size should always be slightly larger than that required, never in great excess. If the calculated size is not available, only the next size larger should be chosen.

FIGURE 22-42 Psychrometric chart showing ADP.

Apparatus Dew Point and Air Quantity. On the psychrometric chart (Figure 22–42), the point where the sensible heat ratio line and the saturation curve intersect is known as the apparatus dew point (ADP). This point represents the lowest temperature at which air can be supplied to the conditioned space and still pick up the required amount of sensible and latent heat. If air at a higher ADP was supplied to the conditioned space, the quantity of air could be adjusted so that the required amount of sensible heat would be picked up.

However, this air quantity would not pick up a sufficient amount of latent heat. The resultant air would then permit a higher relative humidity inside the conditioned space than would be desired. If the ADP was too low, the relative humidity also would be too low. However, if air at any condition which falls on the sensible heat ratio line between condition (point *A*) and ADP (point *B*) is supplied, the quantity of air can be adjusted so that the exact amount of both sensible and latent heat and the design conditions can be met.

The ADP will actually very closely approach the cooling coil surface temperature. Therefore, a very large amount of coil surface would be needed if the conditioned air was to be supplied at the ADP. The air entering the cooling coil will be at its maximum temperature, and the maximum air temperature differential will exist between the air and the coil surface. As the air progresses through the coil, it is cooled so that the temperature differential progressively decreases. The first row of tubing will do the most work and each following row will do less work. Therefore, to cool the air to the ADP would take an infinite number of tubes and the last rows would do very little work because the temperature differential that causes the heat transfer would be very small.

In actuality, the choice of the number of rows of tubing usually falls between a minimum of three rows and a maximum of eight rows. When a large amount of sensible heat is to be removed, a smaller number of rows can be used. Conversely, as the latent heat load increases, more rows of tubing become necessary from both the engineering and economic standpoints.

Bypass Factor. The bypass factor refers to that percentage of the total amount of air that passes through the cooling coil without coming into contact with the coil surfaces and, therefore, is not conditioned. The bypass factor is dependent on the velocity of the air through the coil and the actual design of the coil itself. The number of rows of tubing in the direction of airflow also is a determining factor. Therefore, since coils vary in design from one manufacturer to another, the bypass factor also will be different. However, the average bypass factors of coils presently manufactured and used with normal air velocities are shown in Table 22-10.

It can be seen from the table that if a four-row coil were used, 20% of the total air passing through the coil would not be treated. This amount of air would not be useful in cooling the conditioned space.

TABLE 22-10 Bypass Factor for Different Depths of Coil

Number of Rows	Average Bypass Factor (%)
3	35
4	20
5	10
6	5
7	2.5
8	0.0

□ Duct System Design □

When a new air conditioning system is being considered, the type and design of the duct system are as critical as the load calculation. The location of the outlet, the air velocity, and the cubic feet (or m³) of air per minute are of prime importance in calculating the duct system. The design engineer may make drawings where each pipe, fitting, duct, or other piece of the equipment is shown in detail as it actually appears.

After the heat load has been calculated for the space and the individual rooms within the conditioned space, the ductwork can be calculated and sized.

Air Velocities. Air velocity is the rate of speed at which the air travels through the ducts or openings. There are suggested velocities that should be maintained. Higher velocities will result in increased air noise and electric power consumption (Table 22-11).

Duct Design. Ducts made of metal are preferred because of the smooth surfaces, which do not add resistance to the airflow. Metal ducts also can be formed into compact sizes and shapes, and placed in difficult locations.

Pressure losses that occur in duct systems are caused by the velocity of the airflow, number of fittings, and friction of the air against the walls of the duct. The use of coils, air filters, dampers, and deflectors all add to the resistance to the airflow.

When a more extensive system is to be designed, a good practice is to gradually lower the velocity in both the main duct and the remote branches. This type of design has the following advantages:

1. It enables the air to be distributed in a more uniform manner.
2. It decreases the air friction in the smaller ducts where it would otherwise be the greatest.
3. When the velocity is reduced there is a recovery of velocity pressure that compensates for the duct friction.

Dampers and deflectors should be placed at all points necessary to assure the proper balance of the system. It is difficult to design a duct system that will provide the correct amount of air at each outlet. It is therefore necessary to use dampers and deflectors so that the air supply may be directly proportioned after the system is placed in operation.

Location	Residences	Schools and Public Buildings	Industrial Buildings
Main ducts	700–900 [1000]	1000–1300 [1400]	1200–1800 [2000]
Branch ducts	600 [700]	600–900 [1000]	800–1000 [1200]
Branch risers	500 [650]	600–700 [900]	800 [1000]
Outside air intakes	700 [800]	800 [900]	1000 [1200]
Filters	250 [300]	300 [350]	350 [350]
Heating coils	450 [500]	500 [600]	600 [700]
Air washers	500	500	500
Suction connections	700 [900]	800 [1000]	1000 [1400]
Fan outlets	1000–1600 [1700]	1300–2000 [2200]	1600–2400 [2800]

[a]Maximum duct velocities are given in parentheses.

Duct Calculations. The required duct sizes and velocities can be figured by use of the following general formula:

$$Q = A \times V$$

where

Q = discharge of air [cfm (m³/min)]

A = cross-sectional area of the opening [ft² (m²)]

V = velocity of air [fpm (m/min)]

EXAMPLE 17

Given 1500 cfm (42.45 m³/min) of air to be moved at a velocity of 100 ft (30.48 m) per minute, find the area of duct needed.

Solution

$$Q = A \times V$$
$$1500 = A \times 1000$$
$$A = \frac{1500}{1000} = 1.5 \text{ ft}^2 \ (0.1395 \text{ m}^2)$$

To change these terms to square inches (cm²), multiply A times 144 (929 cm²) as follows:

$$1.5 \times 144 = 216 \text{ in.}^2 \ (0.1395 \text{ m}^2)$$

This gives the area of the duct or opening. This duct can be made in any shape to suit the available space.

The areas of circumferences of circles and the sides of squares of the same area are given in tables such as Table 22–12. To find the equivalent diameter of a circular duct or the side of a square duct note that 216 in.² falls between 213.8 and 226.9 in.². Using the largest value, 226.9, find the diameter of the circular duct to be 15 in.

If the space available for the duct is 18 in. (45.72 cm), then a 12-in. × 18-in. (30.48-cm × 45.72-cm) duct could be used:

$$\frac{216}{18} = 12 \ (30.48 \text{ cm})$$

Friction Loss Calculation. To determine the friction loss of a duct system, it is necessary to know the velocity and size of the duct. The friction loss per hundred feet (30.48 m) of a duct of a given size may be found in Table 22–13.

Allowances also must be made for elbows and bends. These allowances may be calculated by adding to the total length of duct. Add 8 diameters for a round elbow and 5 diameters for a square or rectangular elbow equipped with turning vanes.

EXAMPLE 18

Given a blower moving air through an inlet duct 42 in. (106.68 cm) in diameter and 100 ft (30.48 m) in length, and supplying air through a system of ducts at 4 points, figure the pressure loss (Figure 22–43 and Table 22–13).

AIR CONDITIONING (COOLING)

TABLE 22-12 Areas and Circumferences of Circles and the Sides of Squares of the Same Area

Diameter [in.]	Area of Circle ft²	Area of Circle in.²	Sides of Square of Same Area [in.]	Circumference of Circle [in.]	Diameter [in.]	Area of Circle ft²	Area of Circle in.²	Sides of Square of Same Area [in.]	Circumference of Circle [in.]
1	0.005	0.8	0.9	3.1	32	5.75	804.2	28.3	100.5
1½	0.012	1.7	1.3	4.7	32½	5.85	829.5	28.8	102.1
2	0.022	3.1	1.7	6.2	33	5.93	855.3	29.2	103.6
2½	0.034	4.9	2.2	7.8	33½	6.11	881.4	29.6	105.2
3	0.049	7.0	2.6	9.4	34	6.31	907.9	30.1	106.8
3½	0.067	9.6	3.1	10.9	34½	6.47	934.8	30.5	108.3
4	0.087	12.5	3.5	12.5	35	6.67	962.1	31.0	109.9
4½	0.110	15.9	3.9	14.1	35½	6.87	989.8	31.4	111.5
5	0.136	19.6	4.4	15.7	36	7.08	1017.8	31.9	113.0
5½	0.165	23.7	4.8	17.2	36½	7.28	1046.3	32.3	114.6
6	0.196	28.2	5.3	18.8	37	7.46	1075.2	32.7	116.2
6½	0.230	33.1	5.7	20.4	37½	7.69	1104.4	33.2	117.8
7	0.268	38.4	6.2	21.9	38	7.88	1134.1	33.6	119.3
7½	0.306	44.1	6.6	23.5	38½	8.10	1164.1	34.1	120.9
8	0.350	50.2	7.0	25.1	39	8.30	1194.5	34.5	122.5
8½	0.393	56.7	7.5	26.7	39½	8.52	1225.4	35.0	124.0
9	0.442	63.6	7.9	28.2	40	8.72	1256.6	35.4	125.6
9½	0.493	70.8	8.4	29.8	40½	8.95	1288.2	35.8	127.2
10	0.545	78.5	8.8	31.4	41	9.17	1320.2	36.3	128.8
10½	0.600	86.5	9.3	32.9	41½	9.40	1352.6	36.8	130.3
11	0.660	95.0	9.7	34.5	42	9.64	1385.4	37.2	131.9
11½	0.722	103.8	10.1	36.1	42½	9.85	1418.6	37.6	133.5
12	0.785	113.1	10.6	37.6	43	10.10	1452.2	38.1	135.0
12½	0.852	122.7	11.0	39.2	43½	10.31	1486.1	38.5	136.6
13	0.922	132.7	11.5	40.8	44	10.55	1520.5	38.9	138.2
13½	0.994	143.1	11.9	42.4	44½	10.80	1555.2	39.4	139.8
14	1.07	153.9	12.4	43.9	45	11.05	1590.4	39.8	141.3
14½	1.15	165.1	12.8	45.5	45½	11.30	1625.9	40.3	142.9
15	1.23	176.7	13.2	47.1	46	11.55	1661.9	40.7	144.5
15½	1.31	188.6	13.7	48.6	46½	11.80	1698.2	41.2	146.0
16	1.40	201.0	14.1	50.2	47	12.10	1734.9	41.6	147.6
16½	1.48	213.8	14.6	51.8	47½	12.31	1772.0	42.1	149.2

TABLE 22–12 *continued*

Diameter (in.)	Area of Circle ft²	Area of Circle in.²	Sides of Square of Same Area (in.)	Circumference of Circle (in.)	Diameter (in.)	Area of Circle ft²	Area of Circle in.²	Sides of Square of Same Area (in.)	Circumference of Circle (in.)
17	1.58	226.9	15.0	53.4	48	12.55	1809.5	42.5	150.7
17½	1.67	240.5	15.5	54.9	48½	12.85	1847.4	42.9	152.3
18	1.77	254.4	15.9	56.5	49	13.10	1885.7	43.4	153.9
18½	1.87	268.8	16.4	58.1	49½	13.40	1924.4	43.8	155.5
19	1.97	283.5	16.8	59.6	50	13.65	1963.5	44.3	157.0
19½	2.07	298.6	17.2	61.2	50½	13.95	2002.9	44.7	158.6
20	2.18	314.1	17.7	62.8	51	14.20	2042.8	45.2	160.2
20½	2.29	330.0	18.1	64.4	51½	14.45	2083.0	45.6	161.7
21	2.41	346.3	18.6	65.9	52	14.75	2123.7	46.0	163.3
21½	2.52	363.0	19.0	67.5	52½	15.05	2164.7	46.5	164.9
22	2.64	380.1	19.5	69.1	53	15.35	2206.1	46.9	166.5
22½	2.76	397.6	19.9	70.6	53½	15.62	2248.0	47.4	168.0
23	2.84	415.4	20.3	72.2	54	15.92	2290.2	47.8	169.6
23½	3.01	433.7	20.8	73.8	54½	16.20	2332.8	48.3	171.2
24	3.14	452.3	21.2	75.3	55	16.45	2375.8	48.7	172.7
24½	3.28	471.4	21.7	76.9	55½	16.80	2419.2	49.1	174.3
25	3.41	490.8	22.1	78.5	56	17.10	2463.0	49.6	175.9
25½	3.54	510.7	22.6	80.1	56½	17.45	2507.1	50.0	177.5
26	3.69	530.9	23.0	81.6	57	17.72	2551.7	50.5	179.0
26½	3.83	551.5	23.4	83.2	57½	18.00	2596.7	50.9	180.6
27	3.98	572.5	23.9	84.8	58	18.35	2642.0	51.4	182.2
27½	4.13	593.9	24.3	86.3	58½	18.65	2687.8	51.8	183.7
28	4.28	615.7	24.8	87.9	59	18.95	2733.9	52.2	185.3
28½	4.42	637.9	25.2	89.5	59½	19.35	2780.5	52.7	186.9
29	4.59	660.5	25.7	91.1	60	19.65	2827.7	53.1	188.4
29½	4.74	683.4	26.1	92.6	60½	19.95	2874.7	53.6	190.0
30	4.92	706.8	26.5	94.2	61	20.39	2922.5	54.0	191.6
30½	5.07	730.6	27.0	95.8	61½	20.62	2970.6	54.5	193.2
31	5.23	754.7	27.4	97.3	62	20.95	3019.1	54.9	194.8
31½	5.42	779.3	27.9	98.9	62½	21.32	3068.0	55.4	196.4

TABLE 22–13 Pressure Loss (inches of water per 100 ft of pipe) due to Friction of Air[a]

Diameter of Pipe (in.)	Area (in.²)	Velocity of Air (fpm)														
		200	400	600	800	900	1000	1200	1400	1600	1800	2000	2400	3000	4500	6000
3	7.0	0.023	0.092	0.208	0.370	0.468	0.581	0.832	1.14	1.48	1.87	2.32	3.33	5.20	11.70	20.80
4	12.5	0.017	0.070	0.157	0.280	0.350	0.440	0.624	0.862	1.12	1.40	1.76	2.50	3.94	8.85	15.76
5	19.6	0.014	0.056	0.125	0.226	0.281	0.348	0.498	0.721	0.902	1.12	1.39	1.99	3.12	7.03	12.48
6	28.2	0.011	0.046	0.104	0.185	0.234	0.290	0.417	0.568	0.741	0.935	1.16	1.67	2.60	5.84	10.40
7	38.4	0.010	0.040	0.089	0.158	0.201	0.249	0.357	0.487	0.633	0.805	1.00	1.43	2.22	5.01	8.88
8	50.2	0.009	0.034	0.078	0.138	0.176	0.218	0.312	0.427	0.554	0.703	0.872	1.25	1.95	4.38	7.80
9	63.6	0.008	0.031	0.069	0.123	0.156	0.194	0.277	0.379	0.493	0.625	0.776	1.11	1.73	3.89	6.92
10	78.5	0.007	0.028	0.063	0.111	0.141	0.175	0.250	0.342	0.445	0.562	0.700	1.00	1.56	3.51	6.24
12	113.1	0.006	0.023	0.052	0.092	0.117	0.145	0.208	0.284	0.369	0.467	0.580	0.832	1.30	2.92	5.20
14	153.9	0.005	0.021	0.049	0.086	0.109	0.136	0.194	0.266	0.345	0.437	0.544	0.776	1.21	2.74	4.84
16	201.0	0.004	0.017	0.039	0.070	0.088	0.109	0.156	0.214	0.278	0.352	0.436	0.624	0.977	2.20	3.91
18	254.4	0.004	0.015	0.035	0.060	0.078	0.097	0.138	0.190	0.242	0.312	0.387	0.552	0.867	1.95	3.47
20	314.1	0.004	0.014	0.031	0.055	0.070	0.087	0.124	0.170	0.221	0.280	0.348	0.496	0.778	1.75	3.11
22	380.1	0.003	0.013	0.028	0.050	0.064	0.079	0.113	0.156	0.202	0.256	0.317	0.452	0.711	1.60	2.84
24	452.3	0.003	0.012	0.026	0.046	0.058	0.073	0.104	0.142	0.164	0.234	0.290	0.416	0.650	1.46	2.60
28	615.7	0.003	0.010	0.022	0.040	0.052	0.062	0.089	0.122	0.159	0.202	0.249	0.356	0.558	1.26	2.23
30	706.8	0.003	0.009	0.020	0.037	0.047	0.058	0.083	0.114	0.148	0.187	0.232	0.332	0.520	1.17	2.08
36	1017	0.002	0.008	0.017	0.031	0.039	0.048	0.069	0.095	0.123	0.156	0.193	0.276	0.433	0.976	1.73
42	1385	0.002	0.007	0.015	0.026	0.033	0.041	0.059	0.081	0.106	0.134	0.166	0.236	0.371	0.836	1.48
48	1809	0.002	0.006	0.013	0.023	0.029	0.036	0.052	0.071	0.092	0.117	0.144	0.208	0.325	0.731	1.30
54	2290	0.001	0.005	0.012	0.020	0.026	0.032	0.046	0.063	0.082	0.104	0.129	0.184	0.288	0.650	1.15
60	2827	0.001	0.005	0.010	0.018	0.023	0.029	0.042	0.057	0.074	0.094	0.116	0.168	0.260	0.585	1.04

[a]For other lengths of pipe the loss is directly proportional to the length.

R = 1½ D

R = 1 D

FIGURE 22-43 **Figuring friction loss of a system.**

Solution

Section A: 100 ft (30.48 m) of straight duct 36 in. (91.44 cm) in diameter, an air velocity of 1200 ft (364.8 m) per minute and a friction loss of 0.069 in. (0.175 cm).

Section B: 80 ft (24.32 m) of straight duct, 30 in. (76.2 cm) in diameter, an air velocity of 1000 ft (304.8 m) per minute. For a run of 100 ft (30.48 m) it is found that the loss is 0.058 in. (0.147 cm). Because this run is for 80 ft (24.32 m) take 8/10 of 0.058 or 0.046.

Section C: One 90° bend, 2 ft (0.609 m), the center line radius is equal to 1½ times the diameter. The friction loss is equal to 8 diameters or 16 ft (4.86 m) of straight duct. This length must be added to the straight run of 50 ft (15.2 m), making a total of 66 ft (20.11 m) at a velocity of 900 ft/min (273.6 mpm). Therefore, take 66/100 of 0.058 in. (0.147 cm) or 0.037 in. (0.099 cm).

Section D: In this section there is one 90° bend. The duct is 18 in. (45.72 cm) in diameter. The friction loss is equal to 8 diameters or 12 ft (3.65 m) of straight duct. Add this to the straight run of 40 ft (12.19 m) for a total of 52 ft (15.84 m). Therefore, take 52/100 of 0.060 in. (0.15 cm) and the result is 0.031 in. (0.08 cm).

Section E: This duct is 100 ft (30.48 m) long, and 42 in. (106.68 cm) in diameter. Air travels through this duct at a velocity of 1000 ft (304.8 m) per minute. It is found that the pressure loss is 0.041 in. (0.104 cm).

To find the total friction losses, add the loss in each section, which amounts to a total of 0.224 in. (0.569 cm). In this case the blower selected must be capable of delivering the required volume at ¼ in. (6.5 mm) static pressure.

The example was based on round ducts. If the ducts are to be square or rectangular, however, it would only be necessary to reduce the rectangular ducts to round ducts of equivalent area and proceed with the calculations just described.

☐ Air Conditioning Equipment ☐

The refrigeration equipment used in air conditioning work is much like that used in refrigeration work. The major difference is the duct work. The fact that higher temperatures are encountered necessitates some different design characteristics. Otherwise, the refrigeration cycle is the same as that used in refrigeration applications.

Compressors. The compressors used in air conditioning units are basically the same as those discussed in Chapter 4. They may be either the reciprocating, rotary, or centrifugal type. However, the reciprocating type is most commonly used (Figure 22-44). When used on air-cooled condenser systems, it is mounted outside in the condensing unit. This moves the compressor noise outside and allows more air to flow over it to aid in cooling.

Most equipment manufacturers either use or recommend the use of crankcase heaters to provide compressor protection from refrigerant migration or liquid slugging. Some manufacturers even recommend the use of a pumpdown cycle to prevent the

FIGURE 22-44 Cutaway of hermetic compressor. (Courtesy of Sundstrand Corp.)

possibility of liquid slugging the compressor (Figure 22-45). The major problem with a pumpdown cycle is that the service and installation technicians set the low-pressure control to open at about 10 psig (68.948 kPa) and close at 55 psig (379.212 kPa) so that a good pump-out will be possible to reduce short cycling of the compressor. However, when R-22 is the refrigerant, the compressor would be operating for a part of the cycle at the suction temperature of $-20\,°F$ ($-28.9\,°C$), which is completely out of the operating range of the compressor. This could result in serious damage to the compressor, if not a complete breakdown, especially if the system was operated with a partial loss of charge.

It would be better to set the low-pressure control to open at about 25 psig (172.369 kPa) and close at about 60 psig (413.685 kPa) on systems using R-22 as the refrigerant. At a 60-psig suction pressure the evaporating coil would be completely defrosted and the unit would still have protection if a part of the charge were lost.

Condensers. The function of an air conditioning condenser is to remove heat from the refrigerant and cause the vapor to change to a liquid. Most air conditioning condensers are of the air-cooled type, but there also are water-cooled and evaporative condensers in use. The efficiency of the condenser determines to a great extent the energy efficiency ratio (EER) of the unit. The EER is determined by dividing the Btu (cal) rating of the unit by the wattage drawn by the unit. The higher the EER, the more efficient the unit. There are several units on the market now with an EER of 8.5. The EER is certified by the Air Conditioning and Refrigeration Institute (ARI), and is known as the ARI rating.

Evaporators. Air conditioning evaporators (Figure 22-46) are manufactured in three types: (1) upflow, (2) downflow, and (3) horizontal. This terminology refers to the direction of airflow through the coil. The purpose of the evaporator is to remove heat and moisture from the air delivered to the conditioned space. A drain pan is incorporated in the manufacturer of the coil unit to collect and dispose of the condensation due to cooling the air below the dew-point temperature. The drain connection must be

FIGURE 22-45 Pump down system schematic.

A

B

FIGURE 22–46 Air conditioning evaporators. (Courtesy of Dearborn Stove Co.)

connected to a drain to dispose of the condensate properly.

Flow Control Device. The flow control device used on air conditioning installations may be either the capillary tube or a thermostatic expansion valve. Capillary tubes are used on systems up to approximately 3 tons in capacity. The larger units require a thermostatic expansion valve to control the flow of refrigerant properly.

Air-Handling Unit. The air-handling unit contains the blower and either the heating equipment, the cooling equipment, or both the heating and cooling equipment. In most residences this is the heating furnace. In any case, the blower must be capable of delivering the required amount of air at the proper

velocity with the design static pressure of the duct system.

Types of Air Conditioning Systems. Air conditioning systems are generally classified as being either self-contained or remote systems. Both of these types can be installed as a central unit.

Self-Contained Systems. Self-contained systems have all the components installed on one base (Figure 22–47), with the possible exception of the condenser. Water-cooled systems have a water tower placed at some location outside the building with the water piped to the condenser. Air-cooled systems may be installed on an outside wall with the necessary provisions made to allow the condenser air to circulate outdoors freely (Figure 22–48).

Self-contained units may be larger commercial units known as roof-mounted units (Figure 22–49), window units, or the smaller commercial units (Figure 22–47). Ductwork may be connected to the supply and return air openings and ducted to and from the space.

Remote Split Systems. These systems are in more common use on the residential and small commercial installations. They have the evaporator, blower, and furnace all located in a closet in an equipment room with the condensing unit (Figure 22–50). The refrigerant is piped from the condensing unit to the indoor coil through copper tubing. The air is ducted from the air conditioning equipment to the space to be conditioned. Most residential units have the indoor section centrally located within the space, which eliminates the need for a return air duct system.

In order to obtain the proper and most efficient operation of a system, all of the pieces must be matched as far as Btu (cal) ratings. Most manufacturers build equipment to fit almost any need and the designated combination of fan, furnace, coil, an condensing unit must be used. After the heat load has been calculated, tables can be consulted to determine the desired combination. Care should be exercised in choosing the equipment to obtain the most satisfactory operation.

☐ Safety Procedures ☐

When working on air conditioning units, it is important to remember that the conditioned air must contain enough oxygen to sustain life. Working on a complete air conditioning system involves many

AIR CONDITIONING (COOLING)

FEATURES:

1. FRONT RETURN — Air grille may be interchanged with lower back panel for rear return.

2. AMPLE SPACE — Provided above evaporator coil for field installation of steam, or hot water coil.

3. BLOWER — Full size for maximum air movement at low noise level.

4. MOTOR — has integral overload protection. Variable speed sheave and V-belt drive.

5. CABINET — Heavy gauge, electrolytic zinc coated steel, finished in attractive two-tone acrylic enamel.

6. INSULATION — Thermal and acoustical, spun glass insulation.

7. EXPANSION VALVE — Thermostatic non-adjustable type. Mounted at the factory.

8. COIL — Aluminum finned, copper tubing, manifolded and circuited for maximum heat transfer.

9. FILTERS — Chemically treated, disposable type, easy to change.

FIGURE 22–47 **Cutaway of self-contained unit. (Courtesy of Southwest Manufacturing Company.)**

FIGURE 22–48 **Self-contained condenser connections.**

FIGURE 22–49 **Roof-mounted self-contained unit cutaway. (Courtesy of Lennox Industries, Inc.)**

FIGURE 22–50 Remote split system. (Courtesy of Southwest Manufacturing Company.)

of the dangers involved in refrigeration systems and heating systems.

1. Always wear goggles when working with the refrigerant in these systems.
2. Always wear gloves around ductwork to prevent cutting the hands and fingers.
3. Provide protective shields around operating motors and fans.
4. Be sure that the electric power is off before servicing electrical components.
5. Keep hands, feet, and clothing clear of belts, fans, and rotating parts.
6. Use only safe ladders when working on the roofs of buildings.
7. Never carry equipment, tools, or supplies up or down a ladder. Transport these items with a rope or a hoist.
8. Avoid contact with electric circuits when the area around the equipment is wet.
9. Do not leave loose, light objects on a roof during high winds.
10. Be sure that refrigerant being charged into a system is the same as what the unit is rated to use.

▣ SUMMARY ▣

☐ Air conditioning is the simultaneous mechanical control of temperature, humidity, air purity, and air motion.

☐ The control of humidity can mean either humidifying or dehumidifying.

☐ The two primary reasons for using air conditioning are: (1) to improve the control of an industrial process, and (2) to maintain human comfort.

☐ Human comfort depends on how fast the body loses heat.

☐ The human body is basically a constant-temperature machine.

☐ The human body gives off heat in three ways: (1) convection, (2) radiation, and (3) evaporation.

☐ The three conditions that affect the ability of the body to give off heat are: (1) temperature, (2) relative humidity, and (3) air motion.

☐ The air should be distributed into the room to provide only minor temperature differences within the occupied zone.

☐ The two general types of fans used on air conditioning installations are propeller fan and centrifugal blower.

☐ The air noise caused by air movement is the result of successive air waves caused by the fan blades.

☐ The heating chamber is the place where the heat may be added to the air during the heating season.

☐ The cooling chamber is the place where the cooling and dehumidifying of the air occurs.

☐ The humidifier adds moisture to the air during the heating season.

☐ The supply air ducts direct the conditioned air to the conditioned space.

☐ Supply air outlets are the grills on the end of the supply air ducts that distribute the air within the conditioned space.

☐ The conditioned space is one of the most important parts of the air distribution system.

☐ The return outlets are the openings through which the air from the conditioned space is allowed to enter the return duct system.

☐ The return ducts are used to connect the conditioned space and the air-handling equipment together.

☐ Filters are located on the air inlet side of the fan to prevent the circulation of dust particles through the system.

☐ The science that deals with the relationships that exist within a mixture of air and water vapor is known as psychrometics.

☐ A psychrometric chart is a diagram that represents the various relationships that exist between the heat and moisture content of the air and water vapor.

☐ Any two of the air properties located on the psychrometric chart will definitely determine the condition of the air.

☐ Sensible heat is the heat that can be added to or removed from a substance without a change of state.

☐ The sensible heat content of air is indicated by the dry-bulb temperature.

☐ The dew-point temperature is always an indication of the moisture content of the air.

☐ The moisture content scale used on most psychrometric charts indicates grains of moisture per pounds (453.6 g) of dry air.

- [ ] Sensible heat ratio is the sensible heat removed in Btu (cal) divided by the total heat removed in Btu (cal), usually about 50% in most comfort air conditioning installations.
- [ ] The calculation of the heat load is similar to the calculation of a refrigeration heat load, with the major difference that more sources of heat are involved in air conditioning calculations.
- [ ] Heat sources are listed under two general classifications: (1) internal heat load, and (2) external heat load.
- [ ] The amount of heat that will flow through one square foot per degree temperature difference through a substance is known as the *U* factor.
- [ ] The design temperatures are the average high temperatures for a given locality and not the highest temperatures in that locality.
- [ ] A solar heat load is gained from the direct rays of the sun; therefore, solar heat gain occurs only while the sun is shining.
- [ ] The peak heat load may not occur until hours after the sun is at its peak because of the resistance to heat flow of the construction materials.
- [ ] People who are active create a greater heat load for the cooling equipment than people who are at rest.
- [ ] Infiltration air enters the conditioned space through windows, doors, and cracks and is a source of internal heat load for the cooling equipment.
- [ ] The outdoor wind velocity affects the infiltration of air into the conditioned space.
- [ ] Ventilation air is air introduced into the equipment before it enters the conditioned space and is considered to be an external heat load.
- [ ] Electric appliances and lighting introduce a heat load of 3.41 Btu (859.3 cal) per watt into the conditioned space.
- [ ] Electric motors also are a source of heat in varying amounts depending on the motor horsepower.
- [ ] Steam appliances supply a heat load of 960 Btu/lb (528 kcal/kg) of steam condensed.
- [ ] Appliances heated by gas emit the total amount of heat generated by the burning gas.
- [ ] When ducts are installed in hot places, an additional heat load is placed upon the cooling equipment.
- [ ] The motor used to drive the fan also emits heat that must be removed by the cooling equipment. The amount of heat transferred is in direct proportion to the amount of work done.
- [ ] It is advisable to apply a safety factor of about 10% to the calculated heat load because of the many variables involved.
- [ ] The capacity of the chosen equipment must be as close as possible to the calculated value but never less than the calculated value.
- [ ] On the psychrometric chart, the point where the sensible heat ratio line and the saturation curve intersect is known as the apparatus dew point (ADP).
- [ ] The ADP is the lowest temperature at which air supplied to the conditioned space can still pick up the required amount of both sensible and latent heat.
- [ ] The bypass factor refers to that percentage of the total amount of air that passes through the cooling coil without coming into contact with the coil surfaces and being conditioned.
- [ ] The duct system design is as critical as the heat load calculation.
- [ ] Air velocity is the rate of speed that the air travels through the ducts or openings.
- [ ] Air ducts made of metal are preferred because of the smooth surfaces, which do not add resistance to the airflow.
- [ ] To determine the friction loss of a duct system, it is necessary to know the velocity and size of duct.
- [ ] The major differences in the equipment used in air conditioning work and that used in refrigeration work are the design characteristics used to compensate for the higher temperatures encountered.
- [ ] Air conditioning equipment is generally classified as being either a self-contained or a remote split system.

▣ EXERCISES ▣

1. What are the four basic functions of a true air conditioning system?
2. What are the two primary reasons for using air conditioning?
3. Name the three ways a human body gives off heat.
4. What are the three conditions that affect the ability of the human body to give off heat?
5. Why should the conditioned air be distributed into the space?
6. Name the general types of fans used in air conditioning work.
7. What is the cause of air noise in an air conditioning system?
8. What is the purpose of a humidifier?
9. Where is the heat added to the air during the heating season?
10. What device carries out the dehumidification process in a cooling unit?
11. What part of the air conditioning system directs the air to the conditioned space?
12. Why is the conditioned space one of the most important parts of the air distribution system?
13. What is the purpose of an air filter?
14. What is psychrometrics?
15. What device is used to determine the various relationships that exist between the heat and moisture content of air and water vapor?

16. How many of the air properties must be known before the rest can be determined?
17. What does the dry-bulb temperature indicate?
18. What indicates the moisture content of the air?
19. How is the moisture content indicated on a psychrometric chart?
20. What is the sensible heat ratio in most comfort conditioning installations?
21. How are the heat sources on an air conditioning unit classified?
22. What are design temperatures?
23. From where does solar heat gain come?
24. Does the peak heat load occur at the same time the sun is the hottest?
25. What is infiltration air?
26. Do people who are active create more or less heat load than people who are at rest?
27. Where is ventilation air introduced into the system?
28. Is ventilation air an internal or external heat load?
29. What is the Btu (cal) rating of electric appliances?
30. Why is an additional heat load placed on the equipment when the ducts are located in a hot place?
31. What safety factor percentage is generally used on an average job?
32. Should the capacity of the chosen equipment be less than the calculated capacity?
33. What is the apparatus dew point?
34. To what does the bypass factor refer?
35. What are the two classifications of air conditioning equipment?

23 Cooling and Heating with Water

Performance Objectives

After completion of this chapter, the student should:

■ Be able to name the various methods used in cooling refrigeration condensers with water.

■ Describe the operating principles involved in the use of water chillers.

■ Explain the purpose of air handling units being used when cooling with water.

■ Complete the Exercises with at least 80 percent accuracy.

■ Complete the Laboratory Exercises with at least 80 percent accuracy.

The use of water to transfer heat in cooling and heating systems is more energy efficient than the use of air as the transfer medium. There are several ways to use water in the cooling and heating process: water-cooled refrigeration condensers, water chillers, and hot water boilers. (For a discussion of boilers, see Chapter 20.) A combination of these methods or combining one of the water heat transfer methods with air is also used. One example is a water-cooled air conditioning condenser combined with a direct expansion evaporator where the cooled air is passed through the refrigeration coil. Regardless of which method is used, if water is used on one of the heat transfer surfaces, the efficiency will be increased. This increased efficiency is possible because of the more intimate contact between the water and the refrigeration.

◻ Condenser Cooling ◻

There are several methods used in the cooling of refrigeration condensers with water.

Wastewater Systems. Wastewater systems take water from the city water main, pass it through the condenser, and then dump it into the sewage system. These devices were introduced in Chapter 5. In this chapter we will study the operation of these units in more detail.

There are certain requirements that must be met when using water as a condenser cooling medium. The most basic one is that a certain amount of water must flow through the condenser to obtain the required cooling and condensing of the refrigerant. A good rule is 3 gal (11.355 liters) of water per ton of refrigeration per minute (3 gal/min/ton). Therefore, a 3-ton unit should have a flow of 9 gal (30.06 liters) of water per minute. This volume of water will give a water temperature rise of approximately 10 °F (5.56 °C) and a refrigerant condensing temperature of approximately 105 °F (40.58 °C) (Figure 23–1). A flow of less than 3 gal/min/ton will have a higher temperature rise and a higher than 105 °F refrigerant condensing temperature. Likewise, a flow of more than this will allow a smaller temperature rise and a lower than 105 °F condensing temperature.

The information above can be used in diagnosing troubles in a water-cooled condenser. For example, the compressor discharge pressure of an R-22 unit was checked and found to be 235 psig (1715.44 kPa), and

FIGURE 23-1 Checking water temperature rise through a water-cooled condenser.

FIGURE 23-2 Location of condenser water bleed-off.

the condenser water temperature rise was 6°F (3.33°C). The flow of water was found to be approximately 3 gal/min/ton as required. What would the problem be?

The correct amount of water is being pumped through the condenser; however, a smaller-than-normal temperature rise is found as well as a higher-than-normal discharge pressure. It can now be determined that the condenser has an accumulation of scale on the heat transfer surface, which must be removed. This scale is formed in much the same manner as that formed in a tea kettle. There are two popular ways to remove this scale: (1) reaming the condenser tubes, and (2) using a chemical to dissolve the scale. In some cases the tubes cannot be reamed because of the design of the condenser. Care must be exercised when using chemicals because of possible damage to the equipment and/or the surroundings, grass, shrubs, and the like. Always follow the chemical manufacturers' recommendations when using any chemicals.

The formation of scale in water-cooled condensers can be reduced by using water treatment chemicals to keep the scale from forming particles in suspension in the water. A bleed-off, connected to the water pump discharge used either in conjunction with water treatment or alone, will also help in reducing scale formation (Figure 23-2). This bleed-off should remove about 1 gal/hr/ton. To determine the proper water treatment to be used in each application, a local water treatment analyst should be consulted.

Cooling Towers. The purpose of a cooling tower is to cool the condenser water by removing the heat picked up in the condenser. There are three basic types of cooling towers: (1) forced draft, (2) natural draft, and (3) evaporative condenser. All of these operate on the principle of removing heat from the

water by evaporating a part of the water (Figure 23-3).

A *natural-draft tower* must be located so that nothing will restrict the flow of air through the cooling tower, which uses sprayers to atomize the water so that the air can come into better contact with it and remove the heat more effectively (Figure 23-4).

The temperature drop of the water as it flows through the cooling tower should be approximately 10°F (5.56°C). This can be determined by measuring the water temperature at the spray nozzle discharge and the pump inlet (Figure 23-5). These spray nozzles must produce a full spray cone of water to atomize the water properly for cooling. Wooden slats are placed around the tower to prevent the wind and spray from taking water from the tower. The tower sump must be cleaned periodically to remove the accumulation of dirt and silt that collects there.

A *forced-draft tower* has the same functions as a natural-draft tower. The main difference in operation is that a fan of some type is used in forced-draft towers (Figure 23-6). These towers usually do not use spray headers but use a top water pan with holes in the bottom. The holes allow the water to flow in small streams down through the webbing (fill) of wooden slats. These slats cause the water to break up, allowing the air to come into better contact with the water (Figure 23-7). The temperature drop of the water flowing through the tower should be approximately 10°F (5.56°C), taken where the water enters the top pan and at the pump inlet (Figure 23-8). The water in forced-draft towers is treated in the same manner as that in natural draft towers.

Evaporative condensers flow both water and air over a refrigerant condensing coil (Figure 23-9). The refrigerant is condensed by this combination. The same checkout procedures apply here as with the

Natural draft   Forced draft   Evaporative condenser

FIGURE 23–3   **Natural-draft, forced-draft towers, and evaporative condenser.**

FIGURE 23–4   **Spray nozzle operation.**

FIGURE 23–5   **Checking water temperature drop on a cooling tower.**

FIGURE 23–6   **Forced-draft tower.**

FIGURE 23–7   **Top pan and webbing (fill).**

FIGURE 23–8   **Checking water temperature drop on a forced-draft tower.**

FIGURE 23-9 Evaporative condenser cutaway.

FIGURE 23-10 Cycling pump with a pressure control.

FIGURE 23-11 Three-way valve to bypass condenser water.

FIGURE 23-12 Cycling fan on a forced-draft tower.

other types of towers. A buildup of scale is more easily detected in these units because they can usually be inspected. One advantage that evaporative condensers have over the other towers is that the water circuit can be shut off and drained during freezing weather, and the unit is still operable.

The humidity content of the ambient air has an effect on all water cooling towers. A high wet-bulb temperature will reduce the capacity while a low wet-bulb temperature will increase their capacity. This is because the water will not evaporate as fast during high wet-bulb temperatures and it will evaporate faster during a low wet-bulb temperature.

Basic Cooling Tower Controls. During periods of low outdoor air temperatures it is usually possible to lower the tower water temperature enough so that the compressor discharge pressure will be reduced below the desired operating pressure. There are several controls used to combat this problem. When natural-draft towers are used, the pump may be stopped for short periods of time by using a pressure switch that opens on a fall in pressure (Figure 23-10). An alternative method is to use a modulating three-way valve to divert some of the tower water past the condenser (Figure 23-11). This method may also be used with forced-draft towers.

When forced-draft towers are used, the fan can be stopped to allow the water to heat up without much air flowing through the tower. This is accomplished by use of a thermostat sensing the water temperature in the sump (Figure 23-12). When the water temperature is reduced to the point that the compressor discharge becomes undesirably low, the thermostat stops the fan. Then when the water temperature is increased sufficiently to cause a

satisfactory discharge pressure, the fan will start again, cooling the water. In some extreme cases the condenser bypass method, described above, may also be used.

When evaporative condensers are used during cold weather, it is usually best to shut down the water circuit completely thus cooling the condenser like an air-cooled unit.

Cooling Tower Maintenance. The cooling towers and evaporative condensers used in air conditioning and refrigeration systems require three types of maintenance: (1) mechanical, (2) electrical, and (3) water treatment. Generally, when a complete tower maintenance program is considered, these three are considered together. In most instances all three types are performed at the same time.

Mechanical. Mechanical maintenance includes removing the dirt and other foreign materials that get into the tower water pan during operation. These particles are washed into the pan by the water spray which removes the particles from the air passing through the spray. A thorough spring cleaning is an absolute necessity; otherwise, these particles will remain in the system and will eventually attach themselves to the tubing in the condenser and other parts of the water circulating system.

The float ball must be checked and properly adjusted to prevent water being wasted down the drain from overflow conditions. The proper water level is about 1 in. (25.40 mm) below the top of the overflow drain. When the system is operating and the water is spraying into the tower pan, there should be no overflow of water through the overflow pipe.

Often, a sump overflow will be used as a type of bleed-off to reduce the concentration of solid particles and chemicals in the tower sump, preventing them from entering the system and keeping the tower water solids and chemicals within specified limits. However, it is not a desired type of bleed system because so much of the water is wasted down the drain. The best form of bleed is a regulated bleed line which bleeds off water at the proper rate only when the system is operating rather than all the time, as with the overflow type of bleed.

The complete water system should be inspected monthly to assure that the spray nozzles, pump strainer, and the tower pan are clean and operating properly. The spray nozzles must be producing a full spray cone. The pump strainer must be clean to allow water to flow freely through the system. The tower pan must be clean to prevent the entrance of solid particles into the system. At this time the bleed line must be checked to make certain that it is allowing the proper amount of water to bleed from the system. The discharge pressure of the compressor should be checked to see if there is scaling of the condenser. The phosphate bags, if used, should be washed clean of any sediment that has collected in them so that the water will have better contact with the plates. If algae or slime is present, it should be removed by draining and cleaning the tower pan. Then, to prevent it from returning, an algicide placed in the tower according to the manufacturer's instructions.

Electrical. Any electric motors used on the tower should be inspected and lubricated if required. The belts should be checked and replaced if they appear to be worn. The fuse box should be checked and any deterioration repaired to prevent an electrical short.

The wiring connections to the motor and box should be inspected and repaired. Since the motor operates in a high-humidity atmosphere it should be checked annually for excessive rust and replaced if the housing is in a badly deteriorated condition. The electrical conduit should be checked regularly for deterioration and replaced or repaired when excessive deterioration is found.

Water Treatment. The water treatment program is designed to protect the system from scale and other debris that can get into the system through the water or through atmospheric conditions. The water is treated for scale, corrosion, algae, and slime. An adequate water treatment program involves more than just adding chemicals to the system. It should be approached using the following three steps:

1. Clean the equipment thoroughly during the spring checkup.
2. Start a complete water treatment program. This may involve the use of a water analysis and treatment company.
3. Have a complete and regular maintenance program started.

At the time of the spring cleaning, a scale inhibitor should be placed in the tower. Periodic use of slime/algicides is recommended to prevent their growth in the system. These inhibitors should be used according to their manufacturers' instructions for effective use.

It should be remembered that an effective scale inhibition program includes the proper amount of bleed-off from the tower. This is necessary to prevent excessive buildup of mineral salts in the system because of the evaporation of the water from the tower during operation.

The amount of water treatment to be used should be determined by an accurate water analysis. The amount and type of chemicals to be used will be determined at the time of the analysis. These recommendations should be followed for an effective water treatment program.

□ Water Chillers □

These components are used to cool water by passing it over the refrigerant evaporator on its way to the individual coils. The air is passed over the coils

FIGURE 23-13 Checking water temperature drop through a chiller.

FIGURE 23-14 Chiller thermostat location.

FIGURE 23-15 Flow switch location.

and cooled, or heated as the season requires. The water flowing through the chiller should have a temperature drop of approximately 10 °F (5.56 °C) (Figure 23-13). For best efficiency, the water and refrigerant should flow opposite to one another—the counterflow principle. Water chillers are generally provided with a thermostatic expansion valve refrigerant flow control device. These units are not as apt to scale up as water-cooled condensers; however, the water should be treated to prevent the pipes and tubes from corroding and causing water leaks. A local water treatment analysis company should be consulted for best results. A water flow of approximately 3 gal/min/ton is usually standard for chilled-water units.

Basic Water Chiller Controls. The purpose of water chiller controls is to provide for automatic operation of the equipment and to prevent freezing of the internal tubes. The temperature control is a thermostat whose sensing bulb is mounted in the water outlet pipe (Figure 23-14). These controls are usually set to maintain a leaving water temperature of approximately 40 °F (4.4 °C). A temperature lower than this increases the operating costs while a temperature higher than 45 to 50 °F (7.2 to 10 °C) will not provide sufficient cooling.

A freeze-stat is also used on these units. Its purpose is to prevent freezing and perhaps bursting of the chiller tubes. Bursted chiller tubes will allow water to enter the refrigerant circuit and cause much damage to the refrigeration system. The freeze-stat is also a thermostat. The sensing bulb may be mounted on the outlet water tube or special provisions may be made to mount the bulb in the tube nest. It is generally set not lower than 32 °F (0 °C). When this temperature is reached at the sensing bulb, the compressor will be turned off to prevent damage to the unit.

A flow switch is sometimes incorporated to prevent operation of the refrigeration compressor un-

til a flow of water through the chiller has been proven. This also protects the chiller tubes in case the water circulating pump fails or the system becomes low of water. The flow switch may be mounted at any point in the system; however, it is generally located close to the chiller (Figure 23-15).

These controls are provided in addition to the other controls, which allow safe, automatic operation of the compressor. These include high- and low-pressure switches, oil failure switches, contactors, and others, which were covered in Chapter 14.

□ Air-Handling Units □

These units are made up of a water coil, a blower and motor, a filter rack, and the necessary controls to provide automatic operation (Figure 23-16). Their purpose is to provide heating or cooling, as the season requires, by forcing air over the coil and into the space to be treated. The temperature drop through these units during the cooling season should be approximately 20 °F (11.12 °C) (Figure 23-17). The temperature rise should be approximately 65 °F (18.3 °C), depending on the water temperature. During the heating season, the water temperature will generally be approximately 150 °F (66 °C). This hot water is supplied by a boiler, which was covered in Chapter 20.

FIGURE 23-16 Air-handling unit.

FIGURE 23-18 Three-way valve location.

FIGURE 23-17 Checking temperature drop on an air handler.

Basic Air Handler Controls. The purpose of these controls is to provide automatic operation of the air handler and comfort of the treated space. The type of controls used will depend greatly on the degree of comfort desired in the space. The basic types are modulating and on–off control.

The *modulating system* uses a three-way valve to divert some or all of the treated water around the coil (Figure 23-18). These valves are operated by a modulating type thermostat that signals the valve as the space requirements change (Figure 23-19). A changeover thermostat is sometimes used to reverse the thermostat action from cooling to heating and vice versa. This control is mounted on the water line to the air handler coil. As the water temperature changes, the changeover thermostat senses this change and reverses its contacts.

The *on–off system* uses a standard single-pole double-throw thermostat. The thermostat may be used to control either a water solenoid or a fan relay. The fan cycling system is the most popular because it does not affect the water flow through the system. A changeover thermostat can also be used with this system to provide automatic seasonal changeover operation (Figure 23-20). This control system allows more variation in the space temperature than the modulating system.

Air Vents. Air vents are devices placed in the highest part of the water circuit and automatically expel any air that enters the water circuit (Figure 23-21). Any

FIGURE 23-19 Modulating thermostat and valve motor wiring diagram.

FIGURE 23-20 On-off system diagram using fan relay or solenoid valve.

FIGURE 23-21 Air vent location.

FIGURE 23-22 **Water regulator location.**

air that gets into the system must be removed or the water flow through the coil will be slowed down or sometimes completely stopped thus affecting system operation. Air in the system can be detected by a gurgling sound coming from the water pipes.

Water Regulators

Water regulators are used to control the water pressure in the water circuit and to allow for any water lost or evaporated (Figure 23–22). These controls can be adjusted to give the proper water height to fill the entire water circuit. To determine how much pressure is required in a given system, divide the distance to the highest point in the system by 2.33. This will give the amount of pressure required to force the water to the highest point. [One pound of pressure will raise a column of water 2.33 ft (0.708 m).]

Safety Procedures

The water circuit in hydronic heating and cooling systems sometimes holds a tremendous amount of water. Therefore, special attention must be exercised when working on any of the piping to prevent flooding the equipment room and possible personal injury to the service technician.

1. Always be certain that the water pumps are operating before starting the refrigeration compressor.
2. Be certain that a sufficient amount of water is in both the tower and the chilled hot water circuit before starting the compressor or boiler.
3. Do not set the freeze-stat below 32°F (0°C).
4. Do not use an excessive amount of chemicals when descaling a water-cooled condenser.

5. Be certain that the chemical solution used in descaling a water-cooled condenser is properly neutralized before draining it from the system.
6. Do not jumper the controls used on a chiller or boiler because of possible damage to the system.

☐ SUMMARY ☐

☐ Using water to transfer heat in heating and cooling systems is more efficient than using air.

☐ Some examples of using water in heating and cooling applications are water-cooled refrigeration condensers, water chillers, and hot water boilers.

☐ Some of the methods used in cooling condensers with water are waste water, water tower, and evaporative condensers.

☐ Water-cooled condensers require a 3-gal/min/ton water flow.

☐ The water flowing through a water-cooled condenser will have a temperature rise of approximately 10°F (5.56°C).

☐ Two methods used in removing scale from water-cooled condenser tubes are: (1) reaming the condenser tubes, and (2) dissolving the scale with a chemical.

☐ Follow the chemical manufacturers' recommendations when descaling a condenser by this method.

☐ Water treatment chemicals and a bleed-off can reduce the amount of scale deposited on the condenser tubes.

☐ A cooling tower cools the condenser water by evaporating part of the water.

☐ Forced-draft towers use a fan to force the air through the water.

☐ Water flowing through a tower should be cooled about 10°F (5.56°C).

☐ Evaporative condensers use both air and water to condense the refrigerant.

☐ A high-humidity condition will reduce the tower capacity.

☐ A condenser water bypass can be used to combat low head pressure conditions.

☐ Water chillers are used to cool water by passing it over a refrigeration evaporator.

☐ A temperature drop of 10°F (5.56°C) is normal for water chillers.

☐ Water chillers generally use a thermostatic expansion valve as the refrigerant flow control device.

☐ A chiller freeze-stat should not be set lower than 32°F (0°C).

☐ The chiller leaving water temperature should be approximately 40°F (4.4°C).

☐ A flow switch is used to prevent compressor operation until water flow is proven.

☐ Air-handling units are made up of a water coil, a blower and motor, a filter rack, and the necessary controls to provide automatic operation.

☐ Two basic types of air handler control systems are modulating and on–off.

☐ Air vents are placed in the highest part of the system to expel any air from the system.

☐ Water regulators are used to control the water pressure and the water makeup in the system.

☐ EXERCISES ☐

1. Why is system efficiency increased when cooling or heating with water?
2. Name the methods used for condenser cooling discussed in this text.
3. How much water should be flowing through a water-cooled condenser for proper cooling?
4. What water temperature rise should be expected through a water-cooled condenser?
5. What should be done to the chemical solution used to clean a scaled condenser before dumping it after use?
6. What two methods are used to reduce scaling of water-cooled condensers?
7. What is the purpose of a cooling tower?
8. Name three types of cooling towers.
9. What is the normal water temperature drop as it flows through a tower?
10. Why are spray nozzles used in a cooling tower?
11. What is the purpose of the slats in a forced-draft tower?
12. What type of condenser uses both air and water to cool the refrigerant?
13. What effect will a high relative humidity have on cooling towers?
14. What is the purpose of cooling tower controls?
15. What is normally done to evaporative condensers required to operate during periods of low outdoor ambient temperatures?
16. What is a normal temperature drop of water flowing through a water chiller?
17. What is the normal temperature of water leaving a chiller?
18. What is the purpose of the freeze-stat?
19. Name the control that prevents operation of the compressor until a flow of water is established.
20. Name the control used to switch the air handler and thermostat from one season to another.
21. Which type of control system allows the most variation in temperature (modulating or on–off)?
22. What is the purpose of air vents?
23. At what point in the system are air vents installed?
24. What is the purpose of a water regulator?
25. How high will 1 lb of pressure raise a column of water?

24 Commercial Air Conditioning

Performance Objectives

After completion of this chapter, the student should:

■ Be able to explain where commercial air conditioning systems are used.

■ Demonstrate a better understanding of the types of controls used on commercial air conditioning units.

■ Describe the procedures used in the startup of commercial air conditioning systems.

■ Understand the charging procedures used in charging commercial air conditioning systems.

■ Understand the procedures used in checking the heating system of a commercial air conditioning system.

■ Explain the purpose of induced draft blowers.

■ Be able to name the two basic types of thermal storage units.

■ Complete the Exercises with at least 80 percent accuracy.

■ Complete the Laboratory Exercises with at least 80 percent accuracy.

Commercial air conditioning is, generally, more complicated and larger in size than that used for residential purposes. However, in the larger residential and smaller commercial applications the same equipment is used. The major difference occurs when the equipment reaches about 7½ tons in capacity. At this point in size there are more requirements placed on the equipment for commercial purposes.

Definition. *Commercial air conditioning* may be defined as that which is used in such applications as cafeterias, office buildings, bowling centers, supermarkets, warehouses, and retail establishments, just to name a few.

▢ Commercial Units ▢

When a unit is used for commercial applications the refrigerant circuit is the same as that for any other type of system, with a few refinements, such as: the flow control device is usually a thermostatic expansion valve rather than a capillary tube; the refrigerant lines are usually soldered in rather than connected with quick-connectors; the control system is usually more complicated because of the different requirements; the duct system has provisions for taking in part or all outside air, depending on the building requirements and the economy required; the unit may be equipped to use gas, steam, fuel oil, electricity, or hot water for winter operation; and in most instances a humidifier is used to add humidity in the building.

Capacity control is also popular on commercial units. The capacity may be changed by cycling one or more of the compressors, by cylinder unloading devices, or by the hot-gas bypass method. The method chosen is a matter of the manufacturer's design.

To make certain that the equipment is applied properly, there are certain things that should be known before the installation is started, such as: the type of building construction, how it is used, and the average occupancy; the time of day that the greatest occupancy occurs; the physical characteristics of the building; and the layout of the lighting system.

Other things that must be considered are: the size

of the existing electric service; the type of heating medium available, such as gas, oil, electricity, hot water, or steam; whether chilled water is available (will well water be cold enough to satisfy the needs?); whether there is room for rigging to be used in setting the equipment; whether there will be enough access for proper servicing of the unit after installation; and how much outdoor air will be required. The building orientation will also be a factor, especially if large glass windows are used in the store front. It should also be determined whether a multizone unit will serve the area better or whether a single-zone unit will do the job properly.

Commercial air conditioning equipment is available in split-systems, self-contained systems, and thermal storage systems. The condenser may be air-cooled, water-cooled, or cooled by an evaporative condenser.

☐ Split Systems ☐

Split systems have the air-handling units located on the inside of the building, and the condensing unit, or the water chiller, located outside the building. The two units are connected either by the refrigerant lines or the chilled water lines and the control system electrical wiring. The air handlers may have either direct expansion coils, water coils, or steam coils and a blower with enough capacity to meet the air delivery requirements (Figure 24-1).

The air handler may be located so that it will blow the treated air directly into the space to be conditioned, or it may be connected to a duct system so that the proper distribution and volume requirements to separate areas are met. The conditioning process may require that more than one air handling unit be used. In this application an air handler may be located in each space or there may be several located in a single space so that the desired capacity and air delivery can be met within the economical structure of the application (Figure 24-2).

The air handler generally includes an air filter located in the airstream, upstream of the blower and coils, to protect them from dust and dirt in the airstream (Figure 24-3). The filters used may be either the throwaway or the cleanable type. In either case, these filters must be changed, or cleaned, on a regular basis so that the efficiency of the unit can be maintained.

FIGURE 24-1 **Split system with single air handler.**

FIGURE 24-2 **Split system with multiple air handlers.**

FIGURE 24-3 **Location of air filter in air handler.**

There is also a condensate drain line connected to the air-handling unit to carry away the condensate during the cooling season. This drain line must be installed properly so that the condensate can be disposed of without damage to the property or any products that may be stored in the area.

The condensing unit is generally located outside the building. It may be installed either on the ground near the building or on top of the building. The location will depend on the area available or the desires of the building owner. However, if the condensing unit is to be installed on top of the building, care must be taken to ensure that the building structure will support the unit adequately.

The condensing unit, or outdoor unit, will be equipped with the proper-size compressor or several compressors, to provide the desired capacity. A single compressor may be equipped with cylinder unloaders so that the compressor will not operate at full capacity unless needed. In applications when more than one compressor is used, a compressor or compressors may be cycled off to reduce the capacity as the load drops. There will also be one or more fans to provide the required amount of condenser cooling for the required capacity. Since some installations are required to operate when the outdoor air temperature is too low to maintain the desired discharge pressure, one or more of the fans may be cycled off, or face-and-bypass dampers may be used to modulate the amount of condenser cooling on air-cooled condensers.

To maintain the discharge pressure on a water-cooled type of condenser there may be some means of bypassing the condenser with some of the water, or a means of cycling the tower fan can be used. When evaporative condensers are used, the tower fan may be cycled off to maintain the desired discharge pressure. This will allow the water temperature to rise and maintain a higher discharge pressure.

Water-cooled or evaporative condensers may be located on the ground near the building or on top of the building. Again the building structure must be checked to make certain that it will hold the weight of the tower along with the added weight of the water.

Water-cooled and evaporative condensers may be protected from freezing during cold weather. If the water in one of these units froze and a refrigerant tube ruptured, water would enter the refrigerant system and cause serious damage. It would be almost impossible to remove all the water from the refrigerant system, which would endanger the compressor motor windings as well as other parts of the system.

All of the lines connecting these two units should be insulated, to keep heat transfer to the surrounding ambient air to a minimum and to prevent condensate from dripping on the building, its occupants, and the materials stored inside. Also, if there are water lines connected to the unit, they must be protected from freezing when the unit is shut down for a weekend or for other long periods during cold weather.

Self-Contained Units

Self-contained units have all of the required components inside one cabinet, with the exception of the condenser on some air-cooled models and the cooling tower or evaporative condenser on water-cooled models (Figure 24–4). In this instance the condenser may be located out-of-doors while the remainder of the unit is placed inside. These units may be of the vertical or the horizontal types, depending on the building requirements, design, and equipment space available.

The larger sizes of these units are generally installed on the roof with the necessary ductwork connected to them. They may be of either the down-flow or the horizontal-flow type, depending on the installation requirements. The type used will depend on the design requirements and the building structure. When the down-flow type is used, the ductwork is stubbed up through the roof and the unit is placed on special mountings with special ductwork. The placement of the ductwork and the mountings is very critical or placement of the unit will be very difficult.

When a horizontal flow unit is used, these fixtures are not quite as critical. The external ductwork to fit between the stubbed up ductwork and the unit connections can be made in the shop. The controls and their wiring are located inside the cabinet, with the exception of the thermostat and the electric power wiring. This wiring is installed in the field.

If steam or hot water is available on these types of installations, different temperature zones are very easy to accommodate. Face and bypass dampers are used to direct the circulating air through the cooling coil and through the heating coil—or a part of it through each coil so that the desired temperature air can be delivered to each space being treated (Figure 24–5).

Startup and Operation

Important: If the electrical power to the unit has been off for several hours, the compressors should not be started; the crankcase heaters must be energized for 24 hr before attempting to start the compressors. Set the thermostat temperature levers so

CONDENSER FANS
(1 on -953 models)

EVAPORATOR COIL

FILTERS (4)

BLOWER

HEAT EXCHANGER TUBES

INDUCED
DRAFT
BLOWER

VENT

BURNERS

HEATING
CONTROLS
SECTION

GAS VALVE

CONDENSATE DRAIN

ALTERNATE POWER ENTRY
(requires additional kit)

BLOWER MOTOR

MAIN
POWER
ENTRY

CONTROL BOX

CONDENSER COIL

COMPRESSORS

FIGURE 24-4 Self contained unit and component location. (Courtesy of Lennox Industries, Inc.)

FIGURE 24-5 Air handler with face and bypass dampers.

that there is no demand, to keep the compressors from cycling on.

Preliminary Steps. Use the following preliminary steps in starting up a new unit:

1. Make sure that the unit is installed in accordance with the installation instructions and local codes.
2. Inspect all electrical wiring, both field and factory installed, for loose connections. Tighten as required.
3. Check to ensure that the refrigerant lines are in good condition and do not rub against the cabinet or other refrigerant lines.
4. Check the electrical voltage at the disconnect switch. The voltage must be within the range listed on the unit nameplate. If not, consult the electric company and have the voltage condition corrected before starting the unit.

Startup. Use the following steps in the initial start-up. *Note:* The following is a generalized procedure and does not apply to all thermostat-controlled systems.

1. Set the fan switch to AUTO or ON and move the

system selection switch to the cool position. Adjust the thermostat to a temperature setting far enough below room temperature to bring on both compressors. The compressors should start and cycle on demand from the thermostat.

2. Each refrigerant circuit is charged separately with the correct type of refrigerant. See the unit rating plate for the correct type and amount of charge.
3. Refer to the instructions on cooling operation and maintenance for the proper method to check the refrigerant charge.

Heating Startup. *Caution:* Some of these units are equipped with direct spark ignition systems. Do not attempt to light these systems manually.

Placing the Unit in Operation. Use the following steps to place the unit in operation:

1. Set the thermostat to the OFF position. Close the manual knob on the gas valve.
2. Wait 5 min.
3. Open the manual knob on the gas valve, replace the burner access door, and turn on the unit electrical supply.
4. Set the fan switch to AUTO or ON and move the system selection switch to HEAT. Adjust the thermostat to a setting above room temperature.
5. The induced-draft blower immediately starts. The burners should light within 40 sec.
6. If the unit does not light the first time, it will attempt up to four more times before locking out if it has automatic ignition installed.
7. If lockout occurs, repeat steps 1, 2, 3, and 4. If automatic ignition is not used, the gas line will need to be purged of all air before it can be lighted.

Safety or Emergency Shutdown. Turn off the electrical power to the unit. Close the manual and/or main gas valves.

Extended Period Shutdown. Turn off the thermostat or set it to the "unoccupied" mode, if it is equipped with this feature. Close all gas valves, both internal and external to the unit, to guarantee that no gas will leak into the combustion chamber. Turn off the electric power to the unit. All access panels, covers, and vent caps must be in place and secured.

☐ Refrigeration System ☐

The following steps are general and the procedures for the specific unit should be followed.

Refrigerant Charge and Check. *Important:* All manufacturers do not recommend the same charging and checking procedure. The recommended procedure for the unit being serviced should be followed.

The following procedure is the recommended procedure for the Lennox GCS16 unit. It should be used for this unit only. *Important:* The charging pressures and curves do not apply to this unit. The pressure gauges connect to new service ports located at the thermometer wells (Figure 24-6). Read the following procedure for detailed charging instructions. *Warning:* Do not exceed the nameplate charge under any conditions. The charge should be checked during startup using the "approach" method outlined below. The approach method compares the actual liquid temperature with the outdoor ambient temperature. Thermometer wells have been provided to allow accurate liquid temperature measurement.

1. The indoor temperature should be between 70 and 80°F (21 and 27°C) and the outdoor temperature should be above 60°F (15.56°C).
2. Fill the thermometer well with oil (to assure an accurate reading) and insert the thermometer.

FIGURE 24-6 Charging connections. (Courtesy of Lennox Industries, Inc.)

3. Install the high-pressure gauge to the thermometer well service port (*do not use the compressor discharge port*) and the low-pressure gauge to the suction line service port.
4. Operate *both* cooling stages until the system stabilizes (approximately 5 min).
5. Check each stage separately with both stages operating.
6. Check the ambient (outdoor) temperature.
7. Check the liquid line temperature.
8. Approach temperature equals liquid temperature minus ambient temperature.
9. The approach temperature of each cooling stage should match Table 24–1 (and the unit label).
10. An approach temperature greater than the value shown in the table indicates an undercharge. An approach temperature less than the value shown in the table indicates an overcharge.

When this unit is properly charged, the system pressure should approximate the pressures given in the normal operating pressure table (Table 24–2).

TABLE 24–1 Approach Temperature

Model	Liquid Temperature Minus Ambient Temperature [°F (°C)]
GCS16-953	8 ± 1 (4.5 ± 0.5)
GCS16-1353 two-row condenser coil	5 ± 1 (2.7 ± 0.5)
GCS16-1353 three-row condenser coil	6 ± 1 (3.3 ± 0.5)

Source: Lennox Industries, Inc.

Important: Most GCS16-1353's are equipped with enhanced fin two-row condenser coils. Some early production units may be equipped with standard three-row condenser coils. The operating pressure and approach temperature of each type of coil is different.

Before checking the charge, be sure to determine whether your unit has two- or three-row condenser coils. Table 24–2 is provided to assist in determining the normal operating conditions and is not to be used as a charging procedure. Due to the many differences that exist between installations such as indoor air volume, humidity, and load, this table may be used only as a guide and minor differences should be expected. Significant differences could indicate the malfunction of a component or an improper charge.

Charging. *If the system is completely void of refrigerant, the recommended and most accurate method of charging is to weigh the refrigerant into the unit according to the amount shown on the unit nameplate and in the specifications table (Table 24–3). If weighing facilities are not available or if the unit is just low on charge, use the procedure outlined above.*

□ Heating System □

Use the following procedures when checking out the heating section of the unit:

TABLE 24–2 Normal Operating Pressures (psig)

Type of Coil	Outdoor Coil Entering Air Temp. (°F)	GCS16-953				GCS16-1353			
		Stage 1		Stage 2		Stage 1		Stage 2	
		Liq. ±10	Suct. ±5	Liq. ±10	Suct. ±5	Liq. ±10	Suct. ±5	Liq. ±10	Suct. ±5
Two-row condenser coil	65	160	60	180	63	155	58	155	58
	85	225	70	245	73	250	73	250	73
	95	260	72	275	75	260	75	260	75
	105	285	74	300	78	305	78	305	78
Three-row condenser coil	65					155	58	155	58
	85					255	73	225	73
	95					260	75	260	75
	105					305	78	305	78

Source: Lennox Industries, Inc.

TABLE 24-3   Refrigeration Charge

Model	Stage 1	Stage 2
GCS16-953	6 lb 4 oz	5 lb 6 oz
GCS16-1353 two-row condenser coil	7 lb 2 oz	7 lb 2 oz
GCS16-1353 three-row condenser coil	8 lb 6 oz	8 lb 6 oz

Source: Lennox Industries, Inc.

AGA/CGA Applications and Requirements. All GCS16's are AGA and CGA design certified without modification. Refer to the GCS16 operation and installation instruction manual for more information.

Gas Piping. The gas supply piping must not allow more than 0.5 in. water column (wc) drop in pressure between the gas meter and the unit. The supply pipe must not be smaller than the unit gas connection. The compounds used on the threaded joints of gas piping should be resistant to the action of liquefied petroleum gases.

Testing Gas Piping Pressure. When pressure testing the gas lines, the gas valve must be disconnected and isolated (Figure 24-7). Gas valves can be damaged if subjected to more than 0.5 psig (14 in. wc, 3.45 kPa). If the test pressure is equal to or less than 0.5 psig, use the main manual shutoff valve before pressure testing to isolate the furnace from the gas supply system.

MANUAL MAIN
SHUT-OFF VALVE

WILL NOT HOLD
TEST PRESSURE IN
EXCESS OF 0.5 PSIG (14″ W.C.)

GAS VALVE

CAP

Field Provided
Line Pressure Tap

FIGURE 24-7   Isolating gas valve during pressure test. (Courtesy of Lennox Industries, Inc.)

When checking gas piping connections for gas leaks, use a soap solution or other preferred means. Do not use matches, candles, flame, or other source of ignition to check for leaks. *Note:* In case an emergency shutdown is required, turn off the main manual shutoff valve and disconnect the main power to the unit. These controls should be properly labeled by the installer.

Testing Gas Supply Pressure. When testing the gas supply pressure, connect a test gauge to the inlet pressure tap (field provided; see Figure 24-7). Test the gas supply pressure. Make sure that the reading falls within the ranges of values listed below. A low gas supply pressure may result in erratic operation, or "underfire." A high gas pressure can result in permanent damage to the gas valve, or "overfire." For natural gas units, the operating gas pressure at the unit gas connection must be between 5.5 and 13.5 in. wc. For LP gas units the operating gas pressure should be between 11 and 13.5 in. wc.

Checking and Adjusting the Manifold Pressure. Use the following procedures when checking and adjusting the manifold pressure:

1. The gas valve is factory set and requires no adjustment. All gas valves are factory regulated (Table 24-4). *Note:* The low fire operating pressure is fixed and is not adjustable.
2. The gas valve should cycle off immediately and completely in the event of gas or power failure. The manual shutoff knob can be used to immediately shut off the gas supply.
3. After the gas supply pressure has been checked and adjusted, the manifold pressure may be tested during high-fire operation. See Figures 24-8 and 24-9 for the location of the pressure tap on the gas valve. *Caution:* For safety, connect a shutoff valve between the manometer and the gas tap to permit shutoff of the gas pressure to the manometer.

TABLE 24-4   Unit Input Rating

Unit	Low Fire Input (ft³)	High Fire Input (ft³)
GCS16-953	126 ± 6	200 + 5 − 15
GCS16-1353	170 ± 8	270 + 7 − 20

Source: Lennox Industries, Inc.

FIGURE 24-8 Location of manifold pressure tap for White-Rodgers gas valve. (Courtesy of Lennox Industries, Inc.)

FIGURE 24-9 Location of manifold pressure tap for Honeywell gas valve. (Courtesy of Lennox Industries, Inc.)

4. Connect the test gauge to the outlet pressure tap on the gas valve. Start the unit on high fire and allow 5 min for the unit to reach high-fire steady state.
5. While waiting for the unit to stabilize, notice the flame. The flame should be stable without

flashback and should not lift from the burner heads. Natural gas should burn blue; LP gas should burn mostly blue with some orange streaks.

6. After allowing the unit to stabilize for 5 min, record the manifold gas pressure and compare it to the values given in Table 24-5. *Caution:* Disconnect the heating demand as soon as an accurate reading has been obtained.
7. The maximum input may be reduced by up to 20% on AGA units equipped with adjustable gas valves and operating on natural gas. The high-fire manifold pressure may be set from 3.7 in. wc (maximum) input down to 2.9 in. wc for a maximum derate (Table 24-5). *Note:* Some GCS16 units may be equipped with a White-Rodgers gas valve, which has no high-fire adjustment. These GCS16's may not be derated.

Reducing the Maximum Input. Use the following procedures to reduce the maximum input:

1. Check the manifold pressure at the gas valve pressure tap.
2. Turn the regulator adjusting screw counterclockwise (Figure 24-9).
3. Recheck the manifold gas pressure.

Inshot Burners/Heat Exchangers. Burners, heat exchanger tubes, the spark electrode, the flame sensor, and the ignition circuitry are all factory set and require not adjustment or regular maintenance. Air shutters should always be fully open. A peephole with a cover is furnished in the cabinet access panel for flame viewing. Unless checking the manifold gas pressure, always operate the unit with the access panel in place. The burners can be removed individually for service.

Each burner can be removed by lifting out the adjacent retaining clips (Figure 24-10), then sliding the burner off the orifice. Orifices are sized precisely to

TABLE 24-5 GCS16 Gas Valve Regulation

Fuel	Maximum Inlet Pressure (in. wc)	High Fire Pressure (in. wc)	Low Fire Pressure (in. wc)
Natural	13.0	3.7 ± 0.2	1.6 ± 0.1
LP	13.0	10.5 ± 0.5	5.5 ± 0.3

Source: Lennox Industries, Inc.

FIGURE 24-10 Removing GCS16 burners. (Courtesy of Lennox Industries, Inc.)

3/32" to 7/64" SPARK ELECTRODE GAP
3/32" to 5/32" FLAME SENSOR GAP

FIGURE 24-11 Settings for spark electrode and flame sensor. (Courtesy of Lennox Industries, Inc.)

each burner. The 953 models use six identical orifices and the 1353 models use eight identical orifices. Refer to repair parts for the proper sizing of these parts. *Note:* When replacing a sensor or electrode, be sure to set the gap before installation (Figure 24-11).

The flame-sensing electrode gap may be checked with appropriately sized twist drills or feeler gauges. Remove the right burner by lifting out the retaining clips and sliding the burner off the orifice. The gap on the flame sensing electrode should be between 3/32 and 5/32 in. (2.38 and 3.97 mm). Replace the burner and clips after checking the gap.

The spark electrode gap may also be checked with appropriately sized twist drills or feeler gauges. Remove the left burner by lifting out the retaining

clips and sliding the burner off the orifice. The gap on the spark electrode should be between 3/32 and 7/64 in. (2.38 and 2.78 mm). Replace the burner and the retaining clips after checking the gap. The orifices are sized precisely to each burner. Refer to the Lennox repair parts for correct sizing.

Danger: Shock-hazard, spark-related components contain high voltage. Disconnect the electric power before servicing the unit.

Ignition (Burner) Control (A3). The ignition (burner) control (A3) is factory set and should require no adjustment. The control makes three to five (depending on the control) attempts at ignition and then locks out the system if ignition is not obtained after the third or fifth trial. Reset after lockout only requires breaking and then remaking the thermostat demand. The gas control shuts off the gas flow immediately in the event of a gas or power failure. Upon restoration of the gas or power, the control will restart the ignition process and continue until a flame is established or lockout occurs.

Danger: Shock-hazard, spark-related components contain high voltage. Disconnect the electric power before servicing. The ignition control is not field repairable. Unsafe operation may result.

The gas energy control uses an externally mounted integral fuse to protect the internal circuitry from accidental shorting while servicing the unit. The fuse, mounted on the lower right-hand cover of the control, is a 125-V 3-A slow-blow type.

Induced-Draft Blower (B6). All GCS16's use induced-draft blowers with prelubricated stainless-steel ball bearings. An induced-draft blower is used to draw fresh air into the combustion chamber while simultaneously expelling exhaust gases. The blower operates throughout the heating cycle. On a heating demand, the induced-draft blower energizes immediately, but the ignition control circuit does not. Once the induced-draft blower is energized and moving air through the heat exchanger, the centrifugal switch located near the rear of the induced-draft blower closes to energize the ignition control. The ignition control then begins attempting ignition after 30 to 40 sec.

If the induced-draft blower does not reach full speed, the centrifugal "prove" switch (S18) will not close. If this happens, the ignition control circuit will be locked out until the problem is found and corrected.

COMMERCIAL AIR CONDITIONING

☐ Blower Operation and Adjustment ☐

Use the following procedures when checking and adjusting the blower. *Note:* The following procedures are generalized and does not apply to all thermostat control systems.

Blower Operation

1. Blower operation is dependent on the thermostat control system option that has been chosen for the GCS16. Refer to the discussion of control system options for detailed descriptions of fan operation.
2. Generally, blower operation is set at the thermostat subbase fan switch. With the fan switch in the ON position, the blower operates continuously. With the fan switch in the AUTO position, the blower cycles with demand or runs continuously while the heating or cooling circuits cycle.
3. In all cases, the blower and entire unit will be off when the system switch is in the OFF position.

Determining Unit CFM. Use the following steps to determine the unit cfm:

1. The following measurements must be made with a dry indoor coil. Run the blower without the cooling demand. Air filters must be in place when the measurements are taken.
2. Measure the static pressure external to the unit (from the supply to the return). To measure discharge static pressure:
   (a) Measure the tap locations shown in Figure 24–12.
   (b) Punch a ¼-in. (6.35-mm)-diameter hole. Insert the manometer hose flush with the inside edge of the hole or insulation. Seal around the hole with Permagum. Connect the zero end of the manometer to the discharge (supply)-air side of the system. Connect the other end of the manometer to the return duct as above.
   (c) With only the blower motor running, observe the manometer reading.
   (d) Seal around the hole when the check is complete.
3. Measure the indoor blower motor rpm.
4. Referring to the proper tables, use the static pressure and the rpm to determine the unit cfm.

1/4" DIAMETER HOLE CENTERED ACROSS SUPPLY AND RETURN PLENUMS. CONNECT MANOMETER TO PLENUMS AS SHOWN.

FIGURE 24–12 Measuring static pressure. (Courtesy of Lennox Industries, Inc.)

5. The cfm can be adjusted at the motor pulley (sheave). Loosen the Allen screw and turn the adjustable sheave clockwise to increase the cfm. Turn it counterclockwise to decrease the cfm (Figure 24–13).

Blower Pulley Adjustment. Maximum life and wear can be obtained from belts only if proper pulley alignment and belt tension are maintained. Tension a new belt after the first 24 to 48 hr of operation. This allows the belt to seat in the grooves. To increase the belt tension, loosen the locking bolt and pull the motor mounting plate. Tighten the two bolts so that the motor mounting plate is in a vertical position (Figure 24–13).

☐ Maintenance ☐

Caution: Turn off the gas and electrical power to the unit before performing any maintenance or servicing operations. Remember to follow the lighting instructions attached to the unit when putting it back into operation after service or maintenance. Take care when servicing the unit to avoid accidental contact with any sharp metallic edges, which may cause injury.

Lubrication. *Note:* Always lubricate motors according to the manufacturer's lubrication instructions on each motor. If no instructions are provided, use the following guide:

TO INCREASE CFM
LOOSEN ALLEN SCREW &
TURN PULLEY CLOCKWISE

TO DECREASE CFM
TURN PULLEY COUNTER-
CLOCKWISE

MOTOR & PULLEY

MOTOR MOUNTING
PLATE

TO INCREASE BELT TENSION
1 - LOOSEN LOCKING BOLT.
2 - PULL MOTOR MOUNTING
PLATE.
3 - TIGHTEN MOTOR MOUNTING
PLATE SO PLATE IS VERTI-
CAL.

BELT

BLOWER WHEEL

LOCKING BOLT

CFM ADJUSTMENT

FIGURE 24-13 CFM adjustment. (Courtesy of Lennox Industries, Inc.)

1. *Supply-air blower motor bearings.* These bearings are prelubricated; no further lubrication is required for 10 years of normal operation. Thereafter, clean and repack the bearings with a suitable grease every two years.
2. *Blower shaft bearings.* These bearings are prelubricated and sealed; no further lubrication is necessary.
3. *Induced-draft blower motor bearings.* These bearings are prelubricated. For extended life, lubricate each bearing through the oiling ports provided with a few drops of electric motor oil or SAE10 or SAE20 nondetergent motor oil every two years.

Filters. *Note:* GCS16 units are equipped with either pleated 2-in. (50.80-mm) throwaway filters or 1-in. (25.40-mm) permanent foam filters. The filter rack is adjustable to accommodate either 1- or 2-in. filters. Do not replace 1-in. permanent filters with 1-in. throwaway filters.

1. See Note above.
2. The filters should be checked periodically and cleaned when necessary.
3. The washable filters supplied with the GCS16 units can be cleaned with water and a mild detergent. They should be sprayed with filter Handicoater when dry prior to reinstallation. Filter Handicoater is R.P. Products coating number 418 and Lennox part number P-8-5069.

4. Take note of the airflow direction marking on the filter frame when reinstalling.
5. If the filters must be replaced, filters of a like kind and size must be used. Do not replace permanent filters with throwaway filters. *Note:* Filters must be ULC certified or equivalent for use in Canada.

Burners. Use the following steps when performing maintenance operations:

1. Examine the burner flames for proper appearance periodically during the heating season. Use the inspection port in front of the burner compartment access panel.
2. Before each heating season examine the burners and flue for any deposits or blockages (rodent nest, wasp nest, etc.) which may have occurred.
3. Clean the burners as follows:
   (a) Turn off both the electrical power and the gas supply to the unit.
   (b) Remove the access panel to the burner compartment.
   (c) Remove the burner retaining clips and lift the burners from the orifices.
   (d) Clean as necessary. Check the spark electrode and flame-sensing electrode gaps and replace the burners (refer to the heating adjustment section). Refit the retaining clips and ensure that the burner heads line up correctly. Replace the access panel.

(e) Restore the electrical power and the gas supply. Follow the lighting instructions attached to the unit and use the inspection port in the access panel to check the burner flame. If any burner does not appear to be operating correctly, stop the unit, disconnect the electric power and the gas supply, and adjust as necessary.

Induced-Draft Blower. A centrifugal switch (S18) inside the induced-draft blower closes during blower operation (to prove combustion air operation) to allow power to the ignition (burner) control (A3). If the blower is obstructed, the blower cannot reach full speed and the switch will not close. The ignition control will be locked out.

Under normal operating conditions, the induced-draft blower wheel should be inspected and cleaned prior to the heating season. It should also be examined periodically during the heating season to establish an ideal cleaning schedule. With the power supply disconnected, the condition of the blower wheel can be determined by looking through the vent opening.

1. Shut off the electrical power supply and the gas to the unit.
2. Remove the four screws retaining the vent cap and combustion air blower to the end panel. Clean the vent cap as necessary.
3. Remove the screws holding the blower housing to the flue box cover plate and four wires attached to the motor. Take care not to damage or lose the vent screen.
4. Clean the blower blades with a small brush and wipe off any dust from the housing (Figure 24–14). Clean any accumulated dust from inside the flue box cover. *Caution:* Use care when cleaning the induced-draft blower wheel. The wheel is made of aluminum and may be distorted if too much pressure is applied.
5. Replace the induced-draft blower motor by reversing this procedure. It is recommended that the induced-draft blower gasket be replaced during reassembly.
6. Clean the louvers in the induced-draft air supply (right of vestibule panel) using a small brush.

Evaporator Coil. *Caution:* Disconnect the electric power to the unit before cleaning.

Inspect and clean the coil at the beginning of each

FIGURE 24–14 Cleaning the combustion air blower. (Courtesy of Lennox Industries, Inc.)

cooling and heating season. Clean using a mild detergent or commercial coil cleanser. Flush the coil and condensate drain with water, taking care not to wet the insulation, filters, return ducts, etc.

Condenser Coil. *Caution:* Disconnect the electric power to the unit before cleaning.

Clean the condenser coil annually with detergent or a commercial coil cleaner and inspect monthly during the cooling season. The condenser coil is made of individual coil slabs matched back to face. Dirt and debris may become trapped between the slabs. Carefully separate and clean between the slab sections (Figure 24–15). Flush the coil with water following the cleaning procedure. *Note:* Remove all screws and gaskets before starting the cleaning procedure, and replace afterward.

Supply-Air Blower Wheel. Annually inspect the supply-air blower wheel for accumulated dirt or dust. Turn off the electric power before removing the access panel or attempting to clean the blower wheel.

Replacement Parts. A complete list of replacement parts may be obtained from your local dealer.

☐ Thermal Storage ☐

Thermal storage is the temporary storage of either a high- or a low-temperature medium to be used at a later time. Thermal storage takes advantage of the period of time between when the energy is available and when the unit is to be used. The energy is then usually available at a lower rate. The majority of thermal systems operate on a 24-hr storage cycle; however, other time rates, such as weekly, seasonal, or annual, are sometimes available.

CLEANING PROCEDURE

① Remove unit top panel.

② Remove mullion holding coil.

③ Remove screws connecting coil slabs and separate 3 — 4″ (76 — 102mm).

④ Clean coils according to procedures listed.

⑤ Rinse thoroughly with water and reassemble.

FIGURE 24–15 **Condenser coil cleaning procedure. (Courtesy of Lennox Industries, Inc.)**

When the period of time that the energy is available is longer than the period of time that it is used, thermal storage will allow smaller equipment to be installed to handle the same job. In some installations, both heating and cooling are available at the same time or at different operating seasons. This type of unit requires special design considerations.

Thermal storage is very useful in applications when the load is high for a short period, electric demand charges are high, low-cost energy is available during the off-peak period, when a need for backup heating or cooling is required for refrigeration plant failure, and for many other types of applications that require heating or cooling during periods of off-peak electrical demand.

There are basically two types of thermal storage systems:

1. *Sensible heat storage.* Sensible heat storage is possible by raising the temperature of a storage medium, such as water, rock beds, bricks, sand, or soil. The heat is then removed when needed during periods of high power use.

2. *Latent heat storage.* Latent heat storage is usually accomplished by causing a change in the physical state of a substance. This change is usually from a liquid to a solid. This process takes advantage of the heat of fusion. Typical materials used for this purpose are water/ice, certain polymers, and salt hydrates. The amount of heat stored when latent heat is used is much greater per unit volume of stored material than it is when sensible heat is stored.

For refrigeration and air conditioning purposes, water and other such phase-change materials (PCMs) are the most popular. However, rock and soil are sometimes used. Water has the greatest advantage because it is universally available and is usually low in cost.

Efficiency. Well-designed thermal storage units have an efficiency of about 90% or more when fully charged and discharged on a daily basis and between 80 and 90% when partially charged and discharged on a daily basis.

Packed Rock Beds. A packed bed of rocks is a bed of loosely stored pebbles through which the air is circulated and either gives up heat to or removes heat from the rocks (Figure 24–16). Rocks between the sizes of ¾ and 2 in. (2 and 5 cm) are the most popular for this use.

When heat is being added to the rocks the air flows in one direction, and when they are being cooled the air flows in the opposite direction. To prevent channeling of the air through the rocks, both a discharge and a supply-air plenum should be used, and the inlet and outlet should be on opposite ends of the storage container.

Ice Storage. The thermal energy is stored in the latent heat of fusion of the ice. Water has the highest

FIGURE 24-16 Rock bed.

latent heat of fusion per unit of weight of all the materials in common use. This is an advantage over sensible heat storage systems because the temperature of the medium must be changed during the charging or discharging process. With ice this is not necessary; the ice only has to melt to provide the necessary energy change.

Water/Ice Storage on Refrigerant Coils. There are several types of ice building units available. However, the most popular type and the one that has been in use longest is the direct expansion (DX) type. This type of system has the refrigeration evaporator located in a tank of water. The compressor and the other components are located at a convenient place. The water is frozen on the outside of the evaporating coil to a thickness of up to 2½ in. (63 mm). The water that cools the building is circulated over the ice, causing it to melt and give up the latent heat of fusion (Figure 24-17).

Some units have a method of keeping the water moving inside the tank so that an even buildup of ice is accomplished. There are sensors located in the tank to cycle the compressor when the unit is fully charged. In most systems the compressor is cycled off when approximately one-half of the water is frozen.

There are other variations of this type of system, such as: water/ice storage on brine coils, brine-type solid ice builders, plate ice makers, and ice slurry systems. Each has advantages and disadvantages which must be considered when designing one of these types of systems.

Controls. The controls used on full storage systems are much simpler than those used on partial storage systems. The full storage system is either full on or full off and always operates at full capacity. When the partial storage plant is used, both the chilling

FIGURE 24-17 **Water/ice storage system.**

plant and the storage unit supply a part of the load, at the same time requiring a more complicated control system. In most installations, electronic timers are used to cycle the heating/cooling plant off during peak load hours to prevent operation at a much higher electric rate and increasing the cost of operation.

Sensors. The sensors used in these applications require regular maintenance and calibration. The sensors used in thermal storage applications must perform more accurately than those used in regular air conditioning and refrigeration systems to make certain that the compressor does not operate during the peak use period and to keep the ice buildup at the desired level. Too much ice on the coil will cause the compressor to use excessive energy for the same amount of cooling. Also, the unit could completely freeze up, stopping the flow of water and preventing a discharge of the unit.

Water Treatment. Water treatment is necessary in these types of systems to prevent corrosion. The ice storage tanks and refrigeration coils require proper cleaning and water treatment to keep the system operating efficiently, especially if the tank is drained during the off season.

☐ Safety Procedures ☐

Safety in Working with Hoisting Equipment. Often, air conditioning and refrigeration personnel require the use of cranes and other hoisting equipment to get equipment onto the roof of a building or to move the equipment from one place to another. Although this type of equipment is very useful, there are many safety procedures that must be followed to prevent accidents. The following is a general list to be followed. This list should in no way replace any instructions that are provided by the manufacturer of the equipment. Always follow any safety instructions provided by the manufacturer.

1. Never use a crane under electric power lines. To do so may allow them to come into contact with the lines, causing a high-voltage electrical short through the equipment which could possibly electrocute anyone touching the equipment while being grounded.
2. Always check for the center of gravity before hoisting any piece of equipment. If the weight is not centered on the hoist, it may slip and fall, damaging the equipment and possibly injuring persons standing nearby.
3. Always check the total equipment weight before attempting to hoist it to make certain that the hoisting equipment can lift it adequately.
4. Never move a loaded hoist, crane, or chain fall until everyone is clear of the area underneath the equipment and the path through which the hoisting equipment is to travel.
5. Do not use eyebolts to hoist an entire piece of equipment. The eyebolt may be designed for lifting only a portion of the assembly and may not be adequate for the entire assembly.
6. Do not use rigging equipment that needs repair or is not designed to lift the objects being lifted.
7. Do not use rigging equipment on slippery or otherwise hazardous surfaces where slipping may be a possibility.
8. Do not use ladders that do not have the proper angle from top to bottom.
9. Use only approved methods to rig equipment to be hoisted.
10. Never climb over a unit. Use the proper catwalk or platforms when crossing over equipment.
11. Always be aware of where fellow workers are when moving equipment with hoisting equipment.
12. Use lifting lugs only in accordance with the accompanying instructions.
13. Always post signs indicating that heavy equipment is being moved or hoisted so that other people will stay clear of the area.
14. Always make a final check of all connections and fastenings before hoisting equipment.

Safety in Working with Reciprocating Compressors. When working on reciprocating compressors there are certain precautions that should be taken to prevent personal injury, damage to the equipment, or damage to the property. The following is a basic list of the procedures that should be taken to prevent an accident when working with these units.

1. Never work on the electrical wiring to a compressor until all electrical power is off.
2. Never attach an ohmmeter to any electrical wiring, connections, or terminals until the electric power is turned off. To do so will ruin the meter and possibly cause an injury to the user.
3. Do not purge the refrigerant from the system through a cut, loosened connection, or a broken refrigerant pipe. Always use the gauge manifold so that the flow rate can be controlled to prevent purging oil from the system and possible property damage.
4. Do not turn on the electric power to the compressor until the terminal box cover is in place. This is to prevent possible injury if a compressor terminal should be blown out.
5. Pump the system down to about 2 psig (13.79 kPa) before loosening any bolts or connections to the compressor. This is to prevent possible injury or damage from a refrigerant blowout.
6. Do not attempt to operate a compressor with the discharge service valve closed. To do so could possibly cause damage to the compressor.
7. Open, tag, and lock all electrical disconnect switches while working on any of the electrical circuits, components, or connections.
8. All the compressors in a multiple-compressor system must be valved off before a single compressor can be purged.
9. Never use a hermetic or semihermetic compressor for a vacuum pump. To do so may cause the motor winding to become overheated and short out or cause a terminal to blow out, causing serious injury or property damage.
10. Never weld on a compressor having pressure inside it. To do so may cause a refrigerant and oil blowout, causing serious injury or property damage.
11. Never use oxygen for pressure testing a refrigeration system. The oxygen and compressor oil may combine to cause an explosive mixture.

Safety in Working with Air-Handling Equipment. Air conditioning and refrigeration personnel are constantly involved in working with air-handling equipment. There are certain procedures that should be followed to prevent personal injury, property damage, or damage to the equipment. The following is a list of these safety procedures. This list does not replace the safety instructions provided by the equipment manufacturer.

1. Do not enter the air-handling section while the unit is running. The fan sometimes has a great amount of force that could pull a person into the blades.
2. Do not work on electric circuits, heating elements, or the connections until the electricity is turned off to the unit. The disconnect switch should be tagged, locked, and the fuses removed to prevent turning on equipment accidentally.
3. Never operate belt-driven equipment without the belt guards in place. This is to prevent the possibility of getting clothing or limbs caught in the belt.
4. Make certain that all personnel are clear of the equipment before steam cleaning the coil to prevent possible personal injury to the personnel.
5. Do not work on large fans without securing the sheave with a rope or a strap to prevent it from freewheeling and causing personal injury.
6. Do not reach into a fan or equipment cabinet while the fan is running. The hand or fingers may be caught in the blades.
7. To prevent accidental crushing of the hands or fingers in the dampers, do not work on the air control dampers until the operators are disconnected.
8. To prevent using all the oxygen, always make certain that there is plenty of ventilation when welding or cutting inside an air-handling unit.
9. Do not attempt to remove large access doors of roof-mounted units during high winds without sufficient help.
10. Make certain that rooftop units are properly grounded to prevent accidental shock or property damage should an electric short occur.

Personal Safety Protection. In addition to the safety precautions listed previously, there are other personal precautions that should be observed in special situations that may be encountered in everyday operations involving air-conditioning and refrigeration equipment. Use the following rule when in doubt, if something looks dangerous it probably is.

1. Always wear a hard hat or other head protection when there is the possibility of falling objects.
2. Do not work on live electrical circuits on a wet surface, while wearing wet shoes or when barefooted.
3. Never touch electrical connections with wet or sweaty hands.
4. Always wear gloves when handling components from a system having a burned-out compressor motor winding. The refrigerant and oil from this type of system contains acid, which could possibly cause acid burns to the one handling them.
5. Always wear safety shoes when working around, lifting, or moving heavy objects.
6. Always wear hearing protection equipment when working in areas having noise levels greater than 90 dB.
7. Never wear rings or other jewelry when working around moving belts, moving equipment, electrical wiring, or electrical equipment.
8. Never wear rings, jewelry, long ties, loose clothing, or gloves while working around moving belts and machinery.
9. Always wear safety goggles equipped with side shields when working in manufacturing plants or at construction sites.
10. Always wear goggles and gloves when handling chemicals.
11. Always keep the work area clean of debris and free of liquid spills.

☐ SUMMARY ☐

☐ Commercial air conditioning may be defined as that which is used in such applications as cafeterias, office buildings, bowling centers, supermarkets, warehouses, and retail establishments.

☐ When a unit is used for commercial applications the refrigerant circuit is the same as that in any other type of system, with a few refinements.

☐ The capacity may be changed on commercial units by cycling one or more compressors, by cylinder unloading devices, or by the hot-gas bypass method.

☐ To make certain that the equipment is applied properly, there are certain things that should be known before the installation is started, such as the type of building construction, how it is used, and the average occupancy.

☐ Commercial air conditioning is available in split-systems, self-contained systems, and thermal storage systems.

☐ In split systems the air-handling units is located inside the building and the condensing unit, or water chiller, is located outside the building.

☐ The condensing unit, or outdoor unit, will be equipped with the proper-size compressor or enough compressors to provide the capacity desired. There will also be one or more fans to provide the amount of condenser cooling required.

☐ To maintain the discharge pressure on a water-cooled condenser there may be some means of bypassing the condenser with some of the water, or a means of cycling the tower fan off to maintain the desired discharge pressure.

☐ Water-cooled and evaporative condensers must be protected from freezing during cold weather.

☐ Self-contained units have all the required components inside one cabinet, with the exception of the condenser on some air-cooled units and the cooling tower or evaporative condenser on water-cooled models.

☐ The larger models of these units are generally installed on the roof, with the necessary ductwork connected to them.

☐ The controls and their wiring are located inside the cabinet, with the exception of the thermostat and the electric power wiring. This wiring is installed in the field.

☐ If the electrical power to the unit has been off for several hours, the compressors should not be started; the crankcase heaters must be energized for 24 hr before attempting to start the compressors.

☐ When a safety or emergency shutdown is required, turn off the electrical power to the unit and close the manual and/or main gas valve.

☐ When charging a unit with refrigerant, do not exceed the nameplate charge under any conditions.

☐ If the system is completely void of refrigerant, the most accurate and recommended method of charging is to weigh the refrigerant into the unit according to the amount shown on the unit nameplate and in the specifications.

☐ The flame-sensing electrode gap may be checked with appropriately sized twist drills or feeler gauges.

☐ The gas energy control uses an externally mounted integral fuse to protect the internal circuitry from accidental shorting while servicing the unit. The fuse, mounted on the lower right-hand cover of the control, is a 125-V 3-A slow-blow type.

☐ Thermal storage is the temporary storage of either a high- or a low-temperature medium to be used at a later time.

☐ Thermal storage takes advantage of the period between when the energy is available and when the unit is to be used.

☐ When the period that the energy is available is longer than the period that it is used, thermal storage will allow smaller equipment to be installed to handle the same job.

☐ Thermal storage is useful in applications when the load is high for a short period, electric demand charges are high, low-cost energy is available during the off-peak period, when a need for backup heating or cooling is required for refrigeration plant failure, and for many other types of applications that require heating or cooling during periods of off-peak electrical demand.

☐ There are basically two types of thermal storage systems: sensible heat storage and latent heat storage.

☐ For refrigeration and air conditioning purposes, water and other such phase-change materials (PCMs) are the most popular heat storage materials.

☐ Well-designed thermal storage systems have an efficiency rate of about 90% or more when fully charged and discharged on a daily basis and between 80 and 90% when partially charged and discharged on a daily basis.

☐ When phase-change materials are used, the thermal energy is stored in the latent heat of the material.

☐ There are several types of ice building units available. However, the most popular type and the one that has been in use longest is the direct expansion (DX) type.

☐ The water is frozen on the outside of the evaporating coil to a thickness of up to 2½ in. (63 mm).

☐ The controls used on full storage systems are simpler than those used on partial storage systems.

☐ In most installations, electronic timers are used to cycle the heating/cooling plant off during peak load hours to prevent operation at a much higher electric rate and thus increasing the cost of operation.

☐ Water treatment is necessary in these types of systems to prevent corrosion.

☐ EXERCISES ☐

1. Name the differences between an air conditioning unit used for residential purposes and one used for commercial purposes.
2. What are the types of commercial air conditioning units available?
3. Where is the condensing unit generally located on commercial air conditioning systems?
4. What are two ways to maintain the discharge pressure on air-cooled units?

5. What must be done to water-cooled condensing units during freezing weather?

6. Where are the larger self-contained units generally located?

7. What should be done before starting a unit that has had the electricity turned off for several hours?

8. What two things must be done for a safety or emergency shutdown on heating?

9. Describe the "approach" method of charging a unit.

10. What does an approach temperature less than that shown in the table indicate?

11. What is the best charging method to use if a system is completely out of refrigerant?

12. What is the maximum pressure drop allowed in the gas pipe between the meter and the unit?

13. What should be done to the main gas valve when pressure testing the gas piping?

14. What is the purpose of the induced-draft blower?

15. What type of oil should be used to lubricate motor bearings?

16. What precaution must be taken when cleaning an induced-draft blower wheel?

17. How often should the condenser coil be cleaned during the cooling season?

18. What are the two basic types of thermal storage systems?

19. What are the efficiencies of well-designed thermal storage systems?

20. What devices are used to cycle the plant off during periods of high energy cost?

25 Automotive Air Conditioning

Performance Objectives

After completion of this chapter, the student should:

■ Be able to explain the purpose of an automotive air conditioning system.

■ List the basic components used on an automotive air conditioning system.

■ Describe the various methods of controlling the suction pressure in an automotive air conditioning unit.

■ Know what the compressor controls are used for in an automotive air conditioning system.

■ Demonstrate a better understanding of the temperature controls that are used on automotive air conditioning systems.

■ Demonstrate a better understanding of the various service procedures used on automotive air conditioning systems.

■ Complete the Exercises with at least 80 percent accuracy.

■ Complete the Laboratory Exercises with at least 80 percent accuracy.

Automotive air conditioning systems use the same basic refrigeration system as residential and commercial air conditioning systems. The major differences between an automotive and any other air conditioning system are: (1) the manner used to combine the different components, (2) the manner used to drive the compressor, (3) the fast temperature change inside the automobile, and (4) the complex system of air control dampers used on an automobile.

Automotive heating is accomplished in much the same manner as heating with a hot water boiler. In the automobile, the hot water is taken from the radiator and passed through a heating coil where air is blown over it and heated.

Automotive air conditioning units use doors and dampers to direct the air through the desired function. They also are used to control the volume of air that passes through the desired coil and the amount that bypasses the coil. This bypass is done to aid in temperature control. Each manufacturer will use a different arrangement and method of control for these dampers.

☐ Operation ☐

The automotive air conditioning system functions much the same as any other type of air conditioning system. Its purposes are to remove heat and humidity from inside the automobile and deposit this heat and humidity in the outside air, and to clean the air.

The principles of heat transfer used are the same as in other systems. The refrigeration system is divided into a high-pressure side and a low-pressure side (Figure 25-1). The high-pressure side contains the following components: (1) discharge side of the compressor and the high-side service valve, (2) condenser, (3) receiver-drier, (4) inlet side of the expansion valve, and (5) connecting hoses to the high-side components.

The low side contains the following components: (1) outlet side of the expansion valve, (2) evaporator, (3) suction side of the compressor and its low-side service valve, and (4) connecting hoses to the low-side components.

Cold air
34°
Evaporator

Expansion
valve
28°

Warm air
85°

Suction
line
16°

A

Liquid
line
110°

Compressor

High pressure
line (discharge)
130°

Condenser   Hot air

Sight
glass

Receiver/dryer

RAM air
100°

Primary
liquid
line

HIGH PRESSURE **GAS**   HIGH PRESSURE **LIQUID**   LOW PRESSURE   LOW PRESSURE **GAS**

FIGURE 25–1 Basic refrigeration system.

☐ Basic Components ☐

The refrigeration system used in automotive air conditioning consists of the following components: (1) compressor, (2) magnetic clutch, (3) condenser, (4) receiver-drier, (5) expansion valve, (6) evaporator, (7) suction throttling valve, and (8) high-pressure relief valve. The refrigerant used in automotive air conditioning is R-12, which is presently the standard

type of refrigerant. In most cases the evaporator and expansion valve are installed inside the automobile while the other components are located under the hood. The following is a description of the various components.

Compressor. The compressors used in automotive air conditioning units are of the reciprocating type. There are three styles of compressors in use: (1) the two cylinder, (2) the six cylinder, and (3) the five

A

Sankyo Typical Mount/Drive

2-Cylinder Typical Mount

FIGURE 25-3 Compressor drive pulley arrangement.

B

FIGURE 25-2 (A) Sankyo compressor; (B) cutaway view. (Courtesy of Sankyo International [USA], Inc.)

cylinder (Figure 25-2). They are all driven by a belt, or belts, from the engine crankshaft (Figure 25-3). The functions of the compressor are to create the pressure required for condensing the refrigerant, to circulate the refrigerant through the system, and to create the low pressure required for evaporation of the refrigerant in the evaporator.

Automotive air conditioning compressors are of the open type. That is, they use a shaft seal to prevent the leakage of refrigerant and oil around the shaft where it extends through the compressor body. Since these seals require oil to prevent leakage, they must be operated periodically so that oil will keep the seal faces wet.

There are two service valves located on the compressor for use in servicing the refrigeration system. One is the discharge service valve and the other is the suction service valve. Some service valves are equipped with schrader valves that allow pressure to pass when the service hose is connected to it, much like the valve stem in an automobile tire (Figure 25-4). Others use a stem-operated type service valve to aid in servicing. The easiest way to determine which is which is by checking the refrigerant lines. The discharge line goes to the condenser, while the suction line goes to the evaporator.

Magnetic Clutch. All automotive air conditioning compressors in use today use a magnetic clutch to engage and disengage the compressor on demand from the thermostat inside the automobile. A magnetic clutch is also used when a defrost cycle is needed and is manually turned off when the unit is not in use.

FIGURE 25-4 Schrader valve in-service valve.

FIGURE 25-5 Stationary field magnetic clutch.

All clutches operate basically on the same principle of electromagnetism. They are made in two general types: the stationary field and the rotating field.

Stationary Field Clutch. The stationary field clutch is probably the most desirable clutch because it has fewer parts to wear out (Figure 25-5). The magnetic field is mounted on the compressor body by some mechanical means depending on the type of field and the type of compressor. The clutch rotor is held on the armature by a bearing and snap rings. The armature is mounted on the compressor crankshaft.

When there is no electric current being applied to the clutch field, there is no magnetic field and the rotor is free to rotate on the armature. The armature does not rotate when electric current is not applied.

As the temperature inside the automobile rises, the thermostat contacts close and electric current is applied to the clutch field. A magnetic attraction is then set up between the field and the armature that attracts the armature to the rotor. This magnetic attraction causes the armature and rotor to act as a single piece, and the complete assembly turns while the field remains stationary. The compressor crankshaft begins turning and the refrigeration process is started.

When the temperature inside the automobile is satisfied, the thermostat contacts open, interrupting the electric current to the clutch. The magnetic field fades away and the armature snaps away from the rotor and stops. The rotor continues to turn, but the compressor stops until electric current is again applied to the field.

Rotating Field Clutch. The rotating field clutch operates the same as the stationary field clutch except for the difference in placement of the field. The field is incorporated into the rotor and turns with it. The electric current is fed to the field through brushes mounted on the compressor.

The electric current is fed through the brushes and sets up a magnetic field that attracts the armature to the rotor. The complete assembly turns, including the field, which causes the compressor to turn.

Slots are machined in both the armature and rotor, in both types of clutches, that concentrate the magnetic field and increase the magnetic attraction between them.

Because the clutch is engaged and disengaged at high speeds as required for proper temperature control inside the automobile, considerable scoring of both the armature and rotor will occur. This scoring is normal and should not cause undue alarm.

Condenser. The purpose of the condenser is to receive the hot high-pressure refrigerant vapor from the compressor and cool the vapor until it condenses into a liquid. This is accomplished because heat always flows from a warm substance to a cooler substance. The air passes through the condenser coil and removes the heat.

The condenser is usually mounted in front of the radiator, and, in fact, resembles it (Figure 25-6). Air is forced through the condenser in two ways: (1) it is pulled through the condenser by the radiator fan and (2) it is forced through the condenser as the car moves along the road. This is known as *ram air*.

The compressor increases the pressure over the refrigerant vapor in the condenser. The heat intensity is actually concentrated into a smaller space, thus

Basic Components

High-Capacity Radiator

High-Pressure Radiator Cap

Compressor

Receiver-Drier

Sight Glass

High-Performance Fan

Hot Water Shut-Off Valve

Condenser

Expansion Valve

Blower

Passenger Compartment Unit

Evaporator

FIGURE 25-6 Automotive air conditioning system. (Courtesy of Chrysler Corporation)

raising the refrigerant temperature higher than the temperature of the ambient air. Conditions such as a dirty condenser, loose belts, and improper condenser location will result in poor operation of the condenser and a decrease in overall system efficiency and must be avoided.

Receiver-Drier. Automotive air conditioning units are more susceptible to leaks than other units because of the open-type compressor and the greater amount of vibration to which they are exposed. Over a period of time small leaks will occur requiring the addition of refrigerant. Also, the evaporator refrigerant requirements vary because of the changing heat load, the efficiency of the condenser, and the compressor speed.

A small receiver tank is used in the system to compensate for these variables. Refrigerant is stored in the receiver until it is needed by the evaporator. When a receiver tank is added to the system approximately 1 to 1½ lb (453.6 to 680.4 g) of additional refrigerant is needed (Figure 25-7).

A drying agent, termed a desiccant, such as silica gel or molecular sieve is placed inside the receiver during the manufacturing process. This is the only point in the entire refrigeration system where damaging moisture and acid can be removed (Figure 25-8). If the desiccant reaches its saturation point, that is, when it has absorbed all the moisture it can hold, moisture and acid will be released into the refrigerant circuit. This moisture will cause the expansion valve to freeze up and cause compressor damage. If the

FIGURE 25-7 Receiver-drier. (Courtesy of Refrigeration and Air-Conditioning Division, Parker Hannifin.)

expansion valve freezes up, the flow of refrigerant will stop and a decrease in cooling capacity will result. Because of this, the receiver-drier should be replaced each time the system is opened for repair or service.

When a refrigeration system is opened for repairs or service, foreign particles can enter the system. Moisture and air also will enter and will cause the refrigerant to deteriorate. This deteriorating action will cause a hydrolyzing action resulting in corrosion inside the system. This corrosion is released into the system in the form of solid particles that will eventually stop the flow of refrigerant through the expansion valve. A filter screen is placed inside the receiver-drier to catch these particles. If this screen catches enough of these foreign particles to reduce

FIGURE 25-8 Cutaway of receiver-drier.

FIGURE 25-9 Sight glass in receiver-drier.

the flow of refrigerant, the cooling will be reduced inside the automobile.

A sight glass is usually incorporated into the receiver-drier to aid in servicing the unit (Figure 25-9). A quick check of the sight glass will reveal several conditions to the service technician as follows:

1. A clear sight glass (indicated by a solid stream of refrigerant) indicates that the system has a sufficient amount of refrigerant or possibly an overcharge. An overcharge will be indicated by higher than normal gauge readings.
2. Foam or a steady stream of bubbles indicates that the system is short on refrigerant.
3. Oil streaks on the sight glass indicate that the system is completely out of refrigerant.
4. Occasional bubbles indicate that the system is slightly low on refrigerant or that the receiver-drier is saturated and is releasing moisture into the system.
5. A clouded sight glass indicates that the desiccant is breaking down and is circulating through the system.

Expansion Valve. The thermostatic expansion valve is the most popular flow control device used on automotive air conditioning systems. This device reduces the high-pressure refrigerant liquid to a low-pressure liquid refrigerant (Figure 25-10). The orifice in the expansion valve is the point where the pressure reduction take place. The liquid refrigerant has the lowest temperature upon leaving the expansion valve and entering the evaporator. The expansion valve is controlled by the sensing bulb fastened to the tubing at the evaporator outlet. The expansion and contraction of the charge inside the bulb, in response

FIGURE 25-10 Internal view of thermostatic expansion valve.

Basic Components

External
Equalizer
Line

Inlet

Outlet

Externally Equalized

Screen

Internal
Equalizer
Passage

Inlet

Orifice

Seat

Spring

Outlet

Internally Equalized

FIGURE 25-11 Typical thermostatic expansion valve.

TUBE—EXPANSION

Contains a fixed orifice
through which liquid refrigerant passes from the high
pressure liquid line into the
evaporator. Located in the
inlet tube of the evaporator
core.

**FIGURE 25-12 Expansion tube. (Courtesy of
Chevrolet Division, General Motors Corp.)**

to the evaporator outlet temperature, causes the
valve to open and close as needed.

The thermostatic expansion valve has three functions. It controls the flow of refrigerant into the
evaporator, it provides a quick refrigerant pressure
pulldown on startup, and it overcomes high discharge pressures during idle and low-speed
operation.

There are two types of thermostatic expansion
valves used on automotive air conditioning systems:
(1) internally equalized, which is the most common,
and (2) externally equalized, which is used where
special control is needed (Figure 25-11).

Expansion Tube. Some General Motors units use
an expansion tube in place of the expansion valve
(Figure 25-12). The flow rate through the expansion
tube depends primarily on the amount of subcooling
that takes place in the condenser. Subcooling is the
cooling of the liquid refrigerant after it has been con-

densed to a liquid. The expansion tube is located at
the inlet of the evaporator like the expansion valve.

Evaporator. The evaporator is the point in the system that absorbs the heat from the interior of the
automobile. It is the forced-air finned type of coil.

It is in the evaporator where the refrigerant, which
is under low pressure, evaporates and absorbs heat.
The heat is absorbed from the air as the blower
moves the air through the coil (Figure 25-13). This
flow of air is essential to the evaporation of the liquid
refrigerant. The amount of heat exchanged depends
on the temperature difference of the air and the
refrigerant. The greater the differential, the greater
the amount of heat that will be transferred from the
air to the refrigerant. A high heat load will allow
for the rapid transfer of heat to the refrigerant.
When the blower is on high speed, it will deliver the
greatest volume of air through the evaporator and
produce rapid evaporation of the refrigerant. As the
inside of the automobile is cooled, a temperature will
be reached where very little extra cooling will be experienced if the fan is allowed to continue operating
on high speed. A decrease in air volume will result
when the fan speed is reduced. However, the lower
volume of air will allow the air to remain in contact
with the evaporator coil longer. More heat will be
transferred to the refrigerant, resulting in colder air
being discharged inside the automobile.

The changing of the state of the refrigerant inside the evaporator is as important to the overall ef-

FIGURE 25-13 **Automotive air conditioning evaporator. (Courtesy of Chevrolet Division, General Motors Corp.)**

ficiency of the unit as the airflow over the coil. The liquid refrigerant that is admitted to the evaporator by the expansion valve changes to a vapor as it absorbs heat from the air. Some liquid must be present throughout the entire length of the evaporator in order to obtain full capacity of the unit.

A flooded evaporator is the result of too much refrigerant being fed into the evaporator by the expansion valve. When this condition occurs, the excess of refrigerant is allowed to pass through the evaporator and into the suction line and eventually into the compressor. This condition can result in serious damage to the compressor. There also will be insufficient cooling inside the automobile.

When too little refrigerant is fed into the evaporator, a starved evaporator is the result. When this condition occurs, liquid refrigerant does not reach through the entire length of the coil. Therefore, heat transfer is not accomplished through the complete length of the evaporator, resulting in reduced capacity.

Suction Throttling Valves. Any device used to regulate the flow of refrigerant from the evaporator to the compressor is termed a suction throttling valve or regulator (Figure 25-14). These devices are located in the suction line between the tail end of the evaporator and the compressor suction service valve.

There are several different types of suction throttling valves, and it may be difficult to determine exactly which type is used on a particular unit. However, on inspection of the unit, if no valves are found in the suction line, it may be assumed that a thermostat is used to operate the clutch when an

FIGURE 25-14 **Suction throttling valve cutaway. (Courtesy of Chevrolet Division, General Motors Corp.)**

Basic Components

Altitude
Compensating
Assembly

Adjustment   Lock
Nut   Nut   Diaphragm

Inlet

Adjustment Spring   Spring Cup   Piston   Outlet   Screen

FIGURE 25-15 **POA valve. (Courtesy of Ford Motor Corp.)**

icing condition occurs. The various types of suction throttling valves are discussed below.

Pilot-Operated Absolute (POA) Valve. These valves were used on General Motors units during the period of 1966–1973. Ford Motor Corporation is presently using them on its systems, but calls them STV valves. They are a unique type of valve because of the "straight through" refrigerant flow and the sealed housing. They are basically spring-loaded valves controlled by an evacuated bellows and needle valve assembly located within the valve body. Operation does not depend on atmospheric pressure and, therefore, is not affected by any changes in altitude (Figure 25–15). If the evaporator pressure is above approximately 28.5 psi (296.78 kPa) with the compressor running, the pressure differential between the valve inlet and valve outlet pressures will cause the bellows to be compressed, which will open the valve and permit a free flow of refrigerant to the compressor. If the evaporator pressure drops below 28.5 psi, however, the bellows will expand and close the needle valve. This bellows action tends to equalize the pressure on each side of the valve and permit the valve spring to move the valve needle toward the closed position. When the flow of refrigerant is slowed down by the valve closing, the evaporator outlet pressure will increase and prevent freezing of the evaporator coil. When the evaporator pressure

increases above 28.5 psi, the valve will open, again permitting the full flow of refrigerant to the compressor. This cycling of the valve is constantly repeated throughout the operation of the unit.

Evaporator Pressure Regulator (EPR). These valves are used on 1961 and later Chrysler Corporation automobiles except Dart, Valiant, and automobiles equipped with hang-on type air conditioning units. EPR valves are suction throttling valves located at the compressor (Figure 25–16). These valves are designed to prevent evaporator icing by controlling the suction pressure and keeping it above the freezing temperature. They operate the same as the earlier external type EPR valve with the exception that the valve piston is controlled by a gas-filled bellows. The EPR valve maintains the evaporator pressure between 22 and 26 psi (114.73 and 279.61 kPa) (Figure 25–17). This valve cannot be adjusted like the older type. Replacement of the complete valve is required when a malfunction occurs.

The compressor and EPR are both equipped with an oil passage that allows oil to pass from the suction line into the compressor crankcase. This oil passage has two functions: (1) it allows any oil which has left the compressor to return to it and (2) it is used to pressurize the compressor crankcase and prevent it from falling below atmospheric pressure. This is necessary because air and moisture could be drawn

AUTOMOTIVE AIR CONDITIONING

FIGURE 25–16 **EPR valve location. (Courtesy of Chrysler Corporation.)**

FIGURE 25–17   EPR valve. (Courtesy of Chrysler Corporation.)

FIGURE 25–18   ETR valve location. (Courtesy of Chrysler Corporation.)

FIGURE 25–19   ETR valve. (Courtesy of Chrysler Corporation.)

into the crankcase through the shaft seal if the crankcase pressure was allowed to fall below atmospheric pressure.

Evaporator Temperature Regulator (ETR). In the Chrysler Corporation Auto-Temp System of 1967 and later, the EPR valve has been replaced with the ETR (Figure 25–18). The EPR and the ETR valves are designed to accomplish the same thing. They control the flow of refrigerant from the evaporator to prevent freezing of the evaporator. The major difference between the two valves is that the EPR valve is designed to modulate the flow of refrigerant while the ETR is controlled by a solenoid built into the valve. Therefore, it is either fully open or fully closed. It is a normally open valve (Figure 25–19).

An ETR switch is a temperature-sensitive control that actuates the solenoid. It is located in the rear of the evaporator case and has a sensing capillary tube that extends through the fins on the evaporator coil to sense the evaporator temperature. When the evaporator falls to a point at which the evaporator might freeze up, the ETR switch contacts close, thus closing the ETR valve. When the evaporator temperature increases sufficiently to prevent freezing, the switch contacts open and allow the full flow of refrigerant to pass to the compressor.

High-Pressure Relief Valves

Valves in Receiver (VIR). With the 1973 model, all General Motors automobiles were equipped with

FIGURE 25-20 Cutaway of VIR assembly. (Courtesy of Chrysler Corporation.)

a valves-in-receiver (VIR) assembly. The VIR assembly combines the thermostatic expansion valve, the POA valve, the receiver-drier, and the sight glass into a single unit. The VIR is mounted next to the evaporator and has eliminated the external equalizer connection between the expansion valve and the POA valve. This was accomplished by drilling a hole in the divider between the POA valve and the thermostatic expansion valve cavities in the valve body (Figure 25-20).

The valve was redesigned in 1975 and called the evaporator equalized valves in receiver (EEVIR). This new design reduced the fuel consumption of the automobile. These new valves, the EEVIR, are identified by their gold finish as opposed to the silver finish of the VIR assembly. The valve housing and the expansion valve are two components of the EEVIR that must not be interchanged with the components of the older VIR assembly.

In operation, the liquid refrigerant flows from the condenser into the receiver-drier assembly when it is dehydrated by the desiccant. It then flows from the receiver section through the filter screen at the inlet of the pick-up tube, through the pick-up tube to the expansion valve cavity.

The expansion valve controls the flow of refrigerant into the evaporator by sensing both the temperature and pressure of the refrigerant as it flows through the VIR assembly. This control is accomplished by the power diaphragm of the thermostatic expansion valve. The diaphragm pressure is affected by the temperature of the refrigerant from the evaporator on its way through the VIR inlet to the POA valve. The end of the power diaphragm is now in

the refrigerant vapor stream from the evaporator and any increase in temperature of the refrigerant vapor will cause the power diaphragm to expand. This expansion of the diaphragm will move the needle away from its seat allowing more refrigerant to flow. A drop in temperature of the refrigerant vapor will reverse the action of the power diaphragm and allow the valve to decrease the flow of refrigerant.

The pressure of the refrigerant vapor from the evaporator will also affect the operation of the power diaphragm. The pressure is sensed through the equalizer port between the power diaphragm section of the expansion valve cavity and the POA valve cavity, thereby eliminating the need for an external equalizer line.

The POA valve controls the flow of refrigerant to maintain an evaporator pressure of 30 psi (307.09 kPa). A temperature of 32 °F (0 °C) is maintained with this pressure. This pressure and temperature combination prevents freezing up of the evaporator, which would block the flow of air through the coil.

In operation, the evaporator pressure is applied to the inlet end of the valve piston. The refrigerant then flows through the piston screen and a hole which is drilled through the piston to the bellows cavity. When the pressure from the evaporator increases, the force of the POA valve piston spring and the bellows cavity pressure is overcome, allowing the piston to move and open the main valve port. The refrigerant then flows through the main valve port back to the compressor suction. The opening of the valve allows the refrigerant to flow to the compressor. The evaporator pressure is lowered to 30 psi (307.09 kPa). The force of the piston spring and the bellows cavity pressure overcomes the evaporator pressure, and the valve blocks the valve port and reduces the flow of refrigerant to the compressor.

When the evaporator pressure is greater than the pressure in the bellows cavity, refrigerant flows through an orifice in the piston to maintain the control pressure in the bellows cavity (Figure 25-21). This control pressure affects the piston valve movement and the resulting movement of the valve needle. Refrigerant is allowed to pass from the piston spring area of the bellows cavity through two small holes in the bellows retainer in the cavity surrounding the bellows. With an increase in pressure around the evacuated bellows, it will contract, raising the needle from the seat. The refrigerant will then flow from the bellows cavity through the needle seat orifice and into the suction line. When the flow of refrigerant

AUTOMOTIVE AIR CONDITIONING

FIGURE 25-21 **Refrigerant flow through EEVIR. (Courtesy of Chevrolet Division, General Motors Corp.)**

Labels in figure: From Evaporator; To Compressor Oil Bleed Line; To Evaporator; Sight Glass; From Condenser

through the needle seat orifice is greater than the refrigerant flow through the piston orifice into the bellows cavity, the bellows cavity pressure will decrease. The bellows will expand and move toward the seat, closing off the orifice. The balance of forces acting upon the valve and the controlled flow of refrigerant tend to keep the POA valve piston and needle valve modulating the flow of refrigerant to the compressor.

These POA valves are set at the factory and cannot be adjusted or repaired in the field. If the POA valve is not functioning properly, the complete assembly must be replaced.

EEVIR Equalizer Functions. The equalizer port is designed to speed up the opening of the expansion valve when certain conditions exist. When the POA valve is modulating toward the closed position and slowing down the cooling, the first part of the evaporator coil would eventually become warm enough to open the expansion valve because of the pressure of the R-22 refrigerant vapor in the power diaphragm. When the automobile is moving at high speed, this time lag would be too great to allow an even outlet air temperature because of the time required to cool the evaporator coil after the expansion valve has opened. This fluctuation in temperature has been eliminated by use of the equalizer feature, which helps the expansion valve power diaphragm to open the valve without waiting for the expansion valve to warm up.

A pressure drop in the equalizer port and the resulting drop in pressure under the diaphragm would cause the valve to operate as if the expansion valve sensing element had been heated and pressure exerted on the top of the diaphragm. Thus, the expansion valve will open and allow refrigerant to flow to the evaporator. This is possible because of the reduced pressure under the diaphragm, not the increase in pressure on top of the diaphragm.

☐ Compressor Controls ☐

The controls that operate the compressor are electrical switches. They are installed in series with the magnetic clutch so that the desired function can be obtained by any one of the controls. They are designed to regulate the operation of the compressor. The controls are as follows:

Thermostat Switch. This control is placed inside the automobile and senses the evaporator temperature. It is connected into the compressor clutch electric circuit and when the evaporator approaches the freezing temperature, the thermostat switch will interrupt the electric circuit to the clutch and stop operation of the compressor. When the evaporator temperature rises to a predetermined temperature, the electric circuit is completed, again energizing the magnetic clutch and compressor operation is resumed. Some of these controls have a knob that allows the operator to change the cut-in and cut-out points of the control. Others have an adjustment screw that is used to regulate the switch operation. There are two types of thermostatic switches in use. They are: (1) bimetal and (2) bellows (Figure 25-22).

Bimetal Thermostatic Switch. The actuating device on these controls is a bimetal arm that expands and contracts with a change in temperature. This control is mounted inside the evaporator housing, which allows the discharge air from the evaporator to flow directly over the bimetal arm. This cold air causes the bimetal arm to contract and open the switch contacts. The evaporator discharge air becomes warmer and causes the bimetal arm to expand and close the switch contacts, starting operation of the compressor. Cooling is resumed until the discharge air temperature drops to the switch cutout point and the cycle is repeated.

Bellows Thermostatic Switch. These switches consist of a gas-filled bellows and capillary tube as-

FIGURE 25–22 Thermostatic switches. (Courtesy of Chevrolet Division, General Motors Corp.)

FIGURE 25–23 Superheat switch. (Courtesy of Chevrolet Division, General Motors Corp.)

FIGURE 25–24 Thermal fuse. (Courtesy of Chevrolet Division, General Motors Corp.)

sembly. The capillary tube is the sensing device and is usually inserted into the fins of the evaporator to obtain a close temperature control of the evaporator. An increase in evaporator temperature will cause an increase in pressure inside the bellows assembly. This increase in pressure will cause the switch contacts to close, completing the electric circuit to the compressor clutch, which starts operation of the compressor. As the evaporator temperature is lowered, the pressure within the bellows assembly will drop causing the bellows to contract. Thus, the electric circuit to the clutch is interrupted and the cooling process is stopped. There is a definite differential between the cut-in and the cut-out temperatures of the switch. This differential allows the evaporator to warm before the compressor is started again. An adjustment screw is preset at the factory to maintain a difference of about 12 °F (6.67 °C). However, field adjustments may be made when necessary. A control knob is provided so that the operator can adjust the switch to obtain the desired temperature.

Ambient Switch. Compressor operation when air conditioning is not required may result in compressor damage. Ambient switches are designed to prevent operation of the compressor when the outside air temperature is too low. The switch will normally close at approximately 32 °F (0 °C). However, some manufacturers may require a different setting. When this switch is closed, the electric circuit to the compressor clutch will be completed, which will allow operation of the compressor. Temperatures below this setting will interrupt the electric circuit to the compressor clutch and prevent operation of the compressor.

Superheat Switch. These switches include a diaphragm and sensing tube assembly charged with R-114 refrigerant. The sensing tube is fitted into the suction cavity of the rear compressor head and senses

the temperature of the suction vapor. The suction vapor temperature influences the internal pressure of the diaphragm and sensing tube assembly. The suction pressure exerts an external influence on the diaphragm (Figure 25–23).

The switch is designed so that only a low-pressure high-temperature condition inside the compressor will cause an electric contact, which is welded to the diaphragm, to touch the terminal pin, which may cause a fuse to blow. Refrigerant conditions of low-pressure low-temperature or high-pressure high-temperature will not cause the contacts to close.

Thermal Fuse. The thermal fuse consists of a temperature-sensitive fuse link, a wire-wound resistor, and the necessary electrical terminals. These are all mounted in a plastic casing (Figure 25–24).

672

FIGURE 25-25 Low-refrigerant-charge protection system. (Courtesy of Chevrolet Division, General Motors Corp.)

The thermal fuse allows a time delay before the fuse link is blown. Conditions such as a low refrigerant charge, complete loss of refrigerant charge, a defective POA valve, a defective expansion valve, or an improperly located thermal fuse can cause the thermal fuse to blow and interrupt the electric circuit to the compressor clutch. The thermal fuse prevents possible damage to the compressor due to these conditions.

Low Refrigerant Charge Protection System. Many of the later model General Motors automobiles include a low-refrigerant charge protection system. This system consists of the superheat switch, the thermal fuse, the air conditioning control switch, and the ambient switch. During operation, electric current flows through the air conditioning control switch, the ambient switch, and the thermal fuse to actuate the compressor clutch coil (Figure 25-25). If a partial or total refrigerant charge is lost, the superheat switch, mounted in the rear compressor head, will sense the high suction gas temperature and its contacts will close. When the superheat switch contacts close, electric current flows through the resistance heater in the thermal fuse. The heater provides heat to the fuse link, which warms it to its specific melting point. When the fuse opens, the electric current is interrupted to the compressor clutch. The clutch will disengage, preventing damage to the compressor due to a loss of refrigerant. The cause for the shortage of refrigerant must be found and repaired and the unit recharged with refrigerant or the replacement fuse will be blown.

Automotive Temperature Control Systems

During recent years there has been an increased demand for passenger comfort in automobiles. This demand has led to a greater sophistication of automotive air conditioning systems. Many manufacturers have combined the control of heating and cooling systems into one temperature control package.

Most of the current control systems are similar in their theory of operation. The reheat principle is used almost exclusively. This system directs the air through the evaporator coil where it is cooled and dehumidified, then routed either through or around the heater core and into the air distribution system (Figure 25-26).

Manual Control Systems. The basic difference between the systems being used is the method of moving the various components. Some control systems use mechanical levers and cables, some use vacuum-actuated components, while other systems use a combination of both of these methods. The following is a brief discussion of the various components involved in the control of an automotive air conditioning system.

Control Assembly. This is the first component used in an automobile air conditioning system (Figure 25-27). This assembly consists of levers or a series of switches located on the control panel. The levers or buttons are used to control the mode of operation. The passenger can control whether the system is heating, cooling, or defrosting by changing the control settings on the control panel.

The system selection switch controls the source and the flow of air. For example, when the selector is placed in a position that calls for cooling, an electric circuit is completed to the compressor clutch through the control panel switch and the ambient switch. If the ambient air temperature is above the setting of the ambient switch, the switch contacts will be closed and the compressor will operate. If the control switch or lever were placed in any of the other positions, the compressor would not usually operate.

The system selector switches or levers also determine the direction of airflow. Changing the mode of operation changes the position of a rotary vacuum valve, or switch, which is located on the control

FIGURE 25-26 Typical automotive air conditioning airflow schematic. (Courtesy of Chevrolet Division, General Motors Corp.)

FIGURE 25-27 Control assembly. (Courtesy of Chevrolet Division, General Motors Corp.)

panel (Figure 25-28). The vacuum valve, or switch, controls the position of the air doors located in the duct system. These doors determine the routing of the flow of air. In a pushbutton-type control system, the buttons act as a switch valve that directs the vacuum to one side or the other of a vacuum-actuator diaphragm. The actuator diaphragm is connected to an air door by a rod and linkage. Movement of the diaphragm will either open or close the air door. The same function could be accomplished by use of levers and cables without the vacuum actuator.

The vacuum in these systems is created in the engine manifold. A vacuum storage tank is sometimes used, which tends to eliminate fluctuations in the control system. The vacuum is usually routed to the various vacuum motors by use of a vacuum valve, or switch, which is usually located on the rear of the control panel.

Temperature Control. The temperature lever normally determines the temperature of the discharge air by positioning an air door, which is located in the air ducts. This is called an air temperature door and its position is usually controlled by use of a bowden cable, which connects the temperature lever to the air temperature door. In some later models, the temperature lever is connected to a vacuum valve. This vacuum valve directs the vacuum to an air inlet diaphragm through the system selector rotary vacuum valve (Figure 25-26). This control system will determine the air temperature that is being recirculated inside the passenger compartment. Fast recirculation of the air inside will give a fast cool-down on hot, humid days.

Water Valve. The water valve is used to control the flow of engine coolant through the heater core. When more coolant is allowed to flow through the heater core, more heat is provided in the passenger compartment. When the valve is closed, less coolant is allowed to flow, which reduces the amount of heat provided for the passenger compartment.

Heater Core. The heater core is where the heat is transferred from the engine coolant to the air. The coolant flow through the core begins at the front of the engine intake manifold and flows to the inlet port of the heater core, through the heater core, out

FIGURE 25-28 Auto air conditioning vacuum control system. (Courtesy of Chevrolet Division, General Motors Corp.)

of the outlet port of the heater core, and to the suction of the engine water pump.

Blower Speed. Normally, a blower switch is used to control the airflow by regulating the speed of the blower motor. Several different blower speeds may be available with the control numbers varying with the make and model of the particular automobile.

In checking and repairing automotive air conditioning systems, many malfunctions are traced to conditions that exist in the vacuum system. Therefore, it is necessary that all vacuum connections, vacuum hoses, and check valves be checked if a vacuum problem is suspected. An ohmmeter and a circuit diagram are usually all that is needed to find electrical shorts, grounds, and loose connections.

Automatic Temperature Control Systems. Automatic temperature control systems control both the heating and cooling modes. Therefore, the amount

of each is automatically supplied to produce the desired temperature in the passenger compartment. The temperature is selected by use of a temperature dial similar to a room thermostat. This control may be set to maintain a passenger compartment temperature between 65 and 85°F (18.3 and 29.4°C). A sliding control is usually used to select the specific mode of operation. When the temperature selection has been made, the system operates automatically regardless of any changes in weather conditions (Figure 25-29).

Individual control systems may vary in design, but they will all have the following four basic components: (1) sensor string, (2) control panel, (3) transducer, and (4) power servo.

Sensor String. The sensor is a thermistor type of resistor. The resistance value of a thermistor varies inversely as the temperature changes. That is, a rise in temperature causes its resistance value to decrease. Conversely, a fall in temperature of a thermistor

FIGURE 25-29 **Typical automatic temperature control system.
[Courtesy of Chevrolet Division, General Motors Corp.]**

causes its resistance value to increase. A sensor string is a series of thermistors placed at strategic points throughout the system to perform a particular function.

The most popular locations for these sensors are:

1. The in-car sensor is located on the control panel to sense the air temperature within the passenger compartment.
2. The ambient sensor is generally located in the fresh air duct so that it can sense the incoming air temperature. This sensor also may sense the sunlight effect on the passenger compartment.
3. The duct sensor is located in the discharge air duct plenum to sense the discharge air temperature.

These sensors are connected in series with the temperature selector and with each other. This is done to provide an electrical signal that will reflect the various temperatures of the sensors and temperature selector. This electrical signal is fed to the system amplifier.

Control Panel. The control panel is located on the instrument panel. It contains the amplifier and the temperature dial. The amplifier provides an output voltage from the voltage received from the sensors and the selector switch. The amplifier can be mounted on the control panel or it may be mounted separately.

Transducer. The transducer is an electromechanical device that changes the electrical signal from the amplifier into a vacuum signal which controls operation of the power servo.

Power Servo. The power servo is the control component of the system. It operates in response to the signal from the sensor to program for the desired operation. It is connected to the air temperature door and positions the door to provide either heating or cooling to the passenger compartment. It also regulates the vacuum control and the blower motor speeds to maintain passenger compartment temperature in response to the temperature dial setting.

☐ Universal Automatic Temperature Control Analyzer II ☐

The analyzer is basically a device used to substitute resistances when connected into the wiring harness of the automobile. When properly connected, it bypasses the sensors and the temperature selector and allows for manual control of the system. By switching in preset resistance values, the complete system operation can be checked.

The tester vacuum gauge samples the vacuum at the blower master switch to indicate changes in the transfer valve and mode changes. The voltmeter is used to check operation of the amplifier and servo as well as operation of the sensors and the selector dial potentiometer.

☐ Installation Procedures ☐

These procedures do not refer to any particular make or model of automobile air conditioner. They are merely a guide for use in supplementing the schematic drawings included in each manufacturer's mount and drive kit.

Universal-type auto air conditioning units are fairly easy to install. All of the components are prefabricated by the manufacturer and are accompanied by detailed instructions and schematics. The automotive air conditioning unit consists of a belt-driven compressor mounted on the engine, a condenser mounted between the radiator and grill, a receiver-drier mounted on the fender well, and an evaporator case assembly mounted under the dash panel.

The following steps are suggested by the manufacturer to save time, money, and labor. The directions given (right and left) are given from the driver's seat.

Important: All refrigerant lines and fittings should remain capped or plugged until connected. This is to eliminate the possibility of dirt and moisture contamination of the system. Use one or two drops of clean refrigeration oil on all flare fittings to ensure leakproof connections. Do not use excessive torque when tightening the fittings. Be sure to disconnect the battery negative cable before starting the following procedure.

1. Remove the radiator from the automobile if the crankshaft pulley is not readily accessible. To do this, first drain the radiator, then disconnect the radiator hoses and transmission cooler lines, remove the radiator shroud, remove the support bolts, and lift the radiator out carefully.
2. The compressor mount and drive assembly should be mounted according to the instructions included in each kit. In some cases the compressor mounting bolts may not be accessible. In such cases the compressor should be installed on the bracket before installation of the bracket on the engine.
3. Install the clutch on the compressor crankshaft. Follow the instructions in the clutch box. Be sure to check the alignment of the Woodruff key on the compressor crankshaft with the keyway in the clutch hub.* After the clutch field coil has been installed on the compressor, place the clutch rotor-pulley assembly on the shaft. Rotate the pulley manually to be certain that there is no interference between the clutch field and the rotor. Secure the rotor assembly to the compressor crankshaft with the washer and cap screw provided. Tighten the cap screw again after the installation is completed. A loose cap screw on the clutch may ruin the Woodruff key, the clutch hub, or the tapered compressor crankshaft.
4. Install the condenser between the radiator and grill with the connections on the same side of the automobile as the compressor. Locate the condenser about ½ to 1½ in. (12.7 to 38.1 mm) away from the radiator and as high as possible. Use the sheet metal screws and the brackets provided to secure the condenser to the radiator yoke. Do not cross-thread or twist the condenser fittings.
5. If necessary, cut two holes in the upper part of the radiator yoke on the same side as the compressor. Use a 1¼-in. (3.19-mm) hole saw to make the holes. Run the refrigerant lines into the engine compartment. Put the provided grommets (line protectors) in the holes to prevent damage to the refrigerant lines. It may be necessary to relocate some of the parts to make room for the condenser.
6. The receiver-drier should be mounted as nearly as possible upright on the fender well. Use the sheet metal screws and clamp provided. The inlet fitting is connected to the ⅜-in. (9.525-mm) fitting on the condenser.
7. The evaporator is to be mounted underneath the dash. Attach the hanger brackets to each end of the evaporator using the screws and washers provided. Leave all bolts and nuts loose so that the evaporator can be properly positioned under the dash.
8. Hold the evaporator case under the dash to determine the best location. Be sure that the brake and accelerator are not crowded. With the evaporator parallel to the underside of the dash, mark the hole locations for the hanger brackets, drain holes in the transmission tunnel, and the

---

*A Woodruff key is a crescent-shaped piece of metal that fits in a slot on the compressor shaft and a matching slot in the clutch hub to prevent rotation of the hub.

locations on the firewall for routing the refrigerant lines. Be sure that the hole locations in the firewall clear the engine and its components. Drill the holes as marked. Do not attempt to cut the carpet with drills or hole saws.

9. Route the refrigerant lines from the engine compartment, through the holes in the firewall and secure them to the proper fittings on the evaporator assembly. Be sure the provided grommets are installed in the holes in the firewall. Wrap the exposed portion of the suction line connection on the evaporator with insulating tape.

10. Secure the evaporator assembly underneath the dash. If a long firewall brace is used, extreme care must be used when drilling into the end of the evaporator case or blower housing. Cut one end of the drain hose at an angle to prevent its butting against the transmission beneath the floorboard. Push the drain hose through the hole in the floor, then slip the other end over the aluminum drain tube in back of the evaporator drain pan.

11. Complete the electrical wiring according to the provided diagram. Be sure that the wire with an in-line fuse holder is connected to the ignition switch accessory terminal.

12. Pull the two refrigerant lines through the firewall toward the compressor to remove any slack from the passenger compartment. Connect the suction hose to the suction service valve on the compressor, the liquid line to the outlet fitting on the receiver-drier, and the discharge line to the discharge service valve on the compressor. All refrigerant hoses should be secured away from exhaust manifolds and pipes, hot sections of the engine, and any sharp edges.

☐ Procedure for Evacuating and Charging an Automotive Air Conditioner ☐

It is necessary to evacuate an air conditioning system after installation or any time that the system has been serviced so that it is purged of all refrigerant. Evacuation is necessary to rid the system of all air and moisture that has entered the system during the installation or servicing operations. As the pressure is lowered in the system, the temperature of any moisture present in the system also will be lowered.

FIGURE 25–30 System hookup for evacuation.

The moisture will boil, and it is possible to remove it in the form of vapor.

Following is an outline of the procedure used to connect the vacuum pump to the system to be evacuated:

1. Connect the manifold gauge set to the system (Figure 25–30).
2. Close the gauge manifold hand valves.
3. Remove the protective caps from the inlet and exhaust of the vacuum pump. Be sure that the port cap is removed from the exhaust port to avoid damage to the vacuum pump.
4. Connect the center gauge manifold line to the inlet of the vacuum pump. Connect the high-pressure manifold hose to the discharge service valve on the compressor. The compressor head will be marked with the letter D or DISC. Connect the low-pressure manifold hose to the suction service valve on the compressor. The compressor will be marked with the letter S or SUCTION.

System Evacuation. The following steps outline the procedure for evacuating an air conditioning system after the proper connections have been made as outlined under the above procedure:

1. Start the vacuum pump.
2. Open the low-pressure manifold hand valve and observe the compound gauge needle. It should be indicating a vacuum inside the system. After the vacuum pump has operated about 5 min, the compound gauge should indicate at least a 20-in. (34.35-kPa) vacuum. The high-pressure gauge should be slightly below zero.
3. If the high-pressure gauge needle does not drop below zero, a system blockage is indicated. If the system is blocked, discontinue evacuation. Remove the obstruction. Continue with the evacuation procedure.
4. Operate the vacuum pump for approximately 15 min and observe the gauges. The compound gauge should indicate a reading of 24 to 26 in. (20.61 to 13.74 kPa) of vacuum if there is no leakage in the system. If the desired vacuum has not been reached, close the low-pressure hand valve and observe the compound gauge. If the compound gauge indicates an increase in pressure, indicating a loss of vacuum, there is a leak, which must be repaired before continuing the evacuation procedure.
5. Operate the vacuum pump for at least 30 min. After the evacuation process is completed, close the low-side pressure manifold hand valve. Shut off the vacuum pump and disconnect the center manifold hose from the vacuum pump. Replace the protective caps on the inlet ports of the vacuum pump.
6. Note the compound gauge. It should indicate 26 to 29 in. (13.74 to 0.5 kPa) of vacuum. The compound gauge should remain at this reading. If the system fails to meet this requirement, a partial charge must be put in the system and the system leak checked. After the leak has been found and repaired, the system must be purged of refrigerant and completely reevacuated. If the system holds the specified vacuum (Figure 25-31), it may be charged with Freon.

System Charging Procedure Using One-Pound Cans of R-12. Following is an outline of the charging procedure used on automotive air conditioning units:

1. The gauge manifold should still be connected for the evacuation procedure, but the center hose should be loose.

FIGURE 25-31 **Gauge readings with system under vacuum.**

FIGURE 25-32 **System hookup for charging.**

2. Attach the refrigerant dispenser valve to the refrigerant can. Connect the center manifold hose to the dispenser valve and pierce the can by turning the shutoff valve in a clockwise direction. After piercing the can, back the shutoff valve out in a clockwise direction. The center hose is now charged with both refrigerant and air. Loosen the center hose connector at the manifold until a hiss can be heard. Allow the refrigerant to escape for a few seconds and then retighten the connector. The center hose is now completely free of air (see Figure 25-32).
3. Open the high-pressure gauge manifold hand valve and allow as much refrigerant to enter the system as will flow in freely. This will vary from ½ to 1 lb (226.8 to 453.6 g) of refrigerant. After

TABLE 25-1 Gauge Readings (in.) on a Properly Charged System[a]

Ambient Temperature (°F)	Low-Pressure Gauge Reading	High-Pressure Gauge Reading
80	16	150–170
85	18	165–185
90	20	175–195
95	22	185–205
100	24	210–230
105	26	230–250
110	28	250–270
115	30	265–285

[a]Engine operating 1500 to 1750 rpm (equivalent to 30 mph) and correct refrigerant charge.

this partial charging, both gauges should indicate approximately the same reading. Close the high-pressure manifold hand valve.

4. Start the engine and adjust the speed to maintain approximately 1250 rpm. Turn the blower switch to the high-speed position and set the thermostat to the coldest setting. Open the low-pressure manifold hand valve to allow refrigerant to enter the system. Add refrigerant until the sight glass in the top of the receiver-drier is clear and no bubbles are visible.

5. Turn the low-pressure manifold hand valve off and check the readings of both gauges. The proper readings should be approximately 10:1; that

A. CONDENSER
B. COMPRESSOR
C. DISCHARGE HOSE (COMPRESSOR TO CONDENSER)
D. LIQUID HOSE (CONDENSER TO RECEIVER-DRIER)
E. CONDENSER MOUNTING BRACKETS (4)
F. RECEIVER-DRIER
G. LIQUID HOSE (RECEIVER-DRIER TO EXPANSION VALVE)
H. SUCTION HOSE (EVAPORATOR TO COMPRESSOR)
I. EVAPORATOR HANGER BRACKETS (1 LEFT - 1 RIGHT)
J. EVAPORATOR
K. DRAIN HOSE (2)
L. CLAMP FOR RECEIVER-DRIER

FIGURE 25-33 Typical underdash air conditioner layout.

AUTOMOTIVE AIR CONDITIONING

TY-RAP plastic
strap

5700023 drive
belt

Clutch

Idler

Fan

Alt.

P/S

D/P

9555160 drive
pulley

5/16-18 X 7/8 bolt

5013060 pilot
adapter

Reuse
O.E.M. bolt

3/8-16 X 7/8 bolt

7834206
compressor
strap

3/8-16 X 7/8 bolt

7734201
compressor
mount

3/8-16 X 7/8 bolt

7834205 idler
brace

3/8-16 nut
1/2-20 X 1
bolt

3/8-16 X 1
bolt

9522050
idler pulley

6477008 spacer

8240115 eccentric

5112410
shoulder
bolt

3/8-16 X 5-1/2 bolt (2)

3/8-16 X 1-1/2 bolt

3/8-16 X 1-1/4 bolt (4)

**FIGURE 25-34 Typical eight-cylinder mount and drive pulley
installation.**

is, the high-pressure gauge should indicate approximately 10 times the reading of the compound gauge. However, these readings will vary with the ambient temperature. The gauge readings must be taken with the engine running (Table 25-1). If the gauge readings are approximately 10:1, and the sight glass is clear, the unit will be properly charged and producing cold air.

A typical refrigerant piping schematic of an underdash unit is shown in Figure 25-33. A typical eight-cylinder mount and drive pulley installation is shown in Figure 25-34.

When malfunctions occur in an auto air conditioning system, there are several standard service procedures that can be followed. Table 25-2 should be helpful in solving any problems.

TABLE 25-2 Auto Air Conditioning Service Guide

When you find conditions of no cooling; or insufficient cooling; First: make sure that you connect all gauges properly. At this time find out what is causing the EXISTING TROUBLE. Check the POSSIBLE CAUSES and correct with the ANSWER.

A. REFRIGERATION PROBLEMS

1. LOW HEAD PRESSURE [Existing Trouble]

POSSIBLE CAUSE	ANSWER
a. Leak in system	Repair leak
b. Defective expansion valve	Replace valve
c. Suction valve closed	Open valve
d. Freon shortage	Add freon
e. Plugged receiver drier	Replace receiver drier
f. Compressor suction-valve leaking	Replace valve
g. Bad reed valves in compressor	Replace reed valve

2. HIGH HEAD PRESSURE [Existing Trouble]

POSSIBLE CAUSE	ANSWER
a. Air in system	Recharge system
b. Clogged condenser	Clean condenser
c. Discharge valve closed	Open valve
d. Overcharged system	Remove some refrigerant
e. Insufficient condenser air	Install large fan
f. Loose fan belt	Tighten fan belt
g. Condenser not centered on fan or too close to radiator	Center and check distance from radiator

3. LOW SUCTION PRESSURE [Existing Trouble]

POSSIBLE CAUSE	ANSWER
a. Refrigerant shortage	Add refrigerant
b. Worn compressor piston	Replace compressor
c. Compressor head gasket leaking	Replace head gasket
d. Kinked or flattened hoses	Replace hose
e. Compressor suction valve leaking	Change valve plate
f. Moisture in system	Replace drier
g. Trash in expansion valve screen	Replace drier

4. HIGH SUCTION PRESSURE [Existing Trouble]

POSSIBLE CAUSE	ANSWER
a. Loose expansion bulb	Tighten bulb clamp
b. Overcharged system	Remove some refrigerant
c. Expansion valve stuck open	Replace valve
d. Compressor reed valves	Replace valves in Compressor
e. Leaking head gasket on compressor	Replace head gasket

5. COMPRESSOR NOT WORKING [Existing Trouble]

POSSIBLE CAUSE	ANSWER
a. Broken drive belt	Replace belt
b. Broken clutch wire	Repair wire
c. Broken compressor piston	Replace compressor
d. Bad thermostat switch	Replace thermostat
e. Bad clutch coil	Replace clutch coil

6. ENGINE OVERHEATING [Existing Trouble]

POSSIBLE CAUSE	ANSWER
a. Fan belt slipping	Tighten belt
b. Engine out of time	Tune engine
c. Leaky radiator cap	Replace cap
d. Radiator water flow	Fill radiator
e. Clogged condenser coil fins	Clean fins
f. Engine cooling system clogged	Flush radiator and engine block
g. Not enough air over radiator	Install large fan

TABLE 25-2 Auto Air Conditioning Service Guide

7. EVAPORATOR NOT COOLING [Existing Trouble]

POSSIBLE CAUSE	ANSWER
a. Frozen coil, switch set too high	Defrost coil by turning thermostat down
b. Faulty clutch	Check clutch wire and thermostat
c. Drive belt slipping	Tighten belt
d. Hot air leaks into car	Close heater vents or air vents
e. Plugged receiver drier	Replace receiver drier
f. Capillary tube broken	Replace expansion valve
g. Shortage of freon	Add freon
h. High head pressure	See #2 problem
i. Low suction pressure	See #3 problem
j. High suction pressure	See #4 problem
k. Frozen expansion valve	Evacuate system and replace receiver drier
l. Frozen expansion valve	Replace valve

8. FROZEN EVAPORATOR COIL [Existing Trouble]

POSSIBLE CAUSE	ANSWER
a. Faulty thermostat	Replace thermostat
b. Thermostat not set properly	Set to driving conditions
c. Insufficient evaporator air	Turns switch to higher setting

B. MECHANICAL PROBLEMS

9. BELT TROUBLE [Existing Trouble]

POSSIBLE CAUSE	ANSWER
a. Pulley's not in line	Aline pulley's
b. Belt too tight or too loose	Adjust correctly
c. Wrong belt	Replace belt
d. Overcharged system or excessive head pressure	Discharge some freon
e. Bad bearing in idler pulley	Replace bearing

10. EXCESSIVE VIBRATION OF COMPRESSOR AND MOUNT [Existing Trouble]

POSSIBLE CAUSE	ANSWER
a. Head pressure too high	Check procedure #2
b. Loose or broken bolts in mount	Replace or tighten bolt
c. No lock washers or bolts	Install lock washers
d. Crankshaft pulley not on straight and tight	Tighten all bolts
e. Clutch not tight on compressor	Tighten bolts
f. Too much freon	Discharge some freon
g. Worn or frozen bearings in idler pulley	Replace bearings
h. Loose or defective belt	Tighten or replace

11. NOISY CLUTCH [Existing Trouble]

POSSIBLE CAUSE	ANSWER
a. Be sure coil properly installed	Center clutch and coil
b. Be sure pulley is tightly bolted to compressor shaft	Tighten bolt
c. Check key on shaft of compressor	Line key with clutch

MECHANICAL PROBLEM

12. CLUTCH DOES NOT WORK [Existing Trouble]

POSSIBLE CAUSE	ANSWER
a. Check fuse	Replace if bad
b. Check for broken or loose wires to clutch	Replace or repair wire
c. Check for short in clutch coil	Replace coil
d. Check voltage at clutch	Check connections and insulation on wires
e. Check thermostat	Replace if bad
f. Check blower switch at all positions	Replace if inner brass race is burned
g. Defective compressor [Frozen]	Replace compressor

TABLE 25-2 Auto Air Conditioning Service Guide

13. BLOWER DOES NOT WORK [Existing Trouble]

POSSIBLE CAUSE	ANSWER
a. Check fuse	Replace fuse
b. Check for broken or loose connections	Repair connection
c. Check control switch	Replace if defective
d. Check fan motor and be sure it is not locked or dragging	Center motor and adjust blower clearance
e. Check voltage at motor	Check wires and replace motor if defective
f. Check ground wire	Correct problem

14. BLOWER RUNS TOO SLOWLY [Existing Trouble]

POSSIBLE CAUSE	ANSWER
a. Check for loose wires or shorts	Correct problem
b. Check for binding shaft	Center motor and blower
c. Check for burned out resistors in controls	Replace switch
d. Check voltage at motor	
e. Check Allen Set screw on blower wheel	Tighten set screw

15. LEAKING COMPRESSOR SEAL OR GASKETS [Existing Trouble]

POSSIBLE CAUSE	ANSWER
a. Check for low freon charge and leaks around compressor seal with leak detector	Replace seal, or gasket
b. If oil is on compressor body clutch and underside of car hood above compressor	Check oil level Replace drier if moisture present

16. LEAKING HOSES OR FITTINGS [Existing Trouble]

POSSIBLE CAUSE	ANSWER
a. Check distance from exhaust manifold	Moves hoses
b. Check grommets in fire wall	Install grommets
c. Check wiper cables and arms	Move hose
d. Check flare seats	Replace hose or fittings
e. Check hose to fitting connection	Replace hose

□ Safety Procedures □

Because the engine must be running during the servicing operations on an automotive air conditioning unit, there are several important safety precautions that must be observed.

1. Be sure that the exhaust fumes are properly channeled to the outside of the building.
2. Do not touch the exhaust manifold or serious burns may result.
3. Do not leave loose tools lying on the radiator or other areas where they may fall into the fan.
4. Keep hands, feet, and loose clothing away from fan belts and fans.
5. Keep tools and hands away from the spark plugs because of the high voltage.
6. Do not attempt to remove the radiator cap from a warm radiator.
7. Keep the gauge service lines clear of pulleys, belts, and fans.
8. Be sure to block the wheels of the car when the engine is running.
9. Be sure to wear goggles when servicing the system with refrigerant.
10. Use a protective shield over the fan blade.

□ The major differences between automotive air conditioning and any other air conditioning system are: (1) the manner used to drive the compressor, (2) the manner used to combine the different components, (3) the fast temperature change inside the automobile, and (4) the complex system of air control dampers used in an automobile.

□ Automotive heating is accomplished in much the same manner as heating a home with a hot water boiler.

□ The refrigeration system used on automotive air conditioning units is the same as that used on any air conditioning system.

- The automotive air conditioning system consists of the following components: (1) compressor, (2) magnetic clutch, (3) condenser, (4) receiver-drier, (5) expansion valve, (6) evaporator, (7) suction throttling valve, and (8) high-pressure relief valve.
- The compressors used on automotive air conditioning units are open type.
- The magnetic clutch is used to engage and disengage the compressor on demand from the thermostat.
- The two types of magnetic clutches are: (1) stationary field, and (2) rotating field.
- The condenser is usually mounted in front of the automobile radiator.
- Automotive air conditioning units are more susceptible to leaks than other types of installations because of the vibration encountered.
- The thermostatic expansion valve is the most popular type of flow control device used on automotive air conditioning systems.
- The thermostatic expansion valve has three functions: (1) it controls the flow of refrigerant into the evaporator, (2) it provides a quick refrigerant pressure pulldown on startup, and (3) it overcomes high discharge pressures during idle and low-speed operation.
- Some General Motors units use an expansion tube in place of an expansion valve which operates on the amount of subcooling provided by the condenser.
- POA and STV valves are basically spring-loaded valves controlled by an evacuated bellows and needle-valve assembly located within the valve body.
- EPR valves are suction throttling valves located at the compressor. They are designed to prevent icing of the evaporator by controlling the suction pressure.
- ETR valves are controlled by a solenoid built into the valve. It is either fully open or fully closed.
- The ETR valve is controlled by the ETR switch.
- The VIR assembly combines the thermostatic expansion valve, the POA valve, the receiver-drier, and the sight glass into a single unit.
- The VIR assembly was redesigned and called the EEVIR (evaporator equalized valves in receiver) and reduced the fuel consumption of the automobile.
- The equalizer port on the EEVIR is designed to speed up the opening of the expansion valve under certain conditions.
- The thermostat switch is placed inside the automobile and senses the evaporator temperature.
- The bellows thermostatic switch has its capillary tube inserted into the fins of the evaporator to maintain close temperature control of the evaporator.
- Ambient switches are designed to prevent compressor operation when the outside air temperature is too low.
- Superheat switches are fitted into the compressor cavity on the rear compressor head and sense the temperature of the suction gas and cause a fuse to the compressor clutch to be blown.
- The thermal fuse allows a time delay before the fuse link is blown when overheating conditions occur inside the compressor.
- The low-refrigerant charge protection system consists of the superheat switch, the thermal fuse, the air conditioning control switch, and the ambient switch.
- The control assembly is mounted inside the passenger compartment and is designed to select the mode of operation.
- The temperature lever normally determines the temperature of the discharge air by positioning an air door which is located in the air ducts.
- Normally, a blower switch is used to control the air flow by regulating the speed of the blower motor.
- Automatic temperature control systems control both the heating and cooling modes.
- The sensor string consists of thermistors connected in series with each other and in series with the temperature selection.
- The amplifier provides an output voltage from the voltage received from the sensors and the selector switch.
- The transducer is an electromechanical device that changes the electrical signal from the amplifier into a vacuum signal that controls operation of the power servo.
- The power servo is the control component of the system. It operates in response to the signal from the sensor to program for the desired operation.
- The universal automatic temperature control analyzer is basically a device used to substitute resistances when connected into the wiring harness of the automobile.

▣ EXERCISES ▣

1. List the major differences between automotive air conditioning and other types of air conditioning units.
2. What is the purpose of the doors and dampers on automotive air conditioning units?
3. What type of compressors are used on automotive air conditioning units?
4. Name the types of magnetic clutches used on automotive air conditioning units.
5. What causes the scoring of a magnetic clutch?
6. How is the condenser cooled on an automotive air conditioning unit?
7. How much additional refrigerant is needed in a system when a receiver-drier is used?
8. What is the purpose of the desiccant in a receiver-drier?

9. What is the most popular flow control device used on automotive air conditioning units?
10. Why does the air become colder when the volume of air is reduced?
11. What is meant by the term *flooded evaporator?*
12. What is a suction throttling valve?
13. What operates the ETR valve?
14. What components are incorporated into the VIR assembly?
15. How is an EEVIR valve identified?
16. What pressure does the POA valve maintain in the evaporator?
17. What must be done if the POA valve in an EEVIR assembly is not working properly?
18. What is the function of the EEVIR equalizer?
19. What two types of thermostatic switches are most commonly used in automotive air conditioning units?
20. What is the normal operating temperature of an ambient switch?
21. What is the purpose of a superheat switch?
22. What conditions will cause the fuse to blow in a thermal fuse?
23. What component will cause the low-refrigerant charge protection system to function?
24. What are the three methods of controlling the various components on an automotive air conditioning system?
25. What is a sensor?
26. What is the function of the transducer?
27. What does the power servo control?
28. What components does the automatic temperature control analyzer replace?
29. What vacuum should be obtained when evacuating an air conditioning system?
30. What means can be used to determine whether or not a system has a sufficient refrigerant charge?

26 Solar Energy

Performance Objectives

After completion of this chapter, the student should:
- ■ Be able to explain the basic theory of solar energy.
- ■ Describe how to calculate the amount of solar collector surface needed.
- ■ Know how a basic solar system operates.
- ■ Describe the basic steps involved in designing a solar heating system.
- ■ Complete the Exercises with at least 80 percent accuracy.
- ■ Complete the Laboratory Exercises with at least 80 percent accuracy.

The average efficiency of use of coal and oil to the present time has reached no more than 20% of the demand. The demand for energy has only been about 20% of what is potentially recoverable by using these fuels. Wood, peat, and other vegetable matter provide approximately 15% of present-day fuel requirements. Hydroelectric and nuclear energy provide approximately 2% of the total demand.

In the United States the average energy demand per person is approximately 18,000 kWh/person/year (kilowatt hours per person per year). The future energy demand is estimated at 20,000 kWh/person/yr. When this is combined with the estimate of total population in a steady state to be 10 billion, we can arrive at an ultimate energy demand of approximately 200 trillion kWh/yr. This is approximately 16 times the present level of usage.

This increase arises from an estimated increase in population of three times the present population and an increase in average energy demand per person by more than five times the present demand.

Fuel shortages and the increase in their costs have rekindled man's interest in harnessing the sun's energy. In the past few years, millions of dollars have been spent in learning about the properties and potentials of the star nearest the earth.

☐ Brief History of Solar Energy ☐

The use of the sun's energy to heat buildings and water is not new. It has been traced back several thousand years to when the ancient Greeks used it. A large majority of the principles of solar heat were worked out in the early days of this century. The solar technology developed then has undergone very few changes except in the materials used. Solar heating has not been developed into wide use because of several factors. As long as the conventional fuels were apparently abundant and cheap, there was no incentive for the development of solar heating on a large scale. Solar heating is now coming into the competition of the marketplace because the conventional fuels are in short supply.

Every year the amount of solar energy that falls in this country is almost one thousand times greater than the total yearly energy consumption of the country. When this is added to the facts that solar energy is nonpolluting, inexhaustible, and could significantly reduce the dependence of the United States on imported fuels, it is easy to understand the interest in this source of energy.

TABLE 26–1 Daily Insolation [Btu/hr/ft²] on a South-Facing Roof for Various Angles with Horizontal

	24° N Latitude							40° N Latitude					
	0°	14°	24°	34°	54°	90°		0°	30°	40°	50°	60°	90°
Sept. 21	2194	2342	2366	2322	2212	992	Sept. 21	1788	2210	2228	2182	2074	1416
Oct. 21	1928	2198	2314	2364	2346	1442	Oct. 21	1348	1962	2060	2098	2074	1654
Nov. 21	1610	1962	2146	2268	2324	1730	Nov. 21	942	1636	1778	1870	1908	1686
Dec. 21	1474	1852	2058	2204	2286	1808	Dec. 21	782	1480	1634	1740	1796	1646
Jan. 21	1622	1984	2174	2300	2360	1766	Jan. 21	948	1660	1810	1906	1944	1726
Feb. 21	1998	2276	2396	2446	2424	1476	Feb. 21	1414	2060	2162	2202	2176	1730
Mar. 21	2270	2428	2456	2412	2298	1022	Mar. 21	1852	2308	2330	2284	2174	1484
Apr. 21	2454	2458	2374	2228	2016	488	Apr. 21	2274	2412	2320	2168	1956	1022
May 21	2556	2447	2286	2072	1800	246	May 21	2552	2442	2264	2040	1760	724
June 21	2574	2422	2230	1992	1700	204	June 21	2648	2434	2224	1974	1670	610
July 21	2526	2412	2250	2036	1766	246	July 21	2534	2409	2230	2006	1728	702
Aug. 21	2408	2402	2316	2168	1958	470	Aug. 21	2244	2354	2258	2104	1894	978

	32° N Latitude							48° N Latitude					
	0°	22°	32°	42°	52°	90°		0°	38°	48°	58°	68°	90°
Sept. 21	2014	2288	2308	2264	2154	1226	Sept. 21	1522	2102	2118	2070	1966	1546
Oct. 21	1654	2100	2208	2252	2232	1588	Oct. 21	1022	1774	1860	1890	1866	1626
Nov. 21	1280	1816	1980	2084	2130	1742	Nov. 21	596	1336	1448	1518	1544	1442
Dec. 21	1136	1704	1888	2016	2086	1794	Dec. 21	446	1136	1250	1326	1364	1304
Jan. 21	1288	1839	2008	2118	2166	1779	Jan. 21	596	1360	1478	1550	1578	1478
Feb. 21	1724	2188	2300	2345	2322	1644	Feb. 21	1080	1880	1972	2024	1978	1720
Mar. 21	2084	2378	2403	2358	2246	1276	Mar. 21	1578	2208	2228	2182	2074	1632
Apr. 21	2390	2444	2356	2206	1994	764	Apr. 21	2106	2358	2266	2114	1902	1262
May 21	2582	2454	2284	2064	1788	469	May 21	2482	2418	2234	2010	1728	982
June 21	2634	2436	2234	1990	1690	370	June 21	2626	2420	2204	1950	1644	874
July 21	2558	2422	2250	2030	1754	458	July 21	2474	2386	2200	1974	1694	956
Aug. 21	2352	2388	2296	2144	1934	736	Aug. 21	2086	2300	2200	2046	1836	1208

Source: Sol-R-Tech.

☐ Basic Solar Theory ☐

The earth is 93 million miles from the sun. The sun, sometimes called a solar fusion furnace, provides 442 Btu (111.4 kcal) per hour for every square foot (0.09 m²) of upper atmosphere. By the time this heat reaches the earth, however, the average heat value has dropped to between 250 and 300 Btu (63 to 75.6 kcal) per hour per square foot. This is equal to approximately one horsepower per square yard (0.84 m²). This is one reason why we are not likely ever to have solar-powered automobiles.

There is, however, a variation in the availability of solar energy in different geographic areas and seasons of the year (Table 26–1). The values are expressed in Btu/ft²/hr as measured by a device called a pyroheliometer.

There are a few areas in the United States that receive less than 50% of the normal sunshine dur-

ing the months of January and February, the coldest part of the year (Table 26–2). This will be more than sufficient for home heating and domestic hot water heating.

In practical solar engineering, there are two basic approaches to collecting thermal energy from the sun: concentrating collectors, which focus the energy to a point for collection (Figure 26–1), and using flat plate collectors, which capture the solar output directly as it falls on the surface (Figure 26–2). A basic example of a concentrating collector is the use of a magnifying glass to set fire to a piece of paper. Either a lens or a curved mirror can be used to concentrate the energy into a smaller area. The major advantage of the concentrating collector is its ability to generate very high temperatures, much higher than the flat plate collector. The disadvantage of the concentrating collector is that as the sun moves, so does the spot where the energy is focused. Therefore, the collector must move constantly to follow the sun.

TABLE 26-2 Mean Percentage of Possible Sunshine for Selected Locations

State and Station	Years	Jan.	Feb.	Mar.	Apr.	May	June	July	Aug.	Sept.	Oct.	Nov.	Dec.	Annual
Ala. Birmingham	56	43	49	56	63	66	67	62	65	66	67	58	44	59
Montgomery	49	51	53	61	69	73	72	66	69	69	71	64	48	64
Alaska Anchorage	19	39	46	56	58	50	51	45	39	35	32	33	29	45
Fairbanks	20	34	50	61	68	55	53	45	35	31	28	38	29	44
Juneau	14	30	32	39	37	34	35	28	30	25	18	21	18	30
Nome	29	44	46	48	53	51	48	32	26	34	35	36	30	41
Ariz. Phoenix	64	76	79	83	88	93	94	84	84	89	88	84	77	85
Yuma	52	83	87	91	94	97	98	92	91	93	93	90	83	91
Ark. Little Rock	66	44	53	57	62	67	72	71	73	71	74	58	47	62
Calif. Eureka	49	40	44	50	53	54	56	51	46	52	48	42	39	49
Fresno	55	46	63	72	83	89	94	97	97	93	87	73	47	78
Los Angeles	63	70	69	70	67	68	69	80	81	80	76	79	72	73
Red Bluff	39	50	60	65	75	79	86	95	94	89	77	64	50	75
Sacramento	48	44	57	67	76	82	90	96	95	92	82	65	44	77
San Diego	68	68	67	66	66	60	60	67	70	70	70	76	71	68
San Francisco	64	53	57	63	69	70	75	68	63	70	70	62	54	68
Colo. Denver	64	67	67	65	63	61	69	68	71	71	67	65	63	67
Grand Junction	57	58	62	64	67	71	79	76	72	77	74	67	58	69
Conn. Hartford	48	46	55	56	54	57	60	62	60	57	55	46	46	56
D.C. Washington	66	46	53	56	57	61	64	64	62	62	61	54	47	58
Fla. Apalachicola	26	59	62	62	71	77	70	64	63	62	74	66	53	65
Jacksonville	60	58	59	66	71	71	63	62	63	58	58	61	53	62
Key West	45	68	75	78	78	76	70	69	71	65	65	69	66	71
Miami Beach	48	66	72	73	73	68	62	65	67	62	62	65	65	67
Tampa	63	63	67	71	74	75	66	61	64	64	67	67	61	68
Ga. Atlanta	65	48	53	57	65	68	68	62	63	65	67	60	47	60
Hawaii. Hilo	9	48	42	41	34	31	41	44	38	42	41	34	36	39
Honolulu	53	62	64	60	62	64	66	67	70	70	68	63	60	65
Lihue	9	48	48	48	46	51	60	58	59	67	58	51	49	54
Idaho Boise	20	40	48	59	67	68	75	89	86	81	66	46	37	66
Pocatello	21	37	47	58	64	66	72	82	81	78	66	48	36	64
Ill. Cairo	30	46	53	59	65	71	77	82	79	75	73	56	46	65
Chicago	66	44	49	53	56	63	69	73	70	65	61	47	41	59
Springfield	59	47	51	54	58	64	69	76	72	73	64	53	45	60
Ind. Evansville	48	42	49	55	61	67	73	78	73	67	67	52	42	64
Ft. Wayne	48	38	44	51	55	62	69	74	69	64	58	41	38	57
Indianapolis	63	41	47	49	55	62	69	74	70	68	64	48	39	59
Iowa Des Moines	66	56	56	56	59	62	66	75	70	64	64	53	48	62
Dubuque	54	48	52	52	58	60	63	73	67	61	55	44	40	57
Sioux City	52	55	58	58	59	63	67	75	72	67	65	53	50	63
Kans. Concordia	52	60	60	62	63	65	73	79	76	72	70	64	58	67
Dodge City	70	67	66	68	68	68	74	78	78	76	75	70	67	71
Wichita	46	61	63	64	64	66	73	80	77	73	69	67	59	69
Ky. Louisville	59	41	47	52	57	64	68	72	69	68	64	51	39	59
La. New Orleans	89	49	50	57	63	66	64	58	60	64	70	60	46	59
Shreveport	18	48	54	58	60	69	78	79	80	79	77	65	60	69
Maine Eastport	58	45	51	52	52	51	53	55	57	54	50	37	40	50
Mass. Boston	67	47	56	57	56	59	62	64	63	61	58	48	47	57
Mich. Alpena	45	29	43	52	56	59	64	70	64	52	44	24	22	51
Detroit	69	34	42	48	52	58	65	69	66	61	54	35	29	53
Grand Rapids	56	26	37	48	54	60	66	72	67	58	50	31	22	49
Marquette	55	31	40	47	52	53	56	63	57	47	38	24	24	47
S. Ste. Marie	60	28	44	50	54	54	59	63	58	45	36	21	22	47
Minn. Duluth	49	47	55	60	58	58	60	68	63	53	47	36	40	55
Minneapolis	45	49	54	55	57	60	64	72	69	60	54	40	40	56
Miss. Vicksburg	66	46	50	57	64	69	73	69	72	74	71	60	45	64
Mo. Kansas City	69	55	57	59	60	64	70	76	73	70	67	59	52	65
St. Louis	68	48	49	56	59	64	68	72	68	67	65	54	44	61
Springfield	45	48	54	57	60	63	69	77	72	71	65	58	48	63
Mont. Havre	55	49	58	61	63	63	65	78	75	64	57	48	46	62
Helena	65	46	55	58	60	59	63	77	74	63	57	48	43	60
Kalispell	50	28	40	49	57	58	60	77	73	61	50	28	20	53
Nebr. Lincoln	55	57	59	60	60	63	69	76	71	67	66	59	55	64
North Platte	53	63	63	64	62	64	72	78	74	72	70	62	58	68
Nev. Ely	21	61	64	68	65	67	79	79	81	81	73	67	62	72
Las Vegas	19	74	77	78	81	85	91	84	86	92	84	83	75	82
Reno	51	59	64	69	75	77	82	90	89	86	76	68	56	76
Winnemucca	53	52	60	64	70	76	83	90	90	90	81	65	52	74
N.H. Concord	44	48	53	55	53	51	56	57	58	55	50	43	43	52
N.J. Atlantic City	62	51	57	58	59	62	65	67	66	65	54	58	52	60

State and Station	Years	Jan.	Feb.	Mar.	Apr.	May	June	July	Aug.	Sept.	Oct.	Nov.	Dec.	Annual
N. Mex. Albuquerque	28	70	72	72	76	79	84	76	75	81	80	79	70	76
Roswell	47	69	72	75	77	76	80	74	74	74	74	74	69	74
N.Y. Albany	63	43	51	53	53	57	62	63	61	58	54	39	38	53
Binghamton	63	31	39	41	44	50	56	54	51	57	43	29	26	44
Buffalo	49	32	41	49	51	59	67	70	67	60	51	31	28	53
Canton	43	37	47	50	48	54	61	63	61	54	45	30	31	49
New York	83	49	56	57	69	62	65	66	64	64	61	53	50	59
Syracuse	49	31	38	45	50	58	64	67	63	56	47	29	26	50
N.C. Asheville	57	48	53	56	61	64	63	59	59	62	64	59	48	58
Raleigh	61	50	56	59	64	65	62	62	63	64	62	62	52	61
N. Dak. Bismarck	65	52	58	66	57	68	61	73	69	62	59	49	48	69
Devils Lake	55	53	60	59	60	59	62	71	67	59	58	44	45	58
Fargo	39	47	55	56	58	62	63	73	69	60	57	39	46	59
Williston	43	51	59	60	63	66	66	78	75	65	60	48	43	63
Ohio Cincinnati	44	41	46	52	56	62	69	72	68	68	60	46	39	57
Cleveland	65	29	36	45	52	61	67	71	68	62	54	32	25	50
Columbus	65	36	44	49	54	63	68	71	68	66	60	44	36	55
Okla. Oklahoma City	62	57	60	63	64	65	74	78	78	74	68	64	57	68
Oreg. Baker	46	41	49	56	61	63	67	83	81	74	62	44	37	60
Portland	69	27	34	41	49	52	55	70	65	55	42	28	23	48
Roseburg	29	24	32	40	51	57	59	79	77	68	42	28	18	51
Pa. Harrisburg	60	43	52	55	57	61	65	68	63	82	58	47	43	57
Philadelphia	66	45	56	62	58	61	62	64	61	62	61	53	49	57
Pittsburgh	63	32	39	45	50	57	62	64	61	62	54	39	30	51
R.I. Block Island	48	45	54	47	56	60	62	62	60	59	50	44	56	
S. C. Charleston	61	58	60	65	72	73	70	66	66	67	68	68	57	66
Columbia	55	53	57	62	68	69	69	63	65	64	68	64	51	63
S. Dak. Huron	62	66	62	60	62	65	68	76	72	66	61	52	49	63
Rapid City	53	58	62	63	62	61	66	73	73	69	66	58	54	64
Tenn. Knoxville	62	42	49	53	59	64	66	64	59	64	64	53	41	57
Memphis	55	44	51	57	64	68	74	73	74	70	69	58	45	64
Nashville	63	42	47	54	60	65	69	73	69	68	64	53	42	59
Tex. Abilene	14	64	68	73	66	73	86	83	85	73	71	72	66	73
Amarillo	54	71	71	75	75	75	83	81	81	74	70	76	70	76
Austin	33	46	50	57	60	62	72	76	79	70	70	57	49	63
Brownsville	37	44	49	51	57	65	73	78	78	67	70	54	44	61
Del Rio	36	53	55	61	63	60	66	75	80	69	66	58	52	63
El Paso	53	74	77	81	85	87	87	78	78	80	82	80	73	80
Ft. Worth	33	56	57	65	66	67	75	78	78	74	70	63	58	63
Galveston	56	50	50	55	61	69	76	72	71	70	74	62	49	63
San Antonio	57	48	51	56	58	60	69	74	75	69	67	55	49	62
Utah Salt Lake City	22	48	53	61	68	73	78	82	82	84	73	56	49	69
Vt. Burlington	54	34	43	48	47	53	59	62	59	51	43	25	24	46
Va. Norfolk	60	50	57	60	63	67	66	66	63	64	60	51	62	
Richmond	56	49	55	59	63	67	66	65	62	63	64	58	50	61
Wash. North Head	44	28	37	42	48	48	48	50	46	48	41	31	27	41
Seattle	26	27	34	42	48	53	48	62	56	53	36	24	25	45
Spokane	62	26	41	53	63	64	68	82	79	68	53	28	22	58
Tatoosh Island	49	26	36	39	45	47	48	44	47	38	26	23	40	
Walla Walla	44	24	35	51	63	67	72	86	84	72	59	33	20	60
Yakima	18	34	49	67	70	72	74	86	88	74	61	38	29	63
W. Va. Elkins	55	33	37	42	47	55	55	56	53	55	51	41	33	48
Parkersburg	62	30	36	42	49	60	60	60	53	60	53	37	29	48
Wis. Green Bay	57	44	51	55	56	58	64	70	65	58	52	40	40	55
Madison	59	44	49	52	53	58	64	70	66	56	41	38	56	
Milwaukee	59	44	48	53	56	60	65	73	67	62	56	44	39	57
Wyo. Cheyenne	63	65	66	66	61	59	68	70	68	69	69	65	63	66
Lander	57	66	70	71	66	65	74	76	75	72	67	61	62	69
Sheridan	52	56	61	62	61	61	67	76	74	67	60	53	52	64
Yellowstone Park	35	39	51	55	57	56	63	73	71	65	57	45	38	56
P.R. San Juan	57	64	69	71	66	59	62	65	67	61	63	63	65	65

* Courtesy of Sol-R-Tech

Based on period of record through December 1959, except in a few instances. These charts and tabulation derived from "Normals, Means, and Extremes" table in U.S. Weather Bureau publication Local Climatological Data.

This method requires expensive equipment, which has eliminated it as a practical means of home heating.

The flat plate collector has been accepted almost universally as the best device for gathering solar energy for low-temperature applications, such as home heating (Figure 26-3). The basic principle of operation is very simple. The flat plate collector consists of a flat metal plate coated with a surface that absorbs sunlight and is heated by the rays from the sun. There are two covers over the flat plate to keep heat from being transferred to the ambient air. A

Basic Solar Theory

working fluid is used to carry the heat from the plate to the place where it is to be used for heating the building or hot water tank.

There are two types of flat plate collectors. One type uses air and the other uses liquid as the working fluid. Although systems using air can be made more simple than systems that use a liquid, air does not have the ability to absorb and carry the heat that water and many other fluids do. Although both types of systems are in use, the liquid systems outnumber the air type by a large margin.

The material used for the absorber plate must conduct heat readily. Copper and aluminum are usually the first choices. Even though copper has definite advantages as far as corrosion resistance, the substantially lower cost of aluminum will usually make it the choice of the two.

The coating used for the absorber plate is of great

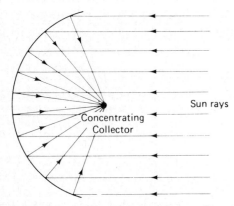

FIGURE 26-1 Concentrating collector. (Courtesy of Sol-R-Tech.)

FIGURE 26-2 Flat plate collector. (Courtesy of Sol-R-Tech.)

FIGURE 26-3 Exploded view of a flat plate collector. (Courtesy of Sol-R-Tech.)

SOLAR ENERGY

importance. It should, ideally, absorb all the energy that falls on it, while reemitting none of it. It is an unfortunate fact of nature that most objects emit radiation most easily at the same frequencies that they absorb them best. Because of this, the best approach to the ideal surface is one that absorbs the visible sunlight but does not emit the very low frequency radiation associated with heat. These selective surfaces are available, but their present cost is very high, and their durability is doubtful.

In actual outdoor tests, it was found that the best coating is one of the simplest, that is, a layer of flat black paint. Almost all collectors presently on the market use this type of surface.

The latest major component of the flat plate collector is the cover plate. Glass is an excellent material optically because it is transparent to visible light but opaque to infrared. This is the reason a greenhouse retains heat. However, it is heavy and is subject to breakage. There are a number of plastic materials available that offer the optical properties of glass without the disadvantages.

The number of cover plates is important to the performance of the collector. If too few are used, sufficient insulation from the cold ambient air will not be provided. Too many will reduce the amount of sunlight reaching the absorber plate. Two cover plates have been found to be the optimum number for most areas.

It is not enough to collect the required amount of energy while the sun is shining. Some of the energy must be stored for use at night and during cloudy periods. The material used for the heat storage medium is crucial to the performance of the system. Presently, the main choices are water or rock in the form of crushed stone or concrete.

The specific heat and heat capacity of common materials used for thermal storage in solar heating systems vary (Table 26–3). Specific heat is a measure of the amount of heat a pound of a material can hold for every degree its temperature is raised. Water

has a specific heat of 1 because the Btu (cal) is defined as the amount of heat required to raise the temperature of 1 pound of water 1 degree.

As seen in Table 26–3, water will store five times as much heat as stone and four times as much heat as concrete. There are no readily available, inexpensive materials that come close to the heat storage ability of water. Eutectic salts are being developed that store heat by melting rather than by getting hotter. However, these heat fusion materials are still in the experimental stage and have not been proven.

☐ Domestic Hot Water ☐

This is probably the most simple and most easily installed system of all solar heating systems. It is generally recommended for existing homes to decrease or eliminate utility bills. The required equipment consists of a solar collector array on the roof, a heat exchanger, a water storage tank, circulator pumps, controls, and an auxiliary water heater, which could be the one in use.

When sunlight strikes a collector panel, several things take place.

1. Solar radiation passes through the outer, fiberglass-reinforced polyester skin. This layer acts as a weather conduction and infrared barrier.
2. The solar radiation then passes through a ½-in. (12.7-mm) desiccated air space that eliminates condensation and acts as a conductive barrier.
3. Next, it passes through an interior layer of transparent temperature-resistant polyvinyl fluoride (Tedlar) that acts as a thermal and infrared barrier.
4. It then passes through a second ½-in. (12.7-mm) air space that acts as a convective barrier.
5. Finally, the solar radiation is absorbed by the blackened tube-in-sheet collector plate of 1100 aluminum alloy.

TABLE 26-3 Specific Heat and Heat Capacity of Common Storage Materials Used in Solar Heating Systems

Material	Specific Heat (Btu/lb-°F)	Density (lb/ft³)	Heat Capacity (Btu/ft³-deg)
Water	1.00	62.5	62.5
Stone	0.21	180	36
Iron ore (magnetite)	0.165	320	53
Concrete	0.27	140	38

6. The working fuid within the collector, inhibited ethylene glycol, captures the heat generated by the absorbed radiation and transfers it to the material being heated (either domestic hot water or space heating) by means of the appropriate plumbing.

7. Conductive losses are kept to a minimum by incorporating 1 in. (25.4 mm) of semirigid fiberglass insulation on back of the panel.

8. All layers are then compressed and sealed by a rigid aluminum T edge to ensure panel stability and aesthetic appearance.

Changing Solar Energy into Hot Water. The heat transfer fluid (ethylene glycol antifreeze) is pumped through a heat exchanger and gives up its energy to the water being pumped from the solar hot water supply tank (Figure 26–4). The electric consumption of these two pumps is minimal. Even in the highest electric rate areas, the pumps will not cost more than about $1.50 per month to run. When there is a demand for hot water, the replacement water is already solar heated, so the thermostat will not signal for more gas, oil, or electricity to raise its temperature. When cloudy days or extreme cold interfere with solar hot water production, the standard hot water heater will take over the load. With a properly designed system, 75 to 95% of the domestic hot water

will be free from utility rates. The pumps are controlled by a solar differential thermostat that turns them on when the temperature of the collector exceeds the solar hot water supply tank temperature. The thermostat also turns off the water pumps when the temperature of the solar collector falls below the solar hot water supply tank temperature. Everything operates automatically. When there has not been enough sunlight to keep the water heated, the auxiliary heater will take over and provide hot water automatically.

□ Solar Collector Efficiency □

The amount of heat captured by a solar collector is dependent on five variables, the first three of which are materials used in fabrication, ambient temperature, and solar intensity (Figure 26–5). All are inherently related. As the outside temperature falls, the collector efficiency will decrease. Clouds, haze, smog, etc. also decrease the efficiency, as do dirty plates or shaded panels.

The fourth factor is concerned with the orientation of the collector. South-facing roofs are best for the collection of solar energy. Southeast and southwest are also acceptable. Even east-west align-

Expansion Tank

Heat Exchanger

Sensor

Pump

Differential Thermostat

Hot

H₂O

Sensor

Standard Hot Water Heater

Solar Storage Tank

——— Piping
– – – Wiring

Cold H₂O

Pump

FIGURE 26–4 Residential solar hot water system. (Courtesy of Sol-R-Tech.)

FIGURE 26-5 **Solar collector efficiency. (Courtesy of Sol-R-Tech.)**

ment will work for solar hot water if the appropriate zoning valves are used. For example, a home at 40° North latitude with a 40° roof pitch and with collectors facing east and west will collect (if plumbed with a differential valve so that only one side operates at a time) 80% of the energy of south facing panels in the month of January and 130% of that collected during the month of July.

The fifth factor that affects solar collector performance is the roof pitch on the building. For the greatest efficiency, the sun should be perpendicular to the collector. For North America, a roof pitch of 15° plus the local latitude is ideal. For example, the city of Washington, D.C., is located at 39° North latitude. A solar-heated house should have a roof pitch of 54° although anything between 39° and 54° would be adequate for heating purposes. However, a flatter roof would have a tendency to produce more hot water and reduce the water bills.

□ Collector Surface Area Needed □

Because most people are interested in how they can save money with solar water heating, all the variables have been taken into account, such as ambient temperature, efficiency, roof orientation, solar radiation, and cloud cover for 80 American cities (Table 26-4). For example, a family with four members in Columbus, Ohio, would require 4 times 0.7 or 2.8 collectors in the summertime and 4 times 3.1 or 12.4 collectors in the wintertime to furnish 90 to 100% of their hot water load. A similar family in Newark, New Jersey, would use the panel requirements for Seabrook because Newark is not listed.

When determining how many collectors are needed, always round off the panel requirement. For example, if 3.2 panels are required, order 3. If 4.5 panels are needed, order 5. Also, remember that the number of panels that deliver 90 to 100% of the midwinter demand will also supply 90 to 100% of the yearly hot water supply.

□ Space Heating with Solar Energy □

There are several important criteria relating to the basic structure of the house. Obviously, to heat with the sun, sunlight must be available. The ideal location of the collectors should be a clear southern exposure, with no objects around to shade the collector panels. Local conditions, such as frequent fogginess, can present problems.

The trees should be trimmed to provide a clear path for the winter sun. If the system is designed for space heating only, deciduous trees can usually be ignored, because they will have lost their leaves by the time the heat is needed.

The ideal roof orientation is due south (Figure 26-6). The only stipulation to the south-facing collector is that the sky remain clear throughout the day. In many areas, however, there is a tendency for clouds to develop at about the same time each day during some periods of the year. For example, in Vermont the winter days often start out sunny, then become cloudy in the afternoon. Because of this the solar collectors in this geographical area are faced about 15 to 25° to the east of south in order to make more use of the morning sun. Since there is presently no source of weather data that would show this kind of trend, personal judgment and observation will be needed to determine if a correction of this sort is needed, and in which direction to make it.

The best roof pitch for any space heating application corresponds to the latitude (measured in degrees) plus 15°. It is possible to compensate for much lower or higher angles by adding more panels. However, this is not recommended if it can be avoided.

In general, the roof should have as many square feet (square meters) of solar panels as equal one-third to one-half the floor space in the home.

Storage Location. The heat storage tank should be located under a heated portion of the structure. Any heat lost from the tank will then go toward heating

TABLE 26-4 Panel Requirements per Person

City	Approximate Latitude	Summer Panel Requirement	Winter Panel Requirement	City	Approximate Latitude	Summer Panel Requirement	Winter Panel Requirement
Albuquerque, New Mexico	35°	0.5	1.4	Lander, Wyoming	30°	0.7	2.2
Annette Island, Alaska	55°	1.2	4.0	Las Vegas, Nevada	36°	0.5	1.2
Apalachicola, Florida	30°	0.7	1.2	Lemont, Illinois	41°	0.8	2.6
Astoria, Oregon	46°	0.9	2.8	Lexington, Kentucky	38°	0.7	2.0
Atlanta, Georgia	33°	0.7	1.8	Lincoln, Nebraska	41°	0.8	2.1
Barrow, Alaska	71°	1.3	175.4	Little Rock, Arkansas	34°	0.7	1.8
Bethel, Alaska	60°	1.2	1.12	Los Angeles, California	34°	0.7	1.2
Bismarck, North Dakota	47°	0.7	2.9	Madison, Wisconsin	43°	0.8	2.8
Blue Hill, Massachusetts	42°	0.9	2.9	Matanuska, Alaska	61°	1.1	14.2
Boise, Idaho	43°	0.6	2.9	Medford, Oregon	42°	0.6	3.1
Boston, Massachusetts	42°	0.8	2.9	Miami, Florida	26°	0.7	0.9
Brownsville, Texas	26°	0.6	1.2	Midland, Texas	32°	0.6	1.6
Caribou, Maine	47°	0.9	4.0	Nashville, Tennessee	36°	0.7	2.1
Charleston, South Carolina	33°	0.7	1.5	Newport, Rhode Island	41°	0.8	2.4
Cleveland, Ohio	41°	0.8	3.2	New York, New York	41°	0.8	2.5
Columbia, Missouri	39°	0.7	2.3	Oak Ridge, Tennessee	36°	0.7	2.1
Columbus, Ohio	40°	0.7	3.1	Oklahoma City, Oklahoma	35°	0.6	1.4
Davis, California	38°	0.6	2.1	Ottawa, Ontario	45°	0.8	3.6
Dodge City, Kansas	38°	0.6	1.6	Phoenix, Arizona	33°	0.5	1.2
East Lansing, Michigan	42°	0.8	3.7	Portland, Maine	43°	0.8	2.3
East Wareham, Massachusetts	42°	0.9	2.4	Rapid City, South Dakota	44°	0.7	2.2
Edmonton, Alberta	53°	0.9	4.6	Riverside, California	34°	0.6	1.2
El Paso, Texas	32°	0.5	1.3	St. Cloud, Minnesota	45°	0.7	3.4
Ely, Nevada	39°	0.6	1.8	Salt Lake City, Utah	41°	No data	2.3
Fairbanks, Alaska	65°	0.9	83.3	San Antonio, Texas	29°	0.6	1.3
Fort Worth, Texas	33°	0.6	1.4	Santa Maria, California	35°	0.7	1.4
Fresno, California	37°	0.6	1.9	Sault Ste. Marie, Michigan	46°	0.8	3.8
Gainesville, Florida	29°	0.7	1.2	Sayville, New York	40°	0.8	2.4
Glasgow, Montana	48°	0.6	3.4	Schenectady, New York	43°	0.9	3.7
Grand Junction, Colorado	39°	0.6	1.8	Seattle, Washington	47°	0.9	3.7
Grand Lake, Colorado	40°	0.8	2.6	Seabrook, New Jersey	39°	0.8	2.4
Great Falls, Montana	47°	0.7	2.8	Spokane, Washington	47°	0.6	4.0
Greensboro, N.C.	36°	0.7	1.9	State College, Pennsylvania	41°	0.8	3.1
Griffin, Georgia	33°	0.7	1.7	Stillwater, Oklahoma	36°	0.7	1.6
Hatteras, North Carolina	35°	0.6	1.5	Tampa, Florida	28°	0.7	1.0
Indianapolis, Indiana	40°	0.7	2.9	Toronto, Ontario	43°	0.8	3.6
Inyokern, California	35°	0.5	1.1	Tucson, Arizona	32°	0.6	1.1
Ithaca, New York	42°	0.8	3.6	Upton, New York	41°	0.8	2.2
Lake Charles, Louisiana	30°	0.7	1.4	Washington, D.C.	39°	0.7	2.1
				Winnipeg, Manitoba	50°	0.8	5.0

Source: Sol-R-Tech.

FIGURE 26–6 **Roof orientation. (Courtesy of Sol-R-Tech.)**

the house. In a new home this is best accomplished by burying the tank under the cellar or ground floor. In an existing home, the basement or garage is usually the only available location.

General Considerations. It should go without saying that when a solar heating system is being designed, the designer should first be sure that the home is well insulated and is free of major air leaks. Insulation should be the heaviest available, normally R-13 in the walls and R-22 in the attic. The windows should be double-glazed (thermopane) or have storm windows. A tight, well-insulated house is important with any heating system, and an absolute necessity for the best performance with solar heat.

☐ Solar Heat Operation ☐

When a system is to be used for solar space heating the heat exchanger can be eliminated (Figure 26–7). Since no one will be drinking the water in the heat storage tanks, it is possible to add a corrosion inhibitor that allows the water to be circulated directly through the solar panels. With this one change, the collector and storage parts of the system work

FIGURE 26–7 **Solar system without heat exchanger. (Courtesy of Sol-R-Tech.)**

in much the same way as in the domestic water heating system. A heat pump is added to distribute the heat to the house.

The use of a heat pump in a solar heating system allows the collectors to operate at a much lower temperature than would otherwise be possible, thereby doubling the efficiency of the collector. Also, during the summer season the refrigeration cycle of the heat pump can be reversed to provide central air conditioning with no extra equipment.

There are several types of solar-assisted heat pumps presently manufactured. Probably the most simple type uses a hot water coil between the outdoor coil and the outdoor fan (Figure 26-8). The hot water from the heat storage tank is pumped through the water coil to provide extra heat when the ambient temperature is low and the efficiency of the heat pump has dropped. When this type is used, a conventional-type heat pump with the added coil will operate satisfactorily.

The balanced flow pump is operated on an internal pressure control to maintain water pressure in the lines connected to it. A solenoid valve is wired to each heat pump to shut off the flow of water when the heat pump is not running (Figure 26-9). This will cause the water pump to shut off automatically when no water is being used.

Heat Pump Cutoff. Some heat pump units must be equipped with a low-temperature cutoff control to prevent operation of the compressor when the water temperature is below 40°F (4.4°C). This is accomplished by wiring a thermostat in series with the compressor contactor coil (Figure 26-10).

Heat Pump Supply Pump. This is usually the smallest available balanced flow pump unit. It must be equipped with an internal pressure control to turn the pump off when no water is being used. A good example of this type of pump is one used for shallow well installations. There are many ways to provide this assistance, but this is the most simple and economical method.

Heat Pump versus Fin-Tube Radiation. In areas where solar radiation is scarce, the heat pump is the best system to choose because it can still extract heat from the solar storage tank when the temperature is at 40°F (4.4°C).

Fin-tube coils can be installed in sunnier and warmer localities. In a fin-tube system the solar-heated water is pumped directly through the heat exchanger tubes located inside the system ductwork (Figure 26-11). Air is blown across the tubes, warmed, and forced into the rooms requiring heating. Obviously, the storage tank temperature

FIGURE 26-8 Solar-assisted heat pump.

FIGURE 26-9 Balanced flow pump for solar-assisted heat pump. (Courtesy of Sol-R-Tech.)

SOLAR ENERGY

Note: On DW-40H Contactor Leads are Top and Bottom

Water Inlet

Tape in Place
Warp with Insulating
Loam Tape (Arma-Flex)

FIGURE 26–10 Heat pump freeze protector. (Courtesy of Sol-R-Tech.)

Heated air to space

Heated air to space

Water coil

Pump

Water storage

FIGURE 26–11 Direct solar heating system.

must be higher for such a system to operate efficiently.

☐ Water Treatment ☐

Even though the getter column is used, it is not possible to circulate untreated water through the solar panels. To do so would cause severe corrosion problems. The following steps are recommended for water treatment:

1. When filling the water storage tank, pass the water through a water softener designed to remove the minerals present in the water supply in that locality. Contact a loal water treatment company for advice.
2. Add 50 parts per million (ppm) of a sodium chromate corrosion inhibitor, such as Barclay Chemical Inhibitor C. An inexpensive color-test

kit should be purchased to allow for the measurement of concentration of the inhibitor.
3. The water is now ready for use.

☐ Heat Storage ☐

A typical heat storage tank is usually very simple (Figure 26–12). The dimensions can be varied to fit the space without affecting the performance of the system as long as the thickness of insulation and relative positions of the pipes are maintained.

The most simple method of waterproofing the heat storage tank is with Thoro-Seal, a cement waterproofing compound available at most local building supply stores. Apply the Thoro-Seal to the inside of the tank according to the directions on the package.

As has already been mentioned, the heat storage tank should be located under the heated portion of the home. If this is not possible, the tank insulation should be increased to 4 in. (10.16 cm) on the sides and 6 in. (15.24 cm) on the top.

☐ Distribution of Heat ☐

Any new home being designed with solar heat in mind should be planned around forced warm air heating. This type of system will accommodate both heat pumps and direct heat exchanger systems. Existing homes with this type of heating system can readily be adapted to solar heating.

To panel pump

Deko labs sensor taped on

4" concrete slab floor

To heat pump

2'-0" square access hole with plywood cover

Pipe access

2" styrofoam

9'

2" styrofoam

8" steel-reinforced concrete walls

2" styrofoam

8'

8'

4" concrete slab

FIGURE 26–12 Storage tank piping arrangement and panel pipe sensor. (Courtesy of Sol-R-Tech.)

Houses that use steam or hot water heating present a more difficult problem, especially if baseboard heaters or radiators are used. Unfortunately, the average temperature required in a hot water system is 180 to 210°F (82 to 99°C), while the temperature that can be reached by the solar panels is usually approximately 150°F (66°C) during the winter season. This could mean that almost twice as much baseboard heating would be needed to heat the building, making such a system uneconomical.

□ Panel Water Supply Pump □

Any ⅓-hp pump can be used to supply the solar collector array, as long as it is self-priming. If self-priming pumps are not available, a foot valve must be installed in the suction line to aid in keeping the pump primed. A check valve should also be installed just beyond the pump discharge.

□ Optional Direct Heat Exchanger □

By adding a hot water coil and providing a three-way solenoid valve to divert the flow of water from the heat pump to the coil, it is possible to provide direct solar heating during some seasons of the year (Figure 26–13). The three-way solenoid should be actuated by a thermostat in the storage tank so that when the water temperature exceeds 100°F (37.8°C), the direct heat exchanger coil is used. The coil uses

the blower already present in the heat pump, preventing the needless duplication of equipment (Figure 26–14).

□ Space Heating Existing Homes □

The equipping of already existing homes with solar heat is a topic of interest to many who wish to remain in their present home while still enjoying the benefits of solar heating. However, the addition of solar heat to an existing home is a difficult and expensive operation. Many of the installation procedures that are easy during construction later become almost impossible.

The optimum time for adding solar heating to a home is during remodeling, or when putting on an addition to the house. Additions are especially helpful since they provide an area where a storage tank can be buried under the house, and the roof can be oriented for optimum collector performance.

Some of the questions that should be answered when determining how feasible it would be to add solar heating to an already existing home are:

1. What kind of heating system is presently used? If it is warm air, fine. If any other type is used, would it be possible to install a warm air system? If not, then solar heating should not be considered. However, domestic hot water from solar energy can still be used.
2. Is there a large, south-facing roof area at an angle of 45° to 60°?

FIGURE 26-13 Optional direct heat exchanger solar heating system. (Courtesy of Sol-R-Tech.)

Note: Field-Supplied Water Solenoid Valve for Heating and Cooling

FIGURE 26-14 Typical wiring diagram for field-supplied water solenoid. (Courtesy of Sol-R-Tech.)

3. Is there an area, preferably underneath a heated space, where a storage tank could be buried?
4. Is the building well insulated?

If all these questions were answered yes, then the home is a good candidate for solar heating. If the answer to question 1, 2, or 3 was no, then the addition of solar heating would probably not be feasible. If the answer to question 4 was no, add to the insulation and proceed with the installation.

☐ Steps in Designing a Solar Heating System ☐

The recommended major steps that should be used in designing a solar heating system may be summarized as follows:

1. *Total net heat loss of the house*
   (a) Select the design outdoor weather conditions,

such as temperature, wind direction, and wind velocity.

(b) Select the indoor air temperature that is to be maintained in each room during the coldest weather.

(c) Estimate the temperature in any adjacent unheated spaces.

(d) Select or compute the heat transfer coefficients for outside walls and glass, and for inside walls adjacent to unheated spaces.

(e) Determine the net wall, roof, ceiling, glass, etc., areas.

(f) Compute the heat transfer loss for each surface.

(g) Compute the heat equivalent of the infiltration losses and the total heat required for positive ventilation, if used.

(h) Sum up the heat losses.

(i) Determine the internal heat generated (range, lights, etc.). This item should be subtracted from the sum of the heat losses.

(j) Additional heating capacity may be required for intermittently heated buildings.

2. *Heating requirement for the design period.* A decision must be made concerning the maximum number of hours during which the required heat will be supplied either by the storage system or by the auxiliary heating system. Assume, for instance, that 48 hr is chosen. By using the sum obtained in section 1h above, compute the 48-hr requirement, Q_D (design load). Part of this heat will be supplied by the auxiliary heater.

The capacities of the solar collector, heat storage unit, and the auxiliary heater are interrelated. In the cold climate zone a full-capacity auxiliary heating unit must be provided to supplement the solar heating system. Otherwise, cost of the collector and storage system will be too high. The auxiliary heating unit will seldom be called upon for heat delivery over an extended period; but the capacity is nevertheless required. The cost of an additional furnace capacity beyond average needs is rather low.

3. *Solar collector size.* The component selection procedure for the solar heating of a residential home may be somewhat simplified. In fact, the size of the solar collector is usually limited by the maximum roof area available facing the southern direction. Therefore, the approximate solar collector surface area is known. This area, however, should be large enough so that the collector will

FIGURE 26–15 Heat balance for sunlit glazing material.

be capable of providing the major portion of the design load (Q_D).

4. *Selection of design parameters.* Choose the type of solar collector, flow rate of the heat transfer fluid (water), and assume a reasonable collector efficiency (η). Determine the design daily total radiation (Q_H) that is incident on the collector (Figure 26–15). The total useful heat supplied by the collector will be given by the relation

$$Q_U = \eta Q_H$$

where

Q_U = Btu (cal) per day heat loss

η = collector efficiency

Q_H = total collector incident radiation

The difference between Q_U and Q_D should be supplied by the auxiliary heater (Figure 26–16).

5. *Design temperatures of the collector and fluid.* By using the following equations, compute the average collector temperature (T), and the water temperature rise (T_f) in the collector. If the water temperature at the collector inlet and the exit are T_1 and T_2, respectively, then

$$T = 0.5(T_1 + T_2)$$
$$T_f = T_2 - T_1$$

Thus T_1 and T_2 are determined. These are the daily average water temperatures. In reality, these temperatures are changing with time.

6. *Head storage capacity.* In its simplest form, it may be assumed that the temperature rise in the storage is equal to the temperature rise in the collector (T_c). Then the heat capacity (C) of the unit is found from

FIGURE 26–16 Residential solar heating system.

$$Q_U = C(T_f)$$

7. *Auxiliary heater capacity.* The auxiliary heating unit capacity should be adequate to supply at least the difference between Q_D and Q_U during the assumed design period.

8. *Airflow rate.* If air is used, for example, to remove heat from the storage to the rooms, compute the required airflow rate and the blower capacity.

Note: Since many of these procedures were covered in Chapter 22, it may be helpful to review that chapter before actually attempting to estimate these calculations.

☐ Safety Procedures ☐

The use of solar energy is comparatively new in the refrigeration and air conditioning industry. However, the safety practices used in the other areas also apply in solar use.

1. Obtain help when moving heavy objects.
2. When cleaning collectors wear sunglasses to protect the eyes from the rays of the sun.
3. Do not touch the reflective plate with the bare hands, since personal burns and etching of the collector surface may result.
4. Always use a safe ladder when working on rooftops.
5. Be careful not to cause separation of the tube and the collector plate.
6. Do not allow the flow of working fluid to stop with some of it still in the collector when the sun is shining on it. Excessive temperatures may result.
7. Be sure that the electric power is off before attempting work on the solution pumps.
8. Do not tamper with the controls for the tracking system or the efficiency of the collector may be decreased.
9. Do not set the working fluid temperature control too high because excessive pressure may result and rupture the storage tank.
10. Be sure that the working fluid drains from the collector when the pump is off or that there is sufficient antifreeze to prevent freezing.
11. When making line connections, use two wrenches to prevent twisting of the tubing.

☐ SUMMARY ☐

☐ Solar energy has been traced back to the ancient Greeks.
☐ Solar energy has not been developed into wide usage because of economic factors.

□ The annual solar energy falling in this country is almost 1000 times greater than the total yearly consumption of the country.

□ The sun provides between 250 and 300 Btu (63 and 75.6 kcal) per hour per square foot (0.09 m²) of the earth's surface.

□ The two types of practical solar collectors are: concentrating collector and flat plate collector.

□ The major advantage of the concentrating collector is its ability to generate very high temperatures.

□ Concentrating collectors must move with the sun.

□ The flat collector is considered the best device for gathering solar energy for low-temperature applications, such as home heating.

□ A "working fluid" is used to carry the heat from the collector to the storage area.

□ The coating used for the absorber plate is of great importance and is usually a flat black paint.

□ If too few cover plates are used, an insufficient amount of insulation from the cold ambient air will be provided.

□ If too many cover plates are used, the amount of sunlight reaching the absorber plate will be reduced.

□ Water will store five times as much heat as stone, and four times as much as concrete.

□ The required equipment for solar heating of domestic hot water is a solar collector array, a heat exchanger, a water storage tank, circulator pumps, controls, and an auxiliary water heater.

□ With a properly designed system 75 to 95% of the domestic hot water will be free of utility rates when heated with solar energy.

□ The amount of heat collected by a solar collector is dependent on materials used in fabrication, ambient temperature, solar intensity, orientation of the collector, and the roof pitch.

□ The ideal orientation of the roof that will carry the collectors is due south.

□ The best roof pitch for a space heating application corresponds to the latitude (measured in degrees) plus 15°.

□ When considering solar heating, the home must be well insulated and be free of any major air leaks.

□ Solar heating can be applied to heat pump systems to increase their efficiency even in low ambient temperatures.

□ On heat pump systems solar-heated water is pumped through an auxiliary coil that adds heat to the refrigerant during the heating season.

□ The water circulating through the collector plates should be treated to prevent corrosion.

□ A solar space heating system should be planned around a forced warm air system.

□ Baseboard heaters and radiators cannot be used with solar heating.

□ Heat pump systems can use an optional direct heat exchanger to supply the total heat when the storage water is above 100°F (37.8°C).

□ The addition of solar heating to an existing home is difficult and expensive.

□ EXERCISES □

1. Why is solar energy not in wide use today?
2. How much solar heat strikes the earth's surface per hour?
3. What are the two major types of solar collectors?
4. Which type of collector will provide the hottest water?
5. Which type of collector is most popular?
6. What type of protective coating is best suited for solar collectors?
7. At present, what is the best heat storage medium for solar heat storage?
8. Why is glass not used for collector cover plates?
9. What is generally used as the heat transfer fluid?
10. What are the five variables that determine how much heat can be collected by a solar collector?
11. What is the ideal orientation of a solar collector?
12. What is the best roof pitch for a roof that is to hold the collectors?
13. In general, how many collector plates are needed for space heating?
14. What type of insulation is recommended for the walls and attic of a home to be heated by solar energy?
15. At what storage temperature should some heat pump compressors be shut off?
16. What type of heat distribution system should be used with solar heating?
17. What are two major considerations that must be given a home when designing a solar heating system?
18. How is the efficiency of a heat pump increased when assisted by solar heating?
19. Are baseboard heaters and radiators suitable for use with solar heating?
20. At what storage temperature can an optional direct heat exchanger be used?

27 Starting and Operating a Business

Performance Objectives

After completion of this chapter, the student should:

- ■ Be familiar with the basic procedures used when purchasing a business.
- ■ Be able to name the four basic forms of businesses.
- ■ Understand the importance of human relations.
- ■ Be able to describe how to pick a company name.
- ■ Demonstrate better understanding of the art of selling.
- ■ Understand the basic procedures used when collecting for the job.
- ■ Discuss the proper procedures used when making a bid.
- ■ Be able to list some of the steps to use in avoiding legal problems.
- ■ Complete the Exercises with at least 80 percent accuracy.

When contemplating starting a new business, there are many things to consider. There are no hard-and-fast rules to guide you through the process—except those imposed by the government agencies: local, state, and federal. Many people who have been in business for several years or have had prior business experience have rules and regulations for their company that they go by—most of the time. Under varying circumstances they will make concessions depending on the customer,

the circumstances, and what advantage each will have for the company in the future.

It takes a special kind of person to own and manage a small business. First, the person must get along with people exceptionally well, must have a good sense of business practices, must know something about handling people, and must be expert in the chosen field, especially if it is going to be a one-person operation for a while. The person must also be neat in appearance, have personal confidence, and above all, be honest. Those who are not completely honest usually do not have a steady list of customers and often do not last long in business. For a long and happy relationship with customers you must be honest.

Anyone who is thinking of buying or starting a new business should give much thought to some basic considerations, such as:

1. Do you have a sufficient amount of capital to buy the tools and equipment necessary to operate the business properly and to see you through several months of slow business with little or no income?
2. Will the geographical location support the kind of business that you plan to establish?
3. What legal requisites must be met before opening and after opening the business? Are licenses required? What types of insurance are required?

When these requirements have all been met or accounted for, there are other legal matters that must be studied in great detail, such as: the types of insurance required and those that are desirable, selecting a name if it is a new business, how to choose an attorney and an accountant, choosing a good bookkeeper, and if it would be better to buy an existing business or start a new one.

☐ Insurance ☐

The owner of any type of business should have a good liability insurance policy. This is in case an accident occurs that damages another person or another's property. Many state laws require that liability insurance be purchased before a license will be issued. Other types of insurance that the owner should consider if not required by law are:

1. Worker's compensation insurance for all employees
2. Hospitalization insurance
3. Comprehensive motor vehicle insurance (required by many state laws)
4. Surety bonds
5. Life insurance to protect the family in case of your accidental death
6. Accidental death and disability insurance

☐ Purchasing a Business ☐

Many times the employees of a business will purchase the business where they work. A lot of effort must be put forth by all involved to make the company profitable. In some instances an outsider will purchase a going business. This also requires that a lot of investigating be done so that the buyer will know whether he or she is making a wise investment. To make this type of purchase, a lawyer and an accountant must be hired to make the necessary investigation. It should be remembered that you can buy a company but you cannot buy the customers.

☐ Forms of Conducting a Business ☐

There are four major methods of conducting a business:

1. *Sole proprietorship.* This is a business enterprise that is owned by one person. It is the simplest form of business. All of the assets and liabilities belong to the proprietor of the business.

2. *General partnership.* In this organizational style, two or more persons are co-owners of a business. In most cases, the co-owners share in the profits of the business in a ratio that is equal to their investment. In this case each partner is individually liable to the extent of his or her personal assets for all liabilities incurred by the partnership while conducting the business.
3. *Limited partnership.* This type of partnership has several general partners and a number of limited partners. A general partner is a member who is responsible for the operation of the business. His or her liabilities are unlimited. A limited partner is not involved in the operation of the business; therefore, his or her liabilities are limited to the amount of the person's investment.
4. *Corporation.* A corporation is a legal entity that is separate from its owners and shareholders. In this type of operation, a number of people put their money together and share in the profits, with only the risk of losing their investment. A corporation is possible only under the laws of state statute.

☐ Human Relations ☐

As an air conditioning and refrigeration service and installation person, your job will involve working with people, equipment, and materials. All of these factors are necessary for you to do your job; the customers are by far the most important. Your success will depend on them and their attitude toward you. It is sometimes difficult to make friends with difficult customers, but your success depends on how you handled difficult situations.

Many potential technicians think an ability to use the proper tools is all that is necessary to get the job done. This idea is far from the truth because customers have basic needs that they want satisfied. Therefore, as a technician you must know how to provide these needs and how to get along with customers. Because of this, one of your top priority jobs is to continually improve the relationships between you and your customers. It must be remembered that your personal success as a technician is, to a great deal, dependent on how well you get along with the customers.

Many technicians who are not the best at working with equipment and materials are very good at getting along with customers and hold higher paying positions within a company. This is, for a major part, because the customers call for them to do the needed work.

The Meaning of Human Relations. There are several schools of thought concerning human relations. Some think that the practice of good human relations is applying the golden rule to relationships. Some think that it is applying psychology to make friends and influence people. Others think that human relations makes use of an ethical approach to personal problems. The fact is that human relations is made up of all these factors.

Good human relations means people getting along with other people in a harmonious manner, reaching satisfactory production, and cooperatively achieving an economic as well as some social satisfaction. The key part to any human relations is the motivation of people. Therefore, the practice of good human relations is more than backslapping, more than being a nice person, and more than glad-handing.

Things to Remember When Working with Others. As a technician you will be working directly with customers. Because of this fact, there are several things you should remember about them as individuals.

1. We are all different. We have thousands of differences. Each of our personalities is different, just as our fingerprints are different. We develop differently from the day we are born. We each have our own mind and think differently. We have our own ideas about life and our wants as an individual.

   Due to these individual differences, any study of human relations must start with the individual. It stands, therefore, with you. You are always a part of a team, but you must remember that the team consists of individuals just like yourself. The team is an entity and has a certain amount of power because of its individual members. The team, however, cannot make decisions. It is the individual members who make the decisions.

2. You must remember that when you are working with a person, you are working with the complete person. You may wish to enjoy only a certain part of that person's personality, but this cannot be done. When any part of a person is considered, the whole person must be considered: mind,

thoughts, and actions. The person's desires and motivations must also be considered, along with his or her home life, which has an effect on how the person responds to different situations. As a company representative, you must remember that each person is different and that in your dealings you must take into consideration the complete person.

3. All the normal behavior of an individual is caused behavior—behavior caused by individual needs and wants. The person is motivated to do things through the belief that doing so will help achieve a goal that he or she feels is worth working for. Remember—the customer is generally not considering what you think is right but what he or she thinks is right. All of us sometimes think that another person's wants, and the reasons behind them, are foolish; but to him or her they are very important and real needs. These are the needs with which you will be dealing when working with customers or employees.

4. Remember that customers are not something that can be programmed for use when we desire and discarded when we are finished with them. We are all human beings and we all need and should be treated with dignity and due respect. This should be done regardless of what we think of a person. Regardless of station in life, a customer should be shown the proper respect for his or her personal choices and abilities.

When dealing with people, you should always keep these four aspects clearly in mind. When doing this, you will find that your understanding of other people and your ability to work with them will improve with experience. As an example, when you realize that all people are different from each other, you will find that categorizing them is most difficult. You will soon stop trying to handle all people in the same way because you will begin to recognize that you are dealing with the complete person as an individual who is unique.

Affecting the Customers' Attitude. Remember, your job is getting things done for people, and that your effectiveness as a technician will be measured by how the customer relates to you and your company. Your success depends on how the customer sees you. Because of this fact, your customers are a very real asset to you and should be treated with respect and consideration. Your customers, in most cases, want to get along with you and help to get the job done as efficiently as is possible. Remember that the work you are doing has an impact on their lives and that they hope to gain some personal satisfaction from its successful completion. Whether or not this satisfaction is realized depends on how well you do your job and your style of presenting yourself. A technician who is generally happy and has self-confidence will usually be seen as a good technician.

Understanding Yourself. Someone once said that if you want to understand other people, you must first understand yourself. When a person begins to know himself or herself, he or she has taken the first step toward understanding others. For example, when you realize that you have certain attitudes about your dress, how fast you should work, and how you should behave, you should also realize that other people have their own ideas about these things and they may not be the same as yours. Therefore, understanding yourself will help you to understand others.

This is especially true of the relationship between the technician and the customer. Any time you become critical about what someone has said or done, you should ask yourself if you have done something to cause the person to react in this manner and what you can do to correct the situation. In most cases, some of the blame is on your shoulders. When none of the blame is yours, change your point of view and have a look at the situation from the other person's point of view. In most cases this will help you gain a better understanding of the problem. As a result, you will generally feel differently toward the other person and the problem that exists, because you will have greater understanding, compassion, and integrity because you know how you would like to be treated under the same circumstances.

Attempting to understand yourself means a good look at both your strong points and weaknesses. Your good points are easy to look at but your weaknesses are a bit difficult to see. You should realize that you have limitations and hangups just like any other person; that you may have a short temper, that you may be prejudiced against certain individuals, and have other such weaknesses. When these shortcomings about yourself are recognized, you are in a better position to control them and do a much better job at human relations.

Earlier we mentioned putting yourself in another person's shoes to help understand his or her behavior. When you do this, you are empathizing; that is, you are trying to see the problem from the other person's point of view. This does not mean that you are going to agree with the person and do things his or her way. It means just the opposite. Once you have an understanding of the problem from the other person's point of view, you are just in a better position to deal with it fairly and effectively. In this manner you are in a better position to appreciate the other person's point of view and feelings without getting involved in his or her personal life. At this point you can truthfully say that you understand why the person feels the way he or she does. When you have truly achieved this ability, you will be in a much better position to solve the problem with fairness to all concerned.

We learned earlier that all behavior is the result of something, that there is a reason for that behavior. As a technician, you should attempt to recognize and understand the cause for the behavior of your customers. The specific reason that a customer does something may not have a logical explanation; it may not even be reasonable, and it may even be ridiculous to you, but it is important to the customer and it has caused his or her actions. A major task facing you as a technician is to understand why a particular customer or fellow employee behaves the way he or she does, and not turn off the person's actions as absurd or ridiculous. As a matter of fact, when you state that the other person's actions are absurd, you are actually admitting your own inability to see things from the other person's point of view and to understand the reasons for his or her actions. If you are not able to see these reasons, then you are going to have a tough time as a technician.

You should make empathizing so much a part of your actions that it comes as easily as shaking hands with a friend. It should be so easy for you to use that solving disputes will be easily done in most cases.

Knowing Your Customers and Fellow Employees. The better a person knows the people with which he or she is associated, the easier it will be to understand them, their viewpoints, and their problems. As a technician, therefore, you should make every effort to understand the complete person, whether a customer or a person at work. That person's life is a part of him and will influence his or her behavior and actions.

To be good at human relations, it is almost impossible for you to know too much about the people with whom you associate. When the facts about the complete person are known, it is much easier to see his or her point of view and to empathize.

What to Expect from Good Human Relations. The total concept of human relations is that something takes place between one person and another. When good human relations are practiced, you will be more able to help satisfy some of the needs of those around you. After all, this is what human relations is all about. For example, we all feel a need to belong to something, a group, or someone. Anything that you can do to fill this need will enhance you in that person's eyes and allow you to do a better job.

However, practicing good human relations does not necessarily mean that everyone will be all smiles and that happiness will abound in all corners. We all have our own set of problems, but practicing good human relations will allow us to work around the difficulties that most problems present.

Achieving good human relations will pay off in a big way. The rewards are sometimes difficult to measure, but they are most certainly there. The practice of good human relations often means that a difficult problem can be avoided, and usually stopped before it gets started. Because of this, good human relations will show in many different ways and will give you the satisfaction of a job well done.

☐ Picking a Company Name ☐

The name that the company is given should be descriptive of what it does. Just to name it Bill's Service Company probably will not get the desired response. The name should tell what the company does. If you plan to service air conditioning equipment only, do not include refrigeration, heating, electrical, or any other type of service that is not provided by the company. Most contractors like to use the terms ''air conditioning'' and ''heating'' in their names because most people do not know that air conditioning includes heating as well as cooling. But whatever you do, pick a name that identifies the services that you plan to provide.

The name should be registered with the local or state agency for protection against its use by someone else. Suppose that the company grew into a large profitable company and someone started using the name as though it were their own. This would take customers away from you as well as possibly causing legal problems with debtors or other people who were dissatisfied with their service.

☐ Licensing ☐

Some states and cities require that air conditioning and refrigeration contractors be licensed. When these are required, they are an added expense. Also, the license must be obtained before the business can operate legally.

☐ Advertising ☐

There are as many ways to advertise as there are ways to spend money. Some people use the telephone Yellow Pages almost exclusively. Others use more sophisticated methods. You should check into those that are available and choose the one that works for your company. You can spend a lot of money on advertising that will not get much response. Be sure to check out an advertising firm to see what their track record is. Direct mail is another way of reaching customers, especially when offering seasonal startup specials or some other type of specials that you think will bring in business. When using direct-mail advertising, be sure to use good paper and name the person to whom you are sending the advertising piece. Do not send it to the building owner or occupant. How do you feel when your mailbox is stuffed with mail addressed to ''occupant''? Also, it is a good idea to follow up on a direct mailing with a phone call. When talking to the people, ask if they have any questions that need to be answered. This is the time to answer them. Each letter should be signed by a company executive. This all costs money but is usually worth the trouble, and often this type of work can be done during the off-season. It not only creates business, but also creates goodwill with customers.

Business cards are a good way of advertising. Leave a card everywhere you go. It is impossible to know when a person may pick up the card and give you a chance to service their equipment.

A person who chooses not to advertise reminds me of an old adage that goes: ''The fellow who winks at a girl in the dark will not accomplish anything because he is the only one that knows what he is doing.''

☐ Purchasing ☐

Purchasing is always a dilemma for the business owner. It is always nice to get a lower price on an item so that you can make some money and the customer can save some. However, quantity purchasing should be approached with caution. You could end up with a truckload or a warehouse full of supplies and parts that are only costing you money because they could not be sold fast enough to cover the billing for them. This way the special price did you no good; it actually cost you money because your money is tied up and not available for use on more profitable projects. It is always a good idea to make friends with the people who own and operate the supply houses where you do business. Sometimes they will actually have a special that will help you. Also, if you are bidding on a job that is very competitive, they can sometimes help by selling you the needed equipment at a lower price, so that you will have a better chance of getting the job. Also, if collections fall off one month, the supplier may give you a few extra days to pay your bill.

☐ Selling ☐

Selling is an art in itself. To be a good salesperson, you must be a good negotiator. The object of selling is to get the most for your product while making certain that the customer also gets a good deal.

Many salespeople say that it is best to start with a price that will allow some negotiation. This is so that the buyer will get some satisfaction out of gaining concessions from you. Concessions should not be given readily. If the customer must negotiate for the best price that he or she can get, the person's feeling of winning will help you to make the sale while giving you the profit that you wanted in the first place.

Rather than giving free gifts, trade. Say that you will lower your price to something nearer that of the customer, but only if he or she will do something for you, such as paying within a certain number of days. Do not give an answer when asked for your bottom-line figure. Merely ask what the customer

needs or remain silent. Keep the climate positive. Do not intimidate any person with whom you are dealing. To do so will probably cost you the sale.

☐ Collecting for the Job ☐

Different people have different methods of collecting money owed for a job. If there were one way that was always effective, there would be very few problems in this area. However, it is one of the greatest loss factors in a small business. If one person does not pay his or her bill, it may take several extra service calls to make up what was lost.

Some collection agencies are very intimidating to the customer. When such a method is used, the customer is probably gone forever. However, this may not be all bad—the time spent collecting from such a person could be better spent doing something more profitable. By the way, ''profit'' is not a dirty word—everyone depends on it for a living.

There are several methods that may be used to make the collection a bit less painful:

1. When a new customer calls for the first time, make arrangements for someone to be at the job site to make payment when the job is done.
2. Have the service technician collect while on the job. This eliminates the possibility of trouble later due to the customer finding fault with the service performed. It also reduces the cost of doing business, as it eliminates the need to send out end-of-month statements. Also, you do not have to wait for the money.
3. Customers who are always late in paying their bills should be asked to make payment upon completion of a job and to pay any outstanding bills at the same time. If they have no good reason for not paying their bills promptly, you do not need them.
4. If there is a problem with the charges, make every effort to settle the problem immediately.
5. If a bill is not paid within 10 days after billing, have someone phone them to determine why it has not been paid. They should try to solve any problems at this time. If they still do not make payment within a given period, say 10 days more, make another phone call. Be more aggressive this time; make sure they realize that their bill will be turned over to a collection agency if it is not paid within 30 days, and then do so.

6. When the contract is signed on large jobs, get a down payment large enough to pay for the equipment and supplies. This should be required of both old and new customers, because when a job takes more than 30 days to complete, the suppliers will want their money. If you have not collected it from the customer, it will have to come from your own pocket, thus reducing your operating capital.

☐ Choosing a Bank ☐

A bank is a vital part of any business. There will be times when you will need to make loans. Even after a company has grown to a substantial size, it will probably still need at least some short-term loans. If you have a good credit record at the bank, there will probably be very little problem getting what you need.

When you are first starting your business, go to the bank you have been dealing with, or one you have chosen, and talk to them about what you are doing, what you plan to do, and where you plan to be in the next few months or years. In other words, give them your game plan. Visit the bank periodically and inform them how things are going. If there are times when you cannot make a loan payment, give the bank a call, or better yet, go and see them, and explain the reason for being late and say when you expect to make the payment. They will respect you much more for this than if you try to ignore them.

☐ Bidding ☐

Bidding for a job is something to which few new business owners give much thought. When bidding on a job there are several things that must be taken into consideration or there will be no profit, and possibly a loss. Overhead and profit margin are two such things.

Overhead. How much does it cost you to operate the company, such as phone bills, rent, utilities, truck insurance, liability insurance, licenses, labor, tools, medical insurance, mailing, advertising, and miscellaneous items such as office supplies—just to name a few. These things must be taken into account and figured as percentages of the total job.

Profit Margin. The company must make a profit to stay in business. There are those who will try to get you to make a bid that does not include any margin of profit. When they do this they are asking you, in truth, to give up your company. Even in slow times a company must make some profit or it will not survive until the time when business picks up. It is better to pass up a job with no profit and wait for one that will make you some money. Usually when you take a no-profit job, a better-paying one will come along soon after you are committed to do the former job.

The method used to figure a profit margin will depend to a great extent on the geographical area in which the business is located. If the economy is slow, the profit margin will be down. If the economy is going strong, the profit margin will be greater. In any case, the charges can only be what the traffic will bear. In other words, you almost must charge what the competition is charging. However, there is one exception to this. If a company has a reputation of doing excellent work at a fair price and at the customer's convenience, a somewhat higher price can be charged for their work.

Those who do the bidding must have a good knowledge of how much labor it will take to do a certain job and must figure the materials accurately. The bidder must then take this information and convert it to dollars and cents. Then the overhead and profit are added to obtain the total cost of the job.

☐ Avoiding Installation Legal Problems ☐

The following steps should be used to aid in avoiding legal problems that may occur with an installation:

1. Provide the customer with a written proposal and obtain a written contract before starting the installation. Check to make certain that all the provisions are specifically spelled out in the contract and that it is signed by a company representative and the customer. Be sure to leave a copy of the contract with the customer.
2. Be sure to use only certified equipment. It should be certified by one or more of the accepted certifying agencies, such as ARI, AGA, or ASME. Seldom will anyone challenge the reputation of such nationally known agencies. Even if someone does, you are protected by their literature.

3. Be sure to follow all the local, state, and national codes and standards. Follow manufacturers' recommendations concerning installation procedures.
4. Be sure to install the equipment as it was intended to be installed. Do not alter the equipment. To do so will void the UL rating, for example.
5. Do not install any equipment that does not comply with the codes in effect, that is, in violation of the rating agency or the equipment listing. If a problem should occur with such an installation, you would be the one to lose.
6. Be sure that only qualified people are involved in the installation. Instruct all your people to advise you of any problems or deviations that they encounter so that you can take care of any small problems before they become large.

☐ Avoiding Service Legal Problems ☐

Use the following steps to help avoid legal problems with a service operation:

1. Be sure that all the personnel on the job have the required current certification for the job that they are doing.
2. Use numbered work orders or service orders for all jobs. Make certain that the ticket is filled out, listing all the work done, and is signed by the person who accepted the service.
3. Use only approved or original-equipment parts on the job. If a problem occurs, the investigators are going to be looking for any deviations from normal and could possibly fault the parts that were used.
4. In no circumstances should any safety device be jumpered and left jumpered. To do so leaves you open to a legal battle.
5. Be sure that any problems with the equipment are pointed out to the customer. It is a good idea to list them and make a statement on the service invoice that the customer did not want to have them repaired at this time and have the customer sign it. This will certainly help if trouble occurs later and you are taken to court for not making the repairs.
6. Always practice safety.

☐ What to Do In Case of an Accident ☐

The following steps are listed as an aid to prevent further problems in case of an accident:

1. Do not make any statement other than that the equipment must be left untouched until the inspectors have completed their investigation of the problem.
2. If the law requires you to do so, inform the local inspector of the problem and what you have done.
3. Notify any architects or engineers involved in the job.
4. Notify the equipment distributor or the supplier about the problem. They may have some suggestions or will help with the investigation.

☐ Choosing a Truck ☐

There are many types of trucks available to serve as a service and/or installation truck. The type chosen depends on the desires of the purchaser. In most instances, a van has a few advantages over a pickup with cover or toolboxes. However, it is more difficult to carry large bulky loads in a van. The van will usually keep tools and supplies drier than will a pickup, although with the toolboxes currently available, this is not always true.

Regardless of the type of truck chosen, it should always be kept clean and in good repair. It is not necessary to have a new truck as long as it is dependable. There are few things that will upset a customer more than having a service truck in his or her driveway that will not start or one that leaves a puddle of oil where it was setting. Sometimes it is necessary to drive a truck that needs repairs. In such a case the truck should be parked in the street.

☐ Stocking the Truck ☐

Since this is a seasonal business, the type of stock carried on the truck will depend largely on the season. It would not be wise to carry heating supplies during the cooling season, or vice versa, but it is a good idea to carry a permanent stock that can be used during every season, such as transformers, fuses, filters, and other items that are needed throughout the year. At the start of each season replace the parts used in the preceding season with parts that will be needed during the current season.

The type of stock carried will depend largely on the type of equipment on which you work. If you specialize in only one brand of equipment, this will reduce the amount of truck inventory that you have. However, this will also limit the amount of work that you will get. It is better to carry a wide variety of supplies that can be used on most brands and make the necessary trips to the supply house for the seldom-used items. The truck stock is money that you have tied up and cannot use until a part is sold. Therefore, reducing the truck inventory as much as possible, as long as you don't have to make several trips a day to the supply house, will increase the amount of money available to you.

☐ SUMMARY ☐

- ☐ There are no hard-and-fast rules in starting a new business.
- ☐ It takes a special kind of person to be the owner of a small business.
- ☐ A business owner must get along well with people, have a good sense of business practices, know how to handle people, and be expert in the field, especially if it is to be a one-person operation for a while.
- ☐ The owner of a small business must be honest in all dealings with customers.
- ☐ A person who is starting a new business or buying an established one must have enough capital to buy the required tools and equipment and to see him or her through several months of low, or no, income.
- ☐ Will the place chosen to locate the business support the establishment?
- ☐ There are many legal requisites for starting or purchasing a business that must be considered and completed.
- ☐ The owner of any type of business should have a good liability insurance policy.
- ☐ Purchasing an existing business requires much investigation to assure that the purchase is a wise one.
- ☐ There are four major methods of conducting a business: (1) sole proprietorship, (2) general partnership, (3) limited partnership, and (4) corporation.
- ☐ Good human relations involves people getting along with each other harmoniously, attaining a satisfac-

tory level of production, and achieving both economic and social goals.
- ☐ The key to good human relations is proper motivation.
- ☐ Remember, your job is getting things done for people. Your effectiveness will be measured by how customers relate to you and your company. Your success depends on how they judge you.
- ☐ A person who is generally happy and self-confident will probably be seen as a good business manager.
- ☐ When a person begins to know himself or herself, he has taken the first step toward understanding others.
- ☐ Attempting to understand yourself involves taking a good look at both your strong points and weaknesses.
- ☐ A major task facing you as a business owner is to understand why a particular customer or employee behaves the way he or she does, and not turn off the person's actions as absurd or ridiculous.
- ☐ To be good at human relations, it is almost impossible for you to know too much about the people with whom you associate.
- ☐ The total concept of human relations is that something takes place between two or more people.
- ☐ Practicing good human relations does not necessarily mean that everyone will be all smiles and that happiness will abound in all corners.
- ☐ The practice of good human relations often means that a difficult problem can be detected, and usually avoided, before it really starts.
- ☐ The name given to a new business should be descriptive of what the company does.
- ☐ The name should be registered with the local or state agency for protection against its use by someone else.
- ☐ Some states require that air conditioning and refrigeration contractors be licensed.
- ☐ A lot of money can be spent on advertising without getting much response. Pick the method that works for you.
- ☐ Quantity purchasing should be approached with caution. Otherwise, you could end up with a lot of equipment and supplies that are difficult to move out of inventory.
- ☐ To be a good salesperson, you must be a good negotiator.
- ☐ An unpaid bill is one of the greatest losses a small business owner has. One unpaid bill may take several other jobs to make up for.
- ☐ A good bank is a necessity to any business. Regardless of how large a company is, there will be times when a loan is needed to keep the company going.
- ☐ When bidding for a job, be sure that all the equipment cost is figured into the job along with the overhead and the profit. Otherwise, a loss could oc-

cur which might have a very bad effect on the company.

☐ There are certain steps that should be used to avoid legal problems when doing both installation and service work.

☐ Choose the type of truck that fits your needs best. Keep the truck clean and in good repair. Stock the truck with only what is needed.

☐ EXERCISES ☐

1. What type of person does it take to be the owner of a successful small business?
2. How much capital should a person contemplating going into business have in hand?
3. Where should a new business be located?
4. What kind of insurance should a small business operator have?
5. What should be done before an existing business is purchased?
6. Name the four types of forms for conducting a business.
7. Why is just getting the job done not always all that is necessary?
8. What do some people think good human relations consist of?
9. What does good human relations mean?
10. Name the four things that as a technician you should remember about people.
11. How will your effectiveness as a businessperson be measured?
12. What must you do if you want to understand other people?
13. What is empathizing?
14. What is a major task facing you as a technician?
15. What is necessary for a person to be good at human relations?
16. What is the total concept of human relations?
17. Of what should a company name be descriptive?
18. What should be done with the company name after it is chosen?
19. What type of advertising is best for a company?
20. Is quantity purchasing always the best way to obtain supplies?
21. To be a good salesperson, what else should you be?
22. What is the best way to collect for a job?
23. Why is it cheaper to have the technician collect while on the job?
24. What should be done when the contract for a large job is signed?
25. How should a bank be approached?
26. What are four major things that a person making a bid should be sure to include?
27. In large part, what determines the price of the job?
28. What should be given to the customer when he or she signs a contract?
29. What type of equipment should be used on the job?
30. What type of statement should be made to the customer concerning an accident on the job site?

Glossary

Absolute humidity: The weight in grains of water vapor actually contained in 1 ft³ (0.0283 m³) of the air and moisture mixture.

Absolute pressure: The sum of gauge pressure plus atmospheric pressure.

Absolute temperature: The temperature at which all molecular motion of a substance stops. At this temperature the substance theoretically contains no heat.

Absorbent: A substance that has the ability to absorb another substance.

Absorptivity: The ratio of radiant energy absorbed by an actual surface at a given temperature to that absorbed by a blackbody at the same temperature.

Accessible hermetic: A single unit containing the motor and compressor; this unit may be field serviceable.

Accumulator: A shell placed in the suction line to prevent liquid refrigerant from entering the compressor by vaporizing the liquid.

Acid condition: A condition in which the refrigerant and/or oil in the system is contaminated with other fluids that are acidic in nature.

ACR tubing: Tubing used in the refrigeration and air conditioning industry. The ends are sealed and the tubing is dehydrated.

Activated alumina: A chemical made up of aluminum oxide used as a desiccant for refrigeration driers.

Activated carbon: A specially processed carbon commonly used to clean air.

Actuator: That part of a regulating valve that changes fluid or thermal or electric energy into mechanical motion to open or close valves, dampers, etc.

Adiabatic compression: The compression of a gas without the addition or removal of heat.

Aeration: The combining of a substance with air.

Agitator: A device used to induce motion in a confined fluid.

Air: An elastic gas. Air is a mechanical mixture of oxygen and nitrogen and slight traces of other gases; it also may contain moisture known as humidity. Dry air weighs 0.075 lb/ft³ (0.0012 g/m³). One Btu (252 cal) will raise the temperature of 55 ft³ (1.56 m³) of air 1 °F (0.56 °C).

Air change: The number of times in an hour the air in a room is changed either by mechanical means or by the infiltration of outside air leaking into the room through cracks around doors, windows, and other openings.

Air cleaner: A device designed to remove airborne impurities such as dust, fumes, and smoke. Air cleaners include air washers and air filters.

Air coil: The coils on refrigeration systems and air conditioning units through which air is blown.

Air conditioning: The simultaneous control of the temperature, humidity, air motion, and air distribution within an enclosure. Where human comfort and health are involved, a reasonable air purity with regard to dust, bacteria, and odors is also included. The primary requirement of a good air conditioning system is a good heating system.

Air conditioning unit: A device designed for the treatment of air; it consists of a means for ventilation, air circulation, air cleaning, and heat transfer with a control to maintain the temperature within the prescribed limits.

Air diffuser: An air distribution element or outlet designed to direct the flow of air in the desired patterns.

Air handler: A unit consisting of the fan or blower, heat transfer element, filter, and housing components of an air distribution system.

Air infiltration: The leakage of air into a house through cracks and crevices and through doors, windows, and other openings by wind pressure and/or temperature difference.

Air-sensing thermostat: A thermostat that operates as a result of air temperature. The sensing bulb is located in the airstream.

Air, standard: Air whose temperature is 68°F (20°C), whose relative humidity is 36%, and whose pressure is 14.7 psi (101.32 kPa).

Algae: A low form of plant life found in water. It is especially troublesome in water towers.

Alternating current: Electric current in which the direction of current reverses periodically, usually 60 times per second. Also called 60-cycle current.

Altitude correction: The change in atmospheric pressure at higher altitudes has the same effect on a vapor pressure bellows as does a barometric change. The effects of altitude are greater and of a permanent nature so that adjustments may be made in the control settings to correct for the effect of altitude.

Ambient temperature: The temperature of the air that surrounds an object.

Ammeter: An electric meter calibrated in amperes that is used to measure current flow in a circuit.

Amperage: The unit of electric current equivalent to a flow of one coulomb per second.

Amplifier: An electric device that increases the electron flow in a circuit.

Analyzer: A device used to check the condition of electrical components.

Anemometer: An instrument used to measure the velocity of air motion.

Anneal: To heat treat metal in order to obtain the desired softness and ductility.

Anode: The positive terminal of an electrolytic cell or battery.

Apparatus dew point: The dew point of the air leaving the air conditioning coil.

Armature: Part of an electric motor or generator that is moved by magnetism.

ASTM standards: Standards set by the American Society for Testing and Materials.

Atmospheric pressure: The pressure exerted by the atmosphere in all directions, as indicated by a barometer. Standard atmospheric pressure at sea level is considered to be 14.7 lb/in.² (101.32 kPa).

Atom: The smallest part of a substance that can exist alone or in combination with other elements.

Atomize: To change a liquid to small particles or fine spray.

Automatic control: The control of the various functions of a piece of equipment without manual adjustment.

Automatic defrost: A method of removing ice and frost from the evaporator automatically.

Automatic expansion valve (AEV): A pressure-controlled valve that reduces the high-pressure liquid refrigerant to a low-pressure liquid and vapor refrigerant.

Automatic ice maker: A refrigeration unit designed to produce a quantity of ice automatically.

Back pressure: The pressure in the low-pressure side of the refrigeration system.

Back seating: The wide-open position of a service valve that usually closes off the gauge connection fitting.

Baffle: A plate or wall of deflecting gases or fluids.

Ball check valve: A valve assembly that allows the flow of fluid in only one direction.

Barometer: An instrument that measures atmospheric pressure.

Battery: An electricity-producing device that uses the interaction of metals and chemicals to create a flow of electric current.

Bearing: A low-friction device used for supporting and aligning moving parts.

Bellows: A corrugated cylindrical container that expands and contracts as the pressure inside changes to provide a seal during the movement of parts.

Bellows seal: A means of preventing leakage of a fluid by rotating parts. The sealing material expands and contracts with a change in pressure.

Bending spring: A spring put on the inside or the outside of a tube to prevent collapse during the bending process.

Bimetal strip: A temperature-regulating device that works on the principle that two dissimilar metals have unequal expansion rates upon a change in temperature.

Bleed: To release pressure slowly from a system or cylinder by opening a valve slightly.

Bleed valve: A valve with a small opening through which pressure is permitted to escape at a slow rate when the valve is closed.

Boiler: A closed vessel in which steam is generated or in which water is heated.

Boiler heating surface: The area of the heat transfer surfaces in contact with the water (or steam) in the boiler on one side and the fire or hot gases on the other.

Boiling point: The boiling temperature of a liquid under a pressure of 14.7 psi (101.32 kPa).

Bourdon tube: A thin-walled tube of elastic metal used in pressure gauges. It is flattened and bent into a circular shape and tends to straighten as the pressure inside is increased.

Boyle's law: A law of physics that deals with the volume of gases as the pressure varies. If the temperature remains constant, the volume varies. If the pressure is increased, the volume is decreased; if the pressure is reduced, the volume is increased.

Braze: To join metals with a nonferrous filler using heat between the temperature of 800°F (427°C) and the melting point of the base metal.

Breaker strip: A strip of plastic used to cover the opening between the case and the liner of a refrigerator or freezer.

British thermal unit (Btu): The quantity of heat required to raise the temperature of 1 lb (453.6 g) of water 1°F

(0.56 °C). (This measurement is approximate but is sufficiently accurate for any work discussed in this book.)

Burner: A device in which the mixing of gas and air occurs.

Butane: A liquid hydrocarbon commonly used as a heating fuel.

Bypass: A pipe or duct usually controlled by a valve or a damper for short circuiting the flow of a fluid.

Calibrate: To position the pointers; also, needles used to determine accurate measurements.

Calorie: The calorie, as used in engineering, is a large heat unit, and is equal to the amount of heat required to raise 1 kg of water 1 °C.

Capacitance: The property of an electric current that permits the storage of electric energy in an electrostatic field and the release of that energy at a later time.

Capacitor: An electric device capable of storing electric energy.

Capacitor-start motor: A motor that uses a capacitor in the starting circuit to increase the starting torque.

Capacity: The capacity of a refrigeration unit is the heat-absorbing capacity per unit of time. Usually measured in Btu/hr.

Capillary tube: A tube of small diameter used to regulate the flow of liquid refrigerant into the evaporator.

Carbon dioxide: A compound made up of carbon and oxygen that is sometimes used as a refrigerant. Its refrigerant number is R-744.

Carbon monoxide: A colorless, odorless, and poisonous gas produced when hydrocarbon fuels are burned with too little air.

Carbon tetrachloride: A colorless, nonflammable, and very toxic liquid sometimes used as a solvent. It should never be allowed to touch the skin and the fumes should not be inhaled.

Cathode: The negative terminal of an electrical device. The electrons leave the device through this terminal.

Cavitation: A gaseous condition found in low-pressure places of a liquid stream.

Celsius (Centigrade): A type of thermometer on which the freezing temperature is 0 ° and the boiling temperature is 100 ° when water is the medium tested.

Centimeter: A metric unit of linear measurement equal to 0.3937 in.

Central fan system: A mechanical indirect system of heating, ventilating, or air conditioning consisting of a central plant where the air is conditioned and then circulated by fans or blowers through a system of distributing ducts.

Centrifugal compressor: A pump that uses centrifugal force to compress vaporous refrigerants.

Change of state: The changing of a substance from one state to another, such as from a liquid to a vapor or a solid to a liquid.

Charge: The amount of refrigerant in a system. Also, to put the refrigerant into a system.

Charging board: A specially designed panel fitted with gauges, valves, and refrigerant cylinders used for evacuating and charging refrigerant and oil into a system.

Charles' law: States that the volume of a given mass of gas at a constant pressure varies according to its temperature.

Check valve: A device that allows the flow of a fluid in only one direction.

Chill factor: A calculated figure based on the dry-bulb temperature and wind velocity.

Chimney: A vertical shaft that carries vent gases from gas heating equipment to the atmosphere.

Chimney effects: The tendency in a duct or other vertical air passage for air to rise when heated due to its decrease in density.

Circuit: The tubing, piping, or electrical wiring that permits flow from the energy source through the circuit and back to the energy source.

Circuit breaker: A safety device used to protect an electric circuit from overload conditions.

Circuit, parallel: The arrangement of electrical devices so that the current is equally divided between each circuit.

Circuit, pilot: A secondary circuit used to control or signal a device in the main circuit. Usually of a different voltage.

Circuit, series: The arrangement of electrical devices in such a way that the current passes through each of the devices one after the other.

Clearance pocket: The space in the cylinder above the piston at the end of the compression stroke.

Closed circuit: A completed electric circuit through which the electrons are flowing.

Closed cycle: Any cycle in which the refrigerant is used over and over again.

Clutch, magnetic: A clutch used on an automotive compressor to start and stop the compressor on demand from the thermostat.

CO_2 indicator: An instrument used to indicate the percentage of carbon dioxide present in the flue gases.

Coefficient of heat transmission (U): The amount of heat transmitted from air to air in 1 hr/ft² of the wall, floor, roof, or ceiling for a difference in temperature of 1 °F (0.56 °C) between the air inside and the air outside of the wall, floor, roof, or ceiling.

Coefficient of performance (COP): The ratio of work performed to energy used.

Cold: The absence of heat.

Cold junction: The part of a thermocouple that absorbs heat from a flame or ambient air.

Cold storage: The process of preserving perishables on a large scale by means of refrigeration.

Cold wall: A type of refrigerator construction in which the inner liner is used as the cooling surface.

Combustible liquid: A liquid whose flash point is at or

above 140 °F (60 °C); usually considered to be a Class 3 liquid.

Comfort cooling: Refrigeration for comfort as opposed to refrigeration for food storage.

Comfort zone: The range of effective temperatures over which the majority of adults feel comfortable.

Commercial refrigeration: A reach-in or service refrigerator of commercial size with or without the means of cooling.

Commutator: The part of a rotor in an electric motor that conveys the electric current to the motor windings.

Compound gauge: An instrument used to indicate the pressure both above and below atmospheric pressure. Commonly used to measure low-side pressures.

Compression ratio: The ratio of the volume of the clearance space to the total volume of the cylinder. It is also used as the ratio of the absolute suction pressure to the absolute discharge pressure.

Compression system: A refrigeration system in which the pressure-imposing element is mechanically operated. Distinguished from an absorption system that uses no compressor.

Compressor: The device used for increasing the pressure on the refrigerant.

Compressor displacement: The volume in inches (mm) represented by the area of the piston head multiplied by the length of the piston stroke.

Compressor seal: The device that prevents leakage between the crankshaft and the compressor body on an open-type compressor.

Condensate: The moisture resulting from the removal of heat from a vapor to bring it below the dew-point temperature.

Condenser: The vessel or arrangement of pipes in which the vaporized refrigerant is liquefied by the removal of heat.

Condenser fan: The fan that forces air over the condenser coil.

Condensing pressure: The pressure inside the condenser at which the refrigerant vapor gives up latent heat of vaporization and changes to a liquid. This pressure varies with the ambient temperature.

Condensing temperature: The temperature inside the condenser at which the vaporous refrigerant gives up latent heat of vaporization and becomes a liquid. This temperature varies with the pressure.

Condensing unit: Usually considered as the high side; consists of the compressor, condenser, receiver, fan, and motor, all on a suitable frame, with the necessary accessories.

Conduction: The transmission of heat through and by means of matter.

Conductivity: The amount of heat in Btu (calories) transmitted in 1 hr through 1 ft² (0.09 m²) of a homogeneous material 1 in. (25.4 mm) thick for a dif-

ference in temperature of 1 °F (0.56 °C) between the two surfaces of the material.

Conductor: A substance or body capable of conducting electric or heat energy.

Constrictor: A device used to restrict the flow of a gas or a liquid.

Contaminant: A substance such as dirt, moisture, or some other material that is foreign to the refrigerant or the oil in the system.

Control: Any device used for the regulation of a machine in normal operation—manual or automatic. It is usually responsive to temperature or pressure, but not to both at the same time.

Control module: Contains the PC card, motor, wiring harness, and switches to receive signals from inputs, including several thermistors, interpreting them and reacting properly to allow the cuber to operate.

Control system: All of the components used for the automatic control of a given process.

Control valve: A valve that regulates the flow of a medium that affects the controlled process. The valve is controlled by a remote signal from other devices which use as power pneumatic, electric, or electrohydraulic devices.

Convection: The transmission of heat by the circulation of a liquid or a gas such as air. When it occurs naturally, it is caused by the difference in weight of hotter and colder fluids.

Convector: A surface designed to transfer its heat to the surrounding air by means of convection.

Cooler: A heat exchanger that transfers heat from one substance to another.

Cooling tower: A device that cools air to the wet-bulb temperature by means of evaporation of water.

Copper plating: An abnormal condition that exists when moisture is present in a refrigeration system. It is a result of the copper being electrolytically deposited on the steel parts of the system.

Corrosion: The deterioration of a metal by chemical action.

Coulomb: The amount of electric energy transferred by an electric current at the rate of one ampere per second.

Counter emf: The tendency for an electric current in an induction coil to reverse its flow as the magnetic field changes.

Counterflow: The opposing direction of flow of fluids, the coldest portion of one meeting with the warmest portion of the other.

Coupling: The mechanical device used to join pipes.

Crank throw: The distance between the center line of a main bearing journal and the center of the crankpin or eccentric.

Crisper: A drawer or compartment used to store vegetables. It helps to maintain the desired humidity for the storage of vegetables.

Critical pressure: The compressed condition of a refrigerant at which the vapor and liquid have the same properties.

Critical temperature: The temperature above which a vapor cannot be liquefied.

Cross-charged: The combining of two fluids to create a desired pressure–temperature curve.

Cuber analyzer: A device used for checking the control module and probes. Thermistor probe signals are substituted by operating switches in the analyzer, which send all combinations of signals to the card. Thus a defective component can be identified by the process of elimination. This can also be used to dry cycle an electronic cuber for demonstration or schools.

Current: The flow of electric energy in a conductor that occurs when the electrons change positions.

Current relay: A device used to start an electric motor. It is caused to function because of the change in current flow in the circuit.

Cut-in: The temperature or pressure at which a set of contacts completes an electric circuit.

Cut-out: The temperature or pressure at which a set of electrical contacts opens an electric circuit.

Cycle: The complete course of operations of a refrigeration system. This includes the four major functions of compression, condensation, expansion, and evaporation.

Cylinder: The chamber in a compressor in which the piston travels to compress the refrigerant vapor.

Cylinder head: A cap or plate that encloses the open end of a cylinder.

Dalton's law: States that a vapor pressure exerted in a container by a mixture of gases is equal to the sum of the individual vapor pressures of the gases contained in the mixture.

Damper: A device used to control the flow of air.

Deaeration: The act of separating air from a substance.

Decibel: A unit commonly used to express sound or noise intensity.

Decomposition: Spoilage.

Defrost: To remove accumulated ice from a cooling coil.

Defrost cycle: The refrigeration cycle in which the ice accumulation is melted from the coil.

Defrost timer: A device connected into the electric circuit to start the defrost cycle and keep it on until the ice has melted from the evaporator.

Degree: The unit of measure on a temperature scale.

Degree day: A unit that represents one degree of difference between the inside temperature and the average outdoor temperature for one day. Normally used for estimating the fuel requirements of a building.

Degree of superheat: The difference between the boiling point of the refrigerant and the actual temperature above the boiling point.

Dehumidifier: A device used to lower the moisture content of the air passing through it.

Dehumidify: To remove water or moisture from the atmosphere; to remove water vapor or moisture from any material.

Dehydrated oil: A lubricant from which the moisture has been removed to an acceptable level.

Dehydrator-receiver: A small tank that serves as a liquid receiver. It also contains a desiccant for removing moisture from the refrigerant and is used on automobile air conditioning systems.

Deice control: A device used to control the compressor and allow for the melting of any accumulation of frost on the evaporator.

Density: The weight per unit volume of a substance.

Deodorizer: A device that absorbs or adsorbs odors. Activated charcoal is usually used for this purpose.

Desiccant: A substance used to collect and hold moisture in a refrigeration system. The most commonly used desiccants are activated alumina and silica gel.

Design pressure: The highest pressure expected to be reached during normal operation. It is usually the operating pressure plus a safety factor.

Dew point: The temperature at which a vapor begins to condense. Usually 100% relative humidity.

Diaphragm: A flexible material, usually a thin metal or rubber, used to separate chambers.

Dichlorodifluoromethane: A halocarbon refrigerant commonly known as R-12.

Differential: The difference between the cut-in and cut-out of a pressure or temperature control.

Direct expansion evaporator: An evaporator that uses either an automatic or a thermostatic expansion valve.

Distributor tester: Also known as a field analyzer, this tester will test individual components, the control module, status indicator, and all four probes individually without being mounted on the cuber. This tester is designed as a bench tester for a shop, or a countertop tester for large dealers and distributors handling many spare parts. This tester can also be used to dry cycle an electronic cuber for demonstration or school.

Double-duty case: A commercial refrigerator in which part of the space is used for storage and part is used for display of goods.

Double pole: The term used to designate the contact arrangement that includes two separate contact forms, i.e., two single-pole contact assemblies.

Double throw: A term applied to a contact arrangement to denote that each contact form included is a break-make, i.e., one contact opens its connection to another contact and then closes its connection to a third contact.

Draft gauge: An instrument used to measure air movement by measuring the air pressure differences.

Draft indicator: An instrument used to measure chimney draft or combustion gas movement.

Draft regulator: A device that keeps the draft in a vent pipe at the desired volume.

Drier: A device used to remove moisture from a refrigerant.

Drip pan: A pan or trough used to collect condensate from an evaporator coil.

Dry bulb: A thermometer used to measure the ambient air temperature.

Dry-bulb temperature: The actual temperature of the air, as opposed to wet-bulb temperature. The temperature of the air indicated by any thermometer not affected by the moisture content of the air.

Dry ice: Compressed carbon dioxide.

Dry system: A refrigeration system in which only droplets are present in the evaporator.

Duct: A tube or channel through which air is forced in a forced air system.

Eccentric: A circle or disc mounted off center on a shaft; used to produce a reciprocating motion.

Eddy currents: The induced currents flowing in a laminated core.

Effective area: The gross area of the grill minus the area of the vanes or bars. The actual flow area of an air inlet or outlet.

Effective temperature: The overall effect on a human being of the air temperature, humidity, and air movement.

Electric defrosting: A method of defrosting an evaporator by using electric current to heat the surface.

Electric heating: A heating system in which the source of heat is electricity.

Electrolysis: The chemical reaction of two substances due to the flow of electricity through them.

Electrolytic capacitor: A plate or surface capable of storing small electrical charges.

Electromagnet: A coil of wire wound around a soft shaft or core. When electric current flows through the coil an electromagnet is formed.

Electromotive force (emf): The voltage or electrical force that causes the free electrons to flow in a conductor.

Electron: A basic component of an atom; it carries a negative electrical charge.

Electronic leak detector: An electronic instrument that senses refrigerant vapor in the atmosphere. An electronic flow change indicates a leak.

End bell: The end structure of an electric motor; it usually holds the bearings, etc.

End play: The slight movement of a shaft along its center line.

Energy: The ability to do work.

Enthalpy: The actual or total heat contained in a substance. It is usually calculated from a base. In refrigeration work the base is accepted as $-40\,°F$ $(-40\,°C)$.

Entropy: The energy in a system. A mathematical factor used in engineering calculations.

Environment: The conditions of the surroundings.

Epoxy: An adhesive formed by mixing two resins.

Eutectic: That certain mixture of two substances that provides the lowest melting temperature of all the various mixes of the two substances.

Evacuation: The removal of air and moisture from a refrigeration system.

Evaporation: The change of state of a liquid to a gas. The greatest amount of heat is absorbed in this process.

Evaporative condenser: A specially designed condenser used to remove heat from refrigerant gas by using the cooling effect of evaporating water.

Evaporator: The part of a refrigeration system where the refrigerant boils and picks up the heat.

Evaporator fan: The fan that forces the air through the evaporator.

Exhaust valve: The outlet port that allows the compressed vapor to escape from the cylinder; also called the discharge valve.

Expansion valve: The device in a refrigeration system that reduces a high-pressure refrigerant to a low-pressure refrigerant.

External equalizer: The tube connected to the low-pressure side of the diaphragm in the thermostatic expansion valve. It lets the evaporator outlet pressure help control the operation of the valve.

Fahrenheit scale: The scale on a standard thermometer. It has the boiling point of water at $212\,°F$ and the freezing point at $32\,°F$.

Fail-safe control: A device that opens an electric current when the sensing element senses an abnormal condition.

Fan: An enclosed propeller that produces motion in air. Commonly designates anything that causes air motion.

Farad: The unit of electric capacity of a capacitor.

Female thread: The internal thread on valves, fittings, etc.

Field pole: That part of the motor stator that concentrates the magnetic field in the field winding.

Filter: A device used to remove the solid particles from a fluid by straining.

Fin: The sheet metal extension on evaporator or condenser tubes.

Finned tube: A tube built up with an extended surface in the form of fins.

Flammability: The ability to burn.

Flammable liquids: Liquids whose flash points are below $140\,°F$ $(60\,°C)$, and whose vapor pressure are less than 40 psi (375.79 kPa) at $100\,°F$ $(37.8\,°C)$.

Flapper valve: A thin metal valve used as the suction and discharge valves in refrigeration compressors.

Flare: An enlargement on the end of a piece of copper tubing used to connect the tubing to a fitting or another piece of tubing. It is usually formed at about a 45° angle.

Flare fitting: A type of soft tubing connector that requires that the tube be flared to make a mechanical seal.

Flare nut: The fitting placed over the tubing to clamp the flared tubing against another fitting.

Flash gas: The gas that is the result of the instantaneous evaporation of refrigerant in a pressure-reducing device. It cools the remaining refrigerant to the evaporation temperature, which exists at the reduced pressure.

Flash point: The temperature of a combustible material at which there is enough vaporization to ignite the vapor, but not enough to support combustion.

Float valve: A valve that is actuated by a float in a liquid chamber.

Flood: To allow liquid refrigerant to flow into a part of the system.

Flooded system: A system in which the refrigerant enters the evaporator from a pressure-reducing valve and in which the evaporator is partly filled with liquid refrigerant.

Flow meter: An instrument used to measure the velocity or volume of fluid movement.

Flue: A gas or air passage through which the products of combustion escape to the atmosphere.

Fluid: A gas or liquid. Commonly defined as matter in any state that flows and takes the shape of its container.

Flush: To remove foreign materials or fluids from a refrigeration system through the use of refrigerant or other fluids.

Flux: A substance applied to the surfaces to be joined by brazing or soldering to keep oxides from forming during the joining process.

Foaming: The formation of foam in the crankcase as a result of liquid refrigerant being absorbed in the oil and rapidly evaporating. This is most likely to occur on compressor startup.

Foam leak detector: Soap bubbles or a special foaming liquid brushed over suspected areas to locate leaks.

Force: Accumulated pressure, expressed in pounds (grams).

Forced convection: The movement of a fluid by a force such as a fan or pump.

Forced-feed oiling: A lubrication system that uses a pump to force the oil to the surfaces of the moving parts.

Freezer: Any device for freezing perishable foods.

Freezer alarm: A bell or buzzer used on freezers to indicate when the storage temperature rises above a safe temperature.

Freeze-up: The formation of ice in the flow control device as a result of the presence of moisture in the system. Also, a frost formation on an evaporator which reduces the flow of air through the fins.

Freezing: The change of state from a liquid to a solid.

Freezing point: The temperature at which a liquid will change to a solid when heat is removed. The temperature at which a given substance freezes.

Freon: The tradename for a family of fluorocarbon refrigerants manufactured by E.I. DuPont de Nemours & Co., Inc.

Frostback: The floodng of liquid refrigerant from the evaporator into the suction line. This condition may or may not be accompanied by frost.

Frost-free refrigerator: A domestic refrigerator that operates with a predetermined defrost period.

Frozen: Term applied to water in the solid state. Also, seized due to the lack of lubrication, as frozen bearings, etc.

Fuel oil: A hydrocarbon oil as specified by U.S. Department of Commerce Commercial Standard CS12 or ASTM D296 whose flash point is not less than 100 °F (37.8 °C).

Furnace: That part of a heating plant in which the combustion process takes place.

Fuse: An electrical safety device consisting of a strip of fusible metal used to prevent overloading of an electric circuit.

Fusible: Capable of being melted.

Fusible plug: A safety plug used in refrigerant containers that melts at a high temperature to prevent excessive pressure from bursting the container.

Gas: The vapor state of a substance.

Gasket: A resilient or flexible material used between mating surfaces to provide a leak-proof seal.

Gas valve: A device in a pipeline that controls the flow of gas to a burner on demand from the thermostat.

Gauge: An instrument used for measuring pressures both above and below atmospheric pressure.

Gauge manifold: A manifold that holds both the pressure and compound gauges, the valves which control the flow of fluids through the manifold ports, and charging hose connections.

Gauge port: The opening or connection provided for the installation of gauges.

Gauge pressure: A measure of pressure taken with a gauge. Pressure measured from atmospheric pressure as opposed to absolute pressure.

Grain: A unit of weight equal to 1/7000 lb (0.06480 g) used to indicate the amount of moisture in the air.

Grill: A perforated covering for an air inlet or outlet usually made of wire screen, processed steel, or cast iron.

Grommet: A plastic or rubber doughnut-shaped protector placed in holes where wires or lines pass through panels to prevent damage to the wires or lines.

Ground coil: A heat exchanger that is buried in the ground; usually the outdoor coil on a heat pump system.

Ground wire: An electrical wire that will safely conduct electricity from a structure or piece of equipment to the ground in case of an electrical short.

Halide refrigerant: The family of synthetic refrigerants that contain halogen chemicals.

Halide torch: A device used to detect refrigerant leaks in a system. A burner equipped with a source of fuel, a mixing chamber, a reactor plate, and an exploring tube. The reactor plate surrounds the flame. When the open end of the exploring tube is held near a refrigerant leak, some of the refrigerant is drawn into the mixing chamber where its presence changes the color of the flame.

Halogens: Substances that contain fluorine, chlorine, bromine, and iodine.

Hanger: A device used to support lines by attachment to a wall.

Head: Unit of pressure usually expressed in feet of water.

Header: A length of pipe large enough to carry the total volume to which several pipes are connected; used to carry fluid to the various points of use.

Head pressure: The pressure against which the compressor must deliver the gas.

Head pressure control: A pressure-operated control that opens the electric circuit when the head pressure exceeds preset limits.

Head velocity: The height of a flowing fluid that is equivalent to its velocity pressure.

Heat: A form of energy produced through the expenditure of another form of energy.

Heat content: The amount of heat, usually stated in Btu (calories) per pound, absorbed by a refrigerant in raising its temperature from a predetermined level to a final condition and temperature. Where change of state is encountered, the latent heat necessary for the change is included.

Heat exchanger: Any device that removes heat from one substance and adds it to another.

Heating coil: A coil of piping used to transfer heat from a liquid.

Heating control: Any device that controls the transfer of heat.

Heating medium: A substance used to convey heat from the boiler, furnace, or other source of heat to the heating unit where the heat is dissipated.

Heating surface: The exterior surface of a heating unit.

Heating value: The amount of heat that may be released by the expenditure of energy. It is usually expressed in Btu/lb, Btu/gal, or cal/g.

Heat intensity: The heat concentration in a substance; indicated by the temperature of the substance through the use of a thermometer.

Heat lag: The amount of time required for heat to travel through a substance heated on one side and not on the other.

Heat leakage: The flow of heat through a substance when a difference in temperature exists.

Heat load: The amount of heat, measured in Btu (calories) or watts, that must be added or removed by a piece of equipment during a 24-hr period.

Heat of compression: The heat developed within a compressor when a gas is compressed as in a refrigeration system.

Heat of fusion: The amount of heat required to change a solid to a liquid or a liquid to a solid with no change in temperature; also called latent heat of fusion.

Heat of the liquid: The heat content of a liquid or the heat necessary to raise the temperature of a liquid from a predetermined level to a final temperature.

Heat of the vapor: The heat content of a gas or the heat necessary to raise the temperature of a liquid from a predetermined level to the boiling point plus the latent heat of vaporization necessary to convert a liquid to a gas.

Heat pump: A compression cycle system used to supply heat to a space by reversing the flow of refrigerant from the cooling cycle.

Heat sink: A relatively cold surface that is capable of absorbing heat usually used as control points.

Heat transfer: The movement of heat from one body to another. The heat may be transferred by radiation, conduction or convection.

Heat unit: Usually refers to a Btu (calorie).

Hermetic compressor: A unit in which the compressor and motor are sealed inside a housing. The motor operates in an atmosphere of refrigerant.

Hertz (Hz): The correct terminology for designating electrical cycles per second.

Hg (mercury): A silver-white heavy liquid metal. It is the only metal that is a liquid at room temperature.

High-pressure cut-out: An electrical control switch operated by the pressure in the high-pressure side of the system that automatically opens an electric circuit when a predetermined pressure is reached.

High side: The part of the refrigeration system that contains the high-pressure refrigerant. Also refers to the condensing unit, which consists of the motor, compressor, condenser, and receiver mounted on one base.

High-side charging: The process of introducing liquid refrigerant into the high side of a refrigeration system. The acceptable manner for placing the refrigerant into the system.

High-vacuum pump: A vacuum pump capable of creating a vacuum in the range of 1000 to 1 microns.

Holding charge: A partial charge of refrigerant placed in a piece of refrigeration equipment after dehydration and evacuation either for shipping or testing purposes.

Holding coil: That part of a magnetic starter or relay that causes the device to operate when energized.

Horsepower: A unit of power. The effort necessary to raise 33,000 lb (14,969 kg), a distance of 1 ft (0.304 m) in one minute.

Hot gas: The refrigerant gas leaving the compressor.

Hot-gas bypass: A connection from the compressor discharge directly into the suction side of a compressor. Sometimes used as a capacity control.

Hot-gas defrost: A method of evaporator defrosting that uses the hot discharge gas to remove frost from the evaporator.

Hot-gas line: The line that carries the hot compressed vapor from the compressor to the condenser.

Hot junction: The part of a thermocouple that releases heat.

Hot water heating system: A heating system in which water is the medium by which heat is carried through pipes from the boiler to the heating units.

Hot wire: A resistance wire used in starting relay that expands when heated and contracts when cooled. Also, an electrical lead whose voltage differs from that of the ground.

Humidifier: A device used to add vapor to the air or to any material.

Humidistat: An automatic control that is sensitive to humidity and is used for the automatic control of relative humidity.

Humidity: Moisture in the air.

Humidity, absolute: The weight of water vapor per unit volume of space occupied, expressed in grains of moisture per cubic foot of dry air.

Humidity, relative: The amount of moisture in the air expressed in terms of percentage of total saturation of the existing dry-bulb temperature.

Hunting: The fluctuation caused by the controls attempting to establish an equilibrium against difficult conditions.

Hydraulics: A branch of physics dealing with the mechanical properties of water and other liquids in motion.

Hydrocarbon: An organic compound containing only hydrogen and carbon atoms in various combinations.

Hydrometer: An instrument used to measure the specific gravity of a liquid by use of floating elements.

Hydronic: Term applied to the process of heating and cooling using water as the heat transfer medium.

Hygrometer: An instrument used to measure relative humidity.

Hygroscopic: Term describing the ability of a substance to absorb and release moisture and to change its physical dimensions as its moisture content changes.

ICC (Interstate Commerce Commission): A governmental body that controls the design, construction, and shipping of pressure containers.

Ice cream cabinet: A commercial refrigerator that maintains approximately 0 °F (-17.8 °C) inside temperature and is used for the storage of ice cream.

Ice harvest switch: Located at the end of the water plate, it resets the timer on the status indicator card after each ice harvest. It is used only with a status indicator.

Ice water thickness: A commercial classification applied to molded cork covering and similar insulation for refrigerant and chilled water lines. Somewhat thinner than "brine thickness."

Idler: A pulley used on some belt drives to allow belt adjustment and to eliminate belt vibration.

Ignition transformer: A transformer designed to provide a high voltage. Normally used to ignite the gas at a pilot burner or to ignite fuel oil at the burner by providing a high intensity spark.

IME (Ice melting equivalent): The amount of heat absorbed in melting ice at 32 °F (0 °C) per pound of ice. It is equal to 144 Btu (36.29 kcal) per pound or 288,000 Btu (72,576 kcal) per day.

Impedance: The opposition in an electric circuit to the flow of an alternating current; similar to electrical resistance in a direct current circuit.

Impeller: The rotating part of a pump that causes the water to flow.

Induced-draft cooling tower: A cooling tower in which the flow of air is created by one or more fans drawing the saturated air out of the tower.

Induced magnetism: Magnetism caused in a metal by magnetic induction.

Inductive reactance: The electromagnetic induction in a circuit that creates a counter emf as the applied current changes. It is in opposition to the flow of alternating current.

Infiltration: The leakage of air into a building or space.

Inhibitor: A substance that prevents a chemical reaction such as corrosion or oxidation to metals.

Instrument: A term broadly used to designate a device that is used for measuring, recording, indicating, and controlling equipment.

Insulation, electrical: A substance that has almost no free electrons.

Insulation, thermal: A material that has a high resistance to heat flow.

Interlock: A device that prevents certain parts of an air conditioning or refrigeration system from operating when other parts of that system are not operating.

Ion: A group of atoms or an atom that is either positively or negatively charged electrically.

IR drop: An electrical term indicating the voltage drop in a circuit. It is expressed as amperes times resistance ($I \times R$).

Isothermal: Term describing a change in volume or pressure under constant-temperature condition.

Isothermal expansion and contraction: Expansion and contraction that takes place without a change in temperature.

Isotron: The refrigerants made by the Pennsalt Chemical Corp.

Jacket: To surround with a bath for temperature control or heat absorption, as in water jacketing.

Joint: The connecting point between two surfaces, as between two pipes.

Journal, crankshaft: That part of a shaft on which the bearing is in contact with the shaft.

Junction box: A box or container that houses a group of electrical terminals or connections.

Kelvin scale: A thermometer scale that is equal to the Celsius scale, and according to which absolute zero is 0 degrees, the equivalent of −273.16°C. Water freezes at 273.16°K and boils at 373.16°K on this scale.

Kilometer (km): A metric unit of linear measurement equal to 1,000 meters.

Kilowatt (kW): A unit of electric power indicating 1000 watts.

Kilowatt hour (kWh): A unit of electrical energy equal to 1000 watt hours.

King valve: A service valve on the liquid receiver outlet.

Lag: A delay in the response to some demand.

Lamp, sterile: A lamp with a high-intensity ultraviolet ray used in food storage cabinets and air ducts to kill bacteria.

Lantern gland (packing): A packing ring inside a stuffing box that has perforations for the introduction or removal of oil.

Lap: To smooth a metal surface to a high polish or accuracy using a fine abrasive. Usually done on compressor valves and shaft seals.

Latent heat: The heat added to or removed from a substance to change its state but which cannot be measured by a change in temperature.

Latent heat of condensation: The heat removed from a vapor to change it to a liquid with no change in temperature.

Latent heat of vaporization: The quantity of heat required to change a liquid to a gas with no change in temperature.

Leak detector: A device used to detect leaks; usually a halide torch, soap bubbles, or an electronic leak detector.

Light-emitting diode (LED): A solid-state semiconducting device that emits colored light when electric power is applied to it.

Limit control: A safety device used in heating equipment to interrupt an electric circuit to the main gas valve when an over temperature condition occurs.

Liquid: A substance in which the molecules move freely among themselves, but do not tend to separate as in a vapor.

Liquid absorbent: A liquid chemical that has the ability to take on or absorb other fluids.

Liquid charge: Usually refers to the power element of temperature controls and thermostatic expansion valves. The power element and remote bulb are sometimes charged with a liquid rather than a gas.

Liquid filter: A very fine strainer used to remove foreign matter from the refrigerant.

Liquid indicator: A device located in the liquid line with a glass window through which the flow of liquid may be observed.

Liquid line: The line carrying the liquid refrigerant from the receiver or condenser-receiver to the evaporator.

Liquid receiver: A cylinder connected to the condenser outlet used to store liquid refrigerant.

Liquid receiver service valve: A two- or three-way manually operated valve located at the receiver outlet used for installation and service operations; also called a king valve.

Liquid sight glass: A glass bull's-eye installed in the liquid line to permit visual inspection of the liquid refrigerant. Used primarily to detect bubbles in the liquid, indicating a shortage of refrigerant in the system. Also, a liquid indicator.

Liquid stop valve: A magnetically operated valve generally used to control the flow of liquid to an evaporator, but may also be used wherever on–off control is permissible. A solenoid electrical winding controls the action of the valve.

Liquid strainer: *See* Liquid filter.

Liquid–vapor valve: A dual hand valve used on refrigerant cylinders to release either gas or liquid from the cylinder.

Liter: A metric unit of volume measurement equal to 61.03 in.^{3}.

Load: The required rate of heat removal. Heat per unit of time that is imposed on the system by a particular job.

Locked-rotor amperes (LRA): The amperage flowing in the circuit to a motor-driven apparatus when the rotor of the motor is locked to prevent its movement. This amperage is typically as much as six times the full-load current of the motor and four to five times the full-load current in hermetic compressors.

Low-pressure control: A pressure-operated switch in the suction side of a refrigeration system that opens its contacts to stop the compressor at a given cut-out setting.

Low side: Composed of the parts of a refrigeration system in which the refrigerant pressure corresponds to the evaporator pressure.

Low-side charging: The process of introducing refrigerants into the low side of the system. Usually reserved for the addition of a small amount of refrigerant after repairs.

Low-side pressure: The pressure in the low side of the system.

LP fuel: Liquefied petroleum used as a fuel gas for heating purposes.

Machine: A piece of equipment, sometimes the total unit.

Machine room: An area where the refrigeration equipment, with the exception of the evaporator, is installed.

Magnetic across-the-line starter: A motor starter or switch that allows full line voltage to the motor windings when engaging.

Magnetic clutch: A pulley operated by electromagnetism to connect or disconnect a source of driving power to a compressor.

Magnetic field: An area in which magnetic lines of force exist.

Magnetic gasket: A rubber-type material with small magnets inside the gasket that keeps the door tightly closed.

Magnetism: The attraction between a magnet and materials made of iron or nickel-coated.

Male thread: The external thread on a pipe, fitting, or valve that makes screw connections.

Manifold: The portion of the refrigerant main in which several branch lines are joined together. Also, a single piece in which there are several fluid paths.

Manifold, discharge: A device used to collect compressed refrigerant from the various cylinders of a compressor.

Manifold, service: A chamber equipped with gauges, manual valves, and charging hoses used in servicing refrigeration units.

Manometer: An instrument for measuring small pressure. Also, a U-shaped tube partly filled with liquid.

Manual shut-off valve: A hand-operated device that stops the flow of fluids in a piping system.

Manual starter: A hand-operated motor switch equipped with an overload trip mechanism.

Mass: A quantity of matter held together to form one body.

Master switch: The main switch that controls the starting and stopping of the entire system.

Mean effective pressure (MEP): The average pressure on a surface exposed to a varying pressure.

Mechanical efficiency: The ratio of work done by a machine to the energy used to do it.

Mechanical refrigeration: A term usually used to distinguish a compression system from an absorption system.

Megohm: A measure of electric resistance. One megohm equals one million ohms.

Megohmmeter: An instrument used to measure extremely high resistances.

Melt: To change state from a solid to a liquid.

Melting point: The temperature, measured at atmospheric pressure, at which a substance will melt.

Mercury bulb: A glass tube that uses a small amount of mercury to make or break an electric circuit.

Met: A measure of the heat released from a human at rest. It is equal to 18.4 Btu/ft²/hr (50 kcal/m²/hr).

Meter: An instrument used for measuring. Also, a unit of length in the metric system.

Metric system: The decimal system of measurement.

Micro: One-millionth of a specified unit.

Microfarad: A unit of electric capacitance equal to 1/1,000,000 of a farad.

Micrometer: A precision instrument for measuring with an accuracy of 0.001 to 0.0001 in (0.0254 to 0.00254 mm).

Micron: A unit of length in the metric system which is equal to 1/1000 mm.

Micron gauge: An accurate instrument used for measuring vacuum that is very close to a perfect vacuum.

Milli-: One-thousandth of a specified unit. For example, millivolt means 1/1000 volt.

Miscibility: The ability of several substances to be mixed together.

Modulating: Refers to a device that tends to adjust by small increments rather than being full-on or full-off.

Modulating control: A type of control system characterized by a control valve or motor that regulates the flow of air, steam, or water in response to a change in conditions at the controller. It operates upon partial degree variations in the medium to which the controller is exposed.

Mol: The unit of weight for mass.

Molecular: Pertaining to or consisting of molecules.

Molecular weight: The average weight of a molecule of a substance; measured against an oxygen atom.

Molecule: The smallest particle of a substance that can exist by itself.

Mollier's diagram: A graph indicating refrigerant pressure, heat, and temperature properties.

Monochlorodifluoromethane: A refrigerant that is better known as Freon-22 or R-22. Its chemical formula is $CHCLF_2$; the cylinder color code is green.

Motor: A device that converts electrical energy to mechanical energy.

Motor burnout: A shorted condition in which the insulation of an electric motor has deteriorated by overheating.

Motor, capacitor: A single-phase induction-type motor that utilizes an auxiliary starting winding (a phase winding) connected in series with a capacitor to provide better starting or running characteristics.

Motor, capacitor-start: An induction motor having separate starting windings. Similar to the split-phase motor except that the capacitor-start motor has an electrical capacitor connected into the starting winding for added starting torque.

Motor, capacitor-start and run: A motor similar to the capacitor-start motor except that the capacitor and start winding are designed to remain in the circuit at all times, thus eliminating the switch used to disconnect the winding.

Motor control: A device used to start and/or stop an electric motor or a hermetic motor compressor at certain temperature and/or pressure conditions.

Motor, shaded pole: A small induction motor with a shading pole used for starting; has a very low starting torque.

Motor, split phase: An induction motor with a separate winding for starting.

Motor starter: A series of electrical switches normally operated by electromagnetism.

Movable contact: The member of a contact pair that is moved directly by the actuating system.

Muffler: A device used in the hot-gas line to silence the compressor discharge surges.

Mullion: A stationary frame member located between two doors.

Mullion heater: An electric heating element mounted in the mullion to keep it from sweating.

Multiple system: A refrigeration system that has several evaporators connected to one condensing unit.

Natural convections: The movement of a fluid caused by temperature differences.

Natural-draft cooling tower: A cooling tower in which the flow of air depends on natural air currents or a breeze. Generally applied where the spray water is relatively hot and will cause some convection currents.

Neoprene: A synthetic rubber that is resistant to refrigerants and oils.

Neutralizer: A substance used to counteract the action of acids.

Neutron: The core of an atom; it has no electrical potential (electrically neutral).

No-frost refrigerator: A low-temperature refrigerator cabinet in which no frost or ice collects on the evaporator surfaces or the materials stored in the cabinet.

Nominal size tubing: Tubing whose inside diameter is the same as iron pipe of the same size.

Noncondensable gas: Any gas, usually in a refrigeration system, that cannot be condensed at the temperature and pressure at which the refrigerant will condense, and therefore requires a higher head pressure.

Nonferrous: Metals and metal alloys that contain no iron.

Nonfrosting evaporator: An evaporator that never collects frost or ice on its surface.

Normal charge: A charge that is part liquid and part gas under all operating conditions.

Normally closed contacts: A contact pair that is closed when the device is in the deenergized condition.

Normally open contacts: A contact pair that is open when the device is in the deenergized condition.

North pole, magnetic: The end of a magnet from which the magnetic lines of force flow.

Odor: The contaminants in the air that affect the sense of smell.

Off cycle: The period when equipment, specifically a refrigeration system, is not in operation.

Ohm (R): A unit of measurement of electrical resistance. One ohm exists when one volt causes one ampere to flow through a circuit.

Ohmmeter: An instrument used to measure electrical resistance in ohms.

Ohm's law: A mathematical relationship between voltage, current, and resistance in an electric circuit. It is stated as follows: voltage (E) = amperes (I) $\times$ ohms (R).

Oil binding: A condition in which a layer of oil on top of liquid refrigerant may prevent it from evaporating at its normal pressure and temperature.

Oil check valve: A check valve installed between the suction manifold and the crankcase of a compressor to permit oil to return to the crankcase but to prevent the exit of the oil from the crankcase upon starting.

Oil, compressor lubricating: A highly refined lubricant made especially for refrigeration compressors.

Oil, entrained: Oil droplets carried by high-velocity refrigerant gas.

Oil equalizer: A pipe connection between two or more pieces of equipment made in such a way that the pressure, or fluid level, in each piece is maintained equally.

Oil filter: A device in the compressor used to remove foreign material from the crankcase oil before it reaches the bearing surfaces.

Oil level: The level in a compressor crankcase at which oil must be carried for proper lubrication.

Oil loop: A loop placed at the bottom of a riser to force oil to travel up the riser.

Oil pressure failure control: A device that acts to shut off a compressor whenever the oil pressure falls below a predetermined point.

Oil pressure gauge: A device used to show the pressure of oil developed by the pump within a refrigeration compressor.

Oil pump: A device that provides the source of power for force-feed lubrication systems in refrigeration compressors.

Oil return line: The line that carries the oil collected by the oil separator back to the compressor crankcase.

Oil separator: A device that separates oil from the refrigerant and returns it to the compressor crankcase.

Oil sight glass: A glass "bull's-eye" in the compressor crankcase that permits visual inspection of the compressor oil level.

Oil sludge: Usually a thick, slushy substance formed by contaminated oils.

Oil trap: A low spot, sag in the refrigerant lines, or space where oil will collect. Also, a mechanical device for removing entrained oil.

On cycle: The period when the equipment, specifically refrigeration equipment, is in operation.

Open circuit: An electric circuit that has been interrupted to stop the flow of electricity.

Open compressor: A compressor that uses an external drive.

Open display case: A commercial refrigerator cabinet designed to maintain its contents at the desired temperatures even though the cabinet is not closed.

Open-type system: A refrigeration system that uses a belt-driven or direct-coupling-driven compressor.

Operating cycle: A sequence of operations under automatic control intended to maintain the desired conditions at all times.

Operating pressure: The actual pressure at which the system normally operates.

Orifice: A precision opening used to control fluid flow.

Output: The amount of energy that a machine is able to produce in a given period of time.

Outside air, fresh air: Air from outside of the conditioned space.

Overload: A load greater than that for which the system or machine was designed.

Overload protector: A device designed to stop the motor should a dangerous overload condition occur.

Overload relay: A thermal device that opens its contacts when the current through a heater coil exceeds the specified value for a specified time.

Oxidize: To burn, corrode, or rust.

Oxygen: An element in the air that is essential to animal life.

Package units: A complete set of refrigeration or air conditioning components located in the refrigerated space.

Packing: A resilient impervious material placed around the stems of certain types of valves to prevent leakage. Also, the slats or surfaces in cooling towers designed to increase the water to air contact.

Packless valve: A valve that does not use a packing to seal around the valve stem.

Partial pressure: A condition occurring when two or more gases occupy a given space and each gas creates a part of the total pressure.

Pascal's law: States that a pressure imposed upon a fluid is transmitted equally in all directions.

Performance: A term frequently used to mean output or capacity, as performance data.

Performance factor: The ratio of heat removed by a refrigeration system to the heat equivalent of the energy required to do the job.

Permanent magnet: A material whose molecules are aligned to enable it to produce its own magnetic field. Also, a piece of metal that has been permanently magnetized.

Phase: A definite part of an operation during a cycle.

Photoelectricity: A physical condition by which an electrical flow is generated by light waves.

Pilot control: A valve arrangement used in an evaporator pressure regulator to sense the pressure in the suction line and to regulate the action of the main valve.

Pilot control, external: A method by which the internal connection of the pilot is plugged and an external connection is provided to make it possible to use an evaporator pressure regulator as a suction stop valve as well.

Pilot tube: A device that measures air velocities.

Piston: A disc that slides in a cylinder and is connected with a rod to exert pressure upon a fluid inside the cylinder.

Piston displacement: The volume displaced in a cylinder by the piston as it travels the full length of its stroke.

Pitch: The slope of a pipe line used to enhance drainage.

Plenum chamber: An air compartment to which one or more pressurized distributing ducts are connected.

Polystyrene: A plastic used as insulation in some refrigeration cabinets.

Potential, electric: An electrical force that causes electrons to move along a conductor or through a resistance.

Potential relay: A starting relay that opens on high voltage on its coil and closes on low voltage.

Potentiometer: A wire-wound coil used for measuring or controlling by sensing small changes in electrical resistances.

Power: The time rate of doing work.

Power element: The sensitive element of a temperature-operated control.

Power factor: The correction factor for the changing of current and voltage values of alternating current electricity.

ppm (parts per million): A unit of concentration in solutions.

Precooler: A cooler used to remove sensible heat before shipping, storing, or processing.

Pressure: The force exerted per unit of area.

Pressure, absolute: The pressure measured above an absolute vacuum.

Pressure, atmospheric: The pressure exerted by the earth's atmosphere. Under standard conditions at sea level, atmospheric pressure is 14.7 psia or 0 psig (101.32 kPa).

Pressure, condensing: *See* Pressure, discharge.

Pressure, crankcase: The pressure in the crankcase of a reciprocating compressor.

Pressure, discharge: The pressure against which the compressor must deliver the refrigerant vapor.

Pressure drop: The loss of pressure due to friction of lift.

Pressure gauge: An instrument that measures the pressure exerted by the contents of a container.

Pressure, gauge: The pressure existing above atmospheric pressure. Gauge pressure is, therefore, 14.7 psi (101.32 kPa) less than the corresponding absolute pressure.

Pressure, heat diagram: A graph representing a refrigerant's pressure, heat, and temperature properties. Also, a Mollier's diagram.

Pressure, limiter: A device that remains closed until a predetermined pressure is reached and then opens to release fluid to another part of the system or opens an electric circuit.

Pressure motor control: A control that opens and closes an electric circuit as the system pressures change.

Pressure-operated altitude valve (POA): A device used to maintain a constant low-side pressure regardless of the altitude of operation.

Pressure-regulating valve: A valve that maintains a constant pressure on its outlet side regardless of how much the pressure varies on the supply side of the valve.

Pressure regulator, evaporator: An automatic pressure-regulating valve installed in the suction line to maintain a predetermined pressure and temperature in the evaporator.

Pressure, saturation: The pressure at which gas at any specific temperature is saturated.

Pressure, suction: The pressure forcing the gas to enter the suction inlet of a compressor.

Pressure switch: A switch operated by a rise or fall in pressure.

Pressure tube: A small line carrying pressure to the sensitive element of a pressure controller.

Pressure water valve: A device that controls water flow through a water-cooled condenser in response to the heat pressure.

Primary control: A device that directly controls the operation of a fuel oil burner.

Prime surface: A heating surface whose heating medium is on one side and whose air or extended surface is on the other side.

Printed circuit card (PC): A circuit card in which the interconnecting wires have been replaced by conductive strips painted or etched onto an insulating board similar to a phonograph.

PROBE: A solid-state sensor housed in an environmentally protective tube or holder with a quick disconnect that plugs into the control module. The sensors used are thermistors, which are sensitive to temperature. Four probes used in the cuber are bin probe, evaporator probe, low-water probe, and high-water probe.

Process tube: A length of tubing fastened to the dome of a hermetic compressor; used when servicing the unit.

Propane: A volatile hydrocarbon that may be used as a refrigerant, but is used primarily as a heating fuel.

Protector, circuit: An electrical device that opens a circuit when an excessive electric load occurs.

Proton: The particle of an atom that has a positive charge.

Psychrometer or wet-bulb hygrometer: An instrument used to measure the relative humidity of air.

Psychrometric chart: A graphic representation of the properties of mixtures and water vapor.

Psychrometric measurement: The measurement of temperature, pressure, and humidity by means of a psychrometric chart.

Psychrometry: The study of water vapor and air mixtures.

Pump: Any one of various machines used to force fluids through pipes from one place to another.

Pump, centrifugal: A pump that produces fluid velocity by centrifugal force.

Pumpdown: The reduction of pressure within a refrigeration system.

Purge: The discharge of impurities and noncondensable gases to the atmosphere.

Pyrometer: An instrument used to measure high temperatures.

Quench: To submerge a hot, solid object in a cooling fluid such as oil.

Quick connect coupling: A device that permits fast and easy connection of two refrigerant lines by use of compression fittings.

Radiant heating: A heating system in which the heating surfaces radiate heat directly into the space to be conditioned.

Radiation: The transmission of heat through space by wave motion.

Ram air: The air forced through the radiator and condenser by the movement of an automobile along the road.

Range: A change within limits of the settings of pressure or temperature control.

Rankine scale: The name given to the absolute Fahrenheit scale. The zero point on this scale is $-460\,°F$.

Rating: The assignment of capacity.

Reactance: The part of impedance of an alternating current due to capacitance or inductance or both.

Receiver, auxiliary: An extra vessel used to supplement the capacity of the receiver when additional storage volume is necessary.

Receiver-drier: A cylinder in an automotive air conditioning system used for storing liquid refrigerant; it also contains a quantity of desiccant and acts as a drier.

Reciprocating: Term describing back-and-forth motion in a straight line.

Recirculated air: Return air that is passed through the conditioning unit before being returned to the conditioned space.

Recording ammeter: An instrument that uses a pen to record on a piece of moving paper the amount of current flowing to a unit.

Recording thermometer: A temperature-sensing instrument that uses a pen to record on a piece of moving paper the temperature of a refrigerated space.

Rectifier, electric: An electrical device that converts alternating current to direct current.

Reed valve: A piece of thin, flat, tempered steel plate fastened to the valve plate.

Refrigerant: A substance that absorbs heat as it expands or evaporates. In general, any substance used as a medium to extract heat from another body.

Refrigerant control: A control that meters the flow of refrigerant from the high side to the low side of the refrigeration system. It also maintains a pressure difference between the high-pressure and low-pressure sides of the system while the unit is running.

Refrigerant tables: Tables that show the properties of saturated refrigerants at various temperatures.

Refrigerant velocity: The movement of gaseous refrigerant required to entrain oil mist and carry it back to the compressor.

Refrigerating capacity: The rate at which a system can remove heat. Usually stated in tons or Btu per hour.

Refrigerating effect: The amount of heat a given quantity of refrigerant will absorb in changing from a liquid to a gas at a given evaporating pressure.

Refrigeration: In general, the process of removing heat from an enclosed space and maintaining that space at a temperature lower than its surroundings.

Refrigeration cycle: The complete operation involved in providing refrigeration.

Register: A combination grill and damper assembly placed over the opening at the end of an air duct to direct the air.

Reheat: To heat air after dehumidification if the temperature is too low.

Relay: A device that is made operative by a variation in the condition of one electric circuit to effect the operation of other devices in the same or another circuit.

Relay, control: An electromagnetic device that opens or closes contacts when its coil is energized.

Relay, thermal overload: A thermal device that opens its contacts when the current through a heater coil exceeds the specified value for a given time.

Relief valve: A valve designed to open at excessively high pressures to allow the refrigerant to escape safely.

Reluctance: A force that works as resistance to the passage of magnetic lines of force.

Remote bulb: A part of the expansion valve. The remote bulb assumes the temperature of the suction gas at the point where the bulb is secured to the suction line. Any change in the suction gas superheat at the point of bulb application tends to operate the valve in a compensating direction to restore the superheat to a predetermined valve setting.

Remote-bulb thermostat: A controller that is sensitive to changes in temperature. A thermostat.

Remote power element control: A control whose sensing element is located separate from the mechanism it controls.

Remote system: A refrigeration system whose condensing unit is located away from the conditioned space.

Resistance: Opposition to the flow of electric current. Measured in ohms.

Resistor: A device offering electrical resistance in an electric circuit for protection or control.

Restrictor: A reduced cross-sectional area of pipe that produces resistance or a pressure drop in a refrigeration system.

Return air: Air taken from the conditioned space and brought back to the conditioning equipment.

Reverse-cycle defrost: A method of reversing the flow of refrigerant through an evaporator for defrosting purposes.

Reversing valve: A valve that reverses the direction of refrigerant flow depending on whether heating or cooling is needed.

Riser: A vertical tube or pipe that carries refrigerant in any form from a lower to a higher level.

Riser valve: A valve used for the manual control of the flow of refrigerant in a vertical pipe.

Rotary blade compressor: A rotary compressor that uses moving blades as cylinders.

Rotary compressor: A mechanism that pumps fluids by means of a rotating motion.

Rotor: The rotating or turning part of an electric motor or generator.

Running time: The amount of time that a condensing unit operates per hour.

Running winding: The electrical winding in a motor through which current flows during normal operation of the motor.

Saddle valve: A valve equipped with a body that can be connected to a refrigerant line. Also, a tap-a-line valve.

Safety control: Any device that allows the refrigeration or air conditioning unit to stop when unsafe pressures or temperatures exist.

Safety factor: The ratio of extra strength or capacity to the calculated requirements to ensure freedom from breakdown and ample capacity.

Safety plug: A device that releases the contents of a container to prevent rupturing when unsafe pressures or temperatures exist.

Safety valve: A quick opening safety valve used for the fast relief of excessive pressure in a container.

Saturation: A condition existing when a substance contains the maximum amount of another substance that it can hold at that particular pressure and temperature.

Schrader valve: A spring-loaded valve that permits fluid to flow in only one direction when the center pin is depressed.

Scotch yoke: A mechanism that changes rotary motion into reciprocating motion; it is used to connect a crankshaft to a piston in some types of refrigeration compressors.

Screw pump: A compressor that is constructed of two mating revolving screws.

Seat: The portion of a valve against which the valve button presses to effect a shut-off.

Seat, front: The part of refrigeration valve that forms

a seal with the valve button when the valve is in the closed position.

Second law of thermodynamics: States that heat will flow from an object at a higher temperature to an object at a lower temperature.

Self-contained air conditioning unit: An air conditioner containing a condensing unit, evaporator, fan assembly, and complete set of operating controls within its casing.

Self-inductance: A magnetic field induced in a conductor carrying a current.

Semiconductor: A class of solids that have the ability to conduct electricity. This class lies between that of conductors and that of insulators.

Semihermetic compressor: A hermetic compressor on which minor field service operations can be performed.

Sensible heat: Heat that causes a change in temperature of a substance but not a change of state.

Sensible heat ratio: The percentage of total heat removed that is sensible heat. It is usually expressed as a decimal and is the quotient of sensible heat removed divided by total heat removed.

Sensor: An electronic device that undergoes a physical change or a characteristic change as surrounding conditions change.

Sequence controls: A group of controls that act in a series or in a timed order.

Serpentining: The arrangement of tubes in a coil to provide circuits of the desired length. Intended to keep pressure drop and velocity of the substance passing through the tubes within the desired limits.

Service valve: A shut-off valve intended for use only during shipment, installation, or service procedures.

Servo: A servomechanism in a low-power device, either electrical, hydraulic, or pneumatic, that puts into operation and controls a more complex or more powerful system. A servomotor supplies power to a servomechanism.

Shell and coil: A designation for heat exchangers, condensers, and chillers consisting of a tube coil within a shell or housing.

Shell and tube: A designation for heat exchangers, condensers, and chillers consisting of a tube bundle within a shell of casing.

Short circuit: An electrical condition occurring when part of the circuit is in contact with another part and causes all or part of the current to take a wrong path.

Short cycle: Starting and stopping that occurs too frequently. A short on cycle and a short off cycle.

Shroud: A housing over a condenser, evaporator, or fan to increase airflow.

Silica gel: An absorbent chemical compound used as a drying agent. When heat is applied, the moisture is released and the compound may be reused.

Silver brazing: A brazing process in which the brazing alloy contains some silver.

Sine wave: A wave form of a single-frequency alternating current. The wave that has this displacement is the sine of an angle which is proportional to time or distance.

Single-phase motor: An electric motor that operates on a single-phase alternating current.

Single-pole, double-throw switch: An electric switch with one blade and two sets of contact points (SPDT).

Single-pole, single-throw switch: An electric switch with one blade and one set of contact points (SPST).

Skin condenser: A condenser that uses the outer shell of a cabinet as the heat radiating medium. Usually used in domestic refrigeration.

Sling psychrometer: A device with a dry-bulb and a wet-bulb thermometer that is moved rapidly through the air to measure humidity.

Slugging: A condition in which a quantity of liquid enters the compressor cylinder, causing a hammering noise.

Smoke test: A test to determine how complete the combustion process is.

Solar heat: Heat created by energy waves, or rays, from the sun.

Solder: To join two metals by the adhesion process using a melting temperature less than 800°F (427°C).

Solenoid valve: An electromagnetic coil with a moving core which operates a valve.

South pole, magnetic: The part of a magnet that receives the magnetic lines of force flow.

Specific gravity: The weight of a liquid as compared to water, which is assigned the value of 1.

Specific heat: The heat required to raise the temperature of 1 lb of a substance 1°F (0.56°C).

Specific volume: The volume per unit of mass. Usually expressed as cubic feet per pound (m³/g).

Splash system, lubrication: The method of lubricating moving parts by agitating or splashing the oil around in the crankcase.

Split-phase motor: A motor with two windings. Both windings are used in the starting of the motor. One is disconnected by a centrifugal switch after a given speed is reached by the motor. The motor then operates on only one winding.

Split system: A refrigeration or air conditioning system whose condensing unit is located apart from the evaporator.

Squirrel cage: A centrifugal fan.

Standard atmosphere: A condition existing when the air is at 14.7 psia (101.32 kPa) pressure, 68°F (20°C) temperature, and 36% relative humidity.

Standard conditions: Conditions used as a basis for air conditioning calculations. They are temperature of 68°F (20°C), pressure of 29.92 inches of mercury (Hg) (101.32) kPa), and a relative humidity of 30%.

Starting relay: An electrically operated switch used to connect or disconnect the starting winding of an electric motor.

Starting winding: A winding in an electric motor used

only briefly during the starting period to provide the extra torque required during this period.

Static head: The pressure due to the weight of a fluid in a vertical column or, more generally, the resistance due to lift.

Status indicator: A PC card, which is optional, whose function it is to give a visual readout of the operating or failure modes of a cuber at a glance.

Status indicator, bin full: Indicated by the lighted yellow LED that the bin is full.

Status indicator, operational: Indicated by a flashing green LED that the machine is operating properly and making ice.

Status indicator, overhead evaporator: Indicated by the lighted green and yellow LEDs that the evaporator is above 140 °F (60 °C).

Status indicator, service required: (Due to a malfunction or a cuber left in the wash cycle.) Indicated by the lighted red and yellow LEDs that the ice harvest sensor switch has not been triggered in a set period of time. (Adjustable on the status indicator card.) This indicates that the ice was not harvested in the set amount of time.

Steam: Water heated to the vapor state.

Steam heating: A heating system that uses steam as the heating medium.

Steam trap: A device that allows the passage of condensed water vapor but prevents the passage of steam.

Strainer: A device, such as a screen or filter, used to remove foreign particles and dirt from the refrigerant.

Stratification of air: A condition existing when there is little or no air movement in the room.

Subcooling: The cooling of a liquid refrigerant below its condensing temperature.

Subcooling coils: A supplementary coil in an evaporative condenser, usually a coil or loop immersed in the spray water tank, that reduces the temperature of the liquid leaving the condenser.

Sublimation: The change of state from a solid to a vapor without the intermediate liquid state.

Suction line: The pipe that carries the refrigerant vapor from the evaporator to the compressor.

Suction pressure: Low-side pressure. Same as evaporator pressure.

Suction riser: A vertical tube or pipe that carries suction gas from an evaporator on a lower level to a compressor on a higher level.

Suction service valve: A two-way manually operated valve installed on the compressor inlet and used during service operations.

Suction side: The low-pressure side of the system, extending from the flow control device through the evaporator and to the compressor suction valve.

Suction temperature: The boiling temperature of the refrigerant corresponding to the suction pressure of the refrigerant.

Sump: The reservoir of an evaporative condenser in which spray water is collected before recirculation.

Superheat: A temperature increase above the saturation temperature or above the boiling point.

Superheated gas: A gas whose temperature is higher than the evaporation temperature at the existing pressure.

Swage: To enlarge one end of a tube so that the end of another tube of the same size will fit inside.

Switch, auxiliary: An accessory switch available for most damper motors and control operators that can be arranged to open or close an electric circuit whenever the control motor reaches a certain position.

Switch, disconnect: A switch usually provided for a motor that completely disconnects the motor from the source of electric power.

Tailpipe: The outlet tube from an evaporator.

Temperature: The measure of heat intensity.

Temperature, ambient: The temperature of the air around the object under consideration.

Temperature, condensing: The temperature of the fluid in the condenser at the time of condensation.

Temperature, discharge: The temperature of the gas leaving the compressor.

Temperature, dry bulb: The temperature of the air measured with an ordinary thermometer. It indicates only sensible heat changes.

Temperature, entering: The temperature of a substance as it enters a piece of apparatus.

Temperature, evaporating: The temperature at which a fluid boils under the existing pressure.

Temperature, final: The temperature of a substance as it leaves a piece of apparatus.

Temperature, saturation: The boiling point of a refrigerant at a given pressure. It is considered to be the evaporator temperature in refrigeration.

Temperature, suction: The temperature of the gas as it enters the compressor.

Temperature, wet bulb: The temperature of the air measured with a thermometer having a bulb covered with a moistened wick.

Test charge: An amount of refrigerant vapor forced into a refrigeration system to test for leaks.

Test light: A light provided with test leads that is used to test electric circuits to determine if electricity is present.

Therm: A symbol used in the gas industry; represents 100,000 Btu (25,200 kcal).

Thermal overload element: The alloy piece used for holding an overload relay closed; it melts when the current draw is too great.

Thermal relay: A heat-operated relay that opens or closes the starting circuit to an electric motor; also called hot-wire relay.

Thermistor: Basically, a semiconductor whose electrical resistance varies with the temperature.

Thermocouple: A device that generates electricity from

heat using two pieces of dissimilar metals that are welded together. When the welded junction is heated, a voltage is produced.

Thermodisc defrost control: An electrical bimetal disc switch that is controlled by changes in temperature.

Thermometer: An instrument used for measuring sensible temperature.

Thermometer well: A small pocket or recess in a pipe or tube designed to provide good thermal contact with a test thermometer.

Thermopile: A number of thermocouples connected in series to produce a higher voltage than can be produced by a single thermocouple.

Thermostat: A device used to control equipment in response to temperature change. A temperature-sensitive controller.

Thermostatic expansion valve: A flow control device operated by both the temperature and pressure in an evaporator.

Thermostat, modulating: A temperature controller that employs a potentiometer winding instead of switch contacts.

Thermostat, on-and-off: A thermostat designed to open or close an electric circuit in response to a temperature change.

Three phase: Term designating operation by means of a combination of three alternating current circuits which differ in phase by ⅓ cycle.

Three-way valve: a multiorifice flow control valve with a common and two alternate paths.

Time delay relay: A relay actuated after a predetermined time from the point of impulse.

Timeout: On a status indicator this is a condition caused by a timing circuit that will cut the cuber off, if the ice harvest switch is not triggered during an ice harvest, in the amount of time determined by the preset timer adjustment.

Timer: A clock-operated mechanism used to open and close an electric circuit on a predetermined schedule.

Ton of refrigeration: A unit of refrigeration capacity measuring 200 Btu (50.4 kcal) per minute, 12,000 Btu (3,024 kcal) per hour, or 288,000 Btu (72,576 kcal) per day. It is so named because it is equivalent in cooling effect to melting one ton of ice in 24 hr.

Torque: A turning or twisting force.

Torque, starting: The amount of torque available to start and accelerate a loaded motor.

Total heat: The sum of sensible and latent heat contained in a substance.

Transducer: A device activated by a change in power from one source for the purpose of supplying power in another form to a second system.

Transformer: An electrical device that transforms electrical energy from one circuit to another by electrical induction.

Triple point: A pressure–temperature condition at which a substance is in equilibrium. Occurs in solid, liquid, and vapor states.

Tube-within-a-tube: Heat exchange surfaces or condensers constructed of two concentric tubes.

Tubing: A pipe with a thin wall.

Two-temperature valve: A pressure-operated valve used in the suction line on multiple refrigeration installations to maintain the evaporators at different temperatures.

Ultraviolet: Invisible radiation waves whose frequencies are shorter than visible wavelengths and longer than X-ray wavelengths.

Unitary system: A combination heating and cooling system that is factory assembled in one package and is usually designed for conditioning one room or space.

Universal motor: An electric motor designed to operate on either ac or dc current.

Urethane foam: A type of insulation that is foamed between the shell and the liner of cabinets.

Useful oil pressure: The difference in pressure between the discharge and suctions sides of the compressor oil pump.

Vacuum: A reduction in pressure below atmospheric pressure. Usually stated in inches of mercury or microns.

Vacuum pump: A pump used to exhaust a system. Also, a pump designed to produce a vacuum in a closed system or vessel.

Valve: A device that controls the flow of fluid.

Valve, cap seal: A manual valve whose stem is protected by a tightly fitting cap.

Valve, charging: A valve located on the liquid line, usually near the receiver, through which refrigerant may be charged into the system.

Valve, check: A valve that permits flow in one direction only; it is designed to close against backward flow.

Valve, condenser shutoff: A valve located in the hot-gas or discharge line at the inlet of the condenser.

Valve, cylinder discharge: The valve in a compressor through which the gas leaves the cylinder.

Valve, cylinder suction: The valve in a compressor through which the gas enters the cylinder.

Valve, expansion: A type of refrigerant flow control device that maintains a constant pressure or temperature in the low side of the system.

Valve plate: The part of the compressor that contains the compressor valves and ports and is located between the compressor body and cylinder head.

Valve port: The passage in a valve that opens and closes to control the flow of a fluid in accordance with the relative position of the valve button to the valve seat.

Valve, purge: A valve through which noncondensable gases may be purged from the condenser or receiver.

Vapor: A fluid in the gaseous state formed by evaporation of the liquid.

Vapor barrier: A sheet of thin plastic or metal foil used in an air-conditioned structure to prevent water vapor from penetrating the insulation.

Vaporization: The changing of a liquid to a gas.

Vapor pressure: The pressure imposed by a vapor.

Vapor, saturated: A vapor in a condition that will result in the condensation into droplets of liquid if the vapor temperature is reduced.

Variable pitch pulley: A pulley that can be adjusted to provide different drive pulley ratios.

V-belt: A type of belt commonly used in refrigeration work to transfer power from the motor to the driven apparatus.

Velocimeter: An instrument used to measure air speeds on a direct reading scale.

Velocity: The speed or rapidity of motion.

Vent: A port or opening through which pressure is relieved.

Voltmeter: An instrument used to measure electrical voltage.

Volumetric efficiency: The relationship between the actual performance of a compressor and the calculated performance of the compressor based on its displacement.

Walk-in cooler: A large commercial refrigerated space often found in supermarkets or places for wholesale meat distribution.

Water-cooled condenser: A condenser cooled by water rather than air.

Water treatment: The treatment of water with chemicals to reduce its scale-forming properties or to change other undesirable characteristics.

Watt: A unit of electric power.

Window unit: A self-contained air conditioner that is installed in a window to cool a single space.

Index